This book is dedicated to my father, who passed on to me an abiding interest in farming and a practical knowledge of the art of cultivation and crop husbandry.

# AGRI INFO

# AGRI INFO

## Guidelines for World Crop and Livestock Production

*Compiled by*

### JOHN FARNWORTH

**JOHN WILEY & SONS**

Chichester • New York • Weinheim • Brisbane • Singapore • Toronto

Published 1997 by John Wiley & Sons Ltd,
            Baffins Lane, Chichester,
            West Sussex PO19 1UD, England

            National      01243 779777
            International (+44) 1243 779777
            e-mail (for orders and customer service enquiries): cs-books@wiley.co.uk
            Visit our Home Page on http://www.wiley.co.uk
                         or http://www.wiley.com

*Other Wiley Editorial Offices*

John Wiley & Sons, Inc., 605 Third Avenue,
New York, NY 10158-0012, USA

VCH Verlagsgesellschaft mbh, Pappelallee 3,
D-69469 Weinheim, Germany

Jacaranda Wiley Ltd, 33 Park Road, Milton,
Queensland 4064, Australia

John Wiley & Sons (Asia) Pte Ltd, 2 Clementi Loop #02-01,
Jin Xing Distripark, Singapore 129809

John Wiley & Sons (Canada) Ltd, 22 Worcester Road,
Rexdale, Ontario M9W 1L1, Canada

**Library of Congress Cataloging-in-Publication Data**

Farnworth, J.
    Agri info 1 : guidelines for world crop and livestock production /
    compiled by John Farnworth.
        p.   cm.
    Includes index.
    ISBN   0-471-97246-0
    1. Agriculture — Statistics. I. Title.
    S439.F38  1997
    630' .2'011 — dc21                                        96-40103
                                                                 CIP

**British Library Cataloguing in Publication Data**

A catalogue record for this book is available from the British Library

ISBN   0-471-97246-0

A PPL book typeset in Garamond, Helvetica and Times from the author's manuscript and collated
material by EverGreen Graphics, Craigweil On Sea, West Sussex. Design by Cecil Smith.
Printed and bound in Great Britain from PostScript files by Biddles Ltd, Guildford and King's Lynn.
This book is printed on acid-free paper responsibly manufactured from sustainable forestation,
for which at least two trees are planted for each one used for paper production.

# Contents

T = Table   F = Figure   N = Note

The contents of the sections and the appendices are as follows :

## Section 15  FARM BUILDINGS

### Crop Storage

### Machinery and Materials Storage

### Livestock Housing

# Foreword

John Farnworth is an agricultural scientist with a strong practical bent and wide experience. After graduating in Agriculture at the University College of North Wales, he worked on arable farms in Lancashire before returning to take a doctorate on aspects of hill land improvement. Immediately after that he became the founder leader of a team appointed to research and develop irrigated forage and livestock production systems for the Saudi Arabian Ministry of Agriculture and Water. The change in environment was a considerable challenge.

With his colleagues he devised improved irrigation systems, imported and modified field machinery, investigated new and traditional crops, built and equipped soil and nutrition laboratories, and upgraded existing herds of cattle and sheep, as well as importing and testing other exotic breeds.

After leaving Saudi Arabia, he led with equal success other research and development groups in the Middle East, lectured in Wales, and has been a freelance consultant in recent years. To my knowledge he has worked in at least twenty countries.

As a suitcase can hold only a limited number of books, he has realized the need for a compendium of facts and figures from a variety of sources to support the preparation of reports and projects as well as field work. Many other people, including myself, have had the same experience.

To meet that need a wide range of summary-type agricultural data has been collected, mostly from the published literature, and assembled in one volume. Inevitably there are differences between tables in presentation, overlaps and minor discrepancies, but these underline the variability of biological data and their presentation and interpretation by different authors.

This volume is essentially a reference book providing a wealth of quantitative facts. It also provides valuable guidance notes and points to some key explanatory books for further reading.

*AGRI INFO* will be invaluable to those on technical assignments, whether they be agriculturalists or others such as engineers, environmentalists and economists who need agricultural facts and figures. It will surely be equally valuable in a more static role as an information source for advisers, students, lecturers, researchers and practitioners in most countries of the world.

*Professor Ian Lucas*

# Introduction

This handbook arose out of a need perceived by the Author during research, development, management, educational and planning work over some twenty-eight years in the UK and developing countries.

Frequently situations were encountered where the necessary texts to provide baseline figures and guidance were not available either in the field or even in the country being visited. Further, it was not a practical option to carry a sufficient range of texts to be able to cope with most circumstances. Not surprisingly it was realized that a concise handbook of information, likely to be useful to a generality of workers in the broad field of agriculture, was required.

The aim of this handbook, therefore, is to present mostly tabulated data felt to be useful to agricultural development workers worldwide. It is also hoped that the information will prove helpful to teachers and students of agriculture and related subjects. The data are of types that might be used directly, extrapolated from, or simply browsed over for an idea or as a prompt to the memory.

A wide range of already summarized and published baseline-type data have been collated. Where certain types of information had apparently not been published in summary form, the Author has produced the necessary material from a range of sources. Additional or improved information from users will be welcomed (see the address at the end of the acknowledgments, page xxix). The extensive fields of crop protection and agricultural product qualities have not been covered in the present volume.

Explanatory notes have been kept to a minimum to restrict the book to portable size. A detailed and sectioned table of contents complements a short, conventional index, but the titles of tables, figures and notes have been somewhat abbreviated to assist in quick scanning for relevant data.

Each item of text is attributed to its original source. Additionally, 'key' texts have been identified in an appendix for the support of specific subjects. They are given to provide reference to more detailed and/or explanatory data.

While every effort has been made to ensure the accuracy of the text, no responsibility can be accepted for its use. When utilizing or extrapolating from data it is advisable that similar information from different sources be considered together with any local data, and that the source, environment and date(s) of the data be taken into account.

*John Farnworth*

# Acknowledgments

Agreat debt is owed to all those who have so generously allowed me to reproduce their material. Clearly this handbook would not otherwise have been possible. In return I hope that my compilation will draw deserved attention to their work. Where possible each piece of reproduced material is individually acknowledged in the text.

Whilst every effort has been made to trace the owners of copyright material, in a few cases this has proved impossible and I take this opportunity to offer my apologies to any copyright holders whose rights may have been unwittingly infringed.

I am also grateful to friends who have considered the contents and made many useful suggestions for their improvement.

Several people have had a profound and beneficial effect on the development of my career which has led to this compilation. I would like particularly to express my gratitude here to: Roland, my uncle, who taught me much about machinery operation and repair, and practical farm management; Eric Cresswell, who created my first overseas post, guided me in it, and has since generously provided valuable counsel; John Pritchard and Norman Williams, from whom I have learnt much about livestock husbandry; Moira, my wife, who has supported all the ideas and schemes over the years; and, not least, team colleagues on overseas assignments with whom much has been debated, and all the overseas nationals who have placed trust in me for their developments.

*John Farnworth*
*3 Bryn Eglwys*
*St. Anne's, Bethesda*
*Gwynedd, LL57 4BQ*
*North Wales, UK*

# Land Use
# and Soils

# Land Quality Classification

---

**TABLE 1 LAND QUALITY A Checklist of Factors to Consider in Assessing Land Quality Potential for Livestock and Crop Production**

---

### Topographical Factors

Altitude
Latitude
Proximity to coast
Slope: consider mechanization, erosion, water catchment, irrigation, drainage
Microtopography: consider levelling and smoothing requirement
Proximity to all-weather road
Flooding hazard

### Physical Factors

Drainage characteristics and requirements
Physical soil type: consider fertility, structure, compaction, workability
Soil texture
Clearance requirements: trees, shrubs, stones, boulders, structures
Farm road requirements
Water resources. Rivers/underground. For humans, livestock and crops
Soil moisture storage capacity
Trafficability of site
Soil depth to limiting horizon
Soil profile

### Chemical Factors

Soil amelioration requirements: pH, fertility, salinity, sodicity
Soil/water toxicity factors
Soil organic matter
Water resources quality
Presence or absence of weatherable minerals in crop utilizable soil profile
Soil nutrient reserves

### Climatic Factors

Rainfall: total, seasonal, variability, intensity
Temperature regime
Radiation input
Balance of potential evapotranspiration and rainfall
Climatic hazards, e.g. wind, hail, frosts, fire, salt spray, extremes of
    temperature and rainfall
Potential for crop maturation and harvesting periods

---

**TABLE 1** *(continued)* **LAND QUALITY**
**A Checklist of Factors to Consider in Assessing Land Quality**
**Potential for Livestock and Crop Production**

---

**Biological Factors**

Presence of weed species
Crop diseases and pest threats
Livestock/human disease threats
Livestock/crop predators
Desirable soil micro-organism population including nitrogen fixation
Presence of pollinating species
Potential economic yield of required crops and livestock

---

**Economic, Technical and Social Factors**

Farm or area size viability
Labour/skills/expertise availability
Physical inputs availability
Proximity to markets
Suitability of site for wide range of crops and livestock rather than narrow
    range of enterprises
Sustainability of proposed crop/livestock regimes
Availability and cost of funds to overcome development constraints, capital
    costs and recurrent costs
Availability of technical data and local experience to support proposed
    enterprises

**TABLE 2   Land Class General Specifications (United States Bureau of Reclamation)**

| Land characteristics | Class 1 – arable | Class 2 – arable | Class 3 – arable |
|---|---|---|---|
| | **Soils** | | |
| **Texture** | Sandy loam to friable clay loam | Loamy sand to very permeable clay | Loamy sand to permeable clay |
| **Depth** (measurements in cm): To sand, gravel or cobble | 90 plus - good free working soil of fine sandy loam or finer; or 105 of sandy loam | 60 plus - good free working soil of fine sandy loam or finer; or 75-90 of sandy loam to loamy sand | 45 plus - good free working soil of fine sandy loam or finer; or 60 to 75 of coarser-textured soil |
| To shale, raw soil from shale or similar material (15 less in each to rock and similar material) | 150 plus; or 135 with minimum of 15 of gravel overlying impervious material or sandy loam throughout | 120 plus; or 105 with minimum of 15 of gravel overlying impervious material or loamy sand throughout | 105 plus; or 90 with minimum of 15 of gravel overlying impervious material or loamy sand throughout |
| To penetrable lime zone | 45 with 150 penetrable | 35 with 120 penetrable | 25 with 90 penetrable |
| **Alkalinity** | pH 9.0 or less, unless soil is calcareous, total salts are low and evidence of black alkali is absent | pH 9.0 or less, unless soil is calcareous, total salts are low and evidence of black alkali is absent | pH 9.0 or less, unless soil is calcareous, total salts are low and evidence of black alkali is absent |
| **Salinity** | Total salts not to exceed 0.2%. May be higher in open permeable soils and under good drainage conditions | Total salts not to exceed 0.5%. May be higher in open permeable soils and under good drainage conditions | Total salts not to exceed 0.5%. May be higher in open permeable soils and under drainage conditions |
| | **Topography** | | |
| **Slopes** | Smooth slopes up to 4% in general gradient in reasonably large-size bodies sloping in the same plane | Smooth slopes up to 8% in general gradient in reasonably large-size bodies sloping in the same plane; or rougher slopes which are < 4% in general gradient | Smooth slopes up to 12% in general gradient in reasonably large-size bodies sloping in the same plane; or rougher slopes which are < 8% in general gradient |
| **Surface** | Even enough to require only small amount of levelling and no heavy grading | Moderate grading required but in amounts found feasible at reasonable cost in comparable irrigated area | Heavy and expensive grading required in spots but in amounts found feasible in comparable irrigated areas |

**TABLE 2** *(continued)* **Land Class General Specifications (United States Bureau of Reclamation)**

| Land characteristics | Class 1 – arable | Class 2 – arable | Class 3 – arable |
|---|---|---|---|
| **Topography** | | | |
| **Cover**<br>(loose rocks and vegetation) | Insufficient to modify productivity or cultural practices, or clearing cost small | Sufficient to reduce productivity and interfere with cultural practices. Clearing required but at moderate cost | Present in sufficient amounts to require expensive but feasible clearing |
| **Drainage** | | | |
| **Soil and topography** | Soil and topographic conditions such that no specific farm drainage requirement | Soil and topographic conditions such that some farm drainage will probably be required but with reclamation by artificial means appearing feasible at reasonable cost | Soil and topographic conditions such that significant farm drainage will probably be required but with reclamation by artificial means appearing expensive but feasible |

**Class 4 – limited arable**

Includes lands having excessive deficiencies and restricted utility but which special economic and engineering studies have shown to be irrigable

**Class 5 – non-arable**

Includes lands which will require additional economic and engineering studies to determine their irrigability and lands classified as temporarily non-productive pending construction of corrective works and reclamation

**Class 6 – non-arable**

Includes lands which do not meet the minimum requirements of the next higher class mapped in a particular survey and small areas of arable land lying within larger bodies of non-arable land

*Source:* United States Bureau of Reclamation (1953) *Irrigated Land Use. Bureau of Reclamation Manual, Vol 5.*

# Soils – Physical

---

### Note 1   Revised Storie Index Soil Rating

---

The Storie Index is presented as a simple method of rating soils. It was developed in California, and a worked example is given. The method of soil rating is based on soil characteristics that govern the land's potential utilization and productive capacity. It is independent of other physical or economic factors that might determine the desirability of growing certain plants in a given location.

Essentially the present revision sets up a new factor C to evaluate slope; the original factor C is now designated as factor X.

Percentage values are assigned to the characteristics of the soil itself, including the soil profile (factor A); the texture of the surface soil (factor B); the slope (factor C); and conditions of the soil exclusive of profile, surface texture, and slope; for example, drainage, alkali content, nutrient level, erosion, and microrelief (factor X). The most favourable or ideal conditions with respect to each factor are rated at 100 per cent. The percentage values or ratings for the four factors are then multiplied, the result being the Storie Index rating of the soil.

**The characteristics of the soil profile (factor A)** are essentially the features of the subsurface layers. For Californian purposes the soils have been divided into nine profile groups. For example, soils that are deep and readily pervious to roots and water (listed in profile group I in the soil-rating chart) are rated at 100 per cent. Profiles with dense clay subsoils (listed in profile group IV on the soil-rating chart) are rated lower. Primary or residual soils (listed in profile groups VII, VIII, and IX) are rated in accordance with the depth to bedrock.

Next, the soils are rated on the basis of the **texture of the surface soils** (designated as **factor B**). Medium-textured soils, such as the loams and the silt loams, are rated highest; the extremes in texture, such as sands and clays, lower.

**Rating of the slope** of the land is considered in **factor C**. Nearly level or gently sloping land is rated at 100 per cent. As the slope increases, the rating for this factor decreases. As shown in the soil-rating chart, single letters are used to indicate simple slopes, and double letters to indicate compound slopes. The percent slope expresses the number of feet rise or fall for 100 feet horizontal distance.

Conditions exclusive of profile, soil texture, and slope are considered in **factor X** on the soil-rating chart. These conditions consist of drainage, alkali or salt content, general nutrient level, acidity, erosion, and microrelief (surface regularity). If two or more conditions exist that are listed under **factor X**, the ratings for each are treated independently; that is, they are multiplied in order to secure the **factor X** rating.

---

## SOIL RATING CHART
### (Storie Soil Index Rating = factor A x factor B x factor C x factor X)

---

### FACTOR A
#### Rating on Character of Physical Profile

per cent

**I**  Soils on recent alluvial fans, flood plains, or other secondary deposits having undeveloped profiles........100
x - shallow phases (on consolidated material), 2 feet deep................................50 - 60
x - shallow phases (on consolidated material), 3 feet deep.........................................70
g - extremely gravelly subsoils...............80 - 95
s - stratified clay subsoils.........................80 - 95

**II**  Soils on young alluvial fans, flood plains,or other secondary deposits having slightly developed profiles........95 - 100
x - shallow phases (on consolidated material), 2 feet deep................................50 - 60
x - shallow phases (on consolidated material), 3 feet deep.........................................70
g - extremely gravelly subsoils...............80 - 95
s - stratified clay subsoils.........................80 - 95

**III**  Soils on older alluvial fans, alluvial plains, or terraces having moderately developed profiles (moderately dense subsoils).......................................................80 - 95
x - shallow phases (on consolidated material), 2 feet deep................................40 - 60
x - shallow phases (on consolidated material), 3 feet deep................................60 - 70
g - extremely gravelly subsoils ...............60 - 90

**IV**  Soils on older plains or terraces having strongly developed profiles (dense clay subsoils)................. 40 - 80

**V**  Soils on older plains or terraces having hardpan subsoil layers at less than 1 foot .......................5 - 20

per cent

at 1 to 2 feet..................20 - 30
at 2 to 3 feet..................30 - 40
at 3 to 4 feet..................40 - 50
at 4 to 6 feet..................50 - 80

**VI**  Soils on older terraces and upland areas having dense clay subsoils resting on moderately consolidated or consolidated material....................................................40 - 80

**VII**  Soils on upland areas underlain by hard igneous bedrock
at less than 1 foot .........10 - 30
at 1 to 2 feet..................30 - 50
at 2 to 3 feet..................50 - 70
at 3 to 4 feet..................70 - 80
at 4 to 6 feet...............80 - 100
at more than 6 feet.............100

**VIII**  Soils on upland areas underlain by consolidated sedimentary rocks
at less than 1 foot .........10 - 30
at 1 to 2 feet..................30 - 50
at 2 to 3 feet..................50 - 70
at 3 to 4 feet..................70 - 80
at 4 to 6 feet...............80 - 100
at more than 6 feet.............100

**IX**  Soils on upland areas underlain by softly consolidated material
at less than 1 foot .........20 - 40
at 1 to 2 feet..................40 - 60
at 2 to 3 feet..................60 - 80
at 3 to 4 feet..................80 - 90
at 4 to 6 feet...............90 - 100
at more than 6 feet.............100

---

### FACTOR B
#### Rating on basis of surface texture

Medium-textured:
fine sandy loam...............................100
loam .................................................100
silt loam ...........................................100

*(continued over)*

**SOIL RATING CHART
FACTOR B** *(continued)*
**Rating on basis of surface texture**

per cent
sandy loam ...........................................95
silty clay loam, calcareous ................95
silty clay loam, non-calcareous ........90
clay loam, calcareous.........................95
clay loam, non-calcareous.........85 - 90

Heavy or fine textured:
silty clay, highly calcareous .....70 - 90
silty clay, non-calcareous .........60 - 70
clay, highly calcareous ..............70 - 80
clay, non-calcareous .................50 - 70

Light or coarse-textured:
coarse sandy loam............................90
loamy sand ........................................80
very fine sand ...................................80
fine sand ............................................65
sand.....................................................60
coarse sand ...............................30 - 60

Gravelly:
gravelly fine sandy loam ..........70 - 80
gravelly loam ............................60 - 80
gravelly silt loam......................60 - 80
gravelly sandy loam..................50 - 70
gravelly clay loam.....................60 - 80
gravelly clay .............................40 - 70
gravelly sand.............................20 - 30

Stony:
stony fine sandy loam...............70 - 80
stony loam..................................60 - 80
stony silt loam ..........................60 - 80
stony sandy loam ......................50 - 70
stony clay loam .........................50 - 80
stony clay...................................40 - 70
stony sand .................................10 - 40

**FACTOR C
Rating on basis of slope**

per cent
A    Nearly level (0 to 2%) ....................100
AA   Gently undulating (0 to 2%) ..95 - 100
B    Gently sloping (3 to 8%)........95 - 100
BB   Undulating (3 to 8%) ..............85 - 100
C    Moderately sloping (9 to 15%)..80 - 95
CC   Rolling (9 to 15%).....................80 - 95
D    Strongly sloping (16 to 30%)....70 - 80
DD   Hilly (16 to 30%)......................70 - 80
E    Steep (30 to 45%) .....................30 - 50
F    Very steep (45% and over) ........5 - 30

**FACTOR X  Rating of conditions other than
those in factors A, B, and C**

Drainage:
well-drained   ................................100
fairly well drained.....................80 - 90
moderately waterlogged...........40 - 80
badly waterlogged ....................10 - 40
subject to overflow..................variable

Alkali:
alkali-free .........................................100
slightly affected.........................60 - 95
moderately affected..................30 - 60
moderately to strongly affected..15 - 30
strongly affected ........................5 - 15

Nutrient (fertility) level:
high ...................................................100
fair............................................95 - 100
poor ...........................................80 - 95
very poor....................................60 - 80

Acidity: according to degree...............80 - 95

Erosion:
none to slight..................................100
detrimental deposition..............75 - 95

**FACTOR X** *(continued)*
**Rating of conditions other than those in factors A, B, and C**

Erosion: *(continued)*

per cent

| | |
|---|---|
| moderate sheet erosion | 80 - 95 |
| occasional shallow gullies | 70 - 90 |
| moderate sheet erosion with shallow gullies | 60 - 80 |
| deep gullies | 10 - 70 |
| moderate sheet erosion with deep gullies | 10 - 60 |
| severe sheet erosion | 50 - 80 |
| severe sheet erosion with shallow gullies | 40 - 50 |
| severe sheet erosion with deep gullies | 10 - 40 |
| very severe erosion | 10 - 40 |
| moderate wind erosion | 80 - 95 |
| severe wind erosion | 30 - 80 |

Microrelief:

| | |
|---|---|
| smooth | 100 |
| channels | 60 - 95 |
| hogwallows | 60 - 95 |
| low hummocks | 80 - 95 |
| high hummocks | 20 - 60 |
| dunes | 10 - 40 |

### Soil Grading

For simplification, six soil grades have been set up in California by combining soils having ranges in index rating as follows:

**Grade 1** (excellent): Soils that rate between 80 and 100 per cent and which are suitable for a wide range of crops, including alfalfa, orchard, truck, and field crops.

**Grade 2** (good): Soils that rate between 60 and 79 per cent and which are suitable for most crops. Yields are generally good to excellent.

**Grade 3** (fair): Soils that rate between 40 and 59 per cent and which are generally of fair quality, with less wide range of suitability than grades 1 and 2. Soils in this grade may give good results with certain specialized crops.

**Grade 4** (poor): Soils that rate between 20 and 39 per cent and which have a narrow range in their agricultural possibilities. For example, a few soils in this grade may be good for rice, but not good for many other uses.

**Grade 5** (very poor): Soils that rate between 10 and 19 per cent are of very limited use except for pasture, because of adverse conditions such as shallowness, roughness, and alkali content.

**Grade 6** (non-agricultural): Soils that rate less than 10 per cent include, for example, tidelands, riverwash, soils of high alkali content, and steep broken land.

### Example of
### Rating the Soil for a Tract of Land

The index for each soil type in the tract is calculated separately, and then a rating for the entire tract is obtained by weighing each soil index according to the proportion of the acreage of that soil in the tract. As an example, using the soil map shown the rating of the tract is determined as follows:

**1** Index for the area Y1-A (Yolo loam, nearly level): this is a recent alluvial soil, deep, smooth, well drained.

Rating in
per cent

| | |
|---|---|
| Factor A: Yolo series, profile group I | 100 |
| Factor B: loam texture | 100 |
| Factor C: slope A, nearly level | 100 |
| Factor X: no other modifying factors | 100 |

*(continued over)*

---

**Example of
Rating the Soil for a Tract of Land
(continued)**

Index rating =
100% x 100% x 100% x 100% = 100%

**2** Index for Ac-BB (Antioch clay loam, undulating): this is a claypan terrace soil with undulating topography.

Rating in
per cent

Factor A: Antioch series, profile group IV..60
Factor B: clay loam texture ........................85
Factor C: undulating topography................95
Factor X: no other modifying factors........100

Index rating =
60% x 85% x 95% x 100% = 48%.

**3** Index for Acl-CC (Altamont clay loam, rolling): this is a brown upland soil from shale parent material; redrock at a depth of 3 feet. Rolling topography, moderate sheet erosion, with occasional gullies.

Rating in
per cent

Factor A: Altamont series,
profile group VIII.........................................70
Factor B: clay loam texture ........................85
Factor C: rolling topography ......................90
Factor X: moderate sheet erosion with
shallow gullies.............................................70

Index rating =
70% x 85% x 90% x 70% = 37%.

**4** The index for the entire tract shown on the map may then be calculated according to the acreage of each soil, as follows:

|  | Index | | Acreage | | |
|---|---|---|---|---|---|
| Yolo loam | 100 | x | 10 | = | 1000 |
| Antioch clay loam | 48 | x | 5 | = | 240 |
| Altamont clay loam | 37 | x | 5 | = | 185 |
| | | | 20 | | 1425 |

Index rating for the tract = $\dfrac{1425}{20}$ = 71%.

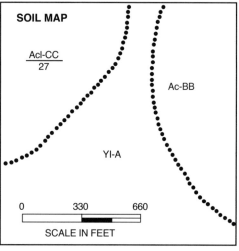

SOIL MAP

Acl-CC
27

Ac-BB

Yl-A

0       330       660

SCALE IN FEET

| MAP SYMBOL | SOILS | ACREAGE | INDEX |
|---|---|---|---|
| Yl-A | YOLO LOAM | 10 | 100 |
| Ac-BB | ANTIOCH CLAY LOAM | 5 | 48 |
| Acl-CC 27 | ALTAMONT CLAY LOAM | 5 | 37 |

*Source:* Storie, R.E. (1978) *Storie Soil Rating.* Division of Agricultural Sciences, University of California. Special Publication No. 3203. Reproduced with kind permission of the University of California.

**TABLE 3  Soil Survey Scale Classifications**

| Descriptive name | Scale | Ratio of ground area to map area 100 mm² | Approximate density of sampling points | Purpose |
|---|---|---|---|---|
| Reconnaissance | 1 : 500 000 to 1 : 250 000 | 25 km² to 6.25 km² | 1 per 50 km² to 1 per 10 km² | Development planning at national scale |
| Detailed Reconnaissance | 1 : 100 000 | 1.0 km² | 1 per 2 km² | Regional planning |
| Semi-detailed | 1 : 50 000 | 25 ha | 1 per 25 ha | Local planning; large agricultural projects, for example, irrigation |
| Detailed | 1 : 25 000 | 6.25 ha | 1 per 5 ha | Farm planning, irrigation or drainage projects |
| Intensive | 1 : 10 000 to 1 : 2500 | 1 ha to 0.06 ha | 1 per 2 ha 1 per 0.5 ha | Detailed farm planning; small projects, urban planning. Design of engineering works, for example, reservoirs, design of agricultural experiments. |

*Source:* Adapted from Hudson, N.W. (1974) *Field Engineering for Agricultural Development.*
Reproduced with kind permission of N.W. Hudson.

**TABLE 4   Soil Particle Size Classifications**

**System**

| INTERNATIONAL | | | USDA | | | |
|---|---|---|---|---|---|---|
| **Fraction** | **Size (mm)** | | **Fraction** | **Size (mm)** | | |
| Gravel | | >2.0 | Gravel | | | >2.0 |
| Coarse Sand | 2.0 – | 0.2 | V. Coarse Sand | 2.0 | – | 1.0 |
| Fine Sand | 0.2 – | 0.02 | Coarse Sand | 1.0 | – | 0.5 |
| Silt | 0.02 – | 0.002 | Medium Sand | 0.5 | – | 0.25 |
| Clay | | <0.002 | Fine Sand | 0.25 | – | 0.1 |
| | | | V. Fine Sand | 0.1 | – | 0.05 |
| | | | Silt | 0.05 | – | 0.002 |
| | | | Clay | | <0.002 | |

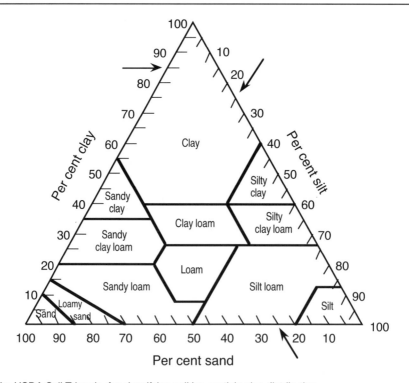

**Fig. 1** The USDA Soil Triangle for classifying soil by particle size distribution

*Source:* USDA Soil Conservation Service (1951) *USDA Soil Survey Manual. Agricultural Handbook No.18.*

**TABLE 4** *(continued)*  **Soil Particle Size Classifications**

## System

| EUROPEAN | | | | SOIL SURVEY OF ENGLAND & WALES | | | |
|----------|---|---|---|---|---|---|---|
| **Fraction** | | **Size (mm)** | | **Fraction** | | **Size  (mm)** | |
| Gravel | | | <2.0 | Stones | | | >2.0 |
| Coarse Sand | 2.0 | – | 0.6 | Coarse Sand | 2.0 | – | 0.6 |
| Medium Sand | 0.6 | – | 0.2 | Medium Sand | 0.6 | – | 0.2 |
| Fine Sand | 0.2 | – | 0.06 | Fine Sand | 0.2 | – | 0.06 |
| Coarse Silt | 0.06 | – | 0.02 | Silt | 0.06 | – | 0.002 |
| Medium Silt | 0.02 | – | 0.006 | Clay | | | < .002 |
| Fine Silt | 0.006 | – | 0.002 | | | | |
| Coarse Clay | 0.002 | – | 0.0006 | | | | |
| Medium Clay | 0.0006 | – | 0.0002 | | | | |
| Fine Clay | | | <0.0002 | | | | |

**TABLE 5**  **Soil Depth Rating**

| Class | Description | Range of depth (mm) | | |
|-------|-------------|---------|---|------|
| 1 | Deep | More than | – | 1500 |
| 2 | Moderately deep | 1000 | – | 1500 |
| 3 | Moderately shallow | 500 | – | 1000 |
| 4 | Shallow | 250 | – | 500 |
| 5 | Very shallow | Less than | – | 250 |

*Source:  USDA Soil Survey Manual. Agricultural Handbook No.18.*

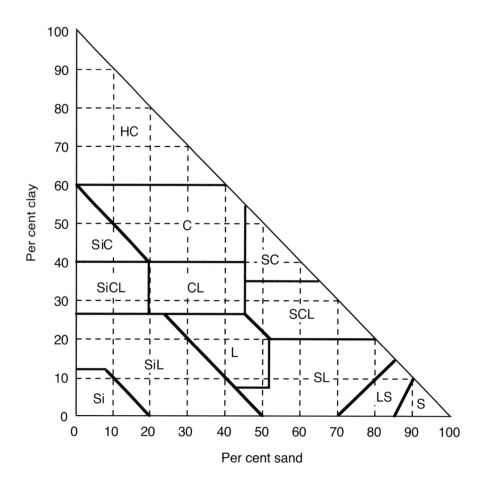

**Fig. 2** Soil texture classes (Canadian System. Percentages of clay and sand in the main textural classes of soil; the remainder of each class is silt)

*Source:* Canada Soil Survey Committee (1978) *Canadian System of Soil Classification.* Research Branch, Canada Dept. of Agriculture. Reproduced with the permission of the Ministry of Supply and Services Canada, 1991.

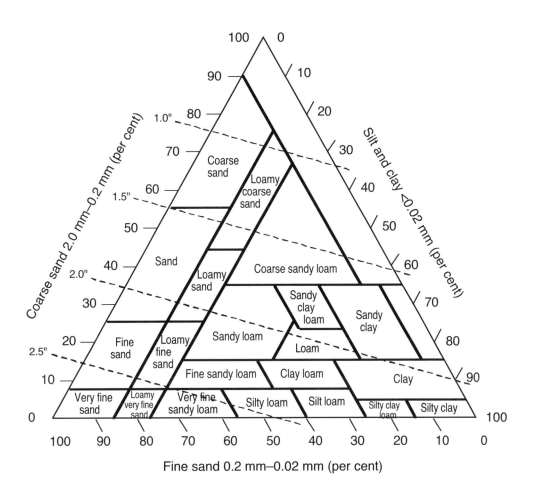

**Fig. 3** Modified triangular co-ordinate diagram (The estimated values of AWC are superimposed as contours (broken lines) at intervals of 0.5 in/ft depth)

*Source:* Salter, P.J. & Williams, J.B. (1967) The Influence of Texture on the Moisture Characteristics of Soils. *J. Soil Sci.* 18, 174. Reproduced with kind permission of MAFF ADAS (UK).

**TABLE 6** Interpretation of Soil Bulk Density Measurements for Mineral (non-peat) soils

|  | Density | Class |
|---|---|---|
|  | 0.9 – 1.2 | Recently cultivated |
|  | 1.1 – 1.4 | Main range uncultivated, uncompact soil |
| Ranges restricting root growth | 1.6 – 1.8 | Sand and loam |
|  | 1.4 – 1.6 | Silts |
|  | Variable | Clays |

*Source:* Landon, J.R., ed. (1984) *Tropical Soil Manual.* Reproduced with kind permission of Booker Tate Ltd.

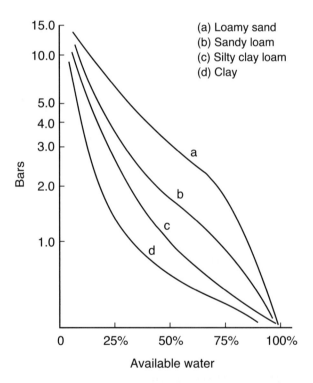

(a) Loamy sand
(b) Sandy loam
(c) Silty clay loam
(d) Clay

**Figs. 4a and 4b** Graphical representations of typical soil moisture and moisture retention characteristics of major soil types

**Fig. 4a** Moisture characteristics for various soil types

*Source:* Withers, B. & Vipond, S. (1974) *Irrigation Design and Practice.* Reproduced with kind permission of B.T. Batsford Ltd.

**TABLE 7  Interpretation of Soil Porosity**

| Porosity % by volume | Class |
|---|---|
| 40 – 70 | Usual range |
| 30 or less | Anaerobic conditions increasingly likely |

*Source:* Landon, J.R., ed. (1984) *Tropical Soil Manual.* Reproduced with kind permission of Booker Tate Ltd.

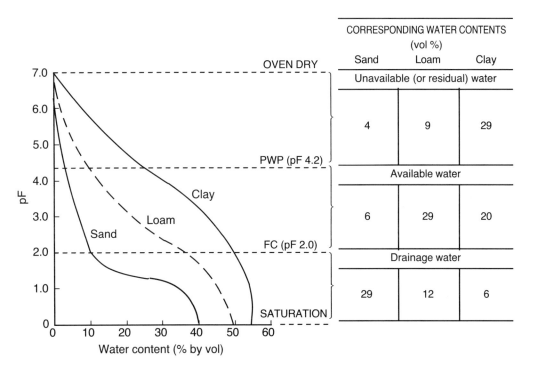

**Fig. 4b**  Indicative soil-water retention curves

*Source:* Landon, J.R., ed. (1984) *Tropical Soil Manual.* Reproduced with kind permission of Booker Tate Ltd.

**TABLE 8  Generalized Moisture Characteristics of Major Soil Types**

| Soil type | Moisture Content % by weight | | Available water |
|---|---|---|---|
| | Field Capacity | Permanent wilting Point | cm/m soil |
| Clay | 45 | 30 | 13.5 |
| Clay loam | 40 | 25 | 15.0 |
| Sandy loam | 28 | 18 | 12.0 |
| Fine sand | 15 | 8 | 8.0 |
| Sand | 8 | 4 | 5.5 |
| Coarse sand and gravel | | | 2 – 6 |
| Sands | | | 4 – 9 |
| Loamy sands | | | 6 – 12 |
| Fine sandy loams | | | 14 – 18 |
| Loams and silt loams | | | 17 – 23 |
| Clay loams and silty clay loams | | | 14 – 21 |
| Silty clays and clays | | | 13 – 18 |
| Clay (Vertisol) | 34 | 17 | 17 |
| Clay (Oxisol) | 44 | 36 | 8 |

*Note a)*  Common practice is to irrigate at 50% AWC depletion adjusting this to 40% for more sandy soils and to 60% for heavier soils.
   *b)*  Readily available water is generally considered to be 66 – 75% of total available water.
   *c)*  For near maximum yields it can be beneficial to some crops not to allow more than 25 – 30% of AWC to be depleted.

*Adapted from:* Withers, B. & Vipond, S. (1974) *Irrigation Design and Practice.*
   Israelsen, O.W. & Hansen, V.E. (1962) *Irrigation Principles and Practices.*
   Hudson, N.W. (1975) *Field Engineering for Agricultural Development.*
   Whiteman, P.C. (1980) *Tropical Pasture Sciences.*
   Landon, J.R., ed. (1984) *Tropical Soil Manual.*

**TABLE 9  Categories of Soil Moisture, Critical Moisture Points and Examples of Measurement Units.**

| Moisture point | Atmospheres | pF | Bar | J kg | Description |
|---|---|---|---|---|---|
| Fully saturated | | 0 | | | Gravitation |
| Field capacity | 0.1 – 0.3* | 2 – 2.5* | 0.1 | -10 | Available |
| Wilting point | 15 | 4.2 | 15 | -1047 | Unavailable |
| Oven dry | | 7 | | | |

* Values for medium and heavy soils; other values for lighter soils.

*Sources:* Whiteman, P.C. (1980) *Tropical Pasture Science.*
Reproduced with kind permission of Oxford University Press.
Landon, J.R., ed. (1984) *Tropical Soil Manual.* Reproduced with kind permission of Booker Tate Ltd.

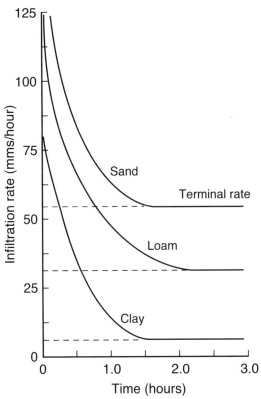

**Fig. 5** Graphical representation of water infiltration characteristics of major soil types

*Source:* Withers, B. & Vipond, S. (1974) *Irrigation Design and Practice.*
Reproduced with kind permission of B.T. Batsford Ltd.

**TABLE 10   Classification of Soil Infiltration Rates**

| Rate  cm/hr | Class |
| --- | --- |
| <0.1 | V. slow |
| 0.1 – 0.5 | Slow |
| 0.5 – 2.0 | Moderately slow |
| 2.0 – 6.0 | Moderate |
| 6.0 – 12.5 | Moderately rapid |
| 12.5 – 25.0 | Rapid |
| >25.0 | Very rapid |

*Source:* Landon, J.R., ed. (1984) *Tropical Soil Manual.*
Reproduced with kind permission of Booker Tate Ltd.

**Table 11   Typical Soil Infiltration Rates According to Soil Texture**

| Soil texture | Representative IR (cm/hr) | Normal range of IR (cm/hr) |
|---|---|---|
| Sand | 5.0 | 2.0 – 25.0 |
| Sandy loam | 2.0 | 1.0 – 8.0 |
| Loam | 1.0 | 0.1 – 2.0 |
| Clay Loam | 1.8 | 0.2 – 1.5 |
| Silty clay | 0.2 | 0.03 – 0.5 |
| Clay | 0.05 | 0.01 – 0.8 |

After Israelsen and Hansen (1962) quoted in FAO (1979).

*Source:* Landon, J.R., ed. (1984) *Tropical Soil Manual.* Reproduced with kind permission of Booker Tate Ltd.

**Table 12   Infiltration Rates According to Two Basic Soil Conditions**

Bare Ground
Good Soil Aggregation
High Organic Content

Bare Ground
Poor Soil Aggregation
Low Organic Content

Open granular structure
and no evidence
of sealing

Uniform Texture

Thin sealed layer at surface
followed by baking
and cracking

Uniform Texture

| | Basic Intake Rate | Reduce for Poor Condition |
|---|---|---|
| Coarse sand | 0.75 – 1.00"/hr. (19 – 25 mm/hr) | 0.35"/hr. (8 mm/hr) |
| Fine sands | 0.50 – 0.75"/hr. (13 – 19 mm/hr) | 0.25"/hr. (6 mm/hr) |
| Fine sandy loams | 0.35 – 0.50"/hr. (9 – 13 mm/hr) | 0.20"/hr. (5 mm/hr) |
| Silt loams | 0.25 – 0.40"/hr. (6 – 10 mm/hr) | 0.12 – 0.15"/hr. (3 – 4 mm/hr) |
| Clay loams | 0.10 – 0.30"/hr. (3 – 8 mm/hr) | 0.05 – 0.10"/hr. (1 – 3 mm/hr) |

*Source:* The Irrigation Association (1983) *Irrigation (5th ed).*
Reproduced with kind permission of the Irrigation Association, Arlington, Virginia.

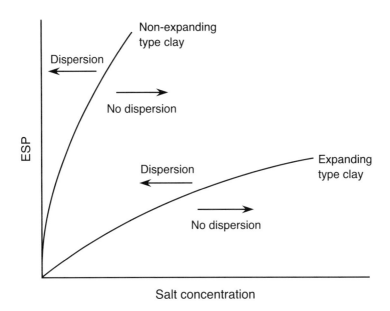

**Fig. 6** Graphical representation of the effect of soil ESP and salt concentration on dispersion of soil according to type of clay mineral

*Note:* An ESP of 15% is a rule and thumb guide to the level above which dispersion is likely to occur. In general coarser soils are tolerant of higher ESP and finer soils may disperse at lower ESP.

*Source:* Smedema, L.K. & Ryecroft, D.W. (1983) *Land Drainage: Planning and Design of Agricultural Systems.* Reproduced with kind permission of D.W. Ryecroft and B.T. Batsford Ltd.

**TABLE 13    Classification of Soil Hydraulic Conductivity**

| Rate m/day | Class |
| --- | --- |
| <0.2 | V. slow |
| 0.2 – 0.5 | Slow |
| 0.5 – 1.4 | Moderate |
| 1.4 – 1.9 | Moderately rapid |
| 1.9 – 3 | Rapid |
| >3 | Very rapid |

*Source:* Landon, J.R., ed. (1984) *Tropical Soil Manual,* (after FAO). Reproduced with kind permission of Booker Tate Ltd.

**TABLE 14   Approximate Relationships Between Texture, Structure and Hydraulic Conductivity**

| Texture | Structure | Indicative hydraulic conductivity (m/day) |
|---|---|---|
| Coarse sand, gravel | Single grain | >12 |
| Medium sand | Single grain | 6 – 12 |
| Loamy sand, fine sand | Medium crumb, single grain | 3 – 6 |
| Fine sandy loam, sandy loam | Coarse, subangular blocky and granular, fine crumb | 1.5 – 3 |
| Light clay loam, silt, silt loam, very fine sandy loam, loam | Medium prismatic and subangular blocky | 0.5 – 1.5 |
| Clay, silty clay, sandy clay, silty clay loam, clay loam, silt loam, silt, sandy clay loam | Fine and medium prismatic, angular blocky, platy | 0.1 – 0.5 |
| Clay, clay loam, silty clay, sandy clay loam | Very fine or fine prismatic, angular blocky, platy | 0.05 – 0.1 |
| Clay, heavy clay | Massive, very fine or columnar | <0.05 |

*Source:* Landon, J.R., ed. (1984) *Tropical Soil Manual.* Reproduced with kind permission of Booker Tate Ltd.

# Soils – Chemical

**TABLE 15  pH Classification**

| pH | Class | Interpretation |
|---|---|---|
| >8.5 | V high | Alkali soil |
| 7 – 8.5 | High | Possibly some nutrient deficiencies |
| 5.5 – 7 | Medium | Optimal range for many crops but too acid for some at lower end |
| <5.5 | Low | Acid soils. Probable nutrient deficiency and toxicity problems |
| 3.6 – 10.3 | | Approximate extreme range for mineral soils |
| 2.8 – 3.2 | | Approximate extreme for acid peat soils |
| 10.1 – 10.7 | | Approximate extreme of alkali mineral soils |
| 5 – 7.2 | | Common range for humid region mineral soils |
| 6.8 – 9 | | Common range for arid region mineral soils |

**TABLE 16  Organic Carbon Rating**

| % Organic carbon in soil* by weight | Class |
|---|---|
| >20 | V. High |
| 10 – 20 | High |
| 4 – 10 | Medium |
| 2 – 4 | Low |
| <2 | V. low |

*NB* 1) Presence of chloride, elemental carbon, ferrous and manganous oxides in soils may give lower values.

2) % Organic Carbon x 1.72 = % Organic Matter.

* Walkley Black Method

*Source:* Landon, J.R., ed. (1984) *Tropical Soil Manual.*
Reproduced with kind permission of Booker Tate Ltd.

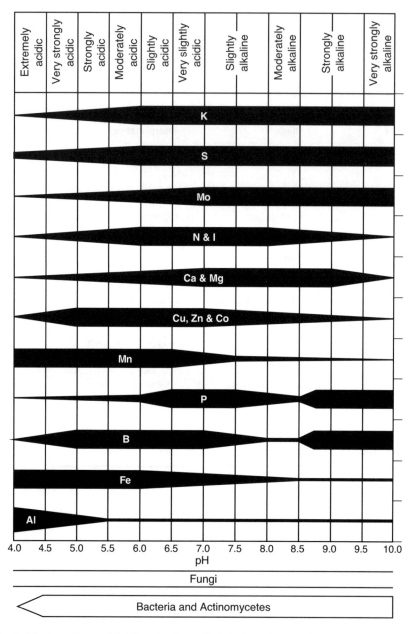

**Fig. 7** Effect of pH on availability of soil supplied plant nutrients and general pH interpretation. (This is principally applicable to mineral soils, see **Fig. 8** for organic soils)

*Source:* Adapted from Landon, J.R., ed. (1984) *Tropical Soil Manual,* and Truog, E. (1947) *Lime in Relation to Plant Nutrients.* Reproduced with kind permission of Booker Tate Ltd.

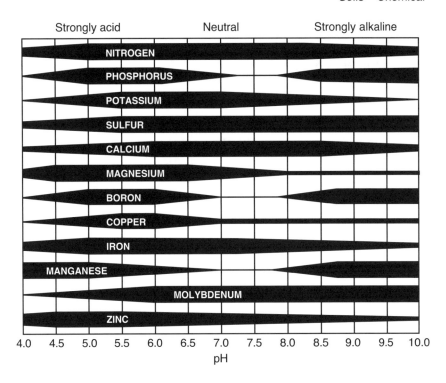

**Fig. 8** Effect of pH on availability of plant nutrients in organic soils

*Source:* Lorenz, O.A., & Maynard, D.N. (1980) *Knott's Handbook for Vegetable Growers.*
© 1980 and reprinted with kind permission of John Wiley and Sons Inc.

**Table 17   Ratings of Organic Matter and its Main Constituents**

| Rating | Total O.M.% | Total C % | Total N % | C/N ratio |
|---|---|---|---|---|
| V. high | >6 | >3.5 | >0.3 | >25 |
| High | 4.3 – 6 | 2.51 – 3.5 | 0.226 – 0.3 | 16 – 25 |
| Medium | 2.1 – 4.2 | 1.26 – 2.5 | 0.126 – 0.225 | 11 – 15 |
| Low | 1 – 2 | 0.6 – 1.25 | 0.050 – 0.125 | 8 – 10 |
| Very low | <1 | <0.6 | <0.050 | <8 |

*NB* Columns must be read independently.

*Source:* ILACO (1981) *Agricultural Compendium.* Reproduced with kind permission of ILACO BV.

**TABLE 18   Soil Nitrogen a) Nitrate b) Ratings**

| % N* in soil by weight | Class | ppm extractable nitrate |
|---|---|---|
| 1 | V. high | >50 |
| >0.5  –  1 | High | 35  –  50 |
| | Moderately high | 25  –  35 |
| 0.2  –  0.5 | Medium | 15  –  25 |
| 0.1  –  0.2 | Low | 5  –  15 |
| <0.1 | V. low | <5 |

\* Kjeldahl Method

*NB* Soil nitrogen content is principally dependent on the previous crop and fertilizer/manure applied.

*Source:* a) Landon, J.R., ed. (1984) *Tropical Soil Manual.* Reproduced with kind permission of Booker Tate Ltd.
   b) ILACO (1981) *Agricultural Compendium.* Reproduced with kind permission of ILACO BV.

**Table 19   Soil P, K, Mg Indices**

The following index system is used by the UK Agricultural Development and Advisory Service.

| | | **mg/l (extractable)** | | |
|---|---|---|---|---|
| Interpretation | Index | Soil P | Soil K | Soil Mg |
| Very low | 0 | 0  –  9 | 0  –  60 | 0  –  25 |
| Low | 1 | 10  –  15 | 61  –  120 | 26  –  50 |
| Moderate | 2 | 16  –  25 | 121  –  240 | 51  –  100 |
| | 3 | 26  –  45 | 241  –  400 | 101  –  175 |
| High | 4 | 46  –  70 | 401  –  600 | 176  –  250 |
| | 5 | 71  –  100 | 601  –  900 | 251  –  350 |
| | 6 | 101  –  140 | 901  –  1500 | 351  –  600 |
| | 7 | 141  –  200 | 1501  –  2400 | 601  –  1000 |
| Excessively | 8 | 205  –  280 | 2401  –  3600 | 1001  –  1500 |
| High | 9 | >280 | >3600 | >1500 |

*Source:* Ministry of Agriculture, Fisheries and Food/ADAS (1980) *Fertilizer Recommendations. Reference Book 209.* Reproduced with kind permission of MAFF/ADAS (UK).

**TABLE 20   Soil Phosphorus Interpretation Guidelines According to Analytical Method**

| Method | Indicative available P values (ppm) | | | Appropriate soils |
| | High | Medium | Low | |
| | Fertilizer response unlikely | Fertilizer response probable | Fertilizer response most likely | |
|---|---|---|---|---|
| **Olsen** 0.5 M NaHCO$_3$ | >15 | 15 – 5 | <5 | All soils especially where pH is >7 |
| **Bray** Dilute HCl/NH$_4$F | >50 | 50 – 15 | <15 | Acid soils |
| **Nelson** Dilute HCl/H$_2$SO$_4$ | >30 | 30 – 10 | <11 | Some acid soils |
| **Truog** Dilute H$_2$SO$_4$ | >40 | 40 – 20 | <20 | Acid soils |
| **Bingham** H$_2$O solution | >2 | 2 – 1 | <1 | All soils |
| **Morgan** Na acetate/acetic acid | >15 | 15 – 5 | <5 | Acid soils |
| **ADAS** NH$_4$ acetate/acetic acid | >40 | 40 – 2 | <2 | Acid soils |

*Source:* Landon, J.R., ed. (1984) *Tropical Soils Manual.* Reproduced with kind permission of Booker Tate Ltd.

**Table 21   Guidelines for Interpretation of Available Soil Phosphorus Determined by Olsen Method**

| Characteristic crop demand | Examples | Indicative available P values (ppm) | | |
| | | Deficient | Questionable | Adequate |
|---|---|---|---|---|
| Low P | Grass, cereals soyabeans, maize | <4 | 5 – 7 | >8 |
| Moderate P | Lucerne, cotton sweetcorn, tomatoes | <7 | 8 – 13 | >14 |
| High P | Sugarbeet, potatoes, celery onions | <11 | 12 – 20 | >21 |

*Source:* Landon, J.R., ed. (1984) *Tropical Soil Manual.* Reproduced with kind permission of Booker Tate Ltd.

**TABLE 22   Interpretation of Available K Values in Soil**

| High values | Medium values | Low values | Comments |
|---|---|---|---|
| **Available K (me/100 g soil) Ammonium acetate extraction** | | | |
| 0.8  –  0.4 | 0.4  –  0.2 | 0.2  –  0.03 | Based on Malawi soils |
| >0.5 | 0.5  –  0.25 | <0.25 | From a general appraisal of USA soils |
| >0.8 | 0.8  –  0.5 | 0.5  –  0.3 | New Zealand |
| >0.6 | 0.6  –  0.15 | <0.15 | UK soils |

*Source:* Landon, J.R., ed. (1984) *Tropical Soil Manual.* Reproduced with kind permission of Booker Tate Ltd.

**TABLE 23   Interpretation of Exchangeable Magnesium Ratings**

| Level (ppm) | Class | Comments |
|---|---|---|
| <30 | Low | Quick-acting Mg fertilizers may be required |
| 30  –  60 | Medium | Use Mg limestone when lime is needed |
| >60 | High | Mg usually sufficient in soil |

*Source:* Landon, J.R., ed. (1984) *Tropical Soil Manual.* Reproduced with kind permission of Booker Tate Ltd.

**TABLE 24   Soil Sulphate Guidelines (UK)**

| $SO_4$ level ppm | | Interpretation |
|---|---|---|
| <3.0 | V. low | Unusual |
| 3.1  –  6.0 | Low | Treatment worthwhile all crops |
| 6.1  –  10 | Moderate | Treatment generally not necessary |
| >10 | High | No treatment required |

*Source:* Atlas Interlates (1989) *Analysis Service User Guide.* Reproduced with kind permission of Atlas Interlates.

**TABLE 25   Interpretation of Sulphur Measurements**

| S measurement | Approximate S level | Effects |
|---|---|---|
| Total S | <200 ppm | Deficiency likely |
| Available S (Morgan's reagent) | <3ppm | Deficiency likely |
| Available S (in saturated extract) | >30 me/l | Excess |
| Extractable S (various methods ) | 6 – 12 ppm | Upper limit for expected response to S |

*Source:* Landon, J.R., ed. (1984) *Tropical Soil Manual.* Reproduced with kind permission of Booker Tate Ltd.

**TABLE 26   Indicative Rating of Soil Boron Levels in the absence of high calcium contents**

| B concentration (ppm) Saturated extract | Hot- water extractable | Category |
|---|---|---|
| - | <1 | Possibly deficient |
| - | 1 – 1.5 | Border line for deficiency |
| ≤0.5 | 1.5 – 3 | Satisfactory for most crops |
| 0.5 – 5 | 3 – 6 | Possibly toxic, depending on crop sensitivity |
| >10 | >6 | Toxic to most crops |

*Source:* Landon, J.R., ed. (1984) *Tropical Soil Manual.* Reproduced with kind permission of Booker Tate Ltd.

**TABLE 27   Indicative Total Trace Elements in Soils**

| Element | Concentration in soil (ppm) Approximate mean | Usual range |
|---|---|---|
| B | 20 | 2 – 270 |
| Cd | 0.35 | 0.01 – 2 |
| Co | 8 | 0.05 – 65 |
| Cr | 70 | 5 – 1500 |
| Cu | 30 | 2 – 250 |
| I | 5 | 0.1 – 25 |
| Mn | 1000 | 20 – 10 000 |
| Mo | 1.2 | 0.1 – 40 |
| Ni | 50 | 2 – 750 |
| Pb | 35 | 2 – 300 |
| S | 700 | 30 – 1600 |
| Zn | 90 | 1 – 900 |

*Source:* Landon, J.R., ed. (1984) *Tropical Soil Manual.* Reproduced with kind permission of Booker Tate Ltd.

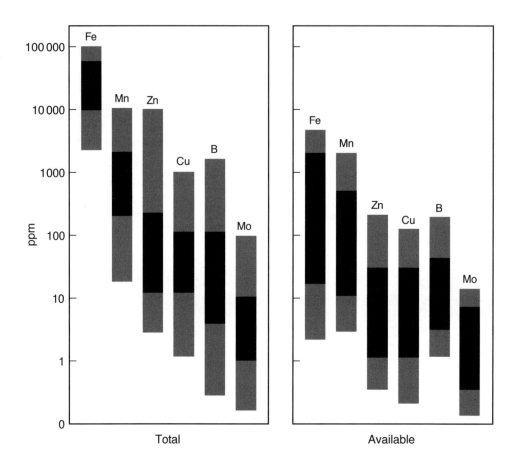

**Fig. 9** Total and Available Levels of Micronutrients in Soils (Dotted areas show unusual values)

*Source:* Blair, G. J. (1980) *Plant Nutrition. The Need for Balance.*
Reproduced with kind permission of the Australian Centre for International Agricultural Research.

**TABLE 28    Indicative Micronutrient Deficiency Levels in Soils**

| Element | Extracting agent | Deficiency levels in soils (ppm available) |
|---------|------------------|--------------------------------------------|
| B | Hot $H_2O$ | 0.1  –  0.7 |
| Cu | Ammonium acetate (pH 4.8) | 0.2 |
| | 0.5 M EDTA | 0.75 |
| | 0.43 M $HNO_3$ | 3.0  –  4.0 |
| | Biological assay | 2.0  –  3.0 |
| | 1 M HCl | 100.0 |
| | 0.1 M HCl | 0.09  –  1.06 |
| Fe | Ammonium acetate (pH 4.8) | 2.0 |
| | *DTPA + $CaCl_2$ (pH 7.3) | 2.5  –  4.5 |
| Mn | 0.05 M HCl +0.025 M $H_2SO_4$ | 5.0  –  9.0 |
| | 0.1 M $H_3PO_4$ and 3M $NH_4H_2PO_4$ | 15.0  –  20.0 |
| | Hydroquinone + ammonium acetate | 25.0  –  65.0 |
| | $H_2O$ | 2.0 |
| Mo | Ammonium oxalate (pH 3.3) | 0.04  –  0.2 |
| Zn | 0.1 M HCl | 1.0  –  7.5 |
| | Dithizone + ammonium acetate | 0.3  –  2.3 |
| | EDTA + $(NH_4)_2 CO_3$ | 1.4  –  3.0 |
| | *DTPA + $CaCl_2$ (pH 7.3) | 0.5  –  1.0 |

*Notes:*  1  Account must also be taken of possible interactive effects from other elements and prevailing environmental and soil conditions.

2  *DTPA is the only suitable method for calcareous arid region soils.

*Source:* Landon, J.R., ed., (1984) *Tropical Soil Manual.*
Reproduced with kind permission of Booker Tate Ltd.

**TABLE 29   Guide to Interpretation of Soil Trace Element Levels**

**BORON**

| Boron level (ppm) | | |
|---|---|---|
| **pH <7.0** | **ph > 7.0** | **Comments** |
| <0.4 | <0.8 | Severe boron deficiency |
| 0.4-0.8 | 0.8-1.2 | Deficiency likely in susceptible crops, (e.g. oilseed rape and sugar beet) |
| 0.8-1.2 | 1.2-4.0 | Satisfactory |
| >1.2 | >4.0 | High |

**COBALT**

| Cobalt level (ppm)* | | |
|---|---|---|
| **pH <6.0** | **pH >6.0** | **Comments** |
| <0.10 | <0.15 | Deficient - low Co levels in herbage likely. |
| 0.10 – 0.25 | 0.15 – 3.0 | Low - low Co levels in herbage possible |
| >0.25 | >3.0 | Satisfactory |

*for organic matter content not exceeding 5.0%

**COPPER**

| | Copper level (ppm) | | |
|---|---|---|---|
| **Organic matter (%)** | **Deficient** | **Low** | **Satisfactory** |
| <3 | 1.3 | 1.3 – 2.0 | >2.0 |
| 3 – 9 | 1.5 | 1.5 – 2.3 | >2.3 |
| 10 – 20 | 1.8 | 1.8 – 2.5 | >2.5 |
| 21 – 25 | 2.0 | 2.0 – 2.9 | >2.9 |
| 26 – 35 | 2.2 | 2.2 – 3.4 | >3.4 |
| 36 – 45 | 3.0 | 3.0 – 4.8 | >4.8 |
| >45 | 4.0 | 4.0 – 6.0 | >6.0 |

**TABLE 29 *(continued)*  Guide to Interpretation of Soil Trace Element Levels**

## MANGANESE

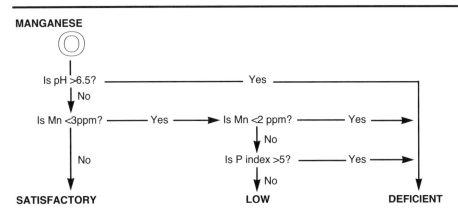

## MOLYBDENUM

| Mo level ppm | Comment |
|---|---|
| <0.1 | Deficiency likely in susceptible crops |
| >0.1 | Deficiency unlikely |

*NB* Mo deficiency can affect susceptible crops on any acid soil

## IRON
Soil analysis for iron is not generally used for diagnostic purposes.

## ZINC
Zn deficiency is likely where the soil level is less than 0.8 ppm.

*Note:* Guidelines based on analysis using analytical methods of the UK Agricultural Development and Advisory Service.

*Source:* Atlas Interlates (1989) *Analysis Service User Guide.*
Reproduced with kind permission of Atlas Interlates.

---

**TABLE 30   Indicative Rating of Exchangeable Aluminium levels**

| Exchangeable Al percentage = Exchangeable Al/CEC % | Effects |
|---|---|
| ≥30 | Sensitive crops may be affected |
| ≤60 | Generally toxic. Only very low Al concentrations expected if no electrolytes present |
| | 60% tolerated by sugarcane |
| 85 | May be tolerated by some crops in some conditions (tea, rubber, cassava, pineapple and some tropical grasses and legumes are notably Al tolerant) |

*Source:* Landon J.R., ed., (1984) *Tropical Soil Manual.*
Reproduced with kind permission of Booker Tate Ltd.

**TABLE 31 Maximum permissible concentrations of potentially toxic elements in soil after application of sewage sludge and maximum annual rates of addition (sampled at 15-25cm)**

| PTE | Maximum permissible concentration of PTE in soil (mg/kg dry solids) | | | | Maximum permissible average annual rate of PTE addition over a 10 year period (kg/ha) (2) |
|---|---|---|---|---|---|
| | pH(1) 5.0 <5.5 | pH(1) 5.5 <6.0 | pH 6.0 - 7.0 | pH(3) >7.0 | |
| Zinc | 200 | 250 | 300 | 450 | 15 |
| Copper | 80 | 100 | 135 | 200 | 7.5 |
| Nickel | 50 | 60 | 75 | 110 | 3 |
| | **For pH 5.0 and above** | | | | |
| Cadium | 3 | | | | 0.15 |
| Lead | 300 | | | | 15 |
| Mercury | 1 | | | | 0.1 |
| Chromium | 400 (Provisional) | | | | 15 (Provisional) |
| * Molybdenum (4) | 4 | | | | 0.2 |
| * Selenium | 3 | | | | 0.15 |
| * Arsenic | 50 | | | | 0.7 |
| * Fluoride | 500 | | | | 20 |

* These parameters are not subject to the provisions of Directive 86/278/EEC

(1) For soils of pH in the ranges of 5.0<5.5 and 5.5<6.0 the permitted concentrations of zinc, copper, nickel and cadmium are provisional and will be reviewed when current research into their effects on certain crops and livestock is completed.

(2) The annual rate of application of PTE to any site shall be determined by averaging over the 10 year period ending with the year of calculation

(3) The increased permissible PTE concentrations in soils of pH greater than 7.0 apply only to soils containing more than 5% calcium carbonate.

(4) The accepted safe level of molybdenum in agricultural soils is 4mg/kg. However there are some areas in UK where, for geological reasons, the natural concentration of this element in the soil exceeds this level. In such cases there may be no additional problems as a result of applying sludge, but this should not be done except in accordance with expert advice. This advice will take account of existing soil molybdenum levels and current arrangements to provide copper supplements to livestock.

*Source:* Dept. of the Environment (1989) *Code of Practice for Agricultural Use of Sewage Sludge.*
Reproduced with kind permission of the Department of the Environment (UK).

**TABLE 32  Maximum permissible concentrations of potentially toxic elements in soil under grass after application of sewage sludge when samples taken to a depth of 7.5cm**

| PTE | Maximum permissible concentration of PTE in soil (mg/kg dry solids) | | | |
|---|---|---|---|---|
| | pH 5.0 <5.5 | pH 5.5 <6.0 | pH 6.0 - 7.0 | pH(3) >7.0 |
| Zinc (1) | 330 | 420 | 500 | 750 |
| Copper (1) | 130 | 170 | 225 | 330 |
| Nickel (1) | 80 | 100 | 125 | 180 |
| | **For pH 5.0 and above** | | | |
| Cadmium (2) | 3/5 | | | |
| Lead | 300 | | | |
| Mercury | 1.5 | | | |
| Chromium | 600 (Provisional) | | | |
| * Molybdenum (4) | 4 | | | |
| * Selenium | 5 | | | |
| * Arsenic | 50 | | | |
| * Fluoride | 500 | | | |

* These parameters are not subject to the provisions of Directive 86/278/EEC

(1) The permitted concentrations of these elements will be subject to review when current research into their effects on the quality of grassland is completed. Until then, in cases where there is doubt about the practicality of ploughing or otherwise cultivating grassland, no sludge applications which would cause these concentrations to exceed the permitted levels specified in Table 31 should be made except in accordance with specialist agricultural advice.

(2) The permitted concentration of cadmium will be subject to review when current research into its effect on grazing animals is completed. Until then, the concentration of this element may be raised to the permitted upper limit of 5mg/kg as a result of sludge applications only under grass which is managed in rotation with arable crops and grown only for conservation. In all cases where grazing is permitted no sludge applications which would cause the concentration of cadmium to exceed the lower limit of 3mg/kg shall be made.

(3)  The increased permissible PTE concentrations in soils of pH greater than 7.0 apply only to soils containing more than 5% calcium carbonate.

(4) The accepted safe level of molybdenum in agricultural soils is 4mg/kg. However there are some areas in UK where, for geological reasons, the natural concentration of this element in the soil exceeds this level. In such cases there may be no additional problems as a result of applying sludge, but this should not be done except in accordance with expert advice. This advice will take account of existing soil molybdenum levels and current arrangements to provide copper supplements to livestock.

*Source:* Dept. of the Environment (1989) *Code of Practice for Agricultural Use of Sewage Sludge.*
Reproduced by kind permission of the Department of the Environment (UK).

**TABLE 33   Classification of Topsoil Cation Exchange Capacity**

| CEC (me/100 g of soil) | Rating |
|---|---|
| >40 | Very high |
| 25 – 40 | High |
| 15 – 25 | Medium |
| 5 – 15 | Low |
| <5 | Very low |

*Source:* Landon, J.R., ed. (1984) *Tropical Soil Manual.* Reproduced with kind permission of Booker Tate Ltd.

**TABLE 34   Exchangeable Cation Ratios: Summary of Critical and Common Values**

| Cation ratio | Approximate value | Effects |
|---|---|---|
| Ca:Mg | ≥5:1 | Mg increasingly unavailable with increasing Ca. With high pH also, P availability may be reduced |
| | 3:1 to 4:1 | Approximate optimum for most crops |
| | <3:1 | P uptake may be inhibited |
| | 1:1 | Suggested lowest acceptable limit. With lower values, Ca availability slightly reduced |

**Ca:Mg ratios commonly decrease with depth, and often with cultivation**

| Cation ratio | Approximate value | Effects |
|---|---|---|
| K:Mg | >2:1 | Mg uptake may be inhibited |
| | <3:2 | Field crops |
| | <1.1 | Vegetables and sugarbeet |
| | <3:5 | Fruit and greenhouse crops |
| K:CEC (EPP) | 2% | Suggested minimum level to avoid K deficiency in humid tropical soils |
| | >25% | K-rich soils (rather rare); similar effects to high Na |
| Na:CEC (ESP) | ≥15% | Sodic soils Effects usually gradual, no sharp change at ESP = 15 |

*Source:* Landon J.R., ed. (1984) *Tropical Soil Manual.* Reproduced with kind permission of Booker Tate Ltd.

**TABLE 35　Classification of Ca:Mg Ratios and Frequently Expected Soil Properties**

| Exch. Ca: Mg | Rating | Physical Soil Property |
|---|---|---|
| >12 | V. high | V. favourable* |
| 12 – 6 | High | V. favourable |
| 6 – 3.5 | Mod. high | Favourable |
| 3.5 – 2.5 | Mod. low | Favourable |
| 2.5 – 1.5 | Low | Unfavourable |
| <1.5 | V. low | V. unfavourable |

* However grazing animals might suffer Mg deficiency

*Source:* ILACO (1981) *Agricultural Compendium.* Reproduced with kind permission of ILACO BV.

**TABLE 36　Ratings of Exchangeable Bases in Soils**

| | Exchangeable Cation me% | | | |
|---|---|---|---|---|
| Rating | Ca | Mg | K | Na |
| V.high | >20 | >8 | >1.2 | >2 |
| High | 10 – 20 | 3 – 8 | 0.6 – 1.2 | 0.7 – 2 |
| Medium | 5 – 10 | 1.5 – 3 | 0.3 – 0.6 | 0.3 – 0.7 |
| Low | 2 – 5 | 0.5 – 1.5 | 0.1 – 0.3 | 0.1 – 0.3 |
| V.low | <2 | <0.5 | <0.1 | <0.1 |

Conversions:　1 me % K = 470 mg $K_2O$/kg soil

1 me % Ca = 280 mg CaO/kg soil

1 me % Mg = 200 mg Mg/kg soil

*Source:* ILACO (1981) *Agricultural Compendium.* Reproduced with kind permission of ILACO BV.

**TABLE 37 Approximate CEC Values of Clay Minerals and Organic Matter**

| Type | Nutrient reserves | Approximate CEC at pH 7 (me/100 g of clay) |
|---|---|---|
| Kaolinite and halloysite | Few nutrient reserves | <10 |
| Illite | Reserves of potassium | 15 – 40 |
| Montmorillonite | Generally with reserves of Mg, K, Fe, etc. | 80 – 100 |
| Vermiculite | Generally with reserves of Mg, K, Fe, etc. | About 100 |
| Organic matter | – | About 200 |

*NB* For soils low in organic matter, the CEC can be usefully expressed as a proportion of the clay, thus:

$$\text{CEC (me/100 g clay)} = \frac{\text{CEC (me/100 g soil)}}{\% \text{ clay}} \times 100$$

*Source:* Landon, J.R., ed. (1984) *Tropical Soil Manual.* Reproduced with kind permission of Booker Tate Ltd.

**TABLE 38 Soil Gypsum Level Classification**

| Gypsum content (% of soil by weight) | Effects on crops |
|---|---|
| <2 | Favours growth |
| 2 – 25 | Little or no adverse effect if in powdery form |
| >25 | Can cause substantial reductions in yield |

*Source:* ILRI (1971) *Gypsiferous Soils Bull. 12.* Reproduced with kind permission of the International Institute for Land Reclamation and Improvement, Wageningen, The Netherlands.

**TABLE 39    Indicative Classification of Salt-affected Soils**

| Soil | $EC_e$ (mS cm$^{-1}$) | ESP | pH (paste) | Description |
|------|------|------|------|------|
| **Saline** | >4 | <15 | Usually <8.5 | Non-sodic soils containing sufficient soluble salts to interfere with growth of most crops |
| **Saline -- sodic** | >4 | >15 | Usually <8.5 | Soils with sufficient exchangeable sodium to interfere with growth of most plants, and containing appreciable quantities of soluble salts |
| **Sodic (Alkali)** | <4 | >15 | Usually >8.5 | Soils with sufficient exchangeable sodium to interfere with growth of most plants, but without appreciable quantities of soluble salts |
| **Degraded Alkali** | <4 | Varying | <8.5 | |

*Source:* Adapted from Landon, J.R., ed. (1984) *Tropical Soil Manual,* after Richards, L.A. *et al.* (1954) *Diagnoses and improvement of saline and alkali soils. USDA Ag. Handbook No. 60.*

**TABLE 40    Soil Salinity Classification (USA)**

| Class | Soil nature | Conductivity mmhos/cm | Total salt Content % | Crop reaction |
|------|------|------|------|------|
| 0 | Salt free | 0.4 | 0 – 0.15 | Negligible salinity effect except most sensitive crops |
| 1 | Slightly saline | 4 – 8 | 0.15 – 0.35 | Restricted yields of many crops |
| 2 | Moderately saline | 8 – 15 | 0.35 – 0.65 | Tolerant crops only |
| 3 | Strongly saline | >15 | >0.65 | Very tolerant crops only |

---

**TABLE 41 Exchangeable Sodium Percentage (ESP) Classification of Soil according to ESP Value and Crop Sensitivity**

---

| Soil ESP | Crop tolerance | Crop type 50% Yield reduction |
|---|---|---|
| 2 – 10 | Very sensitive | Sensitive at ESP <15% |
| 10 – 20 | Sensitive | Semi tolerant at ESP 15 – 25% |
| 20 – 40 | Moderately tolerant | Tolerant at ESP 35% |
| 40 – 60 | Tolerant | |
| >60 | Very tolerant | |

*NB:* 1. ESP is calculated as:

$$ESP = \frac{\text{Exchangeable Na content}}{\text{Cation Exchange Capacity}} \times 100$$

2. Rising ESP levels gradually rather than rapidly cause deterioration of soil conditions.

3. Naturally occurring gypsum in the soil can ameliorate to some extent the effect of high ESP values.

*Source:* Adapted from Landon, J.R., ed. (1984) *Tropical Soil Manual.*
Reproduced with kind permission of Booker Tate Ltd.

# Climate

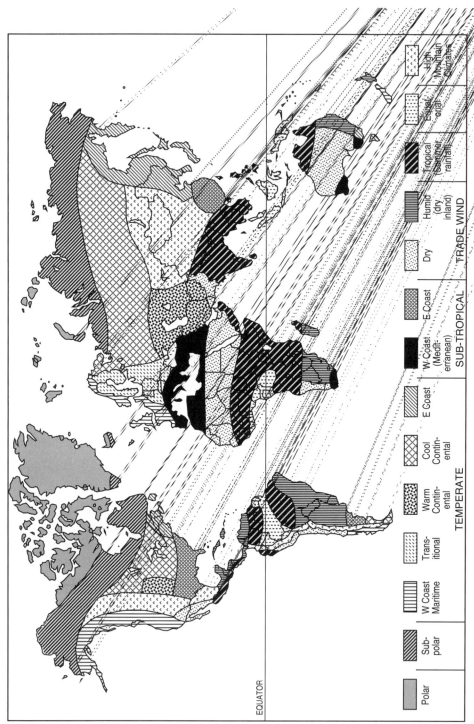

**Fig. 10** Classification of world climates (after E. Neef). *Source: Barry, G. & Chorley, R.J. (1968) Atmosphere, Weather and Climate.*

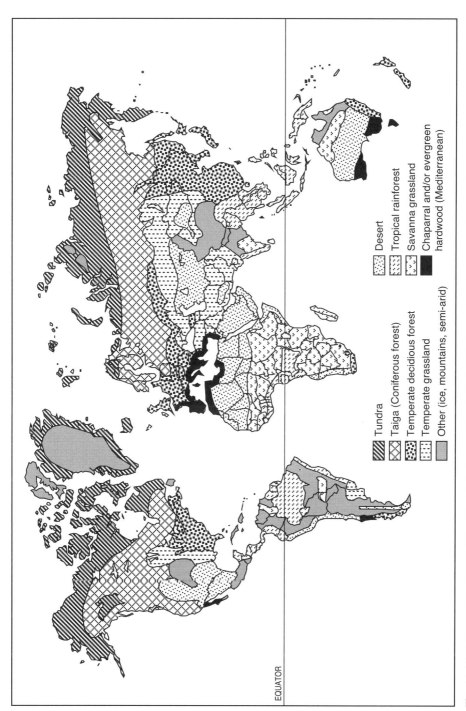

**Fig. 11** World natural vegetation (after Strahler). *Source: Waugh, D. (1990) Geography – An Integrated Approach.*

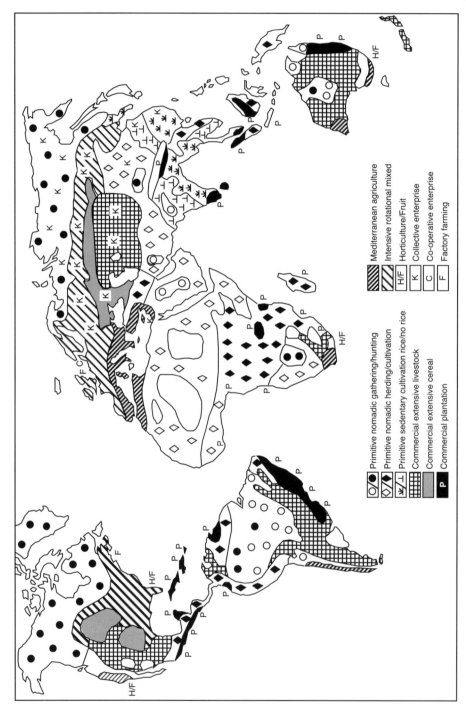

**Fig. 12** The major agricultural systems of the world. *Source:* Newbury, P.A.R. (1980) *A Geography of Agriculture.* Reproduced with kind permission of Pitman Publishing.

Legend:

Primitive nomadic gathering/hunting
Primitive nomadic herding/cultivation
Primitive sedentary cultivation rice/no rice
Commercial extensive livestock
Commercial extensive cereal
Commercial plantation

Mediterranean agriculture
Intensive rotational mixed
H/F  Horticulture/Fruit
K  Collective enterprise
C  Co-operative enterprise
F  Factory farming

**TABLE 42   Simplified Köppen Classification System of Climates**

| Principal types | Subtypes |
|---|---|
| | Characteristics |
| **Polar** | Tundra; very short summer |
| | Perpetual ice and snow |
| | Certain mountainous climates |
| **Cold snow forest** | **Subartic types:** |
| **coldest month<-3°C** | Severe winter; short cool summer; moist all seasons |
| **warmest month >10°C** | Cold winter; short summer; moist all seasons |
| | Severe dry winter; short cool summer |
| | Extremely cold dry winter; short cool summer |
| | **Humid continental types:** |
| | Severe winter; long hot summer; moist all seasons |
| | Severe winter; short warm summer; moist all seasons |
| | Severe dry winter; warm summer |
| **Temperate, mild,** | Marine – mild winter; warm summer; moist all seasons |
| **rainy coldest month** | Marine – mild winter; short, cool summer; moist all seasons |
| **<18°C but >-3°C** | Mediterranean – interior; mild winter; hot dry summer |
| | Mediterranean – coastal; mild winter; short, warm, dry summer |
| | Subtropical monsoon – mild, dry winter; hot summer |
| | Humid subtropical – mild winter; long, hot summer; moist all seasons |
| | Tropical upland – mild, dry winter; short, warm summer |
| **Dry; evaporation exceeds** | Desert – tropical; hot; arid |
| **precipitation** | Desert – mid-latitude; cool or cold; arid |
| | Steppe – tropical; hot; semi-arid |
| | Steppe – mid-latitude; cool or cold semi-arid |
| **Tropical rainy;** | Tropical savannah – hot; dry winter |
| **temperature >18°C** | Tropical monsoon – seasonally excessive rainfall |
| | Tropical rainforest – hot, rainy all seasons |

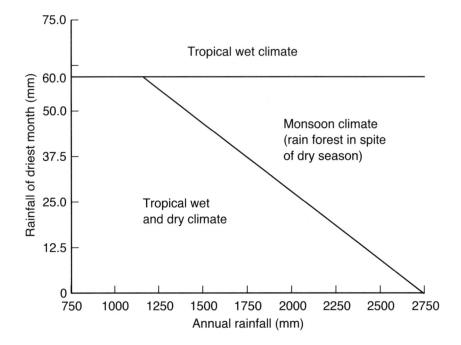

**Fig. 13** A graphical representation of boundaries of Tropical climates (after Köppen)

*Source:* Whiteman, P.C. (1980) *Tropical Pasture Sciences.*
Reproduced with kind permission of Oxford University Press.

**TABLE 43 Global Classification of Precipitation Regimes**

| | |
|---|---|
| **Polar** | Precipitation at all seasons |
| | Very small amounts |
| **Mid-latitudes** | Rain at all seasons |
| **Mediterranean** | |
| Within 30° to 40° N and SW side of continents | Winter rain; Summer dry |
| **Subtropical** | |
| Tropics to 30°N and SW side of continents | Dry desert areas; Negligible rain |
| At the margins – Equatorwards | Summer rain |
| Polewards | Winter rain |
| **Monsoon** | |
| Within and outside the Tropics in the | High summer max. |
| Indian subcontinent, on E side of continents | Long dry season |
| **Tropical** | |
| Between Equator and Tropics of Cancer | Summer rain |
| and Capricorn | Winter dry |
| Outer zone 10°N and S to the Tropics | Single summer max. |
| | Longer dry season |
| Inner zone to 10° N and S | Two summer max. |
| **Equatorial** | |
| Within a few degrees N and S of Equator | Rain all year 2 max. seasons |

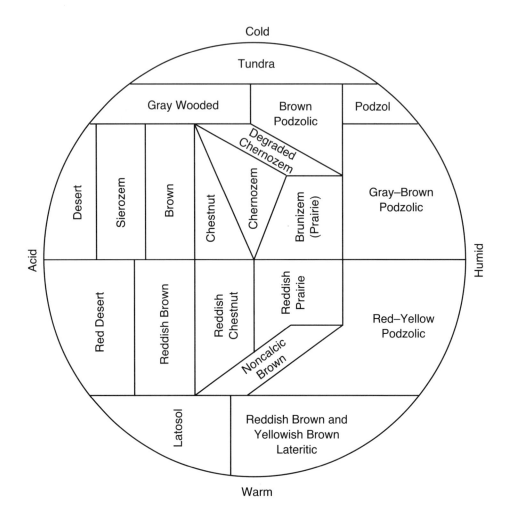

**Fig. 14** A schematic diagram relating the Zonal great soil groups to climate and vegetation

(Names printed vertically are grassland and desert soils; horizontal lettering indicates podzolic and lateritic soils (forested soils) and tundra (shrub vegetation); printed at an angle are names of soils bearing the influence of both grass and trees)

*Source:* Thompson, L.M. & Troeh, F.R. *Soils and Soil Fertility.*
Reproduced with kind permission of F.R. Troeh.

**TABLE 44  Classification of Agro-Climatic Regions according to Reference
Evapotranspiration (ET$_0$ mm (day)**

| Regions | Mean daily temperature °C | | |
| --- | --- | --- | --- |
| | <10 | 20 | >30 |
| | (cool) | (moderate) | (warm) |
| TROPICS | | | |
| humid | 3 – 4 | 4 – 5 | 5 – 6 |
| subhumid | 3 – 5 | 5 – 6 | 7 – 8 |
| semi-arid | 4 – 5 | 6 – 7 | 8 – 9 |
| arid | 4 – 5 | 7 – 8 | 9 – 10 |
| SUBTROPICS | | | |
| Summer rainfall: | | | |
| humid | 3 – 4 | 4 – 5 | 5 – 6 |
| subhumid | 3 – 5 | 5 – 6 | 6 – 7 |
| semi-arid | 4 – 5 | 6 – 7 | 7 – 8 |
| arid | 4 – 5 | 7 – 8 | 10 – 11 |
| Winter rainfall: | | | |
| humid – subhumid | 2 – 3 | 4 – 5 | 5 – 6 |
| semi-arid | 3 – 4 | 5 – 6 | 7 – 8 |
| arid | 3 – 4 | 6 – 7 | 10 – 11 |
| TEMPERATE | | | |
| humid – subhumid | 2 – 3 | 3 – 4 | 5 – 7 |
| semi-arid – arid | 3 – 4 | 5 – 6 | 8 – 9 |

*Note* 1. ET$_O$ represents the evapotranspiration of an extended surface of 8 – 15 cm tall, green, growing grass not short of water and completely shading the ground.

2. Maximum evapotranspiration (ET$_m$) can be calculated by multiplying ET by the appropriate crop coefficient (kc).

3. Wherever possible local data should be used to refine the ET$_O$ figures.

*Source:* FAO (1979) *Yield response to water. Irrigation and Drainage Paper No 33.*
Reproduced with kind permission of the Food and Agriculture Organization of the United Nations.

---

**TABLE 45   Climatological Nomenclature**

---

The following is a general nomenclature for climatic parameters:

**TEMPERATURE**
**General**

$$\text{T mean} = \frac{\text{T max} + \text{T min}}{2}$$

| | | |
|---|---|---|
| hot | T mean >30°C | data collected from max/min |
| cool | T mean <15°C | thermometer or thermograph records. |

**HUMIDITY**
**RH min, minimum relative humidity**

RH min is lowest humidity during daytime and is reached usually at 14.00 to 16.00 hrs. From hygrograph or wet and dry bulb thermometer. For rough estimation purposes when read at 12.00 hrs subtract 5 to 10 for humid climates and up to 30 for desert climates.

| Blaney-Criddle | | Crop coeff. | |
|---|---|---|---|
| low | <20% | dry | <20% |
| medium | 20 – 50% | humid | >70% |
| high | >50% | | |

**RH mean, mean relative humidity**

RH mean is average of maximum and minimum relative humidity or RH mean=(RH max + RH min)/2. Whereas for most climates RH min will vary strongly, RH max equals 90 to 100% for humid climates, equals 80 to 100% for semi-arid and arid climates where T min is 20 – 25°C lower than T max. In arid areas RH max may be 25 – 40% when T min is 15°C lower than T max.

| Radiation method | | Pan method | |
|---|---|---|---|
| low | <40% | low | <40% |
| medium-low | 40 – 55% | medium | 40 – 70% |
| medium-high | 55 – 70% | high | >70% |
| high | >70% | | |

**WIND**
**General**

For rough estimation purposes sum of several wind-speed observations divided by number of readings in m/sec or multiplied by 86.4 to give wind run in km/day.
With 2 m/sec: wind is felt on face and leaves start to rustle
With 5 m/sec: twigs move, paper blows away, flags fly
With 8 m/sec: dust rises, small branches move
With >8 m/sec: small trees start to move, waves form on inland water, etc.

| light | <2 m/sec | <176 km/day |
|---|---|---|
| moderate | 2 – 5 m/sec | 175 – 425 km/day |
| strong | 5 – 8 m/sec | 425 – 700 km/day |
| v.strong | >8 m/sec | >700 km/day |

**TABLE 45 *(continued)*  Climatological Nomenclature**

The following is a general nomenclature for climatic parameters:

| **RADIATION** Blaney-Criddle | | | Ratio between daily actual (n) and daily maximum possible(N) sunshine duration. | |
|---|---|---|---|---|
| **sunshine n/N** | | | n/N > 0.8: | near bright sunshine all day |
| low | <0.6 | | n/N 0.6 – 0.8: | some 40% of daytime hours full cloudiness or partially clouded for 70% of daytime hours. |
| medium | 0.6 – 0.8 | | | |
| high | >0.8 | | | |
| or | | | Mean of several cloudiness observations per day on percentage or segments of sky covered by clouds. | |
| | | | 4 oktas : | 50% of the sky covered all daytime hours by clouds or half of daytime hours the sky is fully clouded |
| **cloudiness** | tenth | oktas | | |
| low | >5 | >4 | 1.5 oktas : | less than 20% of the sky covered all daytime ours by clouds or each day the sky has a full cloud cover for some 2 hours. |
| medium | 2 – 5 | 1.5 – 4 | | |
| high | <2 | <1.5 | | |

*Source:* FAO (1988) *Crop Water Requirements. Irrigation and Drainage Paper No.24.*
Reproduced with kind permission of the Food and Agriculture Organization of the United Nations.

**TABLE 46    Mean Daily Duration of Maximum Possible Sunshine Hours (N) for Different Months and Latitudes**

| Northern Lats | Jan | Feb | Mar | April | May | June |
|---|---|---|---|---|---|---|
| Southern Lats | July | Aug | Sept | Oct | Nov | Dec |
| 50 | 8.5 | 10.1 | 11.8 | 13.8 | 15.4 | 16.3 |
| 48 | 8.8 | 10.2 | 11.8 | 13.6 | 15.2 | 16.0 |
| 46 | 9.1 | 10.4 | 11.9 | 13.5 | 14.9 | 15.7 |
| 44 | 9.3 | 10.5 | 11.9 | 13.4 | 14.7 | 15.4 |
| 42 | 9.4 | 10.6 | 11.9 | 13.4 | 14.6 | 15.2 |
| 40 | 9.6 | 10.7 | 11.9 | 13.3 | 14.4 | 15.0 |
| 35 | 10.1 | 11.0 | 11.9 | 13.1 | 14.0 | 14.5 |
| 30 | 10.4 | 11.1 | 12.0 | 12.9 | 13.6 | 14.0 |
| 25 | 10.7 | 11.3 | 12.0 | 12.7 | 13.3 | 13.7 |
| 20 | 11.0 | 11.5 | 12.0 | 12.6 | 13.1 | 13.3 |
| 15 | 11.3 | 11.6 | 12.0 | 12.5 | 12.8 | 13.0 |
| 10 | 11.6 | 11.8 | 12.0 | 12.3 | 12.6 | 12.7 |
| 5 | 11.8 | 11.9 | 12.0 | 12.2 | 12.3 | 12.4 |
| 0 | 12.1 | 12.1 | 12.1 | 12.1 | 12.1 | 12.1 |

*Source:* FAO (1988) *Crop Water Requirements. Irrigation and Drainage Paper No. 24.*

**TABLE 47    Maximum Active Incoming Shortwave Radiation (cal/cm$^2$/day) * and Gross Dry Matter Production on clear and overcast days (kg/ha/day) for a Standard Crop (after De Wit, 1965) according to latitude**

| North | | Jan | Feb | Mar | Apr | May |
|---|---|---|---|---|---|---|
| South | | July | Aug | Sept | Oct | Nov |
| 0° | Radiation | 343 | 360 | 369 | 364 | 349 |
| | DM clear | 413 | 424 | 429 | 426 | 417 |
| | DM o'cast | 219 | 226 | 230 | 228 | 221 |
| 10° | Radiation | 299 | 332 | 359 | 375 | 377 |
| | DM clear | 376 | 401 | 422 | 437 | 440 |
| | DM o'cast | 197 | 212 | 225 | 234 | 236 |
| 20° | Radiation | 249 | 293 | 337 | 375 | 394 |
| | DM clear | 334 | 371 | 407 | 439 | 460 |
| | DM o'cast | 170 | 193 | 215 | 235 | 246 |
| 30° | Radiation | 191 | 245 | 303 | 363 | 400 |
| | DM clear | 281 | 333 | 385 | 437 | 471 |
| | DM o'cast | 137 | 168 | 200 | 232 | 251 |
| 40° | Radiation | 131 | 190 | 260 | 339 | 396 |
| | DM clear | 219 | 283 | 353 | 427 | 480 |
| | DM o'cast | 99 | 137 | 178 | 223 | 253 |

* That fraction which is useful for photosynthesis

*Source:* FAO (1979) *Yield Response to Water. Irrigation and Drainage Paper No. 33.*

**TABLE 46** *(continued)*   **Mean Daily Duration of Maximum Possible Sunshine Hours (N) for Different Months and Latitudes**

| Northern Lats | July | Aug | Sept | Oct | Nov | Dec |
|---|---|---|---|---|---|---|
| Southern Lats | Jan | Feb | Mar | Apr | May | June |
| 50 | 15.9 | 14.5 | 12.7 | 10.8 | 9.1 | 8.1 |
| 48 | 15.6 | 14.3 | 12.6 | 10.9 | 9.3 | 8.3 |
| 46 | 15.4 | 14.2 | 12.6 | 10.9 | 9.5 | 8.7 |
| 44 | 15.2 | 14.0 | 12.6 | 11.0 | 9.7 | 8.9 |
| 42 | 14.9 | 13.9 | 12.6 | 11.1 | 9.8 | 9.1 |
| 40 | 14.7 | 13.7 | 12.5 | 11.2 | 10.0 | 9.3 |
| 35 | 14.3 | 13.5 | 12.4 | 11.3 | 10.3 | 9.8 |
| 30 | 13.9 | 13.2 | 12.4 | 11.5 | 10.6 | 10.2 |
| 25 | 13.5 | 13.0 | 12.3 | 11.6 | 10.9 | 10.6 |
| 20 | 13.2 | 12.8 | 12.3 | 11.7 | 11.2 | 10.9 |
| 15 | 12.9 | 12.6 | 12.2 | 11.8 | 11.4 | 11.2 |
| 10 | 12.6 | 12.4 | 12.1 | 11.8 | 11.6 | 11.5 |
| 5 | 12.3 | 12.3 | 12.1 | 12.0 | 11.9 | 11.8 |
| 0 | 12.1 | 12.1 | 12.1 | 12.1 | 12.1 | 12.1 |

Reproduced with kind permission of the Food and Agriculture Organization of the United Nations.

**TABLE 47** *(continued)*   **Maximum Active Incoming Shortwave Radiation (cal/cm$^2$/day) * and Gross Dry Matter Production on clear and overcast days (kg/ha/day) for a Standard Crop (after De Wit, 1965) according to latitude**

| June Dec | July Jan | Aug Feb | Sept Mar | Oct Apr | Nov May | Dec June |
|---|---|---|---|---|---|---|
| 337 | 343 | 357 | 368 | 365 | 349 | 337 |
| 410 | 413 | 422 | 429 | 427 | 418 | 410 |
| 216 | 218 | 225 | 230 | 228 | 222 | 216 |
| 374 | 375 | 377 | 369 | 345 | 311 | 291 |
| 440 | 440 | 439 | 431 | 411 | 385 | 370 |
| 235 | 236 | 235 | 230 | 218 | 203 | 193 |
| 400 | 399 | 386 | 357 | 313 | 264 | 238 |
| 468 | 465 | 451 | 425 | 387 | 348 | 325 |
| 250 | 249 | 242 | 226 | 203 | 178 | 164 |
| 417 | 411 | 384 | 333 | 270 | 210 | 179 |
| 489 | 483 | 456 | 412 | 356 | 299 | 269 |
| 261 | 258 | 243 | 216 | 182 | 148 | 130 |
| 422 | 413 | 369 | 298 | 220 | 151 | 118 |
| 506 | 497 | 455 | 390 | 314 | 241 | 204 |
| 268 | 263 | 239 | 200 | 155 | 112 | 91 |

* That fraction which is useful for photosynthesis

Reproduced with kind permission of the Food and Agriculture Organization of the United Nations.

**TABLE 48   Incoming radiation (at the top of the atmosphere) expressed in mm/day evaporation of water**

### Northern hemisphere

| Month | 90° | 80° | 70° | 60° | 50° | 40° | 30° | 20° | 10° | 0° |
|-------|-----|-----|-----|-----|-----|-----|-----|-----|-----|-----|
| Jan | — | — | — | 1.3 | 3.6 | 6.0 | 8.5 | 10.8 | 12.8 | 14.5 |
| Feb | — | — | 1.1 | 3.5 | 5.9 | 8.3 | 10.5 | 12.3 | 13.9 | 15.0 |
| Mar | — | 1.8 | 4.3 | 6.8 | 9.1 | 11.0 | 12.7 | 13.9 | 14.8 | 15.2 |
| April | 7.9 | 7.8 | 9.1 | 11.1 | 12.7 | 13.9 | 14.8 | 15.2 | 15.2 | 14.7 |
| May | 14.9 | 14.6 | 13.6 | 14.6 | 15.4 | 15.9 | 16.0 | 15.7 | 15.0 | 13.9 |
| June | 18.1 | 17.8 | 17.0 | 16.5 | 16.7 | 16.7 | 16.5 | 15.8 | 14.8 | 13.4 |
| July | 16.8 | 16.5 | 15.8 | 15.7 | 16.1 | 16.3 | 16.2 | 15.7 | 14.8 | 13.5 |
| Aug | 11.2 | 10.6 | 11.4 | 12.7 | 13.9 | 14.8 | 15.3 | 15.3 | 15.0 | 14.2 |
| Sept | 2.6 | 4.0 | 6.8 | 8.5 | 10.5 | 12.2 | 13.5 | 14.4 | 14.9 | 14.9 |
| Oct | — | 0.2 | 2.4 | 4.7 | 7.1 | 9.3 | 11.3 | 12.9 | 14.1 | 15.0 |
| Nov | — | — | 0.1 | 1.9 | 4.3 | 6.7 | 9.1 | 11.2 | 13.1 | 14.6 |
| Dec | — | — | — | 0.9 | 3.0 | 5.5 | 7.9 | 10.3 | 12.4 | 14.3 |

### Southern hemisphere

| Month | 90° | 80° | 70° | 60° | 50° | 40° | 30° | 20° | 10° | 0° |
|-------|-----|-----|-----|-----|-----|-----|-----|-----|-----|-----|
| Jan | 17.6 | 17.3 | 16.5 | 16.6 | 17.1 | 17.3 | 17.3 | 16.8 | 15.8 | |
| Feb | 10.7 | 10.5 | 11.2 | 12.7 | 14.1 | 15.2 | 15.8 | 16.0 | 15.7 | |
| Mar | 1.9 | 3.6 | 6.1 | 8.4 | 10.5 | 12.2 | 13.6 | 14.6 | 15.1 | |
| April | — | — | 1.9 | 4.3 | 6.6 | 8.8 | 10.8 | 12.5 | 13.8 | |
| May | — | — | 0.1 | 1.9 | 4.1 | 6.4 | 8.7 | 10.7 | 12.4 | |
| June | — | — | — | 0.8 | 2.8 | 5.1 | 7.4 | 9.6 | 11.6 | |
| July | — | — | — | 1.2 | 3.3 | 5.6 | 7.8 | 10.0 | 11.9 | |
| Aug | — | — | 0.8 | 2.9 | 5.2 | 7.5 | 9.6 | 11.5 | 13.0 | |
| Sept | — | 1.3 | 3.8 | 6.2 | 8.5 | 10.5 | 12.1 | 13.5 | 14.4 | |
| Oct | 7.0 | 7.1 | 8.8 | 10.7 | 12.5 | 13.8 | 14.8 | 15.3 | 15.3 | |
| Nov | 15.3 | 15.0 | 14.5 | 15.2 | 16.0 | 16.5 | 16.7 | 16.4 | 15.7 | |
| Dec | 19.3 | 18.9 | 18.1 | 17.5 | 17.8 | 17.8 | 17.6 | 16.9 | 15.8 | |

*Note* 1.   1 cal/cm$^2$/min $\simeq$ 1mm water evaporated/hr.

2.   The overall energy equation is:

$$R_1 = rR_1 + R_B + H + E$$

where

$R_1$  =  Radiation reaching earths surface. cal/cm$^2$/min

$r$  =  reflectance of the surface

$R_B$  =  long wave back radiation cal/cm$^2$/min

$H$  =  Increase in sensible heat of the atmosphere cal/cm$^2$/min

$E$  =  Energy available for evaporation from the surface cal/cm$^2$/min

*Source:* Withers, B. & Vipond, S. (1974) *Irrigation Design and Practice.*
Reproduced with kind permission of B. T. Batsford Ltd.

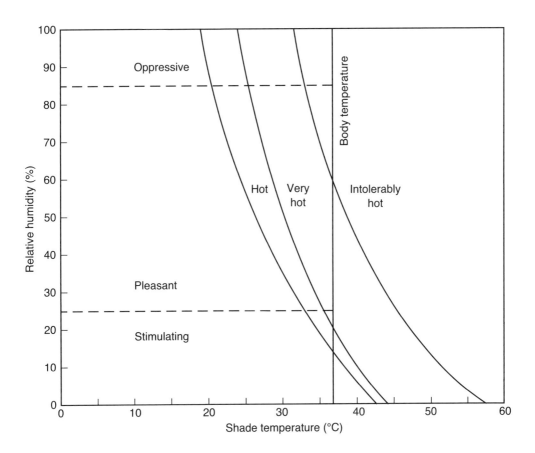

**Fig. 15** A generalized and indicative graphical representation of human climatic comfort according to temperature and humidity

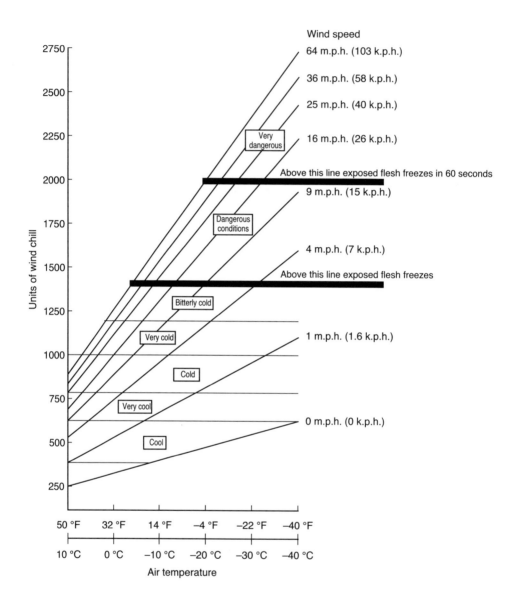

**Fig. 16** Wind chill index for human comfort

*Source:* Penna, E.A. & Smith, C.G. (1984) *The World Weather Guide.* © Hutchinson and Co. Reproduced with kind permission of Sheil Land Associates Ltd.

# Fertilizers and Manures, etc.

# Fertilizers

**TABLE 49   Typical Artificial Fertilizer Analyses**

*NB* Analyses can vary according to regional source of raw materials and manufacturing processes.

| Chemical | N | $P_2O_5$ | $K_2O$ | Mg | Na | S | Ca |
|---|---|---|---|---|---|---|---|
| | | | | | | | |
| Aluminium Calcium phosphates | | 30 | | | | | √ |
| Aluminium Sulphate | | | | | | 14.4 | |
| Ammoniated Phosphates: | | | | | | | |
|     Conc.Super P | 5 | 45 | | 0.7 | | | 12 - 14 |
|     Ord. Super P | 4 | 13 | | 0.3 | | 10 | 16 |
|     Triple Super P | 9 | 41 - 50 | 0.4 | | | 1.4 | 14 |
| Ammonium Phosphate Nitrate | 30 | 9 | | | | | |
| Ammonium Phosphate Sulphate | 13 - 16 | 20 - 48 | | | | 2 - 15 | |
| Ammonium Polyphosphate | 12 - 15 | 55 - 59 | | | | | |
| Ammonium Chloride | 26 | | | | | | |
| Ammonium Nitrate | 35 | | | | | | |
| Ammonium Nitrate Limestone | 20 - 26 | | | 4.4 | | 0.4 | 8.2 |
| Ammonium Nitrate Sulphate | 26 | | | | | 12.1 | |
| Ammonium Phosphate Sulphate | 16 | 20 | | | | | |
| Ammonium Sulphate | 21 | | | | | 24.2 | 0.3 |
| Ammonium Sulphur Solution | 74 | | | | | 10 | |
| Anhydrous Ammonia | 82 | | | | | | |
| Aqueous Ammonia | 21 - 29 | | | | | | |
| Basic Slag | | 16 | | 3 | | 3 | 29 - 32 |
| Blast Furnace Slag | | | | I - 6 | | | √ |
| Bone Meal | 2 - 4 | 2I - 27 | 0.2 | 0.4 | 5.5 | 0.1 | 20 - 25 |
| Burnt Magnesium Lime | | | | 16 | | | √ |
| Calcium Carbonate | | | | | | | 38 |
| Calcium Chloride | | | | | | | 18 |
| Calcined Magnesite | | | | 52 | | | √ |
| Calcium Cyanamide | 21 - 22 | | | | | | 40 |
| Calcium Metaphosphate | | 61 | | | | | 19 |
| Calcium Nitrate | 15.5 | | | | | | 19 |
| Chilean Potash Nitrate | 15 | | 10 | | 20 | | |
| Diammonium Phosphate | 18 - 21 | 46 - 54 | | | | | |
| Dicalcium Phosphate | | 40 - 50 | | | | | 19 - 29 |
| Dolomitic Limestones | | | | varies | | | √ |
| Fused Tri Calcium Phosphate | | 27.5 | | 8 | 8 | 8 | 20 |
| Gas Liquor | 1 - 4 | | | | | | |
| Ground Burnt Magnesian lime | | | | 6 - 20 | | | 20 |
| Ground Rock Phosphate | | 25 - 39 | | | 5 | | 17 - 34 |
| Gypsum | | | | | | 18.6 | 23 |
| Iron Sulphate | | | | | | 11.5 | |
| Kainit | | | 10 - 30 | | | 12.9 | |
| Kieserite | | | | 16 | | | |
| Lime Sulphur Solution | | | | | | 24 | √ |
| Magnesian Limestone | | | | 3 - 12 | | | 20 |
| Magnesite | | | | 27 | | | √ |

**TABLE 49** *(continued)*  **Typical Artificial Fertilizer Analyses**

*NB* Analyses can vary according to regional source of raw materials and manufacturing processes.

| Chemical | N | $P_2O_5$ | $K_2O$ | Mg | Na | S | Ca |
|---|---|---|---|---|---|---|---|
| | | | Approximate Composition % | | | | |
| Magnesium Kainit | | | 6 - 30 | 4 - 14 | 4 - 13 | | √ |
| Magnesium Ammonium Phosphate | 8 | 40 | | 14 | | | √ |
| Magnesium Chloride | | | | 8 - 9 | | | 2 |
| Magnesium Oxide | | | | 42 | | | 3 |
| Magnesium Silicate Glass | | 22 | | 8 - 12 | | | 20 |
| Magnesium Sulphate | | | | 9.7 | | 13 | |
| Monoammonium Phosphate | 11 | 48 - 51 | | | | | 1.5 |
| Nitric Phosphates | 14 - 20 | 13 - 21 | | | | | 6 - 7 |
| Nitrogen solutions | 21 - 29 | (ammonium nitrate + urea + ammonia) | | | | | |
| Phosphoric Acid | | 22 - 26 | | | | | 0.2 |
| Phosphoric Acid -Super | | 73 - 77 | | | | | |
| Potassium Bicarbonate | | | 47 | | | | |
| Potassium Carbonate | | | 67 | | | | |
| Potassium Chloride | | | 60 | 0.1 | | | 0.1 |
| Potassium Metaphosphate | | 58 | 37 | | | | |
| Potassium Nitrate | 13 | | 44 | 0.3 | | | 0.3 |
| Potassium Phosphate | | 41 - 50 | 34 - 54 | | | | |
| Potassium Polyphosphate | | 59 | 39 | | | | |
| Potassium Sulphate | | | 48 - 50 | | | 18 | |
| Rock Phosphate | | 25 - 30 | | | | | 33 - 36 |
| Sodium Chloride | | | | | 37 - 39 | | |
| Sodium Nitrate | 16 | | | | 27 | | 0.1 |
| Sulphuric Acid | | | | | | 32 | |
| Sulphur Dioxide | | | | | 8 | 50 | |
| Superphosphate - ordinary | | 16 - 22 | | 0.2 | | 13 | 13 - 20 |
| Superphosphate - concentrated | 48 | | | | | 0.8 | 12 - 14 |
| Potassium Magnesium Sulphate | | 0 - 5 | 26 | 5 - 11 | | 11 - 18 | 0.1 - 0.7 |
| Triple Superphosphate | | 41 - 50 | 0.3 | 0.3 | | 1.4 | 9 - 14 |
| Urea | 46 | | | | | | |
| Urea Ammonium Phosphate | 23 - 54 | 16 - 34 | | | | | |
| Urea Ammonium Polyphosphate | 22 - 30 | 29 - 43 | | | | | |
| Urea Formaldehyde | 38 | | | | | | |
| Urea Gypsum | 17.3 | | | | | 14.8 | √ |
| Urea Phosphate | 18 | 45 | | | | | |
| Urea Sulphur | 40 | | | | | 10 | |
| Wood Ash | | | 2 - 5 | | | | |

Key: √ = present

*Sources:* Tisdale, S.L., Nelson, W.L. & Beaton, J.D. (1985) *Soil and Soil Fertility.*
Vickar, M.H., Bridger, G.L. & Nelson, L.B. (1963) *Fertiliser Technology and Usage.* Soil. Sci. Socy. America.
Cooke, G.W. (1982) *Fertilizing for Maximum Yield.*
FAO (1984) *Fertilizer and Plant Nutrition Guide. Fertilizer and Plant Nutrition Bulletin No 9.*
Brady, N.C. (1984) *The Nature and Properties of Soils.*
Ministry of Agriculture, Fisheries and Food (UK) (1964) *Fertilizers for the Farm. Bulletin No 195.*
National Research Council (1988) *Nutrient Requirements of Swine.*

**TABLE 50   Equivalent Acidity or Basicity of Common Fertilizers**

| Fertilizer | Equivalent Acid or Base in kg CaCO$_3$ * | |
|---|---|---|
| | ACID | BASE |
| Ammonium Nitrate | 62 | |
| Ammonium Nitrate Sulphate | 68 | |
| Monoammonium Phosphate | 58 | |
| Ammonium Phosphate - Sulphate (13N.39P$_2$O$_5$) | 69 | |
| Ammonium Phosphate - Sulphate (16N.20 P$_2$O$_5$) | 88 | |
| Ammonium Phosphate - Nitrate | 75 | |
| Diammonium Phosphate | 70 | |
| Ammonium Sulphate | 110 | |
| Anhydrous Ammonia | 147 | |
| Aqua Ammonia | 36 | |
| Calcious Ammonium Nitrate Solution | 9 | |
| Calcium Nitrate | | 20 |
| Calcium Cyanamide | | 63 |
| Sodium Nitrate | | 29 |
| Urea | 7l | |
| Urea Formaldehyde | 60 | |
| Urea Ammonium Nitrate Solution | 57 | |
| Single Superphosphate | | NEUTRAL |
| Triple Superphosphate | | NEUTRAL |
| Phosphoric Acid | 110 | |
| Superphosphoric Acid | 160 | |
| Potassium Chloride | | NEUTRAL |
| Potassium Nitrate | 23 | |
| Potassium Sulphate | | NEUTRAL |
| Sulphate of Potash Magnesia | | NEUTRAL |

* Equivalent per 100 kg of each fertilizer

*Source:* FAO (1985) *Water Quality for Agriculture. Irrigation and Drainage Paper No. 29.*
Reproduced with kind permission of the Food and Agriculture Organization
of the United Nations.

# Organic Manures

---

**TABLE 51   Typical Analyses of Organic Manures**

---

*NB* Organic manures are much more variable in analysis than inorganic fertilizers and the following table is therefore very indicative. It is also noted that the availability of nutrients in organic manures is usually much less immediate than in inorganic fertilizers.

| Manure | N | $P_2O_5$ | $K_2O$ | |
|---|---|---|---|---|
| | | % in DM or according to Moisture Content | | |
| Ash, Cottonseed hull | | 5.5 | 27 | |
| Ash, Sunflower stalk | | 2.5 | 36 | |
| Ash, Wood | | 2 | 2 - 5 | |
| Bark | 0.3 | 0.09 | 0.7 | |
| Blood - Dried | 12 - 14 | 2 | 1 | |
| Bone Meal | 2 - 5 | 22 | | |
| Bracken Ash | | | 40 | |
| Brewery sludge, digested | 1.5 | 0.9 | 0.3 | |
| Castor Meal | 5 | 3.4 | | |
| Cattle Manures: | | | | |
| Farm Yard Manure | 0.6 | 0.2 | 0.6 | @ 76% MC |
| Fresh Faeces | 0.4 | 0.2 | 0.1 | @ 85% MC |
| Faeces + Urine | 1.3 | 0.2 | 0.8 | @ 89%MC |
| Cereal Straw | 0.6 | 0.2 | 1.4 | |
| Cocoa Meal | 4 | 2 | 2.5 | |
| Cottonseed Meal | 7 | 3 | 2 | |
| Dissolved Bone | 2 - 3 | 14 - 16 | | |
| Duck manure with litter | 10 | 13 | 5 | @ 61% MC |
| Groundnut Meal | 7 | 1.5 | 1.5 | |
| Farm Yard Manure (typical Values) | 0.6 | 0.3 | 0.6 | |
| Fish Meal | 5 - 10 | 20 - 36 | | |
| Fur, Hair, Skins, Feathers | 8 - 12 | | | |
| Goat Manure | 2.8 | 1.4 | 2.9 | @ 40% MC |
| Goose Manure with litter | 10 | 5 | 5 | @ 67% MC |
| Guano | 10 - 15 | 9 - 13 | 2 | |
| Hoof and Horn | 13 | | | |
| Horse Manure | 0.7 | 0.2 | 0.7 | @ 60% MC |
| Leather Waste | 12 | | | |
| Lignite, ground | 1 | | | |
| Malt Culms and Dust | 3 - 4 | | | |
| Meat and Bone Meal | 5 - 10 | 16 | | |
| Meat Meal | 10 | | | |
| Mushroom Compost, spent | 0.8 | 0.7 | 0.9 | |
| Peat | 0.1 - 2.7 | | | |
| Peanut Meal | 7.2 | 1.4 | 1.2 | |
| Peanut Hull Meal | 1.2 | 0.4 | 0.8 | |
| Pig Slurry | 0.2 - 4 | 0.2 | 0.2 | @97% MC |

*(continued over)*

**TABLE 51** *(continued)*  **Typical Analyses of Organic Manures**

*NB* Organic manures are much more variable in analysis than inorganic fertilizers and the following table is therefore very indicative. It is also noted that the availability of nutrients in organic manures is usually much less immediate than in inorganic fertilizers.

| Manure | N | $P_2O_5$ | $K_2O$ | % in DM or according to Moisture Content |
|---|---|---|---|---|
| Poultry Manures: | | | | |
|     Deep Litter | 1.7 | 2.0 | 1.3 | @ 32%MC |
|     Broiler Litter | 2.3 | 2.0 | 1.3 | @ 32% MC |
|     Battery | 1.5 | 1.1 | 0.7 | @ 66 % MC |
|     Off Dropping Boards | 1.6 | 0.9 | 0.5 | @ 54% MC |
|     Machine dried | 5.0 | 3.0 | 2.0 | @ 14% MC |
| Precipitated Bone Phosphate | | 30 - 35 | | |
| Rabbit Manure | 2 | 1.4 | 1.2 | @ 50% MC |
| Rape Cake | 5 | | | |
| Refuse - Municipal | 0.5 | 0.5 | 0.4 | @ 35% MC |
| Sawdust | 0.2 | 0.02 | 0.15 | |
| Seaweed, dried | 1.3 | 0.2 | 1.8 | |
| Seaweed Ash | | | 12-16 | |
| Sewage Sludge: | | | | |
|     Liquid, undigested | 0.2 | 0.15 | | @ 95% MC |
|     Liquid, digested | 0.2 | 0.15 | | @ 96% MC |
|     Cake, undigested | 0.8 | 0.7 | | @ 75% MC |
|     Cake, Digested | 0.8 | 0.9 | | @ 75% MC |
| Sheep Manure | 1.4 | 0.5 | 1.2 | @ 65% MC |
| Shoddy | 3 - 15 | | | |
| Silage Effluent | 0.2 | 0.1 | 0.4 | |
| Soyabean Meal | 7 | 1.5 | 2. | |
| Spent Bone Charcoal | | 35 - 40 | | |
| Soot | 1 - 6 | | | |
| Steamed Bone Flour | 1 | 28 | | |
| Straw | 0.5 | 0.1 | 0.8 | |
| Tobacco stems | 1.5 | 0.4 | 5 | |
| Turkey Manure | 1.2 | 1.4 | 0.8 | @ 55% MC |
| Woodchips | 0.2 | 0.02 | 0.1 | |

*Note:*  Where straw, hay, green manures and crop residues are used as manurial materials, indicative N, $P_2O_5$, and $K_2O$ values can in some cases  be derived from the feedstuff value tables in this text.

*Sources:* Landon J.R., ed. (1984) *Tropical Soil Manual.* ADAS (UK) (1987) *Use of Sewage Sludge on Agricultural Land.* Brady, N.C. (1984) *Nature and Properties of Soils.* ILACO (1981) *Agricultural Compendium.* Robinson, D.H., ed. (1962) *Fream's Elements of Agriculture.* Cooke, G.W. (1982) *Fertilizing for Maximum Yield.* MAFF (UK) (1964) *Fertilizers for the Farm.* Bull 195. Welsh Development Agency (1987) *Working with Nature. Low Cost Land Reclamation Techniques.*

**TABLE 52   Indicative Concentration of N, P and K in Solid and Liquid Portions of Animal Manures**

| Animal | Fraction | %N | %P | %K |
|--------|----------|----|----|----|
| **Cattle** | Urine | 52 | 100 | 85 |
| | Faeces | 46 | Tr | 15 |
| **Pig** | Urine | 33 | 57 | 43 |
| | Faeces | 67 | 43 | 57 |
| **Sheep** | Urine | 37 | 6 | 70 |
| | Faeces | 63 | 94 | 30 |

*Sources:* Adapted from Brady, N.C. (1984) *Nature and Properties of Soils,* and Landon, J.R., ed. (1984) *Tropical Soil Manual.*

# Trace Elements

**TABLE 53   Indicative Range of Trace Element Levels in Animal Manures**

| Element | kg / t |
|---------|--------|
| Boron | 0.02 - 0.12 |
| Calcium | 2.4 - 74 |
| Copper | 0.01 - 0.03 |
| Iron | 0.08 - 0.93 |
| Magnesium | 1.6 - 5.8 |
| Manganese | 0.01 - 0.18 |
| Molybdenum | 0.0001 - 0.011 |

*Sources:* Adapted from Brady, N.C. (1984) *Nature and Properties of Soils,* and Landon, J.R., ed. (1984) *Tropical Soil Manual.*

**TABLE 54   Indicative Rates of Application and Source Materials for correcting Trace Element Deficiencies**

| BORON | Material | %B |
|---|---|---|
| Borax | $Na_2 B_4 O_7 . 10 H_2O$ | 11 |
| Boric Acid | $H_3 BO_3$ | 17 |
| Boron Frits | - | 2 - 6 |
| Colemanite | $Ca_2 B_6 O_{11} . 5H_2O$ | 10 |
| Sodium Tetraborate | $Na_2 B_4 O_7 . 5H_2O$ | 14 |
| | $Na_2 B_4 O_7$ | 20 |
| Solubor | $Na_2 B_4 O_7 . 5H2O+$ | 20 |
| | $Na_2 B_{10} O_{16} . 10H_2O$ | |

**Soil applications**    1.2 - 3.2 kg B/ha high demand crops
0.6 - 1.2 kg B/ha  low demand crops

**Foliar application**    0.2 - 0.5% Solubor in water

| COBALT | Material | %Co |
|---|---|---|
| Cobalt Sulphate | $CoSO_4 . 7H_2O$ | 21 |

Note: Cobalt is usually required for animal nutrition purposes rather than crop plant nutrition purposes.

**Application rate**    0.06 - 0.24 kg Co/ha

| COPPER | | | Soil application kg / ha Cu | |
|---|---|---|---|---|
| | Material | %Cu | B'cast | Band |
| Copper Sulphate | $CuSO_4 . 5H_2O$ | 25 | 4.5 | 3 |
| Copper Sulphate | $CuSO_4 . H_2O$ | 35 | 4.5 | 3 |
| Cuperous Oxide | $Cu_2O$ | 89 | 4.5 | 2.8 |
| Cuperic Oxide | $CuO$ | 75 | 4.5 | 2.8 |
| Copper chelates | $Na_2$ - Cu EDTA | 13 | 1.6 | 0.5 |
| | Na - Cu EDTA | 9 | 1.6 | 0.5 |

**Foliar applications**   100 g Cu/ha as $CuSO_4$ or 30 g Cu/ha as Cu - EDTA
$CuSO_4$ solution needs neutralizing with lime to avoid scorching

| IRON | Material | %Fe |
|---|---|---|
| Ferrous ammonium sulphate | $( NH_4 )_2 SO_4 . Fe SO_4 . 6H_2O$ | 14 |
| Ferrous carbonate | $Fe CO_3$ | 42 |
| Ferric sulphate | $Fe_2( SO_4 )_3$ | 20 |
| Ferrous sulphate | $FeSO_4 . 7H_2O$ | 20.5 |
| Chelates | Fe-DTPA | 10 |
| | Fe-EDTA | 9 - 12 |
| | Fe-EDDHA | 6 |
| | Fe-HEDTA | 5 - 9 |
| Organic complexes | Iron Frits | 40 ± |
| | Lignin sulphonate | 6 |
| | Methoxy phenyl propane complex | 5 |
| | Poly flavonoid | 6 - 9.6 |

**TABLE 54 *(continued)*  Indicative Rates of Application and source materials for correcting Trace Element Deficiencies**

**IRON**

| | |
|---|---|
| **Soil applications** | Inorganic compounds generally not practicable, high rates required and mostly immobilized<br>1 kg Fe/ha as chelate |
| **Foliar applications** | 1 - 3% Iron Sulphate in water or chelate spray<br>(or 0.3% Fe as $FeSO_4$ with 0.2% silicone based wetting agent can be more effective)<br>2 - 4 kg Fe/ha via irrigation water<br>5 g/tree/yr via trickle irrigation |

| **MANGANESE** | **Material** | **%Mn** |
|---|---|---|
| Manganese carbonate | $Mn\ CO_3$ | 31 |
| Manganese chelate | Mn-EDTA | 12 |
| Manganese chloride | $MnCl_2$ | 17 |
| Manganese frits | – | 10 - 25 |
| Manganese methoxy phenyl propane | Mn MPP | 10 - 12 |
| Manganous oxide | MnO | 41 - 68 |
| Manganese oxide | $Mn\ O_2$ | 63 |
| Manganese sulphate | $MnSO_4 . 3H_2O$ | 26 - 28 |

| | |
|---|---|
| **Soil applications** | 20 - 130 kg Mn /ha Inorganic sources, broadcast<br>6 -11  kg Mn/ha Inorganic sources, banded |
| **Foliar applications** | 0.1 -  0.5 kg Mn/ha Mn EDTA<br>0.5 -  2 kg Mn/ha Inorganic Mn |

| **MOLYBDENUM** | **Material** | **%M0** |
|---|---|---|
| Ammonium Molybdate | $(NH_4)_6\ Mo_7\ O_{24} . 4H_2O$ | 54 |
| Molybdenite | $MoS_2$ | 60 |
| Molybdenum frits | – | 2 - 3 |
| Molybdenum trioxide | $Mo\ O_3$ | 66 |
| Sodium Molybdate | $NaMoO_4 . 2H_2O$ | 39 |

| | |
|---|---|
| **Soil application** | 70 - 400 g  Mo/ha |
| **Foliar application** | 0.1 - 0.3% solution of soluble Mo salt |
| **Seed treatment** | 50 - 100 g/ha Mo in liquid or slurry form and can treat seedbeds to produce Mo 'enriched' seedlings for transplanting to Mo deficient areas |

*(continued over)*

---

**TABLE 54 *(continued)*   Indicative Rates of Application and Source Materials for correcting Trace Element Deficiencies**

---

| ZINC | Material | %Zn |
|---|---|---|
| Chelates | $Na_2$-Zn EDTA | 14 |
| | Na-Zn HEDTA | 8 |
| | Na-Zn NTA | 13 |
| Natural | Zn-lignin sulphonate | 5 |
| | Zn-polyflavonoid | 10 |
| Sphalerite | ZnS | 60 |
| Zinc Ammonium phosphate | $Zn(NH_4)PO_4$ | 37 |
| Zinc Carbonate | $Zn\ CO_3$ | 56 |
| Zinc Chloride | $Zn\ Cl_2$ | 45 - 62 |
| Zinc Dust | – | 99 |
| Zinc Frits | – | 4 ± |
| Zinc Oxide | Zn O | 60 - 80 |
| Zinc Oxide-Sulphate | $ZnO$ - $ZnSO_4$ | 55 |
| Zinc Sulphate | $ZnSO_4$ . $H_2O$ | 36 |
| Zinc Sulphate | $ZnSO_4$ . $7H_2O$ | 23 |

**Soil applications**     5 - 20 kg Zn/ha as inorganic salts broadcast
3 - 5 kg Zn/ha as inorganic salts banded
0.5 - 1 kg Zn/ha as chelates banded
0.5 - 4 kg Zn/ha as polyflavonoid banded

**Foliar application** 15 - 250g Zn/ha as inorganic salt
Root dipping of transplants in 2 - 4% ZnO suspension

---

**Cautions**

• Over zealous applications of micronutrients are always to be avoided as excessive applications can lead to crop damage, crop toxicity and toxic residue levels in soil.

• A single soil application may last for several years; several foliar applications per season may be required.

• Micronutrient treatment should be undertaken with a soil and plant tissue monitoring programme.

It should be noted that organic farmyard manures are a good source of trace elements; however, in some circumstances, their addition to soil can depress trace element availability.

*Source:* Adapted from FAO (1983) *Micronutrients. Fertilizer and Plant Nutrition Bulletin No. 7.* Reproduced with kind permission of the Food and Agriculture Organization of the United Nations.

**TABLE 55    Indicative Levels of Micronutrients in Common Fertilizers and Organic Manures**

(*NB* These are highly variable according to specific material and source)

**FERTILIZERS** ppm (unless indicated)

| | B | Mn | Cu | Zn | Co | Mo |
|---|---|---|---|---|---|---|
| Ammonium sulphate | 6 | 6 | 2 | 0 | 0 | |
| Ammonium Phosphate | | 100 - 220 | 3 - 4 | 80 | | 2 |
| Basic Slag | 10 - 1000 | 1.9 - 5.3% | 5 - 200 | 10 - 30 | 1 - 8 | 5 - 10 |
| Calcium  ammonium nitrate | tr | 10 - 50 | tr - 18 | 8 | | |
| Potassium chloride | 14 | 8 | 3 | 3 | 1 | |
| Potassium sulphate | 4 | 6 | 4 | 2 | 0 | |
| Rock Phosphate | 15 | 13% | 6 - 10 | 25 - 140 | | 6 |
| Sodium nitrate | 0 | 8 | 3 | 1 | 0 | |
| Superphosphate | 11 | 11 | 44 | 150 - 750 | 4 | |
| Triple superphosphate | 530 | 160 - 240 | 2 - 12 | 50 - 100 | | 9 |
| Urea | 0.5 | 0.5 | 0-3.6 | 0.5 | | 0.7 |

**ORGANIC MANURES** ppm in Dry Matter

| | B | Mn | Cu | Zn | Co | Mo |
|---|---|---|---|---|---|---|
| Bonemeal | 715 | 500 | 270 | 660 | | |
| Cattle slurry | | | 57 | 580 | | |
| Compost | 15 | 40 - 60 | 300 - 600 | 3 - 13 | | 2 |
| Farmyard manure | 20 | 410 | 62 | 120 | 6 | |
| Pig slurry | | | 574 | 919 | | |
| Poultry slurry | | | 59 | 495 | | |

*Sources:* Kanwar, J.S. (1979) *Indian Farming. No. 18.*
FAO (1984) *Fertilizer and Plant Nutrition Guide. Fertilizer and Plant Nutrition Bulletin No. 9.*
Cooke, G.W. (1972) *Fertilizing for Maximum Yield. Journ. Aust. Inst. Ag. Sci.* ( Sept-Dec 1977, p. 103. )

**68**

# Lime, Gypsum, etc.

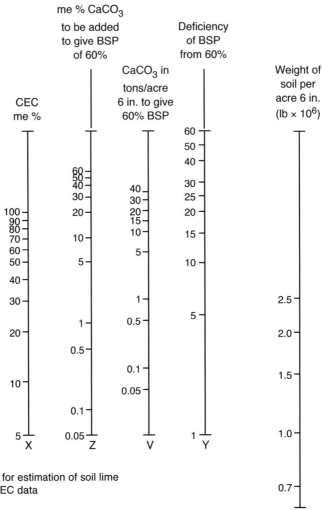

**Fig. 17** Nomogram for estimation of soil lime requirement from CEC data

**Note:** Factors for conversion to metric units are:
  1 lb = 0.45 kg          1 acre 6 in = $4.1 \times 10^3$ m² 15 cm
  1 ton = 1 016 kg                      = $4.1 \times 10^1$ ha 15 cm
      1 ton per acre 6 in = 2 490 kg per ha 15 cm

**Example:** CEC = 40 me/100g (from ammonium acetate method)
  BSP = 35%
  Deficiency of BSP from 60% = 25%
  Soil density = $2.2 \times 10^6$ lb per acre 6 in (= $2.4 \times 10^6$ kg per ha 15 cm)

Align the number 40 of scale X and the number 25 of scale Y and read the CaCO₃ to be added on scale Z (10 me/100g). Then align this 10 on scale Z with 2.2 on scale W, and read the lime requirement on scale V (5 tons per acre 6 in = 12.4 t per ha 15 cm).

*Note:* In the tropics liming is usually aimed at preventing Al toxicity by raising the pH to about 5.5. Application of 1.65 t/ha of CaCO₃ for each 1me/100g of soil of exchangeable Al is often recommended for very acid tropical soils. But some soils with high Fe and Al oxide or allophane contents may require more. A pH of 5.5 is usually quite adequate for tropical soils and crops.

*Source:* Landon, J.R., ed. (1984) *Tropical Soil Manual.*
Reproduced with kind permission of Booker Tate Ltd.

**TABLE 56   Common Liming Materials**

| Material | Material equivalent to 1 t limestone (t) |
|---|---|
| Burned Lime (CaO) | 0.64 |
| Limestone ($CaCO_3$) | 1.0 |
| Limestone - Dolomite ($CaCO_3$, $Mg\ CO_3$) | 0.86 |
| Hydrated Lime $Ca(OH)_2$ | 0.84 |
| Marl $CaCO_3$ | 1.0 |
| Shell $CaCO_3$ | 1.0 |

**TABLE 57   Amelioration of Alkali Soils by Gypsum and other Amendments**

| Exchangeable Na to be removed Meq/100 g soil | Gypsum t/ha metre |
|---|---|
| 1 | 13.9 |
| 2 | 27.8 |
| 3 | 41.8 |
| 4 | 55.7 |
| 5 | 69.6 |
| 6 | 83.5 |
| 7 | 97.4 |
| 8 | 111.3 |
| 9 | 125.1 |
| 10 | 139.2 |

*Source:* Withers, B. & Vipond, S. (1974) *Irrigation Design and Practice.*
Reproduced with kind permission of B.T. Batsford Ltd.

*NB* Alternative Amendment Equivalents

1 t (Pure) Gypsum   $CaSO_4 . 2H_2O$   $\equiv$

| | |
|---|---|
| 1.29 t | Aluminium Sulphate $(AL_2(SO_4)_3 . 18\ H_2O)$ |
| 0.85 t | Calcium Chloride $(CaCl_2 . 2H_2O)$ |
| 0.58 t | Calcium Carbonate $CaCO_3$ |
| 1.06 t | Calcium Nitrate $(Ca(NO_3) . 2H_2O)$ |
| 0.77 t | Calcium Polysulphide $CaS_5$ |
| 1.62 t | Iron Sulphate |
| 0.19 t | Sulphur |
| 0.57 t | Sulphuric Acid |

---

**TABLE 58   Simple Classification of Temperate Soil pH Requirements**

|  | pH Mineral Soils | Peats |
|---|---|---|
| Ley / Arable | 6.5 | 5.8 |
| Permanent Grass | 6 | 5.5 |

---

**TABLE 59   Indicative Maximum Amounts of Lime ($CaCO_3$) that may be Required to Raise Soil pH to 7 (temperate soils)**

| | Lime Required t / ha Soil Type | | | |
|---|---|---|---|---|
| Initial soil pH | Sand | Loam | Silt | Clay |
| 4 | 12.4 | 19.8 | 24.7 | 30.1 |
| 5 | 7.2 | 10.6 | 12.8 | 16.3 |
| 6 | 3.2 | 4.2 | 5.4 | 7.4 |

*NB* Where large amounts of lime are required to correct pH to depth, some should be ploughed under rather than all surface applied.

---

**TABLE 60   Approximate Limestone Requirements to change Soil pH (USA recommendations)**

| Required change in pH to plough depth | Soil Type – t $CaCO_3$ / ha requirement | | | | | |
|---|---|---|---|---|---|---|
| | Sand | Sandy Loam | Loam | Silt Loam | Clay Loam | Muck |
| 4 - 6.5 | 2.9 | 5.6 | 7.8 | 9.4 | 11.2 | 21.3 |
| 4.5 - 6.5 | 2.5 | 4.7 | 6.5 | 7.8 | 9.4 | 18.1 |
| 5 - 6.5 | 2.0 | 3.8 | 5.2 | 6.3 | 7.4 | 14.1 |
| 5.5 - 6.5 | 1.3 | 2.9 | 3.8 | 4.5 | 5.2 | 9.6 |
| 6 - 6.5 | 0.7 | 1.6 | 2.0 | 2.5 | 2.7 | 4.9 |

*Source:* Adapted from Lorenz, O.A. & Maynard, D.N. (1980)
*Knott's Handbook for Vegetable Growers.*
© 1980 and reprinted with kind permission of John Wiley and Sons Inc.

TABLE 61   Approximate Quantities of Sulphur Required to Lower Soil pH
According to Initial pH level and Soil Type

| Required pH change | Sulphur t / ha | | |
| | Sand | Loam | Clay |
|---|---|---|---|
| 8.5 - 6.5 | 2.2 | 2.8 | 3.4 |
| 8 - 6.5 | 1.3 | 1.7 | 2.2 |
| 7.5 - 6.5 | 0.6 | 0.9 | 1.1 |
| 7 - 6.5 | 0.1 | 0.2 | 0.3 |

TABLE 62   Approximate Solubility of Some Common Macro and
Micro Nutrient Fertilizers (Parts in 100 Parts Cold Water)

**Macro Fertilizers**

| | |
|---|---|
| Ammonium Nitrate | 118 |
| Ammonium Sulphate | 71 |
| Calcium Nitrate | 102 |
| Diammonium Phosphate | 43 |
| Monammonium Phosphate | 23 |
| Orthophosphoric Acid | 550 |
| Potassium Chloride | 35 |
| Potassium Nitrate | 13 |
| Potassium Sulphate | 12 |
| Sodium Nitrate | 73 |
| Superphosphate, single | 2 |
| Superphosphate, double | 4 |
| Urea | 78 |

**Micro Fertilizers**

| | |
|---|---|
| Copper Sulphate | 22 |
| Ferrous Sulphate | 29 |
| Manganese Sulphate | 105 |
| Sodium Borate | 5 |
| Sodium Molybdate | 56 |
| Zinc Sulphate | 75 |

Source:  Adapted from Nakayama, F.S. & Bucks, D. (1986)
*Developments in Agricultural Engineering 9.*

---

**TABLE 63   Hydroponic Nutrient Solution**

---

Various nutrient solutions are in use to suit different crops and cropping systems. The Hoagland solution was one of the original hydroponic solutions and is quoted here.

| Element | Concentration mg/l | Source salts | g/1000 litres of source salt |
|---|---|---|---|
| Nitrogen | 210 | $KNO_3$ | 606 |
|  |  | $Ca(NO_3)_2$ | 656 |
| Phosphorus | 31 | $NH_4H_2PO_4$ | 115 |
| Potassium | 234 | From Nitrogen source |  |
| Magnesium | 48 | $Mg\ SO_47H_2O$ | 490 |
| Calcium | 160 | From Nitrogen source |  |
| Sulphur | 64 | From Magnesium source |  |
| Iron | 2.5 | Fe - EDDHA* | 40 |
| Manganese | 0.5 | $MN\ Cl_2 . 4H_2O$ | 1.81 |
| Boron | 0.5 | $H_3BO_3$ | 2.86 |
| Copper | 0.02 | $Cu\ SO_4 . 5H_2O$ | 0.08 |
| Zinc | 0.05 | $Zn\ SO_4 . 7H_2O$ | 0.22 |
| Molybdenum | 0.01 | $H_2MoO_4 . H_2O$ | 0.02 |

* Originally ferrous sulphate and tartaric acid used 3 times weekly as Fe source.

*NB* The pH of hydroponic solutions should normally be maintained between 5.8 and 6.2. The electrical conductivity should be maintained within the range 2000 - 4000 $\mu s\ cm^{-1}$; yield penalties (which may be acceptable economically) can occur at higher salinities. The $NO_3$ and $NH_4$ balance can be manipulated for various crops and growth stages to ensure that the medium pH remains reasonable.

*Source:* Hoagland, D.R. & Arnon, D.I. (1938) *The Water Culture Method for Growing Plants without Soil. California Agriculture Experimental Station Circular* 347.
Quoted in: FAO (1990) *Soilless Culture for Horticultural Crop Production. Plant Production and Protection Paper* No. 101.

---

# Irrigation
# and
# Water Flow

# Irrigation Methods – Soil / Land Suitability

---

**TABLE 64   Guide for Selecting a Method of Irrigation**

| Irrigation Method | Topography | Crops | Remarks |
|---|---|---|---|
| Widely spaced borders | Land slopes capable of being graded to less than 1% slope and preferably 0.2% | Alfalfa and other deep rooted close-growing crops and orchards | The most desirable surface method for irrigating close-growing crops where topographical conditions are favourable. Even grade in the direction of irrigation is required on flat land and is desirable but not essential on slopes of more than 0.5%. Grade changes should be slight and reverse grades must be avoided. Cross slope is permissible when confined to differences in elevation between border strips of 6-9 cm. Less suited to sandy soils. |
| Closely spaced borders | Land slopes capable of being graded to 4% slope or less and preferably less than 1% | Pastures | Especially adapted to shallow soils underlain by claypan or soils that have a lower water intake rate. Even grade in the direction of irrigation is desirable but not essential. Sharp grade changes and reverse grades should be smoothed out. Cross slope is permissible when confined to differences in elevation between borders of 6-9 cm. Since the border strips may have less width a greater total cross slope is permissible than for border irrigated alfalfa. Less suited to sandy soils. |
| Basins | Dead level | All crops | According to soil type and water flow volume, can be up to 5 ha or as small as a few m$^2$. |
| Check back and cross furrows | Land slopes capable of being graded to 0.2% slope or less | Fruit | This method is especially designed to obtain adequate distribution and penetration of moisture in soils with low water intake rates. |
| Corru-gations | Land slopes capable of being graded to slopes between 0.5% and 12% | Alfalfa pasture and grain | This method is especially adapted to steep land and small irrigation streams. An even grade in the direction of irrigation is desirable but not essential. Sharp grade changes and reverse grades should at least be smoothed out. Due to the tendency of corrugations to clog and overflow and cause serious erosion, cross slopes should be avoided as much as possible. |
| Graded contour furrows | Variable land slopes of 2-25% but preferably less | Row crops and fruit | Especially adapted to row crops on steep land, though hazardous due to possible erosion from heavy rainfall. Unsuitable for rodent-infested fields or soils that crack excessively. Actual grade in the direction of irrigation 0.5-1.5%. No grading required beyond filling gullies and removal of abrupt ridges. |

**TABLE 64** *(continued)*   **Guide for Selecting a Method of Irrigation**

| Irrigation Method | Topography | Crops | Remarks |
|---|---|---|---|
| Contour ditches | Irregular slopes up to 12% | Hay, pasture and grain | Especially adapted to foothill conditions. Requires little or no surface grading |
| Rectangular checks (levees) | Land slopes capable of being graded so single or multiple tree basins will be levelled within 6 cm | Orchards | Especially adapted to soils that have either a relatively high or low water intake rate. May require considerable grading. |
| Contour levee | Slightly irregular land slopes of less than 1 % | Fruit, rice, grain and forage crops | Reduces the need to grade land. Frequently employed to avoid altogether the necessity of grading. Adapted best to soils that have either high or low intake rate. |
| Portable pipes | Irregular slopes up to 12% | Hay pasture and grain | Especially adapted to foothill conditions. Requires little or no surface grading. |
| Sub-irrigation | Smooth-flat | Shallow rooted crops such as potatoes or grass | Requires a water table, very permeable subsoil conditions and precise levelling. Very few areas adapted to this method. |
| Sprinkler | Especially suited to undulating 1-35% slopes | All crops | High operation and maintenance costs. Good for rough or very sandy lands in areas of high production and good markets. Good method where power costs are low. May be the only practical methods in areas of steep or rough topography. Good for high rainfall areas where only a small supplemental water supply is needed. Large moving systems (e.g. Centre Pivot Irrigators) can give very low capital investment costs/ha and low labour requirements and have the greatest potential range of irrigation frequencies and amount applied. |
| Contour bench terraces | Sloping land - best for slopes under 3% but useful to 6% | Any crop but particularly suited to cultivated crops | Considerable loss of productive land due to berms. Requires expensive drop structures for water erosion control. |

*(continued over)*

**TABLE 64 *(continued)*   Guide for Selecting a Method of Irrigation**

| Irrigation Method | Topography | Crops | Remarks |
|---|---|---|---|
| Sub-irrigation (installed pipes) | Flat to uniform slopes up to 1% surface should be smooth | Any crop; usually used for row crops of high value | Requires installation of perforated plastic pipe in root zone at narrow spacings. Some difficulties in roots plugging the perforations. Correct spacing also a problem. Field trials on different soils are needed. This is still in the development stage. Not suitable for use with saline waters. |
| Localized (drip, trickle, etc.) | Any topographic condition-suitable for row crop farming | Row crops or fruit | Perforated pipe on the soil surface drips water at base of individual vegetable plants or around fruit trees. Has been successfully used with saline irrigation water. Also suited to orchard and amenity planting in hilly and stony terrains. |
| Wild Flooding | Any - minimal slopes preferred | Pasture | Wasteful system. Little control over distribution, OK if water and labour cheap and no erosion hazard. |

*Source:* Adapted from FAO (1984) *Irrigation Practice and Water Management. Irrigation and Drainage Paper 1.* Reproduced with kind permission of the Food and Agriculture Organization of the United Nations.

**TABLE 65   Indicative Rating of Soil Suitability for Irrigation according to Available Water Capacity**

| Available Water Capacity mm/m | Suitability Class |
|---|---|
| >180 | Highly |
| 120 - 180 | Medium |
| <120 | Low |

*Note:* With modern irrigation techniques such as trickle and high-frequency sprinkler systems, irrigation of apparently low suitability soils becomes ever more feasible.

**TABLE 66   Indicative Rating of Soil Suitability for Flood Irrigation according to Soil Infiltration Rate**

| Basic infiltration rate (cm h⁻¹) | Suitability for surface irrigation |
|---|---|
| < 0.1 | Unsuitable (too slow) but suitable for rice |
| 0.1 - 0.3 | Marginally suitable (too slow); marginally suitable for rice |
| 0.3 - 0.7 | Suitable; unsuitable for rice |
| 0.7 - 3.5 | Optimum |
| 3.5 - 6.5 | Suitable |
| 6.5 - 12.5 | Marginally suitable (too rapid); small basins required |
| 12.5 - 25.0 | Suitable only under special conditions; very small basins required |
| > 25.0 | Unsuitable (too rapid); recommended for overhead methods only |

*Note:* When considering the infiltration rate of an unreclaimed soil it is important to consider that the infiltration rate may well increase with the onset of cultivations and cropping.

*Source:* Landon, J.R., ed. (1984) *Tropical Soil Manual.*

**TABLE 67   Suitability of Mechanized Irrigation Systems according to Terrain**

| System | Flat | Terrain Rolling | Slope up to 30% |
|---|---|---|---|
| Solid set sprinkler | yes | yes | yes |
| Solid set rainguns | yes | yes | yes |
| Side role sprinkler | yes | no | yes |
| Centre pivot | yes | yes | yes |
| Linear move | yes | no | no |
| Linear boom | yes | yes* | yes* |
| Travelling raingun | yes | yes* | yes* |
| Drip/trickle/bubble | yes | yes | yes |

\* must have inbuilt brake/restrain mechanism to cope with slopes

*Source:* Adapted from Sneed McBride International, (1990) *Irrigation Systems Catalog.* Dallas, Texas.

---

**TABLE 68  Indicative Maximum Sprinkler Irrigation Application Rates for Different Soils**

| Soil texture | Maximum rate of application mm/hour |
|---|---|
| Coarse sand | 20 - 40 |
| Fine sand | 12 - 25 |
| Sandy loam | 12 |
| Silt loam | 10 |
| Clay loam<br>Clay | 5 - 8 |

The above rates are for level or gently sloping soils, with good structure or with cover. For soils with poor structure, or bare soils, reduce rates by 20%. For soils steeper than 5%, reduce rates by 20%.

*NB* In practice higher application rates are now being achieved with high output continuously moving sprinkler irrigators which make use of the short period of high initial infiltration rates on soil infiltration rate curves.

*Source:* Hudson, N.W. (1974) *Field Engineering for Agricultural Development.*
Reproduced with kind permission of N.W. Hudson.

---

**TABLE 69  Indicative Desirable Maximum Field Slopes for Flood Irrigated Agriculture**

| Soil type | Max slope (%) |
|---|---|
| Sand | 0.25 |
| Sandy loam | 0.40 |
| Fine sandy loam | 0.50 |
| Clay | 2.50 |
| Loam | 6.25 |

*NB* On sandy soils a compromise is needed between potential deep percolation losses and erosion; hence higher slopes are frequently used.

*Source:* Withers, B., & Vipond, S. (1974) *Irrigation Design and practice.*
Reproduced with kind permission of B.T. Batsford Ltd.

# Irrigation Water Quality

**TABLE 70  General Classification of Saline Groundwaters**

| Class | TDS (mg / l) |
|-------|--------------|
| Fresh | 0 - 1000 |
| Brackish | 1000 - 10 000 |
| Saline | 10 000 - 100 000 |
| Brine | >100 000 |

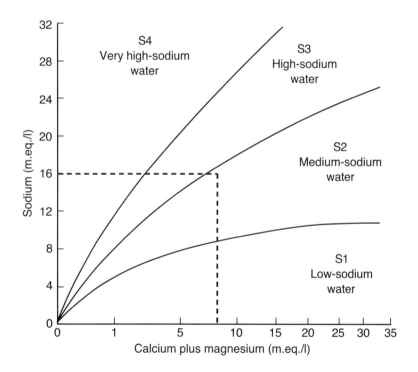

**Fig. 18** Sodium diagram for water classification

*Source:* Wilcox, L.V. (1958) *USDA Information Bulletin 197.*

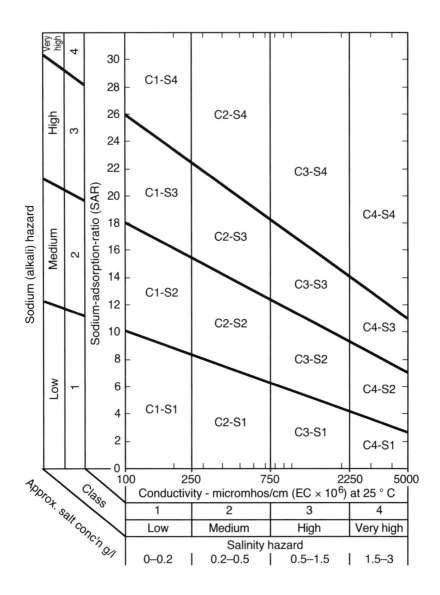

**Fig. 19** Classification of irrigation of water accounting for sodium content and total salt content

*Source:* USDA (1954) *Diagnosis and Improvement of Saline and Alkaline Soils. Ag. Handbook No.60.*

**TABLE 71   Guidelines for Interpretations of Water Quality for Irrigation**

| Potential Irrigation Problems | Units | Degree of Restriction on Use | | |
|---|---|---|---|---|
| | | None | Slight to moderate | Severe |
| **Salinity** (affects crop water availability) | | | | |
| **EC$_w$** | ds/m | <0.7 | 0.7 - 3.0 | >3.0 |
| (or) | | | | |
| **TDS** | mg/l | <450 | 450 -2000 | >2000 |
| **Infiltration** (affects infiltration rate of water into the soil) Evaluate using EC$_w$ and SAR together (1) | | | | |
| **SAR** =   0 - 3 and EC$_w$  = | | >0.7 | 0.7 - 0.2 | <0.2 |
| =   3 - 6        = | | >1.2 | 1.2 - 0.3 | <0.3 |
| =   6 - 12       = | | >1.9 | 1.9 - 0.5 | <0.5 |
| =   12 - 20      = | | >2.9 | 2.9 - 1.3 | <1.3 |
| =   20 - 40      = | | >5.0 | 5.0 - 2.9 | <2.9 |
| **Specific Ion Toxicity** (affects sensitive crops) | | | | |
| **Sodium (Na)** (2) | | | | |
| surface irrigation | SAR | <3 | 3 - 9 | >9 |
| sprinkler irrigation | me/l | <3 | >3 | |
| **Chloride (Cl)** (2) | | | | |
| surface irrigation | me/l | <4 | 4 - 10 | >10 |
| sprinkler irrigation | me/l | <3 | >3 | |
| **Boron (B)** (3) | mg/l | <0.7 | 0.7 - 3.0 | >3.0 |
| **Trace Elements** (4) | | | | |
| **Miscellaneous Effects** (affects susceptible crops) | | | | |
| **Nitrogen NH$_4$ - N** | mg/l | <5 | 5.- 30 | >30 |
| **Nitrogen NO$_3$ - N** (5) | mg/l | <5 | 5 - 30 | >30 |
| **Bicarbonate** (HCO$_3$) overhead sprinkling only | me/l | <1.5 | 1.5 - 8.5 | >8.5 |
| **pH** | | Normal range 6.5 - 8.4 | | |

*Notes:*   1.   Consult Fig. 22
2.   Consult Crop Na & Cl tolerance tables
3.   Consult Crop tolerance tables
4.   Consult Table 77
5.   Include NH$_4$ - N and organic N when testing waste water

*Source:*  FAO (1985) *Water Quality for Agriculture. Irrigation and Drainage Paper No. 24** and University of California (1983) *Soil and Plant Tissue Testing in California.* Bull. 1879.
*Reproduced with kind permission of the Food and Agriculture Organization of the United Nations.

---

**TABLE 72   Salinity and Sodium Classification of Sulphate-free Waters**

---

### 1. Salinity Classification

**C1 -** Low salinity water can be used for irrigation with most crops on most soils, with little likelihood that a salinity problem will develop. Some leaching is required, but this occurs under normal irrigation practices except in soils of extremely low permeability.
EC micromho/cm 0 - 400
Approx TDS ppm 0 - 250

**C2 -** Medium salinity water can be used if a moderate amount of leaching occurs. Plants with moderate salt tolerance can be grown in most instances without special practices for salinity control.
EC 400 - 1200
TDS 250 - 750

**C3 -** High salinity water cannot be used on soil with restricted drainage. Even with adequate drainage special management for salinity control may be required, and plants with good salt tolerance should be selected.

EC 1200 - 2250
TDS 750 - 1450

**C4 -** Very high salinity water is not suitable for irrigation under ordinary conditions but may be used occasionally under very special circumstances. The soil must be permeable, drainage must be adequate irrigation water must be applied in excess to provide considerable leaching, and very salt-tolerant crops should be selected.
EC 2250 - 5000
TDS 1450 - 3200

### 2. Sodium Classification

**S1 -** Low sodium water can be used for irrigation on almost all soils with little danger of the development of a sodium problem. However, sodium-sensitive crops, such as stone-fruit trees and avocados, may accumulate injurious amounts of sodium in the leaves.
SAR <10

**S2 -** Medium sodium water may present a moderate sodium problem in fine textured soils unless there is gypsum in the soil. This water can be used on coarse textured or organic soils that take water well.
SAR 10 - 18

**S3 -** High sodium water may produce troublesome sodium problems in most soils and will require special management, good drainage, high leaching, and additions of organic matter. If there is plenty of gypsum in the soil, a serious problem may not develop for some time. If gypsum is not present, it or some similar material may have to be added.
SAR 18 - 26

**S4 -** Very high sodium water is generally unsatisfactory for irrigation except at low- or medium-salinity levels, where the use of gypsum or some other amendment, or the solution of calcium from the soil, makes it possible to use such water.

SAR >26

### 3. Chloride

<60 ppm

60 - 200 ppm

200 - 600 ppm

>600 ppm

**TABLE 72 *(continued)*   Salinity and Sodium Classification of Sulphate-free Waters**

*Note:* 1. Irrigation water may sometimes dissolve sufficient calcium from calcareous soils to decrease the sodium hazard appreciably, and this should be taken into account in the use of C1-S3 and C1-S4 waters. For calcareous soils with high pH values or for non-calcareous soils, the sodium status of waters in classes C1-S3, C1-S4 may be improved by the addition of gypsum to the water. Similarly, it may be beneficial to add gypsum to the soil periodically when C2-S3 and C3-S2 waters are used.

2. A useful rule of thumb approximation for calculating TDS is:

$$\frac{EC \text{ micromhos}}{1000} \times 640 = TDS \text{ ppm}$$

*Source:* Adapted from Bresler E. *et al.* (1982) *Saline and Sodic Soils.* Reproduced with kind permission of Springer-Verlag. And from Landon, J.R., ed. (1984) *Tropical Soils Manual.* Reproduced with kind permission of Booker Tate Ltd.

**TABLE 73   Residual Sodium Carbonate Classification as expression of Irrigation Water Bicarbonate Hazard**

| RSC Value | Class |
|---|---|
| <1.25 | Probably safe for irrigation |
| 1.25 – 2.5 | Marginally suitable for irrigation |
| >2.5 | Unsuitable for irrigation |

*Note:* RSC = ( $CO_3^{2-}$ + $HCO_3^-$ ) - ( $Ca^{2+}$ + $Mg^{2+}$ ) [ concentrations in me/l ]

**TABLE 74   Sulphate Classification of Irrigation Water**

| Class | Sulphate mg/l |
|---|---|
| Excellent | <192 |
| Good | 192 – 336 |
| Permissible | 336 – 576 |
| Doubtful | 576 – 960 |
| Unsuitable | >960 |

*Source:* McKee, J.E. & Wolf, H.W. (1963) *Water Quality Criteria. Publication No 3.* Resources Agency of California State Water Quality Control Board.

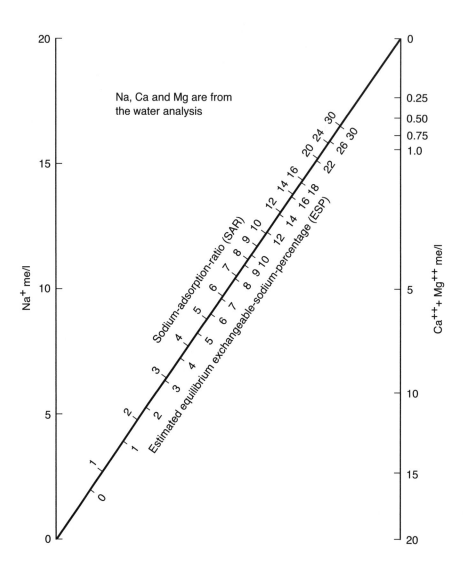

**Fig. 20** Nomogram for determining the SAR value of irrigation water and for estimating the corresponding ESP value of a soil that is at equilibrium with the water

*Source:* Richards, L.A. *et al.* (1954) *Diagnosis and Improvement of Saline and Alkali Soils.* USDA Handbook No. 60.

**TABLE 75   Limit Values for Evaluating the Aggressivity of Water and Soil to Concrete**

| Test | Intensity of attack | | | |
| | None to slight | Mild | Strong | Very Strong |
|---|---|---|---|---|
| **Water** | | | | |
| pH | >6.5 | 6.5 - 5.5 | 5.5 - 4.5 | <4.5 |
| Lime-dissolving carbonic acid ($CO_2$ ), mg/l | <15 | 15 - 3 0 | 30 - 60 | >60 |
| Ammonium ( $NH_4$ ) mg/l | <15 | 15 - 30 | 30 - 60 | >60 |
| Magnesium ( Mg ), mg/l | <100 | 100 - 300 | 300 - 1500 | >1500 |
| Sulphate in Water ( $SO_4$ ),mg/l | <200 | 200 - 600 | 600 - 3000 | >3000 |
| **Soil** | | | | |
| Sulphate in Soil (air-dry) ( $SO_4$ ), mg/kg | <2000 | 2000 - 5000 | >5000 | |

*Source:* FAO (1985) *Water Quality for Agriculture Irrigation and Drainage Paper No. 29.*
Reproduced with kind permission of the Food and Agriculture Organization of the United Nations.

*NB*    1.   $$SAR = \frac{Na^+}{\sqrt{\dfrac{Ca^{++} + Mg^{++}}{2}}}$$

2.   ESP for Soils in Equilibrium with Applied Water
$$ESP = \frac{100 ( 0.01475SAR - 0.0126 )}{0.01475SAR + 0.9874}$$
( For 1 and 2 ionic concentrations me/l )

3.   For high carbonate waters:
adjusted SAR = SAR ( 9.4 - pHc )
and ESP = 2SAR ( 9.4 - pHc )

**Fig. 20 *(continued)***

**TABLE 76  Calculation of New Adjusted SAR (adj R$_{Na}$) for Irrigation Water (now accepted as more accurate and refined calculation)**

Calculation.  $\text{adj } R_{Na} = \dfrac{Na}{\sqrt{\dfrac{Ca_x + Mg}{2}}}$

using 1. Na and Mg me/l in irrigation water
2. Ca$_x$ from table below derived from irrigation water EC and HCO$_3$ and Ca values in me/l

**Calcium concentration (Ca$_x$) expected to remain in near-surface soil-water following irrigation with water of given HCO$_3$/Ca ratio and EC$_w$[1,2]**

| Ratio of HCO$_3$/Ca | Salinity of applied water (EC$_w$) (dS/m) | | | | | | | | | | | |
|---|---|---|---|---|---|---|---|---|---|---|---|---|
| | 0.1 | 0.2 | 0.3 | 0.5 | 0.7 | 1.0 | 1.5 | 2.0 | 3.0 | 4.0 | 6 0 | 8.0 |
| .05 | 13.20 | 13.61 | 13.92 | 14.40 | 14.79 | 15.26 | 15.91 | 16.43 | 17.28 | 17.97 | 19.07 | 19 94 |
| .10 | 8.31 | 8.57 | 8.77 | 9.07 | 9.31 | 9.62 | 10.02 | 10.35 | 10.89 | 11.32 | 12.01 | 12 56 |
| .15 | 6.34 | 6.54 | 6.69 | 6.92 | 7.11 | 7.34 | 7.65 | 7.90 | 8.31 | 8.64 | 9.17 | 9.58 |
| .20 | 5.24 | 5.40 | 5.52 | 5.71 | 5.87 | 6.06 | 6.31 | 6.5 | 6.86 | 7.13 | 7.57 | 7.91 |
| .25 | 4.51 | 4.65 | 4.76 | 4.92 | 5.06 | 5.22 | 5.44 | 5.62 | 5.91 | 6.15 | 6.52 | 6.82 |
| .30 | 4.00 | 4.12 | 4.21 | 4.36 | 4.48 | 4.62 | 4.82 | 4.98 | 5.24 | 5.44 | 5.77 | 6.04 |
| .35 | 3.61 | 3.72 | 3.80 | 3.94 | 4.04 | 4.17 | 4.35 | 4.49 | 4.72 | 4.91 | 5.21 | 5.45 |
| .40 | 3.30 | 3.40 | 3.48 | 3.60 | 3.70 | 3.82 | 3.98 | 4.11 | 4.32 | 4.49 | 4.77 | 4.98 |
| .45 | 3.05 | 3.14 | 3.22 | 3.33 | 3.42 | 3.53 | 3.68 | 3.80 | 4.00 | 4.15 | 4.41 | 4.61 |
| .50 | 2.84 | 2.93 | 3.00 | 3.10 | 3.19 | 3.29 | 3.43 | 3.54 | 3.72 | 3.87 | 4.11 | 4.30 |
| .75 | 2.17 | 2.24 | 2.29 | 2.37 | 2.43 | 2.51 | 2.62 | 2.70 | 2.84 | 2.95 | 3.14 | 3.28 |
| 1.00 | 1.79 | 1.85 | 1.89 | 1.96 | 2.01 | 2.09 | 2.16 | 2.23 | 2.35 | 2.44 | 2.59 | 2.71 |
| 1.25 | 1.54 | 1.59 | 1.63 | 1.68 | 1.73 | 1.78 | 1.86 | 1.92 | 2.02 | 2.10 | 2.23 | 2.33 |
| 1.50 | 1.37 | 1.41 | 1.44 | 1.49 | 1.53 | 1.58 | 1.65 | 1.70 | 1.79 | 1.86 | 1.97 | 2.07 |
| 1.75 | 1.23 | 1.27 | 1.30 | 1.35 | 1.38 | 1.43 | 1.49 | 1.54 | 1.62 | 1.68 | 1.78 | 1.86 |
| 2.00 | 1.13 | 1.16 | 1.19 | 1.23 | 1.26 | 1.31 | 1.36 | 1.40 | 1.48 | 1.54 | 1.63 | 1.70 |
| 2.25 | 1.04 | 1.08 | 1.10 | 1.14 | 1.17 | 1.21 | 1.26 | 1.30 | 1.37 | 1.42 | 1.51 | 1.58 |
| 2.50 | 0.97 | 1.00 | 1.02 | 1.06 | 1.09 | 1.12 | 1.17 | 1.21 | 1.27 | 1.32 | 1.40 | 1.47 |
| 3.00 | 0.85 | 0.89 | 0.91 | 0.94 | 0.96 | 1.00 | 1.04 | 1.07 | 1.13 | 1.17 | 1.24 | 1.30 |
| 3.50 | 0.78 | 0.80 | 0.82 | 0.85 | 0.87 | 0.90 | 0.94 | 0.97 | 1.02 | 1.06 | 1.12 | 1.17 |
| 4.00 | 0.71 | 0.73 | 0.75 | 0.78 | 0.80 | 0.82 | 0.86 | 0.88 | 0.93 | 0.97 | 1.03 | 1.07 |
| 4.50 | 0.66 | 0.68 | 0.69 | 0.72 | 0.74 | 0.76 | 0.79 | 0.82 | 0.86 | 0.90 | 0.95 | 0.99 |
| 5.00 | 0.61 | 0.63 | 0.65 | 0.67 | 0.69 | 0.71 | 0.74 | 0.76 | 0.80 | 0.83 | 0.88 | 0.93 |
| 7.00 | 0.49 | 0.50 | 0.52 | 0.53 | 0.55 | 0.57 | 0.59 | 0.61 | 0.64 | 0.67 | 0.71 | 0.74 |
| 10.00 | 0.39 | 0.40 | 0.41 | 0.42 | 0.43 | 0.45 | 0.47 | 0.48 | 0.51 | 0.53 | 0.56 | 0.58 |
| 20.00 | 0.24 | 0.25 | 0.26 | 0.26 | 0.27 | 0.28 | 0.29 | 0.30 | 0.32 | 0.33 | 0.35 | 0.37 |
| 30.00 | 0.18 | 0.19 | 0.20 | 0.20 | 0.21 | 0.21 | 0.22 | 0.23 | 0.24 | 0.25 | 0.27 | 0.28 |

*Notes:* 1. Assumes a soil source of calcium from lime (CaCO$_3$) or silicates; no precipitation of magnesium, and partial pressure of CO$_2$ near the soil surface (Pco$_2$) is .0007 atmospheres.

2. Ca$_x$, HCO$_3$, Ca are reported in me/l; EC$_w$ is in dS/m.

*Source:* FAO (1985) *Water Quality for Agriculture. Irrigation and Drainage Paper No. 29.*
Reproduced with kind permission of the Food and Agriculture Organization of the United Nations.

## Note 2  Precipitation of Calcium Carbonate from Water

The tendency of calcium carbonate to precipitate out of water can be predicted from the chemical analysis of water. It involves a comparison of the measured pH of the water - pHm - and the theoretical pH of the water, as if it were in equilibrium with $CaCO_3$, calculated from  water analysis - pHc.

The 'saturation index' is pHm-pHc.

If pHm is greater than pHc then a tendency for $CaCO_3$ to precipitate is indicated, i.e. a positive index.

If pHm is less than pHc then a tendency for water to dissolve $CaCO_3$ is indicated, i.e. a negative index.

Calcium carbonate precipitation is primarily important in considerations of calcium supply to saline and sodic soils, and possible precipitates blocking trickle irrigation systems.

pHc is also used in some calculations of adjusted sodium adsorption ratio (old system).

A higher leaching fraction decreases $CaCO_3$ precipitation.

At a pHc greater than 8.4 there is a tendency to dissolve lime from soil through which water moves, then below 8.4 there is a tendency to precipitate lime from the water applied.

At pHc 8.4 there is no precipitation of $CaCO_3$.

At pHm 8.0 water is in close equilibrium to finely ground limestone.

pH 8.4 is the approximate pH of a nonsodic saline soil in equilibrium with $CaCO_3$.

Waters acidified to reduce pH become more temperature dependent and may show increased precipitation at higher temperatures.

Dissolved $CO_2$ reduces precipitation.  Underground waters undergoing de-pressurization lose $CO_2$, increase in pH and tend to show increased $CaCO_3$ precipitation.

The calculation of precipitation hazard is elucidated in such texts as those noted below involving negative logarithm transformation of analytical data.

*Sources:* FAO (1985) *Water Quality for Agriculture. Irrigation and Drainage Paper No.29.* Shainberg, I. & Oster, J.D. (1978) *Quality of Irrigation Water. IIIC Publication No. 2,* © International Irrigation Information Centre. Nakayama, F.S. & Bucks, D. (1986) *Trickle Irrigation for Crop Production.* Wilcox, L.V. (1966) *Tables for Calculating pHc values of Water.* Salinity Laboratory Mimeo, Riverside, California.

**TABLE 77   FAO Recommended Maximum Concentrations of Trace Elements in Irrigation Waters mg/l**

| Element | For water used continuously on all soils | For use up to 20 years on fine textured soils of pH 6.0 - 8.5 |
|---|---|---|
| Al | 5 | 20 |
| As | 0.1 | 2 |
| Be | 0.1 | 0.5 |
| B | 0.7 | 2 (up to 10(3)) |
| Cd | 0.01 | 0.05 |
| Cr | 0.1 | 1 |
| Co | 0.05 | 5 |
| Cu | 0.2 | 5 |
| F | 1 | 15 |
| Fe | 5 | 20 |
| Pb | 5 | 10 |
| Li(1) | 2.5 | 2.5 |
| Mn | 0.2 | 10 |
| Mo | 0.01 | 0.05 (2) |
| Ni | 0.2 | 2 |
| Se | 0.02 | 0.02 |
| V | 0.1 | 1 |
| Zn | 2 | 10 |

These levels will not normally adversely affect plants or soils but specific site conditions may enable ± adjustments to the recommendations.

*Notes:*  1.  Recommended maximum concentration or irrigating citrus is 0.075 mg/l
2.  Only for fine textured acid soils or acid soils with relatively high iron oxide contents.
3.  From Shainberg, I. & Oster, J.D. (1978) *Quality of Irrigation Water. IIIC Publication No. 2,* © International Irrigation Information Centre.

*Source:* Adapted from Landon, J.R., ed. (1984) *Tropical Soil Manual* (after FAO).

**TABLE 78   Classification of Boron-containing Irrigation Waters and Soil (ppm B)**

| Class | Sensitive | Crop Type Semi-Tolerant | Tolerant |
|---|---|---|---|
| **(WATER)** | | | |
| Excellent | <0.33 | <0.67 | <1 |
| Good | 0.33 - 0.67 | 0.67 - 1.33 | 1 - 2 |
| Permissible | 0.67 - 1 | 1.33 - 2 | 2 - 3 |
| Doubtful | 1 - 1.25 | 2 - 2.25 | 3 - 3.75 |
| Unsuitable | >1.25 | >2.5 | >3.75 |
| **(SOIL)** | | | |
| Soil B (sat. ext.) | 0.7 | 1.5 | 1.5 - 2.5 |

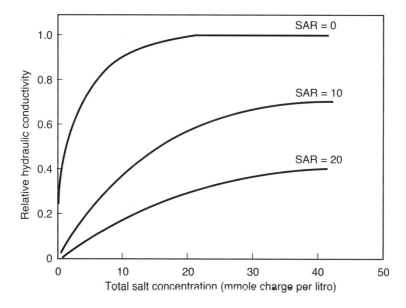

**Fig. 21** Graphical representation of water salinity and SAR effects on hydraulic conductivity of soil

*Source:* Nakayama, F.S. & Bucks, D. (1986) *Trickle Irrigation for Crop Production. Developments in Agricultural Engineering. No. 9.* Reproduced with kind permission of Elsevier Science Publishers.

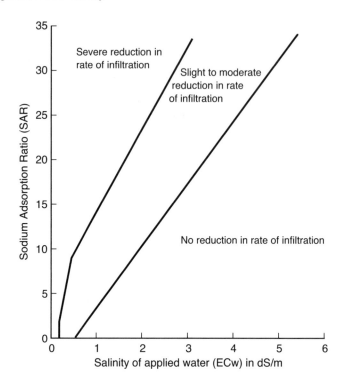

**Fig. 22** Effect of water salinity and sodium adsorption ratio on soil water infiltration rate

*Source:* FAO (1985) *Water Quality for Agriculture. Irrigation and Drainage Paper No. 29.*
Reproduced with kind permission of the Food and Agriculture Organization of the United Nations.

---

**TABLE 79   Influence of Water Quality on the Potential for Clogging Problems in
Localized (Drip) Irrigation Systems**

| Potential Problem | Units | None | Degree of Restriction on Use<br>Slight to Moderate | Severe |
|---|---|---|---|---|
| **Physical** | | | | |
| Suspended Solids | mg/l | <50 | 50 - 100 | >100 |
| **Chemical** | | | | |
| pH | | <7.0 | 7.0 - 8.0 | >8.0 |
| Dissolved Solids | mg/l | <500 | 500 - 2000 | >2000 |
| Manganese | mg/l | <0.1 | 0.1 - 1.5 | >1.5 |
| Iron | mg/l | <0.1 | 0.1 - 1.5 | >1.5 |
| Hydrogen Sulphide | mg/l | <0.5 | 0.5 - 2.0 | >2.0 |
| **Biological** | maximum | | | |
| Bacterial populations | number/ml | <10 000 | 10 000 - 50 000 | >50 000 |

*NB* Mn plant toxicity may occur at these levels.

*Source:* FAO (1985) *Water Quality for Agriculture. Irrigation and Drainage Paper No. 29.*
Reproduced with kind permission of the Food and Agriculture Organization of the United Nations.

**TABLE 80    Indicative Root Zone Leaching Requirement according to
Irrigation Water Salinity, and Leaching Equations**

| Irrigation Water | Fraction of applied irrigation water for leaching as % of applied water | | | |
| --- | --- | --- | --- | --- |
| | Required maximum conductivity of drainage water (millimhos / cm) | | | |
| Salinity micromhos / cm | 4 | 8 | 12 | 16 |
| | % | % | % | % |
| 100 | 2.5 | 1.2 | 0.8 | 0.6 |
| 250 | 6.2 | 3.1 | 2.1 | 1.6 |
| 750 | 18.8 | 9.4 | 6.2 | 4.7 |
| 2250 | 56.2 | 28.1 | 18.8 | 14.1 |
| 5000 | | 62.5 | 41.7 | 31.2 |

*Notes:* 1.  The gross quantity of water applied for leaching purposes must take into account leaching efficiency and irrigation efficiency.

2.  The leaching requirement **for surface irrigation to give 90% potential crop yield** may be calculated as follows:

$$LR = \frac{ECw}{5(ECe) - ECw}$$

where LR = minimum theoretical net leaching requirement as a fraction of the applied water needed to control salts in root zone within the tolerance (ECe) of the crop.

ECw = salinity of applied irrigation in dS/m
ECe = soil saturation extract EC that will result in at least 90% yield potential of crop under investigation.

3.  A general equation for estimating minimum leaching requirement is:

$$\frac{ECi}{ECd} = \frac{Dd}{Di}$$ where   ECi = Ec of irrigation water     Dd =  depth of drainage water

ECd = EC of drainage water     Di =  depth of irrigation water

*NB* This estimates the ADDITIONAL water required for leaching. The maximum permissible ECd value should be inserted in the equation. Alternatively, if ECd is known in a field situation, Dd can be estimated.

4.  For high frequency overhead and drip irrigation:

$$Net\ LR = \frac{ECw}{2\ (Max\ ECe)}$$     LR and ECw as 2 above
Max ECe = ECe corresponding to no yield potential.

*Sources:* Landon, J.R., ed. (1984) *Tropical Soil Manual.* FAO/UNESCO (1973) (table) *Irrigation Drainage and Salinity.* FAO (1985) *Water Quality for Agriculture. Irrigation and Drainage Paper No. 29.* Shainberg, I. & Oster, J.D. (1978) *Quality of Irrigation Water. IIIC Publication No. 2.* Irrigation Association (1983) *Irrigation (5th Ed.).*

**Fig. 23** Graphical representation of leaching requirement according to crop tolerance and applied water salinity

*NB* L is the minimum fraction (%) required for the particular water and crop combination to avoid possible yield loss.

*Source:* Nakayama, F.S. & Bucks, D. (1986) *Trickle Irrigation for Crop Production* and *Developments in Agricultural Engineering No. 9.* Reproduced with kind permission of Elsevier Science Publications.

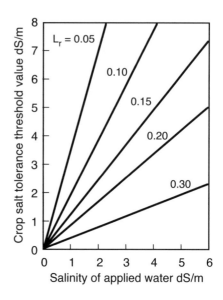

**Fig. 24** Graphical representation of relationship between concentration and EC of several salts commonly found in irrigation waters and irrigated soils

*NB 1.* For the range of EC values that permit plant growth an approximation is: Osmotic Pressure (atm) = 0.36 x EC (mmho/cm).

*NB 2.* EC of aqueous salt solutions increases by approximately 2% per degree centigrade rise in temperature.

*Source:* Shainberg, I. & Oster, J.D. (1978) *Quality of Irrigation Water.* IIIC Publication No.2. © International Irrigation Information Centre.

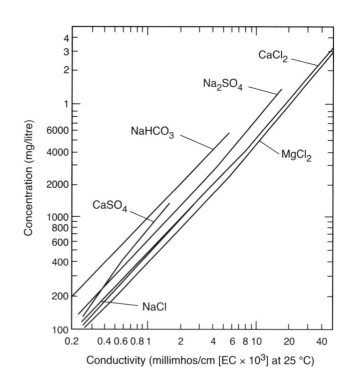

# Estimating Irrigation Requirements

### Note 3  Estimating Evapotranspiration Requirements

The simplest way of estimating crop/water requirements is based on the use of evaporation data obtained from evaporation pans. In its simplest form this is illustrated in **Table 81** and in more refined form in **Table 82** which in turn is related to the tables of crop coefficients. When full climatic data are available computer programmes are now available for calculation of evapotranspiration requirements as typified by the FAO 'cropwat' programme which is based on the Penman approach.

**TABLE 81   Indicative Evaporation Pan Correlation Factors for Estimating Crop Water Use Directly from Pan Evaporation according to Crop Growth Stage**

| Crop Growth Stage | Correlation Factor |
| --- | --- |
| Emergence to early growth | 0.3 - 0.5 |
| During vegetative growth | 0.5 - 1.0 |
| During flowering | 1.0 - 0.8 |
| During wet fruit stage | 0.8 - 0.6 |
| During dry fruit stage | 0.6 - 0 |

*NB*   1.  Crop evapotranspiration $\cong$ pan evaporation x pan factor
      2.  In practice the relationship is highly site-specific and most reliable when the correlation factors are locally derived and not transferred from another site.
      3.  Different designs of evaporation pans are in use; the US Class A pan is most common. Different pan designs generally give different results.

*Source:*  USDA Soil Conservation Service Technical Publication No. 96.

**TABLE 82** Pan Coefficient (kpan) for Class A Pan for different Groundcover and Levels of Mean Relative Humidity and 24-hour Windrun

| | | Pan placed in short green cropped area | | | | Pan placed in dry fallow area | | | |
|---|---|---|---|---|---|---|---|---|---|
| RHmean % | | low <40 | med. 40 -70 | high >70 | | | low <40 | med. 40 - 70 | high >70 |
| Wind km / day | Windward side distance of green crop m | | | | | Windward side distance of dry fallow m | | | |
| **Light** <175 | 1 | .55 | .65 | .75 | | 1 | .7 | .8 | .85 |
| | 10 | .65 | .75 | .85 | | 10 | .6 | .7 | .8 |
| | 100 | .7 | .8 | .85 | | 100 | .55 | .65 | .75 |
| | 1000 | .75 | .85 | .85 | | 1000 | .5 | .6 | .7 |
| **Moderate** 175 - 425 | 1 | .5 | .6 | .65 | | 1 | .65 | .75 | .8 |
| | 10 | .6 | .7 | .75 | | 10 | .55 | .65 | .7 |
| | 100 | .65 | .75 | .8 | | 100 | .5 | .6 | .65 |
| | 1000 | .7 | .8 | .8 | | 1000 | .45 | .55 | .6 |
| **Strong** 425 - 700 | 1 | .45 | .5 | .6 | | 1 | .6 | .65 | .7 |
| | 10 | .55 | .6 | .65 | | 10 | .5 | .55 | .65 |
| | 100 | .6 | .65 | .7 | | 100 | .45 | .5 | .6 |
| | 1000 | .65 | .7 | .75 | | 1000 | .4 | .45 | .55 |
| **Very strong** >700 | 1 | .4 | .45 | .5 | | 1 | .5 | .6 | .65 |
| | 10 | .45 | .55 | .6 | | 10 | .45 | .5 | .55 |
| | 100 | .5 | .6 | .65 | | 100 | .4 | .45 | .5 |
| | 1000 | .55 | .6 | .65 | | 1000 | .35 | .4 | .45 |

*NB* Pan coefficients are used to calculate reference evapotranspiration as follows:
Reference evaporation (ETo)
= Pan coefficient (kpan) x Class A pan evaporation (Epan)

*Source:* FAO (1979) *Yield Response to Water. Irrigation and Drainage Paper No. 33.*
Reproduced with kind permission of the Food and Agriculture Organization of the United Nations.

**TABLE 83** Indicative crop coefficients (kc) for use in relating reference evapotranspiration rate (ETo) to the maximum evapotranspiration rate (ETm) at given crop growth stages

| | | (ETo x kc = ETm) Crop Development stages | | | | Total |
| CROP | Initial | Crop develop- ment | Mid- season | Late season | At harvest | growing period |
|---|---|---|---|---|---|---|
| Banana | | | | | | |
| tropical | 0.4 - 0.5 | 0.7 - 0.85 | 1.0 - 1.1 | 0.9 - 1.0 | 0.75 - 0.85 | 0.7 - 0.8 |
| subtropical | 0.5 - 0.65 | 0.8 - 0.9 | 1.0 - 1.2 | 1.0 - 1.15 | 1.0 - 1.15 | 0.85 - 0.95 |
| Bean | | | | | | |
| green | 0.3 - 0.4 | 0.65 - 0.75 | 0.95 - 1.05 | 0.9 - 0.95 | 0.85 - 0. 95 | 0·85 - 0.9 |
| dry | 0.3 - 0.4 | 0.7 - 0.8 | 1.05 - 1.2 | 0.65 - 0.75 | 0.25 - 0.3 | 0.7 - 0.8 |
| Cabbage | 0.4 - 0.5 | 0.7 - 0.8 | 0.95 - 1.1 | 0.9 - 1.0 | 0.8 - 0.95 | 0.7 - 0.8 |
| Cotton | 0.4 - 0.5 | 0.7 - 0.8 | 1.05 - 1.25 | 0.8 - 0.9 | 0.65 - 0.7 | 0.8 - 0.9 |
| Grape | 0.35 - 0.55 | 0.6 - 0.8 | 0.7 - 0.9 | 0.6 - 0.8 | 0.55 - 0.7 | 0.55 - 0.75 |
| Groundnut | 0.4 - 0.5 | 0.7 - 0.8 | 0. 95 - 1.1 | 0.75 - 0.85 | 0.55 - 0.6 | 0.75 - 0.8 |
| Maize | | | | | | |
| sweet | 0.3 - 0.5 | 0.7 - 0.9 | 1.05 - 1.2 | 1.0 - 1.15 | 0.95 - 1.1 | 0.8 - 0.95 |
| grain | 0.3 - 0.5 | 0.7 - 0.85 | 1.05 - 1.2 | 0.8 - 0.95 | 0. 55 - 0.6 | 0.75 - 0.9 |
| Onion | | | | | | |
| dry | 0.4 - 0.6 | 0.7 - 0.8 | 0.95 - 1.1 | 0.85 - 0.9 | 0.75 - 0.85 | 0.8 - 0.9 |
| green | 0.4 - 0.6 | 0.6 - 0.75 | 0.95 - 1.05 | 0.95 - 1.05 | 0.95 - 1.05 | 0.65 - 0.8 |
| Pea, fresh | 0.4 - 0.5 | 0.7 - 0.85 | 1.05 - 1. 2 | 1.0 - 1.15 | 0.95 - 1.1 | 0.8 - 0.95 |
| Pepper, fresh | 0.3 - 0.4 | 0.6 - 0.75 | 0.95 - 1.1 | 0.85 - 1.0 | 0.8 - 0.9 | 0.7 - 0.8 |
| Potato | 0.4 - 0.5 | 0.7 - 0.8 | 1.05 - 1.2 | 0.85 - 0.95 | 0.7 - 0.75 | 0.75 - 0.9 |
| Rice | 1.1 - 1.15 | 1.1 - 1.5 | 1.1 - 1.3 | 0.95 - 1.05 | 0.95 - 1.05 | 1.05 - 1.2 |
| Safflower | 0.3 - 0.4 | 0.7 - 0.8 | 1.05 - 1.2 | 0.65 - 0.7 | 0.2 - 0.25 | 0.65 - 0.7 |
| Sorghum | 0.3 - 0.4 | 0.7 - 0.75 | 1.0 - 1.15 | 0.75 - 0.8 | 0.5 - 0.55 | 0.75 - 0.85 |
| Soybean | 0.3 - 0.4 | 0.7 - 0.8 | 1.0 - 1.15 | 0.7 - 0.8 | 0.4 - 0.5 | 0.75 - 0.9 |
| Sugarbeet | 0.4 - 0.5 | 0.75 - 0.85 | 1.05 - 1.2 | 0.9 - 1.0 | 0.6 - 0.7 | 0.8 - 0.9 |
| Sugarcane | 0.4 - 0.5 | 0.7 - 1.0 | 1.0 - 1.3 | 0.75 - 0.8 | 0.5 - 0.6 | 0.85 - 1.05 |
| Sunflower | 0.3 - 0.4 | 0.7 - 0.8 | 1.05 - 1.2 | 0.7 - 0.8 | 0.35 - 0.45 | 0.75 - 0.85 |
| Tobacco | 0.3 - 0.4 | 0.7 - 0.8 | 1.0 - 1.2 | 0.9 - 1.0 | 0.75 - 0.85 | 0.85 - 0.95 |
| Tomato | 0.4 - 0.5 | 0.7 - 0.8 | 1.05 - 1.25 | 0.8 - 0.95 | 0.6 - 0.65 | 0.75 - 0.9 |
| Water melon | 0.4 - 0.5 | 0.7 - 0.8 | 0.95 - 1.05 | 0.8 - 0.9 | 0.65 - 0.75 | 0.75 - 0.85 |
| Wheat | 0.3 - 0.4 | 0.7 - 0.8 | 1.05 - 1.2 | 0.65 - 0.75 | 0.2 - 0.25 | 0.8 - 0.9 |
| Alfalfa | 0.3 - 0.4 | | | | 1.05 - 1.2 | 0.85 - 1.05 |
| Citrus | | | | | | |
| clean weeding | | | | | | 0.65 - 0.75 |
| no weed control | | | | | | 0.85 - 0.9 |
| Olive | | | | | | 0.4 - 0.6 |

First figure: Under high humidity (RHmin >70%) and low wind (U <5m/sec).
Second figure: Under low humidity (RHmin <20%) and strong wind (>5m/sec).

*Source:* FAO (1979) *Yield Response to Water. Irrigation and Drainage Paper No. 33.*
Reproduced with kind permission of the Food and Agriculture Organization of the United Nations.

**TABLE 84  kc Values for Full Grown Deciduous Fruit and Nut Trees**

| | Mar | Apr | May | June | July | Aug | Sept | Oct | Nov |
|---|---|---|---|---|---|---|---|---|---|
| With ground cover crop [1] | | | | | | | | | |
| **COLD WINTER WITH KILLING FROST:** | | | | | | | | | |
| **Apples, cherries** | | | | | | | | | |
| humid, light to mod. wind | - | 0.5 | 0.75 | 1.0 | 1.1 | 1.1 | 1.1 | 0.85 | - |
| humid, strong wind | - | 0.5 | 0.75 | 1.1 | 1.2 | 1.2 | 1.15 | 0.9 | - |
| dry, light to mod. wind | - | 0.45 | 0.85 | 1.15 | 1.25 | 1.25 | 1.2 | 0.95 | - |
| dry, strong wind | - | 0.45 | 0.85 | 1.2 | 1.35 | 1.35 | 1.25 | 1.0 | - |
| **Peaches, apricots, pears, plums** | | | | | | | | | |
| humid, light to mod. wind | - | 0.5 | 0.7 | 0.9 | 1.0 | 1.0 | 0.95 | 0.75 | - |
| humid, strong wind | - | 0.5 | 0.7 | 1.0 | 1.05 | 1.1 | 1.0 | 0.8 | - |
| dry, light to mod. wind | - | 0.45 | 0.8 | 1.05 | 1.15 | 1.15 | 1.1 | 0.85 | - |
| dry, strong wind | - | 0.45 | 0.8 | 1.1 | 1.2 | 1.2 | 1.15 | 0.9 | - |
| **COLD WINTER WITH LIGHT FROST:** | | | | | | | | | |
| **Apples, cherries, walnuts[3]** | | | | | | | | | |
| humid, light to mod. wind | 0.8 | 0.9 | 1.0 | 1.1 | 1.1 | 1.1 | 1.05 | 0.85 | 0.8 |
| humid, strong wind | 0.8 | 0.95 | 1.1 | 1.15 | 1.2 | 1.2 | 1.15 | 0.9 | 0.8 |
| dry, light to mod. wind | 0.85 | 1.0 | 1.15 | 1.25 | 1.25 | 1.25 | 1.2 | 0.95 | 0.85 |
| dry, strong wind | 0.85 | 1.05 | 1.2 | 1.35 | 1.35 | 1.35 | 1.25 | 1.0 | 0.85 |
| **Peaches, apricots, pears, plums, almonds, pecans** | | | | | | | | | |
| humid, light to mod. wind | 0.8 | 0.85 | 0.9 | 1.0 | 1.0 | 1.0 | 0.95 | 0.8 | 0.8 |
| humid, strong wind | 0.8 | 0.9 | 0.95 | 1.0 | 1.1 | 1.1 | 1.0 | 0.85 | 0.8 |
| dry, light to mod. wind | 0.85 | 0.95 | 1.05 | 1.15 | 1.15 | 1.15 | 1.1 | 0.9 | 0.85 |
| dry, strong wind | 0.85 | 1.0 | 1.1 | 1.2 | 1.2 | 1.2 | 1.15 | 0.95 | 0.85 |

1.  kc values need to be increased if frequent rain occurs. For young orchards with tree ground cover of 20 and 50%, reduce mid-season kc values by 10 to 15% and 5 to 10% respectively.
3.  For walnuts March-May possibly 10 to 20% lower values due to slower leaf growth.

**Without ground cover crop [2]**
**(clean cultivated, weed free)**

| Mar | Apr | May | June | July | Aug | Sept | Oct | Nov | |
|---|---|---|---|---|---|---|---|---|---|
| **GROUND COVER STARTING IN APRIL** | | | | | | | | | |
| | | | | | | | | | **Apples, cherries** |
| - | 0.45 | 0.55 | 0.75 | 0.85 | 0.85 | 0.8 | 0.6 | - | humid, light to mod. wind |
| - | 0.45 | 0.55 | 0.8 | 0.9 | 0.9 | 0.85 | 0.65 | - | humid, strong wind |
| - | 0.4 | 0.6 | 0.85 | 1.0 | 1.0 | 0.95 | 0.7 | - | dry, light to mod. wind |
| - | 0.4 | 0.65 | 0.9 | 1.05 | 1.05 | 1.0 | 0.75 | - | dry, strong wind |
| | | | | | | | | | **Peaches, apricots, pears, plums** |
| - | 0.45 | 0.5 | 0.65 | 0.75 | 0.75 | 0.7 | 0.55 | - | humid, light to mod. wind |
| - | 0.45 | 0.55 | 0.7 | 0.8 | 0.8 | 0.75 | 0.6 | - | humid, strong wind |
| - | 0.4 | 0.55 | 0.75 | 0.9 | 0.9 | 0.7 | 0.65 | - | dry, light to mod. wind |
| - | 0.4 | 0.6 | 0.8 | 0.95 | 0.95 | 0.9 | 0.65 | - | dry, strong wind |
| **NO DORMANCY IN GRASS COVER CROPS** | | | | | | | | | |
| | | | | | | | | | **Apples, cherries, walnuts[3]** |
| 0.6 | 0.7 | 0.8 | 0.85 | 0.85 | 0.8 | 0.8 | 0.75 | 0.65 | humid, light to mod. wind |
| 0.6 | 0.75 | 0.85 | 0.9 | 0.9 | 0.85 | 0.8 | 0.8 | 0.7 | humid, strong wind |
| 0.5 | 0.75 | 0.95 | 1.0 | 1.0 | 0.95 | 0.9 | 0.85 | 0.7 | dry, light to mod. wind |
| 0.5 | 0.8 | 1.0 | 1.05 | 1.05 | 1.0 | 0.95 | 0.9 | 0.75 | dry, strong wind |
| | | | | | | | | | **Peaches, apricots, pears, plums, almonds, pecans** |
| 0.55 | 0.7 | 0.75 | 0.8 | 0.8 | 0.7 | 0.7 | 0.65 | 0.55 | humid, light to mod. wind |
| 0.55 | 0.7 | 0.75 | 0.8 | 0.8 | 0.8 | 0.75 | 0.7 | 0.6 | humid, strong wind |
| 0.5 | 0.7 | 0.85 | 0.9 | 0.9 | 0.9 | 0.8 | 0.75 | 0.65 | dry, light to mod. wind |
| 0.5 | 0.75 | 0.9 | 0.95 | 0.95 | 0.95 | 0.85 | 0.8 | 0.7 | dry, strong wind |

**2.** kc values assume infrequent wetting by irrigation or rain (every 2 to 4 weeks). In the case of frequent irrigation for March, April and November increase kc and refer to original FAO publication; for May to October use kc values of table 'with ground cover crop'. For young orchards with tree ground cover of 20 and 50% reduce mid-season kc values by 25 to 35% and 10 to 15% respectively.

*Source:* FAO (1984) *Crop Water Requirements. Irrigation and Drainage Paper No. 24.*
Reproduced with kind permission of the Food and Agriculture Organization of the United Nations.

**TABLE 85   Guidelines to Net Amount of Water to Apply per Irrigation according to Soil Type**

| Soil profile | Net amount of moisture to apply - Acre-inches per Acre | | | | | | |
|---|---|---|---|---|---|---|---|
| | 12" | 18" | 24" | 30" for various root depths | 36" | 48" | 72" |
| Coarse sandy soils uniform in texture, 6 ft. | 0.45 | 0.60 | 0.85 | 1.20 | 1.30 | 1.75 | 2.60 |
| Coarse sandy soils over more compact sub-soils | 0.45 | 0.60 | 1.50 | 1.75 | 2.00 | 2.50 | 3.00 |
| Fine sandy loams uniform in texture to 6 ft. | 0.85 | 1.30 | 1.75 | 2.20 | 2.60 | 3.00 | 4.00 |
| Fine sandy loams over more compact sub-soils | 0.85 | 1.50 | 2.00 | 2.40 | 2.80 | 3.25 | 5.00 |
| Silt loams uniform to 6 ft. | 1.10 | 1.70 | 2.25 | 2.75 | 3.00 | 4.00 | 6.00 |
| Silt loams over more compact sub-soils | 1.10 | 1.70 | 2.50 | 3.00 | 3.25 | 4.25 | 6.25 |
| Heavy clay or clay loam soils | 0.90 | 1.40 | 2.00 | 2.40 | 2.85 | 3.85 | 5.50 |

*Source:* Irrigation Association (1983) *Irrigation (5th ed.).*
Reproduced with kind permission of the Irrigation Association, Arlington, Virginia.

**TABLE 86  Guideline Irrigation Water Requirements for a Range of Crops at Four Temperature / Humidity Combinations**

| Crop | degrees C / % relative humidity | | | |
|---|---|---|---|---|
| | 27C/70RH | 32C/60RH | 38C/50RH | 42C/<50RH |
| | l/s/ha | | | |
| Alfalfa | 0.71 | 0.81 | 1.00 | 1.39 |
| Apples | 0.6 | 0.68 | 0.85 | 1.18 |
| Artichokes | 0.67 | 0.76 | 0.94 | 1.32 |
| Barley | 0.74 | 0.85 | 1.05 | 1.46 |
| Beans | 0.71 | 0.81 | 1.00 | 1.39 |
| Beets | 0.71 | 0.81 | 1.00 | 1.39 |
| Broccoli | 0.67 | 0.76 | 0.94 | 1.32 |
| Brussels Sprouts | 0.67 | 0.76 | 0.94 | 1.32 |
| Cabbage | 0.67 | 0.76 | 0.94 | 1.32 |
| Carrots | 0.71 | 0.81 | 1.00 | 1.39 |
| Cauliflower | 0.67 | 0.76 | 0.94 | 1.32 |
| Celery | 0.71 | 0.81 | 1.00 | 1.39 |
| Citrus | 0.42 | 0.48 | 0.59 | 0.83 |
| Coffee | 0.64 | 0.72 | 0.9 | 1.25 |
| Corn | 0.67 | 0.76 | 0.94 | 1.32 |
| Cotton | 0.74 | 0.85 | 1.05 | 1.46 |
| Cucumbers | 0.64 | 0.72 | 0.9 | 1.25 |
| Dates | 0.5 | 0.57 | 0.7 | 0.97 |
| Eggplant | 0.67 | 0.76 | 0.94 | 1.32 |
| Grapes | 0.5 | 0.57 | 0.7 | 0.97 |
| Lettuce | 0.67 | 0.76 | 0.94 | 1.32 |
| Melons | 0.67 | 0.76 | 0.94 | 1.32 |
| Olives | 0.39 | 0.44 | 0.54 | 0.78 |
| Pasture grass | 0.71 | 0.81 | 1.00 | 1.39 |
| Peanuts | 0.67 | 0.76 | 0.94 | 1.32 |
| Peas | 0.74 | 0.85 | 1.05 | 1.46 |
| Peppers | 0.67 | 0.76 | 0.94 | 1.32 |
| Potatoes | 0.74 | 0.85 | 1.05 | 1.46 |
| Radishes | 0.57 | 0.64 | 0.8 | 1.11 |
| Sorghum | 0.67 | 0.76 | 0.94 | 1.31 |
| Soybeans | 0.71 | 0.81 | 1.00 | 1.39 |
| Spinach | 0.67 | 0.76 | 0.94 | 1.31 |
| Stone Fruit | 0.53 | 0.61 | 0.75 | 1.05 |
| Sugar Beets | 0.78 | 0.89 | 1.1 | 1.53 |
| Sugar Cane | 0.74 | 0.85 | 1.05 | 1.45 |
| Tea | 0.64 | 0.72 | 0.9 | 1.25 |
| Tobacco | 0.74 | 0.85 | 1.05 | 1.45 |
| Tomatoes | 0.74 | 0.85 | 1.05 | 1.45 |
| Wheat | 0.74 | 0.85 | 1.05 | 1.46 |

NB  1. Allowances required above these amounts for inefficiency of application and any leaching.
2. Figures are guidelines for crops in active growth with a reasonably complete crop cover for given type of crop.

*Source:* Adapted from Sneed-McBride International (1990) *Irrigation Systems Catalog.* Dallas, Texas.

**TABLE 87   Guide for Determining When to Irrigate Different Soil Types According to 'Feel' of Soil**

| | Feel and appearance of soil | | | |
| | SOIL TYPES | | | |
| Per cent (%) of available moisture remaining in soil | Loamy sands and sandy loams (Light: coarse to moderately coarse) | Very fine sandy loam and silty loams (Medium) | Silty clay loams and clay loams (Heavy) | Indicative action required (depends on rate of evaporation) |
|---|---|---|---|---|
| **100** <br><br> **(field capacity)** | Upon squeezing no free water appears on soil, but wet outline of ball is left on hand. Soil will stick to thumb when rolled between thumb and forefinger <br> [0] | Same as sandy loam <br><br><br><br><br><br> [0] | Same as sandy loam <br><br><br><br><br><br> [0] | Soil very wet Check again |
| **75 to 100** | Forms a weak ball, breaks easily when bounced in the hand, will not slick <br> [0 - 40] | Forms a ball, very pliable, slicks readily <br><br> [0 - 50] | Easily ribbons out between thumb and forefinger, has a slick feeling <br> [0 - 60] | Soil moisture OK Check again |
| **50 - 75** | Tends to ball under pressure but seldom will hold together when bounced in the hand <br><br> [20 - 80] | Forms a ball, somewhat plastic, will slick slightly with pressure <br><br><br> [50 - 100] | Forms a ball, will ribbon out between thumb and forefinger; has a slick feeling <br><br> | Irrigate If you wait any longer you will not be able to irrigate the whole crop until some of it is too dry [60 - 120] |
| **50 or less** | Appears to be dry, will not form a ball with pressure <br><br> [50 - 120] | Somewhat crumbly, but will hold together from pressure <br> [100 - 150] | Somewhat pliable, will ball under pressure <br><br> [120 - 190] | Irrigate, some spots will be too dry already |
| **0** | Dry, loose, flows through fingers <br><br><br><br> [80 - 150] | Powdery, sometimes slightly crusted, but easily broken down into powdery condition [150 - 200] | Hard, baked, cracked, difficult to break down into powdery condition [190 - 250] | Irrigate now Soil is too dry and should have been irrigated earlier |

*NB* Figures in brackets are indicative levels of moisture deficiency in soil - mm/m.

*Sources:* Adapted from *Rhodesion Irrigation Handbook* (1968) and
Israelsen, O.W. & Hansen, V.E. (1962) *Irrigation Practice and Principles.*

---

**TABLE 88   Interpretation of Soil Tensiometer Readings for Irrigation Practice**

---

| Soil State | Tensiometer Reading (atmospheres) | Comment |
| --- | --- | --- |
| **Nearly saturated** | 0 - 0.1 | This state may occur for 1-2 days after irrigation. Poor soil aeration in this phase. |
| **Field capacity** | 0.11 - 0.3 | No irrigation required (lower readings for sands, higher readings for clays). |
| **Irrigation required** | 0.4 - 0.6 | Irrigations usually started in this range. Good soil aeration. Start irrigation of sands at 0.3 - 0.4, loams at 0.4 - 0.5 and clays at 0.5 - 0.6 to ensure available water. |
| **Dry soil** | 0.7 | Crop stress. Reduced production not inevitable, but moisture reserves becoming seriously depleted. |
| | 0.8 | Upper limit of accuracy for tensiometer. |

*NB*  1. Tensiometers are of little value in certain soils where there is a very limited scale deflection between the dry and saturated state.
2. Several tensiometers required in a field to give a reliable average reading.
3. About 75% of available water of coarse soils held up to 0.8, and 25 - 50% for finer soils.
4. Tensiometers are generally more useful in high moisture conditions.

*Source:* Table adapted from FAO (1984) *Irrigation Practice and Management. Irrigation and Drainage Paper No. 1, rev. 1.* (after Stegman, E.C. *et al.* in Jensen, M.E., ed. (1980) *Design and Operation of Farm Irrigation Systems.* ASAE).

# Flood Irrigation

**TABLE 89** Indicative Sizes of Basins for Irrigation according to Stream Flow and Soil Type *

| Stream flow l/s | AREA OF BASIN ha | | | |
| | Sand | Sandy loam | Clay loam | Clay |
| --- | --- | --- | --- | --- |
| 300 | 0.2 | 0.6 | 1.2 | 2.00 |
| 225 | 0.15 | 0.45 | 0.9 | 1.5 |
| 150 | 0.10 | 0.30 | 0.60 | 1.0 |
| 100 | 0.07 | 0.2 | 0.4 | 0.7 |
| 75 | 0.05 | 0.16 | 0.32 | 0.53 |
| 27 | 0.02 | 0.06 | 0.12 | 0.2 |

* Assuming perfect levelling

*Sources:* Combined data from Hudson, N.W. (1974) *Field Engineering for Agricultural Development* & FAO (1988) *Crop Water Requirements. Irrigation and Drainage Paper No. 24.*

**TABLE 90** Maximum Water Flows into Border Strips according to Field Slopes

| Slope % | Max Flow. l/s/m width |
| --- | --- |
| 0.3 | 11.25 |
| 0.4 | 9.0 |
| 0.5 | 7.5 |
| 0.7 | 6.7 |
| 0.9 | 5.0 |
| 1.0 | 4.2 |
| 1.5 | 3.0 |
| 2.0 | 2.5 |
| 3.0 | 2.0 |
| 4.0 | 1.5 |
| 5.0 | 1.25 |

*Source:* Adapted from Withers, B. & Vipond, S. (1974) *Irrigation Design and Practice.* Reproduced with kind permission of B.T.Batsford Ltd.

**TABLE 91    Indicative Border Strip Sizes according to Field Slope, Soil Type, Water Flow and Depth of Water Applied**

| Soil type | Slope % | Depth applied (mm) | Strip width (m) | Strip length (m) | Flow (l/sec) |
|-----------|---------|--------------------|-----------------|------------------|--------------|
|           | 0.25    | 50                 | 15              | 150              | 240          |
|           |         | 100                | 15              | 250              | 210          |
|           |         | 150                | 15              | 400              | 180          |
|           | 1.00    | 50                 | 12              | 100              | 80           |
| Coarse    |         | 100                | 12              | 150              | 70           |
|           |         | 150                | 12              | 250              | 70           |
|           | 2.00    | 50                 | 10              | 60               | 35           |
|           |         | 100                | 10              | 100              | 30           |
|           |         | 150                | 10              | 200              | 30           |
|           | 0.25    | 50                 | 15              | 250              | 210          |
|           |         | 100                | 15              | 400              | 180          |
|           |         | 150                | 15              | 400              | 100          |
|           | 1.00    | 50                 | 12              | 150              | 70           |
| Medium    |         | 100                | 12              | 300              | 70           |
|           |         | 150                | 12              | 400              | 70           |
|           | 2.00    | 50                 | 10              | 100              | 30           |
|           |         | 100                | 10              | 200              | 30           |
|           |         | 150                | 10              | 300              | 30           |
|           | 0.25    | 50                 | 15              | 400              | 120          |
|           |         | 100                | 15              | 400              | 70           |
|           |         | 150                | 15              | 400              | 40           |
|           | 1.00    | 50                 | 12              | 400              | 70           |
| Fine      |         | 100                | 12              | 400              | 35           |
|           |         | 150                | 12              | 400              | 20           |
|           | 2.00    | 50                 | 10              | 320              | 30           |
|           |         | 100                | 10              | 400              | 30           |
|           |         | 150                | 10              | 400              | 20           |

*Source:* USDA (1955) *Yearbook – Water.*

**TABLE 92 Size of Borders and Stream Size for Different Soil Type and Land Slope (Deep Rooted Crops) \***

| Soil type | Slope (%) | Width (m) | Length (m) | Average flow (l/s) |
|---|---|---|---|---|
| Sand | 0.2 - 0.4 | 12 - 30 | 60 - 90 | 10 - 15 |
| | 0.4 - 0.6 | 9 - 12 | 60 - 90 | 8 - 10 |
| | 0.6 - 1.0 | 6 - 9 | 75 | 5 - 8 |
| Loamy sand | 0.2 - 0.4 | 12 - 30 | 75 - 150 | 7 - 10 |
| | 0.4 - 0.6 | 9 - 12 | 75 - 150 | 5 - 8 |
| | 0.6 - 1.0 | 6 - 9 | 75 | 3 - 6 |
| Sandy loam | 0.2 - 0.4 | 12 - 30 | 90 - 250 | 5 - 7 |
| | 0.4 - 0.6 | 6 - 12 | 90 - 180 | 4 - 6 |
| | 0.6 - 1.0 | 6 | 90 | 2 - 4 |
| Clay loam | 0.2 - 0.4 | 12 - 30 | 180 - 300 | 3 - 4 |
| | 0.4 - 0.6 | 6 - 12 | 90 -180 | 2 - 3 |
| | 0.6 - 1.0 | 6 | 90 | 1 - 2 |
| Clay | 0.2 - 0.3 | 12 - 30 | 350+ | 2 - 4 |

\* Under conditions of perfect land grading.

*Source:* FAO (1988) *Crop Water Requirements. Irrigation and Drainage Paper No. 24*
Reproduced with kind permission of the Food and Agriculture Organization of the United Nations.

**TABLE 93 Length of Furrows and Stream Size for Different Soil Type,Land Slope and Depth of Water Application \***

| Slope (%) | Length of furrow (m) | | | | | | | | | | | | average flow (l/s) |
|---|---|---|---|---|---|---|---|---|---|---|---|---|---|
| | heavy texture | | | | medium texture | | | | light texture | | | | |
| 0.05 | 300 | 400 | 400 | 400 | 120 | 270 | 400 | 400 | 60 | 90 | 150 | 190 | 12.0 |
| 0.1 | 340 | 440 | 470 | 500 | 180 | 340 | 440 | 470 | 90 | 120 | 190 | 220 | 6.0 |
| 0.2 | 370 | 470 | 530 | 620 | 220 | 370 | 470 | 530 | 120 | 190 | 250 | 300 | 3.0 |
| 0.3 | 400 | 500 | 620 | 800 | 280 | 400 | 500 | 600 | 150 | 220 | 280 | 400 | 2.0 |
| 0.5 | 400 | 500 | 560 | 750 | 280 | 370 | 470 | 530 | 120 | 190 | 250 | 300 | 1.25 |
| 1.0 | 280 | 400 | 500 | 600 | 250 | 300 | 370 | 470 | 90 | 150 | 220 | 250 | 0.6 |
| 1.5 | 250 | 340 | 430 | 500 | 220 | 280 | 340 | 400 | 80 | 120 | 190 | 220 | 0.4 |
| 2.0 | 220 | 270 | 340 | 400 | 180 | 250 | 300 | 340 | 60 | 90 | 150 | 190 | 0.3 |
| **Application depth (mm)** | 75 | 150 | 225 | 300 | 50 | 100 | 150 | 200 | 50 | 75 | 100 | 125 | |

\* Assuming perfect levelling

*Source:* FAO (1988) *Crop Water Requirements. Irrigation and Drainage Paper No. 24*
Reproduced with kind permission of the Food and Agriculture Organization of the United Nations.

**TABLE 94   Indicative Furrow Lengths according to Field Slope, Soil Type, Depth of Water Applied and Discharge Rate into Furrows**

| Soil type | Application depth (mm) | | Furrow lengths (m) | | | | | |
|---|---|---|---|---|---|---|---|---|
| | | Slope (%) | 0.25 | 0.50 | 1.00 | 1.50 | 2.00 | 3.00 |
| | | Discharge (l/s) | 3.00 | 1.5 | 0.75 | 0.5 | 0.36 | 0.25 |
| **Coarse** | 50 | | 150 | 120 | 70 | 60 | 50 | 25 |
| | 100 | | 210 | 150 | 110 | 90 | 70 | 60 |
| | 150 | | 260 | 180 | 120 | 120 | 90 | 70 |
| **Medium** | 50 | | 250 | 170 | 130 | 100 | 90 | 70 |
| | 100 | | 375 | 240 | 180 | 140 | 120 | 100 |
| | 150 | | 420 | 290 | 220 | 170 | 150 | 120 |
| **Fine** | 50 | | 300 | 220 | 170 | 130 | 120 | 90 |
| | 100 | | 450 | 310 | 250 | 190 | 160 | 130 |
| | 150 | | 530 | 380 | 280 | 250 | 200 | 160 |

*Source:* Adapted from Withers, B. & Vipond, S. (1974) *Irrigation Design and Practice.*
Reproduced with kind permission of B.T. Batsford Ltd.

# Irrigation Efficiency

**TABLE 95   Values for Irrigation Efficiency %**

| Application Method | Light Soils Small Fields | Heavy Soils Large Fields | |
|---|---|---|---|
| Graded border | 60 | 75 | Surface |
| Basin (level border) | 60 | 80 | methods: |
| Contour ditches | 50 | 55 | 40 - 50 |
| Furrows | 55 | 70 | |
| Corrugation | 50 | 70 | |
| **Wind Speed** | **Moderate** | **Low** | |
| Sprinkling, hot dry climate | 60 | 80 | Sprinkler: |
| Sprinkling, moderate climate | 70 | 85 | 65 - 75 |
| Sprinkling, humid climate | 80 | 85 | |
| Drip | 100 | | Drip: 85 - 90 |

*NB*   1.   Nightime irrigation is more efficient especially for sprinkler.

2.   Sprinkler irrigation efficiency improves as greater depth of water is applied per irrigation.

3.   Sprinkler irrigation efficiency decreases slightly as peak use per day rises.

4.   Deep percolation losses incurred in flood irrigation which are returned to a supply aquifer cannot be considered as irretrievable losses like evaporative losses.

*Sources:* ILACO BV (1980) *Agricultural Compendium,* also Landon, J.R., ed. (1984) *Tropical Soil Manual.* Reproduced with kind permission of ILACO BV and Booker Tate Ltd.

TABLE 96    Estimated Deep Percolation Losses as related to Water Application Efficiency, Irrigation Method and Soil Type

| Irrigation Method | Application Practices | Water Application Efficiency $Ea^1$ x 100 | | Average deep percolation as percentage of irrigation water delivered to the field[1] | |
|---|---|---|---|---|---|
| | | soil texture | | soil texture[2] | |
| | | heavy | light | heavy | light |
| **Sprinkler** | - daytime application, moderately strong wind | 60 | 60 | 30 | 30 |
| | - night application | 70 | 70 | 25 | 25 |
| **Localized** | | 80 | 80 | 15 | 15 |
| **Basin** | - poorly levelled and shaped | 60 | 45 | 30 | 40 |
| | - well levelled and shaped | 75 | 60 | 20 | 30 |
| **Furrow,** | - poorly graded and sized | 55 | 40 | 30 | 40 |
| **Border** | - well graded and sized | 65 | 50 | 25 | 35 |

1. $Ea = Et/Id$, where Et is evapotranspiration and $Id$ is irrigation water delivered at the farm gate. Deep percolation losses and losses through surface runoff, specific evaporation effects and tailwater make up the total losses $Id - Et$ according to $Id - Et = (1 - Ea) Id = [ (1 - Ea) / Ea ] Et$.

2. The term 'heavy' is used to refer to a range of finer textured permeable soils. 'Light' refers to a range of coarser soils with a good to fair water holding capacity. The percentages given do not apply to soils having extreme qualities of hydraulic conductivity and infiltration rate.

*Source:* FAO (1984) *Irrigation Practice and Water Management.*
*Irrigation and Drainage Paper No. 1.* Reproduced with kind permission of the Food and Agriculture Organization of the United Nations.

**TABLE 97   Indicative Sprinkler Irrigation Efficiencies**

| | Cool temperatures | | | Moderate temperatures | | | Hot temperatures | | |
| | Rate of application (mm/hr) | | | Rate of application (mm/hr) | | | Rate of application (mm/hr) | | |
| Amount of water applied during each irrigation (mm) | less than 6 | 6 to 10 | more than 10 | less than 8 | 8 to 12 | more than 12 | less than 10 | 10 to 15 | more than 15 |
|---|---|---|---|---|---|---|---|---|---|
| More than 50 | 75 | 80 | 85 | 70 | 75 | 80 | 65 | 70 | 75 |
| 25 - 50 | 70 | 75 | 80 | 65 | 70 | 75 | 60 | 65 | 70 |
| Less than 25 | 65 | 70 | 75 | 60 | 65 | 70 | - | - | - |

*Note:* The above figures are for a dry climate. They can be increased by 5% for humid climates, and still night conditions.

*Source:* Wright Rain Ltd., *Planned Irrigation.* Ringwood, Hants, UK.

# Water Conveyance

**TABLE 98   Guide to Maximum Side Slopes for Water Conveyance Channels**

| Material | Horizontal : Vertical |
|---|---|
| Sand, Soft clay | 3:1 |
| Sandy clay, Silt loam | 2:1 |
| Fine Clay, Clay loam | 1.5:1 |
| Pitching on clay loam | 0.5 - 1:1 |
| Rock | Up to 90% |

*Source:* Withers, B. & Vipond, S. (1974) *Irrigation Design and Practice.*
Reproduced with kind permission of B.T. Batsford Ltd.

TABLE 99   Indicative Water Conveyance Channel Seepage Losses

| Type of soil | Seepage loss $m^3 / m^2 /$ day |
|---|---|
| Self sealing clays | 0 - 0.07 |
| Impervious clay loam | 0.07 - 0.10 |
| Medium clay loam, impervious layer below channel bottom not exceeding 900 mm in depth | 0.10 - 0.15 |
| Clay loam, silty soil | 0.15 - 0.23 |
| Clay loam with gravel, sandy clay loam, gravel cemented with clay particles | 0.23 - 0.30 |
| Sandy loam | 0.30 - 0.45 |
| Sandy soil | 0.45 - 0.55 |
| Sandy soil with gravel | 0.55 - 0.75 |
| Pervious gravelly soil | 0.75 - 0.90 |
| Gravel with some earth | 0.90 - 1.80 |

*Source:* Etchverry, B.A. (1933) *Irrigation Practice and Engineering. Vol.2.*

TABLE 100   Maximum Non-Scouring Water Velocities for Open Channels According to Soil Type and Grass Cover

| Material | Maximum velocity on cover expected after two seasons | | |
|---|---|---|---|
| | Bare | Medium grass cover | Very good grass cover |
| | m/s | m/s | m/s |
| Very light silty sand | 0.3 | 0.75 | 1.5 |
| Light loose sand | 0.5 | 0.9 | 1.5 |
| Coarse sand | 0.75 | 1.25 | 1.7 |
| Sandy soil | 0.75 | 1.5 | 2.0 |
| Firm clay loam | 1.0 | 1.7 | 2.3 |
| Stiff clay or stiff gravelly soil | 1.5 | 1.8 | 2.5 |
| Coarse gravels | 1.5 | 1.8 | Unlikely to form very good grass cover |
| Shale, hardpan, soft rock, etc. | 1.8 | 2.1 | |
| Hard cemented conglomerates | 2.5 | – | – |

*Source:* Hudson, N.W. (1975) *Field Engineering for Agricultural Development.*
Reproduced with kind permission of N. W. Hudson.

**TABLE 101   Design Velocities for Grass Waterways According to Slope and Soil Type**

| | Velocity m/s | | |
|---|---|---|---|
| **Slope** | **0 - 5%** | **5 - 10%** | **10%** |
| Soil Resistant to Erosion | 2 | 1.75 | 1.5 |
| Erodible Soils | 1.75 | 1.5 | 1.25 |

*Source:* Hudson, N.W. (1975) *Field Engineering for Agricultural Development.*
Reproduced with kind permission of N.W.Hudson.

**TABLE 102   Plastic Pipe Pressure Ratings**

| **Class** | **Bar** | **m head (approx)** |
|---|---|---|
| 0 | No pressure | |
| B | 6 | 60 |
| C | 9 | 90 |
| D | 12 | 120 |
| E | 15 | 150 |

**TABLE 103   Roughness Characteristics of Pipes of Various Materials**

| Roughness category | Material | | | | |
|---|---|---|---|---|---|
| | **Plastic** | **Asbestos/ cement** | **Aluminium** | **Galvanized** | **Concrete** |
| **Low friction** | No constriction at joints, for example,taper-fit or solvent-welded | With smooth joints | Permanent | New | – |
| **Medium friction** | Some constriction at joints, for example, butt-welded | – | Portable | – | Spun concrete with gasket joints |
| **High friction** | – | – | – | Old | (a) Precast with gasket joints (b) Cast *in situ* |

*Source:* FAO *Agricultural Development Paper 88.*

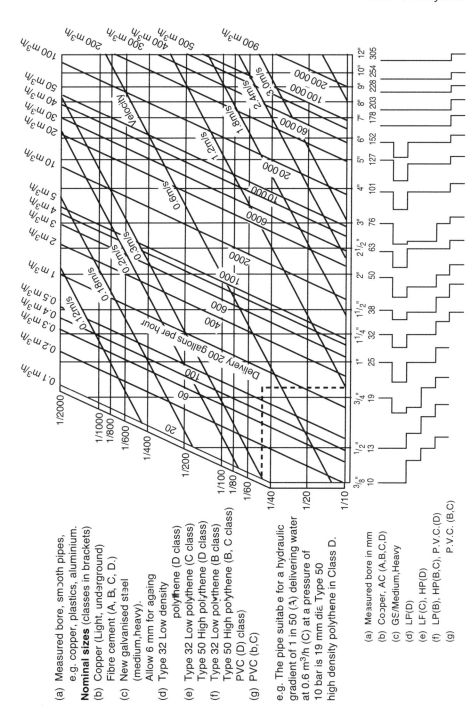

**Fig. 25** Pipe flow chart according to gradient, velocity, pipe bore and pipe material type. *Source:* Electricity Council (UK) (1988) *Pumping and Irrigation. A Farm Electric Handbook.* Reproduced with kind permission of the Electricity Association Technology Ltd.

(a) Measured bore, smooth pipes,
   e.g. copper, plastics, aluminium.
   **Nominal sizes** (classes in brackets)
(b) Copper (Light, underground)
   Fibre cement (A, B, C, D.)
(c) New galvanised steel
   (medium, heavy).
   Allow 6 mm for ageing
(d) Type 32 Low density
       polythene (D class)
(e) Type 32 Low polythene (C class)
   Type 50 High polythene (D class)
(f) Type 32 Low polythene (B class)
   Type 50 High polythene (B, C class)
   PVC (D) class)
(g) PVC (b,C)

e.g. The pipe suitable for a hydraulic
gradient of 1 in 50 (A) delivering water
at 0.6 m³/h (C) at a pressure of
10 bar is 19 mm dia Type 50
high density polythene in Class D.

(a) Measured bore in mm
(b) Copper, AC (A,B,C,D)
(c) GS/Medium,Heavy
(d) LP(D)
(e) LF(C), HP(D)
(f) LP(B), HP(B,C), P.V.C.(D)
(g)              P.V.C. (B,C)

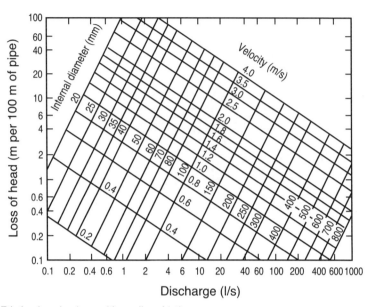

**a.** Friction loss in pipes with medium friction

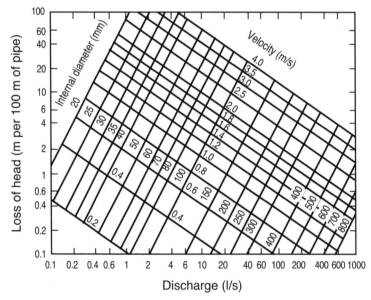

**b.** Friction loss in pipes with high friction

**Figs. 26 a, b, c.** Pipe discharge according to pipe size, velocity and head loss

*Source:* Hudson, N.W. (1974) *Field Engineering for Agricultural Development* (after FAO).
Reproduced with kind permission of N. W. Hudson.

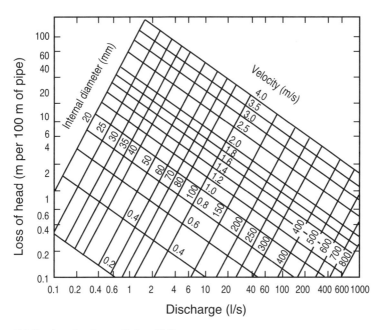

**c.** Friction loss in pipes with low friction

**Fig. 26** *(continued)*

**TABLE 104   Guide for Selecting Irrigation Hose Size for Travelling Sprinkler Systems**

| System | System-capacity | Recommended irrigation hose diameter | Standard full length |
|---|---|---|---|
| ha | l/m | mm | m |
| Up to 8.1 | Up to 568 | 63.5 | 100.6, 152.4 or 201.7 |
| 8.1 to 16.2 | 568 to 1136 | 76.2 | 201.1 |
| 16.2 to 40.5 | 946 to 2271 | 101.6 | 201.1 |
| 24.3 to 48.6 | 1514 to 2839 | 114.3 | 201.1 |
| 32.4to 64.8 | 1892 to 3785 | 127.0 | 201.1 |

*Source:* Irrigation Association (1983) *Irrigation. 5th Ed.*
Reproduced with kind permission of the Irrigation Association, Arlington, Virginia.

**TABLE 105   Indicative Water Flow in Lay Flat Tubing**

| Gradient | Flow l/s<br>138 mm diam | 188 mm diam |
|---|---|---|
| 1 in 25 | 33 | 95 |
| 1 in 50 | 23 | 64 |
| 1 in 100 | 15 | 44 |
| 1 in 250 | 9 | 26 |
| 1 in 500 | 6 | 18 |
| 1 in 1000 | 4 | 12 |

## Water Flow Measurement

**Vertical Case:**

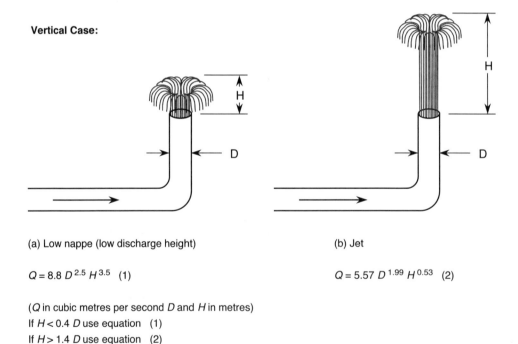

(a) Low nappe (low discharge height)

$$Q = 8.8\ D^{2.5}\ H^{3.5} \quad (1)$$

(b) Jet

$$Q = 5.57\ D^{1.99}\ H^{0.53} \quad (2)$$

($Q$ in cubic metres per second $D$ and $H$ in metres)
If $H < 0.4\ D$ use equation (1)
If $H > 1.4\ D$ use equation (2)
If $0.4D < H < 1.4D$ calculate both equations and take the average

**Fig. 27** Calculation of water flow out of pipes according to nappe characteristics

**Horizontal Case:**

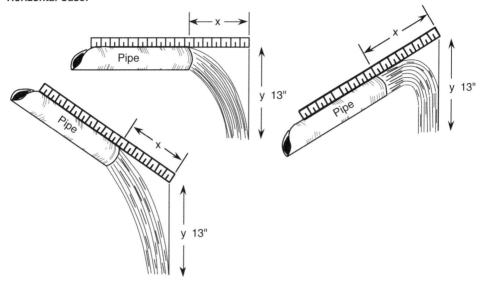

**Discharge (Q) from pipes flowing full in Imperial gallons per minute with vertical drop 'Y' = 13.0 inches**

| Internal pipe dia. (inches) | \multicolumn{13}{c}{Horizontal distance 'X' in inches} |
|---|---|

| Internal pipe dia. (inches) | 12 | 14 | 16 | 18 | 20 | 22 | 24 | 26 | 28 | 30 | 32 | 34 | 36 |
|---|---|---|---|---|---|---|---|---|---|---|---|---|---|
| 2 | 32 | 37 | 42 | 47 | 52 | 57 | 62 | 68 | 73 | 78 | 83 | 89 | 94 |
| 2½ | 49 | 57 | 66 | 74 | 82 | 90 | 99 | 107 | 115 | 123 | 131 | 139 | 148 |
| 3 | 71 | 82 | 94 | 106 | 118 | 130 | 142 | 153 | 165 | 177 | 189 | 200 | 212 |
| 4 | 126 | 146 | 167 | 188 | 209 | 231 | 252 | 272 | 293 | 314 | 335 | 356 | 377 |
| 5 | 197 | 228 | 262 | 295 | 328 | 360 | 392 | 425 | 458 | 491 | 524 | 566 | 589 |
| 6 | 282 | 330 | 377 | 425 | 472 | 518 | 565 | 613 | 660 | 707 | 755 | 800 | 845 |
| 7 | 381 | 449 | 514 | 577 | 642 | 706 | 770 | 834 | 898 | 963 | 1028 | 1090 | 1156 |
| 8 | 504 | 587 | 670 | 755 | 842 | 923 | 1006 | 1090 | 1173 | 1257 | 1341 | 1424 | 1510 |
| 9 | 637 | 743 | 848 | 955 | 1061 | 1168 | 1272 | 1380 | 1485 | 1591 | 1729 | 1805 | 1910 |
| 10 | 785 | 917 | 1048 | 1180 | 1310 | 1440 | 1572 | 1704 | 1836 | 1965 | 2094 | 2222 | 2357 |
| 12 | 1130 | 1320 | 1510 | 1698 | 1888 | 2078 | 2268 | 2456 | 2642 | 2833 | 3021 | 3207 | 3398 |

OR: General Equation $Q_{gpm} = \dfrac{3.007 \times \text{pipe cross sec area sq.ins.} \times X}{\sqrt{Y}}$

*Source: Rhodesion Irrigation Handbook* (1968). Reproduced with kind permission of the Ministry of Lands, Agriculture and Rural Resettlement, Zimbabwe.

**TABLE 106  Discharge Rate for Rectangular Notch Weir-l/s/m Length of Weir Crest**

| Head mm | With end contractions | Suppressed | Head mm | With end contractions | Suppressed |
|---|---|---|---|---|---|
| 25 | 7.2 | 7.27 | 205 | 164 | 171 |
| 30 | 9.5 | 9.56 | 210 | 169.5 | 177 |
| 35 | 11.9 | 12 | 215 | 175 | 183 |
| 40 | 14.6 | 14.7 | 220 | 181.5 | 190 |
| 45 | 17.4 | 17.6 | 225 | 187 | 196 |
| 50 | 20.4 | 20.6 | 230 | 193.5 | 203 |
| 55 | 23.4 | 23.7 | 235 | 200 | 210 |
| 60 | 26.7 | 27 | 240 | 205.5 | 215 |
| 65 | 30.1 | 30.5 | 245 | 212 | 223 |
| 70 | 33.6 | 34.1 | 250 | 218.5 | 230 |
| 75 | 37.2 | 37.8 | 255 | 225 | 237 |
| 80 | 40.9 | 41.6 | 260 | 231 | 244 |
| 85 | 44.8 | 45.6 | 265 | 237.5 | 251 |
| 90 | 48.9 | 49.7 | 270 | 244 | 258 |
| 95 | 52.9 | 53.9 | 275 | 250.5 | 265 |
| 100 | 57.0 | 58.2 | 280 | 257.5 | 273 |
| 105 | 61.3 | 62.6 | 285 | 264 | 280 |
| 110 | 65.6 | 67.1 | 290 | 271 | 288 |
| 115 | 70.1 | 71.8 | 295 | 277.5 | 295 |
| 120 | 74.7 | 76.5 | 300 | 284 | 302 |
| 125 | 79.3 | 81.3 | 305 | 291 | 310 |
| 130 | 84.0 | 86.2 | 310 | 298 | 318 |
| 135 | 88.8 | 91.3 | 315 | 304.5 | 325 |
| 140 | 93.7 | 96.4 | 320 | 311.5 | 333 |
| 145 | 96.4 | 102 | 325 | 319 | 341 |
| 150 | 103.8 | 107 | 330 | 326 | 349 |
| 155 | 108.5 | 112 | 335 | 333 | 357 |
| 160 | 114 | 118 | 340 | 340 | 365 |
| 165 | 119 | 123 | 345 | 347 | 373 |
| 170 | 124.5 | 129 | 350 | 354 | 381 |
| 175 | 130 | 135 | 355 | 361.5 | 389 |
| 180 | 136 | 141 | 360 | 368.5 | 397 |
| 185 | 140.5 | 146 | 365 | 376.5 | 406 |
| 190 | 146 | 152 | 370 | 383.5 | 414 |
| 195 | 152 | 158 | 375 | 390.5 | 422 |
| 200 | 158.5 | 165 | 380 | 398 | 431 |

*Source: Rhodesion Irrigation Handbook* (1968). Reproduced with kind permission of the Ministry of Lands, Agriculture and Rural Settlements, Zimbabwe.

**TABLE 107  Discharge Rates l/s for Cut Throat Flumes**

| Head mm | 0.25 | 0.50 | 0.75 | 1.00 | 2.00 | Head mm | 0.25 | 0.50 | 0.75 | 1.00 | 2.00 |
|---|---|---|---|---|---|---|---|---|---|---|---|
| | | Throat width in metres | | | | | | Throat width in metres | | | |
| 25 | 1.6 | 3.3 | 5.0 | 6.8 | 13.8 | 275 | 68.8 | 140 | 212 | 285 | 580 |
| 30 | 2.2 | 4.4 | 6.7 | 8.9 | 18.2 | 280 | 70.9 | 144 | 219 | 294 | 597 |
| 35 | 2.7 | 5.6 | 8.5 | 11.5 | 23.3 | 285 | 72.8 | 148 | 225 | 302 | 614 |
| 40 | 3.4 | 6.9 | 10.4 | 14.0 | 28.6 | 290 | 74.8 | 152 | 231 | 310 | 630 |
| 45 | 4.1 | 8.3 | 12.7 | 17.0 | 34.6 | 295 | 76.9 | 156 | 237 | 318 | 648 |
| 50 | 4.8 | 9.8 | 14.8 | 19.9 | 40.6 | 300 | 78.9 | 160 | 243 | 326 | 665 |
| 55 | 5.6 | 11.3 | 17.2 | 23.1 | 46.9 | 305 | 81.0 | 165 | 250 | 335 | 683 |
| 60 | 6.4 | 13.0 | 19.8 | 26.6 | 54.1 | 310 | 83.0 | 169 | 256 | 334 | 700 |
| 65 | 7.2 | 14.7 | 22.3 | 30.0 | 61.1 | 315 | 85.1 | 173 | 262 | 352 | 717 |
| 70 | 8.2 | 16.6 | 25.2 | 33.8 | 68.8 | 320 | 87.2 | 178 | 269 | 361 | 736 |
| 75 | 9.1 | 18.4 | 27.9 | 37.5 | 76.4 | 325 | 89.3 | 182 | 276 | 370 | 753 |
| 80 | 10.0 | 20.3 | 30.8 | 41.4 | 84.3 | 330 | 91.6 | 186 | 282 | 379 | 772 |
| 85 | 11.0 | 22.5 | 34.0 | 45.7 | 93 | 335 | 93.7 | 191 | 289 | 388 | 790 |
| 90 | 12.0 | 24.5 | 37.1 | 49.9 | 102 | 340 | 95.8 | 195 | 296 | 397 | 808 |
| 95 | 13.1 | 26.7 | 40.5 | 54.4 | 111 | 345 | 98.1 | 200 | 303 | 406 | 827 |
| 100 | 14.2 | 28.9 | 43.8 | 48.8 | 120 | 350 | 100 | 204 | 309 | 415 | 845 |
| 105 | 15.3 | 31.1 | 47.2 | 63.4 | 129 | 355 | 103 | 209 | 316 | 425 | 865 |
| 110 | 16.5 | 33.6 | 50.9 | 68.3 | 139 | 360 | 105 | 213 | 323 | 434 | 884 |
| 115 | 17.6 | 35.9 | 54.4 | 73.1 | 149 | 365 | 107 | 218 | 331 | 444 | 904 |
| 120 | 18.9 | 38.5 | 58.3 | 78.3 | 159 | 370 | 109 | 223 | 337 | 453 | 922 |
| 125 | 20.1 | 40.9 | 62.0 | 83.3 | 170 | 375 | 112 | 227 | 344 | 463 | 941 |
| 130 | 21.4 | 43.6 | 66.1 | 88.8 | 181 | 380 | 114 | 232 | 352 | 473 | 962 |
| 135 | 22.7 | 46.2 | 70.0 | 94.0 | 191 | 385 | 116 | 237 | 359 | 482 | 981 |
| 140 | 24.0 | 48.8 | 74.0 | 99.4 | 202 | 390 | 119 | 242 | 367 | 492 | 1002 |
| 145 | 25.4 | 51.7 | 78.3 | 105 | 214 | 395 | 121 | 247 | 374 | 502 | 1021 |
| 150 | 26.7 | 54.4 | 82.5 | 111 | 225 | 400 | 124 | 251 | 381 | 512 | 1041 |
| 155 | 28.2 | 57.4 | 86.9 | 117 | 238 | 405 | 126 | 256 | 389 | 522 | 1062 |
| 160 | 29.6 | 60.2 | 91.2 | 123 | 249 | 410 | 128 | 261 | 396 | 532 | 1082 |
| 165 | 31.0 | 63.1 | 95.6 | 128 | 261 | 415 | 131 | 266 | 404 | 542 | 1104 |
| 170 | 32.5 | 66.2 | 100 | 135 | 274 | 420 | 133 | 271 | 611 | 552 | 1124 |
| 175 | 34.0 | 69.2 | 105 | 141 | 287 | 425 | 136 | 276 | 419 | 562 | 1144 |
| 180 | 35.6 | 72.4 | 110 | 147 | 300 | 430 | 138 | 282 | 427 | 573 | 1166 |
| 185 | 37.1 | 75.5 | 114 | 154 | 313 | 435 | 141 | 287 | 434 | 583 | 1187 |
| 190 | 38.7 | 78.6 | 119 | 160 | 326 | 440 | 144 | 292 | 442 | 594 | 1209 |
| 195 | 40.3 | 82.0 | 124 | 167 | 340 | 445 | 146 | 297 | 450 | 604 | 1230 |
| 200 | 41.9 | 85.2 | 129 | 174 | 363 | 450 | 148 | 302 | 458 | 615 | 1251 |
| 205 | 43.6 | 88.7 | 134 | 181 | 367 | 455 | 151 | 307 | 466 | 626 | 1274 |
| 210 | 45.2 | 92.0 | 139 | 187 | 381 | 460 | 154 | 313 | 474 | 636 | 1295 |
| 215 | 46.9 | 95.3 | 145 | 194 | 395 | 465 | 156 | 318 | 482 | 648 | 1318 |
| 220 | 48.6 | 99.0 | 150 | 201 | 410 | 470 | 159 | 323 | 490 | 658 | 1339 |
| 225 | 50.3 | 102 | 155 | 209 | 424 | 475 | 102 | 329 | 496 | 669 | 1361 |
| 230 | 52.2 | 106 | 161 | 216 | 440 | 480 | 164 | 334 | 507 | 680 | 1304 |
| 235 | 53,9 | 110 | 166 | 223 | 454 | 485 | 167 | 340 | 515 | 691 | 1406 |
| 240 | 55.6 | 113 | 172 | 231 | 469 | 490 | 170 | 345 | 523 | 703 | 1430 |
| 245 | 57.5 | 117 | 238 | 238 | 285 | 500 | 175 | 356 | 540 | 725 | 1475 |
| 250 | 59.3 | 121 | 183 | 246 | 500 | | | | | | |
| 255 | 61,2 | 125 | 189 | 254 | 516 | | | | | | |
| 260 | 63.1 | 128 | 195 | 261 | 532 | | | | | | |
| 265 | 65.0 | 132 | 200 | 269 | 548 | | | | | | |
| 270 | 66.9 | 136 | 206 | 277 | 564 | | | | | | |

*Source: Rhodesion Irrigation Handbook* (1968).
Reproduced with kind permission of the
Ministry of Lands, Agriculture and
Rural Resettlement, Zimbabwe.

## TABLE 108 Discharge Rates for Sharp Crested Rectangular Weir Notches with End Contractions - m³/hr

| Head mm | Crest length 0.5 m | 1.0 m | 2.0 m | Head mm | Crest length 0.5 m | 1.0 m | 2.0 m |
|---|---|---|---|---|---|---|---|
| 25 | 13.0 | 25.9 | 52.2 | 255 | 383 | 809 | 1662 |
| 30 | 16.9 | 33.8 | 68.0 | 260 | 393 | 832 | 1709 |
| 35 | 21.2 | 43.2 | 86.4 | 265 | 403 | 854 | 1757 |
| 40 | 25.9 | 52.2 | 105 | 270 | 414 | 879 | 1808 |
| 45 | 31.0 | 62.6 | 126 | 275 | 424 | 902 | 1856 |
| 50 | 36.4 | 73.1 | 147 | 280 | 435 | 926 | 1907 |
| 55 | 41.4 | 84.2 | 169 | 285 | 446 | 949 | 1957 |
| 60 | 47.5 | 96.1 | 194 | 290 | 437 | 973 | 2006 |
| 65 | 53.3 | 108 | 217 | 295 | 467 | 998 | 2059 |
| 70 | 59.8 | 121 | 244 | 300 | 478 | 1022 | 2109 |
| 75 | 65.9 | 134 | 270 | 305 | 490 | 1048 | 2163 |
| 80 | 72.4 | 147 | 296 | 310 | 500 | 1072 | 2214 |
| 85 | 79.2 | 161 | 325 | 315 | 511 | 1096 | 2265 |
| 90 | 86.0 | 175 | 354 | 320 | 522 | 1121 | 2320 |
| 95 | 93.2 | 190 | 384 | 325 | 533 | 1146 | 2372 |
| 100 | 100 | 205 | 414 | 330 | 544 | 1172 | 2427 |
| 105 | 108 | 220 | 445 | 335 | 555 | 1197 | 2480 |
| 110 | 116 | 236 | 478 | 340 | 566 | 1222 | 2533 |
| 115 | 123 | 252 | 510 | 345 | 578 | 1249 | 2591 |
| 120 | 131 | 269 | 544 | 350 | 589 | 1274 | 2644 |
| 125 | 139 | 285 | 577 | 355 | 601 | 1301 | 2701 |
| 130 | 147 | 303 | 613 | 360 | 612 | 1327 | 2756 |
| 135 | 155 | 319 | 648 | 365 | 623 | 1354 | 2814 |
| 140 | 163 | 337 | 683 | 370 | 634 | 1380 | 2869 |
| 145 | 172 | 355 | 721 | 375 | 649 | 1405 | 2925 |
| 150 | 181 | 373 | 757 | 380 | 657 | 1433 | 2984 |
| 155 | 189 | 392 | 796 | 385 | 668 | 1459 | 3040 |
| 160 | 198 | 410 | 834 | 390 | 680 | 1487 | 3100 |
| 165 | 207 | 428 | 872 | 395 | 692 | 1513 | 3157 |
| 170 | 216 | 448 | 913 | 400 | 703 | 1540 | 3214 |
| 175 | 225 | 467 | 952 | 405 | 715 | 1568 | 3275 |
| 180 | 235 | 488 | 994 | 410 | 726 | 1595 | 3332 |
| 185 | 244 | 507 | 1034 | 415 | 738 | 1624 | 3394 |
| 190 | 253 | 527 | 1075 | 420 | 749 | 1650 | 3452 |
| 200 | 272 | 568 | 1160 | 425 | 760 | 1677 | 3510 |
| 205 | 282 | 590 | 1205 | 430 | 773 | 1706 | 3573 |
| 210 | 292 | 610 | 1247 | 435 | 784 | 1734 | 3632 |
| 215 | 301 | 631 | 1290 | 440 | 796 | 1762 | 3695 |
| 200 | 311 | 653 | 1337 | 445 | 807 | 1790 | 3755 |
| 225 | 321 | 674 | 1380 | 450 | 819 | 1819 | 3815 |
| 230 | 332 | 697 | 1428 | 455 | 831 | 1847 | 3878 |
| 235 | 241 | 719 | 1473 | 460 | 842 | 1875 | 3939 |
| 240 | 351 | 741 | 1518 | 465 | 854 | 1904 | 4004 |
| 245 | 362 | 764 | 1566 | 470 | 866 | 1933 | 4065 |
| 250 | 372 | 786 | 1613 | 475 | 877 | 1960 | 4126 |
| | | | | 480 | 889 | 1990 | 4191 |
| | | | | 485 | 901 | 2018 | 4254 |
| | | | | 490 | 912 | 2048 | 4319 |
| | | | | 500 | 935 | 2105 | 4445 |

*Source: Rhodesion Irrigation Handbook* (1968). Reproduced with kind permission of the Ministry of Lands, Agriculture and Rural Resettlement, Zimbabwe.

**TABLE 109    Discharge Rates for 90° 'V' Notch**

| Head mm | Flow l/s | Head mm | Flow l/s |
|---|---|---|---|
| 25 | 0.136 | 205 | 26.4 |
| 30 | 0.215 | 210 | 28.0 |
| 35 | 0.316 | 215 | 29.6 |
| 40 | 0.441 | 220 | 31.4 |
| 45 | 0.592 | 225 | 33.2 |
| 50 | 0.731 | 230 | 35.1 |
| 55 | 0.977 | 235 | 37.0 |
| 60 | 1.21 | 240 | 38.9 |
| 65 | 1.49 | 245 | 41.0 |
| 70 | 1.79 | 250 | 43.1 |
| 75 | 2.11 | 255 | 45.3 |
| 80 | 2.49 | 260 | 47.5 |
| 85 | 2.90 | 265 | 49.8 |
| 90 | 3.34 | 270 | 52.2 |
| 95 | 3.85 | 275 | 54.6 |
| 100 | 4.35 | 280 | 57.2 |
| 105 | 4.92 | 285 | 59.7 |
| 110 | 5.54 | 290 | 62.2 |
| 115 | 6.20 | 295 | 65.0 |
| 120 | 6.91 | 300 | 67.7 |
| 125 | 7.65 | 305 | 70.7 |
| 130 | 8.41 | 310 | 73.5 |
| 135 | 9.27 | 315 | 76.4 |
| 140 | 10.2 | 320 | 79.6 |
| 145 | 11.2 | 325 | 82.6 |
| 150 | 12.1 | 330 | 85.9 |
| 155 | 13.2 | 335 | 89.1 |
| 160 | 14.2 | 340 | 92.3 |
| 165 | 15.3 | 345 | 95.9 |
| 170 | 16.6 | 350 | 100 |
| 175 | 17.8 | 355 | 103 |
| 180 | 19.1 | 360 | 107 |
| 185 | 20.4 | 365 | 110 |
| 190 | 21.8 | 370 | 114 |
| 195 | 23.3 | 375 | 118 |
| 200 | 24.8 | 380 | 122 |

*Source: Rhodesion Irrigation Handbook (1968).*
Reproduced with kind permission of the Ministry of Lands, Agriculture and Rural Resettlement, Zimbabwe.

**TABLE 110    Flow through a Parshall Measuring Flume Cubic Feet Per Second**

| Head in feet | Throat width in inches | | | | |
|---|---|---|---|---|---|
| | 3 | 6 | 9 | 12 | 18 |
| 0.10 | 0.03 | 0.05 | 0.09 | 0.11 | 0.15 |
| 0.15 | 0.05 | 0.10 | 0.17 | 0.20 | 0.30 |
| 0.20 | 0.08 | 0.16 | 0.26 | 0.35 | 0.50 |
| 0.25 | 0.12 | 0.23 | 0.37 | 0.49 | 0.71 |
| 0.30 | 0.15 | 0.31 | 0.49 | 0.64 | 0.94 |
| 0.35 | 0.20 | 0.39 | 0.62 | 0.80 | 1.19 |
| 0.40 | 0.24 | 0.48 | 0.76 | 0.99 | 1.47 |
| 0.45 | 0.29 | 0.58 | 0.90 | 1.19 | 1.76 |
| 0.50 | 0.34 | 0.69 | 1.06 | 1.39 | 2.06 |
| 0.55 | 0.39 | 0.80 | 1.23 | 1.62 | 2.39 |
| 0.60 | 0.45 | 0.92 | 1.40 | 1.84 | 2.73 |
| 0.65 | 0.51 | 1.04 | 1.59 | 2.08 | 3.09 |
| 0.70 | 0.57 | 1.17 | 1.78 | 2.33 | 3.46 |
| 0.75 | – | 1.31 | 1.98 | 2.58 | 3.85 |
| 0.80 | – | 1.45 | 2.18 | 2.85 | 4.26 |
| 0.85 | – | 1.59 | 2.39 | 3.12 | 4.69 |
| 0.90 | – | 1.74 | 2.61 | 3.41 | 5.10 |
| 0.95 | – | 1.90 | 2.84 | 3.70 | 5.55 |
| 1.00 | – | 2.06 | 3.07 | 4.00 | 6.00 |
| 1.05 | – | 2.22 | 3.31 | 4.31 | 6.47 |
| 1.10 | – | 2.40 | 3.55 | 4.62 | 6.95 |
| 1.15 | – | 2.57 | 3.80 | 4.94 | 7.44 |
| 1.20 | – | 2.75 | 4.06 | 5.28 | 7.94 |

*Source: Rhodesian Irrigation Handbook (1968).*
Reproduced with kind permission of the Ministry of Lands, Agriculture and Rural Resettlement, Zimbabwe.

**TABLE 111   Flow Rate in H Flumes**

| Size of flume (maximum depth in millimetres) | Flow in litres per second for depth of flow in millimetres | | | | | | |
|---|---|---|---|---|---|---|---|
| | 25 | 50 | 100 | 250 | 500 | 750 | 1000 |
| 250 | 0.45 | 1.42 | 3.41 | 32.9 | – | – | – |
| 500 | 0.60 | 1.95 | 4.52 | 38.2 | 186.1 | – | – |
| 750 | 0.75 | 2.82 | 5.85 | 42.1 | 201.1 | 526.8 | – |
| 1000 | 0.93 | 3.77 | 8.48 | 48.1 | 208.2 | 546.8 | 1090.3 |

*Source:* USDA *Agricultural Handbook No. 224.*

**TABLE 112   Approximate Water Flow Rate in Half-round, Trapezoidal and Rectangular Open Channels at Different Gradients (155 – 34 900 gall/min)**

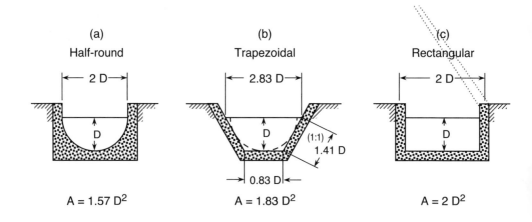

| (a) | (b) | (c) |
|---|---|---|
| Half-round | Trapezoidal | Rectangular |

$A = 1.57 D^2$          $A = 1.83 D^2$          $A = 2 D^2$

**FLOW OF WATER IN OPEN CHANNELS (g/m)**

**TABLE 112** *(continued)*  **Approximate Water Flow Rate in Half-round, Trapezoidal and Rectangular Open Channels at Different Gradients (155 – 34 900 g/m)**

| D inches | Shape | \\ | \\ | \\ | \\ | Gradient, Feet per 100 feet | | | | | | | |
|---|---|---|---|---|---|---|---|---|---|---|---|---|---|
| | | .10 | .15 | .20 | .25 | .30 | .40 | .50 | .60 | .80 | 1.00 | 1.50 | 2.00 |
| | | | | | | | g/m | | | | | | |
| 6 | a | 155 | 190 | 2l5 | 240 | 260 | 305 | 340 | 370 | 450 | 490 | 600 | 675 |
| | b | 185 | 220 | 250 | 280 | 310 | 355 | 410 | 450 | 490 | 560 | 675 | 790 |
| | c | 195 | 235 | 275 | 310 | 340 | 375 | 450 | 490 | 560 | 600 | 750 | 860 |
| 8 | a | 330 | 410 | 450 | 525 | 560 | 675 | 750 | 790 | 940 | 1050 | 1270 | 1460 |
| | b | 375 | 450 | 525 | 600 | 675 | 750 | 860 | 940 | 1090 | 1200 | 1460 | 1720 |
| | c | 410 | 525 | 600 | 675 | 710 | 825 | 940 | 1010 | 1200 | 1310 | 1610 | 1880 |
| 10 | a | 600 | 710 | 825 | 940 | 1010 | 1200 | 1310 | 1460 | 1690 | 1880 | 2290 | 2660 |
| | b | 675 | 860 | 975 | 1090 | 1200 | 1390 | 1540 | 1690 | 1950 | 2180 | 2660 | 3080 |
| | c | 825 | 940 | 1090 | 1200 | 1310 | 1500 | 1690 | 1840 | 2140 | 2400 | 2920 | 3380 |
| 12 | a | 975 | 1200 | 1350 | 1540 | 1690 | 1950 | 2170 | 2360 | 2740 | 3070 | 3750 | 4500 |
| | b | 1160 | 1390 | 1580 | 1800 | 1950 | 2250 | 2510 | 2780 | 3190 | 3560 | 4500 | 4880 |
| | c | 1240 | 1500 | 1760 | 1950 | 2140 | 2480 | 2770 | 3040 | 3490 | 3750 | 4880 | 5620 |
| 15 | a | 1760 | 2140 | 2480 | 2770 | 3040 | 3520 | 3750 | 4120 | 4880 | 5620 | 6750 | 7870 |
| | b | 2020 | 2510 | 2890 | 3220 | 3520 | 4120 | 4500 | 4880 | 5620 | 6370 | 7870 | 9000 |
| | c | 2250 | 2740 | 3150 | 3520 | 3750 | 4500 | 4880 | 5620 | 6370 | 7120 | 8620 | 10 120 |
| 18 | a | 2850 | 3490 | 4120 | 4500 | 4870 | 5630 | 6370 | 7120 | 8250 | 9000 | 11 250 | 12 750 |
| | b | 3340 | 4120 | 4870 | 5250 | 5620 | 6750 | 7500 | 8250 | 9370 | 10 500 | 12 750 | 15 000 |
| | c | 3640 | 4500 | 5250 | 5620 | 6370 | 7130 | 8250 | 9000 | 10 120 | 11 620 | 14 250 | 16 100 |
| 24 | a | 6000 | 7500 | 8620 | 9750 | 10 500 | 12 370 | 13 870 | 15 000 | 17 250 | 19 500 | 23 600 | 27 400 |
| | b | 7120 | 8620 | 10 120 | 11 250 | 12 380 | 14 250 | 16 120 | 17 600 | 20 250 | 22 500 | 27 700 | 31 900 |
| | c | 7870 | 9750 | 11 250 | 12 380 | 13 500 | 15 750 | 17 620 | 19 100 | 22 100 | 24 750 | 30 400 | 34 900 |
| Gradient: 1 in | | 1000 | 660 | 500 | 400 | 330 | 250 | 200 | 166 | 125 | l00 | 66 | 50 |

*Source: Rhodesian Irrigation Handbook* (1968).
Reproduced with kind permission of the Ministry of Lands, Agriculture and Resettlement. Zimbabwe.

**TABLE 113  Water Flow in Several Shapes of Brick-lined Canals (4 – 86 l/s)**

| Section | Grade 1:300 | Grade 1:500 | Grade 1:1000 | No. of bricks per 100 m |
|---|---|---|---|---|
| | | | l/s | |
| 1 | 7.4 | 5.9 | 4.2 | 1760 |
| 2 | 18.1 | 14.7 | 10.1 | 2200 |
| 3 | 29.4 | 24.1 | 16.7 | 2640 |
| 4 | 43.9 | 35.7 | 24.9 | 3080 |
| 5 | 28.3 | 25.2 | 17.6 | 4400 |
| 6 | 15.0 | 12.2 | 8.5 | 3520 |
| 7 | 86.7 | 70.5 | 49.6 | 5280 |
| 8 | 36.5 | 29.7 | 21.0 | 4400 |

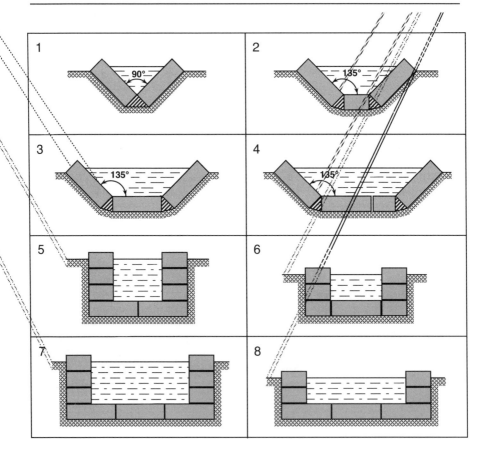

*Note:* The above figures are based on a 25 mm free board, and in practice it is advisable to anticipate a reduced flow due to uneven settlement of the furrow.

*Source: Rhodesian Irrigation Handbook (1968).* Reproduced with kind permission of the Ministry of Lands, Agriculture and Resettlement, Zimbabwe.

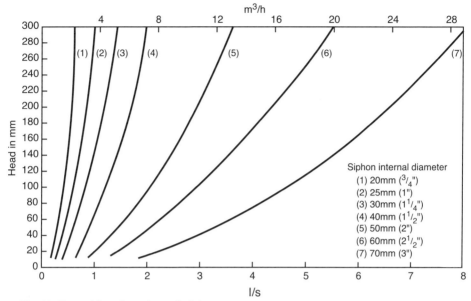

**Fig. 28** Rate of flow through small siphons

*Source: Rhodesian Irrigation Handbook (1968).*
Reproduced with kind permission of the Ministry of Lands, Agriculture and Resettlement, Zimbabwe.

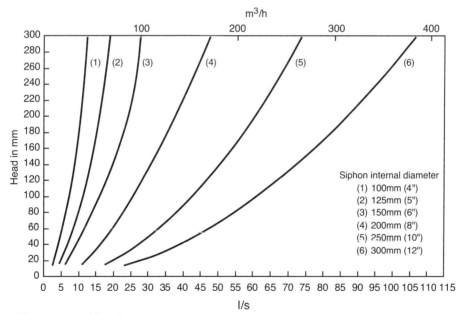

**Fig. 29** Rate of flow through large siphons

*Source: Rhodesian Irrigation Handbook (1968).*
Reproduced with kind permission of the Ministry of Lands, Agriculture and Resettlement, Zimbabwe.

**TABLE 114   Approximate Flow in Unlined Earth Canals (l/s)**

| Flow depth in mm | Base width 450 mm 1½:1 side slopes | | | Base width 600 mm 1½:1 side slopes | | | Base width 900 mm 1:1 side slopes | | | Base width 1.2 m 1:1 side slopes | | |
|---|---|---|---|---|---|---|---|---|---|---|---|---|
| | 1:500 | 1:250 | 1:200 | 1:500 | 1:250 | 1:200 | 1:500 | 1:250 | 1:200 | 1:500 | 1:250 | 1:200 |
| 120 | 11.3 | 17 | 19.8 | 22.6 | 31.1 | 36.8 | - | - | - | - | - | - |
| 180 | 28.3 | 39.6 | 48.1 | 45.3 | 70.8 | 79.3 | - | - | - | - | - | - |
| 240 | 53.8 | 73.6 | 85.0 | 85.0 | 121.8 | 135.9 | - | - | - | - | - | - |
| 305 | 87.8 | 124.6 | 138.7 | 135.9 | 195.4 | 218 | 113 | 161 | 178 | 147 | 209 | 235 |
| 365 | - | - | - | - | - | - | 138 | 343 | 379 | 312 | 439 | 493 |
| 425 | - | - | - | - | - | - | 320 | 456 | 510 | 416 | 586 | 657 |
| 490 | - | - | - | - | - | - | 413 | 568 | 657 | 532 | 753 | 847 |

*Note:* Approximate method for checking flow in canals is as follows:
Select a straight section of channel having a fairly uniform cross-section and measure its average dimensions. Record the time required for a small float or block of wood to pass over a measured length of the channel, making several trials to obtain the average time of travel. Since the velocity of the float on the surface of the water will be greater than the average velocity of the stream, the measurement must be corrected by multiplying by the co-efficient 0.8. Flow is calculated by the formula:

$$Q = 0.8\ A\ V$$

Where
$Q$ = Discharge
$A$ = Area
$V$ = Velocity

*Source: Rhodesian Irrigation Handbook (1968).*
Reproduced with kind permission of the Ministry of Lands, Agriculture and Resettlement, Zimbabwe.

# Pumping Power Requirements

---

**TABLE 115  Indicative Power Requirements for Water Pumping, and Power Losses**

---

In m³/hr :     kW  =   $\dfrac{\text{m}^3/\text{hr x head in metres}}{360 \text{ x efficiency of pump}}$

In l/s :       kW  =   $\dfrac{\text{l/s x head in metres}}{100 \text{ x efficiency of pump}}$

(Efficiency expressed as decimal and obtained from manufacturer's data)


*NB*  Indicative Power Losses incurred in Pumping Water are:

1.    By increase in RPM above maximum pump efficiency RPM:

| % | | | |
|---|---|---|---|
| Increase in RPM | Increase in Discharge | Increase in Head | Increase in Power |
| 5 | 5 | 10 | 15 |
| 10 | 10 | 21 | 33 |
| 15 | 15 | 32 | 54 |
| 20 | 20 | 44 | 73 |
| 25 | 25 | 56 | 96 |

2.    By Transmission losses:
      10% for direct coupled electric motors;
      15% for Vee belt drives from electric motors;
      20% for direct coupled diesel engines;
      25% for Vee belt drives from diesel engines;
      10% shaft losses from tractor PTO to driven unit;
      20% loss in power delivered to Tractor PTO from rated engine power.

3.    By altitude losses:
      3.5%/300 m  above sea level for diesel engines;
      1.5%/300 m  above 1000 m for electric motors.

4.    By temperature losses:
      1% per 5.5 C degree rise in temperature above 15 degrees C
      for diesel engines.

*Source:* Extracted from the *Rhodesian Irrigation Handbook* (1968).

**TABLE 116   Approx. Power Requirements for Water Pumping Schemes (kW)**

| | Head m | | | | | | | | | | | | | |
| | 5 | 10 | 15 | 20 | 25 | 30 | 40 | 50 | 60 | 70 | 80 | 90 | 100 | 200 | 300 |
|---|---|---|---|---|---|---|---|---|---|---|---|---|---|---|---|
| 10 | 0 | 1 | 1 | 1 | 1 | 2 | 2 | 3 | 3 | 4 | 4 | 5 | 6 | 11 | 17 |
| 20 | 1 | 1 | 2 | 2 | 3 | 3 | 4 | 6 | 7 | 8 | 9 | 10 | 11 | 22 | 33 |
| 30 | 1 | 2 | 2 | 3 | 4 | 5 | 7 | 8 | 10 | 12 | 13 | 15 | 17 | 33 | 50 |
| 40 | 1 | 2 | 3 | 4 | 6 | 7 | 9 | 11 | 13 | 16 | 18 | 20 | 22 | 44 | 67 |
| 50 | 1 | 3 | 4 | 6 | 7 | 8 | 11 | 14 | 17 | 19 | 22 | 25 | 28 | 56 | 83 |
| 60 | 2 | 3 | 5 | 7 | 8 | 10 | 13 | 17 | 20 | 23 | 27 | 30 | 33 | 67 | 100 |
| 70 | 2 | 4 | 6 | 8 | 10 | 12 | 16 | 19 | 23 | 27 | 31 | 35 | 39 | 78 | 117 |
| 80 | 2 | 4 | 7 | 9 | 11 | 13 | 18 | 22 | 27 | 31 | 36 | 40 | 44 | 89 | 133 |
| 90 | 3 | 5 | 8 | 10 | 13 | 15 | 20 | 25 | 30 | 35 | 40 | 45 | 50 | 100 | 150 |
| 100 | 3 | 6 | 8 | 11 | 14 | 17 | 22 | 28 | 33 | 39 | 44 | 50 | 56 | 111 | 167 |
| 120 | 3 | 7 | 10 | 13 | 17 | 20 | 27 | 33 | 40 | 47 | 53 | 60 | 67 | 133 | 200 |
| 140 | 4 | 8 | 12 | 16 | 19 | 23 | 31 | 39 | 47 | 54 | 62 | 70 | 78 | 156 | 233 |
| 160 | 4 | 9 | 13 | 18 | 22 | 27 | 36 | 44 | 53 | 62 | 71 | 80 | 89 | 178 | 267 |
| 180 | 5 | 10 | 15 | 20 | 25 | 30 | 40 | 50 | 60 | 70 | 80 | 90 | 100 | 200 | 300 |
| 200 | 6 | 11 | 17 | 22 | 28 | 33 | 44 | 56 | 67 | 78 | 89 | 100 | 111 | 222 | 333 |
| 220 | 6 | 12 | 18 | 24 | 31 | 37 | 49 | 61 | 73 | 86 | 98 | 110 | 122 | 244 | 367 |
| 240 | 7 | 13 | 20 | 27 | 33 | 40 | 53 | 67 | 80 | 93 | 107 | 120 | 133 | 267 | 400 |
| 260 | 7 | 14 | 22 | 29 | 36 | 43 | 58 | 72 | 87 | 101 | 116 | 130 | 144 | 289 | 433 |
| 280 | 8 | 16 | 23 | 31 | 39 | 47 | 62 | 78 | 93 | 109 | 124 | 140 | 156 | 311 | 467 |
| 300 | 8 | 17 | 25 | 33 | 42 | 50 | 67 | 83 | 100 | 117 | 133 | 150 | 167 | 333 | 500 |
| 325 | 9 | 18 | 27 | 36 | 45 | 54 | 72 | 90 | 108 | 126 | 144 | 163 | 181 | 361 | 542 |
| 350 | 10 | 19 | 29 | 39 | 49 | 58 | 78 | 97 | 117 | 136 | 156 | 175 | 194 | 389 | 583 |
| 375 | 10 | 21 | 31 | 42 | 52 | 63 | 83 | 104 | 125 | 146 | 167 | 188 | 208 | 417 | 625 |
| 400 | 11 | 22 | 33 | 44 | 56 | 67 | 89 | 111 | 133 | 156 | 178 | 200 | 222 | 444 | 667 |
| 425 | 12 | 24 | 35 | 47 | 59 | 71 | 94 | 118 | 142 | 165 | 189 | 213 | 236 | 472 | 708 |
| 450 | 13 | 25 | 38 | 50 | 63 | 75 | 100 | 125 | 150 | 175 | 200 | 225 | 250 | 500 | 750 |
| 475 | 13 | 26 | 40 | 53 | 66 | 79 | 106 | 132 | 158 | 185 | 211 | 237 | 264 | 528 | 792 |
| 500 | 14 | 28 | 42 | 56 | 69 | 83 | 111 | 139 | 167 | 194 | 222 | 250 | 278 | 556 | 833 |
| 600 | 17 | 33 | 50 | 67 | 83 | 100 | 133 | 167 | 200 | 233 | 267 | 300 | 333 | 667 | 1000 |
| 700 | 19 | 39 | 58 | 78 | 97 | 117 | 156 | 194 | 233 | 272 | 311 | 350 | 389 | 778 | 1167 |
| 800 | 22 | 44 | 67 | 89 | 111 | 133 | 178 | 222 | 267 | 311 | 356 | 400 | 444 | 889 | 1333 |
| 900 | 25 | 50 | 75 | 100 | 125 | 150 | 200 | 250 | 300 | 350 | 400 | 450 | 500 | 1000 | 1500 |
| 1000 | 28 | 56 | 83 | 111 | 139 | 167 | 222 | 278 | 333 | 389 | 444 | 500 | 556 | 1111 | 1667 |
| 1100 | 31 | 61 | 92 | 122 | 153 | 183 | 244 | 306 | 367 | 428 | 489 | 550 | 611 | 1222 | 1833 |
| 1200 | 33 | 67 | 100 | 133 | 167 | 200 | 267 | 333 | 400 | 467 | 533 | 600 | 667 | 1333 | 2000 |
| 1300 | 36 | 72 | 108 | 144 | 181 | 217 | 289 | 361 | 433 | 506 | 578 | 650 | 722 | 1444 | 2167 |
| 1400 | 39 | 78 | 117 | 156 | 194 | 233 | 311 | 389 | 467 | 544 | 622 | 700 | 778 | 1556 | 2333 |
| 1500 | 42 | 83 | 125 | 167 | 208 | 250 | 333 | 417 | 500 | 583 | 667 | 750 | 833 | 1667 | 2500 |
| 1800 | 50 | 100 | 150 | 200 | 250 | 300 | 400 | 500 | 600 | 700 | 800 | 900 | 1000 | 2000 | 3000 |

(Left axis label: m$^3$/hr)

*Source: Rhodesian Irrigation Handbook* (1968).
Reproduced with kind permission of the Ministry of Lands, Agriculture and Resettlement, Zimbabwe.

**TABLE 117   Electric Motor Power Rating in kW for Various Pumping Duties**
**( based on average pump and motor efficiencies )**

| DISCHARGE m³/hr | Head m | | | | | | | | | | | |
|---|---|---|---|---|---|---|---|---|---|---|---|---|
| | 10 | 20 | 30 | 40 | 50 | 60 | 70 | 80 | 90 | 100 | 110 | 120 |
| 10 | 0.5 | 1.0 | 1.5 | 2.0 | 2.0 | 2.5 | 3.0 | 3.5 | 4.0 | 4.5 | 5.0 | 5.0 |
| 20 | 1.0 | 2.0 | 3.0 | 3.5 | 4.5 | 5.0 | 6.0 | 7.0 | 8.0 | 8.5 | 9.5 | 10 |
| 30 | 1.5 | 2.5 | 4.0 | 5.0 | 6.5 | 8.0 | 9.0 | 10 | 12 | 13 | 14 | 16 |
| 40 | 2.0 | 3.5 | 5.0 | 7.0 | 8.5 | 10 | 12 | 14 | 16 | 17 | 19 | 21 |
| 50 | 2.5 | 4.5 | 6.5 | 9.0 | 11 | 13 | 15 | 17 | 19 | 21 | 24 | 26 |
| 60 | 3.0 | 5.0 | 8.0 | 10 | 13 | 16 | 18 | 21 | 23 | 26 | 28 | 31 |
| 70 | 3.0 | 6.0 | 9.0 | 12 | 15 | 18 | 21 | 24 | 27 | 30 | 33 | 36 |
| 80 | 3.5 | 7.0 | 10 | 14 | 17 | 20 | 24 | 27 | 31 | 34 | 38 | 41 |
| 90 | 4.0 | 8.0 | 12 | 15 | 19 | 23 | 27 | 31 | 35 | 38 | 42 | 46 |
| 100 | 4.5 | 9.0 | 13 | 17 | 21 | 26 | 30 | 34 | 39 | 43 | 47 | 51 |
| 120 | 5.5 | 10 | 15 | 21 | 28 | 31 | 36 | 41 | 46 | 51 | 56 | 62 |
| 140 | 6.0 | 12 | 18 | 24 | 30 | 36 | 42 | 48 | 54 | 60 | 66 | 72 |
| 160 | 7.0 | 14 | 20 | 27 | 34 | 41 | 48 | 55 | 62 | 68 | 75 | 82 |
| 180 | 8.0 | 16 | 23 | 31 | 39 | 46 | 54 | 62 | 69 | 77 | 85 | 92 |
| 200 | 9.0 | 17 | 26 | 34 | 43 | 51 | 60 | 68 | 77 | 85 | 94 | 103 |
| 220 | 9.5 | 19 | 28 | 38 | 47 | 56 | 66 | 75 | 85 | 94 | 103 | 113 |
| 240 | 10 | 20 | 31 | 41 | 51 | 61 | 72 | 82 | 92 | 102 | 113 | 123 |
| 260 | 11 | 22 | 33 | 44 | 56 | 67 | 78 | 89 | 100 | 111 | 122 | 133 |
| 280 | 12 | 24 | 36 | 48 | 60 | 72 | 84 | 96 | 108 | 120 | 132 | 144 |
| 300 | 13 | 26 | 38 | 51 | 64 | 77 | 90 | 103 | 116 | 128 | 141 | 154 |

*Source: Rhodesian Irrigation Handbook* (1968).
Reproduced with kind permission of the Ministry of Lands, Agriculture and Resettlement, Zimbabwe.

**TABLE 118   Diesel Engine Power Rating in kW for Various Pumping Duties**
**( based on average pump and diesel engine efficiencies )**

| DISCHARGE m³/hr | Head m | | | | | | | | | | | |
|---|---|---|---|---|---|---|---|---|---|---|---|---|
| | 10 | 20 | 30 | 40 | 50 | 60 | 70 | 80 | 90 | 100 | 110 | 120 |
| 10 | 1.0 | 1.0 | 2.0 | 2.5 | 3.0 | 3.5 | 4.0 | 4.5 | 5.0 | 5.5 | 6.0 | 7.0 |
| 20 | 1.5 | 2.5 | 3.5 | 4.5 | 6.0 | 7.0 | 8.0 | 9.0 | 10 | 11 | 12 | 14 |
| 30 | 2.0 | 3.5 | 5.0 | 7.0 | 8.5 | 10 | 12 | 14 | 15 | 17 | 19 | 20 |
| 40 | 2.5 | 4.5 | 7.0 | 9.0 | 11 | 14 | 16 | 18 | 20 | 22 | 24 | 27 |
| 50 | 3.0 | 5.5 | 8.5 | 11 | 14 | 17 | 20 | 22 | 25 | 29 | 31 | 34 |
| 60 | 3.5 | 7.0 | 10 | 14 | 17 | 20 | 23 | 27 | 30 | 33 | 37 | 40 |
| 70 | 4.0 | 8.0 | 12 | 16 | 20 | 23 | 27 | 31 | 35 | 39 | 43 | 47 |
| 80 | 4.5 | 9.0 | 14 | 18 | 22 | 27 | 31 | 36 | 40 | 45 | 49 | 54 |
| 90 | 5.0 | 10.0 | 15 | 20 | 25 | 30 | 35 | 40 | 45 | 50 | 55 | 60 |
| 100 | 5.5 | 11.0 | 17 | 22 | 28 | 33 | 39 | 44 | 50 | 56 | 61 | 67 |
| 120 | 7.0 | 14 | 20 | 27 | 33 | 40 | 47 | 53 | 60 | 67 | 73 | 80 |
| 140 | 8.0 | 16 | 23 | 31 | 39 | 47 | 55 | 62 | 70 | 79 | 66 | 94 |
| 160 | 9.0 | 18 | 27 | 36 | 45 | 53 | 62 | 71 | 80 | 89 | 98 | 107 |
| 180 | 10.0 | 20 | 30 | 40 | 50 | 60 | 70 | 80 | 90 | 100 | 110 | 120 |
| 200 | 11.0 | 22 | 33 | 44 | 56 | 67 | 79 | 89 | 100 | 111 | 122 | 134 |
| 220 | 13 | 24 | 37 | 49 | 61 | 73 | 86 | 98 | 110 | 122 | 134 | 147 |
| 240 | 14 | 27 | 40 | 53 | 67 | 80 | 93 | 107 | 120 | 133 | 147 | 161 |
| 260 | 15 | 29 | 43 | 58 | 72 | 87 | 101 | 115 | 130 | 145 | 159 | 174 |
| 280 | 16 | 31 | 47 | 62 | 79 | 93 | 109 | 124 | 140 | 156 | 171 | 187 |
| 300 | 17 | 33 | 50 | 67 | 83 | 100 | 117 | 133 | 150 | 167 | 183 | 201 |

*Source: Rhodesian Irrigation Handbook* (1968).
Reproduced with kind permission of the Ministry of Lands, Agriculture and Resettlement, Zimbabwe.

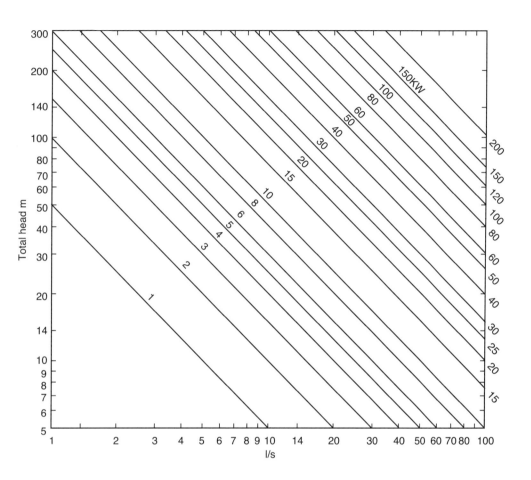

**Fig. 30** Power requirement for pumping water allowing 50% overall efficiency

*Source: Rhodesian Irrigation Handbook* (1968). Reproduced with kind permission of the Ministry of Lands, Agriculture and Resettlement, Zimbabwe.

## Frost Protection

**TABLE 119   Sprinkler Irrigation Application Rate Recommended for Cold Protection Under Different Wind and Temperature Conditions**

| | Wind Speed. km/hr | | |
| | 0. to 1.6 | 3.2 to 6.4 | 8.0 to 12.9 |
|---|---|---|---|
| Minimum Temperature Expected | | Application Rate (mm/hr) | |
| − 2.8C | 2.5 | 2.5 | 2.5 |
| − 3.3C | 2.5 | 2.5 | 3.6 |
| − 4.4C | 2.5 | 4.1 | 7.6 |
| − 5.6C | 3.1 | 6.1 | 12.7 |
| − 6.7C | 4.1 | 7.6 | 15.2 |
| − 7.9C | 5.1 | 10.2 | 17.8 |
| − 9.4C | 6.6 | 12.7 | 22.9 |

*Source:* Adapted from Florida Agricultural Extension Service, *Ext. Circular 287.*

5

# Drainage

## Design Parameters

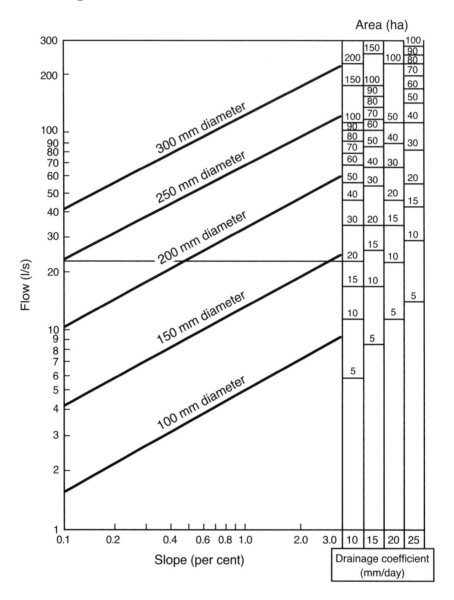

**Fig. 31** Estimating main drain size capacity for land drainage according to flow, slope, drainage coefficient and area

*Example:* 20 ha at 10 mm/day needs 150 mm pipe at 2.5 % or 200 mm pipe at 0.5 % or 250 mm pipe at 0.1 %.

*Source:* Hudson, N. W. (1975) *Field Engineering for Agricultural Development* . Reproduced with kind permission of N.W. Hudson.

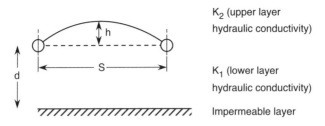

**Fig. 32** Theoretical determination of drain spacing by Houghoudt's equation

Equation: $S = \dfrac{8 K_1 \, dh}{V} + \dfrac{4 K_2 \, h^2}{V}$

where
S = drain spacing
h = hydraulic head
K = hydraulic conductivity
d = depth to impermeable layer
V = rate of water discharge from drain

*Note:* In practice the above formula is imperfect due to soil variability and variable impermeable layers.

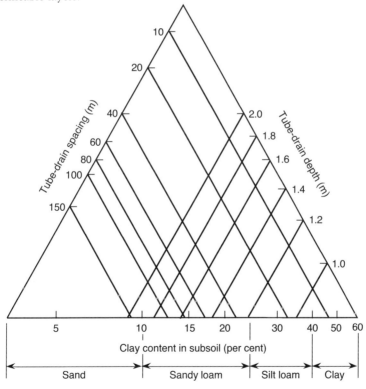

**Fig. 33** American design chart for depth and spacing of lateral field drains

*Note:* USA spacing is usually wider, and depth greater than for European practice.

*Source:* Minnesota Agricultural Experimental Station Tech. Bull. 101.

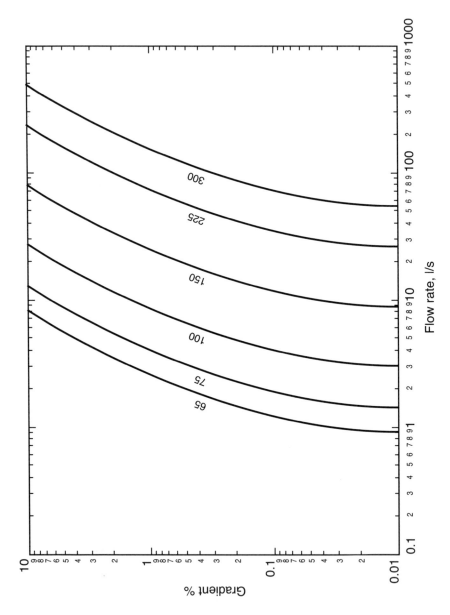

**Fig. 34** Flow rate in field drain pipes according to gradient and diameter. Type: Laterals – clayware pipes. Nominal internal diameter of pipes in mm. *Source:* MAFF/ADAS (1983) *The Design of Field Drainage Pipe Systems.* Reference Book No. 345. Reproduced with kind permission of MAFF ADAS (UK).

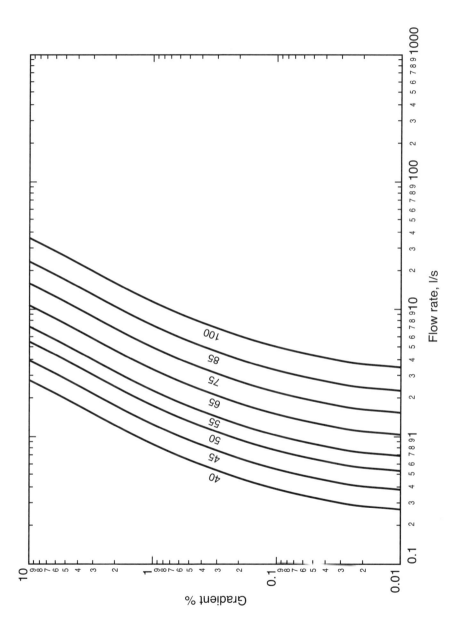

**Fig. 35** Flow rate in field drain pipes according to gradient and diameter. Type: Laterals – smooth plastic pipes with longitudinal slots. Internal diameter of pipes in mm. *Source:* MAFF/ADAS (1983) *The Design of Field Drainage Pipe Systems.* Reference Book No. 745. Reproduced with kind permission of MAFF ADAS (UK).

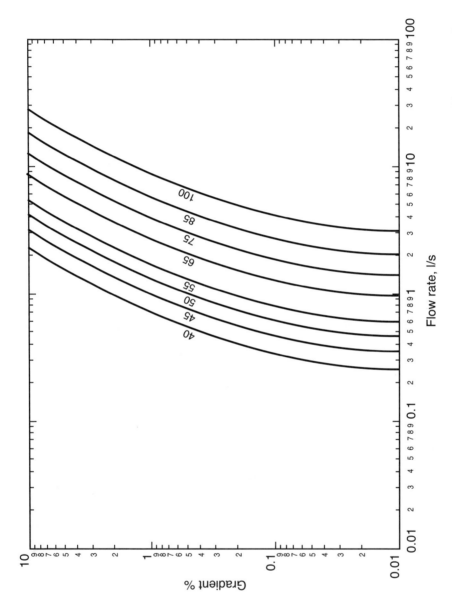

**Fig. 36** Flow rate in field drain pipes according to gradient and diameter. Type: Laterals – smooth plastic pipes with transverse slots. Internal diameter of pipes in mm. *Source:* MAFF/ADAS (1983) *The Design of Field Drainage Pipe Systems.* Reference Book No. 345. Reproduced with kind permission of MAFF ADAS (UK).

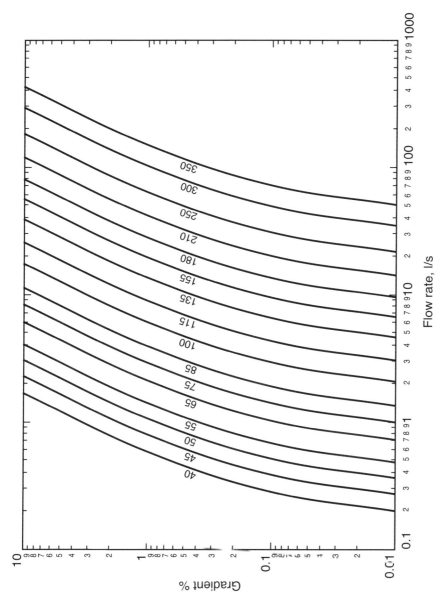

**Fig. 37** Flow rate in field drain pipes according to gradient and diameter. Type: Laterals – corrugated plastic pipes. Internal diameter of pipes in mm. *Source*: MAFF/ADAS (1983) *The Design of Field Drainage Pipe Systems*. Reference Book No. 345. Reproduced with kind permission of MAFF ADAS (UK).

**Table 120  Maximum Safe Velocity in Surface Stormwater Drains**

| Material | Maximum velocity on cover expected after two seasons m/s | | |
| --- | --- | --- | --- |
| | Bare | Medium grass cover | Very good grass cover |
| Very light silty sand | 0.3 | 0.75 | 1.5 |
| Light loose sand | 0.5 | 0.9 | 1.5 |
| Coarse sand | 0.75 | 1.25 | 1.7 |
| Sandy soil | 0.75 | 1.5 | 2.0 |
| Firm clay loam | 1.0 | 1.7 | 2.3 |
| Stiff clay or stiff gravelly soil | 1.5 | 1.8 | 2.5 |
| | | | unlikely to |
| Coarse gravels | 1.5 | 1.8 | form very |
| Shale, hardpan soft rock, etc. | 1.8 | 2.1 | good grass |
| | | | cover |
| Hard cemented conglomerates | 2.5 | - | - |

*Source:* Dept. Conservation and Extension, Government of Rhodesia, Dept. In-service Manual.

**TABLE 121   Indicative Land Drainage Pipe Gradients according to Pipe Diameter and Soil Type**

| | Gradients % | | | | |
| --- | --- | --- | --- | --- | --- |
| | Desirable minimum | Minimum | Maximum | | |
| Pipe diameter | | | Sandy loam | Silt loam | Stiff clay |
| 70 mm | 0.4 | – | 4.0 | 5.0 | 8.3 |
| 100 mm | 0.3 | 0.1 | 1.7 | 3.3 | 6.7 |
| 150 mm | 0.2 | 0.05 | 1.0 | 2.0 | 4.0 |

*NB* Steeper gradients risks pipes being undermined by external flow.

*Source:* Withers, B. & Vipond, S. (1974) *Irrigation Design and Practice.*
Reproduced with kind permission of B.T. Batsford Ltd.

**TABLE 122   Ditch Discharge Capacity in l/s where Flow Width is about 3 x Flow Depth and the Channel is in Moderate Condition**

| | Gradient 1: | | | | | | | | | | | | | |
|---|---|---|---|---|---|---|---|---|---|---|---|---|---|---|
| | 10 000 | 1500 | 1000 | 800 | 600 | 400 | 200 | 100 | 90 | 80 | 70 | 60 | 50 | 40 |
| 0.2 | 6 | 16 | 18 | 21 | 24 | 29 | 41 | 59 | 62 | 66 | 70 | 76 | 83 | 93 |
| 0.3 | 17 | 46 | 55 | 62 | 72 | 87 | 123 | 174 | 182 | 195 | 208 | 225 | 246 | 275 |
| 0.4 | 37 | 99 | 118 | 135 | 153 | 186 | 264 | 373 | 392 | 417 | 446 | 482 | 528 | 590 |
| 0.5 | 68 | 181 | 216 | 246 | 282 | 342 | 483 | 683 | 717 | 764 | 817 | 883 | 966 | 1080 |
| 0.6 | 110 | 292 | 349 | 398 | 455 | 552 | 781 | 1104 | 1158 | 1234 | 1320 | 1427 | 1561 | 1746 |
| 0.7 | 167 | 441 | 527 | 601 | 686 | 833 | 1178 | 1666 | 1747 | 1863 | 1992 | 2153 | 2356 | 2634 |
| 0.8 | 239 | 632 | 756 | 861 | 985 | 1195 | 1810 | 2389 | 2506 | 2671 | 2854 | 3088 | 3379 | 3778 |
| 0.9 | 324 | 857 | 1025 | 1168 | 1336 | 1620 | 2291 | 3240 | 3398 | 3622 | 3874 | 4187 | 4582 | 5123 |
| 1.0 | 433 | 1146 | 1370 | 1562 | 1787 | 2167 | 3064 | 4333 | 4545 | 4845 | 5181 | 5600 | 6128 | 6852 |

Ditch bottom width (m)

*Note:* Not strictly accurate for ditches with vertical sides or for ditches which have a batter greater than 1 in 3. The margin of error increases with discharge and for carrier ditches - an engineer should be consulted.

*Source:* Farr, E. & Henderson, W.C. (1986) *Land Drainage.*
Reproduced with kind permission of Longman Group, UK.

**TABLE 123   Ditch Flow Velocities in m/s Where the Channel is in a Good, Newly-Cleaned Condition**

| | Gradient 1: | | | | | | | | | | | | | |
|---|---|---|---|---|---|---|---|---|---|---|---|---|---|---|
| | 10 000 | 1500 | 1000 | 800 | 600 | 400 | 200 | 100 | 90 | 80 | 70 | 60 | 50 | 40 |
| 0.2 | 0.10 | 0.26 | 0.31 | 0.35 | 0.40 | 0.49 | 0.69 | 0.98 | 1.03 | 1.09 | 1.17 | 1.26 | 1.38 | 1.55 |
| 0.3 | 0.13 | 0.34 | 0.41 | 0.47 | 0.53 | 0.64 | 0.91 | 1.29 | 1.35 | 1.44 | 1.54 | 1.67 | 1.82 | 2.04 |
| 0.4 | 0.16 | 0.41 | 0.49 | 0.56 | 0.64 | 0.78 | 1.10 | 1.56 | 1.63 | 1.74 | 1.86 | 2.01 | 2.20 | 2.46 |
| 0.5 | 0.18 | 0.48 | 0.58 | 0.66 | 0.75 | 0.91 | 1.29 | 1.82 | 1.91 | 2.04 | 2.18 | 2.36 | 2.58 | 2.88 |
| 0.6 | 0.20 | 0.54 | 0.65 | 0.74 | 0.84 | 1.02 | 1.45 | 2.04 | 2.14 | 2.29 | 2.45 | 2.64 | 2.89 | 3.23 |
| 0.7 | 0.23 | 0.60 | 0.72 | 0.82 | 0.94 | 1.13 | 1.60 | 2.27 | 2.38 | 2.54 | 2.71 | 2.93 | 3.21 | 3.58 |
| 0.8 | 0.25 | 0.66 | 0.79 | 0.90 | 1.03 | 1.24 | 1.76 | 2.49 | 2.61 | 2.78 | 2.98 | 3.22 | 3.52 | 3.94 |
| 0.9 | 0.27 | 0.71 | 0.84 | 0.96 | 1.10 | 1.33 | 1.89 | 2.67 | 2.80 | 2.98 | 3.19 | 3.45 | 3.77 | 4.22 |
| 1.0 | 0.29 | 0.76 | 0.91 | 1.04 | 1.19 | 1.44 | 2.04 | 2.89 | 3.03 | 3.23 | 3.46 | 3.73 | 4.09 | 4.57 |

Ditch bottom width (m)

*Source:* Farr, E. & Henderson, W.C. (1986) *Land Drainage.*
Reproduced with kind permission of Longman Group, UK.

**TABLE 124  Ditch Bank Gradients (batter) according to Soil Type**

| | | | |
|---|---|---|---|
| Rock, dry fibrous peat | 1 : 0 | to | 1 : 05 |
| Clay, loess | 1 : 0.5 | to | 1 : 1 |
| Loam, clay loam, silty loam | 1 : 1 | to | 1 : 1.5 |
| Compacted sand or sandy loam | 1 : 1.5 | to | 1 : 2 |
| Loose sand or sandy loam | 1 : 2 | to | 1 : 3 |
| Loose fine sand, soft peat | 1 : 3 | to | 1 : 4 |

*NB* Essentially the gradient of the bank must be less steep than the natural
angle of repose of the material.

*Source:* Farr, E. & Henderson, W.C. (1986) *Land Drainage.*
Reproduced with kind permission of Longman Group, UK.

# Field Systems

**Note 4  Field Drainage**

Drain Spacing depends heavily on drain depth, country, rainfall and intended
crop and animal grazing practices. The following series of tables gives some
examples of practices in different countries.

**TABLE 125  Prevalent and Indicative Depth and Spacing of Lateral Field Drains
according to Soil Type and Permeability (Israel)**

| Soil Type | Hydraulic conductivity cm/day | Drain Spacing (m) | Drain Depth (m) |
|---|---|---|---|
| Clay | 0.15 | 10 - 20 | 1 - 1.5 |
| Clay Loam | 0.15 - 0.5 | 15 - 25 | 1 - 1.5 |
| Loam | 0.5 - 2.0 | 20 - 35 | 1 - 1.5 |
| Fine Sandy Loam | 2.0 - 6.5 | 30 - 40 | 1 - 1.5 |
| Sandy Loam | 6.5 - 12.5 | 30 - 70 | 1 - 2 |
| Peat | 12.5 - 25 | 30 - 100 | 1 - 2 |

**TABLE 126   Suggested Lateral Field Drain Spacings according to Soil Type (USA)**

| Soil Type | Spacing (m) | Permeability |
|-----------|-------------|--------------|
| Clay and Clay Loam | 9 - 18 | Very Slow |
| Silt and Silty Clay Loam | 18 - 30 | Slow - Moderately Slow |
| Sandy Loam | 30 - 90 | Moderately Slow - Rapid |
| Peat and Muck | 15 - 61 | Slow - Rapid |

*Source*: USDA (1955) *Yearbook of Agriculture.*

**TABLE 127   Indicative Lateral Field Drain Spacings and Depths according to Soil Type (UK)**

| | Depth (m) | Spacing (m) |
|---|-----------|-------------|
| Clay | 0.6 - 0.76 | 3.66 - 6.1 |
| Medium Loam | 0.76 - 0.92 | 6.1 - 9.15 |
| Sandy Loam | 0.92 - 1.22 | 9.15 - 12.2 |
| Peat | >1.07 | 5.49 - 6.41 |

**TABLE 128   Drain Space and Depth Recommendations for British Agriculture in the 1940s (m)**

| SPACING | | | Depth |
|---------|---|---|-------|
| **Open soils** | **Medium soils** | **Clay soils** | **Depth** |
| 18.4 - 22.1 | 14.7 - 18.4 | 7.4 - 11.0 | 1.83 |
| 15.6 - 18.4 | 12.0 - 15.6 | 6.4 - 9.2 | 1.52 |
| 12.0 - 14.7 | 10.1 - 12.0 | 4.6 - 7.4 | 1.22 |
| 9.2 - 11.0 | 7.4 - 9.2 | 3.7 - 5.5 | 0.92 |
| 6.4 - 7.4 | 4.6 - 6.4 | 2.8 - 3.7 | 0.61 |
| 4.6 - 5.5 | 3.7 - 4.6 | 1.8 - 2.8 | 0.46 |

*Source:* Adapted from Nicholson, H.H. (1946) *The Principles of Field Drainage.*

**TABLE 129   Depth and Spacing of Lateral Field Drains (European)**

| Soil type | Depth (m) | Spacing (m) |
| --- | --- | --- |
| Sand | 0.6 | up to 60 |
| Sandy Loam | 0.8 - 1 | up to 60 |
| Silt Loam | 0.8 - 1.8 | 20 - 80 |
| Clay Loam | 0.6 - 0.8 | 15 - 100 |
| Peat | 1.2 - 1.5 | 40 - 60 |

*Source:* Hudson, N.W. (1975) *Field Engineering for Agricultural Development.*
Reproduced with kind permission of N.W. Hudson.

# Farm Machinery

# Field Machinery – Outputs / Power Requirements

**TABLE 130   Soil Resistance to Cultivation Implements**

| Operation | Pounds of Draft/Foot of Width | Typical Speed mph | Drawbar Horsepower per Foot of Width |
|---|---|---|---|
| **Ploughing – 8 inches deep** | | | |
| Gumbo | 1250 | 4.0 | 13.3 |
| Clay | 1050 | 4.0 | 11.2 |
| Loam | 950 | 4.5 | 11.4 |
| Sandy Loam | 700 | 5.0 | 9.3 |
| Sand | 350 | 5.0 | 4.7 |
| **Chisel Ploughing – 8 inches deep** | | | |
| Hard, Dry | 800 | 4.0 | 8.5 |
| Medium Clay Loam, Good Moisture | 500 | 5.0 | 6.7 |
| Sand, Sandy Loam | 200 | 6.0 | 3.2 |
| **Field Cultivator** | | | |
| Heavy Clay Soils or | | | |
| Dry and Hard Conditions | 650 | 4.0 | 6.9 |
| Clay Loam | 450 | 5.0 | 6.0 |
| Sandy Loam | 300 | 5.0 | 4.0 |
| Sand | 150 | 6.0 | 2.4 |
| **Tandem Disk Harrow** | | | |
| Heavy Draft | 300 | 4.0 | 3.2 |
| Medium Draft | 200 | 5.0 | 2.7 |
| Light Draft | 100 | 6.0 | 1.6 |
| **Offset or Heavy Tandem Disk** | | | |
| Heavy Draft | 400 | 4.0 | 4.3 |
| Medium Draft | 325 | 5.0 | 4.3 |
| Light Draft | 250 | 6.0 | 4.0 |
| **One-Way Disk** | | | |
| Heavy Draft | 400 | 4.0 | 4.3 |
| Medium Draft | 300 | 5.0 | 4.0 |
| Light Draft | 200 | 6.0 | 3.2 |

*Source:* John Deere & Co. (1981) *Fundamentals of Machinery Operation.*
Reproduced with kind permission of Deere and Company.
© 1981 Deere and Company. All rights reserved.

**TABLE 131    Drawbar Horsepower for Tillage Tools**

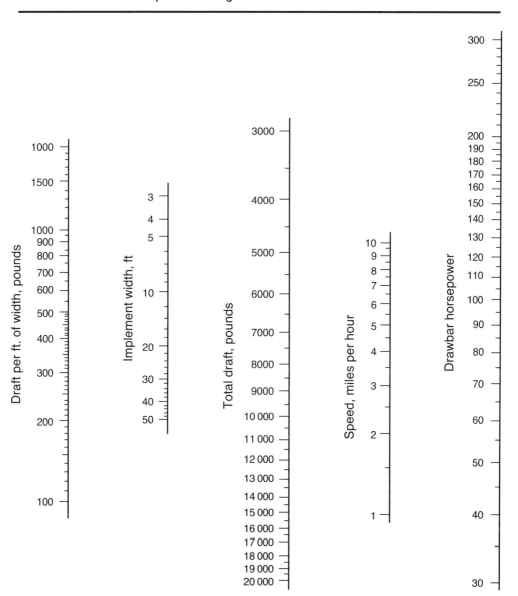

*Source:* John Deere & Co. (1981) *Fundamentals of Machinery Operation.*

---

### Note 5  Field Machinery and Farm Equipment Capacities

---

It is notoriously difficult to predict machinery work rates as they depend critically on such factors as:

- Correct match between tractor power and implement size
- Operator proficiency and tenacity
- Crop volume and condition
- Machine turn around times where filling is involved
- Soil conditions and type
- Field size and shape
- Field topography
- Age of machinery
- For draft operations, tractor wheel slippage
- Specific design of machines

Because of the difficulty in predicting output, examples are given from several regional sources.

---

### TABLE 132  Field Machine Speeds and Power Requirements

---

| Operation | Typical Field Working Speed kph | PTO hp per metre width |
|---|---|---|
| Deep Plough 21 cm deep | | |
| Heavy clay | 6.4 | 80 |
| Clay | 6.4 | 67 |
| Loam | 8 | 54 |
| Sandy loam | 8 | 40 |
| | | |
| Heavy Cultivations 21 cm deep | | |
| Hard and Dry | 6.4 | 50 |
| Medium Damp loam | 8 | 40 |
| Sandy loam | 9.7 | 24 |
| | | |
| Light cultivation S tine | | |
| Heavy and dry | 6.4 | 40 |
| Clay Loam | 8 | 37 |
| Sandy loam | 8 | 24 |
| Sand | 9.7 | 14 |

**TABLE 132** *(continued)*  **Field Machine Speeds and Power Requirements**

| Operation | Typical Field Working Speed kph | PTO hp per metre width |
|---|---|---|
| Tandem Disc Harrow | | |
| Heavy Draft | 6.4 | 24 |
| Medium Draft | 8 | 20 |
| Light Draft | 9.7 | 17 |
| | | |
| Offset Disc Harrow | | |
| Heavy Draft | 6.4 | 30 |
| Medium Draft | 8 | 27 |
| Light Draft | 9.7 | 24 |
| | | |
| Rotovator 10 cm deep | 5 | 40 |
| Power Harrow 15 cm deep. 2bar | 4 | 17 |
| Tooth Harrow 7 cm deep | 6 | 10 |
| Seed Drill <5 cm deep | 6 | 18 |
| Rotary Mower | 5 | 37 |
| | | |
| Subsoiler | | |
| 45 cm deep | 4 | 40 per tine |
| 60 cm deep | | 55 per tine |
| 75 cm deep | | 70 per tine |
| | | |
| Mulch tiller (disc/chisel combination) 10 - 25 cm deep | 8 - 11 | 36 - 60 |

*Sources:* Massey Ferguson (World Export Operations) Ltd. *Field Application Guide,* also ILACO (1985) *Agricultural Compendium* and John Deere brochures. Reproduced with kind permission respectively of Massey Ferguson (UK) Ltd., ILACO BV and Deere & Co.

**TABLE 133    Typical Good Field Machinery Work Rates by Contractors in Large Field Situations in UK using 100 - 120 hp 4WD Tractor for Heavier Cultivations and 80 hp Tractors for Lighter Work as Appropriate**

| Operation | ha / hr | Comments |
|---|---|---|
| Ploughing | 0.75 | |
| Deep Ploughing | 0.6 | |
| Rotovating | 0.75 | Ploughed Land |
| Rotovating | 0.43 | Un-ploughed |
| Subsoiling | 0.75 | |
| Chisel Ploughing | 1.5 | |
| Heavy Cultivating | 2.0 | |
| Heavy Disc | 1.8 | |
| Light Disc | 2 | |
| Power Harrow | 1 | |
| Spring Tine Harrow | 2.25 | |
| Seedbed Harrowing | 1.5 | |
| Ring Rolling | 2.6 | 7.3 m light roller 3 gang |
| Flat Roller | 0.7 | 2.3 m heavy |
| Broadcast Fertilizer | 3.13 | 125 - 375 kg / ha |
| Broadcast Fertilizer | 1.9 | 500 - 1250 kg / ha |
| Pneumatic Spread Fertilizer | 3.75 | 125 - 375 kg / ha |
| Cereal Seeding | 1.25 | Direct Drill |
| Cereal Seeding | 2 | Conventional Drill |
| Cereal Seeding | 1.63 | Combined seed and fertilizer |
| Sugar Beet Seeding | 1.25 | 12 row precision drill |
| Sugar Beet Hoeing | 1 | |
| Potato Planting | 0.2 | 2 row planter. Manual hand fed |
| Potato Planting | 0.35 | 2 row automatic |
| Potato Ridging | 0.7 | 4 row automatic |
| Crop Spraying | 3.13 | Low Volume up to 175 l/ha |
| Crop Spraying | 2.5 | Med Volume 200 - 300 l/ha |
| Crop Spraying | 1.9 | High Volume over 800 l/ha |
| Combine Harvesting | | |
| Cereals | 1.75 | |
| Oilseed rape | 1.5 | Direct cut |
| Beans | 1.25 | |
| Peas | 1 | |
| Combine Harvesting | 1.05 | 1.05 m drum, 3 - 3.7 m table |
| according to drum | 1.3 | 1.3 m drum, 4.3 m table |
| and table width | 1.55 | 1.55 m drum, 5 - 5.55 m table |

**TABLE 133** *(continued)* **Typical Good Field Machinery Work Rates by Contractors in Large Field Situations in UK using 100 - 120 hp 4WD Tractor for Heavier Cultivations and 80 hp Tractors for Lighter Work as Appropriate.**

| Operation | ha / hr | Comments |
|---|---|---|
| Pick-up Baling | 1.25 | Straw |
| Pick-up Baling | 0.8 | Hay |
| Potato Harvesting | 0.125 | 1 row manned complex harvester |
| Potato Harvesting | 0.18 | 2 row digger, 23 casual labourers |
| Potato Harvesting | 0.07 | 1 row manned simple harvester |
| Potato Harvesting | 0.18 | 2 row manned harvester |
| Sugar Beet Harvesting | 0.44 | 3 row harvester |
| Grass Mowing | 1.25 | |
| Grass Tedding/Turning | 1.9 | |
| Grass Seed Sowing | 1.3 | 3.7 m wide drill or barrow |
| Silage Cutting | 0.3 | Direct Cut |
| Silage Pick up and chop | 0.7 | Precision Cut. 60KW tractor |
| Silage Big Round Baling | 0.8 | |
| Forage pick-up wagon | 0.4 | |
| Stone Windrowing | 0.5 | For potatoes, 3 row |
| Ridging for potatoes | 0.7 | 3 row |
| Precision Seeding | 0.3 | 2 row |
| Inter row Cultivation | 0.5 | 3 row, root crops |
| Turnip Harvesting | 0.15 | digger/elevator type machine |
| Windrow oilseed rape | 0.18 | |
| Straw Carting | 1.25 - 1.9 | Two men. Fully mechanized Flat 8/10 bale accumulator system |
| For Tractors over 120 hp | | |
|     Ploughing | 0.88 | |
|     Chisel Ploughing | 1.9 | 3 m |
|     Heavy Cultivating | 3.75 | 7 m |
|     Disc Cultivating | 3.13 | |
|     Power Harrowing | 1.5 | |

Adapted from: Nix, J. (1988 et seq.) *Farm Management Pocketbook.*
      Scottish Agricultural Colleges (1986 et seq.) *Farm Management Handbook.*
Reproduced with kind permission of J. Nix and The Scottish Agricultural Colleges.

*Field Machinery – Outputs / Power Requirements*

**TABLE 134  Large Field Machinery General Specifications and Power Requirements**

| Machine | Specification | | hp Requirement | |
|---|---|---|---|---|
| Big Baler | All Hydraulic. Square bale 95 x 120 cm | | 120 | |
| | Up to 40t/hr capacity | | | |
| Box Scraper | 2 m³ | 1.8 m cut | 65 - 85 | |
| (wheeled | 3 | 1.8 | 75 - 100 | |
| tractors) | 5 | 2 | 100 - 140 | |
| | 6.5 | 2 | 125 - 180 | |
| | 1.5 | 3 | 50 - 100 | |
| | 2 | 4 | 70 - 120 | |
| | 2.5 | 5 | 100 - 160 | |
| (crawler | 5 | 2 | 80 - 105 | |
| tractors) | 6.5 | 2 | 93 - 135 | |
| Chisel Tine cultivators | | | Rigid Tine | Spring Tine |
| for Stubble Mulching | 3.3 m | | 66 - 110 | 49 - 82 |
| | 5.1 | | 102 - 170 | 76 - 127 |
| | 6.3 | | 126 - 200 | 94 - 149 |
| | 7.5 | | 112 - 250 | 83 - 186 |
| | 8.7 | | 130 - 290 | 96 - 216 |
| | 9.9 | | 148 - 330 | 110 - 246 |
| | 11.1 | | 166 - 370 | 126 - 276 |
| | 12.3 | | 184 - 400 | 136 - 298 |
| | 14.1 | | 211 - 450 | |
| | 14.7 | | 220 - 450 | |
| | 15.3 | | 229 - 450 | |
| | 15.9 | | 238 - 450 | |
| | 16.5 | | 247 - 450 | |
| | 17.1 | | 256 - 450 | |
| | 17.7 | | 265 - 450 | |
| Chisel Ripper Ploughs. | 1.5 m | 3 shanks | 90 - 100 | |
| Heavy Duty. 7.5 cm points | 2.4 | 5 | 130 - 160 | |
| | 3.3 | 5 | 150 - 180 | |
| | 4.2 | 7 | 200 - 250 | |
| | 5 | 7 | 220 - 280 | |
| | 5.5 | 9 | 250 - 320 | |
| | 6.9 | 11 | 350 - 400 | |
| | 7.8 | 11 | 420 - 480 | |
| Combine Harvester | 5.4 m cut | | 180 | |
| Disc Mower. Rotary 2 Disc. 3 m cut. Trailed | | | Min 70 | |
| Disc Ridger. Mounted. 2 Ridge, 66 cm discs | | | Min 50 | |
| Disc Ridgers for levees | 8 x 65 cm discs | | 65 - 85 | |
| | 10 x 65 | | 80 - 100 | |
| | 8 x 70 | | 80 - 100 | |
| | 10 x 70 | | 90 - 120 | |

**TABLE 134 *(continued)*   Large Field Machinery General Specifications and
                Power Requirements**

| Machine | Specification | | hp Requirement |
|---|---|---|---|
| Drag Ejector Scraper | 6 m³ | 2.7 m cut | 160 - 180 |
| | 8 | 2.7 | 180 - 230 |
| | 10 | 3.2 | 250 - 300 |
| | 12 | 3.2 | 330 - 350 |
| Harrow: Heavy Duty, | 4 m | | 90 - 110 |
| Rigid Frame, | 6.6 | | 120 - 150 |
| Rigid Vertical Tines, | 8.3 | | 150 - 220 |
| With Crumbler Barrel | 9.3 | | 360 - 400 |
| Harrow: S tine, Rigid Frame, | 3 m | | 50 - 60 |
| With Crumbler Barrel | 4 | | 60 - 80 |
| and Levelling Board | 5 | | 70 - 90 |
| | 6 | | 90 - 120 |
| | 8 | | 140 - 220 |
| | 10 | | 150 - 250 |
| Land Plane | 3 m | | 50 - 100 |
| | 4 | | 70 - 120 |
| | 5 | | 100 - 160 |
| Offset Disc Harrow / | | | |
| Subsoiler / Crumbler Units | 3.1 m | 3 subsoil tines | 160 - 200 |
| (for one pass seed bed | 3.6 | 4 | 250 - 300 |
| preparation) 65 cm discs | 4.2 | 4 | 300 - 350 |
| Offset Disc Plough Harrow: | 3.76 m | | 105 - 135 |
| 65 cm discs, Disc Blade | 4.26 | | 125 - 155 |
| Weighting 40 - 60 kg | 5.11 | | 155 - 190 |
| | 5.58 | | 180 - 220 |
| | 6.26 | | 240 - 270 |
| | 7.32 | | 280 - 320 |
| Offset Disc Plough Harrow: | 1.5 | | 70 - 80 |
| 70 cm discs, Disc Blade | 1.8 | | 90 - 100 |
| Weighting 155 kg | 2.4 | | 110 - 130 |
| | 3 | | 130 - 150 |
| | 3.6 | | 160 - 180 |
| | 4.5 | | 200 - 220 |
| | 5.4 | | 250 - 280 |
| | 6.3 | | 300 - 320 |
| Offset Disc Plough Harrow: | 1.8 | | 100 - 110 |
| 75 cm discs, Disc Blade | 2.4 | | 110 - 130 |
| weighting 200 kg | 3 | | 130 - 160 |
| | 3.6 | | 160 - 180 |
| | 4.5 | | 200 - 240 |
| | 5.4 | | 250 - 300 |
| | 6 | | 300 - 320 |

*( continued over )*

**TABLE 134** *(continued)*  **Large Field Machinery General Specifications and Power Requirements**

| Machine | Specification | | | hp Requirement | |
|---|---|---|---|---|---|
| Offset Disc Plough Harrow: | 1.9 | | | 130 - 160 | |
| 80 cm discs, Disc Blade | 2.28 | | | 170 - 200 | |
| weighting 360 kg | 3.04 | | | 200 - 250 | |
| | 3.8 | | | 260 - 340 | |
| | 4.56 | | | 350 - 420 | |
| | 5.7 | | | 450 - 550 | |
| Offset Disc Plough Harrow: | 1.9 | | | 130 - 170 | |
| 90 cm discs, Disc Blade | 2.28 | | | 180 - 220 | |
| weighting 400 kg | 3.04 | | | 230 - 300 | |
| | 3.8 | | | 300 - 350 | |
| | 4.56 | | | 360 - 430 | |
| | 5.7 | | | 450 - 570 | |
| Rotovators | General Duty | | Heavy Duty | | |
| | m | hp | m | hp | |
| | 3 | 110 - 120 | 2.29 | 120 - 130 | |
| | 3.55 | 120 - 140 | 2.55 | 130 - 140 | |
| | 4.06 | 130 - 150 | 2.8 | 140 - 150 | |
| | 4.57 | 140 - 160 | 3 | 150 - 160 | |
| Sub Surface cultivators: | 1.5 m | 2 Tines | | 60 - 120 | |
| Heavy Duty, Depth Range | 2.4 | 3 | | 120 - 200 | |
| 15 - 60 cm, Winged Shares | 3.3 | 5 | | 200 - 280 | |
| | 4.2 | 7 | | 280 - 360 | |
| | 5 | 9 | | 360 - 400 | |
| Tandem Disc Harrows: | 4 m | | | 80 - 90 | |
| 550 cm discs, Disc Blade | 5 | | | 90 -110 | |
| weighting 40 - 60 kg | 6 | | | 115 - 130 | |
| | 7 | | | 140 - 190 | |
| Subsoiling | | | | crawler tractor | |
|     1 m depth 0.25 ha/hr | | | | 160 | |
|     1.25 m depth 0.25 ha/hr | | | | 300 | |
| Deep Ploughing | | | | crawler tractor | |
|     0.4 m depth | 0.7 m plough width | 0.15 - 0.2 ha/hr | 125 - 140 | | |
|     0.4 m depth | 1 m plough width | 0.15 - 0.2 ha/hr | 250 | | |
|     0.5 m depth | 1.25 m plough width | 0.1 - 0.2 ha/hr | 280 - 300 | | |
|     0.5 m depth | 1.5 m plough width | 0.1 - 0.15 ha/hr | 500 - 550 | | |

*Sources:* Reproduced with kind permission of Simba Machinery Ltd. (Main Source), John Deere (Chisel tine cultivators for Stubble Mulching, Combine), Agrimech Engineering Ltd. (Disc Mower), J.A. Freeman and Son Inc (Big Baler), P.J. Parmiter and Son (Disc Ridger), Agric SA (Rotovators), ILACO (1980) *Agricultural Compendium.*

**TABLE 135    Indicative Outputs of Combine Harvesters, Balers and Forage Harvesters**

**COMBINE HARVESTERS**. Based on 7.4 t/ha standing wheat crop on flat ground

| Engine kw | Header width m | Drum width m | Output t/hr |
|---|---|---|---|
| 227 | 6.1 - 7.3 | 1.56 | 22 |
| 151 | 5.2 - 6.1 | 1.3 | 18 |
| 114 | 4 - 4.6 | 1.3 | 13 |
| 83 | 3.1 - 3.7 | 1.04 | 8 |

**BALERS - Big rectangular**

| Bale Size m | Min hp required | Bales/hr | Bale wt kg | | |
|---|---|---|---|---|---|
| | | | Hay | Straw | Wilted silage |
| 0.6 x 0.9 x 1.2 - 2.5 | 90 | 90 | 325 | 210 | 250 - 425 |

**BALERS - Big round**

| Bale size m width x diameter | hp required | Bales/hr | Bale wt kg | |
|---|---|---|---|---|
| | | | Hay | Straw |
| 1.2 x 1.07 - 1.7 | 50 - 70 | 22 | 370 - 500 | 245 - 330 |
| 1.2 x 1.07 - 1.4 | 45 - 65 | 22 | 250 - 350 | 165 - 230 |

**BALERS - Small rectangular**

| Plunger strokes/min | Bale size m | Bales/hr |
|---|---|---|
| 93 | 0.36 x 0.46 | 220 |
| 80 | 0.36 x 0.46 | 180 |

**BALE PICK-UP WAGON** - Trailed

Automatic for small rectangular bales

1400 - 1600 bales/working day

**FORAGE HARVESTERS**

| Self propelled | 300 hp | 80 t/hr |
|---|---|---|
| | 340 | 100 |

Trailed, double chop, direct cut or pick up and chop

| | 50 - 100 hp | 25 - 30 t/hr |
|---|---|---|

Trailed, pick up and chop

| | 75 - 125 hp | 40 t/hr |
|---|---|---|
| | up to 175 hp | 60 t/hr |

*Source:* Ford-New Holland, Chelmsford, Essex, UK.
Data supplied by and reproduced with kind permission of Ford-New Holland.

**TABLE 136   Indicative Outputs of Forage Harvesting, Raking and Large Baling Machinery**

|  |  |
|---|---|
| **Machine** | **Output** |

Forage Harvester        90 - 140 hp

60 t wet grass/hr
or 30 t 35% DM grass/hr

Single row maize Forage Harvester
                            40 - 100 hp

41 t/hr

Large square baler. Bale size 80 cm x 80 cm
                            Min. 90 hp

Up to 20 t/hr

| | | |
|---|---|---|
| Finger wheel rake. Trailed. Side Delivery | 4.6 m | Up to 6 ha/hr |
| Finger wheel rake. Trailed. V rake | 7.6 m | Up to 10 ha/hr |
| Rake Tedder, double wheel, mounted | 3.6 m | Up to 4 ha/hr |
| Reciprocating tooth rake. Trailed V formation | 7.2 m | Up to 8 ha/hr |

*Source:* Massey Ferguson *Implement Guide.*
Reproduced with kind permission of Massey Ferguson (UK) Ltd.

**TABLE 137   Indicative Tractor Power Requirements according to Cultivation
Implement Type and Width (Zimbabwe)**

| Operation | Working width m | Nominal tractor Engine kw | Operation | Working width m | Nominal tractor Engine kw |
|---|---|---|---|---|---|
| PLOUGHING - MOUNTED | | | DISC HARROW | | |
| 2 disc | | 35 | Trailed.offset, heavy | 2.06 | 56 |
| 3 disc | | 50 | | 2.3 | 65 |
| 4 disc | | 75 | DISC HARROW | | |
| PLOUGHING - TRAILED | | | Rome type | 2.3 | 80 - 100 |
| 5 disc | | 90 | | 2.6 | 100 - 120 |
| 6 disc | | 100 | SPRING TINE CULTIVATOR | | |
| RIPPER | | | 9 tine | 2.74 | 40 |
| 2 tine | | 45 - 60 | 12 tine | 3.66 | 45 |
| 3 tine | | 60 - 80 | GANG TILLER | | |
| 5 tine | | 75 - 90 | 2 row | | 36 |
| 7 tine | | 90 - 100 | 3 row | | 40 |
| DISC HARROW | | | 4row | | 50 |
| Light mounted | 2.9 | 50 | 6 row | | 60 |
| DISC HARROW | | | | | |
| Trailed, single acting | 5.48 | 60 | | | |
| DISC HARROW | | | | | |
| Mounted offset | 1.6 | 40 | | | |
| | 1.83 | 45 | | | |
| | 2.1 | 50 | | | |
| | 2.29 | 60 | | | |

*Source: Rhodesian Farm Management
Handbook* (1970).
Reproduced with kind permission of Ministry of
Lands, Agriculture and Resettlement, Zimbabwe.

**TABLE 138    Typical Field Machinery Work Rates and Fuel Requirements
with 65 kw tractor (Zimbabwe)**

| Operation | ha/hr | l diesel/ha |
|---|---|---|
| **Ploughing** | | |
| 3 disc reversible | 0.43 | 27 |
| 3 disc conventional | 0.37 | 29 |
| 3 furrow mouldboard | 0.37 | 26 |
| 3 disc conventional trailed | 0.29 | 29 |
| **Discing** | | |
| heavy disc 3m Rome type | 0.71 | 9 |
| light disc | 0.88 | 9.5 |
| **Ridging** | | |
| ridge, fertilize, hole, fumigate | | |
| single row | 0.51 | 15 |
| **Landplane** | 1.7 | 4 |
| **Subsoiling and ripping** | | |
| 3 tine ripper after discs | 0.9 | 10 |
| 3 tine ripper | 0.7 | 10 |
| **Planting and Sowing** | | |
| 2 row planter ) | 0.5 | 8 |
| 4 row planter ) manual | 0.9 | 6.5 |
| 6 row planter ) | 1.26 | 5 |
| broadcast spreader | 3 | 2 |
| box type spreader | 1 | 3 |
| Air seed drill | 25.3 | 0.2 |
| **Re-ridging** | | |
| Single row ridger | 0.54 | 11 |
| Double row ridge | 1 | 8 |
| **Inter Row Cultivation** | | |
| Multi row | 1.73 | 6 |
| **Crop spraying** | | |
| Pest and disease - boom sprayer | 2 | 1.5 |
| - mist blower | 4 | 1.5 |
| - air | 25.3 | 1 |
| Herbicide - boom and light disc | 0.88 | 9.5 |
| - boom | 1.8 | 1.5 |
| - air | 25.3 | 1 |
| **Stalk and Stover Destruction** | | |
| Rotary slasher | 0.83 | 7 |
| Ripper with paravane | 0.74 | 10 |
| **Combining** | | |
| Maize, sunflower | 1.6 | 15 |
| **Groundnuts** | | |
| Bed loosen - tractor + blade | 0.78 | 8 |
| Digger/shaker | 0.5 | 10 |

*Source: Rhodesian Farm Management Handbook* (1970).
Reproduced with kind permission of the Ministry of Lands, Agriculture and Resettlement,
Zimbabwe.

**TABLE 139   Small - Medium Sized Field Machinery Power Requirements**

| Machine | Specification | | hp Requirement |
|---|---|---|---|
| Chisel Plough. C Tines, | 5 shank | 1.95 m width | 60 - 80 |
| Spring Loaded | 7 | 2.6 | 80 - 100 |
| | 9 | 3.25 | 100 - 120 |
| Disc Plough | 2 disc | 64 cm width | 30 - 40 |
| | 3 | 90 | 35 - 50 |
| | 3 | 97 | 60 - 70 |
| | 4 | 120 | 90 - 100 |
| | 5 | 150 | 110 - 120 |
| Manure Spreader, | 3.5 t capacity | | 40 - 45 |
| Bed Chain and | 4 | | 45 - 50 |
| Rear Rotor Type | 4.5 | | 50 - 60 |
| | 5.5 | | 60 - 70 |
| Power Harrow | 1.8 m width | | 30 - 60 |
| | 2.3 | | 40 - 70 |
| | 3 | | 70 - 100 |
| Rotary Slasher, | 0.9 m width | | 15 - 20 |
| Blade Type | 1.2 | | 25 - 30 |
| | 1.5 | | 35 - 40 |
| | 2 | | 50 - 60 |
| Rotovators:      to 15 cm depth | | | |
| | 0.76 m width | | 12 - 15 |
| | 1.5 | | 27 - 30 |
| to 20 - 25 cm depth | | | |
| | 1.3 m width | | 40 - 50 |
| | 2.04 | | 70 - 80 |
| Shredder for chopping Straw | 1.23 m width | | 20 - 30 |
| and Maize Stover etc. | 1.95 | | 40 - 60 |
| | 2.55 | | 60 - 80 |
| | 3.2 | | 80 - 100 |
| Stone Picker | 1.6 m width | | 50 (min) |
| | 2.1 | | 65 (min) |
| Tillers, Heavy Duty | 1.85 m width | 7 tine | 25 - 35 |
| spring tine | 2.2 | 9 | 35 - 50 |
| | 2.58 | 11 | 50 - 65 |
| | 3 | 13 | 65 - 90 |
| Tillers, Rigid | 1.3 m width | 5 tine | 30 - 35 |
| Tine Spring Loaded | 2.2 | 9 | 45 - 60 |
| | 3 | 13 | 70 - 90 |
| | 3.4 | 15 | 90 - 120 |

*Source:* Agric SA (Spain) commercial literature.
Reproduced with kind permission of Agric SA, Barcelona, Spain.

**TABLE 140   Indicative Outputs of Single Axle Tractors and Small Farm Machines**

| Machine | Output |
|---|---|
| Power Tiller<br>5 hp petrol engine<br>For paddy fields | Ploughing 0.6 ha/day (mouldboard)<br>Harrowing 0.9 ha/day<br>Reaping 2.5 ha/day |
| Hydro tiller<br>10 hp petrol engine<br>or 6.5 - 9 hp diesel | Rotovator/Puddling action. Incorporates<br>stubble or green manure in paddy fields<br>First pass 1.8 ha/day<br>Second Pass 2 ha/day |
| Cono-Puddler<br>5 hp petrol engine<br>For paddy fields | 1.5 ha/day |
| 1.0 m Reaper<br>3 hp petrol engine | For rice<br>2.4 ha/day |
| Portable Thresher<br>5 hp engine | For rice and some sorghum varieties<br>300 - 600 kg/hr 2 - 3 men required |
| Axial Flow Thresher<br>7 hp engine | For rice<br>Up to 500 kg/hr 3 - 4 men required |
| Axial Flow Thresher<br>10 hp engine | For rice<br>Up to  1000 kg/hr 3 - 4 men required |
| Thresher/Sheller<br>16 hp petrol engine<br>or 11 hp diesel | For rice or Maize (conversion thresher)<br>Up to 5000 kg/hr of corn<br>1000 - 1500 kg rice/hr |
| Portable Grain Cleaner<br>1 hp petrol engine | For rice and other grains<br>Up to 1000 kg/hr 1 - 2 men |
| Two Row Binder 60 cm width<br>3.5 - 5 hp | 10 hr/ha |
| Single Row Binder 30 cm width | 15 hr/ha |
| Soyabean Pod Harvester | 10 hr/ha |
| Mini Combine Harvester<br>60 cm cut (Tracked) | 12.4 hr/ha |
| Mower, 7 hp, 1 m cut | Reciprocating Blade for Grass 8 hr/ha |

*Sources:* Data supplied by and reproduced with kind permission of the International Rice Research Institute, Manila; Willamette Exporting Inc., Oregon (commercial literature); Author's Experience.

**TABLE 141 Field Implement/Tractor Horse Power Compatibility Guide**

| | Massey-Ferguson Tractor Models* | | | 230 |
|---|---|---|---|---|
| **Implement*** | **MF Model*** | **Specification** | **<40** | **40** |
| **PLOUGHS** | | | | |
| f = furrows | | | | |
| sm = semi - mounted | | | | |
| FIXED MOULDBOARD | | | 2,3 f | 2,3 f |
| REVERSIBLE MOULDBOARD | | | | |
| DISC | M-F 765 | Max Depth 240 mm | | 2,3 f |
| | M-F 900 | Max Depth 305 mm | | |
| | M-F 202 | Max Depth 305 mm | | 2,3 f |
| | M-F 204 | Max Depth 305 mm | | |
| | M-F 206 | Max Depth 305 mm | | |
| **DISC HARROWS** | | | | |
| TANDEM | M-F 280 | 24-27 Kg / disc | | 2.3 m |
| | M-F 520 | 43-47 Kg / disc | | |
| OFFSET | M-F 222 | 45-48 Kg / disc | | |
| | M-F 140 | 81-172 Kg / disc | | |
| **CULTIVATORS** | | | | |
| CHISEL / RIPPER | M-F 325-M.DUTY | 510 mm work depth | | |
| | M-F 325-H DUTY | 510 mm work depth | | |
| CHISEL / CULTIVATOR | M-F 750 | 270 mm tine spacing | | |
| SPRING TINE | M-F 305 | C shaped tines with springs | | 2.0 m |
| COIL TINE | M-F 315 | 25 mm coil tines | 2.0 m | 2.0 m |
| FLEXITINE | M-F 754 | 100 mm tine spacing | | |
| SPRING TINE | M-F 755 | 125 mm triple leaf spring tines | | |
| RIGID TINE | M-F 738 | Tiller or adapt to ridger | 1.15 m 1.7 m | 1.7 m 1.88 m |

* *Note* Tractor and Implement Models Nos. as of January 1991
Caution: Guide only referring to 2 and 4 WD models of tractors but soil, climate and topography must be considered.

| 240 | 350 | 355 | 360 | 365 | 375 | 390 | | 398 | 399 | | 3610 | 3630 | 3650 | 3680 | |
|---|---|---|---|---|---|---|---|---|---|---|---|---|---|---|---|
| | | | | | 3050 | 3060 | 3065 | 3070 | 3080 | 3090 | | | | | |
| **Tractor Horse Power** | | | | | | | | | | | | | | | |
| 47 | 52 | 58 | 63 | 68 | 78 | 86 | 92 | 100 | 110 | 116 | 122 | 142 | 158 | 190 | >200 |
| **Compatible Implement Size** | | | | | | | | | | | | | | | |
| | | | | | 5,6 f | 5,6 f | 5,6,7 f | 5,6,7 f | 6,7,8 f | 7,8 f | 8 f | | | | |
| | | | | | sm | sm | sm | sm | sm | sm | sm | | | | |
| 2,3 f | 2,3,4 f | 2,3,4 f | 3,4 f | 3,4,5 f | 3,4,5 f | 4,5 f | 5 f | | | | | | | | |
| | | 2 f | 2,3 f | 2,3 f | 2,3,4 f | 3,4 f | 3,4,5 f | 3,4,5 f | 4,5 f | | | | | | |
| 2,3 f | 2,3 f | 2,3,4 f | 2,3,4 f | 3,4 f | 4 f | 4 f | | | | | | | | | |
| | 2,3 f | 2,3 f | 3 f | 3 f | 3,4 f | 3,4, f | 3,4,5 f | 4,5 f | 4,5,6 f | 5,6 f | 5,6 f | 6 f | 6 f | 6 f | |
| 2,3 f | 2,3 f | 2,3 f | 3 f | 3 f | 3 f | | | | | | | | | | |
| 3 f | 3,4 f | 3,4 f | 3,4 f | 4 f | 4 f | 4 f | | | | | | | | | |
| | 4 f | 4 f | 4,5 f | 4,5 f | 4,5 f | 4,5 f | 5 f | 5 f | | | | | | | |
| 2.3 m | 2.3 m | 2.3 m | 2.7 m | 2.7 m | 3.2 m | 3.2 m | 3.7 m | 3.7 m | | | | | | | |
| | 2.7 m | 3.2 m | 3.2 m | 3.7 m | 3.7 m | | | | | | | | | | |
| | 3.05 m | 3.05 m | 3.05 m | 3.66 m | 3.66 m | 4.27 m | 4.27 m | 5.47 m | 5.47 m | 5.47 m | 6.4 m | 6.4 m | | | |
| | | | 3.66 m | | 4.27 m | | 5.47 m | | | 6.4 m | | | | | |
| 1.8 m | 1.8 m | 1.8 m | 1.8 m | 1.8 m | | | | | | | | | | | |
| | | | | 2.3 m | 2.3 m | 2.3 m | 2.3 m | | | | | | | | |
| | | 2.7 | 3.15 | 3.6 | 4.05 | 4.5 | 4.60 | 4.95 | 5.3 | 5.4 | 5.6 | 6.1 | 6.1 | | |
| | | | | | 2.5 m | 2.5 m | 2.5 m | 2.5 m | 2.5 m | 2.5 m | | | | | |
| | | | | | | | | 2.5 m | 2.5 m | 2.5 m | 2.5 m | 3.0 m | 3.0 m | | |
| | | | | | | | | | | | 3.0 m | 4.2 m | 4.2 m | 4.2 m | 4.2 m |
| 2.0 m | 2.0 m | 2.0 m | 2.0 m | 2.0 m | 2.5 m | 3.0 m | 3.0 m | 3.5 m | 3.5 m | 4.0 m | 4.0 m | | | | |
| | 2.5 m | 2.5 m | 2.5 m | 2.5 m | 3.0 m | 3.5 m | 3.5 m | 4.0 m | 4.0 m | | | | | | |
| | | | 3.0 m | 3.0 m | 3.5 m | 4.0 m | 4.0 m | | | | | | | | |
| 2.0 m | 2.0 m | 2.0 m | 2.0 m | 3.5 m | 3.5 m | 3.5 m | 3.5 m | 3.5 m | 3.5 m | 4.5 m | 4.5 m | | | | |
| | | | | | | 4.5 m | 4.5m | 4.5 m | 6.0 m | 6.0 m | | 6.0 m | 6.0 m | | |
| | | | | | | | 6.0 m | 6.0 m | | | | | | | |
| 2.0 m | 2.0 m | 3.0 m | 3.0 m | 3.5 m | 3.5 m | 3.5 m | 4.5 m | 4.5 m | 4.5 m | 4.5 m | 4.5 m | 6.0 m | 6.0 m | | |
| 3.0 m | 3.0 m | 3.5 m | 3.5 m | 4.5 m | 4.5 m | 4.5 m | 6.0 m | 6.0 m | 6.0 m | 6.0 m | 6.0 m | | | | |
| | 3.5 m | | | | | | | | | | | | | | |
| 3.0 m | 3.0 m | 3.0 m | 3.0 m | 3.5 m | 3.5 m | 4.5 m | 4.5 m | 4.5 m | 5.0 m | | | | | | |
| | 3.5 m | 3.5 m | 4.5 m | 4.5 m | 5.0 m | 5.0 m | 5.0 m | 6.0 m | 6.0 m | 6.0 m | | | | | |
| | | 3.5 m | 3.5 m | 3.5 m | 3.5 m | 3.5 m | 4.5 m | 4.5 m | 6.0 m | 6.0 m | | | | | |
| | | | | 4.5 m | 4.5 m | 4.5 m | 6.0 m | 6.0 m | 6.5 m | 6.5 m | 6.5 m | 6.5 m | 6.5 m | | |
| | | | | | 6.0 m | 6.5 m | 6.5 m | | | | | | | | |
| 1.88 m | 1.88 m | 2.34 m | 2.34 m | 2.79 m | 2.79 m | 2.79 m | 2.79 m | 3.23 m | 3.23 m | | | | | | |
| 2.34 m | 2.34 m | 2.79 m | 2.79 m | 3.23 m | 3.23 m | 3.23 m | 3.23 m | | | | | | | | |
| | 2.79 m | | 3.23 m | | | | | | | | | | | | |

*( continued over )*

**TABLE 141 *(continued)* Field Implement/Tractor Horse Power Compatibility Guide**

| | Massey-Ferguson | | | 230 |
|---|---|---|---|---|
| | Tractor Models* | | | |
| **Implement*** | **MF Model*** | **Specification** | <40 | 40 |
| **CULTIVATORS continued** | | | | |
| TOOLBAR | M-F 352 | Springtine with ducksfoot sweeps | 2.25 m | 2.25 m |
| | M-F 355 | Disc ridger | | |
| RIDGERS | M-F 80 | 4 Lister body ridger | √ | √ |
| | M-F 738 | Spring loaded rigid tiller | 30 MIN | √ |
| BUNDFORMER | M-F 732 | Max (m wide + 0.5 m bund) | | |
| DITCHER / SUBSOILER | M-F 731 | 610 mm max. ditch depth | | |
| ROTARY CULTIVATORS | M-F 762 | 1.3 - 2.5 m width | 35 MIN | √ |
| **PLANTING** | | 0.7 t capacity for 2.9 m | | |
| DRILL END WHEEL | M-F 30 | 1.8 t capacity for 6 m | | |
| CHAIN OR | M-F 500 | 1 t capacity for 3 m | | |
| GRAIN + FERTILIZER | | 1.4 t capacity for 4 m | | |
| PNEUMATIC | M-F 510 | | | |
| | | 1.2 t capacity | | |
| PRECISION PLANTER | M-F 501 | 2,3,4,5,6 row without fertilizer | | |
| | | 2,3,4 row with fertilizer | | |
| TILLER / SEEDER | M-F 738 | 2.3 m width | | |
| WIDE LEVEL DISC SEEDER UNITS | M-F 360 | 4.58 m width / unit capacity:0.15 m | | |
| M = Multiple  S = Single (Units) | | 3 grain & 100 kg fertilizer / 0.3 m width | | |
| FERTILIZER SPREADER | M-F 566 | 900 kg capacity)   } 6 - 15 m | | |
| (mounted,hopper) | M-F 568 | 1000 kg capacity)  } spread | | |
| | M-F 570 | 1100 kg capacity)  } width | | |
| **HAY & FORAGE** | | | | |
| DISC MOWERS | M-F 123 | | | 1.65 m |
| | M-F 125 | | | |
| | M-F 127 | | | |
| DRUM MOWERS | M-F 122 | | | 1.65 m |
| | M-F 124 | | | |
| | M-F 126 | | | |
| MOWER/ | M-F 164 / 165 | | | |
| CONDITIONER | M-F 166 / 167 | | | |
| SLASHERS | M-F 672 | 1 - 2.25 m width | 12 MIN | √ |

* *Note* Tractor and Implement Models Nos. as of January 1991

Caution: Guide only referring to 2 and 4 WD models of tractors but soil, climate and topography must be considered.

| 240 | 350 | 355 | 360 | 365 | 375 | 390 | | 398 | 399 | | 3610 | 3630 | 3650 | 3680 | |
|---|---|---|---|---|---|---|---|---|---|---|---|---|---|---|---|
| | | | | | 3050 | 3060 | 3065 | 3070 | 3080 | 3090 | | | | | |
| **Tractor Horse Power** | | | | | | | | | | | | | | | |
| 47 | 52 | 58 | 63 | 68 | 78 | 86 | 92 | 100 | 110 | 116 | 122 | 142 | 158 | 190 | >200 |
| **Compatible Implement Size** | | | | | | | | | | | | | | | |
| 2.25 m | 2.25 m | 2.25 m | 2.25 m | ALL | ALL | 3.3 m | 3.7 m | 3.7 m | 3.7 m | | | | | | |
| | | | 2.9 m | | | 3.7 m | | | | | | | | | |
| | | 2 | 2 | 2 | 3 | 3 | 3 | | (ridges) | | | | | | |
| √ | √ | √ | √ | √ | √ | √ | | (with | land | wheels) | | | | | |
| √ | √ | √ | √ | √ | √ | √ | (5 - | 13 tines | 20 - | 100h.p.) | | | | | |
| √ | √ | √ | √ | √ | √ | √ | √ | √ | | | | | | | |
| √ | √ | √ | √ | √ | √ | √ | √ | | | | | | | | |
| √ | √ | √ | √ | √ | | | | | | | | | | | |
| 2.4 m | 2.4 m | 2.4 m | 2.4 m | 2.4 m | 2.4 m | 2.4 m | | 6.0 m | 6.0 m | 6.0 m | 6.0 m | 6.0 m | 6.0 m | | |
| | 3 m | 3 m | 3 m | 3 m | 3 m | 3 m | 3 m | 3 m | | | | | | | |
| | | | 4 m | 4 m | 4 m | 4 m | 4 m | 4 m | 4 m | 4 m | | | | | |
| | | | | | | 4.0 m | 4.0 m | 4.0 m | 6.0 m | 6.6 m | 6.6 m | | | | |
| | | | | | | | | 6.0 m | 6.6 m | 8.0 m | 8.0 m | | | | |
| | | | | | | | | 6.6 m | 8.0 m | | | | | | |
| | | √ | √ | √ | √ | √ | √ | √ | √ | | | | | | |
| √ | √ | √ | √ | √ | √ | | | | | | | | | | |
| | | | | | √ | √ | √ | √ | √ | √ | M | M | M | M | |
| | | | | | S | S | S | S | S | S | √ | √ | √ | √ | |
| √ | √ | √ | √ | √ | √ | √ | | | | | | | | | |
| | | √ | √ | √ | √ | √ | √ | √ | | | | | | | |
| | | | √ | √ | √ | √ | √ | √ | | | | | | | |
| 1.65 m | 1.65 m | 1.65 m | 1.65 m | 1.65 m | 1.65 m | 1.65 m | | | | | | | | | |
| 2.16 m | 2.16 m | 2.16 m | 2.16 m | 2.16 m | 2.16 m | 2.16 m | 2.16 m | 2.16 m | | | | | | | |
| | | 2.4 m | 2.4 m | 2.4 m | 2.4 m | 2.4 m | 2.4 m | 2.4 m | 2.4 m | 2.4 m | | | | | |
| 1.65 m | 1.65 m | 1.65 m | 1.65 m | 1.65 m | 1.65 m | 1.65 m | | | | | | | | | |
| 1.85 m | 1.85 m | 1.85 m | 1.85 m | 1.85 m | 1.85 m | 1.85 m | 1.85 | 1.85 m | | | | | | | |
| | | 2.1 m | 2.1 m | 2.1 m | 2.1 m | 2.1 m | 2.1 m | 2.1 m | 2.1 m | 2.1 m | | | | | |
| | | | | | 2.75 m | 2.75 m | 2.75 m | 2.75 m | 2.75 m | 2.75 m | | | | | |
| | | | | | | 3.2 m | 3.2 m | 3.2 m | 3.2 m | 3.2 m | 3.2 m | | | | |
| √ | √ | √ | √ | √ | √ | √ | √ | √ | √ | √ | | √ | √ | | |

*( continued over )*

**TABLE 141** *(continued)* **Field Implement/Tractor Horse Power Compatibility Guide**

| Implement* | MF Model* | Specification | <40 | 40 |
|---|---|---|---|---|
| | Massey-Ferguson | | | 230 |
| | Tractor Models* | | | |
| **HAY & FORAGE Continued** | | | | |
| GRASS FORAGE HARVESTERS | M-F 610 | Max.60 t wet grass or | | (1.75 m |
| | | 30 t 35% DM grass / hr | | |
| MAIZE FORAGE HARVESTERS | M-F 620 | Single row. Up to 41 t / hr | | |
| RAKE | M-F 35 / 36 / 38 / 40 | 3.65 / 4.6 / 5.5 / 7.6 m width | | √ |
| RAKE TEDDERS* / TEDD** | M-F 62*/ 64*/ 72** | 3 / 3 / 5 m width tedding | | |
| RECIPROCATING TINE 'V' RAKE | M-F 104 | 7.2 m width | | |
| BALERS CONVENTIONAL | | Balechamber | | |
| | M-F 1 / 2 / 3 / 4 | Models 1,2: 0.32 & 0.42 m | 30-35 | 1 |
| | | Models 3,4: 0.35 & 0.45 m | min | |
| BALERS ROUND | M-F822 / 828 | Bale dimensions | | |
| | | 822:1.2 m width x 1.3 dia. | | |
| | | 828:1.2 m width x 1.8 dia. | | |
| BALERS LARGE SQUARE | M-F 5 | Up to 20 t / hr | | |
| **MATERIAL HANDLING** | | | | |
| LOADERS | | Capacities at pivot point max height | | |
| | M-F 875 | 1045 kg | | √ |
| | M-F 880 | 1182 kg | | |
| | M-F885 | 2002 kg | | |
| | M-F 890,20 series | Lift capacity at max height | | |
| | | 1020 : 850 kg &#124; 1420 : 1425 kg | | |
| | | 1220 : 1150 kg &#124; 1620 : 1650 kg | | |
| | M-F 890,30 series | 930 : 780 kg | | |
| | | 1430 : 1425 kg | | |
| | M-F 890,40 series | 1640 : 1500 kg | | |
| | | 1840 : 1825 kg | | |
| TRAILERS M-F 700,W/o power brakes - TONS | | | 1/ 2.5 | 5 |
| M-F 700,W power brakes - TONS | | | | 6 |
| BLADES | | | | |
| FRONT | M-F 830 | 1.83 m width | | |
| REAR | M-F 685 - M Duty | | | 1.84 |
| | | | | 2.5 m |
| | M-F 685 - H duty | | | |
| | M-F 721 - H duty | 1.83 m width | | √ |
| POSTHOLE DIGGER | M-F 1 | Max 0.91 m depth x 0.46 m diam | 25 MIN | √ |

* Note Tractor and Implement Models Nos. as of January 1991

Caution: Guide only referring to 2 and 4 WD models of tractors but soil, climate and topography must be considered.

| 240 | 350 | 355 | 360 | 365 | 375 | 390 |  | 398 | 399 |  | 3610 | 3630 | 3650 | 3680 |  |
|---|---|---|---|---|---|---|---|---|---|---|---|---|---|---|---|
|  |  |  |  |  | 3050 | 3060 | 3065 | 3070 | 3080 | 3090 |  |  |  |  |  |
| **Tractor Horse Power** | | | | | | | | | | | | | | | |
| 47 | 52 | 58 | 63 | 68 | 78 | 86 | 92 | 100 | 110 | 116 | 122 | 142 | 158 | 190 | >200 |
| **Compatible Implement Size** | | | | | | | | | | | | | | | |
| pickup | width) |  |  |  | √ | √ | √ | √ | √ | √ | √ |  |  |  |  |
| √ | √ | √ | √ | √ | √ | √ | √ | √ | √ |  |  |  |  |  |  |
| √ | √ | √ | √ | √ | √ | √ | √ | √ | (finger | wheel | type) |  |  |  |  |
| √ | √ | √ | √ | √ | √ | √ | √ | √ | √ |  |  |  |  |  |  |
|  | √ | √ | √ | √ | √ | √ | √ | √ | √ | √ |  |  |  |  |  |
| 1 | 1.2 | 1.2 | 1.2 3 | 1.2 3 | 1.2 3.4 | 2 3.4 | 2 3.4 | 2 3.4 | 3.4 | 3.4 |  |  |  |  |  |
| min | 822 828 min | 822 828 | 822 828 | 822 828 | 822 828 | 822 828 | 822 828 | 828 |  |  |  |  |  |  |  |
|  |  |  |  |  |  | √ | √ | √ | √ | √ | √ | √ | √ |  |  |
| √ | √ | √ | √ |  |  |  |  |  |  |  |  |  |  |  |  |
|  |  |  | √ | √ | √ | √ | √ |  |  |  |  |  |  |  |  |
|  |  |  |  | √ | √ | √ | √ | √ |  |  |  |  |  |  |  |
| 1020 1220 | 1020 1220 | 1020 1220 1420 | 1020 1 220 | ALL 1620 | 1220 1420 | 1420 1620 | 1620 | 1620 | 1620 |  |  |  |  |  |  |
| 930 | 930 | 930 1430 | 930 1430 | 930 1430 | 1430 |  |  |  |  |  |  |  |  |  |  |
|  |  |  |  | 1640 | 1640 | 1640 1840 | 1640 1840 | 1640 1840 | 1640 1840 | 1640 1840 | 1840 | 1840 | 1840 |  |  |
| 5 6 | 5 6 | 5 6 | 5 6 | 6 8 | 6 8 | 6 8 | 6 8 | 8 10 | 8 10 | 8 10 | 8 10 | 8 10 | 8 10 | 8 10 |  |
|  |  |  | √ | √ | √ | √ | √ |  |  |  |  |  |  |  |  |
| 1.84 2.5 m | 1.84 2.5 m | 1.84 2.5 m |  |  |  |  |  |  |  |  |  |  |  |  |  |
|  |  |  | 2.5 m | 2.5 m | 2.5 m | 2.5 m | 2.5 m | 2.5 m |  |  |  |  |  |  |  |
| √ | √ | √ | √ | √ | √ | √ | √ |  |  |  |  |  |  |  |  |
| √ | √ | √ | √ | √ | √ | √ | √ | √ |  |  |  |  |  |  |  |

*Source:* Adapted from *Massey-Ferguson Implement Guide* (1991).

Reproduced with kind permission of Massey-Ferguson (UK) Ltd.

**TABLE 142 a   Effective Capacities of Field Equipment – Distribution and Drilling**

| Equipment | Hopper size cwt | Appli-cation rate cwt / acre | Gang size | Typical implement working width ft | field effici-ency* % | Normal working speed mph | Working Rates acres per hour spot | overall* | Acres per day |
|---|---|---|---|---|---|---|---|---|---|
| Spinner | 6 | 3 | 1 | 20 | 50 | 5 | 12 | 6.0 | 42 |
| Spinner | 6 | 10 | 2 | 20 | 40 | 5 | 12 | 4.8 | 34 |
| Spinner bulk handling | 30 | 3 | 1 | 20 | 75 | 5 | 12 | 9.0 | 63 |
| Spinner bulk handling | 30 | 10 | 1 | 20 | 50 | 5 | 12 | 6.0 | 42 |
| Full width distributor | 6 | 3 | 1 | 8 | 65 | 4 | 3.8 | 2.5 | 18 |
| Full width distributor | 6 | 10 | 2 | 8 | 55 | 4 | 3.8 | 2.0 | 14 |
| Full width distributor | 12 | 3 | 2 | 17 | 65 | 4 | 8.1 | 5.2 | 36 |
| Full width distributor | 12 | 10 | 2 | 17 | 45 | 4 | 8.1 | 3.6 | 25 |
| Combine drill | 4 seed | 1¼ | 2 | 8 | 60 | 5 | 4.8 | 2.9 | 20 |
|  | 5 fert | 3 | - | - | - | - | - | - | - |
| Corn drill | 4 | 1¼ | 1 | 8 | 70 | 5 | 4.8 | 3.4 | 24 |
| Corn drill | 15 | 1¼ | 1 | 13 | 70 | 5 | 8.0 | 5.6 | 40 |
| Spacing drill 5-row | - | - | 1 | 8 | 60 | 2 | 1.9 | 1.1 | 8 |
| Farmyard manure spreader, wheel drive | 40 | 200 | 1 | 7 | + | 3 | 2.5 | + | + |
| Farmyard manure spreader, p.t.o | 80 | 200 | 1 | 8 | + | 5 | 5.0 | + | + |
|  | Tank cap. gal | Gal per acre | | | | | | | |
| Slurry tanker | 300 | 3000 | 1 | - | - | - | - | 0.3 | 2 |
| Slurry tanker | 700 | 3000 | 1 | - | - | - | - | 0.7 | 5 |
| Field crop sprayer | 80 | 20 | 1 | 15 | 50 | 4 | 7.2 | 3.6 | 25 |
| Field crop sprayer | 100 | 20 | 1 | 40 | 40 | 4 | 19.4 | 7.7 | 55 |
| Row crop spray and drill, 5-row | - | - | 1 | 8 | 40 | 2 | 1.9 | 0.8 | 6 |

*   The field efficiency and overall rates allow for carting from store in good
    working conditions at a short (¼ mile) transport distance.
+   Depends on organization.

**TABLE 142 b    Effective Capacities of Field Equipment – Soil Working Implements**

| Implement | Normal working speed mph | Typical Implement working width ft | depth in | draft lb | field effici- ency % | Working Rates acres per hour spot* | overall | Acres per day (\\) (once over) |
|---|---|---|---|---|---|---|---|---|
| Plough 2F G.P + | 3½ | 2 | 6 | 1000 | 80 | 0.8 | 0.6 | 4 |
| Plough 3F G.P.+ | 3½ | 3 | 6 | 1650 | 80 | 1.3 | 1.0 | 7 |
| Plough 4F G.P.+ | 3½ | 4 | 6 | 2200 | 80 | 1.7 | 1.4 | 10 |
| Plough 6F G.P.§ | 3½ | 6 | 6 | 3300 | 80 | 2.6 | 2.0 | 15 |
| Plough 1F Deep+ | 3 | 1¼ | 10 | 1500 | 80 | 0.5 | 0.4 | 3 |
| Plough 2F Deep++ | 3 | 2½ | 10 | 3000 | 80 | 0.9 | 0.7 | 5 |
| Plough 3F Deep++ | 3 | 3¾ | 10 | 4500 | 80 | 1.4 | 1.1 | 8 |
| Plough 4F Deep^ | 3 | 5 | 10 | 6000 | 80 | 1.8 | 1.4 | 10 |
| Plough 5F semi-digger^ | 4 | 5¾ | 10 | 7000 | 80 | 2.8 | 2.2 | 16 |
| Rotary cultivator+ | 2 | 5 | 4 | - | 85 | 1.2 | 1.0 | 7 |
| Rotary cultivator+ | 1½ | 5 | 6 | - | 85 | 0.9 | 0.7 | 5 |
| Rotary cultivator++ | 2½ | 5 | 4 | - | 85 | 1.5 | 1.3 | 9 |
| Rotary cultivator++ | 2 | 5 | 6 | - | 85 | 1.2 | 1.0 | 7 |
| Tine cultivator Heavy++ | 3½ | 7 | 6 | 3000 | 85 | 3.0 | 2.5 | 18 |
| Tine cultivator Heavy§ | 3½ | 10 | 6 | 4500 | 85 | 4.4 | 3.7 | 26 |
| Tine cultivator Heavy^ | 3½ | 12 | 8 | 7000 | 85 | 5.2 | 4.5 | 32 |
| Spring-tine cult harrow+ | 5½ | 9 | 3 | 900 | 85 | 5.5 | 4.6 | 32 |
| Spring-tine cult harrow++ | 5 | 13 | 3 | 1300 | 85 | 8.0 | 6.8 | 48 |
| Spring-tine cult harrow§ | 5 | 20 | 3 | 2000 | 85 | 12.0 | 10.0 | 70 |
| Harrow Light+ | 4 | 10 | 2 | 500 | 85 | 4.2 | 3.5 | 25 |
| Harrow Disc+ | 3½ | 7 | 3 | 750 | 85 | 3.0 | 2.5 | 18 |
| Harrow Disc++ | 3½ | 8 | 4 | 1000 | 85 | 3.4 | 2.8 | 20 |
| Harrow Disc§ | 4 | 10 | 4 | 1300 | 85 | 4.3 | 3.6 | 25 |
| Harrow Disc Heavy^ | 4 | 10½ | 6 | 4500 | 85 | 5.2 | 4.5 | 32 |
| Roll+ | 4 | 16 | - | 600 | 85 | 7.8 | 6.5 | 45 |
| Tractor hoe+ | 2 | 8 | - | - | 80 | 1.9 | 1.54 | 10 |
| Down row thinner+ | 2 | 8 | - | - | 80 | 1.9 | 1.5 | 10 |
| 3 row ridger or scuffler+ | 3 | 7 | - | - | 80 | 2.5 | 2.0 | 14 |

\*   Spot working rate is the number of acres that would be covered by the implement travelling in
     a straight line at a steady speed i.e. it is the product of the forward speed and the working width.
+   Small medium (31-45 hp) tractor.
++  Medium (46-60 hp) tractor.
§   Large-medium (61-80 hp) tractor.
^   Large (over 80 hp) tractor
\\  It is assumed that day consists of 8 working hours, but that only 7 hours on
    average are spent on the actual job, owing to travelling time, etc.

*( continued over )*

**TABLE 142 c   Effective Capacities of Field Equipment – Corn, Hay and Silage Harvesting**

| Machine | Working width ft | Normal working speed mph | Field efficiency % | Working Rates acres per hour spot | overall | Gang size | Acres per day |
|---|---|---|---|---|---|---|---|
| Combine harvester* may be p.t.o. | 6 | 1½ - 3 | 75 | 1 - 2.1 | ¾ - 1½ | 2 | 8+ |
| Combine harvester* medium capacity | 10 | 2 - 4 | 75 | 2.4 - 4.8 | 1¾ - 3½ | 2 | 16+ |
| Combine harvester* high capacity | 12 | 2 - 4 | 75 | 2.9 - 5.7 | 1¾ - 1½ | 2 | 20+ |
| Combine harvester* high capacity | 14 | 2 - 4 | 75 | 3.3 - 6.7 | 2½ - 5 | 2 - 3 | 25+ |
| Combine harvester* giant | 20 | 2 - 4 | 75 | 4.8 - 9.6 | 3½ - 7 | 3 | 35+ |
| Pick-up baler, straw* | 10 | 4 | 60 | 4.8 | 2.9 | 1 | 20+ |
| Pick-up baler, hay | 10 | 4 | 50 | 4.8 | 2.4 | 1 | 15 |
| Mower, finger bar | 5 | 3¼ | 75 | 2.1 | 1.5 | 1 | 10 |
| Mower, flail | 5 | 4 | 85 | 2.4 | 2.0 | 1 | 15 |
| Mower, rotary | 5 | 6 | 80 | 3.6 | 2.9 | 1 | 20 |
| Tedder, 1-row | 5 | 5 | 85 | 3.0 | 2.5 | 1 | 15§ |
| Tedder, 2-row | 10 | 5 | 85 | 6.0 | 5.0 | 1 | 30§ |
| Tedder, 3-row | 15 | 5 | 85 | 9.0 | 7.5 | 1 | 45§ |
| Swath turner/siderake | 10 | 5 | 85 | 6.0 | 5.0 | 1 | 30§ |
| Forage harvester, flail | 4 | 3 | 65++ | 1.4 | 0.9 | 2 | 6 |
| F.H. Full-chop (pick-up) | 10 | 2½ | 65++ | 3.0 | 2.0 | 2 | 12 |

\*   Assume 20-day harvest for normal cropping.
+   Reduce by 25 per cent in North and West.
++  Depends on team.
§   Effective day length for operation reckoned at 6 hours.

**TABLE 142 d   Effective Capacities of Field Equipment – Specialist Potato and Sugar Beet Machinery**

| Machine | Working width ft | Normal working speed mph | Field efficiency % | Working rates acres per hour spot | overall | Gang size | Acres per day |
|---|---|---|---|---|---|---|---|
| **Potato Planters** | | | | | | | |
| Hand-fed 2-row | 5 | 1¼ | 60 | 0.8 | 0.5 | 4 | 3¼* |
| Hand-fed 3-row | 7½ | 1¼ | 60 | 1.2 | 0.7 | 5 | 5* |
| Hand-fed 4-row | 10 | 1¼ | 60 | 1.6 | 1.0 | 6 | 7* |
| Automatic 2-row with | | | | | | | |
| fertilizer | 5 | 3 | 60 | 1.8 | 1.0 | 2 | 7+ |
| oscillating feed | 6 | 5 | 60 | 3.6 | 2.2 | 2 | 15 |
| **Potato Elevator Digger** | | | | | | | |
| 1-row | 2½ | 2 | 70 | 0.6 | 0.4 | 6+12 | 2½++ |
| 2-row | 5 | 2 | 80 | 1.2 | 0.9 | 1** | **§ |
| **Potato Harvester** | | | | | | | |
| (main crop) 1 row | 2½ | 1¼ | 70 | 0.4 | 0.28 | 6-8 | 2^ |
| (unmanned) 2-row | 6 | 1½ | 70 | 1.0 | 0.7 | 4.6 | 5 |
| **Sugar-Beet Harvester** | | | | | | | |
| 1-row side elevator | 1⅔ | 3 | 75 | 0.6 | 0.45 | 3 | 3¼\\ |
| 1-row tanker | 1⅔ | 3 | 75 | 0.6 | 0.45 | 2 | 3¼ |
| 2-stage,3-row | 5 | 3 | 70 | 1.8 | 1.25 | 4 | 8½ |
| 3-stage,5-row | 8⅓ | 2½ | 70 | 2.5 | 1.75 | 6 | 12\\ |

| | |
|---|---|
| * Rate of work reduced by 33% when planted chitted seed. | § Sufficient for day lifted in 2-3 hours. |
| + Some makes not suitable chitted seed. | ^ Partial mechanical separation. |
| ++ Working 6/12 hours a day. | \\ Less late in season. |
| ** Digger works independently. | |

**TABLE 142 e   Effective Capacities of Field Equipment – Transplanters**

| Crop | Typical Spacing In row in | Between rows in | Typical Performance, 1 man / unit Forward speed mph | Plants per hour per unit | Acres per hour (overall for) 2-row planter |
|---|---|---|---|---|---|
| Leeks | 4 | 20 | 0.2 | 2500 | 0.15 |
| Celery | 7½ | 60 | 0.3 | 2700 * | 0.30 |
| Cabbage | 14 | 21 | 0.45 | — | 0.15 |
| Brussels Sprouts | 32 | 36 | 0.9 | — | 0.50 |

* One extra man per 2 rows

*Source:* Culpin, C. (1980) *Profitable Farm Mechanization.*
Reproduced with kind permission of Blackwell Scientific Publications Ltd.

**TABLE 143   Indicative Classification of Volume Rates for Farm Spraying and Spray Droplet Sizes**

| | Volumes l/ha | | Droplet Classification | |
| --- | --- | --- | --- | --- |
| | Field Crops | Trees, Bushes | Median Diameter of Droplets (um) | Classification |
| High volume | ≤600 | ≤-1000 | ≥50 | Aerosol |
| Medium volume | 200 - 600 | 500 - 1000 | 50 - 100 | Mist Spray |
| Low volume | 50 - 200 | 200 - 500 | 101 - 200 | Fine Spray |
| Very low volume | 5 - 50 | 50 - 200 | 201 - 400 | Medium Spray |
| Ultra low volume | ≤5 | ≤-50 | ≥400 | Coarse Spray |

Droplets of a diameter of less than 50 micrometres are of limited use in crop spraying as they tend to drift upwards into the atmosphere and evaporate.

**TABLE 144   Indicative Soil Tilth Conditions resulting from Use of Various Types of Chisels and Tillers**

| Tool | Soil firmness | | | Moisture Conservation | | | Finished Field Soil Ridge Profile | | | Remaining % of Original Field Residue | | Typical Application | | |
| --- | --- | --- | --- | --- | --- | --- | --- | --- | --- | --- | --- | --- | --- | --- |
| | Hard | Medium | Soft | Very good | Good | Fair | Minimum | Medium | Maximum | Heavy Crop Residue | Light Crop Residue | Summerfallow | Weed Control | Deep Tillage |
| Low Crown Sweep | | ● | ● | ● | ● | | ● | | | 55 - 85 | 30 - 55 | ● | ● | ● |
| Medium Crown Sweep | ● | ● | ● | ● | ● | | ● | | | 50 - 80 | 25 - 50 | ● | ● | ● |
| High Crown Sweep | ● | ● | ● | | ● | ● | | ● | | 45 - 65 | 15 - 40 | ● | ● | |
| Chisel | ● | ● | ● | | ● | ● | | ● | | 45 - 65 | 15 - 40 | | | ● |
| Spike | ● | ● | ● | | ● | ● | | ● | | 45 - 65 | 20 - 45 | | | ● |
| Flat Twisted Shovel 3-in. | ● | ● | ● | | ● | ● | | | ● | 40 - 60 | 15 - 35 | | ● | ● |
| Flat Twisted Shovel 4-in. | | ● | ● | | ● | ● | | | ● | 35 - 50 | 10 - 30 | | ● | ● |
| Concave Twisted Shovel 3-in. | ● | ● | ● | | ● | ● | | | ● | 40 - 60 | 15 - 35 | | ● | ● |
| Concave Twisted Shovel 4-in. | | ● | ● | | ● | ● | | | ● | 35 - 50 | 10 - 30 | | ● | ● |

This chart is intended as a general guide for tool equipment used on chisel ploughs and mulch tillers. Specific soil types, climates and operating guidelines may yield results other than specified.

*Source:* John Deere Sales Brochure, *Chisel ploughs, Mulch Tillers and V rippers.*
Reproduced with kind permission of John Deere Ltd. (UK).

# Static Machinery

**TABLE 145  Electric Motors: Speeds and Power Requirements**

| Electricity supply | | Common speeds (rpm) | | |
|---|---|---|---|---|
| 50 cycles per second (Hz) | 725 | 960 | 1450 † | 2900 |
| 60 cycles per second (Hz) | 871 | 1160 | 1750 † | |

| Power output of motor | Electricity supply required | |
|---|---|---|
| Up to 5 kw (7hp) | 110 V or 220 V | Single phase |
| 5 - 20 kw (7 - 25 hp) | 220 V | Three-phase |
| More than 20 kw (25 hp) | 440 V | Three-phase |

† Most common speeds.

*Source:* Hudson, N.W. (1975) *Field Engineering for Agricultural Development.*
Reproduced with kind permission of N.W. Hudson.

**TABLE 146  Approximate Weekly and Annual Outputs of Barley from Different Sizes of Hammer
and Roller Mills based on 40 hours use per Week (8hr x 5 days) and 50 Weeks per Year**

| Mill motor | | t/week | | | t/year | | |
|---|---|---|---|---|---|---|---|
| kw output | (hp) | Hammer mill for pigs/poultry | | Roller mill for cattle | Hammer mills | | Roller mills |
| | | *P | *A | | *P | *A | |
| 2.2 | (3) | 4 | 6 | 14 | 200 | 300 | 700 |
| 3.7 | (5) | 8 | 10 | 20 | 400 | 500 | 1000 |
| 5.6 | (7.5) | 11 | 16 | 36 | 550 | 800 | 1800 |
| 7.5 | (10) | 15 | 21 | 51 | 750 | 1050 | 2550 |
| 11.2 | (15) | 24 | – | 60 | 1200 | – | 3000 |
| 15.0 | (20) | 32 | – | 70 | 1600 | – | 3500 |
| 18.7 | (25) | 40 | 47 | – | 2000 | 2350 | – |

* P - with Pneumatic conveying and 3mm screen
* A - with auger conveying and 3mm screen

*Note:* Throughput can vary significantly with screen size, type of material and its moisture
content. Check manufacturers' specification for each individual machine.

*Source:* Farm Electric Publications.
Reproduced with kind permission of Electricity Association Technology Ltd.

**TABLE 147  Approximate Weekly and Annual Outputs from Different Sizes of Cuber based on 40 hours use per Week (8hr x 5 days) and 50 Week per Year**

| Cuber motor rating | | t/week | | |
|---|---|---|---|---|
| kw output | (hp) | 3.9 mm die | 5.6 mm die | 11 mm die |
| 3.7 | (5) | 6 | 8 | 9 |
| 5.6 | (7.5) | 8 | 10 | 11 |
| 7.5 | (10) | 8 | 12 | 16 |
| 18.7 | (25) | 20 | 30 | 35 |

*Source:* Farm Electric Publications.
Reproduced with kind permission of Electricity Association Technology Ltd.

**TABLE 148  Auger Conveyor Outputs at Various Angles (Metric)**

**Throughputs of clean barley at 17 per cent moisture content**
**Inclination**

| Diameter | 0° | 15° | 30° | 45° t/hr | 69° | 75° | 90° |
|---|---|---|---|---|---|---|---|
| 102 mm | 16.0 | 12.7 | 11.2 | 9.9 | 8.7 | 6.4 | 5.2 |
| 152 mm | 41.1 | 36.8 | 32.5 | 27.9 | 23.1 | 17.9 | 12.4 |

*Source:* Farm Electric (1984) *Grain Drying and Storage.*
Reproduced with kind permission of Electricity Association Technology Ltd.

# Machinery Life

**TABLE 149   Estimated Life of Field Machinery -- Yuma, USA**

| Type | Hours to wear out |
|---|---|
| **TRACTOR AND TRUCKS** | |
| Wheeled tractors | 12 000 |
| Crawler tractors | 12 000 |
| ½ - 1 ton road trucks | 3000 |
| Grain truck | 6000 |
| Feeder truck | 6000 |
| **MATERIAL TRANSPORT** | |
| Self propelled bale wagons | 2500 |
| Cuber/wafer wagon, towed | 4000 |
| Bale wagon, towed | 2500 |
| Cotton trailer | 2000 |
| Forage trailer, PTO unloading | 2000 |
| Grain Trailer | 2000 |
| Hesston stackhand | 2000 |
| Hesston stackmover | 2000 |
| **HARVESTING MACHINERY** | |
| Combine harvester | 2000 |
| Cotton picker/stripper | 2000 |
| Self Propelled forage harvester | 2000 |
| Self propelled swather | 2500 |
| Forage harvester, PTO drive | 2000 |
| Mower,7 foot | 2000 |
| Side rakes | 2500 |
| Balers: PTO and Engine drive | 2500 |
| Stalk cutters: flail and rotary | 2000 |

| Type | Hours to wear out |
|---|---|
| **CULTIVATION AND SEEDING** | |
| Cultivation equipment generally | 2500 |
| Viz.  cultivators, ploughs, discs, rippers, cultipackers, rotary hoes, subsoilers, harrows | |
| Grain drill | 1200 |
| Planter drills | 1200 |
| Broadcast seeder | 1200 |
| Brillion seeder | 1200 |
| Landplanes | 2500 |
| Mulchers | 2500 |
| **FERTILIZING AND SPRAYING** | |
| Fertiliser injector: rowcrop | 1200 |
| Fertiliser broadcaster towed | 1200 |
| Fertiliser side dressing units | 2500 |
| Crop sprayers: mounted and trailed | 1200 |

*Source:* Hathorn, S. (1979) Arizona Field Crop Budgets. Yuma County. Reproduced with kind permission of S. Hathorn Jr., Dept. of Ag. Econ. College of Agriculture, University of Arizona, Tucson.

**TABLE 150  Average Working Life of Tractors and Machinery and Total Repair Costs**

| | Life hr | Repair costs - % of new cost |
|---|---|---|
| Tractor      40 - 50 hp | 7000 | 140 |
| 60 - 70 hp | 8000 | 130 |
| 90 - 100 hp | 10 000 | 110 |
| 3 disc plough | 4000 | 160 |
| 3 furrow mouldboard plough | 4000 | 120 |
| Tandem Disc Harrow | 4000 | 250 |
| Rotovator | 3000 | 150 |
| Tine Harrow | 4000 | 80 |
| Power Harrow | 3000 | 100 |
| Rigid Tine Cultivator | 5000 | 80 |
| Planter and Fertilizer Attachment | 6000 | 100 |
| Seed and Fertilizer Drill | 6000 | 100 |

*Source:* ILACO (1985) *Agricultural Compendium for Rural Development in the Tropics.*
Reproduced with kind permission of ILACO BV.

**TABLE 151  Accumulated Repair Costs\* of Agricultural Machinery as a Percent of Purchase Price**

| Machine | 1/4 LIFE Accumulated Hours - Costs | | 1/2 LIFE Accumulated Hours - Costs | | 3/4 LIFE Accumulated Hours - Costs | | FULL LIFE Accumulated Hours - Costs | |
|---|---|---|---|---|---|---|---|---|
| 2- and 4-Wheel-Drive | | | | | | | | |
| Tractors | 2500 | 9.8% | 5000 | 29.7% | 7500 | 56.8% | 10 000 | 90.0% |
| Crawlers | 4000 | 8.7% | 8000 | 26.4% | 12 000 | 50.5% | 16 000 | 80.0% |
| Combines | 500 | 2.7% | 1000 | 9.5% | 1500 | 19.6% | 2000 | 33.0% |
| Cotton Pickers | | | | | | | | |
| Corn Pickers | 500 | 8.2% | 1000 | 24.7% | 1500 | 47.3% | 2000 | 75.0% |
| Cotton Strippers | | | | | | | | |
| Planters / Drills | 250 | 8.2% | 500 | 24.7% | 750 | 47.3% | 1000 | 75.0% |
| Mowers | 250 | 29.7% | 500 | 73.1% | 750 | 123.7% | 1000 | 180.0% |
| Plows / Swathers | | | | | | | | |
| Balers / Balewagons | 500 | 13.2% | 1000 | 32.5 | 1500 | 55.0% | 2000 | 80.0% |
| Loose Hay Stack Wagons\* | | | | | | | | |
| Forage Harvesters | | | | | | | | |
| Disks / Chisel-Plows\* | 500 | 5.3% | 1000 | 18.7% | 1500 | 38.7% | 2000 | 65.0% |
| Field Cultivators | | | | | | | | |

\* Repair costs estimates in this table do not include the effect of inflation over the period of ownership.

*Source:* John Deere and Co. (1981) *Fundamentals of Machine Operation.* Reproduced with permission of Deere and Company, © Deere and Company. All rights reserved. Also adapted from data from Agricultural Engineering Department, University of Illinois, and the *Agricultural Engineering Yearbook.*

**TABLE 152  Estimated Annual Cost of Spares and Repairs as a Percentage of Purchase Price∗ at Various Levels of Use**

| | Approximate Annual Use (hr) | | | | Additional use per 100 hr ADD |
|---|---|---|---|---|---|
| | 500 | 750 | 1000 | 1500 | |
| | % | % | % | % | % |
| Tractors | 5.0 | 6.7 | 8.0 | 10.5 | 0.5 |

| | Approximate Annual Use (hr) | | | | Additional use per 100 hr ADD |
|---|---|---|---|---|---|
| | 50 | 100 | 150 | 200 | |
| | % | % | % | % | % |
| **Harvesting Machinery:** | | | | | |
| Combine Harvesters, self-propelled and engine-driven | 1.5 | 2.5 | 3.5 | 4.5 | 2.0 |
| Combine Harvesters, p.t.o. driven, metered-chop forage harvesters, pick-up balers, potato harvesters, sugar beet harvesters | 3.0 | 5.0 | 6.0 | 7.0 | 2.0 |
| **Other Implements and Machines:** | | | | | |
| Group 1: | | | | | |
| Ploughs, Cultivators, Toothed harrows, Hoes, Elevator potato diggers    Normal Soils | 4.5 | 8.0 | 11.0 | 14.0 | 6.0 |
| Group 2: | | | | | |
| Rotary cultivators, Mowers, Binders, Pea cutter-windrowers | 4.0 | 7.0 | 9.5 | 12.0 | 5.0 |
| Group 3: | | | | | |
| Disc harrows, Fertilizer distributors, Farmyard manure spreaders, Combine drills, Potato planters with fertilizer attachment, Sprayers, Hedge-cutting machines | 3.0 | 5.5 | 7.5 | 9.5 | 4.0 |
| Group 4: | | | | | |
| Swath turners, Tedders, Side-delivery rakes, Unit drills, Flail forage harvesters, Semi-automatic potato planters and transplanters, Down-the-row thinners | 2.5 | 4.5 | 6.5 | 8.5 | 4.0 |
| Group 5: | | | | | |
| Corn drills, Milking machines, Hydraulic loaders, Simple potato planting attachments | 2.0 | 4.0 | 5.5 | 7.0 | 3.0 |
| Group 6: | | | | | |
| Grain driers, Grain cleaners, Rolls, Hammer mills, Feed mixers, Threshers | 1.5 | 2.0 | 2.5 | 3.0 | 0.5 |

∗ When it is known that a high purchase price is due to high quality and durability or a low price corresponds to a high rate of wear and tear, adjustments to the figures should be made.

*Source:* Nix, J. (1990) *Farm Management Pocketbook.*
Reproduced with kind permission of J. Nix, Wye College, University of London.

# Fuel Consumption

**TABLE 153   Average Energy and Fuel Requirements for Different Field Operations**

| Operation | Energy Required PTO hp-hr per Acre | Gaso-line | Diesel | LP-Gas |
|---|---|---|---|---|
| Shred stalks | 10.5 | 1.0 | 0.72 | 1.20 |
| Plow 8-inches deep | 24.4 | 2.35 | 1.68 | 2.82 |
| Heavy offset disk | 13.8 | 1.33 | 0.95 | 1.60 |
| Chisel plow | 16.0 | 1.54 | 1.10 | 1.85 |
| Tandem disk, stalks | 6.0 | 0.63 | 0.45 | 0.76 |
| Tandem disk, chiseled | 7.2 | 0.77 | 0.55 | 0.92 |
| Tandem disk, plowed | 9.4 | 0.91 | 0.65 | 1.09 |
| Field cultivate | 8.0 | 0.84 | 0.60 | 1.01 |
| Spring-tooth harrow | 5.2 | 0.56 | 0.40 | 0.67 |
| Spike-tooth harrow | 3.4 | 0.42 | 0.30 | 0.50 |
| Rod weeder | 4.0 | 0.42 | 0.30 | 0.50 |
| Sweep plow | 8.7 | 0.84 | 0.60 | 1.01 |
| Cultivate row crops | 6.0 | 0.63 | 0.45 | 0.76 |
| Rolling cultivator | 3.9 | 0.49 | 0.35 | 0.59 |
| Rotary hoe | 2.8 | 0.35 | 0.25 | 0.42 |
| Anhydrous applicator | 9.4 | 0.91 | 0.65 | 1.09 |
| Planting row crops | 6.7 | 0.70 | 0.50 | 0.84 |
| No-till planter | 3.9 | 0.49 | 0.35 | 0.59 |
| Till plant (with sweep) | 4.5 | 0.56 | 0.40 | 0.67 |
| Grain drill | 4.7 | 0.49 | 0.35 | 0.59 |
| Combine (small grains) | 11.0 | 1.40 | 1.00 | 1.68 |
| Combine, beans | 12.0 | 1.54 | 1.10 | 1.85 |
| Combine, corn and grain sorghum | 17.6 | 2.24 | 1.60 | 2.69 |
| Corn picker | 12.6 | 1.61 | 1.15 | 1.93 |
| Mower (cutterbar) | 3.5 | 0.49 | 0.35 | 0.59 |

**TABLE 153** *(continued)*  **Average Energy and Fuel Requirements for Different
Field Operations**

| Operation | Energy Required PTO hp-hr per Acre | Gallons per Acre | | |
|---|---|---|---|---|
| | | Gaso-line | Diesel | LP-Gas |
| Mower conditioner | 7.2 | 0.84 | 0.60 | 1.01 |
| Swather | 6.6 | 0.77 | 0.55 | 0.92 |
| Rake, single | 2.5 | 0.35 | 0.25 | 0.42 |
| Rake, tandem | 1.5 | 0.21 | 0.15 | 0.25 |
| Baler | 5.0 | 0.63 | 0.45 | 0.76 |
| Stack wagon | 6.0 | 0.70 | 0.50 | 0.84 |
| Sprayer | 1.0 | 0.14 | 0.10 | 0.17 |
| Rotary mower | 9.6 | 1.12 | 0.80 | 1.34 |
| Haul small grains | 6.0 | 0.84 | 0.60 | 1.01 |
| Grain drying | 84.0 | 8.40 | 6.00 | 10.08 |
| Forage harvester, green forage | 12.4 | 1.33 | 0.95 | 1.60 |
| Forage harvester, haylage | 16.3 | 1.75 | 1.25 | 2.10 |
| Forage harvester, corn silage | 46.7 | 5.04 | 3.60 | 6.05 |
| Forage blower, green forage | 4.6 | 0.49 | 0.35 | 0.59 |
| Forage blower, haylage | 3.3 | 0.35 | 0.25 | 0.42 |
| Forage blower, corn silage | 18.2 | 1.96 | 1.40 | 2.35 |
| Forage blower, high-moisture ear corn | 5.9 | 0.63 | 0.45 | 0.76 |

*NB* hp outputs per gallon of fuel are approximately:

Diesel  –  13 hp hr/gal

Gasoline  –  9 hp hr/gal

LP Gas  –  7.5 hp hr/gal

*Source.* John Deere (1981) *Fundamentals of Machine Operation.*
Reproduced with kind permission of Deere and Company.

**TABLE 154 Indicative Fuel Consumption for Varying Tractor Horsepower**

| TRACTOR TYPE AND SIZE | AVG DRAWBAR HORSEPOWER | | AVG FUEL CONSUMPTION (l/hr) | |
|---|---|---|---|---|
| | Diesel Tractors | Gasoline Tractors | Diesel Tractors | Gasoline Tractors |
| **Two Wheel Drive PTO hp** | | | | |
| 30 - 39 | – | 31.5 | – | 10.5 |
| 40 - 49 | – | 38.3 | – | 12.3 |
| 50 - 59 | 48.2 | 47.1 | 10.5 | 15.0 |
| 60 - 69 | 56.2 | 55.2 | 13.6 | 20.0 |
| 70 - 79 | 65.5 | 64.5 | 16.4 | 23.2 |
| 80 - 89 | 73.5 | 74.5 | 17.7 | 26.8 |
| 90 - 99 | 84.0 | 82.0 | 20.5 | 28.2 |
| 100 - 119 | 90.3 | – | 22.7 | – |
| 120 - 139 | 111.6 | – | 28.2 | – |
| 140 - 159 | 128.9 | – | 32.3 | – |
| 160 - 179 | 145.8 | – | 38.6 | – |
| Over 180 | 148.9 | – | 39.1 | – |
| **Four Wheel Drive DB hp** | | | | |
| 100 - 149 | 135.0 | – | 33.2 | – |
| 150 - 174 | 159.4 | – | 40.3 | – |
| 175 - 199 | 182.8 | – | 50.0 | – |
| 200 - 224 | 217.4 | – | 53.0 | – |
| 225 - 249 | 235.5 | – | 58.0 | – |
| 250 - 274 | 265.0 | – | 61.4 | – |
| 275 - 299 | 281.3 | – | 62.6 | – |
| 300 - 349 | 305.0 | – | 66.4 | – |
| Over 350 | 385.0 | – | 92.4 | – |

*NB* 75% load factor assumed for these estimates.

*Source:* Alberta Agriculture (1984) *Know Your Farm Machinery Costs.* Agdex 825-6.
Reproduced with kind permission of Alberta Agriculture.

# Manual Work
# and
# Draft Animals

# Manual Work Rates

**TABLE 155 Indicative Outputs of Manually Operated Field Equipment**

| Machine | Performance |
|---|---|
| 6 Row Rice Transplanter | 0.3 - 0.4 ha/day |
| Rolling Injector Planter | 6000 - 16 000 hills/hr |
| Drum seeder | 1 ha/day |
| (For row seeding of germinated rice on puddled soils for subsequent transplanting) | |
| 'Tapak-Tapak' cylinder pump | 3 l/s at 2 m lift |
| for small farm irrigation | 2 l/s at 3m lift |
| Diaphragm Pump | 3 l/s at 1 m head |
| for low lifting pumping | |
| small farm irrigation | 2 l/s at 2 m head |
| Push type Hand Weeder. | 35 - 75 man hr/ha |
| Rotary spike type | |
| for paddy rice in rows | |
| Single Row Cono Weeder. | 40 - 50 man hr/ha |
| Rotary cones with blades | |
| for paddy rice in rows | |
| Double Row Cono Weeder | 25 - 35 man hr/ha |
| Plunger-Auger Fertiliser Injector | |
| 2 row for paddy rice | 16 man hr/ha |

*Source:* Developments of the International Rice Research Institute.
Data supplied by and reproduced with kind permission of the International Rice Research Institute.

**TABLE 156   Indicative Manual Field Work Rate**

| Task | | Work Rate |
|------|---|-----------|
| Weeding paddy rice | | 120 man hr/ha/pass |
| Hand weeding maize | | 50 man hr/ha/pass |
| Hand thinning maize | | 36 man hr/ha |
| Knapsack spraying | | 8 man hr/ha |
| ULV spraying | | 2 man hr/ha |
| Picking maize cobs into sacks or barrows | | 50 man hr/ha |
| Shelling maize | | 85 man hr/ha |
| Cotton | | 35 - 40 kg/day |
| Cabbage transplanting | | 150 - 160 man hr/ha |
| Cabbage harvesting | | 210 - 250 man hr/ha |
| Picking Brussels sprouts | | 320 - 400 man hr/ha |
| Pulling peas | | 475 - 525 man hr/ha |
| Harvesting runner beans | | 625 - 675 man hr/ha |
| Potato planting | chitted seed | 3 men + 16 women 2 ha/day |
| | unchitted seed | 2 men + 12 women 2 ha/day |
| Potato picking after spinner lifting | | 3 - 5 men + 10 - 15 women 65 - 85 ha/day |
| Potato riddling | | 3 - 5 workers 10 - 15 t/day |
| Carting hay bales | | 2 men + 1 or 2 tractors and trailers 7 - 9 t/day |

*Sources: Rhodesia Farm Management Handbook* (1970).
International Rice Research Institute (1991).
Nix, J. (1991) *Farm Management Pocketbook.*

**TABLE 157 Manual Work Rates**

| Activity | Rate |
| --- | --- |
| Digging, loose soil | 35 - 52 day/ha (22.5 - 30 cm deep) |
| Seed broadcasting | 0.6 ha/hr |
| Transplanting | 7500 - 10 000 plants/day (e.g. cabbage) |
| Thinning plants | 0.04 ha/hr (4 - 5 men, e.g. sugar beet) |
| Pitching sheaves | 5000 - 6000/day |
| Cutting with scythe | 0.3 - 0.4 ha/day (grass or cereals) |
| Cutting with sickle | 0.1 ha/day (grass or cereals) |
| Potato raising | 0.4 ha/day (4 - 6 men) |
| Potato raising with spinner | 1 ha/day (with 12 - 14 women picking and working one way only) |
| Potato planting | 0.16 ha/hr (6 - 7 women, place potatoes only) |
| Turnip lifting | 0.4 ha/day (4men, top and tail) |
| Flail threshing | 25 kg/day wheat |
| | 50 kg/day barley |
| | 62 kg/day beans |
| | 100 kg/day oats |
| Dig and load materials on to cart | 6.7 m$^3$ chalk/day |
| | 8.3 m$^3$ clay/day |
| | 10 m$^3$ loam/day |
| | 10.8 m$^3$light loam or sandy clay |
| Sheep shearing | 30 - 60/day (hand shears, according to sheep size) |

*Source:* Primrose McConnell (1897) *The Agricultural Notebook.*
Reproduced with kind permission of Blackwell Scientific Publications Ltd.

# Draft Animal Power

**TABLE 158   Examples of Draft Animal Work Rates and Capacities**

| Animal | Work rate / capacity | |
|---|---|---|
| **In Yemen:(ODA/LRDC)** | | |
| Pair of oxen | 0.26 ha/day ) | Secondary cultivations with |
| Single camel | 0.25 ha/day ) | local single furrow chisel type |
| Pair of donkeys | 0.16 ha/day ) | plough |
| Single donkey | 0.15 ha/day ) | |
| | | |
| **In UK (1897) Horse: (McConnell P)** | | |
| Ploughing | 0.3 - 0.4 ha/day | |
| Hoeing | 0.16 ha/day | |
| Sowing turnips | 0.4 ha/hr with 2 row drill | |
| Harrowing | 0.5 ha/hr | |
| Sowing cereals | 0.5 ha/hr (2 - 3 horses) | |
| Spring tooth cultivator | 0.4 ha/hr (3 horses, 7.5 - 12.5 cm deep) | |
| Mowing grass | 0.4 ha/hr | |
| Reaper/binder | 0.4 ha/hr | |
| Rake or ted hay | 6 - 8 ha/day | |
| | | |
| **General Capacities: (ILACO)** | | |
| Pair of bullocks | Meet power requirements of 6 - 8 ha farm | |
| | Haul 600 kg on cart over 40 km/day | |
| Strong donkey or mule | 150 kg pack for 15 - 20 km/day | |
| Dromedary camel | 150 - 175 kg pack for 30 - 40 km/day | |
| Bactrian camel | 200 - 250 kg pack for 30 - 40 km/day | |
| Pair of buffalo | plough 1 ha/week | |
| | haul 2000 kg at 3 km/hr | |
| Llamas | 50 kg pack at 10 - 20 km/day | |

**Relative Powers: (FAO)**

| Animal | Average weight | Approximate draught | Average speed of work | Power developed |
|---|---|---|---|---|
| | kg | kg | m / s | hp |
| Light horses | 400 - 700 | 60 - 80 | 1.0 | 1.00 |
| Bullocks | 500 - 900 | 60 - 80 | 0.6 - 0.85 | 0.75 |
| Buffaloes | 400 - 900 | 50 - 80 | 0.8 - 0.9 | 0.75 |
| Cows | 400 - 600 | 50 - 60 | 0.7 | 0.45 |
| Mules | 350 - 500 | 50 - 60 | 0 9 - 1.0 | 0.70 |
| Donkeys | 200 - 300 | 30 - 40 | 0.7 | 0.35 |

*NB* FAO suggests that very approximately the pulling power of an animal is
10% of its body weight, but horses usually have slightly more power and up
to 15% of body weight.

*Sources:* Primrose McConnell (1987) *The Agricultural Notebook;* ILACO (1985) *The
Agricultural Compendium;* FAO (1969) Ag. Dev. Paper. No. 91 *Farm Implements for
Tropical and Arid Regions;* Hendy, C. (1981) Yemen Arab Republic Montane Plains and
Wadi Rima Project, *Livestock Production.* Project Record No. 29. ODA/LRDC.

**TABLE 159  Work Performance of Pairs of Draught Animals under Various Conditions (Nepal)**

| Type of animal | Live weight (kg) | Terrain | Job | Hours worked hr/d | EWT (min/hr) | Work output (kJ/team) | Distance travelled km | Average speed (m/s) | Average power output (W) | Comments (d = day) |
|---|---|---|---|---|---|---|---|---|---|---|
| Local oxen | 200 – 290 | Hill terraces | Ploughing with ard | 5.0 | 47 – 54 | 2250 – / 3125 | 4.2 – / 7.25 | 0.22 – / 0.40 | 128 – / 177 | Observation of 5 teams of oxen ploughing on 4 – 6 d each in Nov |
| Local oxen | 173 – 258 | Hill terraces | Ploughing with ard | 5.0 | 46 – 55 | 2228 – / 2641 | 3.9 – / 4.0 | 0.25 – / 0.26 | 149 – / 167 | 2 teams of local oxen ploughing on 6 d each in Jan / Feb. |
| Crossbred oxen | 192 – 220 | | | 5.0 | | 3369 – / 3906 | 5.4 – / 6.1 | 0.35 – / 0.40 | 218 – / 257 | – |
| Large buffalo | 401  464 | Flat land | Carting | *5.0 – | 50 – 56 | 5590 | 16.8 | 0.95 | 317 | Observations of 4 – 6 d per team Dec to Feb:- Load 72 % of live weight |
| Small buffalo | 237  320 | | | *5.0 – | 50 – 60 | 5028 | 16.7 | 0.94 | 285 | Load 68 % of live weight |
| Large oxen | 557  551 | | | 5.0 | 50 – 56 | 5428 | 16.7 | 0.93 | 308 | Load 56 % of live weight |
| Small oxen | 360  424 | | | 5.0 | 50 – 56 | 5664 | 16.9 | 0.94 | 321 | Load 75 % of live weight |
| Healthy buffaloes | 320  237 | Flat land | Carting | 5.1 – / 5.7 | 50 – 56 | 4927 – / 5124 | 16.4 – / 17.1 | 0.87 – / 1.02 | 240 – / 276 | Load 70 % of live weight PCV = 35.37 % |
| 'Anaemic' buffaloes | 233  264 | Flat land | | *5.8 – / 6.1 | 45 – 54 | 4054 – / 4548 | 14.7 / 16.5 | 0.74 / 0.91 | 185 – / 217 | Load 70 % of liveweight PCV = 23.27 % |

* time allowed for wallowing from 19 – 48 minutes in the days work

*Source:* Pearson, R.A. (1991) *Outputs from Draught Animals.* Data supplied by and reproduced with kind permission of R.A. Pearson, Centre for Tropical Veterinary Medicine, Easter Bush, Roslin, Midlothian, EH25 9RG, UK.

A G R I   I N F O

**TABLE 160  Estimated Energy Expenditure on Working Days by Teams of Draught Oxen and Buffaloes Working under Different Conditions**

| Country | Type | Average liveweight (kg) | No Obs | Job | Hours worked | Average energy used as a multiple of maintenance | Comments |
|---------|------|-------------------------|--------|-----|--------------|--------------------------------------------------|----------|
| Costa Rica | Oxen | 630 | 10 | General | 5.5 | 1.44 | Low level of diet |
| | | 470 | 10 | farm | 5.5 | 1.42 | such that animals |
| | | 430 | 5 | work | 5.5 | 1.41 | lost weight |
| Costa Rica | Oxen | 630 | 10 | General | 5.5 | 1.56 | Better diet – |
| | | 470 | 10 | farm | 5.5 | 1.45 | no weight loss |
| | | 430 | 5 | work | 5.5 | 1.51 | |
| Costa Rica | Oxen | 630 | 15 | Ploughing | 5.5 | 1.78 | Animals gained |
| | | 470 | 15 | mowing and | 5.5 | 1.64 | weight - |
| | | 430 | 15 | harrowing 5d | 5.5 | 1.59 | adjustments to |
| | | | | each | | | implements |
| | | | | | | | outside worktime |
| Bangladesh | Oxen | 152 | | Ploughing | 3 | 1.4 | Worked at speeds of |
| | | | | flat land | 4 | 1.5 | about 0.65 m/s |
| | | | | with ard | | | |
| Nepal | Oxen | 289 | 4 | Ploughing | 5 | 1.3 | Animals maintained |
| | | 247 | 5 | hill terraces | 5 | 1.38 | or gained weight |
| | | 268 | 4 | with ard | 5 | 1.32 | when working |
| | | 229 | 4 | | 5 | 1.36 | |
| | | 238 | 6 | | 5 | 1.34 | |
| Nepal | Oxen | 554 | 6 | Carting on | 5 | 1.74 | Animals gained |
| | | 392 | 5 | flat land | 5 | 1.78 | weight when working – |
| | | | | | | | rest stop of 15 - 37 |
| | | | | | | | mins after approx. |
| | | | | | | | 3 - 5 h work |
| Nepal | Buffalo | 432 | 6 | Carting on | 5 | 1.77 | |
| | | 278 | 6 | flat land | 5 | 1.77 | |
| Nepal | Local and crossbred oxen | 173 | 24 | Ploughing hill | 5 | 1.24 | Animals gained weight |
| | | 258 | | terraces | 5 | 1.41 | during working period |

*Source:* Pearson, R.A. (1991) *Outputs from Draught Animals.* Data supplied by and reproduced with kind permission of R.A. Pearson, Centre for Tropical Veterinary Medicine, Easter Bush, Roslin, Midlothian, EH25 9RG,UK.

**TABLE 161  Ox Pair Cultivation Performance for Different Field Operations in Mali**

| Operation | Avg. Daily Work Time. min/day | Days Worked per year | Area covered ha/day | Power Output (W) |
|---|---|---|---|---|
| Harrowing | 250 | 3.4 | 0.49 | 731 |
| Flat ploughing | 171 | 2.7 | 0.17 | 717 |
| Ridging | 273 | 16.2 | 0.32 | 591 |
| Seeding | n/a | 0.9 | n/a | n/a |
| Weeding | 209 | 11.2 | 0.28 | 470 |
| Re-ridging | 282 | 2.7 | 0.36 | 484 |

*Source:* Bartholomew, P. (1992) Personal Communication.
Data supplied by and reproduced with kind permission of P. Bartholomew.

# Crops – General

# General Data

**TABLE 162   Crop Directory**

This directory does not include naturally occurring fodder plants. It attempts to include 'crops' to which man generally makes some contribution to the productive or harvesting process. Therefore naturally occurring pasture and browse species, and many naturally occurring plants whose products are casually harvested by man have generally not been included.

- D/M denotes dicotyledon or monocotyledon.
- A/B/P denotes whether the crops are usually grown as annuals, biennials or perennials.
- The 'principal' products listed do not necessarily indicate the full range of products from a given plant.

The reader is cautioned that synonyms will often be found for the Latin species names in other literature. Similarly two different species can occasionally have the same common name.

| Latin Name | Common Name | D/M | A/B/P | Principal Products |
|---|---|---|---|---|
| **Acacia mearnsii** | Black Wattle | D | P | Tannin (bark), Fuel timber |
| **Acacia nilotica** | Babul | D | P | Tannin (bark) |
| **Acacia senegal** | Sudan Gum Arabic | D | P | Gum |
| **Acca sellowina** | Feijoa Pineapple Guava | D | P | Fruit |
| **Acorus calamus** | Sweet Flag | M | P | Oil (rhizome) |
| **Acroceras macrum** | Nile Grass | M | P | Forage grass |
| **Actinidia chinensis** | Kiwi Fruit Yang Tao Chinese Gooseberry | D | P | Fruit |
| **Adesmia spp.** | | D | A/P | Fodder legumes |
| **Aechmea magdalene** | Pita | M | P | Leaf fibre |
| **Aelanthus gamwelliae** | Nindi | D | P | Oil (inflorescence) |
| **Aeschynomene americana** | American Joint Vetch | D | P | Fodder legume, Green manure |
| **Aeschynomene falcata** | Joint Vetch | D | P | Fodder legume |
| **Agropyron cristatum** | Crested Wheat Grass | M | P | Forage grass |
| **Agropyron desertorum** | Desert Wheat Grass 'Standard' Crested Wheat Grass | M | P | Forage grass |
| **Agropyron elongatum** | Tall Wheat Grass | M | P | Forage grass |
| **Agropyron intermedium** | Intermediate Wheat Grass | M | P | Forage grass |
| **Agropyron semicostatum** | Drooping Wheat Grass | M | P | Forage grass |
| **Agropyron smithii** | Bluestem Western Wheat Grass | M | P | Forage grass |
| **Agropyron spicatum** | Bluebunch Wheat Grass | M | P | Forage grass |
| **Agropyron trachycaulum** | Slender Wheat Grass | M | P | Forage grass |
| **Agropyron trichophorum** | Stiffhair Wheat Grass | M | P | Forage grass |
| **Agrostis gigantea** | Red Top Fiorin | M | P | Forage grass |
| **Agrostis tenuis** | Brown Top Common Bent Colonial Bent | M | P | Forage grass |
| **Aframomum melegueta** | Guinea grains Melegueta Pepper | M | P | Seed spice |
| **Agave amaniensis** | Blue Sisal | M | P | Leaf fibre |
| **Agave angustifolia** | Dwarf Sisal | M | P | Leaf fibre |

| Latin Name | Common Name | D/M | A/B/P | Principal Products |
|---|---|---|---|---|
| **Agave cantala** | Cantala | M | P | Leaf fibre |
| | Maguey | | | |
| **Agave fourcroydes** | Henequeen | M | P | Leaf fibre |
| | White Sisal | | | |
| | Grey Sisal | | | |
| **Agave lecheguilla** | Mexican Fibre | M | P | Leaf fibre |
| | Istle | | | |
| **Agave letonae** | Salvador Henequeen | M | P | Leaf fibre |
| **Agave sisalana** | Sisal | M | P | Leaf fibre |
| **Agrostis stolonifera** | Bent Grass | M | P | Fodder grass |
| **Aleurites montana** | Tung | D | P | Oil |
| | Mu Tree | | | |
| **Albizzia lebbek** | | D | P | Pod and leaf |
| (main Albizzia sp.) | | | | Fodder legume |
| **Allium ampeloprasum** | | M | A/B | |
| var: ampeloprasum | Great Headed Garlic | | | Bulb vegetable |
| kurrat | Kurrat | | | Leaf vegetable |
| porrum | Leek | | | Leaf base vegetable |
| **Allium cepa** | Onion | M | A | Bulb vegetable |
| | (incl: shallots, potato, ever ready, | | | |
| | tree, Egyptian types) | | | |
| **Allium chinense** | Rakkyo | M | A | Bulb vegetable |
| **Allium fistulosum** | Welsh Onion | M | P | Leaf/leaf base |
| | Japanese Bunching Onion | | | vegetable |
| **Allium sativum** | Garlic | M | A/B | Bulb condiment |
| **Allium schoenoprasum** | Chives | M | P | Leaf herb |
| **Allium tuberosum** | Chinese Chives | M | P | Leaf/flower herb |
| **Alocasia macrorrhiza** | | M | P | Stem vegetable |
| **Alocasia indica** | | M | P | Stem vegetable |
| **Alopecurus pratensis** | Meadow Foxtail | M | P | Forage grass |
| | Foxtail | | | |
| **Althaea officinalis** | Marshmallow | D | P | Mucilage |
| **Alysicarpus vaginalis** | Alyce Clover | D | A/P | Fodder legume, |
| | One-leaved Clover | | | Green manure |
| **Amaranthus caudatus** | Amaranths | D | A | Seed |
| cruentus | | | | |
| leucocarpus | | | | |
| **Amaranthus tricolor** | Chinese Spinach | D | A | Leaf vegetable |
| | Tampala | | | |
| **Amorphophallus campanulatus** | | M | P | Tuber vegetable |
| | Elephant Yam | | | |
| **Amorphophallus rivieri** | | M | P | Tuber vegetable |
| **Anacardium occidentale** | Cashew | D | P | Nuts, oil |
| **Ananas comosus** | Pineapple | M | P | Fruit, leaf fibre |
| **Andropogon gayanus** | Gamba grass | M | P | Fodder grass |
| **Andropogon geradi** | Big Bluestem | M | P | Fodder grass |
| **Andropogon hallii** | Sand Bluestem | M | P | Fodder grass |
| **Andropogon scoparius** | Little Bluestem | M | P | Fodder grass, |
| | | | | reclamation |
| **Anethum graveolens** | Dill | D | A/B | Seed spice, |
| | | | | Leaf herb |
| **Angelica archangelica** | | D | A/B/P | Stem |
| **Anthriscus cerefolium** | Salad Chervil | D | A/B | Leaf herb |
| **Anthyllis vulneraria** | Kidney vetch | D | P | Fodder legume |
| **Anthriscus cerefolium** | Chervil | D | A | Leaf herb |
| **Antidesma bunius** | Bignay | D | P | Fruit |
| **Annona cherimolia** | Cherimoya | D | P | Fruit |
| **Annona diversifolia** | Ilama | D | P | Fruit |

| Latin Name | Common Name | D/M | A/B/P | Principal Products |
|---|---|---|---|---|
| **Annona muricata** | Soursop | D | P | Fruit |
| **Annona squamosa** | Sweetsop | D | P | Fruit |
| | Custard Apple, Sugar Apple | | | |
| **Annona reticulata** | Bullock's Heart | D | P | Fruit |
| **Apium graveolens** | Celery | D | A/B | Petiole salad, |
| | Celeriac | | | Root vegetable |
| **Arachis hypogaea** | Groundnut | D | A | Nuts, seed, |
| | Peanut | | | Fodder |
| | Monkeynut | | | |
| **Arbutus unedo** | Strawberry Tree | D | P | Fruit |
| **Areca catechu** | Abeca Palm | M | P | Seed masticatory, |
| | Betel Palm | | | narcotic |
| **Arenga pinnata** | Gomuti Palm | M | P | Sugar/starch |
| | Sugar Palm | | | (peduncles and trunk) |
| **Armorica rusticana** | Horse radish | D | P | Root condiment |
| **Arracacia xanthorrhiza** | Arracacha | D | P | Root vegetable |
| **Arrhenatherum elatius** | Tall Oat Grass | M | P | Forage grass |
| **Artemisia dracunculus** | Tarragon | D | P | Leaf herb |
| **Articum lappa** | Burdock | D | | Root vegetable/ flavouring |
| **Artocarpus altilis** | Breadfruit | D | P | Fruit vegetable |
| | Breadnut | | | |
| **Artocarpus heterophyllus** | Jackfruit | D | P | Fruit, seeds, timber |
| **Arundinaria amabilis** | Bamboo | M | P | Culm poles |
| **Asparagus officinalis** | Asparagus | M | P | Stem vegetable |
| **Astragalus arenarius** | Milk Vetch | D | P | Fodder legume |
| **Astragalus chinensis** | Milk Vetch | D | A | Green manure |
| **Astragalus cicer** | Milk vetch | D | P | Green manure, fodder legume |
| **Astragalus falcatus** | Sickle-pod Vetch | D | P | Fodder legume |
| **Astrocaryum murumuru** | Murumuru Palm | M | P | Oil (fruit) |
| **Atriplex hortensis** | Orache | D | A | Leaf vegetable |
| **Atropa bella-donna** | Deadly Nightshade | D | P | Leaf drug |
| **Avena byzantina** | Red Oat | M | A | Fodder grass |
| | Algerian Oat | | | Seed |
| **Avena sativa** | Oats | M | A/B | Seed, fodder |
| **Avena strigosa** | Bristle Oat | M | A | Fodder grass |
| | Small Oat | | | |
| **Averrhoa bilimbi** | Bilimbi | D | P | Fruit |
| **Averrhoa carambola** | Carambola | D | P | Fruit |
| **Axonopus affinis** | Carpet Grass | M | P | Fodder grass |
| **Axonopus compressus** | Savannah Grass | M | P | Fodder grass |
| | Carpet Grass | | | |
| **Axonopus scoparius** | Micay | M | P | Fodder grass |
| | Gamalote, Maicillo | | | |
| **Baccaurea motleyana** | Rambai | D | P | Fruit |
| **Bambusa vulgaris** | Bamboo | M | P | Culm poles, paper pulp, shoot vegetable |
| **Barbara verna** | Upland Cress | D | A | Leaf salad |
| | American Land Cress | | | |
| **Basella alba** | Indian Spinach | D | P | Leaf vegetable |
| | Malabar Spinach | | | |
| **Beckmannia erucaeformis** | Slough Grass | M | A | Fodder grass |
| **Benincasa hispida** | Wax Gourd | D | A | Fruit vegetable, |
| | White Gourd | | | Seeds |
| | Chinese Squash | | | |

| Latin Name | Common Name | D/M | A/B/P | Principal Products |
|---|---|---|---|---|
| **Bertholletia excelsa** | Brazil Nut | D | P | Nut |
| **Beta vulgaris** | includes: | D | A | |
| | Beetroot | | | Root vegetable |
| | Mangel | | | Root fodder |
| | Fodder Beet | | | Root fodder |
| | Sugar Beet | | | Sugar (root) |
| | Chard | | | Leaf vegetable |
| | Spinach Beet | | | Leaf vegetable |
| **Bixa orellana** | Annatto | D | P | Seeds (Dye) |
| | Bixa | | | |
| **Blighia sapida** | Akee | D | P | Fruit aril |
| **Boehmeria nivea** | Ramie | D | P | Stem fibre |
| | Rhea, China Grass | | | |
| **Boletus edulis** | Cep | | | Mushroom |
| **Borassus aethiopicum** | African Fan Palm | M | P | Fruit |
| **Borassus flabellifer** | Palmyra Palm | M | P | Sugar (sap of inflorescence), leaf stalk fibre |
| **Borago officinalis** | Borage | D | | Leaf herb |
| **Borthriochloa caucasia** | Caucasian Bluestem | M | P | Fodder grass |
| **Borthriochloa insculpta** | Sweet Pitted Grass | M | P | Fodder grass |
| | Creeping Bluegrass | | | |
| **Borthriochloa intermeda** | Australian Bluestem | M | P | Fodder grass |
| **Borthriochloa ischaemum** | Yellow Bluestem | M | P | Fodder grass |
| | Turkestan Bluestem | M | P | |
| **Borthiochloa pertusa** | Hurricane Grass | | | Fodder grass |
| | Seymour Grass, Sour Grass | | | |
| | Indian Bluegrass | | | |
| **Bouea macrophylla** | Gandaria | D | P | Fruit |
| **Bouteloua curtipendula** | Side Oats Grama | M | P | Fodder grass |
| **Bouteloua eriopoda** | Black grama | M | P | Fodder grass |
| **Bouteloua gracilis** | Blue grama | M | P | Fodder grass |
| **Brachiaria brizantha** | Pallisade Grass | M | P | Fodder grass |
| | Signal Grass | | | |
| **Brachiaria decumbens** | Surinam Grass | M | P | Fodder grass |
| | Signal Grass | | | |
| **Brachiaria humidicola** | Koronivia Grass | M | P | Fodder grass |
| | Creeping Signal Grass | | | |
| **Brachiaria mutica** | Para Grass | M | P | Fodder grass |
| **Brachiaria plantaginea** | Marmalade Grass | M | A | Fodder grass |
| **Brachiaria ruzizensis** | Ruzi Grass | M | P | Fodder grass |
| **Brassica alba** | White Mustard | D | A | Seedling salad, oil |
| **Brassica campestris** | Field Mustard | D | A/B | Oil |
| | Turnip Rape | | | |
| **Brassica chinensis** | Pak Choi | D | A | Leaf vegetable |
| | Chinese Cabbage | | | |
| **Brassica juncea** | Indian Mustard | D | A | Oil, leaf salad |
| | Mustard Cabbage, Brown Mustard | | | |
| **Brassica napobrassica** | Swede | D | B | Root vegetable, |
| | Rutabaga | | | Root and leaf fodder |
| **Brassica napus** | Rape Colza | D | A/B | Leaf fodder |
| | Siberian kale | D | A/B | Leaf vegetable |
| | Oilseed rape | D | A/B | Oil seed |
| **Brassica nigra** | Black Mustard | D | A | Oil, condiment |

| Latin Name | Common Name | D/M | A/B/P | Principal Products |
|---|---|---|---|---|
| **Brassica oleracea** | | | | |
| var: alboglabra | Chinese Kale | D | A/B | Leaf vegetable |
| acephala | Borecole | D | B | Leaf/stem fodder |
| | Kale | | | and vegetables |
| | Collard | | | |
| botrytis | Cauliflower | D | A/B | Flower vegetable |
| | Broccoli | | | |
| capitata | Cabbage | D | A/B | Leaf vegetable, |
| | | | | Fodder |
| gemmifera | Brussels Sprouts | D | A/B | Axillary bud |
| | | | | vegetable |
| gongylodes | Kohlrabi | D | A | Stem vegetable |
| italica | Sprouting Broccoli | D | A/B | Flower vegetable |
| **Brassica pikinensis** | Pe Tsai | D | B | Leaf vegetable |
| **Brassica rapa** | Turnip | D | A/B | Tap root vegetable, |
| | | | | leaf/ root fodder |
| **Bromus inermis** | Smooth Brome grass | M | P | Fodder grass |
| **Bromus marginatus** | Mountain Brome Grass | M | P | Fodder grass |
| **Bromus unioloides** | Rescue Grass | M | A/P | Fodder grass |
| | Prairie Grass | M | | |
| **Buchloe dactyloides** | Buffalo Grass | M | P | Fodder grass |
| **Butyrospermum paradoxum** | Shea Butter Tree | D | P | Fat (seeds) |
| **Cajanus cajan** | Pigeon Pea | D | A | Pod vegetable, |
| | Red Gram | | | seed, |
| | Congo Pea | | | fodder legume, |
| | No-eye pea | | | Green manure |
| **Calathea allouia** | Toppe Tambu | M | P | Tuber vegetable |
| **Calocarpum sapota** | Mamm Sapote | D | P | Fruit |
| **Calocarpum viride** | Green Sapote | D | P | Fruit |
| **Calopogonium muconoides** | Calopo | D | P | Fodder legume, |
| | | | | green manure |
| **Callamus spp.** | Rattans | M | P | Canes |
| **Calliandra calothyrsus** | Calliandra | D | P | Fodder legume |
| **Camellia sinensis** | Tea | D | P | Leaf beverage, |
| | | | | Narcotics |
| **Cananga odorata** | Ylang Ylang | D | P | Oil (flower) |
| **Canavalia ensiformis** | Jack Bean | D | A | Fodder, |
| | Horse Bean | | | vegetable, |
| | | | | green manure |
| **Canavalia gladiata** | Sword Bean | D | P | Fodder, |
| | | | | Green manure |
| **Canna edulis** | Edible Canna | M | A | Rhizome |
| | Queensland Arrowroot | | | vegetable, |
| | Purple Arrowroot | | | Rhizome/leaf |
| | Achira | | | fodder |
| **Cannabis sativa** | Hemp | D | A | Stem fibre, oil |
| **Cantharellus cibarius** | Chanterelle | | | Mushroom |
| | Girole | | | |
| **Capparis spinosa** | Caper | D | P | Unopened flowers |
| **Capsicum annum** | Chillies | D | A | Fruits |
| | Red Pepper (sweet/pungent) | | | |
| | Sweet Peppers includes: | | | |
| | Wrinkled Peppers, Cherry Peppers | | | |
| | Cone peppers, Tabasco | | | |
| | Cluster Peppers, Paprika | | | |
| | Long Peppers | | | |
| **Capsicum frutescens** | Bird Chillies | D | P | Fruit (pungent) |
| **Carissa grandiflora** | Natal Plum | D | P | Fruit |

| Latin Name | Common Name | D/M | A/B/P | Principal Products |
|---|---|---|---|---|
| **Carludovica palmata** | Panama Hat Plant | M | P | Leaf fibre |
| **Carthamus tinctorius** | Safflower | D | P | Oil, fodder |
| **Carum carvi** | Caraway | D | B | Fruit spice |
| **Carya illinoensis** | Pecan Nut | D | P | Nut |
| **Caryota urens** | Fish Tail Palm | M | P | Sugar (from |
| | Toddy Palm | | | inflorescence), |
| | | | | Leaf sheath fibre |
| **Casimiroa edulis** | White Sapote | D | P | Fruit |
| **Cassia angustifolia** | Indian Senna | D | P | Senna (pods and |
| | Tinnevelly Senna | | | leaves) |
| **Cassia auriculata** | Avaram | D | P | Tannin (bark) |
| **Cassia didimobotrya** | Candelabra Tree | D | P | Green manure |
| **Cassia hirsuta** | | D | P | Green manure |
| **Cassia multijuga** | | D | P | Green manure |
| | | | | Legume fodder |
| **Cassia pumila** | | D | A | Cover crop |
| **Cassia tora** | Sicke Senna | D | A/P | Green manure, |
| | Wild Senna | | | Fodder legume |
| **Cassia rotundifolia** | Roundleaf Cassia | D | A/P | Fodder legume |
| **Cassia sanna** | Alexandrian Senna | D | P | Senna (pods and |
| | | | | leaves) |
| **Cassia siamea** | | D | P | Fodder legume, |
| | | | | green manure |
| **Castanea sativa** | Sweet Chestnut | D | P | Nut |
| **Carica papya** | Papaya | D | P | Fruit, |
| | Papaw | | | Papain |
| | Pawpaw | | | |
| **Carthamus tinctorius** | Safflower | D | A | Oil, carmin, fodder |
| **Catha edulis** | Khat | D | P | Leaf |
| | Chat | | | (masticatory) |
| | Miraa | | | |
| **Ceiba pentandra** | Kapok | D | P | Inner fruit hair |
| **Cenchrus ciliaris** | Buffel Grass | M | P | Fodder grass |
| | African Foxtail Grass | | | |
| | Rhodesian Foxtail | | | |
| **Cenchrus setigerus** | Birdwood Grass | M | P | Fodder grass |
| **Centrosema pascuorum** | Centurion | D | A | Fodder legume |
| **Centrosema plumieri** | | D | P | Green manure, |
| | | | | fodder legume |
| **Centrosema pubescens** | Centro | D | P | Fodder legume, |
| | | | | Green manure |
| **Ceratonia siliqua** | Carob | D | P | Pod flavouring, |
| | Locust Bean | | | pod fodders |
| **Ceroxylon andicola** | South American Wax palm | M | P | Leaf/trunk |
| | | | | surface wax |
| **Chaerophyllum bulbosum** | Turnip Root Chervil | D | B | Vegetable |
| **Chamaecytisus prolifer** | Tree Lucerne | D | P | Fodder legume |
| | Tagasaste, Escabon | | | |
| **Chamaecytisus stenopetalus** | Gacia | D | P | Fodder legume |
| **Chamaemelum nobile** | Chamomile | D | P | Leaf beverage |
| **Chloris gayana** | Rhodes Grass | M | P | Fodder grass |
| **Chrysanthemum cinerarifolium** | Pyrethrum | D | P | Leaf insecticide |
| **Cicer arietinum** | Chick pea | D | A | Seeds, |
| | Gram | | | Pod/shoot |
| | Garbanzo Bean | | | vegetable, |
| | | | | Fodder legume |
| **Cichorium endiva** | Endive | D | A | Leaf salad |

| Latin Name | Common Name | D/M | A/B/P | Principal Products |
|---|---|---|---|---|
| **Cichorium intybus** | Chicory | D | P | Leaf salad, |
| | Whiteleaf Chicory | | | root beverage, |
| | | | | forage |
| **Chenopodium ambrosioides** | Wormseed | D | A | Oil (hairs) |
| | Mexican Tea | | | |
| **Chenopodium quinoa** | Quinoa | D | A | Seed |
| **Chrysanthemum cinerariaefolium** | Pyrethrum | D | P | Insecticide |
| **Chysanthemum coronarium** | Garland Chrysanthemum | D | A | Shoot vegetable |
| **Chrysobalanus icaco** | | D | P | Fruit |
| **Chrysophyllum cainito** | Star Apple | D | P | Fruit |
| | Caimito | | | |
| **Cinchona calisaya** | Quinine | D | P | Quinine (bark) |
| ledgeriana | | | | |
| officinalis | | | | |
| succirubra + hybrids | | | | |
| **Cinnamomum camphora** | Camphor | D | P | Oil (branches) |
| **Cinnamomum zeylancium** | Cinnamon | D | P | Bark condiment, |
| | | | | oil (bark), oil (leaf) |
| **Citrullus lanatus** | Watermelon | D | A | Fruit, seeds |
| **Citrus aurantifolia** | Sweet Lime | D | P | Fruit |
| **Citrus aurantium** | Sour Orange | D | P | Fruit |
| | Seville Orange | | | |
| **Citrus bergamia** | Bergamot Orange | D | P | Oils |
| **Citrus grandis** | Pummelo | D | P | Fruit |
| | Shaddock | | | |
| **Citrus limon** | Lemon | D | P | Fruit |
| **Citrus medica** | Citron | D | P | Fruit rind |
| **Citrus paradisi** | Grapefruit | D | P | Fruit |
| **Citrus reticulata** | Mandarin | D | P | Fruit |
| | Tangerine | | | |
| **Citrus sinensis** | Sweet Orange | D | P | Fruit |
| **Citrus** hybrids | Various | D | P | Fruit |
| **Clausena lansium** | Wampi | D | P | Fruit |
| **Clitoria cajanifolia** | Butterfly Pea | D | P | Green manure |
| **Clitoria ternatea** | Kordofan Pea | D | P | Fodder legume, |
| | Blue Pea | | | Green manure |
| **Cocos nucifera** | Coconut | M | P | Seed (oil,fruit), |
| | | | | Seed husk fibre, |
| | | | | Endocarp fuel, |
| | | | | Sugar (flower) |
| **Coffea arabica** | Arabic Coffee | D | P | Seed |
| **Coffea canephora** | Robusta Coffee | D | P | Seed |
| **Coffea liberica** | Liberica | D | P | Seed |
| **Coix lachryma jobi** | Job's Tears | M | A | Seed |
| | Adlay | | | |
| **Cola acuminata** | Abata Kola | D | P | Seed (masticatory narcotic) |
| **Cola anomala** | Bamenda Kola | D | P | do. |
| **Cola nitida** | Gbanja Kola | D | P | do. |
| **Cola verticillata** | Owe Kola | D | P | do. |
| **Coleus amboinicus** | Indian Borage | D | P | Leaf herb |
| **Coleus parviflorus** | | D | A | Tuber vegetable |
| **Colocasia esculenta** | | | | |
| var: esculenta | Dasheen | M | A/B | Corm vegetable |
| | Taro | | | Leaf vegetable |
| | Cocoyam | | | |
| antiquorum | Eddoe | M | A | Corm vegetable |
| | | | | Leaf vegetable |

| Latin Name | Common Name | D/M | A/B/P | Principal Products |
|---|---|---|---|---|
| **Corchorus capsuaris** | White Jute | D | A | Stem fibre |
| **Corchorus olitorius** | Jew's Mallow | D | A | Leaf vegetable |
| | Tossa Jute | | | |
| | Melokhia | | | |
| **Coronilla varia** | Crown Vetch | D | P | Soil protection |
| **Cortaderia selloana** | Pampas Grass | M | P | Fodder grass |
| **Copernicia cerifera** | Carnauba Wax Palm | M | P | Leaf surface wax |
| **Cornus mas** | Cornelian cherry | D | P | Fruit |
| **Corylus avellana** | Hazel Nut | D | P | Nut |
| **Corylus colurna** | Turkish Hazel Nut | D | P | Nut |
| **Corylus maxima** | Filibert | D | P | Nut |
| | Kentish Cob | | | |
| **Corypha - umbraculifera** | Talipot Palm | M | P | Sugar (from |
| **- elata** | | | | inflorescence) |
| **Coriandrum sativum** | Coriander | D | A | Fruit spice, |
| | Chinese Parsley | | | Leaf herb |
| **Crambe maritima** | Seakale | D | P | Leaf stalk vegetable |
| **Crataegus azarolus** | Azarole | D | P | Fruit |
| **Craterellus cornucopioides** | | | | Mushroom |
| **Crescentia cujete** | Calabash | D | A | Fruit (utensils) |
| **Crocus sativus** | Saffron | M | P | Stigma flavouring |
| **Crotalaria anagyroides** | | D | P | Green manure |
| **Crotalaria grantiana** | | D | A | Cover crop |
| **Crotalaria incana** | | D | P | Fodder legume, |
| | | | | cover crop |
| **Crotalaria intermedia** | | D | A | Fodder legume |
| **Crotalaria juncea** | Sann | D | A | Stem fibre, |
| | Sun Hemp | | | green manure |
| **Crotalaria lanceolata** | | D | P | Cover crop |
| **Crotalaria usaramoensis** | | D | B/P | Green manure |
| **Croton tiglium** | Croton Oil | D | P | Seed purgative |
| **Cucumis anguria** | W Indian Gherkin | D | A | Fruit vegetable |
| **Cucumis melo** | Melon (various) | D | A | Fruit |
| | incl: Honeydew Melon | | | |
| | Casaba Melon, Muskmelon | | | |
| | Persian Melon | | | |
| | Japanese Pickling Melon, Ogen | | | |
| **Cucumis sativus** | Cucumber | D | A | Fruit |
| | | | | vegetable/salad |
| **Cucurbita maxima** | Pumpkin | D | A | Fruit vegetable |
| | Winter Squash | | | |
| **Cucurbita moschata** | Pumpkin | D | A | Fruit vegetable |
| | Winter Squash | | | |
| | Winter Crookneck Squash | | | |
| **Cucurbita mixta** | Winter Squash | D | A | Fruit vegetable |
| | Pumpkin, Cushaw Squash | | | |
| **Cucurbita pepo** | Winter Squash | D | A | Fruit vegetable |
| | Pumpkin, Acorn Squash | | | |
| | Bush Summer Squash | | | |
| **Cuminum cyminum** | Cumin | D | A | Seed spice |
| **Curcuma amanda** | Mango Ginger | M | P | Rhizome vegetable |
| **Curcuma angustifolia** | Indian Arrowroot | M | P | Rhizome starch |

| Latin Name | Common Name | D/M | A/B/P | Principal Products |
|---|---|---|---|---|
| **Curcuma aromatica** | Wild Turmeric | M | P | Rhizome extracts |
| | Yellow Zedoary | | | |
| **Curcuma caesia** | Black Zedoary | M | P | Rhizome camphor |
| **Curcuma longa** | Turmeric | M | P | Rhizome spice |
| **Curcuma mangga** | | M | P | Rhizome spice |
| **Curcuma xanthorrhiza** | | M | P | Rhizome drug/ vegetable |
| **Curcuma zedoaria** | Zedoary | M | P | Rhizome stimulant/starch |
| **Curcuma domestica** | Turmeric | M | P | Rhizome spice |
| **Cyamopsis tetragonoloba** | Cluster Bean | D | A | Pod vegetable, seed, fodder, Green manure |
| | Guar | | | |
| **Cydonia oblonga** | Quince | D | P | Fruit |
| **Cymbopogon citratus** | Lemon Grass | M | P | Oil (leaf), leaf herb |
| **Cymbopogon flexuosus** | Malabar Grass | M | P | Oil (leaf) |
| **Cymbopogon martini** | Rosha Grass | M | P | Oil (leaf) – |
| var: motia | | | | Palmarosa oil |
| sofia | | | | Gingergrass oil |
| **Cymbopogon nardus** | Citronella Grass | M | P | Citronella oil |
| vars: lenabutu | | | | (leaf) |
| mahapengiri | | | | |
| **Cynara cardunculus** | Cardoon | D | P | Stem vegetable |
| **Cynara scolymus** | Globe Artichoke | D | P | Flower vegetable |
| **Cynodon dactylon** | Bermuda Grass | M | P | Fodder grass |
| | Star Grass, Bahama Grass | | | |
| | Doob, Couch Grass | | | |
| **Cynodon plectostachyum** | Giant Star Grass | M | P | Fodder grass |
| | Naivasha Star Grass | | | |
| **Cynosorus cristatus** | Crested Dogstail | M | P | Fodder grass |
| **Cyperus esculentus** | Chufa | M | P | Tuber vegetable |
| **Cyphomandra betacea** | Tree Tomato | D | B | Fruit |
| | Tamarillo | | | |
| **Cyrtosperma chamissonis** | | M | P | Tuber vegetable |
| **Dactylis glomerata** | Cocksfoot | M | P | Fodder grass |
| | Orchard Grass | | | |
| **Daemonorops spp.** | Rattans | M | P | Canes |
| **Danthonia pilosa** | | M | P | Fodder grass |
| **Daucus carota** | Carrot | D | A/B | Tap root vegetable |
| **Dendrocalamus strictus** | Bamboo | M | P | Culm poles, shoot vegetable |
| **Derris elliptica** | Derris | D | P | Insecticide |
| | Tuba Root | | | (root) |
| **Desmanthus virgatus** | Dwarf Koa | D | P | Fodder legume, green manure |
| **Desmodium spp.** | Tick Clovers | D | A/P | Fodder legumes, green manure |
| | Beggar Weeds | | | |
| important sps: | | | | |
| **Desmodium intortum** | Desmodium Greenleaf | D | P | Fodder legume |
| **Desmodium heterophyllum** | Desmodium Hetero | D | P | Fodder legume |
| **Desmodium uncinatum** | Desmodium Silverleaf | D | P | Fodder legume |
| | Japanese Clover | | | |
| **Dicanthium aristatum** | Angleton Grass | M | P | Fodder grass |
| **Dicanthium caricosum** | Nadi Bluegrass | M | P | Fodder grass |
| | Antigua Hay Grass | | | |
| **Digitalis purpurea** | Common Foxglove | D | P | Leaf pharmaceutical |
| **Digitaria decumbens** | Pangola Grass | M | P | Fodder grass |
| | Pongola Grass | | | |

| Latin Name | Common Name | D/M | A/B/P | Principal Products |
|---|---|---|---|---|
| **Digitaria didactyla** | Queensland Blue Couch | M | P | Fodder grass |
| **Digitaria exilis** | Hungry Rice | M | A | Seed |
| | Fonio, Fundi, Acha | | | |
| **Digitaria pentzii** | Woolly Finger Grass | M | P | Fodder grass |
| **Digitaria smutzi** | Digitaria | M | P | Fodder grass |
| **Digitaria swazilandensis** | Swaziland Finger Grass | M | P | Fodder grass |
| **Dioscorea alata** | Greater Yam | M | A | Tuber vegetable |
| | Water Yam, Winged Yam | | | |
| | Asiatic Yam, White Yam | | | |
| | Chinese Yam | | | |
| **Dioscorea batatas** | Japanes Yam | M | A | Tuber vegetable |
| **Dioscorea bulbifera** | Potato Yam | M | A | Aerial tuber |
| | Aerial Yam | | | vegetable |
| **Dioscorea cayenensis** | Yellow Guinea Yam | M | A | Tuber vegetable |
| | Yellow Yam, Twelve Month Yam | | | |
| **Dioscorea dumetorum** | African Bitter Yam | M | A | Tuber vegetable |
| | Cluster Yam | | | |
| **Dioscorea esculenta** | Lesser Yam | M | A | Tuber vegetable |
| **Dioscorea hispida** | Asiatic Bitter Yam | M | A | Tuber vegetable |
| **Dioscorea nummularia** | | M | A | Tuber vegetable |
| **Dioscorea opposita** | Chinese Yam | M | A | Tuber vegetable |
| | Cinnamon Yam | | | |
| **Dioscorea pentaphylla** | | M | A | Tuber vegetable |
| **Dioscorea rotundata** | White Guinea Yam | M | A | Tuber vegetable |
| | White Yam, Eight Months Yam | | | |
| **Dioscorea trifida** | Cush Cush Yam | M | A | Tuber vegetable |
| **Diospyros - ebenaster** | Black Sapote | D | P | Fruit |
| - **digyna** | | | | |
| **Diospyros discolor** | Velvet Apple | D | P | Fruit |
| **Diospyros kaki** | Chinese Persimmon | D | P | Fruit |
| | Kaki | | | |
| **Diospyros lotus** | Date Plum | D | P | Fruit |
| **Diospyros virginiana** | Persimmon | D | P | Nut |
| **Diplachne fusca** | Kallar Grass | M | P | Fodder |
| **Dipteryx odorata** | Tonka Bean | D | P | Flavourings/scents (beans) |
| **Dolichos bulbosus** | (see Pachyrrhizus) | | | |
| **Dolichos lablab** | (see Lablab) | | | |
| **Dovyalis caffra** | Kei Apple | D | P | Fruit |
| **Doyvalis hebecarpa** | Kitembilla | D | P | Fruit |
| **Durio zibethinus** | Duria | D | P | Fruit |
| **Echinochloa colona** | Jungle Rice | M | A | Fodder, seed |
| **Echinochloa crus - galli** | Barnyard Millet | M | A | Fodder, seed, shoot vegetable |
| **Echinochloa frumentacea** | Japanese Barnyard Millet | M | A | Fodder, seed |
| **Echinochloa pyramidalis** | Antelope Grass | M | P | Fodder grass |
| **Echinochloa stagina** | | M | P | Fodder grass |
| **Ehrharta calycina** | Perennial Veld grass | M | P | Fodder grass |
| **Elaeis guineensis** | Oil Palm | M | P | Palm oil/palm kernel oil from seed mesocarp/ endosperm |
| **Eleagnus angustifolia** | Oleaster | D | P | Fruit |
| **Eleocharis dulcis** | Water Chestnut | M | B | Corm vegetable |
| **Eleusine coracana** | Finger Millet | M | A | Seed |
| | Rapoko | | | |
| **Eleusine indica** | Fowl Foot Grass | M | A | Fodder, seed |
| | Crowsfoot Grass, Rapoka Grass | | | |

| Latin Name | Common Name | D/M | A/B/P | Principal Products |
|---|---|---|---|---|
| **Ellettaria cardomomum** | Cardamom | M | P | Fruit spice |
| **Elymus angustus** | Wildrye Altai | M | P | Fodder grass |
| **Elymus triticoides** | Beardless Wild Rye | M | P | Fodder grass |
| **Elymus canadensis** | Canada Wild Rye | M | P | Fodder grass |
| **Elymus junceus** | Russian Wild Rye | M | P | Fodder grass |
| **Ensete ventricosa** | Enset | M | P | Corm/stem vegetable |
| **Eragrostis curvula** | Weeping Love Grass | M | P | Fodder grass |
| **Eragrostis lehmanniana** | Lehmann Love Grass | M | P | Fodder grass |
| **Eragrostis Tef** | Teff | M | A | Seed, fodder |
| **Eragrostis trichodes** | Sand Love Grass | M | P | Fodder grass |
| **Eriobotrya japonica** | Loquat | D | P | Fruit |
| **Eriochloa polystachya** | Carib Grass | M | P | Fodder grass |
| **Eruca sativa** | Rocket | D | A | Leaf salad |
| **Erythrina abyssinica** | Erythrina | D | P | Fodder legume |
| **Erythroxylum coca** and novogranatense | Coca | D | P | Cocaine(leaf), Masticatory |
| **Eucalyptus spp.** | Eucalyptus Oil | D | P | Oil (leaf) |
| **Euchlaena mexicana** | Teosinte | M | A | Fodder grass |
| **Eugenia caryophyllus** | Clove | D | P | Cloves (flower bud), oil (flower bud and stem, leaf) |
| **Eugenia jambos** | Rose Apple | D | P | Fruit |
| **Eugenia javanica** | Java Apple Wax Apple | D | P | Fruit |
| **Eugenia malaccensis** | Pomerac Malay Apple | D | P | Fruit |
| **Eugenia uniflora** | Surinam Cherry Pitanga | D | P | Fruit |
| **Euphoria longana** | Longan | D | P | Fruit |
| **Fagopyrum esculentum** | Buckwheat | D | A | Seeds |
| **Feijoa sellowiana** | Feioja | D | P | Fruit |
| **Festuca arundinacea** | Tall Fescue | M | P | Fodder grass |
| **Festuca ovina** | Sheep's Fescue | M | P | Fodder grass |
| **Festuca pratensis** | Meadow Fescue | M | P | Fodder grass |
| **Festuca rubra** | Red Fescue | M | P | Fodder grass |
| **Ficus carica** | Fig | D | P | Fruit |
| **Ficus elastica** | India Rubber Fig | D | P | Rubber |
| **Flacourtia jangomas** | Cherry | D | P | Fruit |
| **Foeniculum vulgare** | Fennel Florence Fennel | D | A/P | Petiole salad/ vegetable, leaf herb, oil (flower) |
| **Fortunella spp.** | Kumquat | D | P | Fruit |
| **Fragaria ananassa** | Strawberry | D | P | Fruit |
| **Fragaria moschata** | Hautbois Strawberry | D | P | Fruit |
| **Fragaria vesca** | Wild Strawberry | D | P | Fruit |
| | Alpine Strawberry | D | P | Fruit |
| **Furcraea gigantea** | Mauritius Hemp | M | P | Leaf fibre |
| **Garcinia mangostana** | Mangosteen | D | P | Fruit |
| **Garcinia hanburyi** | Gamboge | D | P | Dye (resin) |
| **Gigantochloa apus** | Bamboo | M | P | Culm poles |
| **Gliricidia maculata** | | D | P | Green manure, fodder legume |
| **Glyceria maxima** | Reed Sweet Grass | M | P | Fodder grass |
| **Glediitschia triacanthos** | Honey Locust Sweet Locust | D | P | Fodder legume, pod fodder |

| Latin Name | Common Name | D/M | A/B/P | Principal Products |
|---|---|---|---|---|
| **Glycine max** | Soybean | D | A | Seed, oil, |
|  | Soya Bean |  |  | Fodder, green manure |
| **Glycine wightii** | Glycine | D | P | Fodder legume |
| **Glycyrrhiza glabra** | Liquorice | D | P | Roots |
|  |  |  |  | Fibre types: |
| **Gossypium herbaceum** | Cotton | D | A | Short staple |
| **arboreum** |  |  |  | Short staple |
| **hirsutum** |  |  |  | Med and med long |
| **barbadense** |  |  |  | Long staple. Also oil |
| **Grewia asiatica** | Phalasa | D | P | Fruit |
| **Guadua angustifolia** | Bamboo | M | P | Culm poles |
| **Guilielma gasipaes** | Peach Palm | M | P | Fruit |
|  | Pejibaye |  |  |  |
| **Guizota abyssinica** | Niger Seed | D | A | Oil |
| **Hedysarum coronarium** | Sulla | D | P | Fodder legume |
| **Helianthus annus** | Sunflower | D | P | Seed, oil, fodder |
| **Helianthus tuberosus** | Jerusalem Artichoke | D | A/P | Tuber vegetable |
| **Hevea brasiliensis** | Para Rubber | D | P | Rubber (latex) |
| **Hibiscus cannabinus** | Kenaf | D | A | Stem fibre |
| **Hibiscus esculentus** | Lady's Finger | D | A | Fruit vegetable |
|  | Okra |  |  |  |
| **Hibiscus sabdariffa** | Roselle | D | A | Acid flavouring |
|  | Jamaican Sorrel |  |  | (Calyces), leaf salad |
| **Hyssopus officinalis** | Hyssop | D | P | Leaf herb |
| **Holcus lanatus** | Yorkshire Fog | M | P | Fodder grass |
|  | Velvet Grass |  |  |  |
| **Hordeum bulbosum** |  | M | P | Fodder grass |
| **Hordeum vulgare** | Barley | M | A/B | Seed, fodder grass |
| **Humulus lupulus** | Hop | D | P | Flowers |
| **Hyparrhenia hirta** |  | M | P | Fodder grass |
| **Hyparrhenia rufa** | Jaruga | M | P | Fodder grass |
|  | Thatching Grass |  |  |  |
| **Hyphaene thebaica** | Doum Palm | M | P | Fruit |
|  | Dum Palm |  |  |  |
| **Hyptis spicigera** |  | D | A | Oil |
| **Ilex paraguensis** | Mate | D | P | Leaf beverage |
|  | Paraguay Tea |  |  |  |
| **Indigofera arrecta** | Indigo | D | P | Green manure |
| **Indigofera endecaphylla** | Trailing Indigo | D | P | Green manure, |
|  |  |  |  | fodder legume |
| **Indigofera hirsuta** | Hairy Indigo | D | A | Green manure, |
|  |  |  |  | fodder legume |
| **Indigofera suffruticosa** | Indigo | D | P | Green manure |
|  |  |  |  | Indigo |
| **Indigofera tinctoria** | Indigo | D | P | Green manure |
|  |  |  |  | Indigo |
| **Inga spp.** |  | D | P | Pod pulp fruit |
| **Inula helenium** | Elecampane | D | P | Root oil |
| **Ipomea batatas** | Sweet Potato | D | P | Tuber vegetable, |
|  |  |  |  | Tuber and tops |
|  |  |  |  | fodder, starch |
| **Ipomea aquatica** | Water Spinach | D | P | Leaf vegetable, |
|  |  |  |  | Vine fodder |
| **Ipomea purga** | Jalap | D | P | Tuber vegetable |
| **Ischaemum indicum** | Batiki Blue Grass | M | P | Fodder grass |
| **Ixophorus unisetus** |  | M | P | Fodder grass |
| **Jasminum officinale** | White Jasmine | D | P | Flower perfume |
| **Juglans rebia** | Walnut | D | P | Nut |

| Latin Name | Common Name | D/M | A/B/P | Principal Products |
|---|---|---|---|---|
| **Juniperus communis** | Common Juniper | M | P | Fruit flavouring |
| **Lablab purpureus** | Hyacinth Bean | D | P | Pod vegetable, |
| | Seim Bean | | | seed, |
| | Lubia Bean | | | green manure, |
| | Indian Bean | | | fodder legume |
| | Lablab Bean | | | |
| | Egyptian Bean | | | |
| | Bovanist Bean | | | |
| **Lactarius deliciosus** | Saffron Milk Cap | | | Mushroom |
| **Lactuca sativa** | Lettuce | D | A | Leaf vegetable |
| **Lagenaria siceraria** | Bottle Gourd | D | A | Fruit shell |
| | Calabash Gourd | | | utensils |
| **Languas galanga** | Greater Galangal | M | P | Rhizome spice |
| **Languas officinarum** | Lesser Galangal | M | P | Rhizome spice |
| **Lansium domesticum** | Duku | D | P | Fruit |
| | Langsat | | | |
| **Lathyrus cicera** | Vetchling | D | A/P | Green manure, |
| | Flat Pod Pea Vine | | | fodder legume |
| **Lathyrus hirsutus** | Rough Pea | D | A | Fodder legume |
| | Caley Pea, Singletary Pea | | | |
| **Lathyrus ochrus** | Cyprus Vetch | D | A | Fodder legume |
| **Lathyrus sativus** | Grass Pea | D | A | Seeds, |
| | Chickling Pea | | | Fodder legume |
| | Wedge Pea | | | |
| **Lathyrus sylvestris** | Flat Pea | D | P | Fodder legume |
| | | | | Erosion control |
| **Laurus nobilis** | Sweet Bay | D | P | Leaf flavouring |
| **Lavendula angustifolia** | Lavender | D | P | Oil (flower) |
| **Lavendula latifolia** | Broad-leaved Lavender | D | P | Oil (flower) |
| **Lecythis zabucajo** | Paradise Nut | D | P | Nut |
| | Sapucaia Nut, Monkey Pot | | | |
| **Leersia hexandra** | | M | P | Fodder grass |
| **Leginaria siceraria** | Chinese squash | D | A | Fruit vegeable |
| **Lens esculenta** | Lentil | D | A | Seeds |
| **Lentinus edodes** | Shiitake Mushroom | | | Mushroom |
| **Leopoldinia piassaba** | Para Piassava Palm | M | P | Leaf base fibres |
| **Lepidium sativum** | Garden Cress | D | A | Seedling salad |
| **Lepiota procera** | Parasol Mushroom | | | Mushroom |
| **Lespedeza cuneata** | Sericea | D | P | Fodder legume, |
| | Lespedeza | | | green manure |
| **Lespedeza stipulacea** | Korean Lespedeza | D | A | Fodder legume, |
| | | | | green manure |
| **Lespedeza striata** | Common Lespedeza | D | A | Fodder legume, |
| | Japanese Lespedeza | | | green manure |
| **Leucaena leucocephala** | White Popinac | D | P | Fodder legume, |
| | Kao Haole | | | green manure |
| | Lamtoro, Leucaena | | | |
| **Leucaena glabrata** | | D | P | Fodder legume |
| **Leucaena pulverulenta** | | D | P | Fodder legume |
| **Levisticum officinale** | Lovage | D | P | Herb |
| **Linum usitatissimum** | var. Linseed | D | A | Oil |
| | var. flax | | | Stem fibre |
| **Lippia citriodora** | Lemon scented Verbena | D | P | Leaf beverage |
| **Lippia micromeria** | Spanish Thyme | D | P | Leaf herb |
| **Litchi chinensis** | Litchi | D | P | Fruit aril |
| | Lychee | | | |

| Latin Name | Common Name | D/M | A/B/P | Principal Products |
|---|---|---|---|---|
| **Lodoicea maldivica** | Double Coconut Coco-De-Mer | M | P | Fruit shell utensil |
| **Lolium multiflorum** | Italian Ryegrass | M | A/B | Fodder grass |
| **Lolium perenne** | Perennial Ryegrass | M | P | Fodder grass |
| **Lolium perenne x multiflorum** | Short rotation Ryegrass Hybrid Ryegrass | M | B/P | Fodder grass |
| **Lolium rigidum** | Wimmera Ryegrass Annual Ryegrass | M | A | Fodder grass |
| **Loofa acutangula** | Angled Loofah Chinese Okra | D | A | Vegetable |
| **Loofa cylindrica** | Smooth Loofah Vegetable sponge | D | A | Sponge (fruit), vegetable |
| **Lotononis bainesii** | Lotononis | D | P | Fodder legume |
| **Lotus corniculatus** | Birdsfoot Trefoil | D | P | Fodder legume |
| **Lotus uliginosus** | Big Trefoil Maku, Greater Lotus | D | P | Fodder legume |
| **Lucuma bifera** | Egg Fruit | D | P | Fruit |
| **Lupinus alba** | White Lupin | D | A | Green manure, seed, fodder legume |
| **Lupinus angustifolius** | Blue Lupin Narrow Leaved Lupin Annual Lupin | D | A | Green manure, seed Fodder legume |
| **Lupinus luteus** | Yellow Lupin | D | A | Green manure, seed, fodder legume |
| **Lupinus spp.** | Lupins | D | A/P | Various as above |
| **Lycopersicon esculentum** | Tomato | D | A | Fruit salad/vegetable |
| **Macadamia ternifolia** | Macadamia Nut Queensland Nut | D | P | Nuts |
| **Macroptilium lathyroides** | Phasey Bean | D | A/B | Fodder legume |
| **Macrotyloma axillare** | Axillaris | D | P | Fodder legume |
| **Macrotyloma uniflorum** | Horsegram Kulthi Bean | D | A | Fodder legume, green manure |
| **Madhuca longifolia** | Mahua | D | P | Fat (fruit) |
| **Malpighia glabra** | Barbados Cherry West Indian Cherry, Acerola | D | P | Fruit |
| **Malus sylvestris** | Crab apple | D | P | Fruit |
| **Malus pumila** | Apples | D | P | Fruit |
| **Mammea americana** | Mammey Apple | D | P | Fruit |
| **Mangifera indica** | Mango | D | P | Fruit |
| **Manihot esculenta** | Cassava Manioc Tapioca | D | P | Tuber vegetable, starch |
| **Manilkara achras** | Sapodila Chku Nispero Naseberry | D | P | Fruit, latex (chicle-chewing gum) |
| **Manilkara bidentata** | Balata | D | P | Latex (balata rubber) |
| **Maranta arundinacea** | Arrowroot West Indian Arrowroot | M | P | Starch (rhizomes) |
| **Marjorana hortensis** | Sweet marjoram Knotted Marjoram | D | A | Leaf herb |
| **Marjorana onites** | Pot Marjoram | D | P | Leaf herb |
| **Matricaria recutita** | Wild/German Chamomile | D | A | Leaf beverage |
| **Medicago arabica** | Spoted Burr Clover Spotted Medic Southern Burr Clover Heart Clover, Heart Trefoil Purple Grass, St Mawes Clover | D | A | Fodder legume |

| Latin Name | Common Name | D/M | A/B/P | Principal Products |
|---|---|---|---|---|
| **Medicago falcata** | Sickle Medic | D | P | Fodder legume |
| **Medicago polymorpha** | Burr Clover | D | A | Fodder legume |
| | Toothed Medic | | | |
| | Toothed Burr Clover | | | |
| **Medicago littoralis** | Strand Medic | D | A | Fodder legume |
| **Medicago lupulina** | Black Medic | D | A/B | Fodder legume |
| **Medicago minima** | Little Burr Clover | D | A | Fodder legume |
| **Medicago orbicularis** | Snail Medic | D | A | Fodder legume |
| **Medicago rigidula** | Tift Burr Clover | D | A | Fodder legume |
| **Medicago scutellata** | Snail Medic | D | A | Fodder legume |
| **Medicago sativa** | Lucerne | D | P | Fodder legume |
| | Alfalfa, Purple Alfalfa | | | |
| | Purple medic, Snail Clover | | | |
| | Burgandy Clover, Chilean Clover | | | |
| **Medicago truncatula** | Barrel Medic | D | A | Fodder legume |
| **Melinis minutiflora** | Molasses Grass | M | P | Fodder grass |
| **Melicocca bijuga** | Genip | D | P | Fruit |
| | Mamoncillo | | | |
| **Melilotus alba** | Sweet Clover | D | A/B | Fodder legume |
| | White Sweet Clover | | | |
| | Bokhara Sweet Clover | | | |
| | Honey Clover, White Melilot | | | |
| | Hubam | | | |
| **Melilotus indica** | Yellow Annual Sweet Clover | D | A | Fodder legume |
| | Sour Clover | | | Reclamation |
| | King Island Clover, Senji | | | |
| | Indian Clover | | | |
| **Melilotus officinalis** | Yellow Melilot | D | B/A | Fodder legume |
| | Biennial Yellow Sweet Clover | | | |
| **Melilotus suaveolens** | Daghestan Sweet Clover | D | B | Fodder legume |
| **Melissa officinallis** | Balm | D | P | Leaf herb |
| **Melocanna baccifera** | Bamboo | M | P | Culm poles, paper pulp |
| **Mentha arvensis** | Japanese Mint | D | P | Menthol oil |
| **Mentha pulegium** | Pennyroyal | D | P | Leaf herb |
| **Mentha rotundifolia** | Round Leaved Mint | D | P | Leaf herb |
| **Mentha spicata** | Spearmint | D | P | Leaf herb |
| | Common Garden Mint | | | |
| **Mentha x piperata** | Peppermint | D | P | Oil (leaf) |
| **Mespilus germanica** | Medlar | D | P | Fruit |
| **Metroxylon spp.** | Sago Palms | M | P | Sago (trunk) |
| **Mimosa invisa** | Mimosa | D | A/P | Green manure |
| **Momordica charantia** | Bitter Gourd | D | A | Fruit vegetable, seed condiment |
| | Bitter Cucumber | | | |
| | Balsam Pear, Bitter Melon | | | |
| **Monstera delicosa** | Ceriman | M | P | Fruit |
| **Morchella esculenta** | Edible Morrel | | | Mushroom |
| **Morus alba** | White Mulberry | D | P | Fruit, silkworm fodder |
| **Morus nigra** | Black Mulberry | D | P | Fruit |
| **Mucuna pruriens** | Velvet Bean | D | A | Fodder legume |
| **Musa cvs** | Banana | M | P | Fruit |
| **Musa fehi** | Fe'i Banana | M | P | Fruit |
| **Musa textilis** | Abaca | M | P | Petiole sheath fibre |
| | Manila Hemp | | | |
| **Musa paradisiaca** | Plantain | M | P | Fruit |
| **Muscari comosum** | Grape Hyacinth | M | P | Bulb vegetable |

| Latin Name | Common Name | D/M | A/B/P | Principal Products |
|---|---|---|---|---|
| **Myrciaria cauliflora** | Jaboticaba | D | P | Fruit |
| **Myristica fragrans** | Nutmeg | D | P | Nutmeg(seed), Mace (husk) |
| **Myrrhis odorata** | Sweet Cicely | D | P | Herb |
| **Nasturtium officinale** | Watercress | D | P | Leaf vegetable |
| **Nelumbo nucifera** | Lotus Root | D | A/P | Rhizome veg. |
| **Neoglaziovia variegata** | Caroa | M | P | Leaf fibre |
| **Neonotonia wightii** | Glycine | D | P | Fodder legume |
| **Nephelium lappaceum** | Rambutan | D | P | Fruit aril |
| **Nephelium mutabile** | Pulasan | D | P | Frut aril |
| **Nicotiana rustica** | Nicotine Tobacco | D | A | Tobacco (leaf), Nicotine |
| **Nicotiana tabacum** | Tobacco | D | A | Tobcco (leaf) |
| **Nigella sativa** | Black Cumin | D | A | Seed spice |
| **Nypa fructicans** | Nipa Palm | M | P | Sugar (peduncle) |
| **Ocimum basilicum** | Basil | D | P | Leaf herb, oil |
| **Olea europaea** | Olive | D | P | Oil, fruit |
| **Onobrychis vicifolia** | Sainfoin Esparcette | D | P | Fodder legume |
| **Opuntia ficus-indica** | Prickly Pear Barbary Fig | D | P | Fruit |
| **Orbignya spp.** | Barbacu Palms | M | P | Oil (nut) |
| **Origanum vulgare** | Oregano Wild Marjoram | D | P | Leaf herb |
| **Ornithopus compressus** | Serradella | D | A | Fodder legume, green manure |
| **Oryza glaberrima** | African Rice | M | A | Seed |
| **Oryza sativa** s.sp. indica japonica javanica | Rice | M | A | Seed |
| **Oryzopsis hymenoides** | Indian Rice Grass | M | P | Fodder grass |
| **Oryzopsis miliacea** | Smilo | M | P | Fodder grass |
| **Oxalis tuberosa** | Oca | D | A | Tuber vegetable |
| **Oxytenanthera abyssinica** | Bamboo | M | P | Culm timber |
| **Pachyrrhizus erosus** | Yam Bean | D | P | Tuber vegetable, pod vegetable |
| **Pachyrrhizus tuberosus** | Yam Bean Potato Bean Manioc Bean | D | P | Tuber vegetable, fodder legume |
| **Palaquium gutta** | Gutta Percha | D | P | Latex |
| **Pandanus spp.** | Screw Pines | M | P | Various incl. fruit, scents, seeds, fibre |
| **Panicum antidotale** | Giant Panic Grass Blue Panic | M | P | Fodder grass |
| **Panicum bulbosum** | Texas Grass | M | P | Fodder grass |
| **Panicum coloratum** | | M | P | Fodder grass |
| **Panicum coloratum** var. makarikariense | Makarakari Grass | M | P | Fodder grass |
| **Panicum maximum** | Guinea Grass | M | P | Fodder grass |
| **Panicum miliaceum** | Common Millet Proso Millet, Russian Millet Hog Millet, Brown Corn Millet Broomcorn Millet | M | A | Seed, fodder |
| **Panicum prolutum** | | M | P | Fodder grass |
| **Panicum sumatrense** | Little Millet | M | A | Seed, fodder |
| **Panicum turgidum** | | M | P | Fodder grass |

| Latin Name | Common Name | D/M | A/B/P | Principal Products |
|---|---|---|---|---|
| **Panicum virgatum** | Switch Grass | M | P | Fodder grass |
| **Papaver somniferum** | Poppy | D | A | |
| **ssp. somniferum** | | | | Opium (latex) |
| **ssp. hortense** | | | | Seeds, oil |
| **Parthenium argentatum** | Guayule | D | P | Fruit |
| **Paspalum conjugatum** | Buffalo Grass | M | P | Fodder grass |
| **Paspalum dilatatum** | Dallis Grass | M | P | Fodder grass |
| **Paspalum guenoarum** | | M | P | Fodder grass |
| **Paspalum notatum** | Bahia Grass | M | P | Fodder grass |
| **Paspalum scrobiculatum** | Kodo Millet | M | A | Grain, fodder |
| | Scrobic | | | |
| **Paspalum plicatulum** | Plicatulum | M | P | Fodder grass |
| **Paspalum urvillei** | Vasey Grass | M | P | Fodder grass |
| | Upright Paspalum | | | |
| **Paspalum virgatum** | Upright Paspalum | M | P | Fodder grass |
| **Paspalum wettsteinii** | Broad Leaf Paspalum | M | P | Fodder grass |
| | | | | |
| **Parmentiera edulis** | Cuachilote | D | P | Fruit |
| **Pastinaca sativa** | Parsnip | D | A/B | Tap root vegetable |
| **Passiflora edulis** | Passion Fruit | D | P | Fruit |
| | Purple Passion Fruit | | | |
| **Passiflora quadrangularis** | Giant Granadilla | D | P | Fruit, Fruit vegetable |
| **Pelargonium graveolens** | Geranium | D | P | Oil (leaf) |
| **Pennisetum clandestinum** | Kikuyu Grass | M | P | Fodder grass |
| **Pennisetum orientale** | | M | P | Fodder grass |
| **Pennisetum purpureum** | Elephant Grass | M | P | Fodder grass |
| | Napier Grass | M | P | |
| | Uganda Grass | M | P | |
| **Pennisetum purpureum** | | M | P | Fodder grass |
| var merkeri | Merker Grass | | | |
| **Pennisetum pedicellatum** | | M | A | Fodder grass |
| **Pennisetum stramineum** | | M | P | Fodder grass |
| **Pennisetum typhoides** | Bulrush Millet | M | A | Seeds, Fodder |
| **Pennisetum villosum** | | M | P | Fodder grass |
| **Persea americana** | Avocado | D | P | Salad fruit |
| | Alligator Pear | | | |
| **Petasites japonicus** | Butterbur | D | A | Petiole vegetable |
| **Petroselinum crispum** | Parsley | D | B | Leaf herb, |
| | Italian Parsley | | | root vegetable |
| | Hamburg Parsley | | | |
| | Turnip Rooted Parsley | | | |
| **Phacelia tanacetifolia** | Phacelia | D | A | Green manure |
| **Phalaris arundinacea** | Reed Canary Grass | M | P | Fodder grass |
| **Phalaris canariensis** | Birdseed | M | A | Seed |
| **Phalaris minor** | Small Canary Grass | M | A | Fodder grass |
| **Phalaris tuberosa** | Toowoomba Canary Grass | M | P | Fodder grass |
| **Phalaris tuberoa var. stenoptera** | | M | P | Fodder grass |
| | Harding Grass | | | |
| **Phaseolus aconitifolius** | Mat Bean | D | A | Pod vegetable, |
| | Moth Bean | | | Seeds, fodder |
| | Philipesera | | | legume |
| **Phaseolus acutifolius** | Tepary Bean | D | A | Seeds, fodder |
| | Texas Bean | | | legume |
| **Phaseolus angularis** | Adzuki Bean | D | A | Seed, fodder |
| | | | | legume |

| Latin Name | Common Name | D/M | A/B/P | Principal Products |
|------------|-------------|-----|-------|--------------------|
| **Phaseolus aureus** | Green Gram | | | |
| | Gold Gram | D | A | Seeds, pod vegetable, fodder legume, green manure |
| **Phaseolus calcaratus** | Rice Bean | D | A | Seeds, pod/leaf vegetable |
| | Red Bean | | | |
| **Phaseolus coccineus** | Scarlet Runner Bean | D | A/P | Seeds, pod/leaf vegetable |
| **Phaseolus lathyroides** | Phasemy Bean | D | A/P | Fodder legume |
| **Phaseolus lunatus** | Lima Bean | D | A | Seeds, pod vegetable, Green manure |
| | Sieva Bean | | | |
| | Butterbean | | | |
| | Madagascar Bean, Burma Bean | | | |
| **Phaseolus mungo** | Black Gram | D | A | Seeds, fodder, green manure, pod vegetable, fodder legume |
| | Urd | | | |
| **Phaseolus radiatus** | Green Gram | D | A | Green manure, seeds, green manure |
| | Green Mung | | | |
| **Phaseolus semi-erectus** | | D | A | Green manure, Fodder legume |
| **Phaseolus vulgaris** | Common Bean | D | A | Pod vegetables, Seeds |
| | Kidney Bean | | | |
| | French Bean, Haricot Bean | | | |
| | Salad Bean, String Bean, Frijoles, | | | |
| | Snap Bean, 'Baked beans' | | | |
| | Garden Bean | | | |
| **Phleum pratense** | Timothy Grass | M | P | Fodder grass |
| **Phoenix dactylifera** | Date Palm | M | P | Fruits, timber |
| **Phoenix reclinata** | Wild Date Palm | M | P | Fruits |
| **Phoenix sylvestris** | | M | P | Sugar (stem) |
| **Phorium tenax** | Hemp | M | P | Leaf fibre |
| | New Zealand Flax | | | |
| **Phyllanthus embilica** | Aonla | D | P | Fruit |
| | Myrobolan | | | |
| **Phyllanthus acidus** | Otaheite Gooseberry | D | P | Fruit |
| **Physalis ixocarpa** | Tomatillo | D | P | Fruit |
| **Physalis peruviana** | Cape Gooseberry | D | P | Fruit |
| **Phytelephas macrocarpa** | Ivory Nut Palm | M | P | Ivory (seeds) |
| **Pimento dioica** | Allspice | D | P | Spice (fruit) |
| | Pimento | | | Oil (spice) |
| **Pimento acris** | Bay Oil | D | P | Oil (leaf) |
| **Pimpinella anisum** | Anise | D | A | Seed spice |
| **Pinus pinea** | Stone Pine | M | P | Pine nuts (kernels) |
| **Piper betle** | Betel Pepper | D | P | Leaf masticatory |
| **Piper methysticum** | Kava | D | P | Root beverage |
| **Piper nigrum** | Pepper | D | P | Black and white pepper spice (fruit) |
| **Pistacia lentiscus** | Mastic Tree | D | P | Resin |
| **Pistacia vera** | Pistachio Nut | D | P | Nut |
| **Pisum sativum** | Pea | D | A | Pod vegetable, seeds, green manure, fodder legume |
| **Plectranthus esculentus** | Hausa Potato | D | P | Tuber vegetable |
| **Pleurotus ostreatus** | Oyster Mushroom | | | Mushroom |

| Latin Name | Common Name | D/M | A/B/P | Principal Products |
|---|---|---|---|---|
| **Poa ampla** | Big Bluegrass | M | P | Fodder grass |
| **Poa compressa** | Canada Bluegrass | M | P | Fodder grass |
| **Poa palustris** | Swamp Meadow Grass | M | P | Fodder grass |
| | Fowl Bluegrass | | | |
| **Poa pratensis** | Smooth Stalked Meadow Grass | M | P | Fodder grass |
| | Kentucky Bluegrass | | | |
| **Poa trivialis** | Rough-stalked Meadow Grass | M | P | Fodder grass |
| **Pogostemon cablin** | Patchouli | D | P | Oil (shoot) |
| **Portulaca oleracea** | Purslane | D | A | Leaf salad/ vegetable |
| **Prosopis juliflora** | Mesquite | D | P | Pod fodder |
| **Prosopis chilensis** | Mesquite | D | P | Pod fodder |
| **Prunus amygdalus** | Almond | D | P | Nut |
| **Prunus armeniaca** | Apricot | D | P | Fruit |
| **Prunus avium** | Gean Cherry | D | P | Fruit |
| | Sweet Cherry | | | |
| **Prunus besseyi** | Sand Cherry | D | P | Fruit |
| **Prunus cerasifera** | Myrobalan | D | P | Fruit |
| | Cherry Plum | | | |
| **Prunus cerasus** | Morello Cherry | D | P | Fruit |
| **Prunus domestica** | Plum, Prune | D | P | Fruit |
| | Damson, Greengage, Bullace | | | |
| **Prunus dulcis** | Almond | D | P | Nut |
| **Prunus spinosa** | Sloe | D | P | Fruit |
| | Blackthorn | | | |
| **Prunus persica** | Peach & Nectarine | D | P | Fruit |
| **Psalliota arvensis** | Horse Mushroom | | P | Mushroom |
| **Psalliota bisporus** | Cultivated Mushroom | | A | Mushroom |
| **Psalliota campestris** | Field Mushroom | | P | Mushroom |
| **Psidium guajava** | Guava | D | P | Fruit |
| **Psophocarpus tetragonolobus** | | D | P | Pod vegetable, |
| | Goa Bean | | | Seeds, tuber |
| | Asparagus Pea | | | root vegetable, |
| | Winged Bean | | | fodder, green |
| | Princess Pea | | | manure |
| | Four Angled Bean | | | |
| | Manila Bean | | | |
| **Puccinela airoides** | Alkaligrass | M | P | Fodder grass |
| | Nuttal | | | |
| **Pueraria phaseoloides** | Puero | D | P | Fodder legume, |
| | Tropical Kudzu | | | Erosion control |
| **Pueraria thunbergiana** | Kudzu | D | P | Fodder legume |
| **Punica granatum** | Pomegranate | D | P | Fruit |
| **Pyrus communis** | Pear | D | P | Fruit |
| **Quercus suber** | Cork Oak | D | P | Bark |
| **Raphanus sativus** | Radish | D | A | Tap root |
| | Daikon | | | vegetable |
| **Rheum rhaponticum** | Rhubarb | D | P | Petiole fruit |
| **Rhynchelytrum repens** | Natal Redtop | M | A/P | Fodder grass |
| | Natal Grass | | | |
| **Ribes uva-crispa** | Gooseberry | D | P | Fruit |
| **Ribes nigrum** | Black currant | D | P | Fruit |
| **Ribes sativum** | White currant | D | P | Fruit |
| | Red currant | | | |
| **Ricinus communis** | Castor | D | A/P | Oil |
| **Rosmarinus officinalis** | Rosemary | D | P | Leaf herb |
| **Rubus caesius** | Dewberry | | | Fruit |

| Latin Name | Common Name | D/M | A/B/P | Principal Products |
|---|---|---|---|---|
| **Rubus chamaemorus** | Cloudberry | D | P | Fruit |
| **Rubus idaeus** | Raspberry | D | P | Fruit |
| **Rubus illecebrosus** | Wineberry | D | P | Fruit |
| **Rubus longanobacus** | Loganberry | D | P | Fruit |
| **Rubus phoenicolasius** | Wineberry | D | P | Fruit |
| **Rubus fruticosus** | Blackberry Bramble | D | P | Fruit |
| **Rubus ursinus** | Boysenberry | D | P | Fruit |
| **Rumex acetosa** | Sorrel | D | P | Leaf vegetable |
| **Rumex patientia** | Patience Dock | D | P | Leaf vegetable |
| **Rumex scutatus** | Round Leaved Sorrel | D | P | Leaf vegetable |
| **Saccharum:** | | | | |
| principal economic sp: | | | | |
| **officinarum** | Sugar Cane Noble Cane | M | P | Sugar (stem) Fodder (tops) |
| lesser spps: | | | | |
| **sinense** | Japanese Cane Uba Cane | | | Sugar (stem) Fodder grass, |
| **barberi** | | | | Sugar (stem) |
| **Saccharum spontaneum** | Wild Cane | M | P | Fodder grass |
| **Sagittaria sagittifolia** | Chinese Arrowhead | M | P | Corm vegetable |
| **Salacca edulis** | Salak Palm | M | P | Fruit |
| **Salvia officinalis** | Sage | D | P | Leaf herb, oil (leaf) |
| **Santalum album** | Sandlewood | D | P | Oil (wood) |
| **Sansevieria guineensis cylindrica liberica** | Bowstring Hemp | M | P | Leaf fibre |
| **Salvia officinalis** | Sage | D | P | Leaf herb |
| **Salvia sclarea** | Clary | D | B | Leaf flavouring |
| **Satureja hortensis** | Summer Savory | D | A | Leaf herb |
| **Satureja montana** | Winter Savory | D | P | Leaf herb |
| **Schleichera oleosa** | Lac Tee | D | P | Fat (seed), host for Lac insect |
| **Scolymus hispanicus** | Spanish Oyster Plant | D | A/B | Root vegetable |
| **Scorzonera hispanica** | Black Salsify | D | A/B | Root vegetable |
| **Secale cereale** | Rye | M | A/B | Seed, fodder grass |
| **Secale montanum** | Mountain Rye | M | A | Fodder grass |
| **Sechium edule** | Choyote Christophine | D | P | Fruit/tuber root vegetable, |
| **Sesame indicum** | Sesame Simsim | D | A | Oil, seeds |
| **Sebania aegyptica** | Sesbania | D | P | Fodder legume, Green manure |
| **Sesbania brachycarpa** | Sesbania | D | P | Fodder legume, Green manure |
| **Sesbania grandiflora** | Sesbania | D | P | Fodder legume, green manure |
| **Sesbania macrocrapa** | Common Sesbania | D | P | Green manure |
| **Sesbania speciosa** | Sesbania | D | P | Green manure |
| **Setaria incrassata** | Purple Pigeon Grass | M | P | Fodder grass |
| **Setaria italica** | Foxtail Millet | M | A | Grain, fodder grass |
| **Setaria sphacelata** | Golden Timothy Grass | M | P | Fodder |
| **Setaria splendida** | | M | P | Fodder grass |
| **Setaria italica** | Foxtail Millet | M | A | Seed, fodder |
| **Simmondsia chinensis** | Jojoba | D | P | Oil (seed) |
| **Smilax spp.** | Sarsaparilla | D | B/P | Root flavouring |

| Latin Name | Common Name | D/M | A/B/P | Principal Products |
|---|---|---|---|---|
| **Solanum melongena** | Egg Plant | D | A | Fruit vegetable |
| | Aubergine, Brinjal, Melongene | | | |
| | Japanese Long Egg Plant | | | |
| **Solanum muricatum** | Pepino Dulce | D | A | Fruit |
| **Solanum quitoense** | Naranjilla | D | A | Fruit |
| **Solanum topiro** | Cocona | D | A | Fruit |
| **Solanum tuberosum** | Potato | D | A | Tuber vegetable |
| **Sorghum x almum** | Columbus Grass | M | P | Fodder |
| **Sorghum arundinaceum** | | | | |
| **var. sudanese** | Sudan Grass | M | A | Fodder |
| **Sorghum bicolor** | Sorghum | M | A | Grain, fodder |
| **(vulgare)** | Kafir Corn, Great Millet | | | |
| | Guinea Corn, Milo, Sorgo, | | | |
| | Kaoliang, Durra, Mtama, Jola, | | | |
| | Jawa, Cholam | | | |
| **Sorghum halepense** | Johnson Grass | M | P | Fodder grass |
| **Sorghum vulgare x Sorghum** | | | | |
| **verticilliflorum** | Kavirondo | M | P | Fodder grass |
| | Perennial Sorghum | | | |
| **Sphaerophysa salsula** | Spherophysa | M | P | Fodder grass |
| **Spinacia oleracea** | Spinach | D | A/B | Leaf vegetable |
| **Spondias cytherea** | Otahaie Apple | A | P | Fruit |
| | Pomme de Cythere | | | |
| **Spondias mombin** | Yellow mombin | D | P | Fruit |
| | Hog Plum | | | |
| **Spondias purpurea** | Red Mombin | D | P | Fruit |
| **Spondias tuberosa** | Umbu | D | P | Fruit |
| **Sporobulus airoides** | Alkali Sacaton | M | P | Fodder grass |
| **Stenotaphrum secundatum** | Buffalo Grass | M | P | Fodder grass |
| | St Augustine Grass | | | |
| | Crab Grass | | | |
| **Stachys affinis** | Chinese Artichoke | D | B | Tuber vegetable |
| **Stilozobium** | Velvet Bean | D | A/P | Seeds, |
| | Florida Bean | | | Fodder legume |
| | Deering Velvet Bean | | | |
| **Stizolobium aterrimum** | Bengal Bean | D | A | Green manure, fodder legume |
| **Stizolobium spp.** | Velvet Beans | D | A/P | Green manure, fodder legume |
| **Stylosanthes gracilis** | Common Stylo | D | P | Fodder legume |
| **Stylosanthes guyanensis** | Fine Stem Stylo | D | P | Fodder legume |
| **Stylosanthes hamata** | Caribbean Stylo | D | P | Fodder legume |
| | Verano Stylo | | | |
| **Stylosanthes humilis** | Townsville Stylo | D | A | Fodder legume |
| **Stylosanthes scabra** | Shrubby Stylo | D | P | Fodder legume |
| **Sambucus nigra** | Elder | D | P | Fruit |
| **Swainsona stipularis** | | D | P | Fodder legume |
| **Symphyytum x uplandicum** | Russian Comfrey | D | P | Fodder |
| **Syzygium cumini** | Jamun | D | P | Fruit |
| | Jambolan | | | |
| **Syzygium jambos** | Rose Apple | D | P | Fruit |
| **Tacca leontopetaloides** | East Indian Arrowroot | M | P | Tuber vegetable, starch |
| **Tamarindus indica** | Tamarind | D | P | Seed pulp, seeds, charcoal |
| **Taraktogenos kurzi** | Chaulmoogra Oil | D | P | Oil |
| **Taraxacum officinale** | Dandelion | D | P | Leaf salad |
| **Telfairia pedata** | Oyster Nut | D | P | Nut |

| Latin Name | Common Name | D/M | A/B/P | Principal Products |
|---|---|---|---|---|
| **Tephrosia spp.** | | D | P | Green manure |
| **Teramnus uncinatus** | | D | P | Fodder legume |
| **Tetragonia tetragonioides** | New Zealand Spinach | D | A | Leaf vegetable |
| **Tilia spp.** | Lime | D | P | Flower beverage |
| | Linden | | | |
| **Tillandsia usneoides** | Spanish Moss | M | A/P | Stem fibre |
| **Themeda triandra** | Red Oat Grass | M | P | Fodder grass |
| **Theobroma cacao** | Cocoa | D | P | Cocoa (seeds) |
| **Thymus serpyllum** | Breckland Thyme | D | P | Leaf herb |
| **Thymus vulgaris** | Garden Thyme | D | P | Leaf herb, oil |
| **Tragopogon porrifolius** | Salisy | D | A/B | Tap root |
| | Oyster Plant | | | vegetable |
| **Trichosanthes cucumerina** | Snake Gourd | D | A | Fruit vegetable |
| **Trifolium alexandrium** | Egyptian Clover | D | A | Fodder legume |
| | Berseem | | | |
| **Trifolium balansae** | Balansa Clover | D | | Fodder legume |
| **Trifolium campestre** | Large Hop Clover | D | A | Fodder legume |
| **Trifolium dubium** | Small Hop Clover | D | A | Fodder legume |
| | Hop Trefoil | | | |
| **Trifolium fragiferum** | Strawberry Headed Clover, Trefoil | D | P | Fodder legume |
| **Trifolium hybridum** | Alsike Clover | D | P | Fodder legume |
| **Trifolium incarnatum** | Crimson clover | D | A | Fodder legume |
| | Carnation Clover, French Clover | | | |
| | Italian Clover | | | |
| **Trifolium pratense** | Red Clover | D | P | Fodder legume |
| | Purple Clover, Meadow Clover, | | | |
| | Cowgrass | | | |
| **Trifolium repens** | White Clover | D | P | Fodder legume |
| **Trifolium resupinatum** | Persian Clover | D | A | Fodder legume |
| | Annual Strawberry Clover | | | |
| | Shaftal, Bird's Eye Clover | | | |
| **Trifolium semipilosom** | Kenya White Clover | D | P | Fodder legume |
| | Safari Clover | | | |
| **Trifolium subterraneum** | Subterranean Clover | D | A | Fodder legume |
| other Trifoliums include: | | | | |
| **T. africanum** | Yellow clover, Hop Clover | | | |
| **T. alpinum** | Kura Clover | | | |
| **T. arvense** | Rabbit Foot Clover | | | |
| **T. badium** | Carolina Clover | | | |
| **T. cheranganiensis** | | | | |
| **T. glomeratum** | Cluster Clover | | | |
| **T. hirtum** | Rose Clover | | | |
| **T. johnstonii** | Kenya Wild White Clover | | | |
| **T. lappaceum** | Lappa Clover | | | |
| **T. macrocephalum** | Ball Clover | | | |
| **T. panonicum** | Hungarian Clover | | | |
| **T. parviflorum** | Small Flowered Clover | | | |
| **T. polymorphum** | Buffalo Clover | | | |
| **T. striatum** | Striata Clover | | | |
| **Trigonella foenum graecum** | Fenugreek | D | A | Leaf herb, fodder legume, seed |
| **Tripsacum dactyloides** | Gama Grass | M | P | Fodder grass |
| **Tripsacum laxum** | Guatemala Grass | M | P | Fodder grass |
| **Trisetum flavescens** | Yellow Oat Grass | M | P | Fodder grass |
| **Triticum aestivum** | Common Wheat | M | A/B | Seed, fodder grass |
| | Bread Wheat | | | |
| **Triticum dicoccum** | Emmer Wheat | M | A | Seed |

| Latin Name | Common Name | D/M | A/B/P | Principal Products |
|---|---|---|---|---|
| **Triticum durum** | Durum Wheat | M | A | Seed |
| **Triticum monococcum** | Einkorn | M | A | Seed |
| | Small spelt | | | |
| **Triticum polonicum** | Polish Wheat | M | A | Seed |
| **Triticum x secale** | Triticale | M | A | Grain, fodder |
| **Triticum spelta** | Spelt Wheat | M | A | Seed |
| **Tropaeolum tuberosum** | Anu | D | A | Tuber vegetable |
| **Tuber melanosporum** | Perigord Truffle | | | Truffle |
| **Tuber magnatum** | Italian White Truffle | | | Truffle |
| **Urena lobata** | Aramina Fibre | D | A/P | Stem fibre |
| | Congo Jute | | | |
| **Urochloa mosambicensis** | Sabi Grass | M | P | Fodder grass |
| **Vaccinium corymbosum** | Highbush Blueberry | D | P | Fruit |
| **Vaccinium macrocarpon** | American Cranberry | D | P | Fruit |
| **Vaccinium myrtillus** | Bilberry | D | P | Fruit |
| **Vaccinium oxycoccus** | Cranberry | D | P | Fruit |
| **Vaccinium vitis idaea** | Cowberry | D | P | Fruit |
| **Valerianella locusta** | Cornsalad | D | A | Leaf salad |
| | Lamb's Lettuce | | | |
| **Vanilla fragrans** | Vanilla | M | P | Flavouring/spice |
| also: **pompona** | | | | (fruit) |
| **tahitensis** | | | | |
| **Verbena officinalis** | Vervain | D | P | Leaf beverage |
| **Vetiveria zizanioides** | Khuskhus | M | P | Oil (roots), anti- |
| | Vetiver | | | erosion |
| **Vicia angustifolia** | Narrow leaved Vetch | D | A | Fodder legume |
| | Augusta Vetch | | | |
| **Vicia articulata** | Monantha | D | A | Fodder legume, |
| | One leaved Vetch | | | Seeds |
| **Vicia atropurpurea** | Purple Vetch | D | A | Fodder legume |
| **Vicia dasycarpa** | Wooly Pod Vetch | D | A/B | Fodder legume |
| **Vicia faba** | Broad Bean | D | A/B | Seeds, pod |
| | Horse Bean | | | vegetable, |
| | Windsor Bean | | | fodder legume |
| | Field Bean, Tick Bean | | | |
| **Vicia grandiflora** | Bigflower Vetch | D | A/B | Fodder legume |
| **Vicia narbonensis** | Narbonne Vetch | D | A/B | Fodder legume, |
| | | | | green manure |
| **Vicia pannonica** | Hungarian Vetch | D | A | Fodder legume |
| **Vicia sativa** | Vetch | D | A | Fodder legume, |
| | Spring Vetch | | | green manure, |
| | Tares | | | seed |
| **Vicia villosa** | Hairy Vetch | D | B | Fodder legume |
| | Winter Vetch, Smooth Vetch | | | |
| **Vigna catjang** | Catjang | D | A | Fodder legume |
| | | | | Seeds |
| **Vigna luteola** | Dalrymple Vigna | D | P | Fodder legume |
| **Vigna oligosperma** | Sarawak Bean | D | A | Green manure |
| **Vigna parkeri** | Creeping Vigna | D | P | Fodder legume |
| **Vigna unguiculata** | Cowpea | D | P | Seeds, pod |
| | Black Eye Pea | | | vegetable, green |
| | China Pea | | | manure, fodder |
| | Marble Pea | | | legume |
| | Southern Bean, Black Eye Bean | | | |
| | Yard Long Pea, Kaffir Pea | | | |

| Latin Name | Common Name | D/M | A/B/P | Principal Products |
|---|---|---|---|---|
| **Vigna vexillata** | | D | P | Root vegetable |
| **Vitis** | Grape | D | P | Fruit |
| **vinifera** | 'European' | | | |
| **labrusca** | 'American' | | | |
| **rotundifolia** | 'Muscadine' | | | |
| **hybrids** | | | | |
| **Voandzeia subterranea** | Bambara Groundnut | D | A | Seeds |
| | Juga Bean, Madagascar Peanut | | | |
| | Earth Pea | | | |
| **Xanthosoma sagittifolium** | Tannia | M | A/P | Corm vegetable, |
| | Tanier | | | leaf vegetable |
| | Yautia, New Cocoyam | | | |
| **Zea mays** | Maize | M | A | Seed, fodder grass |
| | Corn, Indian Corn | | | |
| **Zingiber officinale** | Ginger | M | P | Rhizome spice |
| **Zizania aquatica** | Wild Rice | M | A | Seed |
| **Zizyphus jujuba** | Chinese Jujube | D | P | Fruit |
| **Zizyphus mauritiana** | Indian Jujube | D | A | Fruit |

*Sources:* de Rougemont, G.M. (1989) *Field Guide to the Crops of Britain and Europe.*
Acland, J.D. (1975) *East African Crops.* Whyte, R.O. *et al.* (1959) *Grasses in Agriculture.* FAO of UN.
Hubbard, C.E. (1974) *Grasses.* Whyte, R.O. *et al.* (1953) *Legumes in Agriculture.* FAO of UN.
Weiss, E.A. (1983) *Oilseed Crops.* Purseglove, J.W. (1974) *Tropical Crops. Dicotyledons.*
Purseglove, J.W. (1976) *Tropical Crops. Monocotyledons.* Samson, J.A. (1986) *Tropical Fruits.*

## TABLE 163  Indicative Seed Weights

*NB* Seed weights vary widely between varieties, and lots of the same variety. Seeds of
unimproved local varieties can often be lighter than those of highly bred types.

| Crop | Seeds / kg | Crop | Seeds / kg |
|---|---|---|---|
| American joint vetch | 352 000 | Centrosema plumieri | 5500 |
| Asparagus | 45 000-60 000 | Chervil | 500 000 |
| Artichoke, Jerusalem | 23 760 | Chicory | 620 000-690 000 |
| Alfalfa | 495 000 | Chinese cabbage | 300 000- 400 000 |
| Axillaris | 120 000 | Chive | 800 000 |
| Bahia grass | 748 000 | Climbing French bean | 2 000-4 000 |
| Barley, spring | 30 000-50 000 | Clover, Alsike | 1 485 000 |
| Barley, winter | 35 000-55 000 | Clover, Alyce | 660 000 |
| Basil | 700 000 | Clover, berseem | 455 000 |
| Beetroot | 40 000-90 000 | Clover, burr | 48 510 |
| Borage | 70 000 | Clover, cluster | 2 200 000 |
| Borecole | 200 000-400 000 | Clover,crimson | 326 700 |
| Bean, Adzuki | 8800 | Clover, Daghestan sweet | 550 000 |
| Bean, broad | 800-1000 | Clover,large hop | 5 380 000 |
| Bean, hyacinth | 3080 | Clover, small hop | 2 200 000 |
| Bean, navy, sall | 6000 | Clover, Kenya | 700 000-1 000 000 |
| Bean, navy, large | 4000 | Clover, Ladino | 1 918 000 |
| Bean, mat | 44 000 | Clover,lappa | 1 496 000 |
| Bean, mung | 23 800 | Clover, little burr | 880 000 |
| Bean, phasey | 119 000 | Clover, mamoth | 594 000 |
| Bean, rice | 22 000 | Clover, Persian | 1 402 000 |
| Bean, runner | 800-1200 | Clover,red | 594 000 |
| Bean, Sarawak | 37 400 | Clover, rose | 308 000 |
| Bean, Tepary | 55 000 | Clover, spotted Bur | 462 000 |
| Bean, velvet | 1980 | Clover, subterranean | 155 000 |
| Bermuda grass | 3 900 600 | Clover, striata | 506 000 |
| Birdsfoot trefoil | 805 860 | Clover, strawberry | 790 000 |
| Birdood grass | 176 000 | Clover,sweet | 564 300 |
| Broadleaf Paspalum | 880 000 | Clover, yellow annual sweet | 605 000 |
| Broccoli | 200 000-400 000 | Clover, Tifton Bur | 330 000 |
| Brome grass | 297 000 | Clover,white | 1 485 000 |
| Broom corn | 59 400 | Cocksfoot | 1 427 000 |
| Brussels sprouts | 180 000-350 000 | Columbus Grass | 121 000 |
| Buckwheat | 44 550 | Cotton | 8 800 |
| Buffel grass | 330 000-550 000 | Cowpea | 400-15 000 |
| Burnet | 250 000 | Crested wheatgrass | 594 000 |
| Cabbage, pointed | 200 000-350 000 | Crotalaria anagyroides | 55 000 |
| Cabbage, spring | 170 000-400 000 | Crotalaria grantiana | 330 000 |
| Cabbage, summer | 200 000-350 000 | Crotalaria incana | 187 000 |
| Cabbage, autumn | 200 000-350 000 | Crotalaria intermedia | 220 000 |
| Cabbage, storage | 200 000-350 000 | Crotalaria lanceolata | 374 000 |
| Cabbage, savoy | 200 000-350 000 | Cucumber | 32 000 |
| Cabbage, winter | 200 000-350 000 | Dallis grass | 495 000 |
| Calabrese | 200 000-400 000 | Desmodium, greenleaf | 750 000 |
| Calopo | 66 000-73 000 | Desmodium, silverleaf | 210 000 |
| Caraway | 300 000 | Dill | 700 000 |
| Carpet grass, narrowleaf | 1 100 000-2 450 000 | Dolichos (Archer) | 88 000 |
| Carrot | 600 000-1 200 000 | Dolichos (Leichhardt) | 33 000 |
| Cauliflower | 240 000-450 000 | Dolichos | 4000 |
| Celery | 1 600 000-3 000 000 | Dwarf French bean | 400-6500 |
| Centro | 40 000 | Egg plant | 150 000-250 000 |

**TABLE 163 *(continued)*  Indicative Seed Weights**

*NB* Seed weights vary widely between varieties, and lots of the same variety. Seeds of unimproved local varieties can often be lighter than those of highly bred types.

| Crop | Seeds / kg | Crop | Seeds / kg |
|---|---|---|---|
| Endive | 700 000-800 000 | Onion | 200 000-350 000 |
| Fennel | 200 000-300 000 | Onion, salad | 200 000-350 000 |
| Fennel, common | 120 000 | Panic grass, blue | 1 400 000 |
| Fenugreek | 50 600 | Panic grass, gatton | 1 280 000 |
| Flax | 176 200 | Panic grass, green | 1 280 000 |
| Glycine | 130 000-200 000 | Para grass | 450 000-935 000 |
| Green gram | 22 000 | Parsley | 400 000-700 000 |
| Guinea grass | 2 400 000 | Parsnip | 200 000-300 000 |
| Indigo | 330 000 | Paspalum | 570 000-700 000 |
| Indigo, hairy | 440 000 | Pea | 4000-8000 |
| Indigo, trailing | 440 000 | Pea, butterfly | 28 600 |
| Johnsongrass | 287 100 | Pea, chick | 2200 |
| Kale, thousand head | 311 850 | Pea, flat | 17 600 |
| Kentucky bluegrass | 4 752 000 | Pea, grass | 11 000 |
| Kentucky fescue | 495 000 | Pea, pigeon | 17 600 |
| Kikuyu grass | 400 000 | Pea, rough | 33 000 |
| Kudzu | 88 000 | Peanut | 2000 |
| Lablab | 4000 | Pepper, green | 110 000-190 000 |
| Leek | 325 000-450 000 | Phalaris | 700 000 |
| Lentils | 19 800 | Plicatulum | 530 000-950 000 |
| Lespedeza, common | 814 000 | Prairie grass | 110 000 |
| Lespedeza, kobe | 743 000 | Puero | 81 000 |
| Lespedeza, sericea | 812 200 | Radish | 80 000-180 000 |
| Lespedeza, Japanese | 660 000 | Rape | 342 500 |
| Lespedeza, Korean | 495 000 | Reed canary grass | 1 188 000 |
| Lettuce | 650 000-1100 000 | Rhodes grass | 3 300 000-4 400 000 |
| Leucaena | 26 000 | Rice, rough | 28 600-41 800 |
| Leucaena pulverulenta | 55 000 | Ruzi grass | 270 000 |
| Lotononis | 3 500 000 | Rye | 40 000-47 000 |
| Lupin, blue | 5500 | Ryegrass, perennial | 610 000 |
| Lupin, white | 3300 | Ryegrass , Italian | 550 000 |
| Lupin, yellow | 8800 | Ryegrass, Wimmera | 550 000 |
| Maize | 2970 | Sabi grass | 870 000 |
| Maize, popcorn | 6930 | Safflower | 29 000 |
| Makarikari grass | 1 600 000 | Sage | 150 000 |
| Marjoram, sweet | 3 000 000 | Sainfoin | 55 000 |
| Marrow | 4000-7000 | Sann | 33 000 |
| Meadow fescue | 475 200 | Savory, summer | 1 300 000 |
| Medic, barre | 230 000-330 000 | Scrobic | 370 000-660 000 |
| Medic, black | 580 000 | Sericea | 770 000 |
| Medic, burr | 310 000 | Sesbania | 68 000 |
| Medic, snail (M. orbicularis) | 330 000 | Setaria | 1 300 000-1 900 000 |
| Medic, snail (M. scutellata) | 99 000 | Sickle senna | 48 400 |
| Medic, sickle | 506 000 | Signal grass | 286 000-700 000 |
| Millet, bulrush | 200 000 | Siratro | 79 000 |
| Millet, Japanese | 316 800 | Sorghum, grain | 50 000 |
| Mimosa | 165 000 | Sorghum, sweet | 61 000 |
| Molasses grass | 11 000 000-14 500 000 | Sorghum almum | 160 000 |
| Monantha | 26 400 | Soyabeans | 6000-15 000 |
| Oatgrass, tall meadow | 326 700 | Spinach | 80 000-160 000 |
| Oats | 27 700 | Stylo | 265 000-350 000 |

*(continued over)*

**TABLE 163 *(continued)*   Indicative Seed Weights**

*NB* Seed weights vary widely between varieties, and lots of the same variety. Seeds of unimproved local varieties can often be lighter than those of highly bred types.

| Crop | Seeds / kg | Crop | Seeds / kg |
|---|---|---|---|
| Stylo, Caribbean | 275 000-350 000 | Trefoil, birdsfoot | 715 000 |
| Stylo, shrubby | 425 000 | Triticale | 23 000 |
| Stylo, Townsville | 400 000-485 000 | Turnip | 370 000-520 000 |
| Sudan grass | 118 000 | Urochloa | 850 000 |
| Sugar pea | 400-6 000 | Vetch, kidney | 385 000 |
| Sulla | 220 000 | Vetch, milk | 286 000 |
| Sunflower | 59 400 | Vetch, big flower | 70 400 |
| Swede | 260 000-420 000 | Vetch, Hungarian | 22 000 |
| Seet corn | 4000-7000 | Vetch, narrow leaf | 66 000 |
| Tall fescue | 500 000 | Vetch, purple | 22 000 |
| Tarragon, Russian | 4 000 000 | Vetch, sickle pod | 286 000 |
| Tephrosia candida | 55 000 | Vetch, winter | 35 640 |
| Tephrosia toxicana | 99 000 | Vetch, spring | 18 810 |
| Tephrosia vogelli | 33 000 | Vetchling | 17 600 |
| Thmye | 5 000 000 | Wheat, winter | 25 000-60 000 |
| Timothy | 2 475 000 | Wheat, spring | 32 000 |
| Tomato | 250 000-350 000 | Yellow mellilot | 572 000 |
| Trefoil, big | 2 200 000 | | |

*Sources:* Samuel Yates Ltd. (1990) *Yates Seed Catalogue.* Nickersons Seeds (1990) *Vegetable Seed Catalogue.* Keystone Seeds Information leaflet. FAO (1973) *Legumes in Agriculture.* Humphreys, L.R. (1980) *A Guide to Better Pastures in the Tropics and Sub-Tropics.* Wright Stephenson and Co. ICI Publication, *Growing Cereals.* FAO (1953) *Legumes in Agriculture.* O'Reilly, M.V. (1975) *Better Pastures for the Tropics.* Arthur Yates and Co. Pty. Ltd.

**Note 6  Guideline Crop Seed Rates and Plant Spacings for Principal Crops**

Crop seed rates and final plant spacings vary greatly according to specific circumstances. For high value horticultural crops final plant spacings can be very critical in order that a particular size of product is achieved, and such considerations may take precedence over achieving maximum weight of crop per unit area. The main factors influencing both seed rate and plant population are:

- **environment, particularly water supply**
- **variety**
- **grade, type and quality of required crop product**
- **soil fertility status and crop fertiliser programme**
- **prevalence of insect, fungus and weed problems**
- **mechanization system as it affects row widths and seed spacing within rows; also row arrangements on bed layouts**
- **plant habit**

Other factors which specifically influence seed rate are:

- **soil conditions at sowing, e.g. moisture, tilth**
- **anticipated soil conditions following sowing which may affect crop establishment (particularly soil moisture)**
- **seed viability**
- **seed size, seed weight**
- **sowing method, e.g. drill, broadcast, precision drill, sowing for transplanting, use of pre-germinated seed**
- **use of seed coatings**
- **whether sown in mixture with another specie**

Table 164 is intended as a guideline to seed rates and final plant spacings required to achieve relatively high yields where there are no major constraints to crop productivity. It is noted that particularly in situations of limited moisture supply and low fertility significantly lower seed rates and plant spacings will be required. Advice from commercial seed producers and suppliers to specific locations should always be sought as well as locally available field trial results and local farming practice.

Generally, seed rate should be derived from a consideration of the number of seeds per unit weight and desired plant population taking into account seeding method and climatic considerations.

Unless indicated, seed rate guidelines for tropical fodder species are for rainfed situations. Under intensive irrigated conditions much higher seed rates are usually justified.

For fruit trees it is noted that new varieties and dwarf types, as well as new mechanized harvesting and pruning techniques increasingly dictate higher than traditional populations. Also, fruit trees may be planted close to enhance yields in the early years of an orchard and then later thinned.

---

**TABLE 164    Guideline Field Seed Rates and Final Plant Populations for Principal Crops (drilled unless indicated)**

| CROP | SEED RATE kg / ha (or plants / ha) | | FINAL PLANT POPULATION per ha |
|---|---|---|---|
| Alfalfa | 25 - 40 | Irrigated | |
| | 6 | Dry land | |
| Allspice | 175 | trees / ha | |
| Almonds | 250 | trees / ha | |
| Angleton grass | 4 | | |
| Apricot | 185 | trees / ha | |
| Apple | 250 | trees / ha | |
| Aramina fibre | 75 | | |
| Artichoke, globe | 2500 | root sections per ha | |
| | | (135 - 225 kg / ha) | |
| Asparagus | 2.8 | | 23 900 |
| Asparagus | 21 000 | crowns per ha | |
| Avocado | 150 | trees / ha | |
| | | (traditional) - 750 trees / ha | |
| Axillaris | 5 | | |
| Bahia grass | 5 | | |
| Bambarra groundnut | 34 | | |

*(continued over)*

A G R I   I N F O

**TABLE 164** *(continued)* **Guideline Field Seed Rates and Final Plant Populations for Principal Crops (drilled unless indicated)**

| CROP | SEED RATE kg/ha (or plants / ha) | | FINAL PLANT POPULATION per ha |
|---|---|---|---|
| Banana, tall types | 1100 | plants / ha | |
| dwarf types | 3000 | plants / ha | |
| Barley, spring | 155 | | 400 / sq m |
| Barley, winter | 190 | | 250 / sq m |
| Bean, Adzuki | 22 - 28 | broadcast | |
| Bean, broad | 180 | | 179 000 |
| Bean, cluster | 34 - 45 | broadcast | |
| | 17 | drilled | |
| Bean, dwarf | 59 | French | 403 500 |
| | 30 | Climbing | 80 000 |
| Bean, hyacinth | 60 | | |
| Bean, Jack | 56 | | 60 000 |
| Bean, Lima, bush | 56 | vegetable | |
| Bean, Lima, pole | 39 | vegetable | |
| Bean, Lima, seed | 120 | large seed types | |
| | 67 | small seed types | |
| Bean, mat | 5 - 10 | drilled | |
| | 17 - 45 | broadcast | |
| Bean, mung | 28 | | |
| Bean, navy | 70 | | |
| Bean, phasey | 3 | | |
| Bean, rice | 78 - 90 | broadcast | |
| Bean, runner | 93 | Ground (pinched) | 107 500 |
| | 74 | Maincrop stick | 92 219 |
| Bean, snap, bush | 89 | | 174 000 |
| Bean, snap, pole | 28 | | |
| Bean, sword | 39 - 67 | broadcast | |
| | 8 | drilled | |
| Bean, Tepary | 28 - 34 | broadcast | |
| | 20 | drilled | |
| Bean, velvet | 33.6 | | 63,500 |
| Barley, spring | 185 | | 300 - 350 / sq m |
| winter | 185 | | 280 / sq m |
| Beetroot | 9 | Early bunch | 538 000 |
| | 10.5 | Maincrop | 1 075 900 |
| | 20 | Baby process | 1613900 |
| Bermuda grass | 7 | | |
| Big trefoil | 3.5 - 5.6 | broadcast | |
| Birdsfoot trefoil | 10 | | |
| Birdseed (Phalaris canariensis) | 28 | | |
| Birdwood grass | 4 | | |
| Bitter gourd | | | 50 000 |
| Black Current | 7000 | bushes / ha | |
| Black Gram | 12 | | |
| Black Wattle | 5 | | 1700 |
| Bluegrass, Kentucky | 34 | | |
| Bluegrass, Canada | 23 | | |
| Bluegrass, Creeping | 4 | | |
| Bluegrass, Indian | 4 | | |
| Blue Panic Grass | 4 | | |
| Borage | 22 | | |
| Borecole | 1 | Processing | 193 648 |

**TABLE 164** *(continued)*   Guideline Field Seed Rates and Final Plant Populations for
Principal Crops (drilled unless indicated)

| CROP | SEED RATE kg / ha (or plants / ha) | | FINAL PLANT POPULATION per ha |
|---|---|---|---|
| Borecole | | Fresh | 40 000 |
| Breadfruit | 90 | trees / ha | |
| Broadbean, field, winter | 200 | Dry seed production | 280 000 |
| Broccoli | 1.1 | Sprouting | 22 415 |
| | | Heading | 30 000 |
| Brome Grass | 23 | | |
| Broom Corn | 9 | | |
| Brussels sprouts | 1.1 | Freezing | 40,755 |
| | | Prepack | 32 278 |
| | | General market | 20 518 |
| Buckwheat | 67 | | |
| Buffel grass | 25 | Irrigated | |
| | 5 | Rainfed | |
| Cabbage | 1.1 | Spring green | 172 159 |
| | | Spring heart | 107 593 |
| | | Early summer | 86 075 |
| | | Summer | 86 075 |
| | | Autumn / Winter / Savoy | 26 898 |
| | | Winter white prepack | 40 347 |
| Calabrese | 4 | Small heads | 430 373 |
| | 1.5 | Large heads | 125 525 |
| Calopo | 3.3 | | |
| Canary grass | 10 | | |
| Cantaloupe | 2.5 | | 6500 |
| Cape gooseberries | 30 g | for transplants | 6200 |
| Cardoon | 5 | | |
| Carpet grass | 40 | Narrowleaf | |
| Carrot | 1.9 | Large dicing | 538 460 |
| | 3.1 | Market ware | 1 075 932 |
| | 4.9 | Bunch / medium prepack | 1 614 145 |
| | 8.6 | Can / small prepack | 3 765 515 |
| | 11.1 | Canning | 4 767 100 |
| Cashew | 275 | trees / ha | |
| Cassava | From cuttings | | 6000 |
| Cassia, roundleaf | 4 | | |
| Castor bean | 6 (on hills) | | 24 000 |
| Cauliflower | 1.1 | Summer | 35 864 |
| | 0.5 | Autumn / Winter | 24 564 |
| Celerlac | 1.7 | | 106 666 |
| Celery | 1.7 | Self blanching | 129 106 |
| | | Giant types | 61 312 |
| Centro | 5 | | |
| Centrosema plumieri | 3.3 | | |
| Chicory | 4.5 | | 222 000 |
| Chick pea | 40 | | |
| Chinese Cabbage | 1.7 | | |
| Choyote | | | 4300 |
| Cinnamon | 2200 | trees / ha | |
| Citrus: Grapefruit | 175 | trees / ha | |
| Orange and Lemon | 270 | trees / ha | |
| Lime and Mandarin | 350 | trees / ha | |
| Clove | 120 | | |
| Clover, red | 11.2 | | |

*(continued over)*

**TABLE 164** *(continued)*   Guideline Field Seed Rates and Final Plant Populations for
Principal Crops (drilled unless indicated)

| CROP | SEED RATE kg / ha (or plants / ha) | | FINAL PLANT POPULATION per ha |
|---|---|---|---|
| Clover, Alsike | 2.5 | (in mixtures) | |
| | 7 | (pure stand) | |
| | 7 - 12 | broadcast | |
| Clover, alyce | 11 - 22 | broadcast | |
| Clover, balansa | 4 | | |
| Clover,berseem | 20 | | |
| | 45 - 65 | broadcast | |
| Clover, burr | 44.8 | | |
| Clover, little burr | 22 | broadcast | |
| Clover, cluster | 3.4 - 5 | broadcast | |
| Clover, crimson | 17 | | |
| Clover, daghestan sweet | 11 - 17 | broadcast | |
| Clover, hop | 13.4 | | |
| Clover, small hop | 4.5 - 6 | broadcast | |
| Clover, Kenya white | 3 | | |
| Clover, Ladino | 6.7 | | |
| Clover, lappa | 4.5 - 5.6 | broadcast | |
| Clover, mammoth | 11.2 | | |
| Clover, Persian | 9 | | |
| Clover, red | 9 - 13.4 | broadcast | |
| Clover, rose | 17 - 22 | broadcast | |
| Clover, serradella | 10 | | |
| Clover, shaftal | 4 | | |
| Clover, spotted burr | 22 | broadcast | |
| Clover, striata | 9 - 13.4 | broadcast | |
| Clover, strawberry | 4 | | |
| | 6 - 12 | broadcast | |
| Clover, subterranean | 2 | (in mixtures) | |
| | 20 | (pure stand) | |
| Clover, Tifton burr | 22 | broadcast | |
| Clover, sweet | 17 | | |
| | 11 - 22 | broadcast | |
| Clover, white | 5 | (in mixtures) | |
| | 17 | (pure stand) | |
| Clover, white Dutch | 8.9 | | |
| Clover, yellow | 17 | | |
| Cocksfoot grass | 31.3 | | |
| Cocoa | 750 | trees/ha | |
| Coffee | 1260 | trees/ha | |
| Collard | 3.4 | | |
| Columbus grass | 10 | | |
| Cotton | 35 | (fuzzy seed) | 42 000 |
| Corn salad | 11.2 | | |
| Cowpea | 45 | | 200 000 |
| Crested wheat grass | 13.4 | | |
| Crotalaria anagyoides | 22 - 34 | broadcast | |
| Crotalaria goreensis | 11 | broadcast | |
| Crotalaria grantiana | 11 - 15 | broadcast | |
| Crotalaria incana | 17 - 20 | broadcast | |
| Crotalaria intermedia | 11 - 17 | broadcast | |
| Crotalaria lanceolata | 9 - 13 | broadcast | |
| Crotalaria mucronata | 12 - 25 | broadcast | |
| Crotalaria usaramoensis | 9 - 13 | broadcast | |
| Cucumber | 2.8 | Processing | 104 000 |
| | | Trailing,salad | 20 000 |

**TABLE 164** *(continued)*   Guideline Field Seed Rates and Final Plant Populations for Principal Crops (drilled unless indicated)

| CROP | SEED RATE kg / ha (or plants / ha) | | FINAL PLANT POPULATION per ha |
|---|---|---|---|
| Custard apple | 275 | trees / ha | |
| Dallis grass | 11.2 | | |
| Dandelion | 2.2 | | |
| Dasheen | 1700 | kg / ha of 100 g corms | 15 900 |
| Date | 100 | trees / ha | |
| Derris | 14 000 | rooted cuttings / ha | |
| Desmodiums | 3 | | |
| Desmodium discolor | 12 | | |
| Digitaria | 4 | | |
| Dolichos (lablab) | 25 | | |
| Dolichos, leichardt | 9 | | |
| Durian | 100 | trees / ha | |
| Eggplant | 2.2 | | 18 000 |
| Endive | 3.9 | | 92 699 |
| Evening Primrose | 3.7 | | |
| Fenugreek | 28 - 48 | broadcast | |
| Fescue, alta | 20.1 | | |
| Fescue, meadow | 34 | | |
| Fescue, tall | 39.2 | | |
| Figs | 160 | trees / ha | |
| Flax | 90 | | |
| Florence fennel | 1.9 | | 86 450 |
| Garlic | 1750 | kg / ha of cloves | |
| Gamba grass | 4 | | |
| Giant granadilla | | | 10 750 |
| Glycine wightii | 5 | | |
| Goa Bean | | | 13 000 |
| Grape | 1750 | trees / ha | |
| Grass pea | 45 | | |
| Green gram ( P. aureus) | 67 - 78 | broadcast | |
| Green gram (P. radiatus) | 17 | | |
| Green panic grass | 8 | | |
| Guava | 275 | trees / ha | |
| Guinea grass | 8 | | |
| Hemp | 70 | | |
| Horsegram | 45 | | |
| Horseradish | 10 000 | root cuttings per ha | |
| Indigo, hairy | 9 - 11 | broadcast | |
| Jackfruit | 70 | | |
| Japanese radish | 3.3 | For Fodder roots | 22 000 |
| Jerusalem artichoke | 1400 | kg / ha of 55 g tubers | 55 000 |
| Johnson grass | 34 | | |
| Kale | 3.4 | Vegetable | |
| Kale, thousand head | 5.6 | Fodder | |
| Kapok | 270 | trees / ha | |
| Kenaf | 34 | | 740 000 |
| Kikuyu grass | 8 | | |
| Kiwi fruit | 300 - 400 | plants / ha | |
| Kohlrabi | 4.5 | | 147 000 |
| Kola | 175 | trees / ha | |
| Koronivia grass | 4 | | |
| Kudzu | 1 100 | crowns / ha | |
| | 7 - 12 | unhulled seed | |

*(continued over)*

**TABLE 164 *(continued)*  Guideline Field Seed Rates and Final Plant Populations for Principal Crops (drilled unless indicated)**

| CROP | SEED RATE kg / ha (or plants / ha) | | FINAL PLANT POPULATION per ha |
|---|---|---|---|
| Leek | 2.5 | Prepack | 429 780 |
| | 2 | Fresh market | 143 358 |
| Lentil | 34 - 90 | broadcast | |
| | 12 - 15 | drilled | |
| Lespedeza | 28 | Common,Kobe,Korean, Sericea types | |
| Lespedeza, Japanese | 4.5 | | |
| Lettuce | 2.2 | Butterhead types | 129 106 |
| | | Crisp and Cos | 92 699 |
| Leucaena glauca | 20 | | 25 000 |
| Linseed | 55 | Oil types | |
| Litchi | 70 | trees / ha | |
| Longan | 70 | trees / ha | |
| Loquat | 275 | trees / ha | |
| Loofa | | | 2700 |
| Lotononis | 1.1 | | |
| Lotus peduncukatus | 5 | (in mixture) | |
| Lupins | 90 | Cover or green manure | |
| Lupin, blue | 78 - 112 | broadcast | |
| Lupin, white | 120 - 134 | broadcast | |
| Lupin, yellow | 56 - 90 | broadcast | |
| Macadamia nut | 125 | trees / ha | |
| Makarikari grass | 4 | | |
| Mango | 90 | trees / ha | |
| Mangosteen | 275 | trees / ha | |
| Marrow / Courgette | 9.9 | Early | 32 278 |
| | | Maincrop | 17 784 |
| Maize | 12 | Grain | 60 000 |
| | 40 | Forage | 100 000 |
| | 12 | Sweet Corn | 31 746 |
| Mango | 250 | trees / ha | |
| Mangosteen | 100 - 275 | trees / ha | |
| Medic, barrel | 2 | | |
| Medic, black | 23 | | |
| Medic, snail (M. orbiculata) | 22 | broadcast | |
| Medic, snail (M. scutellata | 112 | broadcast | |
| Medic, strand | 2 | | |
| Millet, pearl | 33.6 | Hay | |
| | 16.8 | Grain | |
| Millet, Hungarian | 54 | Hay | |
| | 34 | Grain | |
| Millet, Japanese | 34 | Irrigated | |
| Mimosa | 7 - 9 | broadcast | |
| Molasses grass | 5 | | |
| Muskmelon | 3.4 | | 4300 |
| Mustard | 4.5 | | |
| Nectarines | 190 | trees / ha | |
| New Zealand hemp | 3000 | plants / ha | |
| New Zealand spinach | 17 | | |
| Niger seed | 8 | | |
| Nindi | | | 7000 |
| Nutmeg | 70 | trees / ha | |
| Oats, winter | 160 | | 320 / sq m |
| Oats, spring | 190 | | |
| Oatgrass, tall meadow | 56 | | |

**TABLE 164** *(continued)*  Guideline Field Seed Rates and Final Plant Populations for Principal Crops (drilled unless indicated)

| CROP | SEED RATE kg / ha (or plants / ha) | | FINAL PLANT POPULATION per ha |
|---|---|---|---|
| Okra | 7.8 | | 35 000 |
| Onion | 4.3 | Spring sown ware | 645 559 |
| | 4.9 | Autumn sown ware | 860 745 |
| | 27.2 | Pickling | 4 000 000 |
| | 17.3 | Spring, salad, bunch | 2 678 962 |
| | 22.2 | Autumn, salad, bunch | 3 744 026 |
| | 8.7 | Japanese bunch | 2 678 962 |
| Onion | 940 | kg / ha of sets | |
| Papya | | | 1000 - 2000 |
| Para grass | 5 | | |
| Parsley | 3.9 | | 193 667 |
| Parsnip | 4 | Prepack | 484 120 |
| | 1.9 | Ware | 114 500 |
| Paspalum, broadleaf | 7 | | |
| Paspalum | 11 | | |
| Passion fruit | 1100 | plants / ha | |
| Pawpaw | 1500 | trees / ha | |
| Pea, field | 247 | Autumn, fresh | 1 150 000 |
| | 185 | Spring, fresh | 1 150 000 |
| | 100 | Sugar | 500 000 |
| | 220 | Dry, combining | 950 000 |
| Pea, flat | 67 - 78 | broadcast | |
| Pea, grass | 78-90 | broadcast | |
| Pea, pigeon | 12 | Seed | |
| | 33.6 | Green manure | |
| Pea, rough | 56 - 67 | broadcast | |
| | 20 | drilled | |
| Southern pea | 34 | | |
| Pea, wedge | 56 | | 150 000 |
| Peaches | 190 | trees / ha | |
| Peanuts | 100 | | 80 000 |
| Pecan nut | 40 | trees / ha | |
| Pepper: hot or sweet | 4.5 | | 25 000 |
| Phalaris aquatica | 4 | | |
| Pigeon Pea | 16.5 | | |
| Pilcatulum | 4 | | |
| Pineapple - cannery | 40 000 | plants / ha | |
| - fresh export | 70 000 | plants / ha | |
| Plum | 185 | trees / ha | |
| Potato | 3000 | kg / ha of tubers or tuber sections | |
| Prairie grass | 40 | | |
| Puero | 3 | | |
| | 6 - 17 | broadcast, unhulled. | |
| Pumpkin | 3 | | |
| Purple pigeon grass | 4 | | |
| Pyrethrum | 1.5 | (for transplanting) | 36 000 |
| Quinine | 6750 | trees / ha | |
| Radish | 74 | Prepack | 8 330 000 |
| | 22 | Bunch | 2 000 000 |
| Rambutan | 70 | trees / ha | |
| Ramie | 2500 | cuttings / ha | |
| Rape | 8 | Broadcast. For fodder | |
| Rape | 5 | Drilled. For fodder | |
| Rape, Oilseed | 7 | Winter types | |

*(continued over)*

*General Data*

TABLE 164 *(continued)* Guideline Field Seed Rates and Final Plant Populations for Principal Crops (drilled unless indicated)

| CROP | SEED RATE kg / ha (or plants / ha) | | FINAL PLANT POPULATION per ha |
|---|---|---|---|
| Rape,Oilseed | 8 | Spring types | • |
| Rapoko grass (Eleusine) | 7 | For grain | |
| Raspberries | 17 000 | canes / ha | |
| Reed canary grass | 13.4 | | |
| Rescue grass | 33.6 | | |
| Rhodes grass | 25 | Irrigated | |
| | 7 | Dryland | |
| Rhubarb | 11 000 | crown divisions per ha | |
| Rice | From 6 kg / ha for transplanting to 100 kg / ha for drilled crops | | 1 500 000 |
| Roselle | 4.5 | | 9000 |
| Rubber(Para) | 300 | trees / ha | |
| Rutabaga | 1.7 | | |
| Ruzi grass | 5 | | |
| Rye | 125 | Grain, spring type | |
| Rye | 225 | Fodder | |
| Ryegrass, perennial | 30 | | |
| Ryegrass, annual and biennials | 35 | Annual, L. Multiflorum, tetraploids and Italians | |
| Sabi grass | 6 | | |
| Sann | 39 - 45 | broadcast | |
| Safflower | 25 | Forage | |
| | 11 | Seed | |
| Salsify | 10.1 | | |
| Sainfoin | 34 - 39 | broadcast | |
| Sapote | 155 | trees / ha | |
| Scrobic | 5.6 | | |
| Serradella | 22.4 | | |
| Sesame | 7 | | |
| Sesbania, common | 22 | broadcast | |
| Setaria | 5 | | |
| Sheeps burnett | 12 | | |
| Sickle senna | 45 - 50 | broadcast | |
| Signal grass | 8 | | |
| Sisal | 2600 | plants / ha Suckers or bulbils | |
| Siratro | 3.5 | | |
| Sorghum | 40 | Forage | |
| | 12 | Grain | |
| Sorrel | 2.8 | | |
| Soybean | 75 | | 450 000 |
| | 135 | broadcast | |
| Spinach | 20 | Fresh | 1 075 392 |
| | 43 | Processing | 3 227 796 |
| Strawberries | 20 - 30 000 | plants / ha | |
| Stylo, Caribbean | 4 | | |
| Stylo, common | 3.5 | | |
| Stylo, fine stem | 4 | | |
| Stylo, shrubby | 6 | | |
| Stylo, Townsville | 6 | | |
| Stylo, verano | 6 | | |
| Sudan grass | 35 | irrigated | |
| | 10 - 15 | dryland | |

**TABLE 164** *(continued)*  Guideline Field Seed Rates and Final Plant Populations for Principal Crops (drilled unless indicated)

| CROP | SEED RATE kg / ha (or plants / ha) | | FINAL PLANT POPULATION per ha |
|---|---|---|---|
| Sugar beet | 10 | rubbed seed | 75 - 100 000 |
| Sugar cane | 22 000 | setts / ha | |
| Sulla | 22 - 39 | broadcast | |
| Sunflower | 9 | | |
| Sunhemp | 33.6 | For green manure or forage | |
| | 78 | For fibre | |
| Swede | 1.9 | | 172 149 |
| Sweet potato | 800 | kg / ha of roots for bedding | 30 000 |
| Swiss chard | 7.8 | | |
| Sweet corn | 17 | | 43 037 |
| Tall fescue | 30 | | |
| Tea | | | 7000 |
| Tephrosia sps. | 6 - 8 | | |
| Timothy grass | 11.2 | | |
| Tobacco, burley | 18 000 | plants / ha (from about 5 g of seed) | |
| Tobacco, flue cured | 13 500 | plants / ha (from about 5 g of seed) | |
| Tobacco, Turkish | 250 000 | plants / ha (from about 30 g of seed) | |
| Tomato | 0.25 | | 12 000 |
| Tonka Bean | 1100 | trees / ha | |
| Tossa jute | 7 | | |
| Triticale | 17 | trees / ha | |
| Turnip | 3 | | 344 293 |
| Uniflorus | 20 | | |
| Urochloa | 2.5 | | |
| Watermelon | 2.2 | | 6666 |
| Wheat, spring | 150 | Mexican types | |
| spring | 155 | Temperate types | 400 / sq m |
| winter | 210 | | 300 / sq m |
| White jute | 12 | | |
| Vetch, American joint | 45 | broadcast | |
| Vetch, big flower | 40 - 45 | broadcast | |
| Vetch, Hungarian | 70 - 80 | broadcast | |
| Vetch, kidney | 17 - 22 | broadcast | |
| Vetch, milk | 22 - 28 | broadcast | |
| Vetch, narrowleaf | 34 - 45 | broadcast | |
| Vetch, purple | 34 - 45 | broadcast | |
| Vetch, spring | 67.2 | | |
| Vetch, winter | 40 | | |
| Vetchling | 67 - 78 | broadcast | |
| Yellow melilot | 11 - 17 | broadcast | |
| Yam bean | | | 35 000 |

*Sources:* The above guidelines are derived principally from:
a range of commercial seed catalogues and,
Lorenz, O.A. & Maynard, D.N. (1980) *Knott's Handbook for Vegetable Growers.*
Samson, J.A. (1986) *Tropical Fruits.*
Purseglove, J.W. (1974) *Tropical Crops. Dicotyledons.*
Purseglove, J.W. (1976) *Tropical Crops. Monocotyledons.*
FAO (1953) *Legumes in Agriculture.*

### Note 7  Guidelines to Field Crop Macro Nutrient Requirements

**TABLE 165** gives guidelines to crop nutrient requirements for high yield levels. The table has in the main been compiled from higher rate recommendations to be found in the literature. Clearly they cannot be site specific and are given as a base from which to extrapolate. Actual site recommendations must depend upon such factors as:

- **local experimental evidence**
- **local farmers' experience**
- **soil nutrient status and nutrient availability**
- **soil pH**
- **climate**
- **moisture regime**
- **desired crop yield**
- **desired crop quality**
- **use or not of farmyard manures**
- **method of fertilizer application**
- **nutrient release rate from applied fertilizers**
- **soil type**
- **fertilizer application method**
- **likely return of nutrients from grazing animals**
- **crop disease potentials**
- **length of growing season**
- **nutrient inputs from irrigation water and rainfall**
- **soil temperatures at critical periods**
- **crop variety**
- **management level**
- **likely leaching of nutrients during the cropping period**
- **nutrient residues from previous crop**
- **whether the land is being reclaimed**
- **organic or non-organic soil**
- **economics of crop production**
- **for legumes, soil rhizobium status may affect response**
- **light intensity**
- **previous cropping history of the site**
- **requirement to build up soil reserves of P and K**
- **desired marketing period**
- **weed competition**
- **offtake of nutrients by the crop (this can often be calculated approximately for many crops by reference to animal feed analyses)**
- **presence or absence of nitrogen-producing soil organisms**

**CAUTIONS**

The following guidelines are intended to denote upper level nutrient rates needed to achieve high yield levels in generally non-limiting environments BUT ASSUMING LOW AVAILABLE SOIL NUTRIENT RESERVES.

Particular caution is required in extrapolating nitrogen requirements from this list to organic soils and moisture limited situations.

Where a crop production programme is started by extrapolation from these guidelines, it should be accompanied by a full soil and plant tissue nutrient level monitoring programme.

As a rule of thumb guide application of phosphate and potassium nutrients can be reduced by half to one-third where levels of these elements in the soil are known to be moderate to good.

Total applications of more than 100 kg/ha of nitrogen must normally be made in split applications to avoid crop damage.

High levels of fertilizer should not be placed in close proximity to seeds, seedling or roots of bare root transplants.

---

**TABLE 165   Guideline Crop Nutrient Requirements**

*NB*  • kg / ha unless stated
    • per year for perennials
    • per crop for annuals
    • for tree crops guidelines are for mature trees

| Crop | N | $P_2O_5$ | $K_2O$ |
|---|---|---|---|
| Apples, dessert | 40 | 80 | 220 |
| Apples, culinary and cider | 100 | 80 | 220 |
| Arecanut | 60 | 60 | 100 |
| Asparagus | 100 | 200 | 200 |
| Avocado | 170 | 100 | 170 |
| Banana | 350 | 150 | 750 |
| Barley | 125 | 75 | 75 |
| Beans, Broad vegetable | 60 | 250 | 250 |
| Beans, Field | 0 | 75 | 120 |
| Beans, French vegetable | 150 | 250 | 275 |
| Beans, Runner vegetable | 150 | 250 | 275 |
| Beetroot | 250 | 100 | 300 |
| Blackberries | 100 | 110 | 220 |
| Broccoli | 175 | 200 | 200 |
| Brussels Sprouts | 300 | 175 | 200 |
| Blackcurrants | 140 | 110 | 250 |
| Cabbage, Chinese and long season | 300 | 200 | 300 |
| Cabbage, over wintered long season | 225 | 200 | 300 |
| Cabbage, over wintered spring type | 400 | 200 | 300 |
| Cabbage, spring / summer short season | 400 | 200 | 300 |
| Calabrese | 250 | 150 | 150 |
| Cardamom | 45 | 45 | 45 |
| Carrots | 60 | 250 | 250 |
| Cashew   kg / tree | 0.25 | 0.15 | 0.15 |
| Castor | 120 | 90 | 60 |
| Cassava | 110 | 65 | 190 |
| Cauliflowers, shorter season types | 250 | 175 | 300 |
| Cauliflowers, overwintered types | 275 | 175 | 300 |
| Celery, winter types | 225 | 200 | 400 |
| Celery, self blanching | 250 | 200 | 400 |
| Cherries | 90 | 80 | 220 |

*(continued over)*

### TABLE 165 *(continued)*  Guideline Crop Nutrient Requirements

*NB*  • kg / ha unless stated
   • per year for perennials
   • per crop for annuals
   • for tree crops guidelines are for mature trees

| Crop | | N | $P_2O_5$ | $K_2O$ |
|---|---|---|---|---|
| Chicory | | 150 | 200 | 200 |
| Cinchona | | 285 | 70 | 70 |
| Citrus | kg / tree | 1.5 | 0.75 | 1.5 |
| Cocoa | | 120 | 120 | 240 |
| Coconut | kg / palm | 0.5 | 0.3 | 1.2 |
| Cocoyam | | 100 | 200 | 200 |
| Coffee, shaded | | 100 | 50 | 100 |
| Coffee, un-shaded | | 250 | 90 | 250 |
| Congo Jute | | 45 | 45 | 65 |
| Cotton | | 250 | 100 | 150 |
| Courgettes | | 100 | 200 | 250 |
| Cucumber | | 125 | 150 | 200 |
| Dates | kg / palm | 2 | 0.5 | 0.5 |
| Egg Plant | | 100 | 200 | 200 |
| Endive | | 90 | 180 | 180 |
| Fodder Beet | | 125 | 100 | 300 |
| Garlic | | 180 | 120 | 100 |
| Ginger | | 60 | 60 | 120 |
| Gooseberries | | 100 | 110 | 250 |
| Grapes | | 160 | 110 | 250 |
| Hops | | 225 | 300 | 450 |
| Jute | | 90 | 60 | 60 |
| Kale | | 125 | 100 | 100 |
| Kenaf | | 100 | 40 | 60 |
| Lettuce | | 200 | 300 | 175 |
| Leeks | | 250 | 300 | 275 |
| Linseed | | 100 | 75 | 75 |
| Loganberries | | 100 | 110 | 250 |
| Mangels | | 125 | 100 | 300 |
| Mango | kg / tree | 1.8 | 0.4 | 1.8 |
| Manilla Hemp | | 200 | 150 | 170 |
| Maize, temperate, forage or grain | | 60 | 75 | 100 |
| Maize, tropical, forage or grain | | 250 | 80 | 100 |
| Maize, sweet corn | | 100 | 150 | 150 |
| Melon, Musk | | 180 | 160 | 220 |
| Melon, Water | | 180 | 160 | 220 |
| Millet, bulrush, pearl | | 160 | 90 | 70 |
| Millet, finger | | 100 | 45 | 45 |
| Marrows | | 100 | 200 | 250 |
| Oats | | 125 | 75 | 75 |
| Oil Palm | kg / palm | 1 | 1.3 | 3.6 |
| Olives, dryland | kg / tree | 0.75 | 0.45 | 0.7 |
| Olives, irrigated | kg / tree | 1.5 | 0.9 | 1.5 |
| Onions, bulb,short season | | 250 | 70 | 210 |
| Onions, bulb,over wintered | | 90 | 300 | 275 |
| Onions, salad bunching | | 200 | 300 | 275 |
| | | 125 | 250 | 125 |
| Papya | | 200 | 135 | 300 |

**TABLE 165 *(continued)*  Guideline Crop Nutrient Requirements**

*NB*  • kg / ha unless stated
   • per year for perennials
   • per crop for annuals
   • for tree crops guidelines are for mature trees

| Crop | | N | $P_2O_5$ | $K_2O$ |
|---|---|---|---|---|
| PASTURES - Grazed, Fodder, Seed | | | | |
| | | | | |
| Grass, perennial, temperate, forage | | 450 | 200 | 280 |
| Grass, perennial, temperate, grazed | | 400 | 60 | 60 |
| Grass, perennial, temperate, seed | | 150 | 100 | 120 |
| Grass, perennial, tropical, forage, irrigated | | 2000 | 600 | 800 |
| Legume, temperate, perennial, forage | | 0 | 250 | 300 |
| Legume, temperate, perennial, seed | | 0 | 100 | 120 |
| Legume, tropical, perennial, fodder | | 0 | 500 | 300 |
| | | | | |
| Parsnips | | 100 | 175 | 225 |
| Peanuts | | 25 | 80 | 80 |
| Pears | | 90 | 80 | 220 |
| Peas, field | | 0 | 50 | 150 |
| Peas, vegetable | | 0 | 75 | 120 |
| Pepper, climbing | kg / plant | 0.25 | 0.36 | 0.15 |
| Pepper, sweet | | 115 | 200 | 200 |
| Pineapple | | 500 | 150 | 700 |
| Plums | | 90 | 80 | 220 |
| Potatoes - early | | 180 | 350 | 180 |
| Potatoes - maincrop | | 220 | 350 | 350 |
| Potatoes - seed, canning | | 180 | 350 | 250 |
| Pyrethrum | | 60 | 115 | 145 |
| Radish | | 60 | 100 | 275 |
| Ramie | | 245 | 100 | 350 |
| Rape, oilseed, spring | | 150 | 40 | 40 |
| Rape, oilseed, winter | | 260 | 40 | 40 |
| Rape, forage | | 100 | 75 | 100 |
| Raspberries | | 100 | 110 | 250 |
| Redcurrants | | 100 | 110 | 250 |
| Rhubarb | | 250 | 100 | 300 |
| Rice, wetland, short | | 160 | 60 | 80 |
| Rice, wetland, tall | | 50 | 35 | 45 |
| Rice, dryland | | 100 | 35 | 45 |
| Roselle | | 80 | 45 | 80 |
| Rubber | kg / tree | 0.15 | 0.2 | 0.18 |
| Rye | | 125 | 75 | 75 |
| Safflower | | 150 | 170 | 140 |
| Sesame | | 80 | 60 | 70 |
| Sisal | | 100/yr | 125/cycle | 250/cycle |
| Sorghum | | 200 | 50 | 75 |
| Soyabeans | | 25 | 70 | 100 |
| Spinach | | 175 | 300 | 300 |
| Strawberries | | 40 | 110 | 220 |
| Sugar Beet - temperate | | 120 | 100 | 200 |
| Sugar Beet - tropical, irrigated | | 200 | 100 | 200 |
| Sugar Cane | | 300 | 120 | 200 |

*(continued over)*

**TABLE 165 *(continued)*   Guideline Crop Nutrient Requirements**

*NB*   • kg / ha unless stated
  • per year for perennials
  • per crop for annuals
  • for tree crops guidelines are for mature trees

| Crop | N | $P_2O_5$ | $K_2O$ |
|------|---|----------|--------|
| Sunflower | 100 | 100 | 150 |
| Swedes, fodder | 75 | 150 | 125 |
| Swedes, vegetable | 100 | 150 | 250 |
| Sweet Potato | 65 | 200 | 300 |
| Tea | 225 | 100 | 150 |
| Tobacco, Turkish and flue cured | 20 | 200 | 240 |
| Tobacco, Burley, sun / air cured and dark fire cured | 120 | 120 | 250 |
| Tobacco, Wrapper leaf | 225 | 115 | 225 |
| Tomatoes, temperate | 100 | 200 | 250 |
| Tomatoes, tropical | 360 | 200 | 500 |
| Tung | 140 | 70 | 170 |
| Turmeric | 100 | 40 | 65 |
| Turnips, stubble | 100 | 75 | 100 |
| Turnips, vegetable | 100 | 150 | 250 |
| Watermelon | 180 | 160 | 220 |
| Wheat - winter | 150 | 75 | 75 |
| Wheat - spring and dwarf | 150 | 75 | 75 |
| Yams | 80 | 85 | 135 |

**Magnesium** is not commonly needed but where a requirement is suspected between 60 - 100 kgs/ha of Mg are recommended as an initial application to be monitored.

**Sodium** applied to certain crops may give response as follows:

In a K deficient situation barley, broccoli, Brussels sprouts, carrots, cotton, millet oats, peas, tomatoes and wheat will respond.

In a K - sufficient situation celery, mangel, sugar beet, swiss chard, beetroot and turnip will normally give a good response.

In a K - sufficient situation cabbage, kale, kohlrabi, mustard, radish and rape may give a limited response.

**Calcium** is rarely a deficient element. Its use as a soil 'fertilizer' is mainly in the context of modifying soil pH. Deficiency usually occurs on acid soils or highly alkali soils low in calcium. In both these situations addition of calcium compounds is usually required for the broader purpose of pH amelioration as well as increasing calcium availability. Soil requirements for calcium ameliorants are given in **TABLES 56 - 60**

**Sulphur** is becoming an increasingly required nutrient particularly in areas of intensive agriculture away from industrial pollution and using relatively sulphur-free concentrated fertilizers. Where a deficiency is indicated 30 kg/ha of sulphur or sulphur equivalent should be applied and monitored.

*Sources:* Russel, E.J. (1961) *Soil Conditions and Plant Growth.* de Geus, J.G. (1973)
*Fertilizer Guide for the Tropics and Subtropics.* Ministry of Agriculture, Fisheries and Food (1978)
*Lime and Fertiliser recommendations. No 1. Arable Crops and Grassland.* Ministry of Agriculture
Fisheries and Food (1985/86) *Lime and Fertiliser Recommendations. No 2. Vegetables and Bulbs.*
Lorenz, O.A. & Maynard, D.N. (1980) *Knott's Vegetable Handbook.* Landon, J.R. ed. (1984)
*Tropical Soil Manual.*

## Note 8  Crop Yields

Yields of the same crop vary widely throughout the world reflecting the influence of a wide range of factors. Management is a prime influence on yields: good and bad management can achieve totally different results when using the same level of inputs. Natural factors such as soil and climate also have a great influence over yield levels achieved. Soil may limit production more in economies where inputs to the soil are not financially possible. Climate also sets the limits for crop production. As well as determining yield it can determine crop quality and year to year consistency of yield and quality. All other factors being equal, the growing of crops in their optimal environments will place growers in an advantageous economic position.

Because of the diverse range of factors which affect crop yields and the diverse range of environments in which they are grown it is not possible to give comprehensive coverage to the range of crop yields likely to be achieved around the world. However the following lists of yields have been prepared from various regional sources to act as guidelines for planning purposes. They are designed to indicate possible yield levels for crops given reasonable inputs and management, and taking into account likely economic environments. Wherever possible local yield data from trials, farmers' records and visual appraisal of crop vigour should be taken into account in trying to estimate or plan yield levels.

**TABLE 166  Indicative Crop Yields in a Temperate Region (UK)**

*NB* These yields are intended to reflect averages and trends over several seasons. Yields can vary widely between seasons. Individual farmers may achieve yield levels higher or lower than those stated. The yields of the principal crop product only are given; extra crop value may come from secondary products such as straw from grain crops or sub-standard vegetable produce used as animal feed. With fodder crops all the yield may not be efficiently utilized by grazing stock.

| Crop | Production Level t / ha | | |
|---|---|---|---|
| | Low | Avg | High |
| Apples, culinary | | 17.5 | |
| Apples, dessert | | 11 | |
| Barley, spring | 3.5 | 4.6 | 5.5 |
| Barley, winter | 4.25 | 5.75 | 6.75 |
| Beans, Broad processing | | 3.1 | |
| Beans, green processing | | 6 | |
| Beans, fresh Runner | 7 - 17 early crops; 24 - 35 main crops | | |
| Beans,winter field | 3 | 3.6 | 4.2 |
| Beans, spring field | 2.8 | 3.4 | 4 |
| Blackcurrants | | 5 | |
| Borage | | | 0.7 |
| Broccoli, heading | | 13 000 heads | |
| Brussels Sprouts | | 12.5 | |
| Cabbage, summer | | 20 | |

*(continued over)*

**TABLE 166** *(continued)*  **Indicative Crop Yields in a Temperate Region (UK)**

| Crop | Production Level t / ha | | |
|---|---|---|---|
| | Low | Avg | High |
| Cabbage, spring | | 15 | |
| Cabbage, winter | | 25 | |
| Calabrese | | 4.5 | |
| Carrots | | 37.5 | |
| Cauliflower | | 15 000 heads | |
| Evening Primrose | 0.15 | 0.4 | 0.75 |
| Flax | 1.5 fibre from 8 straw | | |
| **FODDER CROPS** | | | |
| Arable Silage | | 5 - 9 D.M. | |
| Fodder Beet | | 60 | 90 |
| Forage Pea silage | | 4 - 7 DM | |
| Grass | 2 | 7 | 12 (all as DM) |
| Kale | | 7 - 8 DM | |
| Maize | 9 | 10 | 11 (all as DM) |
| Mangels | | 7.5 - 11.5 DM | |
| Rape | | 5 DM | |
| Rye, early graze | | 2 - 3 DM | |
| Stubble Turnips | | 40 | |
| Swedes | | 75 | 100 |
| Turnips | | 60 | 80 |
| Grapes | | 5 | |
| **HERBAGE SEEDS** | | | |
| Italian Ryegrass | | 1.05 | 1.4 |
| Hybrid Ryegrass | | 1.15 | 1.55 |
| Perennial Ryegrass, early | | 1 | 1.4 |
| Perennial Ryegrass, intermediate | | 1.03 | 1.4 |
| Perennial Ryegrass, late | | 1.1 | 1.5 |
| Timothy | | 0.45 | 0.6 |
| Wild White Clover | | 0.08 | 0.11 |
| Leeks | | 20 | |
| Lettuce | | 12 | |
| Linseed | 1.25 | 1.9 | 2.5 |
| Lupins | 1 | 2 | 3 |
| Oats, winter | 4 | 5.5 | 6.5 |
| Oats, spring | 3.35 | 4.5 | 5.35 |
| Oilseed Rape, winter | 2.25 | 3 | 3.75 |
| Oilseed Rape, spring | 1.5 | 2.1 | 2.7 |
| Onions, dry bulb | | 27.5 | |
| Parsnips | | 10 | |
| Pears, Conference | | 10 | |
| Peas, dried | 2.5 | 3.25 | 4 |
| Peas, freezing | | 4.8 (fresh) | |
| Plums | | 10 | |
| Potatoes, maincrop | 28 | 38 | 48 |
| Potatoes, early | say, 12 - 28 depending on lifting date | | |
| Potatoes, seed | | 22 (plus ware) | |
| Raspberries | | 5.5 - 8.3 (after establishment) | |
| Rhubarb | | 20 | |

**TABLE 166** *(continued)*  **Indicative Crop Yields in a Temperate Region (UK)**

| Crop | Production Level t / ha | | |
|---|---|---|---|
| | Low | Avg | High |
| Rye | 3.5 | 4.5 | 5.5 |
| Strawberries | | 7 (after establishment) | |
| Swedes, domestic | | | 39 |
| Sugar Beet | 30 | 40 | 50 |
| Triticale | 3.5 | 4.75 | 6 |
| Vetch, winter | | | 2.5 |
| Wheat, Durum | 3 | 4 | 5 |
| Wheat, spring | 3.35 | 4.5 | 5.35 |
| Wheat, winter | 5 | 6.75 | 8 |

*Principal sources:* Nix, J. (1990) *Farm Management Pocketbook.* Scottish Agricultural Colleges (1986-87) *Farm Management Handbook.*
Reproduced with kind permission of J. Nix and the Scottish Agricultural Colleges.

**TABLE 167   Indicative Yields for Selected Tropical and Sub-Tropical Crops**

*NB* Yields of crops in these regions will primarily be influenced by climate which varies greatly according to latitude and altitude and thereby permitting the production of a very diverse range of 'tropical' and 'temperate' crops in specific localities and seasons. Where these are not optimal for the crop, yields will be inherently lower.

| Crop | Yield t / ha / yr or t / ha / annual crop | Comments |
|---|---|---|
| Adzuki Bean | 0.4 - 1.1 | |
| Alexandrian Senna | 1.2 | Pods |
| Alfalfa | 7.5 - 10 DM | Rainfed |
| | 10 - 25 DM | Irrigated |
| Allspice | 0.002 - 0.022 | Green berries, 10 - 15 year tree |
| Almonds (in shell nuts) | 1.9 | USA avg yield |
| | 0.8 | Unimproved situations |
| Apples | 0.055 / tree | Five to six year old tree |
| Apricots | 0.055 / tree | Five year old tree |
| Asparagus | 2.5 - 3.7 | |
| Avaram | 1.5 | Green bark |
| Avocado | 5 - 15 | 5 - 10 avg range |
| Bambarra groundnut | 0.9 - 3.5 | |
| Banana | 40 - 60 | Good commercial yield |
| | 15 - 25 | Average farmer rainfed |
| | 35 - 50 | Average farmer, irrigated |
| Barley | 5 - 7 | Avg irrigated yield range |
| Bean (P. vulgaris) | 6 - 8 | Fresh |
| | 1.5 - 2 | Dry beans,commercial,irrigated |
| | 0.5 - 1 | Small farm, rainfed |
| | 1 - 1.5 | Small farm, irrigated |
| Beetroot | 17 - 25 | Average commercial range |
| Berseem | 4 - 12 (DM) | Fodder |
| Birdseed | 0.9 - 1.5 | Rainfed / irrigated range |
| Black Gram | 0.5 | |

**TABLE 167** *(continued)*  **Indicative Yields for Selected Tropical and Sub-Tropical Crops**

| Crop | Yield t / ha / yr or t / ha / annual crop | Comments |
|---|---|---|
| Black Wattle | 10 - 17 | Fresh bark, 8 - 10 year old stand |
| Bowstring Hemp | 2.25 | Fibre yield per year |
| Breadfruit | Up to 700 fruits / tree of 1 - 4 kg each | |
| Broad Bean | 12.3 - 17.3 | As green beans in pods |
| | 1.8 - 3.1 | As dry beans |
| Brussels sprouts | 7.5 - 11.1 | |
| Buckwheat | 1.3 | |
| Cabbage | 25 - 35 | Commercial, rainfed |
| | 80 | Commercial irrigated upper limit |
| | 10 - 20 | Avg rainfed farmer |
| | 20 - 40 | Avg irrigated farmer |
| Cape Gooseberries | 3.7 - 7.5 | As husked fruit |
| Cardomom | 0.1 - 0.3 | Dried capsules |
| Carrots | 30 | |
| Cashew | 1 - 5 | Avg range |
| Castor Bean | 1.4 | Favourable rainfed |
| | 3.9 | Good irrigated |
| Cassava | 5 - 15 | Rainfed smallholder |
| | 30 - 40 | Commercial plantation average |
| | >50 | Commercial improved vars. |
| | 15 - 20 | Avg rainfed farmer |
| | 25 - 35 | Avg irrigated farmer |
| Cauliflower | 12.3 - 25 | |
| Chicory | 16 - 25 | As fresh roots |
| Chick Pea | 0.45 - 1.75 | Dried seeds |
| Chillies | 0.25 - 2.75 | Dried, rainfed - irrigated range |
| Cinnamon | 0.07 - 0.22 | Quill yield, 5 - 10 year old trees |
| **CITRUS (commercial)** | | |
| Orange | 25 - 40 | |
| Grapefruit | 40 - 60 | |
| Lemon | 30 - 45 | |
| Mandarin | 20 - 30 | |
| **CITRUS (average farmer)** | | |
| Orange | 10 - 20 | Rainfed |
| | 20 - 30 | Irrigated |
| Grapefruit | 8 - 15 | Rainfed |
| | 15 - 25 | Irrigated |
| Mandarin | 8 - 15 | Rainfed |
| | 15 - 25 | Irrigated |
| Cloves | 0.003 - 0.018 / tree | Dried cloves from mature trees |
| Cluster Beans | 0.75 - 1.5 | Rainfed - irrigated range |
| Cocoa beans | 0.6 - 2 | Rainfed, good management |
| | 0.8 - 1.5 | Range for estates and smallholders |
| Coconuts (copra) | 0.026 / tree | Minimum for tree selection |
| | 0.4 - 2.5 | Range of yields smallholder to estates |
| | 15 - 30 nuts / palm | Annual production range |

**TABLE 167** *(continued)*  **Indicative Yields for Selected Tropical and Sub-Tropical Crops**

| Crop | Yield t / ha / yr<br>or t / ha / annual crop | Comments |
|---|---|---|
| Cocoyam | < 5 | Low |
| (Colocasia) | 10 | Medium |
| | >20 | High |
| Cocoyam | 5 - 7 | Tropical peasant agriculture |
| (Xanthosoma) | 12.5 - 20 | Good avg yield tropical peasant |
| | agriculture | |
| Coffee-Arabica | 0.5 - 1.2 | Smallholder rainfed |
| | 1 - 2 | Estate |
| Coffee-Robusta | 0.5 - 1.2 | Smallholder rainfed |
| | 0.8 - 1.9 | Estate |
| Cotton | 1.1 - 1.8 | Seed cotton, rainfed |
| | 2.2 - 3.3 | Seed cotton, irrigated |
| | 0.1 - 3.3 | Lint. Poor - v.high range |
| Cowpeas | 1.1 - 2.2 | Seed |
| | 5.5(DM) | Fodder |
| Cucumber | 20 | |
| Custard Apple | 20 | Average improved farmer yield |
| Dates | 5 - 50 kg / palm | Near East / North Africa avg |
| | 150 - 170 kg / palm | High Californian yield |
| Desmodium | 8 (DM) | From four cuts / year |
| Dolichos | 1.1 | Seed |
| | 4.4 (DM) | Fodder |
| Durian | 10 - 18 | Good yield range |
| Egg Plant | 25 | |
| Finger millet | 0.4 - 4.5 | Poor rainfed / best commercial |
| Fig | 0.135 / tree | Good mature tree yield |
| Fodder Beet | 45 - 50 | Much higher yields possible |
| Garlic | 4.5 - 11 | |
| Ginger | 7.5 - 30 | Fresh roots |
| Grape | 15 - 30 | Good avg commercial, sub tropic |
| | 3 - 5 | Avg rainfed farmer |
| | 5 - 10 | Avg irrigated farmer |
| Grass pea | 1.1 | Seed |
| Grasses, tropical | Up to 50 (DM) | Irrigated |
| Green Gram | 0.45 - 1.3 | Dry beans |
| Groundnut (unshelled) | 2 - 3 | Good commercial rainfed |
| | 3.5 - 4.5 | Good commercial irrigated |
| | 0.5 | Unimproved smallholder |
| | 1 - 2 | Avg farmer rainfed |
| | 1.5 - 2 | Avg farmer irrigated |
| Guava | 0.045 / tree | Three year old tree |
| | 0.225 / tree | Good mature tree yield |
| Hemp | 0.3 | Prepared fibre |
| Horsegram | 0.25 - 0.65 | Dry seeds |
| Hungry Rice | 0.6 - 1.1 | |
| Hyacinth Bean | 0.45 - 1.4 | Dry seeds |
| Indian Senna | 0.08 - 0.165 | Pods |
| Jack Bean | 2.2 - 3.8 | Seed |
| | 7 (DM) | Hay |

*(continued over)*

**TABLE 167** *(continued)*  **Indicative Yields for Selected Tropical and Sub-Tropical Crops**

| Crop | Yield t / ha / yr or t / ha / annual crop | Comments |
|---|---|---|
| Japanese Barnyard Millet | 0.7 - 0.8 | Grain |
| | 1 - 1.5 | Straw |
| Japanese Radish | 300 | Good total fodder yield |
| Jute (processed fibre) | 1.0 - 2.5 | Rainfed / Irrigated range |
| Kapok | 0.5 / tree | Clean floss, young tree |
| | 1.8 / tree | Clean floss, 7 - 10 year tree |
| Kenaf | 1.1 - 3.3 | dry fibre |
| Kiwi fruit | 20 | Good yield from mature stand |
| Kudzu | 11 (DM) | Yield from older stand |
| Kurrat | 37 | Yield of tops in Egypt |
| Leek | 20 | Good commercial yield |
| Lentil | 0.2 - 2 | Poor rainfed - irrigated range |
| Lettuce | 10 - 13 | |
| Leucena (seed) | 0.9 | Up to 6t recorded |
| Lima Bean | 1.3 | Dry beans |
| Linseed | 0.5 - 1.2 | Seed |
| | 0.45 | Fibre |
| Litchi | 0.180 / tree | From well grown tree |
| Little Millet | 0.2 - 0.9 | |
| Longan | 6 | Avg Thai yield |
| Loquat | 0.135 / tree | Mature tree yield |
| Loofah | 60 000 fruits | High yield level |
| Macadamia nut | 8 - 10 | normal range from mature trees |
| Maize | 6 - 9 | Good commercial irrigated |
| | >3 | Good rainfed smallholder |
| | 0.5 - 1.5 | Unimproved rainfed smallholder |
| | 1.5 - 3 | Avg rainfed farmer |
| | 4 - 5 | Avg irrigated farmer |
| Mango | 8 - 16 | Avg to high commercial range |
| | 200 - 1000 fruits per tree | |
| Mangolds | 45 - 50 | Much higher yields possible |
| Mangosteen | 200 - 1500 fruits per tree recorded | |
| Marrow | 7.5 - 25 | Depends on size at harvest |
| Mauritius Hemp | 3.7 | Fibre yield per cut |
| Muskmelon | 20 | |
| Millet | 0.4 - 2 | Rainfed range |
| | 3.5 | Good improved level |
| Nectarines | 0.068 / tree | |
| New Zealand Hemp | 1 | Fibre yield per cut |
| Niger Seed | 0.45 | |
| Nindi | 0.001 - 0.002 / plant | Harvested inflorescence |
| Nutmeg | 0.1 - 0.2 | As mace |
| Oil Palm | 3 - 4 (oil) | Good Estate |
| Okra | 5.5 | Fresh fruits |
| Olive | 2 - 2.5 | Oil, rainfed |
| | 3 - 4 | Oil, irrigated |
| Onion | 35 - 45 | Good commercial irrigated |
| | 5 - 10 | Avg rainfed farmer |
| | 10 - 20 | Avg irrigated farmer |
| Oyster Nut | 2.5 | |

**TABLE 167** *(continued)*  **Indicative Yields for Selected Tropical and Sub-Tropical Crops**

| Crop | Yield t / ha / yr or t / ha / annual crop | Comments |
|---|---|---|
| Papya | 100 (yrs 2 + 3 + 4) | High commercial yield |
| | 20 (yrs 2 + 3 + 4) | Unimproved yield |
| | 0.14 - 0.17 | Dried latex |
| Parsnip | 25 | For domestic market |
| Passion fruit | 10 - 20 | Avg yield. Up to 60 recorded. |
| Pepper (fresh fruit) | 10 - 15 | Normal commercial |
| | 20 - 25 | Best commercial |
| Pea | 2.2 | Good yield dry seed |
| Pears | 0.055 / tree | Mature tree |
| Peaches | 0.068 / tree | |
| Pigeon Pea ('Dahl') | 1.1 - 2.2 | Seed yield |
| | 4 - 10 (DM) | Fodder |
| Pineapple | 25 - 35 | Good commercial sub tropics |
| | 15 - 25 | Good commercial tropics |
| | 25 | Avg rainfed farmer |
| | 40 | Avg irrigated farmer |
| Plums | 0.055 / tree | Five year old tree |
| Potato | 15 - 40 | Commercial range |
| | 5 - 7 | Without blight control |
| Pyrethrum | 0.4 - 1.3 | Dried flowers |
| Quince | 14.8 | |
| Quinine | 10 - 18 | Bark, over 8 - 10 year cycle |
| Ramie | 0.8 - 1.2 | De-gummed fibre |
| Rape | 25 - 50 | As green fodder |
| Rapoko | 2.6 - 4.4 | Improved peasant farmer range |
| Raspberry | 11 | |
| Rhubarb | 20 - 30 | |
| Rice (paddy unhusked) | 6 - 8 | Good commercial controlled irrigation |
| | 3 - 4 | Good flood irrigation |
| | 0.5 - 1.5 | Typical unimproved smallholder |
| | 1.5 - 2.5 | Avg rainfed farmer |
| | 4 - 5 | Avg irrigated farmer |
| Rice Bean | 0.2 | |
| Rosella | 1 - 2.5 | Dry processed fibre |
| Rubber | 1 - 2 | Good conditions avg dry rubber yield |
| Rye | 1.8 - 3.3 | |
| Safflower | 1 - 2.5 | Good rainfed |
| | 2 - 4 | Good irrigated |
| | 0.8 - 1.3 | Avg rainfed farmer |
| | 1.5 - 2 | Avg irrigated farmer |
| Sann Hemp | 0.33 - 0.55 | Dry fibre |
| Sapodilla | 0.300 / tree | Tobago yield |
| Sesame | 0.2 - 2.2 | |
| Sisal (fibre) | 2 | Average plantation |
| | 1.5 - 2 | Average smallholder rainfed |

*(continued over)*

**TABLE 167** *(continued)* **Indicative Yields for Selected Tropical and Sub-Tropical Crops**

| Crop | Yield t / ha / yr or t / ha / annual crop | Comments |
|------|------|------|
| Sorghum | 3.5 - 5 | Good commercial irrigated |
| | 0.8 - 1.3 | Traditional spate irrigation |
| | 2 - 3 | Good smallholder irrigated |
| | 0.2 - 0.8 | Unimproved smallholder |
| | 1.3 - 2 | Average rainfed farmer |
| | 4 - 5 | Average irrigated farmer |
| | 7 | High commercial yield |
| | 12 - 75 | Green fodder |
| Soyabean | 1.5 - 2.5 | Good rainfed |
| | 2.5 - 3.5 | Good irrigated |
| | 0.8 - 1.3 | Average rainfed farmer |
| | 1.5 - 2 | Average irrigated farmer |
| Squash | 20 | |
| Strawberry | 7.5 | |
| Sweet Corn | 3 - 8 | As cobs |
| Sweet Potato | 19 - 50 | Tubers |
| | 20 - 60 | Vines |
| Sugar Beet | 40 - 60 | Good commercial |
| | 20 - 30 | Average rainfed farmer |
| | 40 - 45 | Average irrigated farmer |
| | 12 - 25 | Tops |
| Sugar Cane | 70 - 100 | Good rainfed estate humid tropics |
| | 110 - 150 | Good irrigated estate tropics and sub-tropics |
| Sunflower (seed) | 0.8 - 1.5 | Normal rainfed |
| | 2.5 - 3.5 | Commercial irrigated |
| | 1 - 1.5 | Average rainfed farmer |
| | 1.5 - 2 | Average irrigated farmer |
| Sunhemp | 6 (DM) | Fodder |
| | 0.4 - 1.3 | Fibre |
| | 1.3 | Seed |
| Sweet Potato | 2.5 - 50 | Attainable range |
| | 17.5 - 20 | Satisfactory yields |
| | 5 - 10 | Average rainfed farmer |
| | 12 - 18 | Average irrigated farmer |
| Tea | 0.2 | 18 - 30 month bushes made tea |
| | 1.7 | Mature (>10yrs) bushes made tea |
| | 1.5 - 2.5 | Fermented leaf smallholders to estates |
| Teff | 0.3 - 3 | 1/t considered good |
| Tepary Bean | 0.2 - 1.65 | Poor rainfed - irrigated range |
| Tobacco | 2 - 2.5 | Good commercial fresh leaf |
| | 0.5 - 1 | Flue cured avg rainfed farmer |
| | 1 - 1.5 | Air cured avg rainfed farmer |
| | 1 - 2 | Flue cured avg irrigated farmer |
| | 1.5 - 3 | Air cured avg irrigated farmer |

**TABLE 167** *(continued)*  **Indicative Yields for Selected Tropical and Sub-Tropical Crops**

| Crop | Yield t / ha / yr or t / ha / annual crop | Comments |
|---|---|---|
| Tobacco - Turkish | 0.6 - 1.3 | |
| Tobacco - Burley | 1.1 - 1.65 | |
| Tomato | 45 - 65 | Good irrigated commercial |
| | 10 - 20 | Average rainfed farmer |
| | 20 - 40 | Average irrigated farmer |
| Tonka Bean | 0.5 - 1 kg / tree | |
| Tung | 0.6 - 2.5 | Processed oil |
| Turmeric | 13 - 25 | Fresh roots |
| Turnips | 22 | For domestic market |
| Vanilla | 0.5 - 0.8 | Cured beans |
| Velvet Beans | 3.4 | Seed, good yield |
| | 4.4 - 7.7 (DM) | Fodder |
| Vetch | 10 (DM) | Good fodder yield |
| Watermelon | 25 - 35 | Good commercial irrigated |
| | 10 - 20 | Average rainfed farmer |
| | 20 - 40 | Average irrigated farmer |
| Wedge Pea | 1.3 | Seed |
| | 5 (DM) | Fodder |
| West Indian Cherry | 60 | High yield level |
| Wheat | 4 - 8 | Good commercial irrigated subtropics |
| | 0.7 | Unimproved rainfed smallholder |
| | 1.3 - 2 | Average farmer rainfed |
| | 3 - 5 | Average farmer irrigated |
| White Jute | 1.65 - 2.76 | Dry fibre |
| Yam | 7.5 - 17.5 | Rainfed W Africa smallholder |
| | 12.5 - 25 | Rainfed SE Asia smallholder |
| | 20 - 30 | Rainfed W Indies |
| | 12 - 25 | Average rainfed farmer |

*Main Sources:* FAO (1982) *Date Production and Protection. Plant Production and Protection Paper No. 35.* Samson, J.A. (1988) *Tropical Fruits. Rhodesian Farm Management Handbook* (1973). Acland, J.D. (1975) *East African Crops.* Purseglove, J.W. (1974) *Tropical Crops. Dicotyledons.* Purseglove, J.W. (1976) *Tropical Crops. Monocotyledons.* ILACO (1985) *Agricultural Compendium.* Landon, J.R., ed. (1984) *Tropical Soil Manual.*

**TABLE 168  Indicative Yields of Vegetable Crops in USA t / ha (These yields are intended to reflect the diversity of cropping environments found in the USA)**

| Crop | Approx avg yield | Good yield |
|---|---|---|
| Artichoke | 9.26 | 12.35 |
| Asparagus, processing | 3.09 | 4.94 |
| Bean, market | 4.94 | 12.35 |
| Bean, processing | 6.79 | 12.35 |
| Bean, Lima, processing | 3.71 | 4.94 |
| Beets, market | 17.29 | 24.7 |
| Beets, processing | 32.11 | 37.05 |
| Broccoli | 10.5 | 13.59 |
| Brussels Sprouts | 15.43 | 19.76 |
| Cabbage, market | 30.88 | 37.05 |
| Cabbage, processing | 49.4 | 61.75 |
| Carrot, topped | 34.6 | 43.23 |
| Cauliflower | 12.35 | 18.53 |
| Celeriac | | 24.7 |
| Celery | 61.75 | 86.45 |
| Chard, Swiss | | 18.53 |
| Corn, market | 9.58 | 14.82 |
| Corn, processing | 12.35 | 16.06 |
| Cucumbers, market | 13.59 | 30.88 |
| Cucumbers, processing | 12.35 | 14.68 |
| Eggplant | 24.7 | 30.88 |
| Endive | 17.29 | 22.23 |
| Garlic | 16.06 | 19.76 |
| Horseradish | | 9.88 |
| Lettuce | 27.17 | 43.23 |
| Melon, Persian | 14.82 | 18.53 |
| Melon, Honeydew | 22.7 | 30.88 |
| Muskmelon | 17.3 | 24.7 |
| Okra | | 12.35 |
| Onion | 38.3 | 49.4 |
| Pea, market | 4.94 | 7.41 |
| Pea, processing, shelled | 3.46 | 4.94 |
| Pepper, bell | 13.59 | 24.7 |
| Pepper, chilli, dried | 4.94 | 7.41 |
| Pepper, pimento | 4.94 | 7.41 |
| Potato | 30.88 | 43.23 |
| Pumpkin | | 49.4 |
| Rhubarb | | 6.18 |
| Rutabaga | | 49.4 |
| Spinach, market | 8.65 | 18.53 |
| Spinach, processing | 17.29 | 24.7 |
| Squash, summer | | 37.05 |
| Squash, winter | | 49.05 |
| Sweet potato | 14.20 | 24.7 |
| Tomato, market | 21 | 24.7 |
| Tomato, processing | 54.34 | 74.1 |
| Turnip | | 37.25 |
| Watermelon | 14.2 | 24.7 |

*Source:* Lorenz, O.A. & Maynard, D.N. (1980) Adapted from *Knott's Handbook for Vegetable Growers.* © 1980 and reproduced with kind permission of John Wiley and Sons Inc.

# Growing Periods and Development Patterns

**TABLE 169    Indicative, Approximate Growing Periods of Annual Crops**

The principal factors controlling the period from planting to harvest are variety and climate. An increasing range of long and short season varieties become available for more commercialized crops. Exceptionally short season varieties are frequently found in low rainfall areas of the world. In higher latitudes long duration crops are found which may overwinter in a semi-dormant state. Transplanting may add to the production period. Some crops require harvesting immediately they are ready, others will stand for long periods without deterioration.

| Crop | Growing Period (day) | Crop | Growing Period (day) |
|---|---|---|---|
| Barley, spring type | 100 - 210 | Dandelion | 85 |
| Barley, winter type | 300 +/- | Egg Plant | 125 - 150 |
| Bean, broad, winter type | 300 +/- | Endive | 85 - 100 |
| Bean, broad, spring type | 120 - 180 | Flax | 150 - 220 |
| Bean, broad, fresh vegetable | 120 | Florence Fennel | 100 |
| Bean, bush snap, vegetable | 50 - 65 | Gram, green | 60 - 120 |
| Bean, pole snap, vegetable | 65 - 90 | Groundnut | 95 - 140 |
| Bean, common | 60 - 180 | Hemp | 120 - 150 |
| Bean, Lima | 100 - 270 | Jute | 100 - 130 |
| Bean, Lima, fresh vegetable | 65 - 88 | Kale, vegetable | 55 - 180 |
| Beetroot | 45 - 180 | Kale, fodder | 150 - 240 |
| Berseem | 120 - 270 | Kohlrabi | 50 - 150 |
| Broccoli, overwintered | 240 +/- | Landcress | 90 |
| Broccoli, summer type | 90 - 180 | Leeks | 150 - 300 |
| Brussels Sprouts | 140 - 300 | Lentil | 90 - 170 |
| Cabbage | 120 - 270 | Lettuce, butterhead | 55 - 90 |
| Cabbage, Chinese | 75 | Lettuce, Cos | 70 - 75 |
| Cardoon | 120 | Lettuce, Webb types | 70 - 100 |
| Carrot | 50 - 250 | Linseed | 130 - 180 |
| Cassava | 180 - 700 | Maize, grain | 90 - 180 |
| Castor | 100 - 200 | Maize, forage | 90 - 180 |
| Cauliflower, overwintered | 270 +/- | Melon, Casaba type | 110 |
| Cauliflower, summer | 95 - 120 | Melon, Canteloupe | 85 - 150 |
| Celeriac | 110 | Melon, honeydew type | 110 |
| Celery | 125 - 300 | Melon, Persian type | 110 |
| Chard | 50 - 60 | Millet, Bullrush | 60 - 90 |
| Chevril | 60 | Millet, common | 60 - 90 |
| Chickpea | 120 - 180 | Muskmelon | 85 - 95 |
| Chicory | 65 - 150 | New Zealand Spinach | 70 |
| Chillies | 100 | Oats, spring | 180 - 210 |
| Chives | 90 | Oats, winter | 300 +/- |
| Collard | 70 - 120 | Okra | 90 - 140 |
| Cotton | 150 - 270 | Onion, dry bulb | 100 - 180 |
| Cowpea | 60 - 150 | Onion, dry bulb, overwintered | 240+/- |
| Cucumber | 80 - 150 | Onion, green bulb | 90 - 120 |
| Cucumber, pickling | 50 - 60 | Parsnip | 120 - 180 |
| Cucumber, slicing | 62 - 72 | Parsley | 70 - 100 |

*(continued over)*

**TABLE 169** *(continued)* **Indicative, Approximate Growing Periods of Annual Crops**

| Crop | Growing Period (day) | Crop | Growing Period (day) |
|------|------|------|------|
| Pea, fresh | 55 - 90 | Sugar Beet | 160 - 300 |
| Pea, dry | 90 - 140 | Sugar Cane | 270 - 720 |
| Pepper, hot and sweet | 95 - 110 | Sunflower | 70 - 165 |
| Pigeon pea | 150 - 360 | Swede | 120 - 180 |
| Potato | 100 - 240 | Sweet Corn | 55 - 110 |
| Pumpkin | 75 - 120 | Sweet Potato | 100 - 180 |
| Radish | 20 - 75 | Squash, winter pumpkin | 95 - 120 |
| Rice, Upland | 110 - 180 | Squash, Zucchini | 100 +/- |
| Rice, wetland | 110 - 180 | Sweet Potato | 120 - 150 |
| Roselle | 120 - 175 | Tobacco | 90 - 120 |
| Rutabaga | 90 | Tomato | 90 - 180 |
| Safflower | 120 - 190 | Triticale | 140 - 160 |
| Salsify | 150 - 180 | Turnip | 45 - 160 |
| Scolymus | 150 | Watercress | 180 |
| Scorzonera | 150 | Watermelon | 120 - 150 |
| Sorghum | 90 - 140 | Wheat, Durum | 140 |
| Sesame | 40 - 140 | Wheat, spring type | 110 - 210 |
| Soyabean | 95 - 150 | Wheat, winter type | 300 +/- |
| Spinach | 60 - 120 | Yam | 180 |

*Sources:* Author's experience. ILACO (1980) *Agricultural Compendium.* Dalgety/Master seeds, *Linseed Growers Guide.* FAO (1988) *Crop Water requirements. Irrigation and Drainage Paper No. 24.* Lorenz, O.A. & Maynard, D.N. (1980) *Knott's Vegetable Handbook.* Kenya Seed Co., *Planting Guide.* Sunseeds, California, *Vegetable Seeds Guide.*

**TABLE 170** **Length of Growing Season and Crop Development Stages of Selected Field Crops; Some Indications**

Key. 40/40/250/30 and (360) for example, stand respectively for initial, crop development, mid-season and late season crop development stages in day and (360) for total growing period from planting to harvest in days.

**Artichokes**    Perennial, replanted every 4-7 years; example Coastal California with planting in April 40/40/250/30 and (360); subsequent crops with crop growth cutback to ground level in late spring each year at end of harvest or 20/40/220/30 and (310).

**Barley**    Also wheat and oats; varies widely with variety; wheat Central India November planting 15/25/50/30 and (120); early spring sowing, semi-arid, 35°-45° latitudes and November planting Rep. of Korea 20/25/60/30 and (135); wheat sown in July in East African highland at 2500 m altitude and Rep. of Korea 15/30/65/40 and (150).

**TABLE 170** *(continued)*  **Length of Growing Season and Crop Development Stages of Selected Field Crops; Some Indications**

**Beans**
(green)

February and March planting California desert and Mediterranean 20/30/30/10 and (90); August-September planting California desert, Egypt, coastal Lebanon 15/25/25/10 and (75).

**Beans**
(dry)
Pulses

Continental climates late spring planting 20/30/40/20 and (110); June planting Central California and West Pakistan 15/25/35/20 and (95); longer season varieties 15/25/50/20 and (110).

**Beets**
(table)

Spring planting Mediterranean 15/25/20/10 and (70); early spring planting Mediterranean climates and pre-cool season in desert climates 25/30/25/10 and (90).

**Carrots**

Warm season of semi-arid to arid climates 20/30/30/20 and (100); for cool season up to 20/30/80/20 and (150); early spring planting Mediterranean 25/35/40/20 and (120); up to 30/40/60/20 and (150) for late winter planting.

**Castorbeans** Semi-arid and arid climates, spring planting 25/40/65/50 and (180).

**Celery**

Pre-cool season planting semi-arid 25/40/95/20 and (180); cool season 30/55/105/20 and (210); humid Mediterranean mid-season 25/40/45/15 and (125).

**Corn**
(maize)
(sweet)

Philippines, early March planting (late dry season) 20/20/30/10 and (80); late spring planting Mediterranean 20/25/25/10 and (80); late cool season planting desert climate 20/30/30/10 and (90); early cool season planting desert climates 20/30/50/10 and (110).

**Corn**
(maize)
(grains)

Spring planting East African highlands 30/50/60/40 and (180); late cool season planting, warm desert climates 25/40/45/30 and (140); June planting sub-humid-Nigeria, early October India 20/35/40/30 and (125); early April planting Southern Spain 30/40/50/30 and (150).

**Cotton**

March planting Egypt, April-May planting Pakistan, September planting South Arabia 30/50/60/55 and (195); spring planting, machine harvested Texas 30/50/55/45 and (180).

**Crucifers**

Wide range in length of season due to varietal differences; spring planting Mediterranean and continental climates 20/30/20/10 and (80); late winter planting Mediterranean 25/35/25/10 and (95); autumn planting coastal Mediterranean 30/35/90/40 and (195).

**Cucumber**

June planting Egypt, August-October California desert 20/30/40/15 and (105); spring planting semi-arid and cool season arid climates, low desert 25/35/50/20 and (130).

*(continued over)*

**TABLE 170 *(continued)*  Length of Growing Season and Crop Development Stages of Selected Field Crops; Some Indications**

| | |
|---|---|
| **Egg plant** | Warm winter desert climates 30/40/40/20 and (130); late spring-early summer planting Mediterranean 30/45/40/25 and (140). |
| **Flax** | Spring planting cold winter climates 25/35/50/40 and (150); pre-cool season planting Arizona low desert 30/40/100/50 and (220). |
| **Grain,** small | Spring planting Mediterranean 20/30/60/40 and (150); October-November planting warm winter climates; Pakistan and low deserts 25/35/65/40 and (165). |
| **Lentil** | Spring planting in cold winter climates 20/30/60/40 and (150); pre-cool season planting warm winter climates 25/35/70/40 and (170). |
| **Lettuce** | Spring planting Mediterranean climates 20/30/15/10 and (75) and late winter planting 30/40/25/10 and (105); early cool season low desert climates from 25/35/30/10 and (100); late cool season planting, low deserts 35/50/45/10 and (140). |
| **Melons** | Late spring planting Mediterranean climates 25/35/40/20 and (120); mid-winter planting in low desert climates 30/45/65/20 and (160). |
| **Millet** | June planting Pakistan 15/25/40/25 and (105); central plains USA spring planting 20/30/55/35 and (140). |
| **Oats** | See Barley. |
| **Onion** (dry) | Spring planting Mediterranean climates 15/25/70/40 and (150); pre-warm winter planting semi-arid and arid desert climates 20/35/110/45 and (210). |
| (green) | Respectively 25/30/10/5 and (70) and 20/45/20/10 and (95). |
| **Peanuts** (groundnuts) | Dry season planting West Africa 25/35/45/25 and (130); late spring planting coastal plains of Lebanon and Israel 35/45/35/25 and (140). |
| **Peas** | Cool maritime climates early summer planting 15/25/35/15 and (90); Mediterranean early spring and warm winter desert climates planting 20/25/35/15 and (95); late winter Mediterranean planting 25/30/30/15 and (100). |
| **Peppers** | Fresh - Mediterranean early spring and continental early summer planting 30/35/40/20 and (125); cool coastal continental climates mid-spring planting 25/35/40/20 and (120); pre-warm winter planting desert climates 30/40/110/30 and (210). |
| **Potato** (Irish) | Full planting warm winter desert climates 25/30/30/20 and (105); late winter planting arid and semi-arid climates and late spring-early summer planting continental climate 25/30/45/30 and (130); early-mid spring planting central Europe 30/35/50/30 and (145); slow emergence may increase length of initial period by 15 days during cold spring. |

**TABLE 170 *(continued)*   Length of Growing Season and Crop Development Stages of Selected Field Crops; Some Indications**

**Radishes**    Mediterranean early spring and continental summer planting 5/10/15/5 and (35); coastal Mediterranean late winter and warm winter desert climates planting 10/10/15/5 and (40).

**Safflower**    Central California early-mid spring planting 20/35/45/25 and (125) and late winter planting 25/35/55/30 and (145); warm winter desert climates 35/55/60/40 and (190).

**Sorghum**    Warm season desert climates 20/30/40/30 and (120); mid-June planting Pakistan, May in Mid-West USA and Mediterranean 20/35/40/30 and (125); early spring planting warm arid climates 20/35/45/30 and (130).

**Soybeans**    May planting central USA 20/35/60/25 and (140); May-June planting California desert 20/30/60/25 and (135); Philippines late December planting, early dry season - dry: 15/15/40/15 and (85) vegetables 15/15/30/- and (60); early-mid June planting in Japan 20/25/75/30 and (150).

**Spinach**    Spring planting Mediterranean 20/20/15/5 and (60); September-October and late winter planting Mediterranean 20/20/25/5 and (70); warm winter desert climates 20/30/40/10 and (100).

**Squash**    Late winter planting Mediterranean and warm winter desert climates
(winter)    20/30/30/15 and (95); August planting California desert 20/35/30/25 and (110);
pumpkin    early June planting maritime Europe 25/35/35/25 and (120).

**Squash**    Spring planting Mediterranean 25/35/25/15 and (100+); early summer
(zucchini)    Mediterranean and maritime Europe 20/30/25/15 and (90+); winter planting
crookneck    warm desert 25/35/25/15 and (100).

**Sugarbeet**    Coastal Lebanon, mid-November planting 45/75/80/30 and (230); early summer planting 25/35/50/50 and (160); early spring planting Uruguay 30/45/60/45 and (180); late winter planting warm winter desert 35/60/70/40 and (205).

**Sunflower**    Spring planting Mediterranean 25/35/45/25 and (130); early summer planting California desert 20/35/45/25 and (125).

**Tomato**    Warm winter desert climate 30/40/40/25 and (135); and late autumn 35/45/70/30 and (180); spring planting Mediterranean climates 30/40/45/30 and (145).

**Wheat**    See Barley.

*Source:*  FAO (1988) *Crop Water Requirements. Irrigation and Drainage Paper No. 24.*
Reproduced with kind permission of the Food and Agriculture Organization of the United Nations.

**TABLE 171    Developmental Pattern of Principal Cereal Crops**

| | | | Timing of Developmental Stages in Days | | | |
|---|---|---|---|---|---|---|
| Crop | Type[a] | Sowing to floral different'n | Floral differentiation to flower | Flower to maturity | Sowing to maturity | |
| | | | | | Standard[b] | Extremes |
| **Barley** | A | 17 – 32 | 30 – 37 | 30 – 40 | 75 – 110 | 65 – 140 |
| | WA | 8 – 25[c] | 28 – 35 | 30 – 45 | 65 – 105 | 60 – 130 |
| **Maize** | A | 24 – 54[d] | 28 – 75 | 40 – 60 | 90 – 180 | 65 – 330 |
| **Millet-pearl** *Pennisetum typhoides* | A | 22 – 55 | 25 – 60 | 35 – 60 | 100 – 160 | 90 – 190 |
| **Millet-foxtail** *Setaria italica* | A | 12 – 25 | 30 – 40 | 20 – 30 | 65 – 100 | 60 – 120 |
| **Millet-proso** *Panicum miliaceum* | A | 15 – 30 | 30 – 40 | 18 – 30 | 63 – 100 | 60 – 110 |
| **Oat** | A | 24 – 45 | 28 – 40 | 30 – 45 | 80 – 125 | 70 – 160 |
| | WA | 10 – 30 | 28 – 40 | 30 – 45 | 70 – 110 | 65 – 150 |
| **Rice** | AU | 35 – 60 | 28 – 42 | 45 – 75 | 110 – 180 | 80 – 200 |
| | AP | 45 – 70 | 30 – 45 | 50 – 75 | 120 – 190 | 90 – 240 |
| **Rye** | A | 20 – 32 | 32 – 40 | 33 – 45 | 80 – 115 | 70 – 150 |
| | WA | 5 – 25 | 30 – 40 | 30 – 60 | 60 – 110 | 55 – 140 |
| **Sorghum** *Sorghum bicolor* | A | 28 – 60 | 30 – 60 | 35 – 70 | 95 – 190 | 90 – 230 |
| **Triticale** *x Triticosecale* | A | 20 – 40 | 35 – 40 | 40 – 60 | 90 – 130 | 85 – 170 |
| | WA | 10 – 30 | 35 – 45 | 35 – 45 | 85 – 120 | 80 – 180 |
| **Wheat** | A | 20 – 35 | 32 – 40 | 38 – 45 | 85 – 130 | 75 – 160 |
| | WA | 19 – 25 | 32 – 40 | 35 – 45 | 75 – 110 | 73 – 140 |

[a]  A = annual; W = winter; U = upland; P = paddy.
[b]  Standard applies to areas of major production of the crop.
[c]  For winter annuals the starting base is beginning of spring growth instead of sowing date.
[d]  Range encompasses male and female; the male preceded the female as a rule by a matter of 2 to 6 days.

*Source:* Christie, B.R. (1987) *Handbook of Plant Science in Agriculture. Vol 1.*
© CRC Press Inc., Boca Raton, FL. Reprinted with permission.

**TABLE 172    Development of Pulse Crops**

| Crop[a] | Sowing to flowering (days) | Flowering to maturity (days) | Sowing to maturity (days) | |
|---|---|---|---|---|
| | | | Standard | Extremes |
| **Bean**—common dry *Phaseolus vulgaris* | 45 – 60 | 50 – 70 | 100 – 140 | 90 – 160 |
| **Chickpea** | 35 – 55 | 55 – 90 | 90 – 130 | 85 – 180 |
| **Cowpea** | 35 – 70 | 25 – 35 | 80 – 160 | 60 – 210 |
| **Faba bean** *Vicia faba* | 40 – 60 | 45 – 75 | 95 – 135 | 90 – 150 |
| **Lentil** | 45 – 58 | 40 – 50 | 75 – 96 | 70 – 130 |
| **Pea** *Pisum sativum* | 32 – 50 | 25 – 65 | 60 – 100 | 50 – 160 |
| **Peanut** (ground nut) | 25 – 70 | 55 – 75 | 110 – 150 | 100 – 170 |
| **Pigeon pea** | 60 – 110 | 50 – 100 | 120 – 180 | 90 – 240 |

[a] All but pigeon pea are herbaceous annuals. Pigeon pea is a woody shrub normally grown as an annual.

*Source:* Christie, B.R. (1987) *Handbook of Plant Science in Agriculture. Vol 1.*
© CRC Press Inc., Boca Raton, FL. Reprinted with permission.

*Growing Periods and Development Patterns*

---

**TABLE 173  Development of Annual Oilseed Crops**

---

| | | Timing of Developmental Stages in Days | | | |
|---|---|---|---|---|---|
| Crop | Type[a] | Sowing to flowering | Flowering to maturity | Sowing to maturity Standard[b] | Extremes |
| **Castor bean**[c] | A/P | 80 – 110 | 70 – 100 | 160 – 180 | 140 – |
| **Linseed** | A | 45 – 75 | 35 – 55 | 90 – 110 | 85 – 130 |
| **Rapeseed** | WA | 25 – 45[d] | 40 – 70 | 90 – 115 | 80 – 125 |
| *Brassica napus* | A | 45 – 65 | 40 – 55 | 80 – 120 | 70 – 150 |
| *B. campestris* | A | 35 – 50 | 32 – 48 | 70 – 100 | 60 – 115 |
| **Safflower** | A | 65 – 140[c] | 35 – 60 | 110 – 200 | 100 – 250 |
| **Sesame** | A | 70 – 110 | 40 – 80 | 130 – 160 | 120 – 180 |
| **Soybean** | A | 35 – 60 | 40 – 100 | 100 – 160 | 75 – 195 |
| **Sunflower** | A | 65 – 90 | 30 – 60 | 95 – 130 | 70 – 170 |

[a]  A = annual; P = perennial; W = winter.

[b]  Standard applies to areas of major production of the crop.

[c]  Castor bean, usually grown as a woody annual for oil seed, will persist and bear as perennial in suitable environment.

[d]  For this winter annual, the starting base is beginning of growth in spring instead of sowing date.

[e]  Fall seeded in regions of mild winters, the safflower is delayed in early development.

*Source:* Christie, B.R. (1987) *Handbook of Plant Science in Agriculture. Vol 1.*
© CRC Press Inc., Boca Raton, FL. Reprinted with permission.

**TABLE 174    Development of Tree Fruits**

| Crop | Type[a] | Time to bearing (years) | Flower initiation (season)[b] | Flowering (season) | Bloom to maturity (days) | Reach full production (years) | Production life (years) |
|---|---|---|---|---|---|---|---|
| **Apple** | | | | | | | |
| *Malus domestica* | D | 5 – 8 | ESu | MSp | 80 – 150 | 10 – 12 | 20 – 40 |
| *M. bacata* | D | 4 – 6 | ESu | MSp | 65 – 110 | 8 – 18 | 30 – 50 |
| **Apricot,** *Prunus armeniaca* | D | 4 – 6 | LSu | ESp | 60 – 120 | 8 – 10 | 15 – 20 |
| **Avocado,** *Persea americana* | E | 5 – 7 | Su | W[c] | 140 – 400 | 7 – 10 | 30 – 40 |
| **Cherry** | | | | | | | |
| *Prunus avium* (sweet) | D | 5 – 7 | Su[d] | VE Sp | 35 – 65 | 9 – 12 | 25 – 30 |
| *P. cerasus* (sour) | D | 3 – 6 | Su | E Sp | 50 – 80 | 8 – 10 | 20 – 25 |
| **Date,** *Phoenix dactylifera* | E | 5 – 6 | LSu | E Sp[e] | 160 – 240 | 8 – 10 | 50 + |
| **Fig,** *Ficus carica* | D | 3 – 5 | LSu,Sp[f] | Sp; LSu | 70 – 130 | 5 – 7 | – |
| **Grapefruit,** *Citrus paradisi* | E | 4 – 7 | Ir | In[g] | 190 – 380 | 7 – 12 | 30 – 50 |
| **Kiwifruit,** *Actinidia chinensis* | DV | 4 – 5 | VESp | Sp | 160 – 200 | 7 – 8 | – |
| **Lemon,** *C. limon* | E | 4 – 7 | In | In | 220 – 280 | 7 – 12 | 30 – 50 |
| **Lime,** *C. aurantifolia* | E | 4 – 7 | In | In | 140 – 230 | 6 – 10 | 30 – 50 |
| **Mandarin,** *C. reticulata* | E | 4 – 8 | W | ESp | 170 – 330 | 6 – 10 | 20 – 40 |
| **Mango,** *Mangifera indica* | E | 3 – 7 | LA | VESp[h] | 70 – 180 | 8 – 12 | 35 – 40 |
| **Orange,** *C. sinensis* | E | 4 – 7 | In | In | 170 – 370 | 6 – 12 | 20 – 50 |
| **Papaya,** *Carica papaya* | EH | 1 | Ap | MSu[i] | 80 – 220 | – | 4 – 8 |
| **Peach,** *Prunus persica* | D | 3 – 4 | MSu | ESp | 80 – 150 | 6 – 10 | 30 – 160 |
| **Pear,** *Pyrus communis* | D | 6 – 8 | LSp,Su | MSp | 75 – 160 | 8 – 12 | 30 – 58 |
| **Plum** | | | | | | | |
| *Prunus domestica* | D | 4 – 6 | LSu | Esp | 75 – 150 | 7 – 9 | 20 – 30 |
| *P. salicina* | D | 4 – 6 | LSu | VESp | 50 – 120 | 7 – 9 | 20 – 30 |

[a]    D = deciduous; E = evergreen; V = vine; H = herbaceous.

[b]    E = early; M = mid; L = late; Sp = spring; Su = summer; A = autumn; W = winter; In = intermittent.

[c]    Purseglove describes three avocado types with widely different fruit development periods, all of which tend to flower from November to May in the northern hemisphere, but the flowering period may last up to 6 months.

[d]    Sweet cherries initiate flowering after harvest of the preceding crop.

[e]    Tolerant of only mild frosts, the date is a dry climate plant which flowers in February to May in Arizona and California, depending on cultivar, weather, and age.

[f]    The fig may initiate flowers during the growing season, resulting in a second crop; in some environments the latter is the main crop.

[g]    Citrus may flower mainly in spring grown in a climate with a cool winter, and may produce some flowers continuously in a tropical climate; characteristically there is a season (or two or three seasons) when a flush of flowering occurs in response to a dry period or other inducement.

[h]    Mango may depart from the pattern shown, depending on timing of dry seasons.

[i]    Flowering in papaya is seasonal only in year one; thereafter fruits develop in leaf axils, about one per week at maximum.

*Source:* Christie, B.R. (1987) *Handbook of Plant Science in Agriculture. Vol 1.*
© CRC Press Inc., Boca Raton, FL. Reprinted with permission.

**TABLE 175    Development of Small Fruits**

| | Type[a] | Begin to bear (years) | Floral initiation (season)[b] | Time of bloom (season)[b] | Bloom to maturity (days) | Fully productive (years) | Longevity (years) |
|---|---|---|---|---|---|---|---|
| **Blueberry,** *Vaccinium spp.* | | | | | | | |
| Highbush, *V. corymbosum* | DS | 4 – 5 | Au | ESp | 70 – 90 | 5 – 8 | 20 – 30 |
| Lowbush, *V. angustifolium* | DS | 1 – 4 | Au | ESp | 60 – 80 | 2 – 3 | – |
| **Cranberry,** *V. rnacrocarpon* | DV | 3 – 4 | LSu | MSu | 75 – 100 | 5 – 6 | 40 – 50 |
| **Currant,** *Ribes sativum* | DS | 3 – 4 | MSu | ESp | 45 – 80 | 3 – 4 | 12 – 15 |
| **Gooseberry,** *R. hirtellum* | DS | 2 – 4 | LSu | ESp | 45 – 75 | 4 – 5 | 10 – 12 |
| **Grape:** *Vitis spp.* | | | | | | | |
| V. vinifera | DV | 3 – 4 | MSu | LSp | 80 – 190 | 5 – 6 | 30 – 50 |
| V. labrusca | DV | 3 – 4 | MSu | LSp | 80 – 180 | 5 – 6 | 30 – 50 |
| V. rotundifolia | DV | 3 – 4 | MSu | LSp | 100 – 120 | 5 – 6 | 30 – S0 |
| **Raspberry:** *Rubus spp.* | | | | | | | |
| Red: *R. idaeus* | DC | 2 – 3 | LSu | MSp | 40 – 80 | 3 – 4 | 8 – 10 |
| Black: *R. occidentalis* | DC | 2 – 3 | Au | MSp | 60 – 85 | 3 – 4 | 8 – 10 |
| **Strawberry,** *Fragaria vesca* | HP | 1 – 2 | Au | VESp | 20 – 45 | 1 – 2 | 4 – 6 |

[a]    D = deciduous; S = shrub; V = vine; C = cane; H = herbaceous; P = perennial.

[b]    Au = autumn; E = early; M = mid; L = late; Sp = spring; Su = summer; V = very.

*Source:* Christie, B.R. (1987) *Handbook of Plant Science in Agriculture. Vol 1.*
© CRC Press Inc., Boca Raton, FL. Reprinted with permission.

**TABLE 176    Production Cycle of Field-Seeded Vegetables**

| | | Field-Seeding Dates by Frost-Free Period[a] | | | | | | |
|---|---|---|---|---|---|---|---|---|
| Date: month-day | | Long season 2-28 to 12-10[c] | | Medium season 4-10 to 10-10 | | Short season 5-20 to 9-20 | | Ready for use |
| Crop | Type[b] | First | Last | First | Last | First | Last | (in days) |
| **Beans, snap or green,** *Phaseolus vulgaris* | A | 4 – 10 | 9 – 1 | 4 – 25 | 7 – 20 | 5 – 15 | 6 – 1 | 48 – 60 |
| **Beans, lima,** *P. limensis* | A | 4 – 15 | 8 – 20 | 5 – 1 | 6 – 1 | – | – | 65 – 90 |
| **Beans, broad,** *Vicia faba* | A | – | – | 3 – 15 | 5 – 10 | 4 – 20 | 5 – 5 | 70 – 100 |
| **Beets,** *Beta vulgaris* | B | 3 – 1 | 9 – 1 | 3 – 20 | 6 – 15 | 5 – 10 | 6 – 1 | 50 – 80 |
| **Carrots,** *Daucus carota* | B | 3 – 1 | 9 – 10 | 4 – 10 | 6 – 15 | 5 – 1 | 6 – 10 | 70 – 90 |
| **Corn, sweet,** *Zea mays* | A | 4 – 1 | 8 – 1 | 4 – 25 | 6 – 5 | 5 – 10 | 6 – 1 | 70 – 100 |
| **Cucumber,** *Cucumis sativus* | A | 4 – 1 | 8 – 15 | 5 – 10 | 6 – 10 | 5 – 25 | 6 – 10 | 50 – 70 |
| **Endive,** *Cichorium endivia* | A,B | 3 – 1 | 9 – 1 | 4 – 1 | 7 – 1 | 5 – 15 | 6 – 10 | 60 – 90 |
| **Kale,** *Brassica oleracea*[d] | B | 2 – 15 | 9 – 1 | 4 – 1 | 7 – 1 | 5 – 1 | 6 – 1 | 50 – 80 |

**TABLE 176** *(continued)* **Production Cycle of Field-Seeded Vegetables**

| Date: month-day | | Field-Seeding Dates by Frost-Free Period[a] | | | | | | Ready for use (in days) |
|---|---|---|---|---|---|---|---|---|
| | | Long season 2-28 to 12-10[c] | | Medium season 4-10 to 10-10 | | Short season 5-20 to 9-20 | | |
| Crop | Type[b] | First | Last | First | Last | First | Last | |
| Kohlrabi, *B. oleracea*[d] | B | 2 – 20 | 9 – 1 | 4 – 10 | 7 – 1 | 5 – 1 | 6 – 1 | 55 – 80 |
| Leek, *Allium ampelo-prasum* | B | 2 – 15 | 8 – 15 | 4 – 1 | – | – | – | 100 – 150 |
| Lettuce, *Lactuca sativa* | A | 3 – 1 | 9 – 15 | 3 – 15 | 8 – 1 | 4 – 5 | 7 – 10 | 40 – 50 |
| Muskmelon, *Cucumis melo* | A | 4 – 1 | 7 – 1 | 4 – 25 | 6 – 1 | 5 – 5 | 5 – 15 | 85 – 110 |
| Onions, cooking, *Allium cepa* | B | 2 – 1 | 9 – 15 | 3 – 20 | 4 – 15 | 4 – 15 | 4 – 25 | 80 – 125 |
| Onions, bunching, *A. cepa* | B | 2 – 1 | 9 – 20 | 4 – 15 | 7 – 9 | 5 – 1 | 7 – 15 | 35 – 75 |
| Okra, *Hibiscus esculentus* | A | 3 – 25 | 8 – 1 | 6 – 1 | – | – | – | 55 – 65 |
| Parsley, *Petroselinum crispum* | B | 2 – 15 | 9 – 1 | 4 – 10 | 7 – 1 | 5 – 15 | 6 – 10 | 70 – 90 |
| Parsnip, *Pastinaca sativa* | B | 2 – 15 | 9 – 1 | 4 – 1 | 5 – 20 | 5 – 1 | 5 – 15 | 120 – 170 |
| Peas, garden, *Pisum sativum* | A | 3 – 1 | 10 – 1 | 4 – 1 | 6 – 5 | 4 – 25 | 6 – 1 | 55 – 90 |
| Potato, *Solanum tuberosum* | A | 3 – 1 | 7 – 20 | 4 – 20 | 5 – 20 | 5 – 1 | 6 – 1 | 75 – 100 |
| Radish, *Raphanus sativus* | A | 3 – 1 | 8 – 1 | 3 – 20 | 8 – 15 | 4 – 25 | 8 – 1 | 25 – 40 |
| Rutabaga, *Brassica napobrassica* | B | 1 – 20 | 10 – 1 | 6 – 1 | – | 6 – 1 | – | 80 – 100 |
| Spinach, *Spinacea oleracea* | A | 2 – 15 | 10 – 1 | 4 – 1 | 8 – 10 | 4 – 25 | 8 – 1 | 40 – 60 |
| Squash, summer, *Cucurbita* spp.[e] | A | 4 – 1 | 6 – 15 | 5 – 1 | 6 – 1 | 5 – 25 | 6 – 1 | 45 – 60 |
| Squash, winter, *Cucurbita* spp. | A | 7 – 15 | 8 – 1 | 5 – 20 | – | 5 – 15 | – | 85 – 120 |
| Sweet potato, *Ipomoea batatas* | A | 4 – 15 | 6 – 1 | – | – | – | | 110 – 130 |
| Turnip, *Brassica rapa* | B | 2 – 15 | 10 – 1 | 3 – 20 | 6 – 1 | 5 – 25 | 6 – 1 | 40 – 80 |
| Watermelon, *Citrulus lunatus* | A | 3 – 15 | 7 – 1 | 5 – 15 | 6 – 10 | 5 – 25 | – | 80 – 110 |

[a] Range in dates allows for a succession of plantings where appropriate.
[b] A = annual; B = biennial (in terms of seed production; produce harvest is in year of planting).
[c] Average dates of last spring frost and first fall frost.
[d] Cole crops, *Brassica oleracea*; kale = var. *acephala*; kohlrabi = var. *botrytis,* subvar. *gongylodes.*
[e] Squash: *Cucurbita* spp; Summer - *C. pepo*; Winter - *C. moschata*; Pumpkin - *C. mixta.*

*Source:* Christie, B.R. (1987) *Handbook of Plant Science in Agriculture. Vol 1.*
© CRC Press Inc., Boca Raton, FL. Reprinted with permission.

**TABLE 177   Production Cycle of Transplanted Vegetables**

| Crop | Type[a] | Sowing days be- fore field planting | Field planting dates by length of season[b] (month / day) | | | Ready for use (in days)[c] |
|------|---------|------|------|------|------|------|
| | | | Long (>200 days) | Medium (130 – 200 days) | Short (<130 days) | |
| **Cole,** *Brassica oleracea* | | | | | | |
| **Broccoli,** Var. *botrytis,* | | | | | | |
| s. var. *cymosa* | B | 40 – 50 | 2/15 – 7/15 | 3/15 – 7/15 | 5/25 – 6/15 | 60 – 90 |
| **Brussels sprouts,** Var. | | | | | | |
| botrytis, s.var. *gemmifera.* | B | 40 – 55 | 2/15 – 7/1 | 3/15 – 6/20 | 5/20 – 6/1 | 90 – 140 |
| **Cabbage,** Var. *botrytis,* | | | | | | |
| s.var. *capitata* | B | 40 – 50 | 2/1 – 7/10 | 3/15 – 6/20 | 5/20 – 6/1 | 60 – 96 |
| **Cauliflower,** Var. *botrytis,* | | | | | | |
| s.var. *cauliflora* | B | 40 – 55 | 2/20 – 7/10 | 3/25 – 6/15 | 5/25 – 6/5 | 90 – 96 |
| **Collards,** Var. *acephala* | B | 40 – 50 | 2/20 – 7/1 | 3/25 – 6/10 | – | 50 – 80 |
| | B | 55 – 70 | 2/20 – 7/1 | 3/10 – 6/1 | – | 120 – 135 |
| **Celery,** *Apium graveolens* | A | 40 – 55 | 4/10 – 6/20 | 5/15 – 6/10 | 6/1 – 6/10 | 65 – 90 |
| **Eggplant,** | | | | | | |
| *Solanum melongena* | B | 60 – 70 | 4/20 – 6/1 | 5/1 – 6/10 | – | 75 – 90 |
| **Lettuce (head),** | | | | | | |
| *Lactuca sativa* | A | 35 – 45 | 2/15 – 3/10 | 4/1 – 5/15 | 5/10 – 6/1 | 65 – 80 |
| **Onion (Spanish),** | | | | | | |
| *Allium cepa* | B | 50 – 65 | 2/1 – 3/1 | 4/1 – 5/1 | 5/1 – 5/15 | 90 – 140 |
| **Pepper,** *Capsicum annuum* | A | 40 – 55 | 4/2 – 6/20 | 5/10 – 6/15 | 5/15 – 6/10 | 60 – 90 |
| **Tomato,** | | | | | | |
| *Lycopersicon esculentum* | A | 40 – 50 | 4/2 – 5/15 | 5/10 – 6/1 | 5/24 – 6/10 | 65 – 90 |

[a]   A = annual; B = biennial.

[b]   Length of season based on frost-free period.

[c]   Period from field transplanting to harvesting for edible use.

*Source:* Christie, B.R. (1987) *Handbook of Plant Science in Agriculture. Vol 1.*
© CRC Press Inc., Boca Raton, FL. Reprinted with permission.

**TABLE 178**  **Developmental Pattern of Perennial Spice Crops**

| | Type[a] | Time to bearing (years) | Time to full production (years) | Flowering time[b] | Bloom to harvest (days) | Bearing life (years) |
|---|---|---|---|---|---|---|
| **Allspice (pimento),** *Pimenta dioica* | ETD | 6 – 8 | 13 – 16 | sesnl[c] | 85 – 125 | 50 – 100 |
| **Cloves,** *Syzigium aromaticum* | ET | 7 – 9 | 15 – 25 | sesnl | 170 – 190[d] | 50 – 100 |
| **Nutmeg,**[e] *Myristica fragrans* | ETD | 5 – 8 | 15 – 17 | cont | 170 – 250 | 30 – 50 |
| **Pepper,** *Piper nigrum* | EWV | 3 – 4 | 7 – 8 | sesnl[f] | 180 – 240 | 12 – 20 |
| **Vanilla,** *Vanilla planifolia* | HV | 3 – 4 | 7 – 8 | sesnl | 120 – 270 | 10 – 12 |

[a]  E = evergreen, T = tree, D = deciduous, W = woody, H = herbaceous, V = vine.

[b]  cont = continuous; sesnl = seasonal.

[c]  In Jamaica, pimento flowers in March to June.

[d]  The unopen bud is the commercial product.

[e]  Mace is a companion product of nutmeg, forming part of the same fruit.

[f]  A dry period checks vegetative growth and brings on floral initiation; subsequent moisture stimulates flowering which usually lasts about 2 months.

*Source:* Christie, B.R. (1987) *Handbook of Plant Science in Agriculture. Vol 1.* © CRC Press Inc., Boca Raton, FL. Reprinted with permission.

**TABLE 179 Development of Perennial Forage Grasses**

| Crop | Induction[a] | First growth to heading (days) | Heading to bloom (days) | First growth to maturity (days) |
|---|---|---|---|---|
| *Agropyron cristatum*<br>**Crested wheatgrass**<br>**(Fairway)** | V-SD | 48 – 57 | 15 – 20 | 85 – 105 |
| *A. desertorum*<br>**Crested wheatgrass**<br>**(Standard)** | V-SD | 45 – 55 | 15 – 20 | 80 – 100 |
| *A. intermedium*<br>**Intermediate wheatgrass** | V-SD | 60 – 70 | 15 – 25 | 95 – 120 |
| *A. smithii*<br>**Western wheatgrass** | SD | 60 – 70 | 13 – 25 | 95 – 120 |
| *A. trachycaulum*<br>**Slender wheatgrass** | V-SD | 60 – 70 | 15 – 25 | 95 – 115 |
| *Agrostis alba*<br>**Redtop** | SD | 65 – 75 | 15 – 20 | 95 – 120 |
| *Andropogon gayanus*<br>**Gamba grass** | – | – | – | 100 – 150 |
| *Axonopus affinis*<br>**Carpet grass** | – | – | – | 80 – 140[b] |
| *Bouteloua gracilis*<br>**Blue grama** | – | 45 – 60 | 15 – 25 | 85 – 110[c] |
| *Bromus inermis*<br>**Smooth bromegrass** | V-SD | 50 – 70 | 15 – 25 | 90 – 120 |
| *Buchloe dactyloides*<br>**Buffalo grass** | – | 45 – 65 | 10 – 20 | 90 – 130 |
| *Chloris gayana*<br>**Rhodes grass** | – | 120 – 150[d] | 10 – 25 | 150 – 180[b] |
| *Cynodon dactylon*<br>**Bermuda grass** | – | – | – | 95 – 130[b] |
| *Dactylis glomerata*<br>**Orchard grass** | V-SD | 50 – 65 | 8 – 15 | 80 – 110 |
| *Elymus canadensis*<br>**Canadian wild rye** | V-SD | 57 – 75 | 15 – 30 | 95 – 125 |
| *Elymus junceus*<br>**Russian wild rye** | V-SD | 55 – 70 | 15 – 25 | 85 – 120 |
| *Eragrostis curvula*<br>**Weeping lovegrass** | – | 80 – 110 | – | 120 – 160[b] |
| *Eragrostis trichoides*<br>**Sand lovegrass** | – | 150 – 200 | – | 170 – 220[b] |

**TABLE 179 *(continued)* Development of Perennial Forage Grasses**

| Crop | Induction[a] | First growth to heading (days) | Heading to bloom (days) | First growth to maturity (days) |
|---|---|---|---|---|
| *Festuca arundinacea*<br>**Tall fescue** | V-SD | 60 – 75 | 15 – 20 | 95 – 130 |
| *Festuca elatior*<br>**Meadow fescue** | V-SD | 55 – 70 | 20 – 25 | 90 – 120 |
| *Festuca rubra*<br>**Creeping red fescue** | V+-SD | 55 – 70 | 12 – 20 | 80 – 110 |
| *Lolium perenne*<br>**Perennial ryegrass** | OV-SD[e] | 60 – 80 | 20 – 30 | 95 – 135 |
| *Panicum virgatum*<br>**Switchgrass** | O | 60 – 85 | 15 – 35 | 100 – 140 |
| *Paspalum dilatatum*<br>**Dallis grass** | O | 75 – 80 | 15 – 20 | 105 – 120 |
| Phalaris arundinacea<br>**Reed canary grass** | V-SD | 55 – 70 | 15 – 20 | 80 – 105 |
| Phleum pratense<br>**Timothy** | V-SD[f] | 65 – 75 | 20 – 30 | 95 – 140 |
| Poa pratensis<br>**Kentucky bluegrass** | V-SD | 30 – 50 | 15 – 25 | 65 – 85 |
| Setaria anceps<br>**Setaria grass** | – | 60 – 80 | – | 100 – 140[b] |

[a] V = vernalization advantageous or required at least in most cultivars. SD = short day generally required for induction. O = no.

[b] These tropical species usually head and bloom over a long period.

[c] More than one seed crop per year may be harvested in long-season areas.

[d] From emergence of crop grown from seed.

[e] Cultivars differ in requirements for induction.

[f] Most cultivars are induced more fully at low than high temperatures, but some require neither cold nor short days.

*Source:* Christie, B.R. (1987) *Handbook of Plant Science in Agriculture Vol 1*
© CRC Press Inc., Boca Raton, FL. Reprinted with permission.

**TABLE 180  Developmental Pattern of Forage Legumes**

| Crop | | First growth to bloom[a] (days) | Fertilization to mature seed[b] (days) | First growth to harvest (days) |
|---|---|---|---|---|
| *Lespedeza cuneata* | **Sericea lespedeza** | 70 – 100 | 28 – 42 | 120 – 130 |
| *Lotus corniculatus* | **Birdsfoot trefoil** | 40 – 60 | 24 – 47 | 90 – 120 |
| *Medicago sativa* | **Alfalfa** | 45 – 75[c] | 28 – 40 | 110 – 140 |
| *Melilotus alba* | **White sweet clover** | 68 – 80 | 25 – 40 | 105 – 120 |
| *M. officinalis* | **Yellow sweet clover** | 60 – 70 | 18 – 30 | 95 – 110 |
| *Onobrychis viciaefolia* | **Sainfoin** | 50 – 60 | 42 – 49 | 95 – 110 |
| *Trifolium hybridum* | **Alsike clover** | 45 – 70[c] | 21 – 25 | 90 – 120 |
| *T.incarnatum* | **Crimson clover** | 50 – 60 | 26 – 30 | 90 – 115 |
| *T. pratense* | **Red clover** | 50 – 75[c] | 21 – 30 | 95 – 130 |
| *T. repens* | **White clover** | 25 – 40[c] | 20 – 28 | 80 – 110 |
| *T. subterraneum* | **Subterranean clover** | 50 – 90 | 28 – 39 | 95 – 210[d] |

[a]  From first visible spring growth to initial stage of profuse blooms; many species bloom profusely over a long period.

[b]  This refers to developmental time of individual florets, in view of the indeterminate flowering of several of the species.

[c]  The shorter period applies to regrowth after a clipping or grazing usually at bud or early bloom.

[d]  The long growth period refers to early winter planting in an environment such as Melbourne.

*Source:* Christie, B.R. (1987) *Handbook of Plant Science in Agriculture. Vol 1.*
© CRC Press Inc., Boca Raton, FL. Reprinted with permission.

**TABLE 181  Harvested Fraction of High Producing Crop Varieties**

| Crop | Product | % harvested* | Crop | Product | % harvested* |
|---|---|---|---|---|---|
| Afalfa | Hay | 40 – 50 (First year) | Pineapple | Fruit | 20 – 40 |
| | | 80 – 90 (Second Year) | Potato | Tuber | 55 – 65 |
| Bean | Grain | 25 – 35 | Rice | Grain | 40 – 50 |
| Cabbage | Head | 60 – 70 | Sorghum | Grain | 30 – 40 |
| Cotton | Lint | 8 – 12 | Soyabean | Grain | 30 – 40 |
| Groundnut | Grain | 25 – 35 | Sugarbeet | Sugar | 35 – 45 |
| Maize | Grain | 35 – 45 | Sugarcane | Sugar | 20 – 30 |
| Onion | Bulb | 70 – 80 | Tobacco | Leaf | 50 – 60 |
| Pea | Grain | 30 – 40 | Tomato | Fruit | 25 – 35 |
| Pepper | Fruit | 20 – 40 | Wheat | Grain | 35 – 45 |

* %. of total dry matter production

*Source:* FAO (1979) *Yield Response to Water. Irrigation and Drainage Paper No. 33.*
Reproduced with kind permission of the Food and Agriculture Organization of the United Nations.

# Crop
# Tolerances

# Nutrients

---

**TABLE 182   General Description of Mineral Nutrient Deficiencies**

---

*NB* Visual symptoms offer a qualitative approach to defining mineral deficiencies. The following notes summarize common visual symptoms which may occur singly or in combination for a given deficiency with some examples of specific crops. Symptoms vary between species and growth stages, and different parts of the plant may be differently affected. Multiple deficiencies are difficult to recognize. It is noted that certain mineral excesses may induce similar symptoms. Confusion can arise in distinguishing mineral deficiencies from those induced by other factors, e.g. salinity, waterlogging, drought and cold effects.

| Nutrient | Symptoms |
|---|---|
| **Nitrogen** | Chlorosis of whole plant often with reddening. Older leaves generally affected first. Branching reduced. Growth retarded. The most common mineral deficiency. For graminaceous crops drying up of leaves - 'firing' - starts at the bottom of plant and works upwards. Reduced tillering in cereals with reduced ear numbers which are small and poorly filled. Dramatically reduces growth of crops where leaf is the prime product, e.g. brassicas. Stems thin, erect, hard. Leaves smaller than normal. Lower protein content. Flowering reduced in acute deficiency. |
| **Phosphorus** | Dark green foliage, reddening, purpling or bronzing of leaves and petioles (similar to cold effects). Purpling occurs first on under-side of leaves. Scorch may ultimately occur round whole leaf. At first leaves are darker or more blue green than normal. Stunted growth. Deficiency often more easily seen in seedling than in mature plants. Deficiency can markedly affect fruit storage quality, e.g. low temperature breakdown in Bramley apples. Most soils in all parts of world are deficient in phosphorus. Plants have poorly developed root system. Fruiting poor and maturity delayed. Stems thin and shortened; spindly growth. |
| **Potassium** | Older leaves may show yellow, brown, grey or necrotic spots or marginal burn; younger leaves may develop red pigmentation or become interveinally chlorotic and show a shiny surface. Firing of distal leaf margins which later causes death of the distal end of the older leaves. Plants lodge easily. Shrivelled seed or fruit. Terminal or lateral buds may die. Quite common deficiency in many parts of the world. 'Banding' disease - necrotic lesions - in Sisal. Small, dark leaves pointing upwards in citrus - 'dogs ears'. |

| Nutrient | Symptoms |
|---|---|
| **Calcium** | Deficiency not often observed because secondary effects associated with high acidity which limits growth. Growing point dies; stem elongation restricted. In fruit crops disorders of fruit. In leaf crops disorders such as tip burn. Leaves may be cup shaped. Root tips die and root growth restricted. Younger leaves small and roll together in legumes. In grasses distal ends of leaves turn brown, twist and die. Stem structure weakened. Seed set may be reduced. Premature fall of older leaves; also buds and blossoms. Scorched leaf margins in lettuce. 'Cavity spot' in carrot roots. Growing point dies in many root crops and celery or stems and petioles may collapse; celery 'blackheart'. Small depressions on surface of apples - 'Bitter Pit' - with corky spots beneath; develops in store. 'Blossom end rot' of tomato. In cacao chlorosis spreading from leaf edge to midrib in terminal leaves. Unfilled pods - 'pops' - in peanuts. |
| **Magnesium** | Marginal or interveinal chlorosis often quite strongly coloured may be mosaic patterned. Green area of leaf may form 'arrowhead' in woody plants, e.g. tea. Strong reddening may border the chlorotic zone. Symptoms usually occur on older tissue first; with continuing deficiency younger leaves become affected. Chlorosis often on older leaves first as yellowing of leaf between main veins. Leaf withering and fall with prolonged deficiency. Some species show brilliant red patches. Twigs weak and prone to fungus attack; premature leaf drop. Brown to red leaf margins in lettuce and lemon; in lemon spreading into leaf. Brown necrotic margins to terminal leaflets of tomato leaves. Citrus trees take on characteristic bronze appearance as deficiency advances; symptoms more prevalent on older leaves and in autumn; leaves may become fully chlorotic and drop. Leaf yellowing results in parallel banding pattern in cereals; older leaves affected first and die back prematurely. Characteristic 'cigarette burn' necrosis on sugar beet. In oilseed rape yellowing followed by purpling between leaf veins. In potatoes yellowing between leaf veins turning to brown patches characteristic of deficiency in many crops. In cabbage interveinal yellowing then purple discoloration spreading from tips and margins with premature leaf drop. Yellowing between veins in strawberry followed by purpling and ultimately leaf death. Distinct red patches on leaf of apple. In cacao interveinal chlorosis spreading from midrib to edge of leaf. 'Herringbone vein' chlorosis in coffee. 'Sand drown' in tobacco on sandy soils in rainy season. Small chlorotic leaves followed by leaf fall in rubber. Interveinal chlorosis in sweet potato. In cotton lower leaves turn dull red with pale green mid-rib. |

*(continued over)*        A G R I   I N F O

| Nutrient | Symptoms |
| --- | --- |
| **Sulphur** | Chlorosis of the whole plant, often younger leaves affected first. Similar to nitrogen deficiency but whole plant affected more quickly. Plants do not become green when N fertilizers added. Chances of deficiency greater away from industrial centres. Plants may be more rapidly affected than with nitrogen deficiency. Flower production often indeterminate. Stems stiff, woody and small in diameter. In sweet potato paleness and long, thin, spindly vines. Yellow discoloration of young cotton leaves and yield reductions before visual symptoms. Defoliation and dieback in coffee. 'Tea Yellows' in tea. |
| **Copper** | Death of young leaves, chlorosis, failure of fertilization and fruit set. S-shaped growth and fruit gumming in citrus. Young leaves may first become pale to greyish green. Possible marginal necrosis, leaflets dish-shaped with rolled edges, abnormal curvature of leaves and stems. Young shoots die back particularly in fruit trees. Interveinal chlorosis in some species. Growth of internodes depressed. Leaf tips may become white and leaves narrow and twisted. Flowering and fruit formation severely depressed in fruit trees. In grain crops ears often jagged and empty; spiralling of flag leaves and withering of leaf tips – 'Whither Tip'. In oats terminal and new leaves roll at tips, are narrow and become chlorotic, grey spots appear turning yellow white, seed light or shrivelled and ears not fully developed. No chlorisis in alfalfa but terminal petioles show apinastic curvature followed by withering and death of leaflets. No chlorosis in carrots but stunted top growth and restricted root development. In wheat new leaves pale green, lack turgor, and roll and yellow, leaves die to a bleached grey. In apple shoots die back and over time tree assumes a bushy stunted appearance. Similar terminal shoot symptoms in orange but get large dark green leaves on soft, long angular shoots; leaves have irregular contours usually with bowing up of midrib; reddish excrescences from bark and fruits. Dieback of new growth in citrus. In barley awns curled and bent at ear emergence. Irregular shaped ears in wheat; indeterminate tillering; restricted ear production. Onion bulbs soft with thin, pale yellow scales. Yellowing of pinnae at tip of fronds in oil palms. Inhibits fermentation in tea. 'Crookneck' in pineapple is a combined deficiency syndrome of copper and zinc. Copper deficient pineapple has very narrow, waxy and pale leaves. |
| **Zinc** | Deficiency symptoms mostly appear on second and third fully mature leaves from top of plant. Little leaf syndrome particularly- |

| Nutrient | Symptoms |
| --- | --- |

in terminal shoots, rosetting and bunching of leaves,
chlorotic mottle in less severe cases. Light green, yellow or
bleached spots in interveinal areas particularly on older and
lower leaves. Brown rusty spots in some crops. Younger leaves
twisted and necrotic. Visual symptoms may not occur before 50%
yield reduction. Root growth retarded. Dwarfism resulting from
shortened internodes. Uneven crop stand, delayed maturity.
Improperly developed fruits and low yield. Brown rusty appear-
ances to rice fields; starts as small scattered light yellow
spots two or three weeks after transplanting. 'White Bud'
syndrome in maize when deficiency severe; more commonly green
and yellow broad stripes at base of leaves. Older leaves
whither and drop off in beans; small red-brown spots on cotyle-
don leaves. Rosette disease in apples with crinkled narrow
leaves. 'Little leaf' and 'mottle' in citrus, deficiency often
more noticeable on south side of tree; twigs die back. Small
chlorotic leaves in grapes. Tendency to round fruits in avocado.
Bronzing in tung. 'Sickle leaf' in cacao. Reduced leaf width and
twisted leaves in rubber. In wheat longitudinal band of white
and yellow leaf tissue followed by interveinal chlorotic mottling and white to brown necrotic lesions in middle of leaf blade.

**Manganese**  Deficiency usually first seen on new leaves. Interveinal
chlorosis and mottled areas often on younger leaves; but not as
severe as iron deficiency. When severe, necrotic spots or
streaks may form. Often occurs first on middle leaves. Dark
brown or grey specks on upper leaf surface. 'Grey speck' in
oats and similar effect in barley. In some species small dish
shaped leaves. Dark green bands in citrus along midribs and
veins; when severe these turn paler green with areas between
becoming grey or white, foliage becomes thin and affected
branches may die back; symptoms generally more noticeable on
north side of tree and are more pronounced in spring flush of
growth. Sugar beet leaves have upright habit, triangular shape
and margins that curl forwards. Perforating lesions develop in
chlorotic interveinal areas-speckled disease. 'Marsh spot' in
peas (seeds) characterized by slight interveinal chlorosis of
leaves and brown spots internally in peas with cavity in centre;
browning of pod. 'Streak disease' in sugar cane shows as light
green streaks between darker veins. 'Phala blight' in sugar
cane. Grey flecks on lower leaves of strawberries. Chlorosis
often appears in young, tillered cereal plants. Brown mottling
of mature leaves in alfalfa. Uniform interveinal mottling in
tomato. In beetroot foliage becomes intensely red. Onions and

*(continued over)*                                    A G R I   I N F O

| Nutrient | Symptoms |
|----------|----------|
| | corn show narrow striping of yellow; onions pale. In wheat initially pale leaves then rows of interveinal white streaks near mid-point of leaf. In celery strong interveinal yellowing in older leaves near leaf margins. Carrot crops show light green patches in field. Some varieties of potatoes show black-ish-brown spots along veins best seen on leaf undersides. In brassicas fine interveinal mottling - yellow/yellow white on older leaves. Most species of fruit show interveinal yellowing of older leaves. In rubber paling of leaves with bands of green tissue outlining midrib and veins. |
| **Iron** | Distinct interveinal chlorosis which in severe cases may mean yellowing followed by total bleaching of young foliage and necrosis. Occurs first on young leaves; points and margins of leaves keep colour longest. Chlorosis at first interveinal, leaf margins may turn in and die. Reduced tillering. Initially yellow and green stripes along leaf in some species. Growth retarded and in severe cases death of plants. Interveinal chlorosis common in sorghum, maize, citrus, strawberries and can occur in wheat. In citrus single branches or sections of a tree may be affected with great variation between trees in same field; and loss of leaves, die-back of buds/branches/twigs. Complete leaf chlorosis in rice and peaches. Irregular manifestation of problem on field scale with seriously affected areas often immediately adjacent to unaffected areas. Observed in olives as pale hard leaves. 'Herring bone' interveinal yellow or white chlorosis in apple. 'Droopy top' in sugar cane. 'Fishnet' chlorosis in coffee. |
| **Boron** | First symptoms on young part of plant. Death of growing points which may stimulate sprouting of laterals and increase bushy appearance. Axillary buds may burst. Stems shortened and hard. Some species show leaf distortion characteristic of impaired metabolism caused by auxins. Fruit may be distorted or show woody pits or cracking of surface. Petiole cracking in celery and hollowness in some vegetable species. Short internodes in some species. Young leaves much reduced and show wavy margins. Younger leaves may be misshapen, wrinkled, thicker and darker green. Leaves possibly wilting, becoming brittle and chlorotic spotting. Seed formation reduced. Roots appear slimy, thickened and necrotic tips; root growth stunted. Cracked stems, scaly surfaces and formation of external and internal cork. Fruit formation impaired; small, brown flecks, necrosis, cracks, dry rot. In alfalfa death of terminal bud, rosetting, yellow top and sometimes reddening; little flowering and poor seed set. In cotton excessive shedding of squares, young bolls, ruptures at |

| Nutrient | Symptoms |
| --- | --- |

base of bolls, dark internal discoloration of boll, half-opened bolls retain green leaves until frost. 'Crumple top' in cotton. In maize short, bent, barren ears, blank stalks and poor kernel development. In apples death of terminal buds, development of laterals to give bush appearance and cracking, necrosis, puffing and corking of fruit. In grapes terminal buds remain dormant and underdeveloped fruits with some fully developed fruits. Oranges have thickened ring and gum pockets near axis; citrus leaf colour becomes generally yellow-bronze and brittle. Crown and heart rot in beets ('blackheart') with leaves dying in crown and older leaves wilting. In cabbage hollow stem, watery areas yellow buds and stunted. Kale leaves curled and mottled around edges. In turnips terminal bud breaks down; leaves curled, rolled, purplish yellow, scorched, shrivelled and sometimes cracked and internal browning. Spinach leaves small and yellowish with roots dry and dark. Brown curd and white tipped leaves in cauliflower with hollow stem. Mottled, convoluted leaves in oilseed rape. Breaking off of flower heads in sunflower. Reduces pegging in peanuts; hollowed out brown areas in meat of kernel. Malformed fruits in cacao. Terminal shoot death and 'fanning' of growth in coffee. Internal and external 'cork' symptoms in apple. In celery brown cracks along ridges of outer surface of petiole.

**Molybdenum**    Paleness in legumes. Can appear similar to nitrogen deficiency as it is involved in N metabolism. Mottled, pale appearance in non legumes. Legumes bear many ineffective and small nodules so nitrogen fixation severely affected. Chlorosis syndrome sometimes interveinal. Leaf margins tend to curl and roll. Limp leaves typical. Infolding of leaves. Restricted growth. Narrow leaves. Marginal burn of mature leaves in rock melon, maize, sunflower. 'Whiptail' in cauliflower with twisted and elongated laminae, corrugated and cupped; in extreme cases lamina may not form. Pale green leaves in alfalfa, retarded growth, older leaves scorch and die and drop, occurs unevenly in the field. 'Yellow spot' in citrus shows as water-soaked areas on leaves turning to interveinal chlorotic spots with sometimes brown gum forming on under side.

**Chlorine**    Wilting of leaflet tips, chlorosis of leaves followed by bronzing and drying.

*Sources:* Reuter, D.J. & Robinson, J.B. (1986) *Plant Analysis. An Interpretation Manual.* Phosyn Chemicals commercial literature. FAO (1983) *Micronutrients. Fertilizer and Plant Nutrition Bulletin No.7.* de Geus, J.G. (1973) *Fertilizer Guide for the Tropics and Sub-Tropics.* Lorenz, O.A. & Maynard, D.N. (1980) *Knott's Handbook for Vegetable Growers.* Whiteman, P.C. (1980) *Tropical Pasture Science.* FAO (1984) *Fertilizer and Plant Nutrition Guide. Fertilizer and Plant Nutrition Bulletin No.9* . Ministry of Agriculture, Fisheries and Food, UK. (1983) *Trace Element Deficiencies in Field Crops.* Booklet 2197.

**TABLES 183 and 184** are given to indicate crop response to micronutrients under deficiency situations and relative sensitivity to micronutrient deficiencies. Some inconsistencies will be noted in the latter where some crops are listed under two or three categories because of variation in soils, crop varieties or growing conditions; and different sources of information.

---

**TABLE 183    Response of Crops to Micronutrients under Deficiency Situations**

Response: L = low; M = medium; H = high

| Crop | Zn | Fe | Mn | Mo | Cu | B |
|---|---|---|---|---|---|---|
| Alfalfa | L | M | M | M | H | H |
| Apples | H | | H | L | M | H |
| Asparagus | L | M | L | L | L | L |
| Barley | M | H | M | L | H | L |
| Beans | H | H | H | L | L | L |
| Beetroot | M | H | H | M | H | H |
| Blueberries | | H | L | L | M | L |
| Broccoli | | H | M | M | M | M |
| Cabbage | | M | M | M | M | M |
| Carrots | L | | M | L | H | M |
| Cauliflower | | H | M | H | M | H |
| Celery | | | H | L | M | H |
| Clover | M | | M | H | M | M |
| Cucumber | | | H | | M | L |
| Corn | H | M | L | L | M | L |
| Cotton | H | | | L | M | M |
| Grapefruit | H | H | H | M | H | L |
| Grapes | L | H | H | L | | M |
| Lettuce | | | H | H | H | M |
| Oats | L | M | H | M | H | L |
| Onion | H | | H | H | H | H |
| Orange | H | H | H | M | H | L |
| Pea | L | | H | M | L | L |
| Peaches | H | | H | L | M | M |
| Pears | M | | | L | M | M |
| Peppermint | L | L | M | L | L | L |
| Potato | M | | M | L | L | L |
| Radish | | | H | M | M | M |
| Raspberries | | H | H | L | | M |
| Rye | L | | L | L | L | L |
| Soyabean | M | H | H | M | L | L |
| Sorghum | H | H | H | L | M | L |
| Spearmint | L | | M | L | L | L |
| Spinach | | H | H | H | H | M |
| Strawberry | | H | H | | M | M |
| Sudan grass | H | H | H | L | H | L |
| Sugar beet | M | H | H | M | M | H |
| Sweet Corn | H | M | M | L | M | L |
| Rice | M | H | M | L | L | L |
| Tomato | M | H | M | M | M | M |
| Turnip | | | M | M | M | H |
| Wheat | L | L | H | L | H | L |

*Sources: Principally* Mortvedt, J.J. *et al.* (1972) *Micronutrients in Agriculture.* Reproduced with kind permission of the Soil Science Society of America, Inc.; also Lorenz, D.A. & Maynard, D.M. (1980) *Knott's Handbook for Vegetable Growers.*

**TABLE 184    Relative Sensitivity of Crops to Micronutrient Deficiencies**

|  | SENSITIVITY |  |
| --- | --- | --- |
| **Sensitive** | **Moderate** | **Tolerant** |

| | ZINC | |
| --- | --- | --- |
| Citrus | Cotton | Peas |
| Deciduous tree fruits | Potatoes | Small grains |
| Pecans | Tomatoes | Peppermint |
| Grapes | Alfalfa | Asparagus |
| Beans (Phaseolus) | Clovers (Trifolium) | Crucifers |
| Soya beans | Sorghum | Forage grasses |
| Hops | Sudan Grass | Safflower |
| Maize | Sugar beets | Carrots |
| Lima Beans | Peanut | |
| Flax | Mango | |
| Castor Beans | Coffee | |
| Onions | Tea | |
| Pecan | Banana | |
| Tung | Pineapple | |
| Cacao | Wheat | |
| Coffee | Barley | |
| Avocado | | |
| Rice | | |
| Sorghum | | |

| | IRON | |
| --- | --- | --- |
| Berries | Alfalfa | Alfalfa |
| Citrus | Barley | Barley |
| Field beans | Maize | Maize |
| Flax | Cotton | Cotton |
| Sorghum | Field beans | Flax |
| Grapes | Flax | Grasses |
| Mint | Grain Sorghum | Millet |
| Groundnuts | Grasses | Oats |
| Soya beans | Oats | Potatoes |
| Sudan Grass | Rice | Rice |
| Tree fruits | Soya beans | Soya beans |
| Vegetables | Tree fruits | Sugar beets |
| Walnuts | Vegetables | Vegetables |
| Coffee | Wheat | Wheat |
| Sugar Cane | Banana | |
| Coconut | Pineapple | |
| Potato | | |

*(continued over)*

TABLE 184 *(continued)*  **Relative Sensitivity of Crops to Micronutrient Deficiencies**

### SENSITIVITY

| Sensitive | Moderate | Tolerant |
|-----------|----------|----------|

**MANGANESE**

| Sensitive | Moderate | Tolerant |
|-----------|----------|----------|
| Beans | Alfalfa | Asparagus |
| Cucumber | Barley | Maize |
| Lettuce | Cabbage | Rye |
| Oats | Cauliflower | Blueberries |
| Peas | Clover | |
| Radish | Peppermint | |
| Soyabean | Potato | |
| Sorghum | Sweet Corn | |
| Spinach | Rice | |
| Sugar beet | Tomato | |
| Wheat | Turnips | |
| Grapefruit | Oil Palm | |
| Orange | Rubber | |
| Apples | | |
| Peaches | | |
| Grapes | | |
| Raspberry | | |
| Strawberry | | |
| Coconut | | |

**COPPER**

| Sensitive | Moderate | Tolerant |
|-----------|----------|----------|
| Alfalfa | Broccoli | Asparagus |
| Barley | Cabbage | Beans |
| Carrots | Cauliflower | Peas |
| Lettuce | Celery | Peppermint |
| Oats | Clover | Potato |
| Spinach | Cucumber | Rye |
| Sudan grass | Maize | Spearmint |
| Beetroot | Cotton | Soyabean |
| Wheat | Radish | Rice |
| Grapefruit | Sorghum | |
| Orange | Sugar beet | |
| | Sweet Corn | |
| | Tomato | |
| | Turnip | |
| | Apple | |
| | Peaches | |
| | Pears | |
| | Blueberries | |
| | Strawberries | |
| | Tea | |
| | Pineapple | |
| | Banana | |
| | Oil Palm | |
| | Potato | |

**TABLE 184** *(continued)*   **Relative Sensitivity of Crops to Micronutrient Deficiencies**

| | SENSITIVITY | |
| Sensitive | Moderate | Tolerant |
| --- | --- | --- |
| | **BORON** | |
| Alfalfa | Cabbage | Beans |
| Cauliflower | Carrot | Barley |
| Celery | Cotton | Maize |
| Sugar beet | Lettuce | Oats |
| Beetroot | Radish | Peas |
| Turnip | Spinach | Soyabean |
| Apple | Peaches | Potato |
| Sisal | Pears | Rice |
| Coffee | Grapes | Wheat |
| Asparagus | Pecan | Grapefruit |
| Cruciferae | Sweet Potato | Sorghum |
| Mangold | Tobacco | Citrus |
| Sunflower | Tomato | Clover (forage) |
| Groundnuts | Oil Palm | Forage grasses |
| | Clover (seed) | Strawberry |
| | Peanut | |
| | Maize | |
| | Potato | |
| | Bananas | |
| | Coconuts | |
| | **MOLYBDENUM** | |
| Cauliflower | Alfalfa | Barley |
| Clover | Cabbage | Beans |
| Lettuce | Oats | Carrots |
| Spinach | Peas | Celery |
| Pasture legumes | Soya Beans | Maize |
| Soya Bean | Radish | Cotton |
| | Sugar beet | Potato |
| | Beetroot | Sorghum |
| | Tomato | Rice |
| | Turnips | Wheat |
| | Citrus | Apples |
| | Sisal | Peaches |
| | Wheat | Grapes |
| | Field Beans | |
| | Maize | |
| | Groundnuts | |
| | Chick Peas | |
| | Pigeon Peas | |
| | **COBALT** | |

Legumes are sometimes sensitive

*Sources:*  FAO (1983) *Micronutrients. Fertilizer and Plant Nutrition Bulletin No. 7.*
Jones, V.S. (1979) *Fertilizers and Soil Fertility.* Whiteman, P.C. (1980) *Tropical Pasture Science.* de Geus, J.G. (1973) *Fertilizer Guide for the Tropics and Sub Tropics.*
Shorrocks, V.M. (1984) *Micronutrient Deficiencies and Their Correction.*

**TABLE 185  Indicator and Susceptible Crop Species for Nutrient Deficiencies**

| Deficient element | Indicator plants |
|---|---|
| **Nitrogen** | Cereals, mustard, apple, citrus, grass |
| **Phosphorus** | Maize, barley, lettuce, tomato |
| **Potassium** | Potato, clover, lucerne, bean, tobacco, cucurbits, cotton, tomato, maize |
| **Calcium** | Lucerne, other legumes |
| **Magnesium** | Potato, cauliflower, sugar beet |
| **Sulphur** | Lucerne, clover, rapeseed |
| **Iron** | Sorghum, barley, citrus, peach, cauliflower |
| **Zinc** | Maize, onion, citrus, peach |
| **Copper** | Apple, citrus, barley, maize, lettuce, oats, onion, tobacco, tomato, sugar beet, carrots |
| **Manganese** | Apple, apricot, bean, cherry, citrus, cereals, pea, radish |
| **Boron** | Lucerne, turnip, cauliflower, apple, peach, sugar beet, swedes, mangels, kale, celery |
| **Molybdenum** | Cauliflower, other brassica spp., citrus, legumes, oats, spinach |
| **Chlorine** | Lettuce |

*Main Source:* FAO (1984) *Fertilizer and Plant Nutrition Guide, Fertilizer and Plant Protection. Bulletin No. 9;* reproduced with kind permission of the Food and Agriculture Organization of the United Nations; also Ministry of Agriculture Fisheries and Food UK (1978) *Lime and Fertilizer Recommendations. No. 1. Arable Crops and Grassland* and Ministry of Agriculture Fisheries and Food UK (1978) *Fertiliser Recommendations.*

---

### NOTE 9  Plant Tissue Analysis

---

**Use:** Mineral analysis of plant tissue can provide guidance to the status of the mineral nutrition of a crop. It is particularly helpful if used in conjunction with soil analyses, and irrigation water analyses if the crop is irrigated. Average concentration of elements in plants are:

| Nutrients | ppm in dry matter | Nutrients | % in dry matter |
|-----------|-------------------|-----------|-----------------|
| Mo | 0.3 | S | 0.2 |
| Co | 0.3 | P | 0.25 |
| Cu | 10 | Mg | 0.2 |
| Mn | 50 | Ca | 0.5 |
| Zn | 75 | K | 1.0 |
| Fe | 100 | N | 2.0 |
| Cl | 100 | O | 45.0 |
| Si | 100 | C | 45.0 |
| B | 75 | H | 6.0 |
| Na | 350 | | |

*Source:* Blair, G.J. *Plant Nutrition: The need for balance.* University of New England. Armidale, Australia. Reproduced with kind permission of the Australian Centre for International Agricultural Research

### Interpretation:

Tissue analysis can vary widely according to season of growth, environmental factors, maturity of plant part part-sampled and actual plant part-sampled. Analytical technique can also affect values obtained. For general purposes the following sampling procedure may be adopted:

**vegetable and arable crops:**    whole leaves 3rd, 4th and 5th from the growing point;

**fruit trees and bushes:**    three leaves from the middle length of a new non-fruiting shoot in each quarter of the tree or bush;

**grasses and cereals - small:**    complete plant cut just above ground level;

**grasses and cereals - large:**    four young mature leaves from each sampled plant.

*(continued over)*                                                    A G R I   I N F O

In all cases as many plants as possible should be sampled; growth stage and plant parts accurately recorded to aid future interpretation of analyses. Consistency of sampling technique in future sampling should be ensured to enable fair comparison of sequential analyses. Indicative foliar rating for micronutrients are:

## (concentration in recently matured leaves-ppm)

|       | Deficient | Adequate   | Excess or Toxic |
|-------|-----------|------------|-----------------|
| **B**  | <15       | 20 – 100   | >200            |
| **Cu** | <4        | 5 – 20     | >20             |
| **Fe** | <50       | 50 – 250   | ?               |
| **Mn** | <20       | 20 – 500   | >500            |
| **Mo** | <0.1      | 0.5 – ?    | ?               |
| **Zn** | <20       | 25 – 150   | >400            |

*Source:* Mortvedt, J.J. *et al.* (1972) *Micronutrients in Agriculture.*
Reproduced with kind permission of the Soil Science Society of America.

**TABLE 186** gives guidelines for the interpretation of temperate crop tissue analyses for a range of crops in UK based on very specific sampling by time and plant part. It is stressed that these are guidelines only and represent experience in UK. Below these levels growth rate, yield or quality would be expected to decline significantly.

**TABLE 187** is an attempt to give similar guideline levels for tropical and sub-tropical crops. These have been compiled mainly from extensive referenced data from many countries presented by Reuter and Robinson who give values for deficient, marginal, critical, adequate, high and toxic levels. As would be expected values vary according to source and plant part. Critical values are often not available or are for varying yield reduction levels. In order to compile a table for tropical and sub-tropical crops in similar vein to **TABLE 186** lower level 'adequate' figures have been collated and averaged from Reuter and Robinson to provide guidelines for these crops without concern for standard sampling between individual datum. Where 'adequate' values were not available, 'critical' values have been used. Only values for vegetative parts were considered, values for seed or fruits being ignored. Similar criteria were applied to the few data from other sources. BECAUSE OF THE UNAVOIDABLE OVERVIEW NATURE OF **TABLE 187** THE READER IS CAUTIONED TO USE IT ONLY FOR INITIAL GUIDELINE PURPOSES AND NOT DEFINITE ASSESSMENT. Even when using the table in this way the reader should consider the origin and sampling of any tissue analyses under review, possible nutrient interactive effects, varietal effects etc. and wherever possible seek out more definitive local data. Reference to the extensive and more detailed data presented in Reuter and Robinson and others is always recommended. Any local data will always be particularly relevant. The reader's attention is also drawn to mineral analyses to be found in the livestock feedstuff analyses tables in this volume which may have relevance in circumstances where no other analyses are available to the crop nutritionist.

**TABLE 186 Guideline Levels for Leaf Analysis in the UK**

| Crop | Ca % | Mg % | Mn ppm | B ppm | Cu ppm | Mo ppm | Fe ppm | ZN ppm | S % | N % | P % | K % | Recommended Leaf Sampling Position (Optimum timing in brackets) |
|---|---|---|---|---|---|---|---|---|---|---|---|---|---|
| Amenity Grass | 0.30 | 0.12 | 30 | 5 | 7.0 | - | 50 | 20 | 0.2 | 3.0 | 0.26 | 1.80 | Clean, uncontaminated samples (May) |
| Amenity Tree | 1.00 | 0.30 | 30 | 20 | 5.0 | - | 150 | 25 | 0.1 | 1.6 | 0.13 | 0.70 | Mid-third extension growth (mid-August) |
| Apple (Cox) | 1.50 | 0.20 | 30 | 20 | 5.0 | - | 150 | 25 | 0.1 | 2.6 | 0.20 | 1.20 | Mid-third extension growth (mid-August) |
| Apple (Bramley) | 1.50 | 0.20 | 30 | 20 | 5.0 | - | 150 | 25 | 0.1 | 2.4 | 0.18 | 1.20 | Mid-third extension growth (mid-August) |
| Apple | 1.50 | 0.20 | 30 | 20 | 5.0 | - | 150 | 25 | 0.1 | 2.5 | 0.20 | 1.20 | Mid-third extension growth (mid-August) |
| Artichoke | 1.00 | 0.15 | 30 | 20 | 5.0 | 1.0 | 150 | 20 | 0.2 | 1.5 | 0.30 | 2.50 | Youngest fully expanded leaves (June-August) |
| Asparagus | 0.40 | 0.15 | 25 | 40 | 5.0 | 0.3 | 30 | 20 | 0.1 | 3.0 | 0.26 | 1.40 | Take 25 samples from top 12" of Fern (Sept.) |
| Aubergine | 1.50 | 0.20 | 35 | 35 | 4.0 | 0.3 | 200 | 15 | 0.1 | 2.0 | 0.30 | 2.00 | Youngest fully expanded leaves (Pre-fruit) |
| Barley | 0.30 | 0.12 | 30 | 5 | 7.0 | - | 50 | 20 | 0.2 | 3.0 | 0.26 | 1.80 | 2-4 blades from top of plant (April-May) |
| Beans | 0.40 | 0.25 | 35 | 20 | 5.0 | 1.0 | 30 | 15 | 0.2 | 3.0 | 0.16 | 1.25 | Uppermost mature blades (early flowering) |
| Blackcurrant | 1.30 | 0.15 | 30 | 20 | 5.0 | - | 150 | 25 | 0.1 | 2.8 | 0.25 | 1.50 | Fully expanded extension growth (Pre-harvest) |
| Blueberries | 1.30 | 0.15 | 30 | 20 | 5.0 | - | 150 | 25 | 0.1 | 2.8 | 0.25 | 1.50 | Fully expanded extension growth (Pre-harvest) |
| Borage | 1.50 | 0.20 | 40 | 25 | 7.0 | 1.5 | 50 | 30 | 0.2 | 3.0 | 0.30 | 2.50 | Youngest fully expanded leaves (Pre-flower) |
| Broad Beans | 0.40 | 0.25 | 35 | 20 | 5.0 | 1.0 | 30 | 15 | 0.2 | 3.0 | 0.16 | 1.25 | Uppermost mature blades (early flowering) |
| Broccoli | 1.00 | 0.20 | 30 | 25 | 4.0 | 2.0 | 30 | 15 | 0.25 | 3.5 | 0.35 | 3.00 | Youngest fully expanded leaves (Pre-heading) |
| Brussels Sprouts | 1.00 | 0.20 | 30 | 25 | 4.0 | 2.0 | 30 | 15 | 0.25 | 3.5 | 0.35 | 3.00 | Youngest fully expanded leaves (July) |
| Bulbs | 1.00 | 0.15 | 30 | 20 | 5.0 | 1.0 | 150 | 20 | 0.2 | 1.5 | 0.30 | 1.25 | Youngest fully expanded leaves (Pre-flower) |
| Cabbage | 1.00 | 0.20 | 30 | 25 | 4.0 | 2.0 | 30 | 15 | 0.2 | 3.5 | 0.35 | 3.00 | Youngest fully expanded leaves (Pre-heading) |
| Calabrese | 1.00 | 0.20 | 30 | 25 | 4.0 | 2.0 | 30 | 15 | 0.2 | 3.5 | 0.35 | 3.00 | Youngest fully expanded leaves (Pre-heading) |
| Carnation | 1.00 | 0.15 | 30 | 20 | 5.0 | 1.0 | 150 | 20 | 0.2 | 1.5 | 0.30 | 1.25 | Youngest fully expanded leaves (Pre-flower) |
| Carrots | 1.50 | 0.30 | 40 | 25 | 5.0 | 0.5 | 30 | 15 | 0.25 | 2.0 | 0.26 | 2.50 | Mature leaf cut off at crown (Pre-bulking) |
| Cauliflower | 1.00 | 0.20 | 30 | 25 | 4.0 | 2.0 | 30 | 15 | 0.2 | 3.5 | 0.35 | 3.00 | Youngest fully expanded leaves (Pre-heading) |
| Celery | 1.00 | 0.20 | 20 | 15 | 5.0 | 0.3 | 75 | 15 | 0.1 | 2.0 | 0.30 | 1.40 | Youngest fully elongated leaf (10-15" tall) |
| Cereals | 0.30 | 0.12 | 30 | 5 | 7.0 | - | 50 | 20 | 0.2 | 3.0 | 0.26 | 1.80 | 2-4 blades from top of plant (April-May) |
| Cherry | 1.50 | 0.25 | 30 | 20 | 5.0 | - | 150 | 25 | 0.1 | 2.4 | 0.20 | 1.50 | Mid-third extension growth (mid-August) |
| Chick Peas | 0.40 | 0.25 | 35 | 20 | 5.0 | 2.0 | 30 | 15 | 0.1 | 3.0 | 0.16 | 1.25 | Uppermost mature blades (Pre-flowering) |
| Chicory | 0.40 | 0.15 | 25 | 40 | 5.0 | 0.3 | 30 | 20 | 0.1 | 3.0 | 0.26 | 1.40 | Youngest fully elongated leaf (6-10" tall) |
| Chinese Cabbage | 1.00 | 0.30 | 30 | 20 | 4.0 | 2.0 | 50 | 20 | 0.2 | 3.0 | 0.40 | 3.00 | Youngest fully expanded leaves (Pre-heading) |
| Chives | 0.40 | 0.15 | 25 | 40 | 5.0 | 0.3 | 30 | 20 | 0.1 | 3.0 | 0.26 | 1.40 | Youngest fully expanded leaves (Pre-flower) |
| Chrysanthemum | 1.00 | 0.15 | 30 | 20 | 5.0 | 1.0 | 150 | 20 | 0.1 | 1.5 | 0.30 | 1.25 | Youngest fully expanded leaves (Pre-flower) |
| Clover | 0.30 | 0.12 | 30 | 5 | 7.0 | - | 50 | 20 | 0.2 | 3.0 | 0.26 | 1.80 | Clean, uncontaminated samples (May) |
| Courgette | 1.00 | 0.30 | 20 | 20 | 4.0 | 0.3 | 60 | 25 | 0.2 | 2.5 | 0.25 | 2.00 | Youngest fully expanded leaves (Pre-fruit) |
| Cress | 1.00 | 0.20 | 30 | 25 | 4.0 | 2.0 | 30 | 15 | 0.2 | 3.5 | 0.35 | 3.00 | Clean, uncontaminated samples (May) |
| Cucumber | 1.00 | 0.30 | 20 | 20 | 4.0 | 0.3 | 60 | 25 | 0.2 | 2.5 | 0.25 | 2.00 | Youngest fully expanded leaves (Pre-fruit) |
| Durum | 0.30 | 0.12 | 30 | 5 | 7.0 | - | 50 | 20 | 0.2 | 3.0 | 0.26 | 1.80 | 2-4 blades from top of plant (April-May) |
| Evening Primrose | 1.00 | 0.20 | 30 | 25 | 4.0 | 1.0 | 50 | 20 | 0.2 | 3.0 | 0.30 | 3.00 | Youngest fully expanded leaves (Pre-flower) |

*(continued over)*

**TABLE 186** *(continued)* **Guideline Levels for Leaf Analysis in the UK**

| Crop | Ca % | Mg % | Mn ppm | B ppm | Cu ppm | Mo ppm | Fe ppm | ZN ppm | S % | N % | P % | K % | Recommended Leaf Sampling Position (Optimum timing in brackets) |
|---|---|---|---|---|---|---|---|---|---|---|---|---|---|
| Field Bean | 0.40 | 0.25 | 35 | 20 | 5.0 | 1.0 | 30 | 15 | 0.2 | 3.0 | 0.16 | 1.25 | Uppermost mature blades (Early flowering) |
| Fir Trees | 0.30 | 0.10 | 20 | 10 | 3.0 | | 50 | 15 | 0.1 | 1.6 | 0.13 | 0.70 | Mid-third extension growth (mid-August) |
| Fodder Beet | 1.00 | 0.20 | 30 | 30 | 5.0 | 0.5 | 30 | 15 | 0.2 | 2.0 | 0.22 | 1.25 | Youngest fully expanded leaves (June-August) |
| Gherkin | 1.50 | 0.20 | 35 | 35 | 4.0 | 0.3 | 200 | 15 | 0.1 | 2.0 | 0.30 | 2.00 | Youngest fully expanded leaves (Pre-fruit) |
| Gooseberry | 1.30 | 0.20 | 30 | 20 | 5.0 | | 150 | 25 | 0.1 | 2.5 | 0.25 | 1.50 | Fully expanded extension growth (Pre-harvest) |
| Grass (Seed) | 0.30 | 0.12 | 30 | 5 | 7.0 | | 50 | 20 | 0.2 | 3.0 | 0.26 | 1.80 | Clean, uncontaminated samples (May) |
| Green Bean | 0.40 | 0.25 | 35 | 20 | 5.0 | 1.0 | 30 | 15 | 0.2 | 3.0 | 0.16 | 1.25 | Uppermost mature blades (Early flowering) |
| Herbs | 1.00 | 0.15 | 30 | 20 | 5.0 | 1.0 | 150 | 20 | 0.2 | 1.5 | 0.30 | 1.25 | Youngest fully expanded leaves (Pre-flower) |
| Hops | 1.30 | 0.30 | 40 | | 5.0 | 0.5 | 75 | 25 | 0.2 | 2.0 | 0.30 | 1.50 | Leaves adjacent to clusters (Post-bloom) |
| Kale | 1.00 | 0.20 | 30 | 25 | 4.0 | 2.0 | 30 | 15 | 0.25 | 3.5 | 0.35 | 3.00 | Youngest fully elongated leaves (Aug-Sept) |
| Kohlrabi | 1.00 | 0.20 | 30 | 25 | 4.0 | 2.0 | 30 | 15 | 0.2 | 3.5 | 0.35 | 3.00 | Youngest fully expanded leaves (Pre-heading) |
| Kiwi Fruit | 1.30 | 0.20 | 30 | 20 | 5.0 | | 150 | 25 | 0.1 | 2.5 | 0.25 | 1.50 | Mid-third extension growth (mid-August) |
| Leek | 1.00 | 0.20 | 20 | 25 | 4.0 | 0.5 | 30 | 20 | 0.1 | 2.0 | 0.30 | 1.40 | Youngest fully elongated leaves (Aug-Sept) |
| Lentil | 0.40 | 0.25 | 35 | 20 | 5.0 | 1.0 | 30 | 15 | 0.2 | 3.0 | 0.16 | 1.25 | Uppermost mature blades (Early flowering) |
| Lettuce | 1.00 | 0.30 | 30 | 15 | 7.0 | 0.5 | 75 | 25 | 0.2 | 3.0 | 0.40 | 3.50 | Youngest wrapper leaves (Pre-heading) |
| Linseed | 1.00 | 0.20 | 25 | 25 | 7.0 | 1.5 | 50 | 20 | 0.2 | 3.0 | 0.30 | 3.00 | Youngest fully expanded leaves (Pre-flower) |
| Loganberry | 1.30 | 0.20 | 30 | 20 | 5.0 | | 150 | 25 | 0.1 | 2.5 | 0.25 | 1.50 | Fully expanded non-fruiting canes (Fruiting) |
| Lucerne | 1.00 | 0.20 | 25 | 20 | 7.0 | 0.6 | 50 | 15 | 0.1 | 4.0 | 0.15 | 1.40 | Mature leaves one-third down (Early flower) |
| Lupins | 1.00 | 0.15 | 40 | 15 | 5.0 | 1.5 | 75 | 20 | 0.1 | 3.0 | 0.16 | 1.25 | Youngest fully expanded leaves (Pre-flower) |
| Maize | 0.60 | 0.15 | 35 | 5 | 7.0 | | 75 | 25 | 0.1 | 2.0 | 0.20 | 1.80 | Flag leaf (at full tassling) |
| Melon | 1.50 | 0.20 | 35 | 35 | 4.0 | 0.3 | 200 | 15 | 0.1 | 2.0 | 0.30 | 2.00 | Youngest fully expanded leaves (Pre-fruit) |
| Mustard | 1.00 | 0.20 | 30 | 25 | 4.0 | 2.0 | 30 | 15 | 0.25 | 3.5 | 0.35 | 3.00 | Youngest fully expanded leaves (Pre-flower) |
| Oats | 0.30 | 0.12 | 30 | 5 | 7.0 | | 50 | 20 | 0.2 | 3.0 | 0.26 | 1.80 | 2-4 blades from top of plant (April-May) |
| Oilseed Rape | 1.00 | 0.20 | 40 | 30 | 4.0 | 2.0 | 30 | 15 | 0.25 | 3.5 | 0.35 | 3.00 | Youngest fully expanded leaves (Pre-flower) |
| Onions | 1.00 | 0.20 | 30 | 25 | 7.0 | 0.5 | 30 | 25 | 0.2 | 2.0 | 0.30 | 1.40 | Mature leaf cut off at crown (Pre-bulking) |
| Ornamental Trees | 1.00 | 0.30 | 30 | 20 | 5.0 | | 150 | 25 | 0.1 | 3.0 | 0.26 | 1.80 | Mid-third extension growth (mid-August) |
| Parsley | 1.00 | 0.15 | 30 | 20 | 5.0 | 1.0 | 150 | 20 | 0.2 | 1.5 | 0.30 | 1.25 | Clean, uncontaminated samples (May) |
| Parsnip | 1.40 | 0.20 | 30 | 25 | 5.0 | 0.5 | 30 | 15 | 0.2 | 2.0 | 0.26 | 2.50 | Youngest fully expanded leaves (June-July) |
| Peaches | 1.50 | 0.20 | 30 | 20 | 5.0 | | 150 | 20 | 0.1 | 2.5 | 0.15 | 2.00 | Mid-third extension growth (mid-August) |
| Pears (Comice) | 1.30 | 0.20 | 30 | 20 | 5.0 | | 150 | 25 | 0.1 | 1.8 | 0.15 | 1.20 | Mid-third extension growth (mid-August) |
| Pears (Conference) | 1.30 | 0.20 | 30 | 20 | 5.0 | | 150 | 25 | 0.1 | 2.1 | 0.15 | 1.20 | Mid-third extension growth (mid-August) |
| Peas | 0.40 | 0.25 | 35 | 20 | 5.0 | 2.0 | 30 | 15 | 0.2 | 3.0 | 0.16 | 1.25 | Uppermost mature blades (Pre-flowering) |
| Pepper | 1.30 | 0.20 | 30 | 20 | 5.0 | | 150 | 25 | 0.1 | 2.5 | 0.25 | 1.50 | Youngest fully expanded leaves (Pre-fruit) |
| Plum | 1.30 | 0.20 | 30 | 20 | 5.0 | | 150 | 25 | 0.1 | 2.0 | 0.15 | 1.50 | Mid-third extension growth (mid-August) |
| Potatoes | 1.00 | 0.25 | 40 | 15 | 4.0 | | 30 | 15 | 0.1 | 3.0 | 0.35 | 2.50 | 4th leaf from main stalk tip (Jun-Aug) |
| Radish | 1.00 | 0.20 | 30 | 25 | 4.0 | 2.0 | 30 | 15 | 0.25 | 3.5 | 0.35 | 3.00 | Mature leaf cut off at crown (Pre-bulking) |

**TABLE 186 (continued) Guideline Levels for Leaf Analysis in the UK**

| Crop | Ca % | Mg % | Mn ppm | B ppm | Cu ppm | Mo ppm | Fe ppm | ZN ppm | S % | N % | P % | K % | Recommended Leaf Sampling Position (Optimum timing in brackets) |
|---|---|---|---|---|---|---|---|---|---|---|---|---|---|
| Raspberries | 1.30 | 0.30 | 30 | 20 | 5.0 | - | 150 | 25 | 0.1 | 2.4 | 0.20 | 1.50 | Fully expanded non-fruiting canes (Fruiting) |
| Red Beet | 1.00 | 0.20 | 30 | 30 | 5.0 | 0.5 | 30 | 15 | 0.2 | 3.5 | 0.22 | 1.25 | Mature leaf cut off at crown (Pre-bulking) |
| Redcurrant | 1.30 | 0.20 | 30 | 20 | 5.0 | - | 150 | 25 | 0.1 | 2.5 | 0.25 | 1.50 | Fully expanded extension growth (Pre-harvest) |
| Roses | 1.00 | 0.15 | 30 | 20 | 5.0 | 1.0 | 150 | 20 | 0.2 | 1.5 | 0.30 | 1.25 | Youngest fully expanded leaves (Pre-flower) |
| Rye | 0.30 | 0.12 | 30 | 5 | 7.0 | - | 50 | 20 | 0.2 | 3.0 | 0.26 | 1.80 | 2-4 blades from top of plant (April-May) |
| Shallots | 1.00 | 0.20 | 30 | 25 | 7.0 | 0.5 | 30 | 25 | 0.2 | 2.0 | 0.30 | 1.40 | Mature leaf cut off at crown (Pre-bulking) |
| Soya Bean | 0.40 | 0.25 | 35 | 20 | 5.0 | 1.0 | 30 | 15 | 0.2 | 3.0 | 0.16 | 1.25 | Uppermost mature blades (Early flowering) |
| Spinach | 0.60 | 0.35 | 40 | 30 | 7.0 | 0.3 | 30 | 20 | 0.1 | 2.0 | 0.30 | 1.40 | Mature leaf cut off at stalk (May) |
| Strawberries | 1.30 | 0.15 | 30 | 20 | 5.0 | - | 150 | 25 | 0.1 | 2.6 | 0.25 | 1.50 | Lamina of recently matured leaves (Fruiting) |
| Sugar Beet | 1.25 | 0.30 | 40 | 30 | 7.0 | 0.5 | 30 | 15 | 0.2 | 3.5 | 0.22 | 1.25 | Mature leaf cut off at crown (Pre-bulking) |
| Sunflower | 0.80 | 0.20 | 30 | 40 | 5.0 | - | 50 | 25 | 0.2 | 3.0 | 0.25 | 3.00 | Youngest fully expanded leaves (Pre-flower) |
| Swedes | 1.00 | 0.20 | 30 | 25 | 4.0 | 2.0 | 30 | 15 | 0.2 | 3.0 | 0.35 | 3.00 | Mature leaf cut off at crown (Pre-bulking) |
| Tickbean | 0.40 | 0.25 | 35 | 20 | 5.0 | 1.0 | 30 | 15 | 0.2 | 3.0 | 0.16 | 1.25 | Uppermost mature blades (Early flowering) |
| Tomato | 1.00 | 0.35 | 20 | 20 | 4.0 | 0.3 | 60 | 20 | 0.1 | 2.0 | 0.30 | 2.00 | Youngest fully expanded leaves (Pre-fruit) |
| Top Fruit | 1.50 | 0.20 | 30 | 20 | 5.0 | - | 150 | 25 | 0.1 | 2.5 | 0.20 | 1.20 | Mid-third extension growth (mid-August) |
| Triticale | 0.30 | 0.12 | 30 | 5 | 7.0 | - | 50 | 20 | 0.2 | 3.0 | 0.26 | 1.80 | 2-4 blades from top of plant (April-May) |
| Turnips | 1.00 | 0.20 | 30 | 25 | 4.0 | 2.0 | 30 | 15 | 0.2 | 3.5 | 0.35 | 3.00 | Mature leaf cut off at crown (Pre-bulking) |
| Vines | 1.30 | 0.20 | 40 | 30 | 5.0 | 0.5 | 75 | 25 | 0.2 | 2.5 | 0.20 | 1.75 | Leaves adjacent to clusters (Post-bloom) |
| Wheat | 0.30 | 0.12 | 30 | 5 | 7.0 | - | 50 | 20 | 0.2 | 3.0 | 0.26 | 1.80 | 2-4 blades from top of plant (April-May) |

**INTERPRETATION:**

| | |
|---|---|
| Greater than guideline: | Adequate |
| 75% - 99% of guideline: | Slightly low |
| 50% - 74% of guideline: | Low |
| Less than 50% of guideline: | Very low |

*Source:* Data supplied by and reproduced with kind permission of Phosyn Chemicals Ltd., Manor Place, The Airfield, Pocklington, York YO4 2NR

**TABLE 187  Indicative Guidelines for Leaf Analysis of Tropical and Sub-Tropical Crops**

| Crop | Ca % | Mg % | Mn ppm | B ppm | Cu ppm | Mo ppm | Fe ppm | Zn ppm | S % | N % | P % | K % |
|---|---|---|---|---|---|---|---|---|---|---|---|---|
| **FIELD CROPS** | | | | | | | | | | | | |
| Cassava | 0.6 | 0.25 | 95 | 15 | 7 | - | 60 | 40 | 0.3 | 5 | 0.3 | 0.83 |
| Chickpea | - | - | 120 | 40 | 4 | - | - | 12 | 0.15 | 2.3 | 0.27 | - |
| Cotton | 2.1 | 0.46 | 30 | 20 | 9 | 2.4 | 40 | 19 | 0.59 | 3.9 | 0.33 | 3.0 |
| Cowpea | 0.81 | 0.17 | 280 | - | - | - | 100 | 30 | 0.34 | 1.9 | 0.23 | 4.9 |
| Ginger | 1.7 | 0.5 | 438 | 118 | 8 | 0.5 | 130 | 29 | 0.36 | 3.1 | 0.27 | 3.8 |
| Guar | 1.3 | 0.5 | 60 | 60 | 8 | - | - | 90 | - | 2.2 | 0.21 | 1.75 |
| Lima Bean** | - | - | - | - | - | - | - | - | - | 3.0 | 0.25 | 1.88 |
| Mung Bean | - | - | - | - | - | - | - | - | - | 3.6 | 0.27 | - |
| Navy (and other beans) | 1.5 | 0.5 | 50 | 20 | - | 4 | - | 22 | 0.2 | 3.8 | 0.33 | 1.75 |
| Peanuts | 1.25 | 0.3 | 75 | 19.5 | 10 | 1 | 75 | 22 | 0.2 | 3.5 | 0.28 | 1.92 |
| Pigeon Pea | 1.08 | 0.22 | 49 | 52 | 10 | 0.23 | 139 | 26 | 0.16 | 3.6 | 0.21 | 1.45 |
| Rice | 0.24 | 0.14 | 245 | 10 | 5.2 | - | 82 | 7 | 0.17 | 3.15 | 0.36 | 2.19 |
| Rubber* | - | 0.24 | - | 4.2 | - | 0.05 | - | 25 | - | 3.25 | 0.2 | 1.15 |
| Sisal* | - | 0.2 | 30 | - | 3 | 0.2 | - | 15 | - | - | - | - |
| Snap Bean (bushy)** | - | - | - | - | - | - | - | 23 | - | 1.75 | 0.33 | 1.13 |
| Sorghum | 0.51 | 0.24 | 20 | 4 | 7.3 | - | 79 | 21 | 0.2 | 3 | 0.22 | 1.91 |
| Soya Bean | 0.57 | 0.27 | 30 | 24 | 7.3 | 0.5 | 51 | 22 | 0.38 | 4.1 | 0.29 | 1.61 |
| Sugar Cane | 0.16 | 0.11 | 20 | 1.9 | 4.4 | 0.07 | 25 | 15 | 0.25 | 1.68 | 0.21 | 1.44 |
| Sunflower | 1.95 | 0.29 | 51 | 40 | 4.4 | 0.28 | - | 22 | 0.43 | 3.28 | 0.29 | 3.5 |
| Sunhemp | 0.13 | 0.17 | - | 25 | - | - | 120 | - | 0.16 | 2.4 | 0.14 | 0.82 |
| Tea | 0.4 | 0.16 | 100 | 12 | 12 | - | 120 | 5 | 0.1 | 3.5 | 0.3 | 1.6 |
| Tobacco (N. tabacum) | 1.65 | 0.37 | 182 | 20 | 9.7 | 0.38 | 70 | 22.5 | 0.27 | 2.75 | 0.23 | 2.5 |
| **PASTURE CROPS** | | | | | | | | | | | | |
| Barley Grass | - | - | - | - | - | - | - | - | 0.13 | - | - | - |
| Buffel Grass | - | - | - | - | - | - | - | - | 0.12 | - | 0.27 | - |
| Centro | - | - | - | - | - | - | - | 20 | - | - | 0.21 | 1.1 |
| Clover, Berseem | - | - | - | - | 5 | - | - | - | - | - | - | - |
| Clover, Kenya White | - | - | - | - | 5 | - | - | - | 0.2 | - | - | - |
| Clover, Strawberry | - | - | - | - | 5 | - | - | - | - | - | - | 1.2 |
| Clover, Subterranean | - | - | 20 | - | 3.5 | 0.1 | - | 12 | 0.2 | - | 0.1 | 1.6 |
| Clover, White | 0.75 | 0.18 | 25 | 25 | 5.5 | 0.15 | 50 | 16 | 0.24 | 4.8 | 0.3 | 1.43 |

**TABLE 187 (continued)  Indicative Guidelines for Leaf Analysis of Tropical and Sub-Tropical Crops**

| Crop | Ca % | Mg % | Mn ppm | B ppm | Cu ppm | Mo ppm | Fe ppm | Zn ppm | S % | N % | P % | K % |
|---|---|---|---|---|---|---|---|---|---|---|---|---|
| Columbus Grass | - | - | - | - | - | - | - | - | - | - | 0.22 | - |
| Cowpea | - | - | - | - | - | - | - | 17 | - | - | - | 0.8 |
| Creeping Indigo | - | - | - | - | 5 | - | - | - | - | - | - | 0.85 |
| Desmodium, greenleaf | - | - | - | - | - | - | - | 19 | 0.2 | - | 0.24 | 0.9 |
| Desmodium, silverleaf | - | - | - | - | 5 | - | - | - | 0.2 | - | 0.25 | - |
| Glycine | - | - | - | - | - | - | - | 22 | 0.22 | - | 0.25 | - |
| Great Brome Grass | - | - | - | - | - | - | - | - | 0.14 | - | 0.22 | - |
| Green Panic | - | - | - | - | - | - | - | - | 0.12 | - | 0.24 | - |
| Kikuyu | - | - | - | - | - | - | - | - | 0.12 | - | 0.19 | - |
| Lotononis | - | - | - | - | - | - | - | 22 | 0.2 | - | 0.24 | 0.9 |
| Lucerne | 0.67 | 0.23 | 25 | 25 | 5 | 0.18 | 45 | 13.5 | 0.23 | 3.7 | 0.47 | 1.83 |
| Medic, Barrel | - | - | - | - | 5 | - | - | - | 0.2 | - | 0.2 | 1.1 |
| Medic, Burr | - | - | - | - | - | - | - | - | 0.2 | - | - | - |
| Molasses Grass | - | - | - | - | - | - | - | - | - | - | - | - |
| Pangola Grass | - | - | - | - | - | - | - | - | 0.12 | - | 0.18 | - |
| Paspalum | - | - | - | - | - | - | - | - | 0.12 | - | 0.27 | - |
| Phalaris | 0.25 | 0.2 | 60 | 5 | 4 | 0.3 | - | 15 | 0.2 | 2 | 0.25 | 1.5 |
| Phasey Bean | - | - | - | - | 5 | - | - | - | 0.17 | - | 0.2 | 0.8 |
| Rhodes Grass | - | - | - | - | 5 | - | - | - | 0.12 | - | 0.24 | 0.5 |
| Ryegrass, perennial | 0.25 | 0.18 | 50 | 5 | 5.5 | 0.3 | 50 | 14.5 | 0.24 | 3.5 | 0.28 | 1.75 |
| Ryegrass, Wimmera | - | - | - | - | - | - | - | - | 0.14 | - | 0.23 | - |
| Setaria | - | - | - | - | - | - | - | - | 0.12 | - | 0.25 | - |
| Siratro | - | - | - | - | 5 | - | - | 24 | 0.2 | - | 0.19 | 2.5 |
| Stylo | 1.5 | - | - | - | - | - | - | - | - | - | 0.11 | 0.8 |
| Stylo, Capica | 0.7 | - | - | - | - | - | - | - | - | - | 0.13 | 0.9 |
| Stylo, Caribbean | - | - | - | - | - | - | - | - | 0.13 | - | 0.1 | 1 |
| Stylo, Macrocephala | 0.8 | - | - | - | - | - | - | - | - | - | 0.17 | 0.9 |
| Stylo, Townsville | - | - | - | - | - | - | - | 34 | 0.18 | - | 0.27 | 0.65 |
| Vigna luteola | - | - | - | - | - | - | - | - | - | - | 0.27 | - |

*(continued over)*

**TABLE 187** *(continued)* Indicative Guidelines for Leaf Analysis of Tropical and Sub-Tropical Crops

| Crop | Ca % | Mg % | Mn ppm | B ppm | Cu ppm | Mo ppm | Fe ppm | Zn ppm | S % | N % | P % | K % |
|---|---|---|---|---|---|---|---|---|---|---|---|---|
| **FRUITS, VINES AND NUTS** | | | | | | | | | | | | |
| Almond | - | - | - | 20 | 25 | 4 | - | 25 | - | 2 | 0.1 | 1.4 |
| Apple | 1.1 | 0.4 | - | 21 | 6 | 50 | - | 20 | 0.2 | 2 | 0.15 | 1.2 |
| Apricot | 2 | 0.3 | 40 | 20 | 5 | - | 100 | 20 | - | 2.4 | 0.14 | 2 |
| Avocado | 1 | 0.25 | 30 | 50 | 5 | 0.05 | 50 | 30 | 0.2 | 2 | 0.08 | 0.71 |
| Banana | 0.8 | 0.3 | 1 000 | 20 | 7 | 1.5 | 70 | 21 | 0.23 | 2.8 | 0.2 | 3.1 |
| Cacao* | 0.3 | 0.2 | - | 8.5 | 4 | - | 50 | 15 | - | 2.17 | 0.19 | 1.6 |
| Citrus | 3 | 0.26 | 25 | 31 | 5.1 | 0.1 | 60 | 25 | 0.21 | 2.4 | 0.12 | 0.7 |
| Coconut* | 0.5 | 0.25 | - | - | - | - | - | - | - | 1.8 | 0.12 | 0.8 |
| Coffee | 0.75 | 0.25 | 50 | 40 | 16 | - | 70 | 15 | - | 2.5 | 0.15 | 2.1 |
| Custard Apple | 0.6 | 0.35 | - | 15 | - | - | - | 15 | - | 2.5 | 0.16 | 1 |
| Fig | 3 | 0.75 | 20 | - | 4 | - | - | - | - | 2 | 0.1 | 1 |
| Grape | 1.2 | 0.3 | 25 | 30 | 6 | - | - | 26 | - | 425 ppm* NO3-N | 0.2 | 1.5 |
| Guava | 0.9 | 0.25 | - | - | 3 | - | - | - | - | 1.3 | 0.14 | 1.3 |
| Hazel nut | 0.7 | 0.25 | 26 | 31 | 3 | - | 51 | 16 | - | 2.2 | 0.14 | 0.9 |
| Kiwi Fruit | 3.11 | 0.4 | 104 | 31 | 5 | - | - | 15 | 0.33 | 2.37 | 0.17 | 1.54 |
| Lychee | 0.56 | 0.21 | - | - | - | - | - | - | 0.1 | 1.3 | 0.15 | 0.8 |
| Macadamia | - | - | - | - | - | - | - | - | 0.17 | 1.3 | 0.08 | 0.65 |
| Mango | 2.5 | 0.2 | 60 | 50 | 10 | - | 70 | 20 | - | 1.0 | 0.08 | 0.3 |
| Oil Palm | 0.6 | 0.3 | 150 | 10 | 5 | 0.5 | - | 15 | - | 2.7 | 0.18 | 1.3 |
| Olive | 1 | 0.1 | 20 | 19 | 40 | - | 20 | 10 | - | 1.5 | 0.1 | 0.8 |
| Papaw | - | 1 | 25 | 20 | 4 | - | 100 | 10 | 0.3 | 1.3 | 0.2 | 3 |
| Passion Fruit | 0.5 | 0.25 | 50 | - | 5 | - | 100 | 45 | 0.2 | 4.75 | 0.25 | 2 |
| Peach | 1.8 | 0.3 | 40 | 20 | 5 | - | 60 | 20 | 0.2 | 3 | 0.14 | 2 |
| Pear | 1.5 | 0.3 | 60 | 20 | 9 | - | 60 | 20 | 0.17 | 2.3 | 0.14 | 1.2 |
| Pecan | 0.7 | 0.3 | 150 | 20 | 4 | - | 50 | 50 | 0.15 | 2.5 | 0.12 | 0.75 |
| Persimmon | 1.35 | 0.17 | 238 | 48 | 1 | - | 56 | 5 | 0.21 | 1.57 | 0.1 | 2.4 |
| Pineapple | 0.22 | 0.41 | 150 | - | 10 | - | 80 | 15 | 0.07 | 1.5 | 0.1 | 4.3 |
| Pistachio | 1.3 | 0.6 | 30 | 55 | - | - | - | 7 | - | 1.5 | 0.14 | 1 |
| Plum | 1.5 | 0.3 | 40 | 25 | 6 | - | 100 | 20 | - | 2.4 | 0.14 | 1.6 |

**TABLE 187 (continued)   Indicative Guidelines for Leaf Analysis of Tropical and Sub-Tropical Crops**

| Crop | Ca % | Mg % | Mn ppm | B ppm | Cu ppm | Mo ppm | Fe ppm | Zn ppm | S % | N % | P % | K % |
|---|---|---|---|---|---|---|---|---|---|---|---|---|
| Prune** | 1 | 0.25 | - | 27 | - | - | - | 18 | - | 2.2 | - | 1.15 |
| Raspberry | 0.6 | 0.4 | 80 | 25 | 2 | - | - | 34 | - | 2.4 | 0.3 | 1.5 |
| Strawberry | 1 | 0.4 | 50 | 30 | 5 | - | - | 30 | 0.1 | 2.5 | 0.3 | 1.5 |
| Tea* | 0.2 | - | - | - | 12 | 0.5 | 70 | - | - | 4.1 | 0.35 | 2.0 |
| Walnut | 1 | 0.3 | 30 | 35 | 4 | - | - | 20 | - | 2.5 | 0.1 | 1.2 |
| **VEGETABLES** | | | | | | | | | | | | |
| Capsicum | 0.85 | 0.63 | 26 | 35 | 10 | - | - | 35 | - | 2.95 | 0.3 | 4.8 |
| Cucumber | 3.71 | 0.021 | - | 50 | 7 | - | - | 20 | - | 3.43 | 0.45 | 1.15 |
| Garlic** | - | - | - | - | - | - | - | - | - | 3.5 | 0.25 | 2.5 |
| Pumpkin (incl. marrows, squash) | 7.38 | - | - | - | - | - | - | - | - | 2.6 | 0.55 | 2.55 |
| Pepper, sweet** | - | - | - | - | - | - | - | - | - | 2.75 | 0.2 | 2.0 |
| Rockmelon | 5.1 | 0.85 | - | 34 | - | - | - | 30 | - | 2 | 0.3 | 4.28 |
| Sweet Corn | 0.84 | 0.3 | 20 | 45 | 5 | 0.2 | 60 | 70 | - | 2.92 | 0.27 | 2.6 |
| Sweet Potato | 0.73 | 0.56 | 40 | 118 | - | - | - | - | 0.08 | 2.65 | 0.18 | 3.95 |
| Watermelon | 2.35 | 0.5 | 60 | 33 | 4.5 | - | 120 | 20 | - | 2.5 | 0.3 | 2.5 |

*Main Source*: Reuter, D.J. & Robinson, J.B. (1986) *Plant Analysis. An International Manual.*
Reproduced with kind permission of Inkata Press.
Also:
* de Geus, J.G. (1973) *Fertilizer Guide to the Tropics and Sub Tropics.*
Div. of Ag. Sci., Univ. of California. Bull. 1879 and
** Reisenauer, H.M., ed. (1983) *Soil and Plant Tissue Testing in California.*

## TABLE 188   Factors Favouring Mineral Nutrient Deficiencies

*NB* The effects of soil pH, supply and availability of nutrients are noted elsewhere in this text; also some effects of nutrient interactions. Varieties of the same specie may be more tolerant of deficiency situations, as is often the case with Fe.

| Nutrient | Factors favouring deficiency |
|---|---|
| **Nitrogen** | Waterlogging, compaction, excessive leaching on light soils. |
| **Phosphorous** | Cold, wet soils. Water stress. |
| **Potassium** | Excessive leaching on light soils. |
| **Calcium** | Very dry soils. Soils with high potassium levels. Highly leached soils. |
| **Magnesium** | Soils with high potassium levels. Leached light soils. |
| **Copper** | Muck and peat soils. Poor drainage. Sandy soils. High nitrogen and phosphate levels; also Zn. Excessive uptake of other nutrients in acid soils. |
| **Zinc** | Wet soils in cold, early springs. Heavy phosphorus fertilization. Calcareous soils after leaching and erosion. Acid, leached soils. Coarse sands. Subsoils exposed by land levelling. Low organic matter soils AND peat and muck soils. Compacted soils. Restricted root zones. Old orchard, corral and barnyard sites. High nitrogen applications. |
| **Manganese** | Sand, muck and peat soils. Calcareous soils. High iron availability; also Cu and Zn. Dry weather. Excessively limed organic rich acid soils. Cold weather. Poorly drained soils. Dry weather. Excess sodium or potassium. Low light intensity. |
| **Iron** | Cold, wet alkali soils. Calcareous soils; free $CaCo_3$. Excess phosphorus. Low organic matter acid soils. Sandy soils. Low soil temperature. Land levelling exposing subsoil. High moisture. High levels of Cu, Mn, Mo and Zn. Soil moisture extremes. High bicarbonate. Poor soil aeration (high $CO_2$). Heavily manured alkali soils. Root damage. Temperature extremes. Heavy manuring of alkali soils. |
| **Boron** | Droughty conditions. Acid leached soils. Overlimed acid soils. Peat and muck soils AND low organic matter soils. Coarse textured sandy soils. Dry soils. High light intensities. Moderate to heavy rainfall. |
| **Molybdenum** | Low organic matter soils. Sandy soils. Soils high in free $Fe_2O_3$. Soils high in sulphate or copper. |
| **Cobalt** | Sandy soils, soils derived from silicic igneous, parent material. |

*Sources:* Lorenz, O.A. & Maynard, D.N. (1987) *Knott's Handbook for Vegetable Growers.* Reuter, D.J. & Robinson, J.B. (1986) *Plant Analysis. An Interpretation Manual.* Mortvedt, J.J. *et al.* (1972) *Micronutrients in Agriculture.* Brady, N.C. (1984) *The Nature and Properties of Soils.* Jones, V.S. (1979) *Fertilisers and Soil Fertility.* Landon, J. R., ed. (1984) *Tropical Soil Manual.* Kelso, I. (1991) Personal Communication.

**TABLE 189   General Nutrient Toxicity Symptoms and Examples of Indicator Plants**

| Nutrient | Symptoms | Accumulators or indicators of high soil content |
|---|---|---|
| **Nitrate nitrogen** | Edge burn which may be followed by interveinal collapse. | |
| **Ammonium nitrogen** | Initial necrosis, blackening around tips and edges of leaves. Root death may occur. | |
| **Phosphorus** | Interveinal chlorosis in younger leaves, necrosis and tip death may follow in more susceptible species, marginal scorch and shedding of older leaves. | |
| **Sodium** | Marginal chlorosis and burn. Salinity stress generally first indicated by blue colouring of leaves. | Beets, chard |
| **Chlorine** | Bronzing, chlorosis, marginal burn; leaf drop may be premature. In some species the marginal burn is accompanied by upward cupping. | |
| **Manganese** | Yellowing, beginning at the leaf edge of older leaves, sometimes with upward cupping. Interveinal bronze-yellow chlorosis in beans. Orange-yellow marginal and interveinal chlorosis in lemons. Brown 'tar spots' in orange leaves. Necrosis in apple bark. Chlorosis of younger leaves in tropical legumes. Brown, red and black spotting of older leaves in cereals and grasses. 'Stem streak' necrosis in potatoes. | |
| **Aluminium** | Symptoms on shoots may resemble those of phosphorus deficiency. Roots often stunted with short laterals.        Tea | |
| **Boron** | Interveinal necrosis which is often spotty at first. | Cruciferae, Pea |
| **Flourine** | Scorching of leaf tip and margin extending into interveinal areas. | |
| **Molybdenum** | 'Golden shoot' in tomato. | |

*Sources:* Reuter, D.J. & Robinson, J.B. (1986) *Plant Analysis. An Interpretation Manual.* Whiteman, P.C. (1980) *Tropical Pasture Science.* Walsh, L.M. & Beaton, J.D. (1980) *Soil Testing and Plant Analysis.* FAO (1984) *Fertilizer and Plant Nutrition Guide. Fertilizer and Plant Nutrition Bulletin No. 9.*

**TABLE 190  Indicative Summary of Principal Plant-nutrient Interactions**

| Element acting | B | Ca | Cu | Fe | K | Mg | Mn | Mo | N | Na | P | S | Zn |
|---|---|---|---|---|---|---|---|---|---|---|---|---|---|
| | | | | | | **Element affected** | | | | | | | |
| B | / | - | - | - | - | - | - | - | - | - | A | - | - |
| Ca | A | / | - | A | A | A | A | - | - | A | A | - | A |
| Cu | - | E | / | A | - | - | A | A | - | - | A | - | A |
| Fe | - | - | A | / | A | - | A | - | - | - | A | - | - |
| K | A | A | - | E/ | / | A | E | A | A | A | - | - | - |
| Mg | - | A | A | A | A | / | A | E | - | A | E | E | A |
| Mn | - | - | - | A | - | - | / | A | - | - | - | - | - |
| Mo | - | - | A | A | - | - | A | / | - | - | - | A | - |
| N | A | E | A | - | A | E | - | E | / | - | E | - | A |
| Na | - | A | - | - | A | A | - | - | - | / | A | - | A |
| P | - | A | A | A | A | E | A | E | A | - | / | E | A |
| S (as SO₄²⁻) | - | - | - | - | - | - | - | A | - | - | E | / | - |
| Zn | - | - | A | A | - | A | A | - | - | - | A | A | / |

A = Antagonizes action
E = Enhances action
- = Insufficient data or effects too variable to summarize simply

*NB* This is a highly generalized table; interactions vary greatly depending on plant type and soil conditions, the latter including, in particular, pH, temperature, drainage status, relative levels and forms of the elements occurring, and presence or absence of other elements. In particular, where the acting element is present in amounts limiting to crop growth, applications of additional quantities up to a 'sufficient' level may enhance the action of the affected elements which are otherwise antagonized.

An entry in the table means no more than that the potential interaction indicated should be borne in mind during situation assessment.

*Source:* Landon, J.R., ed. (1984) *Tropical Soil Manual.*
Reproduced with kind permission of Booker Tate Ltd.

**TABLE 191 a & b   Planting, Grazing and Harvesting Constraints to Use of Sewage Sludge**

In order to enable appropriate measures to be taken to minimize the risk to health of humans, animals and plants, it is necessary to co-ordinate sludge applications in time with planting, grazing or harvesting operations. Sludge must not be applied to growing fruit and vegetable crops nor used where crops are grown under permanent glass or plastic structures. Untreated sludge must not be used in orchards or on land used for growing nursery stock (including bulbs). Further constraints which must be taken into account are set out below.

## a. Acceptable uses of *treated* sludge in agriculture

| **When applied to growing crops** | **When applied before planting crops** |
|---|---|
| Cereals, oil seed rape | Cereals, grass, fodder, sugar beet, oil seed rape, etc. |
| Grass (1) | Fruit trees |
| Turf (2) | Soft fruit (3) |
| Fruit Trees (3) | Vegetables (4) |
| | Potatoes (4), (5) |
| | Nursery stock (5) |

(1)  No grazing or harvesting within 3 weeks of application
(2)  Not to be applied within 3 months before harvest
(3)  Not to be applied within 10 months before harvest
(4)  Not to be applied within 10 months before harvest if crops are normally in direct contact with soil and may be eaten raw
(5)  Not to be applied to land used or to be used for a cropping rotation that includes the following:
      a)  basic seed potatoes
      b)  seed potatoes for export
(6)  Not to be applied to land used or to be used for a cropping rotation that includes the following:
      a)  basic nursery stock
      b)  nursery stock (including bulbs) for export

## b. Acceptable uses of *untreated* sludge in agriculture

| **When applied to growing crops by injection\*** | **When cultivated or injected\* into the soil before planting crops** |
|---|---|
| Grass (1) | Cereals, grass, fodder, sugar beet, oil seed rape, etc |
| Turf (2) | Fruit trees |
| | Soft fruit |
| | Vegetables (3) |
| | Potatoes (3), (4) |

(1)  No grazing or harvesting within 3 weeks of application
(2)  Not to be applied within 6 months before harvest
(3)  Not to be applied within 10 months before planting if crops are normally in direct contact with soil and may be eaten raw
(4)  Not to be applied to land used or to be used for cropping rotation that includes seed potatoes
\* Injection carried out in accordance with WRC publication FR 008 1989,
*Soil Injection of Sewage Sludge - A Manual of Good Practice (2nd Edition).*

*Source:* Dept. of Environment (1989) *Code of Practice for Agricultural Use of Sewage Sludge.*
Reproduced with kind permission of the Department of the Environment (UK).

TABLE 192   Tolerances of Plant Foliage to Urea Sprays in Pounds per 100 Gallons of Water

| Vegetable crops | | Plantation or tropical crops | | Deciduous tree and small fruit crops | | Field crops | |
|---|---|---|---|---|---|---|---|
| Crop | Tolerance | Crop | Tolerance | Crop | Tolerance | Crop | Tolerance |
| Cucumber | 3 - 5 | Pineapple | 20 - 50 | Grape | 4 - 6 | Potatoes | 20 |
| Bean | 4 - 6 | Cacao | 5 - 10 | Raspberry | 4 - 6 | Sugar beets | 20 |
| Tomato | 4 - 6 | Sugar cane | 10 - 20 | Apple | 4 - 6 | Alfalfa | 20 |
| Pepper | 4 - 6 | Banana | 5 - 10 | Strawberry | 4 - 6 | Corn | 5 - 20 |
| Sweet corn | 4 - 6 | Cotton | 20 - 50 | Plum | 5 - 15 | Wheat | 20 - 800 |
| Lettuce | 4 - 6 | Tobacco | 3 - 10 | Peach | 5 - 20 | Bromegrass | 20 - 800 |
| Cabbage | 6 - 12 | Citrus | 5 - 10 | Cherry | 5 - 20 | | |
| Carrots | 20 | Hops | 40 - 50 | | | | |
| Celery | 20 | | | | | | |
| Onions | 20 | | | | | | |

*Source:* McVickar, M.H. *et al.*, eds. (1983) *Fertilizer Technology and Usages.*
Reproduced with kind permission of the Soil Science Society of America Inc.

# Climatic Effects

TABLE 193   Crop Groups According to Optimum Temperature Range and Photosynthetic Characteristic

| Crop group | Photosynthetic pathway | Maximum leaf photosynthesis rate (1) (kg $CH_2O$ / ha / hr) | Optimum temperature range (deg. C) | Major Crops |
|---|---|---|---|---|
| I | C3 | 20 | 15 - 20 | potato, chickpea, lentil, french bean, rape, cabbage, sunflower, barley, bread wheat, linseed, tomato, oats, rye, grape (TEC), pyrethrum, sugar beet, olive, arabica coffee |
| II | C3 | 38 | 25 - 30 | groundnut, cowpea, soyabean, french bean, tobacco, sunflower, sesame, tomato, safflower, rice, kenaf, fig, cotton, castor bean, sweet potato, cassava, yam, robusta coffee, banana |
| III | C4 | 65 | 30 - 35 | barnyard millet (TRC), foxtail millet (TRC) millet, common millet (TRC), pearl millet, hungry rice, sorghum (TRC), maize (TRC) sugarcane |
| IV | C4 | 65 | 20 - 30 | barnyard millet (TEC, TRHC), foxtail millet (TEC,TRHC), common millet (TEC, TRHC), sorghum (TEC,TRHC), maize (TEC,TRHC) |

NOTES:  1. Maximum photosynthesis rates applicable in the optimum temperature range only
        2. TEC :      temperate cultivar
           TRC :      tropical cultivar
           TRHC :    tropical highland cultivar
NB See also Fig. 39

*Source:* De Pauw, W. (1988) *A Summary of the Agricultural Ecology of Ethiopia.*
Reproduced with kind permission of the Food and Agriculture Organization of the United Nations.

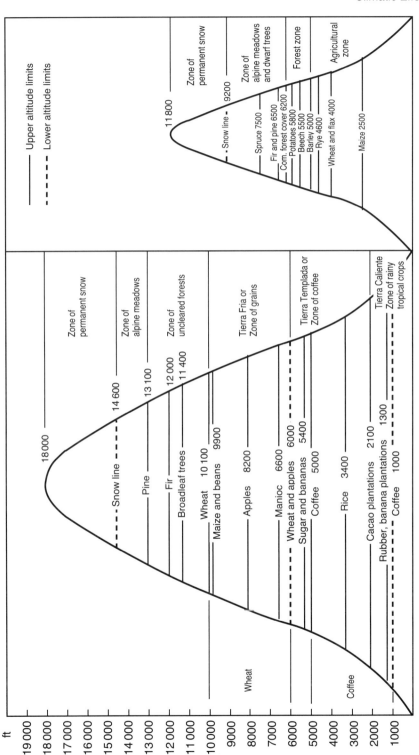

**Fig. 38** Temperature zones as related to altitude in mountainous areas of both the tropics and middle latitudes

*Source:* Wilsie, C.P. (1962) *Crop Adaption and Distribution.* Copyright © (1962) W.H. Freeman. Reprinted with permission.

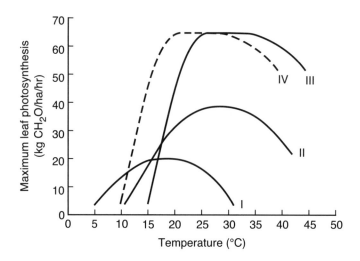

**Fig. 39** Temperature effect on photosynthetic rates of crop groups

*NB* For typical species of each crop group see **TABLE 193.**

*Source:* De Pauw, W. (1988) *A Summary of the Agricultural Ecology of Ethiopia.*
Reproduced with kind permission of the Food and Agriculture Organization of the United Nations.

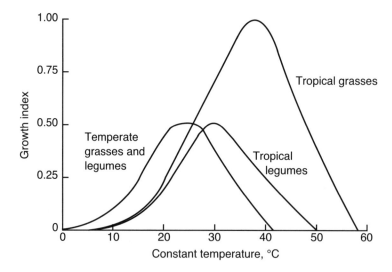

**Fig. 40** Generalized representation of dry matter production responses of temperate and tropical pasture species to temperature. The growth index represents dry matter production relative to maximum dry matter production of tropical grasses at their optimum temperature

*Source:* Whiteman, P.C. (1980) *Tropical Pasture Science.*
Reproduced with kind permission of the Oxford University Press.

Recognizing that varieties of vegetables perform differently in different climates the NUNHEMS ZADEN seed company of Holland has grouped its vegetable varieties as suitable for three main climatic regions across N. America, Europe and N. Africa as shown below.

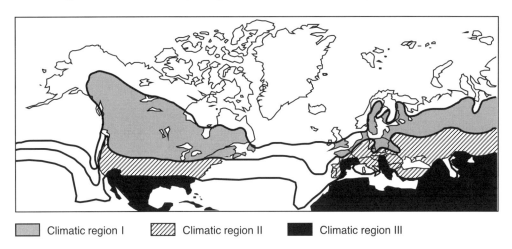

    Climatic region I        Climatic region II        Climatic region III

**Fig. 41** Climatic regions for vegetable varieties

*NB* This map is intended as a general guide only and is dependent on location and altitude.

*Source: Nunhems Specialities. Vegetable Seed Catalogue* (1991). Reproduced with kind permission of Nunhems Zaden BV, PO Box 4005, 6080 AA Haelen, Holland.

---

**TABLE 194   Photoperiodic Classes**

---

| Class of Plants | Characteristics |
| --- | --- |
| **Long Day** | Require short dark periods to initiate floral activity. Critical dark period length - longer than which they will not flower. Flowering may be promoted by continuous light, or a period of light in the night. |
| **Intermediate** | Have both critical minimum and maximum night lengths. Will only flower in this range. |
| **Short Day** | Require relatively long dark periods to initiate floral activity. Critical minimum dark period, shorter than which they will not flower. Can prevent flowering by causing low light intensity period in the middle of the dark period. |
| **Day Neutral** | Insensitive to a wide range of day lengths for floral stimulation. |

**TABLE 195   Photoperiodic Characteristics of Some Crop Plants**

| Crop | Characteristic |
|------|----------------|
| Alfalfa | Day Neutral |
| Banana | Day Neutral |
| Bean | Short Day/Day Neutral |
| Cabbage | Long Day |
| Citrus | Day Neutral |
| Cotton | Short Day/Day Neutral |
| Groundnut | Day Neutral |
| Maize | Short Day/Day Neutral |
| Onion | Long Day/Day Neutral |
| Pea | Day Neutral |
| Pepper | Short Day/Day Neutral |
| Pineapple | Short Day |
| Potato | Long Day/Day Neutral |
| Rice | Short Day/Day Neutral |
| Sorghum | Short Day/Day Neutral |
| Soyabean | Short Day/Day Neutral |
| Sugar beet | Long Day |
| Sugarcane | Short Day/Day Neutral |
| Sunflower | Short Day/Day Neutral |
| Tobacco | Short Day/Day Neutral |
| Tomato | Day Neutral |
| Watermelon | Day Neutral |
| Wheat | Day Neutral/Long Day |

**TABLE 196   Indicative Chilling Requirements for Main Deciduous Fruit Tree Species**

| Fruit | Hrs below $7°C$ |
|-------|----------------|
| Almonds | 0 - 800 |
| Peaches | 100 - 1250 |
| Japanese Plums | 100 - 800 |
| Apples and Pears | 200 - 1400 |
| European Plums | 800 - 1500 |
| Cherries | 800 - 1700 |

*Source:* Philipe, J.M. (1971) *Informe sobre la seleccion de frutas y especialmente de variedades de melocoton.* Rep.UNDP/SF Proj Spain/16 FAO Rome Italy. Reproduced with kind permission of the Food and Agriculture Organization of the United Nations.

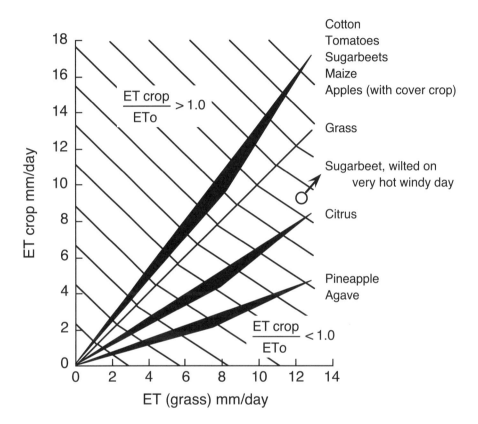

**Fig. 42** Conceptual representation of relationship between ETcrop and ETo for major crop types

*Note:* The figure illustrates the wide variation between crop types due to factors such as variable resistance to evapotranspiration, waxy leaves (e.g. citrus), day closure of stomata (e.g. pineapple), crop height, crop roughness.

*Source:* FAO (1988) *Crop Water Requirements. Irrigation and Drainage Paper No. 24.* Reproduced with kind permission of the Food and Agriculture Organization of the United Nations.

**TABLE 197   Grass Species for Cultivation in the Main Climatic Regions**

| Climatic classification[1] | Geographical areas | Characteristic, natural or cultivated species |
|---|---|---|
| **TEMPERATE** | | |
| Extended rainfall | NW Europe | *Dactylis glomerata* |
| | N Atlantic seaboard of N America | *Lolium italicum* |
| | N Pacific seaboard of N America | *L. perenne* |
| | New Zealand | *Phleum pratense* |
| | SE Australia (part) | *Poa pratensis* |
| Winter rainfall | Mediterranean region | *Dactylis glomerata* |
| (Mediterranean climate) | S Africa (Cape) | *D. hispanica* |
| | Southern Australia | *Ehrharta calycina* |
| | California (part) | *Festuca arundinacea* |
| | Chilean seaboard (part) | *Hordeum bulbosum* |
| | | *Lolium rigidum* |
| | | *Oryzopsis miliacea* |
| | | *Phalaris tuberosa* |
| Summer rainfall | Central Europe and Ukraine | *Agropyron cristatum* |
| | NE Great Plains of N America | *Bromus inermis* |
| | Central China | *Elymus canadensis* |
| Semi-arid | Steppes of USSR and central Asia | *Agripyron cristatum* |
| | N China | *A. smithii* |
| | Patagonia | *A. trachycaulum* |
| | NW Great Plains of N America | *Bromus inermis* |
| **WARM TEMPERATE** | U.S. Gulf region | *Chloris gayana* |
| **OR** | Queensland and New South Wales littoral | *Cynodon dactylon* |
| **SUBTROPICAL** | S China (part) | *Paspalum dilatatatium* |
| Extended rainfall | NE Argentina and Uruguay | *P. notatum* |
| | | *Sorghum halepense* |
| Summer rainfall | SE Great Plains of N America | *Andropogon ischaemum* |
| | S Africa plateau | *Bouteloua gracilis* |
| | Central and W Argentina | *Buchloë dactyloides* |
| | S China (part) | *Centhrus ciliaris* |
| | Australia (sub-coastal Queensland, | *Chloris gayana* |
| | northern coast of New South Wales) | *Eragrostis curvula* |
| | | *Panicum dilatatum* |
| | | *P. Maximum* |
| | | *Pennisetum clandestinum* |
| | | *Sorghum almum* |
| | | Annual sorghum species |

**TABLE 197** *(continued)* **Grass Species for Cultivation to the Main Climatic Regions**

| Climatic classification[1] | Geographical area | Characteristic, natural or cultivated species |
|---|---|---|
| Semi-arid | SW Great Plains of N. America<br>S Africa (part)<br>W Argentina (part)<br>Areas S and E of Mediterranean | *Buchloë dactyloides*<br>*Bouteloua curtipendula*<br>*B. gracilis*<br>*Cenchrus ciliaris*<br>*Eragrostis curvula*<br>*E. trichodes* |
| **TROPICAL**<br>Extended rainfall | Equatorial belt<br>Islands and littorals N and S of the<br>Equator | *Brachiari mutica*<br>*Chloris gayana*<br>*Cynodon dactylon*<br>*Hyparrhenia rufa*<br>*Melinis minutiflora*<br>*Panicum maximum*<br>*Pennisetum clandestinum*<br>(4000 to 7000 ft)<br>*P. purpureum* |
| Restricted rainfall | Monsoon regions of the tropics<br>Summer rainfall regions of central and<br>southern Africa | *Cenchrus ciliaris* and<br>all the grasses of the preceding<br>column in favourable sites<br>Annual sorghum species |
| Semi-arid | Monsoon regions with a more restricted<br>rainfall (under 25 inches) | *Cenchrus ciliaris*<br>Annual sorghum species |

*NB* The inclusion of a species in the third column does not necessarily mean that that particular species is grown or occurs naturally in all the areas or countries listed in the second column.

[1]    It is well known that in mountainous regions climate is modified by altitude. At elevations above 7000 feet in the tropics, grasses from northwestern Europe may be successfully cultivated. The effects of seasonal rainfall can be modified by irrigation, and those species adapted to an extended rainfall can then be grown.

*Source:* FAO (1959) *Grasses in Agriculture. Agricultural Studies No. 42.*
Reproduced with kind permission of the Food and Agricultural Organization of the United Nations.

**TABLE 198  Indicative Yields of Selected Crops in Three Climatic Regions**

Good yields of high-producing varieties adapted to the climatic conditions of the available growing season under adequate water supply and high level of agricultural inputs under irrigated farming conditions.

t/ha
CLIMATIC REGIONS

| CROP | | Tropics [1] < 20° C [4] | > 20° C | Subtropics [2] < 20° C | > 20° C | Temperate [3] < 20° C | > 20° C |
|---|---|---|---|---|---|---|---|
| Alfalfa | hay | 15 | | 25 | | 10 | |
| Banana | fruit | | 40-60 | | 30-40 | | |
| Bean: fresh | pod | 6-8 | | 6-8 | | 6-8 | |
| dry | grain | 1.5-2.5 | | 1.5-2.5 | | 1.5-2.5 | |
| Cabbage | head | 40-60 | | 40-60 | | 40-60 | |
| Citrus: | | | | | | | |
| grapefruit | fruit | | 35-50 | | 40-60 | | |
| lemon | fruit | | 25-30 | | 30-45 | | |
| orange | fruit | | 20-35 | | 25-40 | | |
| Cotton | seed cotton | | 3-4 | | 3-4.5 | | |
| Grape | fruit | 5-10 | | 15-30 | | 15-25 | |
| Groundnut | nut | | 3-4 | | 3.5-4.5 | | 1.5-2 |
| Maize | grain | 7-9 | 6-8 | 9.10 | 7-9 | | 4-6 |
| Olive | fruit | | | 7-10 | | | |
| Onion | bulb | 35-45 | | 35-45 | | 35-45 | |
| Pea: fresh | pod | 2-3 | | 2-3 | | 2-3 | |
| dry | grain | 0.6-0.8 | | 0.6-0.8 | | 0.6-0.8 | |
| Fresh pepper | fruit | 15-20 | | 15-25 | | 15-20 | |
| Pineapple | fruit | | 75-90 | | 65-75 | | |
| Potato | tuber | 15-20 | | 25-35 | | 30-40 | |
| Rice | paddy | | 6-8 | | 5-7 | | 4-6 |
| Safflower | seed | | | 2-4 | | | |
| Sorghum | grain | 3-4 | 3.5-5 | 3-4 | 3.5-5 | | 2-3 |
| Soybean | grain | 2.5-3.5 | | 2.5-3.5 | | | |
| Sugar beet | beet | | | 40-60 | | 35-55 | |
| Sugarcane | cane | | 110-150 | | 100-140 | | |
| Sunflower | seed | 2.5-3.5 | | 2.5-3.5 | | | 2-2.5 |
| Tobacco | leaf | | 2-2.5 | | 2-2.5 | | 1.5-2 |
| Tomato | fruit | 45-65 | | 55-75 | | 45-65 | |
| Water melon | fruit | | 25-35 | | 25-35 | | |
| Wheat | grain | 4-6 | | 4-6 | | 4-6 | |

[1]  Semi-arid and arid areas only
[2]  Summer and winter rainfall areas
[3]  Oceanic and continental areas
[4]  Mean temperature

*Source:* FAO (1979) *Yield Response to Water. Irrigation and Drainage Paper No. 33.* Reproduced with kind permission of the Food and Agriculture Organization of the United Nations.

**TABLE 199   Maximum DM Production Rates (in kg/ha/hour) for Crop Groups and Mean Temperatures**

| | Mean temperature °C | | | | | | | | |
|---|---|---|---|---|---|---|---|---|---|
| Crop group | 5 | 10 | 15 | 20 | 25 | 30 | 35 | 40 | 45 |
| I cool | 5 | 15 | 20 | 20 | 15 | 5 | 0 | 0 | 0 |
| I warm | 0 | 0 | 15 | 32.5 | 35 | 35 | 32.5 | 5 | 0 |
| II cool | 0 | 5 | 45 | 65 | 65 | 65 | 45 | 5 | 0 |
| II warm | 0 | 0 | 5 | 45 | 65 | 65 | 65 | 45 | 5 |

I cool:    alfalfa, bean, cabbage, pea, potato, tomato, sugar beet, wheat
I warm:    alfalfa, citrus, cotton, groundnut, pepper, rice, safflower, soybean, sunflower, tobacco, tomato
II cool:   some maize and sorghum varieties
II warm:   maize, sorghum, sugarcane

*Source:* FAO (1979) *Yield Response to Water, Irrigation & Drainage Paper No. 33.* Reproduced with kind permission of the Food and Agriculture Organization of the United Nations.

## Soil Moisture

**TABLE 200   Relative Tolerance of Certain Crops to Waterlogging and Flooding**

*NB* There has been little systematic research on this subject. Few crops tolerate long periods of waterlogging or flooding; for most crops lack of aeration in the root zone for any prolonged period is undesirable except the classic case of paddy rice. Some crops can withstand a high water table [not the WHOLE soil profile] whilst others cannot. It will be noted from this summary that authorities rank some crops differently. Where an authority ranked a crop overlapping two categories the crop has been placed in the more sensitive category. Where an authority gave four rankings the middle two were placed in the medium category.

**Sensitive**

Alfalfa
Apricot
Barley
Bean, green
Birdwood Grass
Blue Panic Grass
Buffel Grass
Cabbage
Cashew
Cassava
Carribean Stylo
Cherry
Citrus

Clover, Ladino
Clover, Strawberry
Clover, Sweet
Cocoa
Coffee
Cotton
Cowpea
Cucumber
Date Palm
Desmodium Silverleaf
Flax
Glycine
Gram
Green Panic Grass
Groundnut

Hemp
Kenaf
Lettuce
Leucaena
Lubia
Maize
Molasses Grass
Oats
Oil Palm
Olive
Onions
Peach
Peas
Pigeon Pea
Pineapple

*(continued over)*

---

**TABLE 200** *(continued)* **Relative Tolerance of Certain Crops to Waterlogging and Flooding**

---

**Sensitive**

Potato
Pyrethrum
Raspberry
Ruzi Grass
Rye
Sabi Grass
Safflower
Sesame
Shrubby Stylo
Sisal
Sirato
Tapioca
Tea
Tobacco
Tomato
Townsville Stylo
Wheatgrass, Crested
Yam

**Medium**

Apple
Axillaris
Bahia grass
Banana
Barley
Bermuda Grass
Berseem
Blackberry
Bromegrass
Buffel Grass
Calopo
Castor
Centro
Citrus
Cocksfoot
Columbus Grass
Cotton
Crimson Clover
Dallis grass
Desmodium Hetero
Desmodium Silverleaf
Elephant Grass
Fescue, Meadow
Fine Stem Stylo
Guinea Grass
Jute
Kenya White Clover
Kikuyu Grass

Lablab
Mango
Millet, Bulrush,
    Finger, Panicum
Oats
Onion
Pear
Phalaris
Plicatulum
Plum
Prairie Grass
Puero
Red Clover
Rhodes Grass
Rice, Upland
Ronpha Grass
Rosella
Rye
Ryegrass, perennial,
    hybrids, Wimmera
Sabi Grass
Scrobic
Setaria
Signal Grass
Siratro
Sisal
Sorghum
Soyabean
Stylo
Subterranean Clover
Sudan Grass
Sugar Beet
Sugar Cane
Sunflower
Tea
Timothy Grass
Tomato
Trefoil, Birdsfoot
Trefoil, Narrowleaf
Uniflorus
Wheat
Wheatgrass, Slender
White Clover

**Tolerant**

Bahia Grass
Broad Beans
Broadleaf Paspalum
Canarygrass, Reed

Centro
Clover, Alsike
Clover, White
Cocksfoot
Coconut
Cocoyam
Columbus Grass
Dallis Grass
Date Palm
Desmodium Greenleaf
Desmodium Hetero
Fescue, Tall
Greenleaf Desmodium
Kikuyu Grass
Lotononis
Makarakari Grass
Mangolds
Mustard
Oil Palm
Pangola Grass
Para Grass
Pear
Phalaris
Phasey Bean
Plicatulum
Puero
Rape
Rice, Paddy
Rubber
Signal Grass
Strawberry
Strawberry Clover
Sugar Cane
Tall Fescue
Trefoil, Big

*Sources:* FAO/UNESCO.
(1973) *Irrigation, Drainage
and Salinity.* Landon, J.R.,
ed. (1984) *Tropical Soil
Manual.* Whiteman, P.C.
(1980) *Tropical Pasture
Science.* Humphreys, L.R.
(1980) *A Guide to the Better
Pastures for the Tropics and
Sub-Tropics.* Wright
Stevenson Co. *Tropical seed
sowing Guide.*

**TABLE 201 a & b  Sensitivity of Crops to Soil Water Depletion**

## a. Crop Groups

| Crop Group | Typical Crops |
|------------|---------------|
| 1 | onion, pepper, potato |
| 2 | banana, cabbage, grape, pea, tomato |
| 3 | alfalfa, bean, citrus, groundnut, pineapple, sunflower, watermelon, wheat |
| 4 | cotton, maize, olive, safflower, sorghum, soybean, sugarbeet, sugarcane, tobacco |

**Increasing Tolerance** (arrow pointing down alongside groups 1–4)

## b. Indicative Critical Soil Moisture Depletion Fraction for different Crop Groups according to Evapotranspiration Rate

| Crop Group | ETm* mm / day | | | | | | | | |
|------------|------|-------|-------|------|------|-------|-------|------|-------|
|  | 2 | 3 | 4 | 5 | 6 | 7 | 8 | 9 | 10 |
| 1 | 0.50 | 0.425 | 0.35 | 0.30 | 0.25 | 0.225 | 0.20 | 0.20 | 0.175 |
| 2 | 0.675 | 0.575 | 0.475 | 0.40 | 0.35 | 0.325 | 0.275 | 0.25 | 0.225 |
| 3 | 0.80 | 0.70 | 0.60 | 0.50 | 0.45 | 0.425 | 0.375 | 0.35 | 0.30 |
| 4 | 0.875 | 0.80 | 0.70 | 0.60 | 0.55 | 0.50 | 0.45 | 0.425 | 0.40 |

\* ETm = maximum evapotranspiration rate of the crop when soil water is not depleted

*Note:* Irrigation frequency and amount will be further determined by the rooting depth of the individual crops.

*Source:* FAO (1979) *Yield Response to Water. Irrigation and Drainage Paper No. 33.* Reproduced with kind permission of the Food and Agriculture Organization of the United Nations.

**TABLE 202  Soil Water Suction (for soil depth with maximum root activity) at which water should be applied for maximum yields of various crops grown in deep, well drained soil fertilized and managed for maximum production**

| Crop | Soil suction (bars) | Crop | Soil suction (bars) |
|------|---------------------|------|---------------------|
| **Vegetative crops** | | **Seed crops** | |
| Alfalfa | 1.50 | Alfalfa | 2.00 |
| Beans (snap, lima) | 0.75 - 2.00 | Alfalfa (bloom) | 4.00 - 8.00 |
| Broccoli (early) | 0.45 - 0.55 | Alfalfa (ripening) | 8.00 - 15.00 |
| Broccoli (post-bud) | 0.60 - 0.70 | Seed carrots (60 cm depth) | 4.00 - 6.00 |
| Cabbage | 0.60 - 0.70 | Onions (7 cm depth) | 4.00 - 6.00 |
| Canning Peas | 0.30 - 0.50 | Seed onions (15 cm depth) | 1.50 |
| Cauliflower | 0.60 - 0.70 | Lettuce (productive) | 3.00 |
| Celery | 0.20 - 0.30 | Coffee | required short |
| Grass | 0.30 - 1.00 | | periods of |
| Lettuce | 0.40 - 0.60 | | low potential to |
| Sugar cane | 0.25 - 0.30 | | break bud dormancy, |
| Sweet corn | 0.50 - 1.00 | | followed by high |
| Tobacco | 0.30 - 0.80 | | water potential |
| **Root crops** | | **Fruit crops** | |
| Carrots | 0.55 - 0.65 | Avocados | 0.50 |
| Onions (bulbing) | 0.55 - 0.65 | Bananas | 0.30 - 1.50 |
| Onions (early) | 0.45 - 0.55 | Cantaloupe | 0.35 - 0.40 |
| Potatoes | 0.30 - 0.50 | Deciduous fruit | 0.50 - 0.80 |
| Sugar beets | 0.40 - 0.60 | Grapes (early) | 0.40 - 0.50 |
| | | Grapes (mature) | 1.00 |
| **Grain crops** | | Lemons | 0.40 |
| Corn (vegetative) | 0.50 | Oranges | 0.20 - 1.00 |
| Corn (ripening) | 8.00 - 12.00 | Strawberries | 0.20 - 0.30 |
| Small grains(vegetative) | 0.40 - 0.50 | Tomatoes | 0.80 - 1.50 |
| Small grains (ripening) | 8.00 - 12.00 | | |

Where two values for soil water suction are given, the lower suction value is used when evaporative demand is high and the higher value when it is low; intermediate values are used when the atmospheric demand for evapotranspiration is intermediate. These values are subject to revision as additional experimental data become available.

*Source*:  FAO/UNESCO (1973) *Irrigation, Drainage and Salinity*.
Reproduced with kind permission of the Food and Agriculture Organization of the United Nations.

**TABLE 203   Sensitive Moisture Stress Periods for Selected Crops**

| | |
|---|---|
| **Alfalfa** | just after cutting (and for seed production at flowering) |
| **Banana** | throughout but particularly during first part of vegetative period flowering and yield formation |
| **Bean** | flowering and pod filling; vegetative period not sensitive when followed by ample water supply |
| **Cabbage** | during head enlargement and ripening |
| **Citrus** | |
|     **grapefruit** | flowering and fruit set to fruit enlargement |
|     **lemon** | flowering and fruit set to fruit enlargement; heavy flowering may be induced by withholding irrigation just before flowering |
|     **orange** | flowering and fruit set to fruit enlargement |
| **Cotton** | flowering and boll formation |
| **Grape** | vegetative period, particularly during shoot elongation and flowering to fruit filling |
| **Groundnut** | flowering and yield formation, particularly during pod setting |
| **Maize** | flowering to grain filling; flowering very sensitive if no prior water deficit |
| **Olive** | just prior flowering and yield formation, particularly during the period of stone hardening |
| **Onion** | bulb enlargement, particularly during rapid bulb growth to vegetative period (and for seed production at flowering) |
| **Pea** | flowering and yield formation to vegetative, ripening for dry peas |
| **Pepper** | throughout but particularly just prior and at start of flowering |
| **Pineapple** | during period of vegetative growth |
| **Potato** | period of stolonization and tuber initiation, yield formation to early vegetative period and ripening |
| **Rice** | during period of head development and flowering to vegetative period and ripening |
| **Safflower** | seed filling and flowering to vegetative |
| **Sorghum** | flowering yield formation to vegetative; vegetative period less sensitive when followed by ample water supply |
| **Soyabean** | yield formation and flowering; particularly during pod development |
| **Sugar beet** | particularly first month after emergence |
| **Sugarcane** | vegetative period, particularly during period of tillering and stem elongation to yield formation |
| **Sunflower** | flowering to yield formation to late vegetative, particularly period of bud development |
| **Tobacco** | period of rapid growth to yield formation and ripening |
| **Tomato** | flowering to yield formation to vegetative period, particularly during and just after transplanting |
| **Watermelon** | flowering, fruit filling to vegetative period, particularly during vine development |
| **Wheat** | flowering to yield formation to vegetative period; winter wheat less sensitive than spring wheat |

*Source:* FAO (1979) *Yield Response to Water. Irrigation and Drainage Paper No. 33.*
Reproduced with kind permission of the Food and Agriculture Organization of the United Nations.

**TABLE 204   Typical Crop Rooting Depths and Some Indications of Soil Water readily available to Plants**

*NB* Rooting Depth can vary considerably according to such factors as husbandry, soil depth, soil type, cultivations, length of growing season, moisture regime, presence of inhospitable soil layers.

| Crop | Rooting depth cm | Main depth for water and nutrient uptake cm | Soil Water readily available as % of AWC |
|---|---|---|---|
| Alfalfa | >300 | 100 - 200 | 55 |
| Almonds | >150 | | |
| Artichokes | 100 - 300 | | |
| Asparagus | 200 - 300 | | |
| Banana | 90 | 50 - 75 | 35 |
| Barley | 100 - 150 | <100 | 55 |
| Bean (Phaseolus) | 100 - 150 | 50 - 70 | 45 |
| Beetroot | 100 | | |
| Broccoli | 70 | | |
| Brussels Sprouts | 30 - 60 | | |
| Cabbage | 60 | 40 - 50 | 45 |
| Cantaloupe | 150 | | |
| Castor Bean | >150 | | |
| Carrot | 50 - 100 | <60 | 35 |
| Cauliflower | 100 | | |
| Celery | 30 - 50 | | |
| Chard | 120 | | |
| Citrus | 100 - 200 | 120 - 160 | 40 Flower |
| | | | 60 - 70 Fruit |
| Clover | 100 | | |
| Cocoa | 100 | 15 | 20 |
| Cotton | 180 | 100 - 170 | 65 |
| Cucumber | >120 | 70 - 120 | 50 |
| Date | >300 | 150 - 250 | 50 |
| Deciduous Orchards | 200 - 300 | 80 - 100 | |
| Eggplant | 120 | | |
| Endive | 60 | | |
| Flax | 150 | | |
| Garlic | 60 | | |
| Grape | >300 | 100 - 200 | 40 |
| Grass Pasture | 100 - 150 | | |
| Groundnut | 180 | 50 - 100 | 50 |
| Leek | 60 | | |
| Lettuce | >50 | 30 - 50 | 30 |
| Lima Bean | >120 | | |
| Maize | 200 | 80 - 100 | 55 |
| Melon | >150 | 100 - 150 | 35 |
| Muskmelon | 120 | | |
| Mustard | 120 | | |
| Oil Palm | >200 | <45 | 40 |
| Olive | >180 | 120 - 170 | 65 |

**TABLE 204** *(continued)* **Typical Crop Rooting Depths and Some Indications of Soil Water readily available to Plants.**

*NB* Rooting Depth can vary considerably according to such factors as husbandry, soil depth, soil type, cultivations, length of growing season, moisture regime, presence of inhospitable soil layers.

| Crop | Rooting depth cm | Main depth for water and nutrient uptake cm | Soil Water readily available as % of AWC |
|------|------------------|---------------------------------------------|------------------------------------------|
| Onion | 50 | 30 - 50 | 25 |
| Pastures | 50 - 200 | 60 - 80 | |
| Parsley | 60 | | |
| Parsnips | 100 | | |
| Pea | 100 - 150 | 60 - 100 | 40 |
| Pepper | 100 | 50 - 100 | 25 |
| Pineapple | 100 | 30 - 60 | 50 |
| Potato | 60 | 40 - 60 | 25 |
| Pumpkin | >150 | | |
| Radish | 30 - 60 | | |
| Rice | 100 | <50 | 20 |
| Rubber | 300 - 400 | <200 | 55 |
| Rutabaga | 120 | | |
| Safflower | 100 - 200 | 100 - 200 | 60 |
| Sisal | >300 | 50 - 100 | 80 |
| Sorghum | 100 - 200 | 100 - 200 | 55 Growth |
| | | | 80 Ripen |
| Soyabean | >180 | 60 - 130 | 55 |
| Spinach | 30 - 50 | | |
| Squash | 100 | | |
| Strawberries | 100 | | |
| Sugar beet | >150 | 70 - 120 | 50 |
| Sugarcane | >300 | 120 - 200 | 40 - 70 |
| Sunflower | 200 - 300 | 80 - 150 | 50 |
| Sweet Potato | >180 | 100 - 150 | 65 |
| Tobacco | >100 | 50 - 100 | 35 Early Tobacco |
| | | | 65 Late Tobacco |
| Tomato | 150 | 70 - 150 | 40 |
| Turnips | 100 | | |
| Vegetables | 60 | 30 - 60 | 20 |
| Walnuts | >150 | | |
| Watermelon | 150 - 200 | 100 - 150 | 45 |
| Wheat, Spring | 120 - 150 | 90 | 50 Growth |
| | | | 90 Ripen |
| Wheat, Winter | 150 - 200 | 100 - 150 | 50 Growth |
| | | | 90 Ripen |
| Winter Squash | >120 | | |

*Sources:* Adapted from Landon, J.R., ed (1984) *Tropical Soil Manual.* Withers, B. & Vipond, S. (1974) *Irrigation Design and Practice.* FAO (1984) *Irrigation Practice and Management. Irrigation and Drainage Paper No.1.* Lorenz, O.A. & Maynard, D.N. (1980) *Knott's Handbook for Vegetable Growers. Rhodesian Irrigation Handbook* (1968).

# Salinity

---

**TABLE 205   Relative Susceptibility of Crops to Foliar Injury from Saline Sprinkling Waters**

---

## Na or Cl concentrations (mol/m$^3$) causing foliar injury[1]

| <5 | 5 - 10 | 10 - 20 | >20 |
|---|---|---|---|
| Almond | Grape | Alfalfa | Cauliflower |
| Apricot | Pepper | Barley | Cotton |
| Citrus | Potato | Corn | Sugar beet |
| Plum | Tomato | Cucumber | Sunflower |
| | | Safflower | |
| | | Sesame | |
| | | Sorghum | |

[1]   Foliar injury is influenced by cultural and environmental conditions. These data are presented only as general guidelines for daytime sprinkling.

*Source:* Maas, E.V. (1986) *Salt Tolerance of Plants.* Appl. Ag. Res. Vol.1 No.1.

---

**TABLE 206   Relative Salt Tolerance of Various Crops at Emergence and During Growth to Maturity According to Soil Salinity**

---

| Crop | Electrical conductivity of saturated soil extract to give: | |
|---|---|---|
| | 50% Yield dS / m | 50% Emergence dS / m |
| Barley | 18 | 16 - 24 |
| Cotton | 17 | 15 |
| Sugar beet | 15 | 6 - 12 |
| Sorghum | 15 | 13 |
| Safflower | 14 | 12 |
| Wheat | 13 | 14 - 16 |
| Beet, red | 9.6 | 13.8 |
| Cowpea | 9.1 | 16 |
| Alfalfa | 8.9 | 8 - 13 |
| Tomato | 7.6 | 7.6 |
| Cabbage | 7.0 | 13 |
| Corn | 5.9 | 21 - 24 |
| Lettuce | 5.2 | 11 |
| Onion | 4.3 | 5.6 - 7.5 |
| Rice | 3.6 | 18 |
| Bean | 3.6 | 8.0 |

*NB*   This table should be considered as indicative only since so many variables affect germination.

*Source:* Maas, E.V. (1986) *Salt Tolerance of Plants.* Appl. Ag. Res. Vol.1 No.1.

**TABLE 207    Tolerance of Some Fruit Crop Cultivars and Rootstocks to Root Zone and Irrigation Water Chloride Levels**

| Crop | Rootstock or Cultivar | | Maximum Permissible Cl⁻ without Leaf Injury[1] | |
|------|------------------------|--|---|---|
| | | | Root Zone (me/l) | Irrigation Water [2,3] (me/l) |
| **Avocado** | West Indian | | 7.5 | 5.0 |
| | Guatemalan | | 6.0 | 4.0 |
| | Mexican | | 5.0 | 3.3 |
| **Citrus** | Sunki Mandarin | Cleopatra mandarin | 25.0 | 16.6 |
| | Grapefruit | Rangpur lime | | |
| | Sampson tangelo | Sour orange | 15.0 | 10.0 |
| | Rough lemon | Ponkan mandarin | | |
| | Citrumelo 4475 | Sweet orange | 10.0 | 6.7 |
| | Trifoliate orange | Savage citrange | | |
| | Cuban shaddock | Rusk citrange | | |
| | Calamondin | Troyer citrange | | |
| **Grape** | Salt Creek, 1613 – 3 | | 40.0 | 27.0 |
| | Dog Ridge | | 30.0 | 20.0 |
| **Stone Fruits** | Marianna | | 25.0 | 17.0 |
| | Lovell, Shalil | | 10.0 | 6.7 |
| | Yunnan | | 7.5 | 5.0 |
| **Berries** | Boysenberry | | 10.0 | 6.7 |
| | Olallie blackberry | | 10.0 | 6.7 |
| | Indian Summer Raspberry | | 5.0 | 3.3 |
| **Grape** | Thompson seedless | | 20.0 | 13.3 |
| | Perlette | | 20.0 | 13.3 |
| | Cardinal | | 10.0 | 6.7 |
| | Black Rose | | 10.0 | 6.7 |
| **Strawberry** | Lassen | | 7.5 | 5.0 |
| | Shasta | | 5.0 | 3.3 |

1. For some crops, the concentration given may exceed the overall salinity tolerance of that crop and cause some reduction in yield in addition to that caused by chloride ion toxicities.

2. Values given are or the maximum concentration in the irrigation water. The values were derived from saturation extract data ($EC_e$) assuming a 15-20 per cent leaching fraction and $EC_e = 1.5 EC_w$

3. The maximum permissible values apply only to surface irrigated crops. Sprinkler irrigation may cause excessive leaf burn at values far below these.

*Source:* FAO (1985) *Water Quality for Agriculture. Irrigation and Drainage Paper No. 29.*
Reproduced with kind permission of the Food and Agriculture Organization of the United Nations.

**TABLE 208  Chloride Tolerance of Agricultural Crops. Listed in Order of Increasing Tolerance[a]**

| Crop | Maximum Cl⁻ concentration[b] without loss in yield (Threshold) ($mol/m^3$) | Percent decrease in yield at Cl⁻ concentrations[b] above the threshold (Slope) (% per $mol/m^3$) |
|---|---|---|
| Strawberry | 10 | 3.3 |
| Bean | 10 | 1.9 |
| Onion | 10 | 1.6 |
| Carrot | 10 | 1.4 |
| Radish | 10 | 1.3 |
| Lettuce | 10 | 1.3 |
| Turnip | 10 | 0.9 |
| Rice, Paddy[c] | 30[d] | 1.2[d] |
| Pepper | 15 | 1.4 |
| Clover, Strawberry | 15 | 1.2 |
| Clover, Red | 15 | 1.2 |
| Clover, Alsike | 15 | 1.2 |
| Clover, Ladino | 15 | 1.2 |
| Corn | 15 | 1.2 |
| Flax | 15 | 1.2 |
| Potato | 15 | 1.2 |
| Sweet Potato | 15 | 1.1 |
| Broad Bean | 15 | 1.0 |
| Cabbage | 15 | 1.0 |
| Foxtail, Meadow | 15 | 1.0 |
| Celery | 15 | 0.6 |
| Clover, Berseem | 15 | 0.6 |
| Orchardgrass | 15 | 0.6 |
| Sugarcane | 15 | 0.6 |
| Trefoil, Big | 20 | 1.9 |
| Lovegrass | 20 | 0.8 |
| Spinach | 20 | 0.8 |
| Alfalfa | 20 | 0.7 |
| Sesbania[c] | 20 | 0.7 |
| Cucumber | 25 | 1.3 |
| Tomato | 25 | 1.0 |
| Broccoli | 25 | 0.9 |
| Squash, Scallop | 30 | 1.1 |
| Vetch, Common | 30 | 1.1 |
| Wild rye, Beardless | 30 | 0.6 |
| Sudan grass | 30 | 0.4 |

TABLE 208 *(continued)*   Chloride Tolerance of Agricultural Crops. Listed in Order of Increasing Tolerance[a]

| Crop | Maximum Cl⁻ concentration[b] without loss in yield (Threshold) ($mol/m^3$) | Percent decrease in yield at Cl⁻ concentrations[b] above the threshold (Slope) (% per $mol/m^3$) |
|---|---|---|
| Wheat grass, Standard crested | 35 | 0.4 |
| Beet, Red[c] | 40 | 0.9 |
| Fescue, Tall | 40 | 0.5 |
| Squash, zucchini | 45 | 0.9 |
| Hardinggrass | 45 | 0.8 |
| Cowpea | 50 | 1.2 |
| Trefoil, Narrow-leaf bird's foot | 50 | 1.0 |
| Ryegrass, Perennial | 55 | 0.8 |
| Wheat, Durum | 55 | 0.5 |
| Barley (forage)[c] | 60 | 0.7 |
| Wheat[c] | 60 | 0.7 |
| Sorghum | 70 | 1.6 |
| Bermuda Grass | 70 | 0.6 |
| Sugar beet[c] | 70 | 0.6 |
| Wheat Grass, Fairway crested | 75 | 0.7 |
| Cotton | 75 | 0.5 |
| Wheat Grass, Tall | 75 | 0.4 |
| Barley[c] | 80 | 0.5 |

[a]   These data serve only as guidelines to relative tolerances among crops. Absolute tolerances vary, depending upon climate, soil conditions and cultural practices.

[b]   Cl⁻ concentrations in saturated-soil extracts sampled in the rootzone. To convert Cl⁻ concentrations to ppm, multiply threshold values by 35. To convert % yield decreases to % per ppm, divide slope values by 35.

[c]   Less tolerant during emergence and seedling stage.

[d]   Values for paddy rice refer to the Cl⁻ concentration in the soil water during the flooded growing conditions.

*Source:* Maas, E.V. (1990) Crop Salt Tolerance. In Tanji, K.K. ed., *Agricultural Salinity Assessment and Management. ASCE Manuals and Reports on Engineering Practice No. 71.*

**TABLE 209  Boron Tolerance Limits for Agricultural Crops according to Soil Water Boron Contents**

| Common name | Threshold[a] (g/m$^3$) | Common name | Threshold[a] (g/m$^3$) |
|---|---|---|---|
| **Very sensitive** | | Pea[b] | 1.0 - 2.0 |
| Lemon[b] | <0.5 | Carrot | 1.0 - 2.0 |
| Blackberry[b] | <0.5 | Radish | 1.0 - 2.0 |
| | | Potato | 1.0 - 2.0 |
| **Sensitive** | | Cucumber | 1.0 - 2.0 |
| Avocado[b] | 0.5 - 0.75 | Lettuce | 1.3 |
| Grapefruit[b] | 0.5 - 0.75 | | |
| Orange[b] | 0.5 - 0.75 | **Moderately tolerant** | |
| Apricot[b] | 0.5 - 0.75 | Cabbage[b] | 2.0 - 4.0 |
| Peach[b] | 0.5 - 0.75 | Turnip | 2.0 - 4.0 |
| Cherry[b] | 0.5 - 0.75 | Bluegrass, Kentucky[b] | 2.0 - 4.0 |
| Plum[b] | 0.5 - 0.75 | Barley | 3.4 |
| Persimmon[b] | 0.5 - 0.75 | Cowpea | 2.5 |
| Fig, Kadota[b] | 0.5 - 0.75 | Oats | 2.0 - 4.0 |
| Grape[b] | 0.5 - 0.75 | Corn | 2.0 - 4.0 |
| Walnut[b] | 0.5 - 0.75 | Artichoke[b] | 2.0 - 4.0 |
| Pecan[b] | 0.5 - 0.75 | Tobacco[b] | 2.0 - 4.0 |
| Onion | 0.5 - 0.75 | Mustard[b] | 2.0 - 4.0 |
| Garlic | 0.75 - 1.0 | Clover, Sweet[b] | 2.0 - 4.0 |
| Sweet potato | 0.75 - 1.0 | Squash | 2.0 - 4.0 |
| Wheat | 0.75 - 1.0 | Muskmelon[b] | 2.0 - 4.0 |
| Sunflower | 0.75 - 1.0 | Cauliflower | 4.0 |
| Bean, Mung[b] | 0.75 - 1.0 | | |
| Sesame[b] | 0.75 - 1.0 | **Tolerant** | |
| Lupine[b] | 0.75 - 1.0 | Tomato | 5.7 |
| Strawberry[b] | 0.75 - 1.0 | Alfala[b] | 4.0 - 6.0 |
| Artichoke, Jerusalem[b] | 0.75 - 1.0 | Vetch, Purple[b] | 4.0 - 6.0 |
| Bean, Kidney[b] | 0.75 - 1.0 | Parsley[b] | 4.0 - 6.0 |
| Bean, Lima[b] | 0.75 - 1.0 | Beet, Red | 4.0 - 6.0 |
| Bean, Snap | 1.0 | Sugar beet | 4.9 |
| Peanut | 0.75 - 1.0 | | |
| | | **Very tolerant** | |
| **Moderately sensitive** | | Sorghum | 7.4 |
| Broccoli | 1.0 | Cotton | 6.0 - 10.0 |
| Pepper, red | 1.0 - 2.0 | Celery | 9.8 |
| | | Asparagus[b] | 10.0 - 15.0 |

[a]  Maximum permissible concentration in soil water without yield reduction. Boron tolerances may vary, depending upon climate, soil conditions, and crop varieties.

[b]  Tolerance based on reductions in vegetative growth.

*Source:*  Maas, E.V. (1990) *Crop Salt Tolerance.* In Tanji, K.K. ed., *Agricultural Salinity Assessment and Management. ASCE  Manuals and Reports on Engineering Practice No. 71.*

**TABLE 210   Crop Sensitivity to Boron in Irrigation Water based on Toxicity Symptoms Observed on Plants Grown in Sand Culture arranged in Descending Order of Tolerance**

| Sensitive 0.3-1 ppm boron | Semi-tolerant 1-2 ppm boron | Tolerant 2-4 ppm boron |
|---|---|---|
| Citrus | Lima Bean | Carrot |
| Pecan | Sweet Potato | Lettuce |
| Avocado | Bell Pepper | Cabbage |
| Black Walnut | Oat | Turnip |
| Apricot | Sorghum Grain | Onion |
| Peach | Corn | Broad Bean |
| Cherry | Wheat | Alfalfa |
| Persimmon | Barley | Garden Beet |
| Fig | Olive | Mangel |
| Grape | Field Pea | Sugar beet |
| Apple | Radish | Palm |
| Pear | Tomato | Asparagus |
| Plum | Cotton | Date Palm |
| Navy Bean | Potato | |
| Jerusal Artichoke | Pumpkin | |
| Walnut | Sunflower | |
| Thornless Blackberry | | |
| **Corresponding Soil B Content**  0.7 | 1.5 | 1.5 - 2.5  (ppm in sat. extract) |

*Source:* Adapted from *Boron Injury to Plants. USDA Ag. Inf. Bull. 211* (1960).

**TABLE 211   Citrus and Stone-Fruit Rootstocks Ranked in Order of Increasing Boron Accumulation and Transport to Scions**

|  | Common name | |
|---|---|---|
| **Citrus** | **Citrus** | **Stone fruit** |
| Alemow | Savage Citrange | Almond |
| Gajanimma | Cleopatra Mandarin | Myrobalan Plum |
| Chinese Box Orange | Rusk Citrange | Apricot |
| Sour Orange | Sunki Mandarin | Marianna Plum |
| Calamondin | Sweet Lemon | Shalil Peach |
| Sweet Orange | Trifoliate Orange | |
| Yuzu | Citrumelo 4475 | |
| Rough Lemon | Ponkan Mandarin | |
| Grapefruit | Sampson Tangelo | |
| Rangpur Lime | Cuban Shaddock | |
| Troyer Citrange | Sweet Lime | |

*Source:* Maas, E.V. (1986) Salt Tolerance of Plants. *Appl Ag. Res. Vol. 1 No. 1.*

.

---

**TABLE 212   Tolerance of Various Crops to Exchangeable Sodium Percentage (ESP) under Non-saline Conditions**

---

| Tolerance to *ESP* and range at which affected | Crop | Growth response under field conditions |
|---|---|---|
| Extremely sensitive (*ESP* = 2-10) | **Deciduous fruits** **Nuts** **Citrus** **Avocado** | Sodium toxicity symptoms even at low *ESP* values |
| Sensitive (*ESP* = 10-20) | **Beans** | Stunted growth at low *ESP* values, even though the physical condition of the soil may be good |
| Moderately tolerant (*ESP* = 20-40) | **Clover** **Oats** **Tall fescue** | Stunted growth as a result of both nutritional factors and adverse soil conditions |
| Tolerant (*ESP* = 40-60) | **Wheat** **Cotton** **Alfalfa** **Barley** **Tomatoes** **Beets** | Stunted growth, usually due to adverse physical condition of soil |
| Most tolerant (*ESP* = more than 60) | **Crested and Fairway Wheatgrass** **Tall wheatgrass** **Rhodes grass** | Stunted growth, usually due to adverse physical condition of soil |

---

*Source:* Shainberg, I. & Oster, J.D. (1978) *Quality of Irrigation Water* (after Pearson, 1960). IIIC Publication. No. 1. Reproduced with kind permission of the International Irrigation Information Centre.

**TABLE 213    Generalized Classification of the Relative Tolerance of Selected Crops to Exchangeable Sodium**

| Sensitive ESP <15 | Semi-tolerant ESP 15 - 40 | Tolerant ESP >40 |
|---|---|---|
| Avocado | Carrot | Alfalfa |
| Deciduous fruits | Clover, Ladino | Barley |
| Nuts | Dallisgrass | Beet, Garden |
| Bean, Green | Fescue, Tall | Beet, Sugar |
| Cotton (at germination) | Lettuce | Bermuda Grass |
| Maize | Bajara Grass | Cotton |
| Peas | Sugarcane | Paragrass |
| Grapefruit | Berseem | Rhodes Grass |
| Orange | Benji | Wheatgrass, Crested |
| Peach | Raya | Wheatgrass, Fairway |
| Tangerine | Oat | Wheatgrass, Tall |
| Mung | Onion | Karnal grass |
| Mash | Radish | |
| Lentil | Rice | |
| Groundnut (peanut) | Rye | |
| Gram | Ryegrass, Italian | |
| Cowpeas | Sorghum | |
| | Spinach | |
| | Tomato | |
| | Vetch | |
| | Wheat | |

*Source:* FAO (1985) *Water Quality for Agriculture. Irrigation and Drainage Paper No. 29.*
Reproduced with kind permission of the Food and Agriculture Organization of the United Nations.

**TABLE 214   ESP Levels at which Selected Crops show 50% Yield Reduction**

| Sensitive:<br>50% yield reduction at<br>ESP < 15% | Semi-tolerant:<br>50% yield reduction at<br>ESP 15 - 25% | Tolerant:<br>50% yield reduction at<br>ESP 35% |
|---|---|---|
| Avocado | Dwarf Kidney Bean | Alfalfa |
| Green Bean | Ladino Clover | Barley |
| Corn | Carrot | Sugar beet |
| Tall Fescue | Lemon | Cotton |
| Peach | Lettuce | Dallis Grass |
| Sweet Orange | Oats | Onion |
| Grapefruit | Rice | Bermuda Grass |
|  | Sorghum |  |
|  | Wheat |  |
|  | Sugarcane |  |

*Note:* The relationships tabulated above should only be considered as broad guidelines, since local conditions can markedly affect crop response.

For soils that are to be irrigated, the ESP that will develop in equilibrium with the irrigation water is of major importance, rather than values measured in advance from the soil. Given adequate drainage, the ESP values that are likely to develop can be approximately predicted from the SAR of the irrigation water.

*Source:* Landon, J. R., ed. (1984) *Tropical Soil Manual.*
Reproduced with kind permission of Booker Tate Ltd.

**TABLE 215   Crop Germination Tolerance to Salinity**

| Conductivity ($\mu$S) | Index | Comment |
|---|---|---|
| 1900 - 2200 | 0 | Satisfactory |
| 2210 - 2400 | 1 | Satisfactory |
| 2410 - 2600 | 2 | Satisfactory |
| 2610 - 2700 | 3 | Seed germination may be affected Tolerance varies according to specie |
| 2710 - 2800 | 4 | Seed germination may be affected Tolerance varies according to specie |
| 2810 - 3000 | 5 | Seed germination may be affected Tolerance varies according to specie |
| 3010 - 3300 | 6 | Seed germination may be affected Tolerance varies according to specie |
| 3310 - 3700 | 7 | Seed germination may be affected Tolerance varies according to specie |
| 3710 - 4000 | 8 | Seed germination may be affected Tolerance varies according to specie |

*Source:* Atlas Interlates Analysis Service User Guide 1989.
Reproduced with kind permission of Atlas Interlates Ltd.

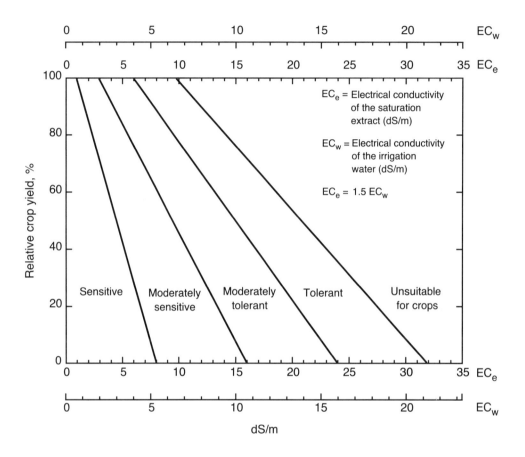

**Fig. 43** Graphical representation of salt tolerance ratings of agricultural crops according to irrigation water quality

*NB* $EC_w$ = EC of Irrigation Water dS/m

$EC_e$ = EC of Soil (saturated extract) dS/m

An EC concentration factor of 1.5 is used to represent a 15-20% leaching fraction

*Source:* Maas, E.V. (1986) *Salt Tolerance of Plants. Appl. Ag. Res.* Vol. 1 No. 1.

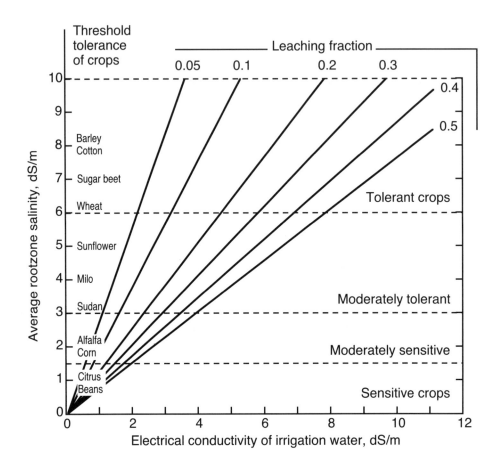

**Fig. 44** Assessing salinity hazards in conventional irrigation.
(Relations between average rootzone salinity (saturation extract basis), electrical conductivity of irrigation water, and leaching fraction to use for conditions of conventional irrigation management)

*Source:* Rhoades, J.D. (1988) *Evidence of the Potential to use Saline Water for Irrigation.* International Seminar on the re-use of low quality water for Irrigation. Water Research Centre Bari Institute, Cairo/Aswan, Egypt. Reproduced with kind permission of J D. Rhoades.

---

**TABLE 216   Crop Tolerance to Soil Salinity (Quantitative)**

---

*NB*  1.  The following table should be considered as guideline only as salinity tolerance can vary according to the following factors:
2.  In gypsiferous soils plants tolerate about 2dS/m higher $EC_e$;
3.  Some of the ratings are estimates gleaned from the literature
4.  Varieties can vary in tolerance;
5.  Some species are less tolerant during the emergence and seedling stage; climate can also affect tolerance;
6.  The value for Paddy rice is of soil in submerged conditions;
7.  The tolerance ratings relate to **Fig. 43**;
8.  General high quality management practice including correct leaching practices and frequent irrigation in irrigated situations, and provision of adequate fertility conditions in general will all enhance crop salt tolerance and productivity.

**EC units are dS/m**

| Crop | Salinity Threshold Soil $EC_e$ at which yield decrease starts (A) | % Productivity decrease per unit increase beyond the threshold soil $EC_e$ (B) | General Tolerance Rating (Qualitative) | Theoretical Zero Yield soil $EC_e$ | Max EC of Irrigation Water at Zero Yield $EC_e$ assuming 15-20% leaching fraction |
|---|---|---|---|---|---|
| | (S = Sensitive, MS = Moderately Sensitive, MT = Moderately Tolerant, T = Tolerant) | | | | |
| Alfalfa | 2 | 7.3 | MS | 16 | 10 |
| Alkaligrass, Nuttal | | | T | | |
| Alkali Sacaton | | | T | | |
| Almond | 1.5 | 19 | S | 6.8 | 4.5 |
| Apple | 1 | | S | | |
| Apricot | 1.6 | 24 | S | 5.8 | 3.8 |
| Artichoke | | | MT | | |
| Asparagus | 4.1 | 2 | T | | |
| Avocado | 1 | | S | | |
| Barley, grain | 8 | 5 | T | 28 | 19 |
| Barley, forage | 6 | 7.1 | MT | 19 | 13 |
| Bean (Phaseolus) | 1 | 19 | S | 6.3 | 4.2 |
| Bean, Broad | 1.6 | 9.6 | MS | | |
| Beetroot | 4 | 9 | MT | 15 | 10 |
| Bentgrass | | | MS | | |
| Bermuda grass | 6.9 | 6.4 | T | 23 | 15 |
| Blackberry | 1.5 | 22 | S | 6 | 4 |
| Blue Gramma | | | MT | | |
| Bluestem, Angleton | | | MS | | |
| Boysenberry | 1.5 | 22 | S | 6 | 4 |
| Broccoli | 2.8 | 9.2 | MS | | |
| Brome, Mountain | | | MT | | |
| Brome, Smooth | | | MS | | |
| Brussels Sprouts | | | MS | | |
| Buffelgrass | | | MS | | |

*(continued over)*

**TABLE 216** *(continued)*  **Crop Tolerance to Soil Salinity (Quantitative)**

EC units are dS/m

| Crop | Salinity Threshold Soil EC$_e$ at which yield decrease starts (A) | % Productivity decrease per mmho/cm increase beyond the threshold soil EC$_e$ (B) | General Tolerance Rating (Qualitative) | Theoretical Zero Yield soil EC$_e$ | Max EC of Irrigation Water at Zero Yield EC$_e$ assuming 15-20% leaching fraction |
|---|---|---|---|---|---|
| | (S = Sensitive, MS = Moderately Sensitive, MT = Moderately Tolerant, T = Tolerant) | | | | |
| Broad Bean | 1.6 | 9.6 | MS | 12 | 8 |
| Burnet | | | MS | | |
| Cabbage | 1.8 | 9.7 | MS | 12 | 8.1 |
| Canarygrass, Reed | | | MT | | |
| Canteloupe melon | | | MS | | |
| Carrot | 1 | 14 | S | 8.1 | 5.4 |
| Castor Bean | | | MS | | |
| Cauliflower | 2.5 | | MS | | |
| Celery | 1.8 | 6.2 | MS | 18 | 12 |
| Cherimoya | | | S | | |
| Cherry, Sweet | | | S | | |
| Cherry, Sand | | | S | | |
| Clover, Alsike | 1.5 | 12 | MS | 9.8 | 6.6 |
| Clover, Berseem | 1.5 | 5.7 | MS | 19 | 13 |
| Clover, Hubam | | | MT | | |
| Clover, Ladino | 1.5 | 12 | MS | 9.8 | 6.6 |
| Clover, Red | 1.5 | 12 | MS | 9.8 | 6.6 |
| Clover, Strawberry | 1.5 | 12 | MS | 9.8 | 6.6 |
| Clover, Subterranean | | | S | | |
| Clover, Sweet | | | MT | | |
| Clover, White Dutch | | | MS | | |
| Cocksfoot | 1.5 | 6.2 | MS | 18 | 12 |
| Corn, Grain | 1.7 | 12 | MS | 10 | 6.7 |
| Corn, Forage | 1.8 | 7.4 | MS | 15 | 10 |
| Corn, Sweet | 1.7 | 12 | MS | 10 | 6.7 |
| Cotton | 7.7 | 5.2 | T | 27 | 18 |
| Cowpea, seed | 4.9 | 12 | MT | 13 | 8.8 |
| Cowpea, forage | 2.5 | 11 | MS | 12 | 7.8 |
| Cucumber | 2.5 | 13 | MS | 10 | 6.8 |
| Currant | | | S | | |
| Dallisgrass | | | MS | | |
| Date Palm | 4 | 3.6 | T | 32 | 21 |
| Eggplant | 1.1 | 6.9 | MS | | |
| Fescue, Tall | 3.9 | 5.3 | MT | 20 | 13 |
| Fescue, Meadow (Alopecurus) | 1.5 | 9.6 | MS | | |
| Fig | 4.2 | | MT | | |
| Flax | 1.7 | 12 | MS | 10 | 6.7 |
| Foxtail | 1.5 | 9.6 | MS | 12 | 7.9 |
| Gooseberry | | | S | | |
| Grama, Blue | | | MS | | |
| Grape | 1.5 | 9.6 | MS | 12 | 7.9 |
| Grapefruit | 1.8 | 16 | S | 8 | 5.4 |

**TABLE 216** *(continued)* **Crop Tolerance to Soil Salinity (Quantitative)**

| | EC units are dS/m | | | | |
|---|---|---|---|---|---|
| | Salinity Threshold Soil $EC_e$ at which yield decrease starts (A) | % Productivity decrease per mmho/cm increase beyond the threshold soil $EC_e$ (B) | General Tolerance Rating (Qualitative) | Theoretical Zero Yield soil $EC_e$ | Max EC of Irrigation Water at Zero Yield $EC_e$ assuming 15-20% leaching fraction |
| Crop | (S = Sensitive, MS = Moderately Sensitive, MT = Moderately Tolerant, T = Tolerant) | | | | |
| Guar | 8.8 | 17 | T | | |
| Guayule | 15 | 13 | T | | |
| Hardingrass (Phalaris) | 4.6 | 7.6 | MT | 18 | 12 |
| Jojoba | | | T | | |
| Jujube | | | MT | | |
| Kale | 6.5 | | MS | | |
| Kallargrass | | | T | | |
| Kenaf | | | MT | | |
| Kohlrabi | | | MS | | |
| Lemon | 1 | | S | | |
| Lettuce | 1.3 | 13 | MS | 9 | 6 |
| Lime | | | S | | |
| Loquat | | | S | | |
| Lovegrass | 2 | 8.4 | MS | 14 | 9.3 |
| Mango | | | S | | |
| Milkvetch, Cicer | | | MS | | |
| Millet, Foxtail | | | MS | | |
| Mulberry | | | MS | | |
| Muskmelon | 2.5 | | MS | | |
| Natal Plum | 6 | | T | | |
| Oats, Forage | | | MS | | |
| Oats, Grain | | | MT | | |
| Oatgrass, Tall | | | MS | | |
| Okra | | | S | | |
| Olive | 4 | | MT | | |
| Onion | 1.2 | 16 | S | 7.4 | 5 |
| Orange | 1.7 | 16 | S | 8 | 5.3 |
| Panicgrass, Blue | | | MT | | |
| Panicgrass, Green | | | MT | | |
| Papaya | | | MT | | |
| Parsnip | | | S | | |
| Passion Fruit | | | S | | |
| Pea | 2.5 | | S | | |
| Peach | 1.7 | 21 | S | 6.5 | 4.3 |
| Peanut | 3.2 | 29 | MS | 6.6 | 4.4 |
| Pear | 1 | | S | | |
| Pepper | 1.5 | 14 | MS | 8.6 | 5.8 |
| Persimmon | | | S | | |
| Pineapple | | | MT | | |
| Pineapple Guava | 1.2 | | S | | |
| Plum | 1.5 | 18 | | | |
| Plum, Prune | 1.5 | 18 | S | 7.1 | 4.7 |

*(continued over)*

**TABLE 216** *(continued)* **Crop Tolerance to Soil Salinity (Quantitative)**

| | EC units are dS/m | | | | |
|---|---|---|---|---|---|
| Crop | Salinity Threshold Soil $EC_e$ at which yield decrease starts (A) | % Productivity decrease per mmho/cm increase beyond the threshold soil $EC_e$ (B) | General Tolerance Rating (Qualitative) | Theoretical Zero Yield soil $EC_e$ | Max EC of Irrigation Water at Zero Yield $EC_e$ assuming 15-20% leaching fraction |
| | (S = Sensitive, MS = Moderately Sensitive, MT = Moderately Tolerant, T = Tolerant) | | | | |
| Pomegranate | 4 | | MT | | |
| Potato | 1.7 | 12 | MS | 10 | 6.7 |
| Pummelo | | | S | | |
| Pumpkin | | | MS | | |
| Radish | 1.2 | 13 | MS | 8.9 | 5.9 |
| Rape, Forage | | | MT | | |
| Raspberry | 1 | | S | | |
| Rescuegrass | | | MT | | |
| Rhodesgrass | | | MT | | |
| Rice, Paddy | 3 | 12 | S | 11 | 7.6 |
| Rose Apple | | | S | | |
| Rosemary | 4.5 | | T | | |
| Rye, Grain | 11.4 | 10.8 | T | | |
| Rye, Forage | | | MS | | |
| Ryegrass, Italian | | | MT | | |
| Ryegrass, Perennial | 5.6 | 7.6 | MT | 19 | 13 |
| Ryegrass, Wimmera | | | MT | | |
| Safflower | 6.5 | | MT | | |
| Saltgrass, Desert | | | T | | |
| Sapote, White | | | S | | |
| Sesame | | | S | | |
| Sesbania | 2.3 | 7 | MS | 17 | 11 |
| Siratro | | | MS | | |
| Sorghum | 6.8 | 16 | MT | 13 | 8.7 |
| Soyabean | 5 | 20 | MT | 10 | 6.7 |
| Sphaerophysa | 2.2 | 7 | MS | 16 | 11 |
| Spinach | 2 | 7.6 | MS | 15 | 10 |
| Squash, Scallop | 3.2 | 16 | MS | 9.4 | 6.3 |
| Squash, Zucchini | 4.7 | 9.4 | MT | 15 | 10 |
| Strawberry | 1 | 33 | S | 4 | 2.7 |
| Sudan grass | 2.8 | 4.3 | MT | 26 | 17 |
| Sugar beet | 7 | 5.9 | T | 24 | 16 |
| Sugarcane | 1.7 | 5.9 | MS | 19 | 12 |
| Sunflower | | | MS | | |
| Sweet Potato | 1.5 | 11 | MS | 11 | 7.1 |
| Tangerine | | | S | | |
| Timothy Grass | | | MS | | |
| Tomato | 2.5 | 9.9 | MS | 13 | 8.4 |
| Trefoil, Big | 2.3 | 19 | MS | 7.6 | 5 |
| Trefoil, Narrowleaf Birdsfoot | 5 | 10 | MT | 15 | 10 |
| Trefoil, Broadleaf Birdsfoot | | | MT | | |

**TABLE 216 *(continued)*  Crop Tolerance to Soil Salinity (Quantitative)**

| | EC units are dS/m | | | | |
|---|---|---|---|---|---|
| | Salinity Threshold Soil $EC_e$ at which yield decrease starts (A) | % Productivity decrease per mmho/cm increase beyond the threshold soil $EC_e$ (B) | General Tolerance Rating (Qualitative) | Theoretical Zero Yield soil $EC_e$ | Max EC of Irrigation Water at Zero Yield $EC_e$ assuming 15-20% leaching fraction |
| Crop | (S = Sensitive, MS = Moderately Sensitive, MT = Moderately Tolerant, T = Tolerant) | | | | |
| Triticale | 6.1 | 2.5 | T | | |
| Turnip | 9 | 9 | MS | 12 | 8 |
| Vetch, Common | 3 | 11 | MS | 12 | 8.1 |
| Walnut | | | MS | | |
| Watermelon | | | MS | | |
| Wheat, Grain | 6 | 7.1 | MT | 20 | 13 |
| Wheat, Forage | 4.5 | 2.6 | MT | | |
| Wheat, semidwarf | 8.6 | 3 | T | | |
| Wheat, Durum | 5.9 | 3.8 | T | 24 | 16 |
| Wheat, Durum forage | 2.1 | 2.5 | MT | | |
| Wheatgrass, Standard Crested | 3.5 | 4 | MT | 28 | 19 |
| Wheatgrass, Fairway Crested | 7.5 | 6.9 | T | 22 | 15 |
| Wheatgrass, Intermediate | | | MT | | |
| Wheatgrass, Slender | | | MT | | |
| Wheatgrass, Tall | 7.5 | 4.2 | T | 31 | 21 |
| Wheatgrass, Western | | | MT | | |
| Wildrye, Altai | | | T | | |
| Wildrye, Beardless | 2.7 | 6 | MT | 19 | 13 |
| Wildrye, Canadian | | | MT | | |
| Wildrye, Russian | | | T | | |

*NB*  1.  Calculation of relative yield for any given salinity is as follows:

$$\text{Relative Yield} = 100 - B(k_e - A)$$

where A = Salinity Threshold

B = % Productivity decrease per unit salinity increase above the threshold

$k_e$ = Salinity of soil saturation extract

2.  It is now increasingly considered that traditional salt tolerance ratings, such as those listed in the above table, are too conservative and that in many circumstances it is possible to use more saline waters economically.

*Adapted from:* Maas, E.V. (1986) Salt Tolerance of Plants. *Appl. Ag. Res.* Vol. 1 No. 1. Bresler, E. *et al.* (1984) *Saline and Sodic Soils.* FAO (1985) *Water Quality for Agriculture. Irrigation and Drainage Paper No. 29.* Landon, J.R., ed. (1984) *Tropical Soil Manual.* Jensen, M.E., ed. (1983) *Design and Operation of Farm Irrigation Systems.* ASAE Monograph American Society of Agronomy. Maas, E.V. (1990) Crop Salt Tolerance. In Tanji, K.K. ed., *Agricultural Salinity Assessment and Management. ASCE Manuals and Reports on Engineering Practice No. 71.*

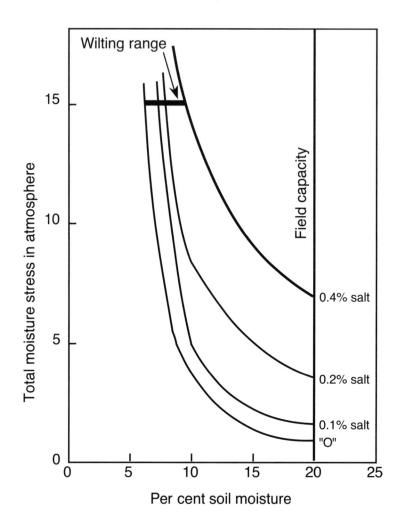

**Fig. 45** Relation of moisture stress to moisture percentage at different levels of soil salinity

*Source:* Irrigation Association (1983) *Irrigation.* 5th edition. (After Waddleigh & Ayers (1945) *Plant Physiology.* Am. Soc. Plant Physiol.) Reproduced with kind permission of the Irrigation Association, Arlington, Virginia.

# Soil pH

**TABLE 217  Relative Tolerance of Vegetable Crops to Soil Acidity**

Vegetables in the slightly tolerant group can be grown successfully on soils that are on the alkaline side of neutrality. They do well up to pH 7.6 if there is no deficiency of essential nutrients. Vegetables in the very tolerant group will grow satisfactorily at a soil pH as low as 5.0. For the most part even the most tolerant crops grow better at pH 6.0 - 6.8 than in more acid soils.

| Slightly Tolerant (pH 6.8 - 6.0) | Moderately Tolerant (pH 6.8 - 5.5) | Very Tolerant (pH 6.8 - 5.0) |
|---|---|---|
| Asparagus | Bean | Chicory |
| Beet | Bean, Lima | Dandelion |
| Broccoli | Brussels Sprouts * | Endive |
| Cabbage | Carrot | Fennel |
| Cauliflower | Collard | Potato ** |
| Celery | Corn | Rhubarb |
| Chard, Swiss | Cucumber | Shallot |
| Cress | Eggplant | Sorrel |
| Chinese Cabbage | Garlic | Sweet potato |
| Leek | Gherkin | Watermelon |
| Lettuce | Horseradish | |
| Muskmelon | Kale * | |
| New Zealand Spinach | Kohlrabi * | |
| Okra | Mustard | |
| Onion | Parsley | |
| Orach | Pea | |
| Parsnip | Pepper | |
| Salsify | Pumpkin | |
| Soybean | Radish | |
| Spinach | Rutabaga | |
| Watercress | Squash | |
| | Tomato | |
| | Turnip | |

*Notes:*    \*   Risk of club root disease at lower pH
         \*\*  Risk of common scab at higher pH

*Source:* Lorenz, O.A., & Maynard, D.N. (1980) *Knott's Handbook for Vegetable Growers.*
© 1980 and reproduced with permission of John Wiley and Sons, Inc.

**TABLE 218   pH Preferences for Crop Plants**

*NB* It is noted that liberal dressings of farmyard manure can somewhat offset the effects of acidity; also that crops are more tolerant of acidity in cool moist climates than warm dry ones. Certain peculiarities of specific soils may also shift the normal pH tolerance of crops.

| Crop | Optimal pH | Tolerable pH range (for satisfactory yield) |
|---|---|---|
| Alfalfa | 6.5 - 7.5 | 6 - 8 |
| Asparagus | | 6 - 7 |
| Avocado | | 6 - 8 |
| Barley | | 6.3 - 7.7 |
| Bean | 6 - 7 | 5.5 - 7.5 |
| Banana | 6 - 7.5 | 4 - 8 |
| Cabbage | 7.2 | 6 - 7.5 |
| Cantaloupe Melon | | 6 - 6.5 |
| Carrot | | 5.5 - 6.5 |
| Cashew | | 5.5 - 7 |
| Cassava | | 5.5 - 6.5 |
| Celery | | 6 - 6.5 |
| Cocoa | 6 - 7 | 4.5 - 8 |
| Coconut | 6 - 7.5 | 5 - 8 |
| Coffee | 5 - 6 | 4.5 - 7 |
| Corn | | 6 - 7 |
| Cotton | 5.2 - 6 | 4.8 - 7.5 |
| Cowpea | 6 - 7 | 5.5 - 7.5 |
| Cucumber | | 6.5 - 7.5 |
| Date | | 6.5 - 8 |
| Eggplant | | 6 - 7 |
| Grape | | 6 - 7 |
| Gram | 6 - 7 | 5.5 - 7.5 |
| Groundnut | 5.3 - 6.6 | 5 - 7 |
| Hemp | | 6 - 7 |
| Kale | 6.3 - 7.2 | 5.7 - 7.2 |
| Kenaf | | 6 - 7 |
| Kudzu | | 5.5 - 6.5 |
| Lemon | | 5.5 - 7 |
| Lettuce | | 6 - 7 |
| Maize | 5.5 - 7 | 5 - 8 |
| Mango | | 5.5 - 7.5 |
| Millet (bulrush, finger, panicum ) | | 5 - 6 |
| Oats | 5.8 - 6.5 | 5 - 6.5 |
| Pol Palm | 5.5 - 6 | 4 - 8 |
| Olive | 7 | |
| Onion | | 6 - 7.5 |
| Orange | | 5 - 7 |
| Pangola Grass | | 6 - 7.5 |

**TABLE 218 *(continued)*  pH Preferences for Crop Plants**

*NB*  It is noted that liberal dressings of farmyard manure can somewhat offset the effects of acidity; also that crops are more tolerant of acidity in cool moist climates than warm dry ones. Certain peculiarities of specific soils may also shift the normal pH tolerance of crops.

| Crop | Optimal pH | Tolerable pH range (for satisfactory yield) |
|---|---|---|
| Papya | 6 - 6.5 | |
| Passion Fruit | | 6 - 8 |
| Pea | | 6 - 8 |
| Pecan | | 6 - 7 |
| Pepper | | 6 - 6.5 |
| Pineapple | | 5 - 6.5 |
| Potato | 5 - 5.8 | 4.5 -7 |
| Pumpkin | | 5.5 - 6.5 |
| Radish | | 6 - 7 |
| Red Clover | 6 - 7 | 5.4 - 7.2 |
| Rice, paddy | 5 - 6.5 | 4 - 8 |
| Rice, upland | | 4.5 - 7.5 |
| Rice, hungry | | 4.5 - 7.5 |
| Rosella | | 6 - 7.5 |
| Rubber | 4 - 6.5 | 3.5 - 8 |
| Rye | 4.8 - 5.8 | 4.8 - 6.3 |
| Ryegrass | 6 - 6.7 | 5.5 - 7.3 |
| Safflower | | 5.5 - 6.5 |
| Sesame | | 5.5 - 7 |
| Sisal | | 6.5 - 8 |
| Sorghum | 5.5 - 6.5 | 5 - 8.5 |
| Soyabean | 6 - 7 | 4.5 - 7.5 |
| Sweet Potato | 5.8 - 6 | 5 - 7 |
| Sugar Cane | 6 - 7.5 | 4.5 - 8.5 |
| Sunflower | | 6 - 7.5 |
| Sugar Beet | 6.7 - 7.5 | 6.2 - 8 |
| Taro | | 5.5 - 6.5 |
| Tea | 4 - 5.5 | 4 - 6.5 |
| Tobacco | 5.5 - 6 | 5 - 7.5 |
| Tomato | | 5 - 7 |
| Turnips | 5.7 - 6.5 | 5.3 - 7 |
| Walnut | | 6 - 8 |
| Watercress | | 6 - 8 |
| Water Melon | | 5 - 5.5 |
| Wheat | 6 - 7 | |
| White Clover | 6 - 6.5 | 5.5 - 7 |
| Yams | | 5.5 - 6.5 |

*Sources:*  Landon, J.R., ed. (1984) *Tropical Soil Manual.* Whiteman, P.C. (1980) *Tropical Pasture Science.* UKF Fertilisers, *Soil Analysis.* Russel, E.W. (1961) *Soil Conditions and Plant Growth.*

**TABLE 219   Critical Lower pH Values for Crop Growth – UK Mineral Soils**

| Crop | pH at which growth will be restricted and at which lime must be applied | Crop | pH at which growth will be restricted and at which lime must be applied |
|---|---|---|---|
| Alfalfa | 6.2 | Mangel | 5.8 |
| Apple | 5 | Mint | 6.6 |
| Asparagus | 5.9 | Mustard | 5.4 |
| Barley | 5.9 | Oats | 5.3 |
| Bean | 6 | Onion | 5.7 |
| Beetroot | 5.9 | Parsley | 5.1 |
| Blackberry | 4.9 | Parsnip | 5.4 |
| Blackberry, Thornless | 5.4 | Pea | 5.9 |
| Brussels Sprouts | 5.7 | Pear | 5.3 |
| Blackcurrant | 6 | Plum | 5.6 |
| Cabbage | 5.4 | Potato | 4.9 |
| Carrot | 5.7 | Rape | 5.6 |
| Cauliflower | 5.6 | Raspberry | 5.5 |
| Celery | 6.3 | Redcurrant | 5.6 |
| Chicory | 5.1 | Rhubarb | 5.4 |
| Clover, Alsike | 5.7 | Rye | 4.9 |
| Clover, Red | 5.9 | Ryegrass | 4.7 |
| Clover, White | 5.6 | Sainfoin | 6.2 |
| Clover, Wild white | 4.7 | Spinach | 5.8 |
| Cocksfoot | 5.3 | Strawberry | 5.1 |
| Cucumber | 5.5 | Sugar Beet | 5.9 |
| Fescues | 4.7 | Swede | 5.4 |
| Hops | 5.5 | Timothy | 5.3 |
| Kale | 5.4 | Tomato, Outdoor | 5.1 |
| Leek | 5.8 | Trefoil | 6.1 |
| Lettuce | 6.1 | Turnip | 5.4 |
| Linseed | 5.4 | Vetches | 5.9 |
| Maize | 5.5 | Wheat | 5.5 |

*Source:* Ministry of Agriculture Fisheries and Food (1979) *Fertiliser Recommendations. GF 1.*
Reproduced with kind permission of MAFF ADAS (UK).

# Sundry Effects

**TABLE 220   Effectiveness Groupings of Legume Crops**

**Legumes that tend to respond similarly when inoculated with the same strain of rhizobia**

| Rhizobium species | Effectiveness groupings | Leguminous species |
|---|---|---|
| ***Rhizobium meliloti*** | 1. | *Medicago sativa, M. falcata, M. minima, M. tribuloides, Melilotus denticulata, M. alba, M. officinalis, M. indica* |
| | 2. | *Medicago arabica, M. hispida, M. lupulina, M. orbicularis, M. praecox, M. truncatula, M. scutellata, M. polymorpha, M. rotata, M. rigidula, Trigonella foenum-graecum* |
| | 3. | *Medicago laciniata* |
| | 4. | *Medicago rugosa* |
| ***Rhizobium trifolii*** | 5. | *Trifolium incarnatum, T. subterraneum, T. alexandrinum, T. hirtum T. arvense, T. angustifolium* |
| | 6. | *Trifolium pratense, T. repens, T. hybridum, T. fragiferum, T. procumbens, T. nigescens, T. glomeratum* |
| | 7. | *Trifloum vesiculosum, T. bertytheum, T. bocconei, T. boissieri, T. compactum, T. leucanthum, T. mutabile, T. vernum, T. physodes, T. dasyurum* |
| | 8. | *Trifolium ruepellianum, T. tembense, T. usambarense, T. steudneri, T. buchellianum var. burchellianum, T. burchellianum var. johnstonii, T. africanum, T. pseudostriatum* |
| | 9. | *Trifolium semipilosum var. kilimanjaricum, T. masaiense, T. cheranganiense, T. ruepellianum var. lanceolatum* |
| | 10. | *T. medium, T. sarosience, T. alpestre* |
| | 11. | *T. ambiguum* |
| | 12. | *T. heldreichianum* |
| | 13. | *T. masaiense* |
| | 14. | *T. reflexum* |
| | 15. | *T. rubens* |
| | 16. | *T. semipilosum* |
| ***Rhizobium leguminosarum*** | 17. | *Pisum sativum, Vicia villosa, V. hirsuta, V. faba, V. tenuifol- ia, V. tetrasperma, Lens esculenta, Lathyrus aphaca, L. cicera, L. hirsutus, L. odoratus, L. sylvestris* |
| | 18. | *Lathyrus ochrus, L. tuberosus, L. szenitzii* |
| | 19. | *Lathyrus sativus, L. clymenum, L. tingitanus* |
| | 20. | *Vicia faba, V. narbonensis* |
| | 21. | *Vicia sativa, V. amphicarpa* |

*(continued over)*

**TABLE 220** *(continued)*   **Effectiveness Groupings of Legume Crops**

**Legumes that tend to respond similarly when inoculated with the same strain of rhizobia**

| Rhizobium species | Effectiveness groupings | Leguminous species |
|---|---|---|
| *Rhizobium phaseoli* | 22. | *Phaseolus vulgaris, P. coccineus, P. angustifolius* |
| *Rhizobium lupini* | 23. | *Lupinus albicaulis, L. albifrons, L. albus, L. angustifolius, L. arboreus, L. argenteus, L. benthamii, L. formosus, L. luteus, L. micranthus, L. perennis, L. sericeus, Lotus uliginosus, L. americanus, L. pedunculatus, L. strictus, L. strigosus* |
| | 24. | *Lupinus densiflorus, L. vallicola* |
| | 25. | *L. nanus* |
| | 26. | *L. polyphyllus* |
| | 27. | *L. subcarnosus* |
| | 28. | *L. succulentus* |
| *Rhizobium japonicum* | 29. | *Glycine max* |
| *Rhizobium spp. (cowpea type)* | 30. | *Vigna unguiculata, V. sesquipedalis, V. luteola, V. cylindrica, V. angularis, V. radiata, V. mungo, Desmodium sp., Alysicarpus vaginalis, Crotalaria sp., Macroptilium lathyroides, M. atropurpureum, Psophocarpus sp., Lespedeza striata, L. stipulacea, Indigofera sp., Cajanus cajan, Cicer arietinum* |
| | 31. | *Phaseolus limensis, P. lunatus, P. aconitifolius, Canavalia ensiformis, C. lineata* |
| | 32. | *Arachis hypogaea, A. glabrata, Cyamopsis tetragonoloba, Lespedeza sericea, L. japonica, L. bicolor* |
| | 33. | *Centrosema pubescens, Galactia sp.* |
| | 34. | *Lotononis bainesii* |
| | 35. | *Lotononis angolensis* |
| *Rhizobium spp. (lotus)* | 36. | *Lotus corniculatus, L. tenuis, L. angustissimus, L. tetragonolobus, L. caucasicus, L. crassifolius, L. creticus, L. edulis, L. frondosus, L. subpinnatus, L. weilleri, Dorycnium hirsutum, D. rectum, D. suffrutocosum, Anthyllis culneraria, A. lotoides* |
| | 37. | *Lotus uliginosus, L. americanus, L. scoparius, L. angustissimus, L. pedunculatus, L. strictus, L. strigosus, Ornithopus sativus, Lupinus angustifolius, L. albus, L. luteus* |
| *Rhizobium spp. (Coronilla, Petalostemon-Onobrychis)* | 38. | *Coronilla varia, Onobrychis viciifolia, Petalostemon purpureum, P. candidum, P. microphyllum, P. multiflorus, P. villosum, Leucaena leucocephala, L. retusa* |

**TABLE 220 (continued)   Effectiveness Groupings of Legume Crops**

**Legumes that tend to respond similarly when inoculated with the same strain of rhizobia**

| Rhizobium species | Effectiveness groupings | Leguminous species |
|---|---|---|
| ***Rhizobium*** | 39. | *Dalea alopecuroides* |
| ***spp.*** | 40. | *Strophostyles helvola* |
| ***(various)*** | 41. | *Robinia pseudoacacia, R. hispida* |
| | 42. | *Amorpha canescens* |
| | 43. | *Caragana arborescens, C. frutescens* |
| | 44. | *Oxytropis sericea* |
| ***Rhizobium spp.*** | | |
| ***Astragalus sp.*** | 45. | *A. cicer, A. falcatus, A. canadensis, A. mexicanus, A. orbiculatus* |

*Source:* FAO (1984) *Legume innoculants and their use.* Reproduced with kind permission of the Food and Agriculture Organization of the United Nations.

**TABLE 221   Legume Rhizobial Requirements**

**Legume Inoculant as recommended by Australian Inoculant Research & Control Service (AIRCS)**

| Group | Legumes Inoculated |
|---|---|
| Lucerne | Lucerne and Barrel, Harbinger and Cyprus Medic |
| Paragosa | Parogosa Medic |
| White Clover | Red, Strawberry, Alsike, Suckling, and all White Clovers |
| | All Subterranean Clovers |
| Sub Clover | Rose, Cupped, Crimson Clover |
| Lotus | Lotus major |
| Pea | Pea, Vetch, Tares, Broad Bean, Tick Bean |
| Bean | French and Climbing Beans |
| | Lupin |
| Lupin | Serradella |
| Soybean | Soybean |
| | Cowpea, Peanut, Velvet Bean, Mung Bean, Poona Bean |
| Cowpea | Siratro, Phasey Bean, Puero, Calopo Glycine, Stylo, |
| | Townsville Stylo, Caribbean Stylo |
| Lablab | Lablab |
| Centro | Centro |
| Desmodium | Desmodium |
| Leucaena | Leucaena |
| Lotononis | Lotononis |

**IN ADDITION: Special strains of rhizobium are available for Paragosa medic, Kenya White Clover, Soybean (var Hardee) and Desmodium hetero.**

*Source:* Humphreys, L.R. (1980) *A guide to better pastures for the tropics and sub-tropics.* Wright Stephenson & Co.

**TABLE 222  Indicative Planting Depth for Seeds of Different Sizes**

| Normal depth of seeding mm | Normal maximum depth for emergence mm | Seeds / kg | Representative crops |
|---|---|---|---|
| 6 - 13 | 25 - 50 | 150 000 to 2 500 000 | Red-top, carpet grass, timothy, fescues, white clover, alsike clover and tobacco |
| 13 - 19 | 50 - 76 | 75 000 to 150 000 | Lucerne, red clover, sweet clover, lespedeza, crimson clover, ryegrass, foxtail millet turnip |
| 19 - 38 | 76 - 100 | 25 000 to 75 000 | Flax, Sudan grass, Crotalaria, millets, bromegrass, sugar beet |
| 38 - 50 | 76 - 120 | 5000 to 25 000 | Wheat, oats, barley, rye, rice, sorghum, buckwheat, hemp, vetch, mung bean |
| 50 - 76 | 100 - 200 | 200 to 5000 | Maize, pea, cotton, broad bean |

*Source:* FAO (1961) *Agricultural and Horticultural Seeds. Agricultural Studies No. 55.*
Reproduced with kind permission of the Food and Agriculture Organization of the United Nations.

# Livestock – General

---

**TABLE 223  Latin Names of Commercially Important Livestock**

---

| | | |
|---|---|---|
| **Cattle** | ***Bos taurus*** | (European types) |
| | ***Bos indicus*** | (Zebu types) |
| | ***Bos grunniens*** | (Yak) |
| | ***Bibos banteng*** | (Bali cattle) |
| **Buffalo** | ***Bubalus bubalis bubalis*** (as found in India, Pakistan, China, Turkey, Egypt, Europe and Brazil) | |
| | ***Bubalus bubalis fluvus*** (NE India and Assam) | |
| | ***Bubalus bubalis kerebau*** (Sri Lanka, Philippines, Indonesia) (In practice the main distinction in buffaloes is between working and milking types) | |
| **Sheep** | ***Ovie aries*** (In practice the multiplicity of sheep breeds are often grouped according to their fleece type) | |
| **Goats** | ***Capra hiscus*** | |
| **Pigs** | ***Sus vittatus*** (Far Eastern types) | |
| | ***Sus scrofa*** (European types) | |
| **Camels** | ***Camelus dromedarius*** (single hump dromedary) | |
| | ***Camelus bactrianus*** (two hump bactrian) | |
| **Llamas** | ***Lama glama*** (Llama) | |
| | ***Lama pacos*** (Alpaca) | |
| **Hens** | ***Gallus*** | |
| **Duck** | ***Anas platyrhynchos*** (Mallard) | |
| | ***Cairina moschata*** (Muscovy) | |
| **Geese** | ***Anser*** | |
| **Turkeys** | ***Meleagris gallopavo*** | |
| **Rabbit** | ***Oryctolagus cuniculus*** | |
| **Horse** | ***Equus caballus*** | |
| **Donkey** | ***Equus asinus*** | |

*NB* Some authorities consider that there should be further division of some livestock 'breeds' or 'types' into separate species - this is particularly so for pigs.

**TABLE 224**    Some Recognized Breeds of Cattle and Sheep Developed from Simple Crossbred Foundations, and Pigs Developed from Inbred Lines

| Breed | Foundation Breeds/Species | Origin |
|---|---|---|
| **CATTLE** | | |
| **Ankole** | *Bos taurus, B. indicus* | Africa |
| **Australian Braford** | 5/8 Hereford, 3/8 Brahman | Australia |
| **Australian Brangus** | 5/8 Angus, 3/8 Brahman | Australia |
| **Australian Milking Zebu** | Sahiwal, Sindhi, Jersey | Australia |
| **Australian Friesian Sahiwal** | 1/2 Friesian, 1/2 Sahiwal | Australia |
| **Barzona** | Africander, Hereford, S. Gertrudis | USA |
| **Beefalo** | 3/8 Buffalo, 3/8 Charolais, 1/4 Hereford | USA |
| **Beefmaster** | 1/2 Zebu, 1/4 Hereford, 1/4 Shorthorn | USA |
| **Belmont Red** | 1/2 Africander, 1/4 Hereford + Shorthorn | Australia |
| **Bonsmara** | 5/8 Africander, 3/8 Hereford + Shorthorn | S. Africa |
| **\* Botswana Beef Synthetic** | Sanga, Zebu and European | Botswana |
| **Brangus** | Brahman, Angus | USA |
| **Brayford** | Brahman, Hereford | USA |
| **Canchim** | 5/8 Charolais, 3/8 Nelore | Brazil |
| **Charbray** | 3/8 Brahman, 5/8 Charolais | USA and Brazil |
| **Damascus** | *B. taurus, B. indicus* | Asia |
| **Drakensberger** | *B. taurus, B. indicus* | Africa |
| **Droughtmaster** | 3/8-1/2 Brahman x 1/2-5/8 Hereford, Shorthorn and Devon-Shorthorn | Australia |
| **Grati** | Holstein-Friesian x Javanese and Madura | Indonesia |
| **Hays Converter** | Hereford, Brown Swiss, Holstein | USA |
| **Indubrasil** | Two breeds of Zebu | Brazil |
| **Jamaica Black** | Angus, Zebu | Jamaica |
| **Jamaica Hope** | 80% Jersey, 5% Friesian, 15% Sahiwal | Jamaica |
| **Jamaica Red** | 90% Red Poll, 10% Zebu | Jamaica |
| **Karan Fries** | 5/8 Friesian, 3/8 Tharparker | India |
| **Karan Swiss** | 1/2-3/4 Brown Swiss, 1/4-1/2 Sahiwal | India |
| **\* Lavinia** | Brown Swiss x Gujera | Brazil |
| **\* Mandalong Special** | Continental European 58%, British 28%, Zebu 17% | Australia |
| **Mpwapwa** | 3/4-7/8 Tanzanian Zebu + Boran 1/8-1/4 Ayrshire and Sahiwal | Tanzania |
| **Murray Grey** | Shorthorn, Angus | Australia |
| **Nganda** | *B. taurus, B. indicus* | Africa |
| **Nguni** | *B. taurus, B. indicus* | Africa |
| **\* Nuras** | Africander x (Simmental x Hereford) | Namibia |
| **Occampo** | Holstein, Zebu | West Indies |
| **Pitanguerias** | 5/8 Red Poll, 3/8 Zebu | Brazil |
| **Rana** | B. taurus, B. indicus | Africa |
| **Renitelo** | 1/2-5/8 Limousin, 1/2-3/8 Africander and Madagascar Zebu | Madagascar |
| **Romagna Red** | 1/2 Criollo, 1/2 Zebu | Dominican Rep. |
| **Santa Gertrudis** | 5/8 Shorthorn, 3/8 Zebu | USA |
| **Senepol** | 3/4-7/8 Red Poll,  1/8-1/4 N'Dama | West Indies |

\* These are still in the formative stage    *(continued over)*

**TABLE 224** *(continued)* **Some Recognized Breeds of Cattle and Sheep Developed from Simple Crossbred Foundations, and Pigs Developed from Inbred Lines**

| Breed | Foundation Breeds/Species | Origin |
|---|---|---|
| **Siboney** | 5/8 Holstein Friesian, 3/8 Zebu | Cuba |
| **Siri** | *B. taurus, B. indicus* | Asia |
| * **Suisbu** | Brown Swis x Zebu | Argentina |
| **Sunandini** | 5/8 Brown Swiss, 3/8 local Indian Zebu | India |
| **Tuli** | *B. taurus, B. indicus* | Africa |
| **Tuni** | *B. taurus, B. indicus* | Africa |
| * **Quasar** | 1/2 *B. indicus*, 1/2 *B. taurus* | Australia |

* These are still in the formative stage

**SHEEP**

| Breed | Foundation Breeds/Species | Origin |
|---|---|---|
| **Columbia** | Lincoln, Rambouillet | USA |
| **Corriedale** | Lincoln, Leicester, Merino | USA |
| **Debouillet** | Delaine Merino, Rambouillet | USA |
| **Dormer** | Dorset Horn, German Merino | S. Africa |
| **Dorper** | Dorset Horn, Persian Blackhead | S. Africa |
| **Montadale** | Cheviot, Columbia | USA |
| **Panama** | Rambouillet, Lincoln | USA |
| **Polwarth** | 3/4 Merino, 1/4 Lincoln | Australia |
| **Romedale** | Romney, Rambouillet | USA |
| **Soviet Corriedale** | Lincoln, Rambouillet | Russia |
| **Targhee** | Rambouillet, Corriedale, Lincoln | USA |
| **Thribblecross** | Cotswold, Spanish Merino | USA |

**PIGS (hogs) developed in USA**

| Breed | Foundation Breeds/Species |
|---|---|
| **Beltsville No 1** | 75% Landrace, 25% Poland China |
| **Beltsville No 2** | 58% Yorkshire, 30% Duroc, 6% Landrace, 6% Hampshire |
| **Maryland No 1** | 38% Berkshire, 62% Landrace |
| **CPF No 1** | Beltsville No. 1, San Pierre |
| **CPF No 2** | 25% Beltsville No. 1, 50% Maryland No. 1, 25% Yorkshire |
| **Minnesota No 1** | 45% Tamworth, 55% Landrace |
| **Minnesota No 2** | 40% Yorkshire, 60% Poland China |
| **Minnesota No 3** | Gloucester Old Spot, Welsh Pig, English Large White, Beltsville No. 2 and others |
| **Montana No 1** | 45% Hampshire, 55% Landrace |
| **Palouse** | Chester White, Landrace |
| **San Pierre** | Berkshire, Chester White |

*Sources:* Chamberlain, A. (1989) *Milk Production in the Tropics.* McDowell, R.E. (1972) *Improvement of Livestock in Warm Climates.* Smith, A.J., ed. (1974) *Beef Cattle Production in Developing Countries.* Blakely, J. & Bade, D. (1976) *The Science of Animal Husbandry.* Maule, J.P. (1990) *The Cattle of the Tropics.* CTVM, University of Edinburgh.

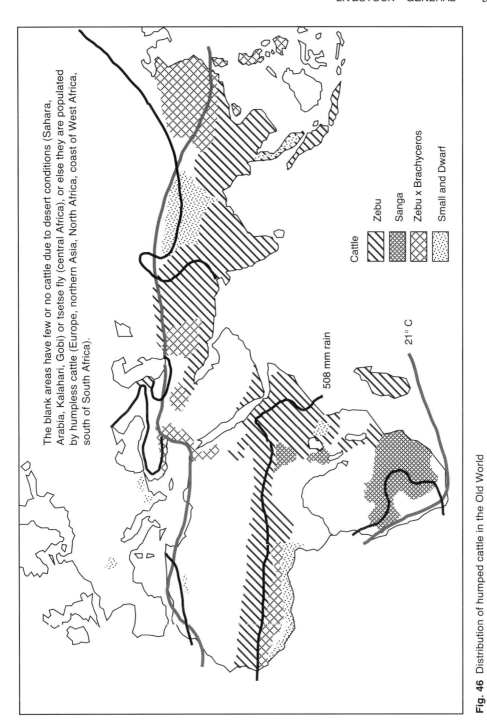

**Fig. 46** Distribution of humped cattle in the Old World

*Source:* Smith, A.J., ed. (1974) *Beef Cattle Distribution in Developing Countries.* Reproduced with kind permission of A. J. Smith.

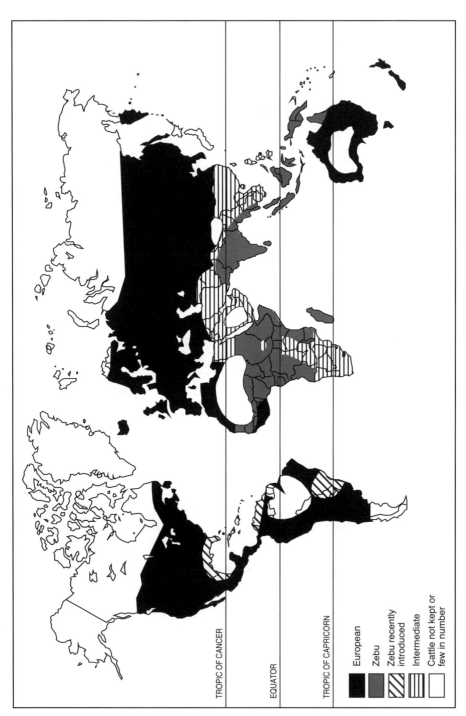

**Fig. 47** World distribution of the two basic types of cattle (USDA)

TROPIC OF CANCER

EQUATOR

TROPIC OF CAPRICORN

European

Zebu

Zebu recently introduced

Intermediate

Cattle not kept or few in number

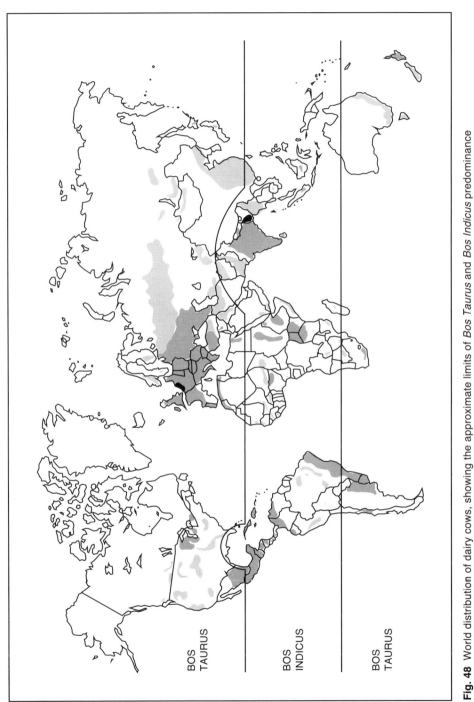

**Fig. 48** World distribution of dairy cows, showing the approximate limits of *Bos Taurus* and *Bos Indicus* predominance

*Source:* FAO (1987) *Crossbreeding Bos Indicus and Bos Taurus for milk production in the Tropics. Animal Production and Health Paper No. 68.* Reproduced with kind permission of the Food and Agriculture Organization of the United Nations.

BOS TAURUS

BOS INDICUS

BOS TAURUS

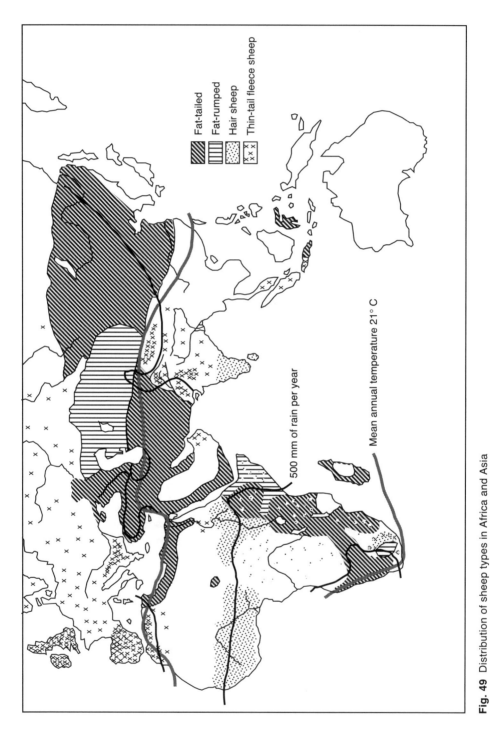

**Fig. 49** Distribution of sheep types in Africa and Asia

*Source:* Gatesby, N. M. (1986) *Sheep Production in the Tropics and Sub Tropics*. Reproduced with kind permission of Longman Group UK Ltd.

**TABLE 225   Drinking Water Requirements for Various Livestock**

| Animal | | l / head / day |
|---|---|---|
| Camel | | 40 - 90 |
| **600kg cow** | 10kg milk / day, <16°C | 81 |
| | 10kg milk / day, >20°C | 105 |
| | 40kg milk / day, <16°C | 113 |
| | 40kg milk / day, >20°C | 147 |
| **Young cattle** | | |
| | up to 6 months | 15 - 25 |
| | 6 - 12 months | 25 - 35 |
| | 12 - 18 months | 35 - 40 |
| **Ewes** | | 2.4 - 4.5 |
| **Pregnant pigs** | | 4.5 - 9 |
| **Lactating pigs** | | 18 - 23 |
| **Pork pigs** | | 4 - 5 |
| **Bacon/heavy pigs** | | 4.5 - 9 |
| **Horse** | | 30 - 40 |
| **Rabbits** | | |
| | Non pregnant / early pregnant does | 0.28 |
| | Adult bucks | 0.28 |
| | Late pregnant does | 0.57 |
| | Suckling does post weaning | 0.60 |
| | Doe plus 7 x 6 week young | 2.3 |
| | Doe plus 7 x 8 week young | 4.5 |
| **Poultry (general)** | | 0.2 - 0.3 |
| **Broilers** | | l / 1000 / day |
| | Age (days)          7 | 53 - 59 |
| | 14 | 95 - 106 |
| | 21 | 138 - 155 |
| | 28 | 176 - 198 |
| | 35 | 210 - 234 |
| | 42 | 245 - 275 |
| | 49 | 272 - 306 |
| | 56 | 291 - 328 |

*NB*   1. In temperatures above 25°C water consumption of broilers is increased and above 30°C additional drinkers should be provided.

2. Water requirements for all stock will vary with environmental temperature.

3. Grazing livestock may obtain a considerable proportion of their water requirements from forage specially when it is wet with rain or dew.

4. Housed livestock may obtain some of their water from that contained in feed.

*Sources:* Agro Business Consultants (Nov. 1990) *The Agricultural Budgeting and Costing Book No. 31.* ILACO (1985) *The Agricultural Compendium.* Portsmouth, J. (1979) *Commercial Rabbit Meat Production.* McDonald, P., Edwards, R.A. & Greenhalgh, J.F.D. (1990) *Animal Nutrition.* The Cobb Breeding Co. (1991) *Broiler Management Guide.*

**TABLE 226   Livestock and Poultry Drinking Water Salinity Rating**

| Water Salinity (dS/m) | Class |
|---|---|
| <1.5 | Excellent |
| 1.5 - 5 | Very satisfactory |
| 5 - 8 | Unfit for poultry |
|  | Satisfactory for livestock |
| 8 - 11 | Unfit for poultry |
|  | Limited use for livestock |
| 11 - 16 | Unfit for poultry and probably pigs |
|  | Very limited use for livestock |
| >16 | Not recommended |

*NB* As water salinity increases, so the risk of its use on pregnant or lactating animals increases. Poultry are particularly susceptible to poor quality.

*Source:* FAO (1985) *Water Quality for Agriculture. Irrigation and Drainage Paper No. 29 Rev. 1.* Reproduced with kind permission of the Food and Agriculture Organization of the United Nations.

**TABLE 227   Guideline Levels of Upper Limits of Toxic Elements for Livestock Drinking Water**

| Element | Upper limit (mg/l) | Element | Upper limit (mg/l) |
|---|---|---|---|
| Al | 5 | Pb | 0.1 |
| As | 0.2 | Mn * | 0.05 |
| Be * | 0.1 | Hg | 0.01 |
| B | 5 | Nitrate N + Nitrite N | 100 |
| Cd | 0.05 | Nitrite N | 10 |
| Cr | 1 | Se | 0.05 |
| Co | 1 | V | 0.1 |
| Cu | 0.5 | Zn | 24 |
| F | 2 | TDS | 10 000 |
| Fe | not needed | | |

*NB*   1 As Pb is accumulative, problems may start at .05 mg/l
      2 *Insufficient livestock data. Values for marine life and humans, respectively, used.

*Source:* FAO (1985) *Water Quality for Agriculture. Irrigation and Drainage Paper No. 29.* Reproduced with kind permission of the Food and Agriculture Organization of the United Nations.

**TABLE 228   Suggested Limits for Magnesium in Livestock Drinking Water**

| Stock | Mg (mg / l) |
|---|---|
| Poultry | <250 |
| Pigs | <250 |
| Horses | 250 |
| Cows - lactating | 250 |
| Ewes - suckling | 250 |
| Beef cattle | 400 |
| Adult sheep, dry fed | 500 |

*Source:* FAO (1985) *Water Quality for Agriculture. Irrigation and Drainage Paper No.29.*
Reproduced with kind permission of the Food and Agriculture Organization of the United Nations.

**TABLE 229   Effluent Production based on National Research Council of Canada (1970)**

| Class of stock | | Age | Weight (kg) | Output * (urine + faeces) (m³ / head per week) |
|---|---|---|---|---|
| **Cattle** | Young stock | 1 - 3 months | 40 - 100 | 0.04 |
| | | 3 - 6 months | 101 - 170 | 0.05 |
| | | 6 - 15 months | 171 - 390 | 0.10 |
| | | 15 - 24 months | 391 - 540 | 0.15 |
| | Dairy cow | adult | 540 | 0.32 |
| | Beef cow | adult | 540 | 0.20 |
| **Pigs** | Young stock | 3 - 6 weeks | 4 - 11 | 0.008 |
| | | 6 - 9 weeks | 12 - 23 | 0.015 |
| | | 9 - 12 weeks | 24 - 34 | 0.025 |
| | | 12 - 16 weeks | 35 - 57 | 0.035 |
| | Baconers | 16 - 20 weeks | 58 - 79 | 0.050 |
| | | 20 - 22 weeks | 80 - 91 | 0.060 |
| | Sow | adult | 110 | 0.80 |
| **Poultry** | Broilers | 0 - 12 weeks | 0 - 1.8 | 0.000 56 |
| | Layers | 6 - 24 months | 2.25 | 0.001 |
| **Sheep** | Ewe | adult | 75 | 0.002 |

* Output quantities vary with different data sources

*NB*  1.  Mature buffalo produce about 19 kg faeces + 13 kg urine/day (FAO)
  2.  Mature rabbits (nursing doe) produce about 0.15 - 0.2 kg faeces +
     0.25 - 0.3 kg urine/day (FAO)

*Source:* Weller, J.B. & Willets, S.C. (1977) *Farm Wastes Management.*
Reproduced with kind permission of Blackwell Scientific Publications.

**TABLE 230   Moisture Content of Animal Manures and Faeces to Urine Ratios**

| Animal | Faeces / Urine | % Water |
|---|---|---|
| Cattle | 80 : 20 | 85 |
| Goat | | 40 |
| Poultry | 100 : 0 | 62 |
| Pigs | 60 : 40 | 85 |
| Rabbit | | 50 |
| Sheep | 67 : 33 | 66 |
| Horse | 80 : 20 | 66 |

*Source:* Adapted from Brady, N.C. (1984) *Nature and Properties of Soils,* and Landon, J.R., ed. (1984) *Tropical Soil Manual.*

**TABLE 231   Approximate Daily Excreta Volume from Different Animals**

| Animal | Wt kg | Faeces + Urine, or droppings l / day |
|---|---|---|
| Dairy Cow | 500 | 41 |
| Fattening Bullock | 400 | 27 |
| Laying Hen | 2 | .114 |
| Pig   Dry Feed | 50 | 4 |
| Liquid Feed | 50 | 7 |
| Whey Feed | 50 | 14 |

*Source:* Adapted from Cooke, G.W. (1984) *Fertilizing for Maximum Yields.* Reproduced with kind permission of Blackwell Scientific Publications.

**TABLE 232   a & b Livestock Units**

'Livestock Units' (LSU) is a concept developed for equating in general terms the feed requirements of a range of livestock species for overall feed supply and/or land stocking purposes. Several systems have been developed such as that by the UK Ministry of Agriculture Fisheries and Food.

### a. LSU VALUES

| Type of Livestock | Liveweight | Age in Months or as Stated | Calculated Recommended Allowance of Dietary ME(MJ) | LSU Livestock Unit |
|---|---|---|---|---|
| **Dairy Cow** (except Channel Island) | 600 kg | Mature animal for a complete year | 48 000 | 1.0 |
| Standard yield of 4500 litres (36 g/kg BF, 86 g/kg SNF) | | | | |
| Adjust for liveweight and milk yield 50 kg weight change ± 0.03 LSU 500 litres yield change ± 0.05 LSU | | | | |
| **Channel Island Breeds** | 450 kg | Mature animal for a complete year | 41 689 | 0.87 |
| Standard yield of 3600 litres (48 g/kg BF, 90 g/kg SNF) | | | | |
| Adjust for liveweight and milk yield 50 kg weight change ± 0.03 LSU 500 litres yield change ± 0.06 LSU | | | | |
| **Beef Cows** (with calf) | 550 kg | Mature animal for a complete year | 35 656 | 0.75 |
| Standard yield of 2300 litres | | | | |
| Adjust for liveweight and milk yield 50 kg weight change ± 0.03 LSU 100 litres yield change ± 0.01 LSU | | | | |
| (assuming dairy and beef cows breeding regularly) | | | | |

*(continued over)*

**TABLE 232 a & b** *(continued)*  **Livestock Units**

### a. LSU VALUES

| Type of Livestock | Liveweight | Age in Months or as Stated | Calculated Recommended Allowance of Dietary ME(MJ) | LSU Livestock Unit |
|---|---|---|---|---|
| **Bulls** Any breed | 750 kg | Mature animal for a complete year | 31 500 | 0.65 |
| **Dairy / Beef Replacements** | | | | |
| Birth to calving (approx 2 years) | Birth - 270 kg | 0 - 12 | 14 553 | 0.30 |
| | 270 - 500 kg | 12 - 24 | 25 986 | 0.54 |
| **Growing Cattle** | | | | |
| Finishing at 12 months (Barley beef or similar intensive) | Birth - 400 kg | 0 - 12 | 19 807 | 0.41 |
| Finishing at 18 months | Birth - 310 kg | 0 - 12 | 16 632 | 0.35 |
| | 310 - 450 kg | 12 - 18 | 13 669 | 0.28 |
| Finishing at 24 months | Birth - 270 kg | 0 - 12 | 14 364 | 0.30 |
| | 270 - 550 kg | 12 - 24 | 28 555 | 0.59 |
| **Ewes** (including ewe replacements) | | | | |
| Lightweight breeds (100% lambing) | 40 kg | Mature animal for a complete year | 3411 | 0.07 |
| Mediumweight breeds (130% lambing) | 60 kg | Mature animal for a complete year | 4328 | 0.09 |
| Heavyweight breeds (170% lambing) | 80 kg | Mature animal for a complete year | 5328 | 0.11 |
| **Rams** | | | | |
| Mediumweight | 80 kg | Mature animal for a complete year | 3942 | 0.08 |
| **Lambs** | | | | |
| Store lamb | Birth - 30 kg | 0 - 6 | 1684 | 0.04 |
| Fat lamb | Birth - 40 kg | 0 - 4½ | 1881 | 0.04 |
| Hogget | Birth - 50 kg | 0 - 10 | 3472 | 0.07 |
| Purchased stores | 25 - 50 kg | 3 - 4 months | 1882 | 0.04 |

**TABLE 232 a & b** *(continued)* **Livestock Units**

## b. LSU COEFFICIENTS - NON-RUMINANTS

| Type of Stock | Livestock Unit Coefficient |
|---|---|
| **Pigs:** | |
| Sows (including litters to weaning) | 0.44 |
| Boars | 0.35 |
| Pigs Fattened | |
| per pig fattened during the year | 0.09 |
| per pig on the farm at any one time | 0.17 |
| **Poultry:** | |
| Over 6 months | 0.017 |
| Under 6 months (excluding broilers) | 0.0044 |
| Broilers | 0.0017 |
| Turkeys (breeding stock only) | 0.017 |

*Source:* MAFF/ADAS (UK) *Livestock Units Handbook. Booklet 2267.*
Reproduced with kind permission of MAFF ADAS (UK).

**a.** Incisor dentition of sheep according to age

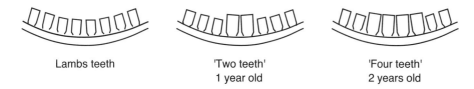

| Lambs teeth | 'Two teeth'<br>1 year old | 'Four teeth'<br>2 years old |

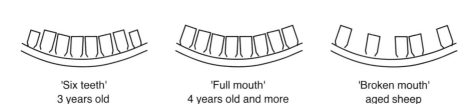

| 'Six teeth'<br>3 years old | 'Full mouth'<br>4 years old and more | 'Broken mouth'<br>aged sheep |

**Fig. 50  a, b, c, d, e** Dentition of sheep, cattle and buffalo

(*NB* Level of nutrition can have an effect on time of emergence of permanent teeth)

**b.** Indicative ages for teeth emergence in sheep

| Approx. date of tooth emergence (months) | Incisors | | | | Molars & pre-molars | | | | | | Comments |
|---|---|---|---|---|---|---|---|---|---|---|---|
| | 1 | 2 | 3 | 4 | 1 | 2 | 3 | 4 | 5 | 6 | |
| 0 - 1 | ○ | ○ | ○ | ○ | ○ | ○ | ○ | | | | Incisors in lower jaw only |
| 3 | ○ | ○ | ○ | ○ | ○ | ○ | ○ | ● | | | |
| 9 | ○ | ○ | ○ | ○ | ○ | ○ | ○ | ● | | | |
| 12 - 15 | ● | ○ | ○ | ○ | ○ | ○ | ○ | ● | ● | | |
| 18 | ● | ○ | ○ | ○ | ○ | ○ | ○ | ● | ● | ● | |
| 21 | ● | ● | ○ | ○ | ● | ● | ○ | ● | ● | ● | |
| 24 | ● | ● | ○ | ○ | ● | ● | ● | ● | ● | ● | |
| 27 | ● | ● | ● | ○ | ● | ● | ● | ● | ● | ● | |
| 31 - 36 | ● | ● | ● | ● | ● | ● | ● | ● | ● | ● | |

○ = temporary teeth          ● = permanent teeth

**c.** Lower jaw dentition of mature cattle

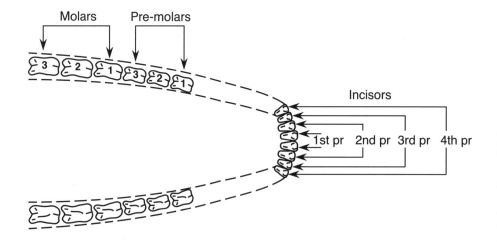

**Fig. 50** *(continued)*

**d.** Indicatives ages for teeth emergence in cattle

| Approx. date of tooth emergence (months) | Incisors | | | | Molars & pre-molars | | | | | | Comments |
|---|---|---|---|---|---|---|---|---|---|---|---|
| | 1 | 2 | 3 | 4 | 1 | 2 | 3 | 4 | 5 | 6 | |
| 0 - 1 | ○ | ○ | ○ | ○ | ○ | ○ | ○ | | | | Incisors in lower jaw only |
| 6 | ○ | ○ | ○ | ○ | ○ | ○ | ○ | ● | | | |
| 12 - 15 | ○ | ○ | ○ | ○ | ○ | ○ | ○ | ● | ● | | |
| 15 - 18 | ● | ○ | ○ | ○ | ○ | ○ | ○ | ● | ● | ● | |
| 24 | ● | ○ | ○ | ○ | ● | ● | ○ | ● | ● | ● | |
| 27 | ● | ● | ○ | ○ | ● | ● | ○ | ● | ● | ● | |
| 31 | ● | ● | ● | ○ | ● | ● | ◑ | ● | ● | ● | |
| 39 + | ● | ● | ● | ● | ● | ● | ● | ● | ● | ● | |

○ = temporary teeth      ● = permanent teeth

**e.** Indicative ages for emergence of permanent teeth in buffalo (months)

| | Incisors | Pre Molars | Molars |
|---|---|---|---|
| 1st pair | 30 - 36 | 24 | 15 |
| 2nd pair | 42 - 48 | 47 | 17 |
| 3rd pair | 48 - 60 | 48 | 32 |

*NB* Temporary incisors appear at 2 - 5 months

**f.** Goats

Goats show a similar eruption pattern to sheep but tending to be slightly earlier.

**Fig. 50** *(continued)*

A G R I   I N F O

**Handling point for assessing body condition**
(Grip loin halfway between the hip (hook) bone and last rib)

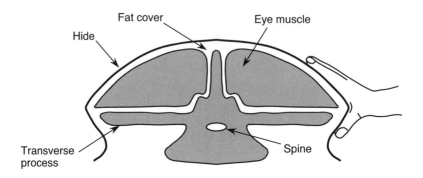

**Score**

Body condition is scored on a six-point scale from 0 (severe emaciation) to 5 (grossly overfat) as follows:

**0**     Spine is prominent, and transverse processes feel very sharp with no detectable fat cover.

**1**     Spine is still prominent, but transverse processes feel less sharp.

**2**     Transverse processes can still be felt by the thumb, but they are rounded with a covering of fat.

**(2.5   Desirable condition at mating)**

**3**     Separate transverse processes can only be detected with firm pressure from the thumb.

**4**     Transverse processes cannot be felt even with firm pressure.

**5**     Transverse processes cannot be felt and are obviously covered by a thick layer of fat.

**Fig. 51** Condition scoring in cattle *(Bos taurus)*

*Source:* Meat and Livestock Commission (1991) *Suckler Beef.*
Reproduced with kind permission of the Meat and Livestock Commission, UK.

---

### Note 10  Body Condition Scoring of Ewes

---

Condition scoring of ewes is used to assess the nutritional status of flocks as follows:

| Condition Score | Description |
|---|---|
| **0** | Extremely emaciated and on the point of death. It is not possible to detect any muscular or fatty tissue between the skin and the bone. |
| **1** | The spinous processes are prominent and sharp; the transverse processes are also sharp, the fingers pass easily under the ends, and it is possible to feel between each process; the loin muscles are shallow with no fat cover. |
| **2** | The spinous processes are prominent but smooth, individual processes can be felt only as fine corrugations; the transverse processes are smooth and rounder, and it is possible to pass the fingers under the ends with a little pressure; the loin muscles are of moderate depth, but have little fat cover. |
| **3** | The spinous processes have only a small elevation, are smooth and rounded, and individual bones can be felt only with pressure; the transverse processes are smooth and well covered, and firm pressure is required to feel over the ends; the loin muscles are full, and have a moderate degree of fat cover. |
| **4** | The spinous processes can just be detected with pressure as a hard line; the ends of the transverse processes cannot be felt; the loin muscles are full and have a thick covering of fat. |
| **5** | The spinous processes cannot be detected even with firm pressure; there is a depression between the layers of fat where the spinous processes would normally be felt; the transverse processes cannot be detected; the loin muscles are very full with very thick fat cover. |

*Source:* Meat and Livestock Commission (1983) *Body Condition Scoring of Ewes. Sheep Improvement Services.* Reproduced with kind permission of Meat and Livestock Commission, UK.

---

**Note 11  Condition Assessment of Lambs for Carcass Production (UK system for UK breeds)**

---

| FAT CLASS | DOCK (root of tail) | LOIN (lumber region) |
|---|---|---|
| 1 | Fat cover very thin. Individual bones very easy to detect. | Spinous processes – very prominent. Individual processes felt very easily. Transverse processes – prominent. Very easy to feel between each process. |
| 2 | Fat cover thin. Individual bones detected easily with light pressure. | Spinous processes prominent. Each process felt easily. Transverse process – each process felt easily. |
| 3 | Individual bones detected with light pressure. | Spinous and transverse processes – tips rounded. With light pressure individual bones felt as corrugations. |
| 4 | Fat cover quite thick. Individual bones detected only with firm pressure | Spinous processes – tips of individual bones felt as corrugation with moderate pressure. Transverse processes – tips detected only with firm pressure. |
| 5 | Fat cover thick. Individual bones cannot be detected even with firm pressure. | Spinous and transverse processes – individual bones cannot be detected even with firm pressure. |

*NB*  Practice in handling live lambs to assess individual levels of fatness is essential, whether selling lambs to a particular deadweight contract or making up even batches of lambs for presentation at an auction market. Live handling combined with a visit to examine the carcasses in the abattoir is the best way to gain experience and improve the accuracy of selection.

*Source:* Meat and Livestock Commission (1989) *Lamb Carcass Production. Planning to Meet Your Market.* Reproduced with kind permission of Meat and Livestock Commission, UK.

**TABLE 233    Pulse and Respiration Rates, and Temperatures**

| Specie | Respiratory Rate / min | Pulse Rate / min | Temp °C |
|---|---|---|---|
| Cow | 12 - 16 | 60 - 70 | 38.3 - 38.8 |
| Dog | 15 - 25 | 80 - 90 | 38.3 - 38.8 |
| Goat | 15 - 30 | 70 - 90 | 38.6 - 40.6 |
| Horse | 8 - 12 | 38 - 43 | 37.7 - 38.3 |
| Pig | 20 - 30 | 70 - 80 | 38.8 - 39.4 |
| Sheep | 20 - 30 | 75 - 80 | 39.4 - 40 |
| Fowl | 12 - 28 | 130 - 140 | 41.5 |
| Rabbit | | | 38.3 |

**TABLE 234    Oestrus Period etc. (days)**

| Specie | Duration | Return after parturition | Return if not impregnated |
|---|---|---|---|
| Buffalo | 1 | | 21 - 22 |
| Camel | 3 - 4 | 28 | 21 - 28 |
| Cow | 1 | 21 - 28 | 21 - 28 or more |
| Dog | 7 - 21 | 150 - 180 | 150 - 180 |
| Donkey | 8 - 10 | | 21 - 28 |
| Goat * | 1 | | 18 - 23 |
| Horse | 5 - 7 | 7 - 10 | 14 - 21 or more |
| Pig | 2 - 4 | 35 - 42 | 20 - 22 |
| | | (often 5 - 6 days after weaning) | |
| Rabbit | Not definable | 1 (re-mating after parturition commercially) | |
| Sheep * | 1 - 2 | 120 - 180 | 14 - 20 |

* In temperate latitudes sheep and goats normally breed only seasonally whereas in the tropics all year round breading is common.

*Sources: Rhodesian Farm Management Handbook* (1970). Wilkinson, J.M. & Stark, B.A. (1987) *Commercial Goat Production.* ILACO (1985) *Agricultural Compendium. Collins Farmers Diary* (1992). FAO (1982) *Camel and Camels Milk. Animal Production and Health Paper No. 26.* Hylyne Rabbits Ltd., Lymn, Cheshire, UK (1991).

**TABLE 235  Time of Ovulation in Relation to Oestrus**

| Specie | Time of Ovulation (hr) |
|---|---|
| Buffalo | 18 - 24 from end |
| Camel | Not known |
| Cattle | 9 - 14 from end |
| Goat | 30 - 40 after start |
| Horse | 24 - 36 from end |
| Pig | 35 after start |
| Sheep | 24 - 48 after start |

**TABLE 236  Typical Gestation and Incubation Periods**

| Specie | Gestation / Incubation days | Specie | Gestation / Incubation days |
|---|---|---|---|
| Alpacas | 340 | Guinea fowl | 26 - 28 |
| Ass | 380 | Hen | 21 |
| Buffalo | 320 | Hen - Bantam types | 19 - 21 |
| Cat | 63 - 64 | Horse | 340 - 345 |
| Cattle - Temperate types | 281 - 285 | Japanese Quail | 17 |
| Cattle - Zebu types | 290 | Llama | 340 |
| Camel - Dromedary | 365 - 395 | Ostrich | 40 - 42 |
| Camel - Bactrian | 402 | Partridge | 21 - 25 |
| Deer | 230 - 235 | Peafowl | 28 - 30 |
| Dog | 63 - 65 | Pheasant | 25 |
| Donkey | 365 | Pigeon | 17 - 19 |
| Duck | 26 - 28 | Rabbit | 30 |
| Duck - Muscovey | 33 - 35 | Sheep | 140 - 147 |
| Elephant - Indian | 630 | Turkey | 26 - 30 |
| Goat | 140 - 147 | Quail - Bobwhite | 22 - 24 |
| Goose | 28 - 35 | Quail - Coturnix | 16 - 18 |

*Sources:* FAO (1982) *Camel and Camel Milk. Animal Production and Health Paper No. 26. Collins Farmers Diary* (1992). *Rhodesian Farm Management Handbook* (1970). ILACO (1985) *Agricultural Compendium.* Oluyemi, J.A. & Roberts, F. (1979) *Poultry Production in Warm Wet Climates.*

# Livestock Nutrition

# General Notes

---

**Note 12 Livestock Nutrition**

---

Several systems of allocating nutrients to livestock and predicting subsequent performance are in use around the world.

The representative feed requirement and allowance tables in this text are in the main, but not exclusively, of UK origin. In using them extensively it is advised that full reference to the original texts be made where much important qualifying and explanatory information is to be found.

It is stressed that livestock performance predictions given in the tables, and nutritional formulae, are constantly under review. As such they represent the best estimates of performance currently available on the basis of relatively recent summarized research.

Livestock nutrition research in the tropics is limited by comparison with that in more temperate climates. Therefore the prediction of animal performance in the tropics continues to rely heavily on extrapolation from temperate experience. Equally the analysis and understanding of tropical feedstuffs is not as advanced.

In considering any predictions of animal performance from the data provided in this text it is important to appreciate that significant deviations will occur due to a range of factors including:

- **feed palatability,**
- **physical nature of feed, e.g. meal v. crumbs v. pellets etc.,**
- **climate (macro and micro),**
- **standard of housing,**
- **genetic potential of livestock including variation between individuals,**
- **specific strains of livestock (particularly hybrid poultry),**
- **age of livestock,**
- **history of livestock,**
- **transportation standard (if stock purchased from outside farm),**
- **consistency and constancy of feed supply,**
- **consistency and constancy of water supply,**
- **mineral imbalances (either in the feed or in the water supply), or**
  **caused by secondary factors such as soil type or interactive effects,**
- **non-mineral nutrient imbalances or deficiencies (e.g. protein, vitamins),**
- **livestock disease,**
- **fungal, soil, rodent or insect contamination of feed,**

- method and efficiency of feed preservation,
- prevailing management standards,
- climatic degradation of feed in storage,
- variation in analysis of different batches of the same feed type,
- inaccurate laboratory analyses (note also that different laboratories may use different analytical methods which do not give strictly comparable results),
- wastage of allocated feed either during grazing or at the trough,
- energy expenditure by animals in obtaining feed.

With particular reference to Zebu type cattle the following relevant points are noted:

- Zebu cattle are considered to have a lower fasting metabolism than temperate cattle. This is associated with lower growth rates under better feed conditions.
- Zebus may have lower energy requirements for maintenance when kept for a long period at the same weight.
- Levels of protein required by cattle in the tropics may be expected to be less than those found in temperate areas.
- Growing Zebu cattle deposit proportionately less fat than temperate types; hence less energy is considered to be required per unit of weight gain.
- Requirements for milk production in Zebus is considered to be essentially the same as for temperate types, but genetic yield potentials are inherently lower.

*Source:* Butterworth, M. H. (1985) *Beef Cattle Nutrition and Tropical Pastures.*

More emphasis is given to the nutrition of ruminants in this section. It is considered that with the ever-expanding use of hybrid poultry and pigs around the world rearers of such stock should in the first instance, and whenever possible, consult the stock suppliers for guidelines on diet composition since new hybrids can benefit from specific rations fed according to age and environment.

It is considered that a knowledge of nutritional constraints and limitations is important for the proper use of the tables in this section; also a knowledge of the wider characteristics of individual feedstuffs.

To take full advantage of this section the reader should use it together with such standard reference texts as *Energy Allowances and Feeding Systems for Ruminants* by the UK MAFF Agricultural Development and Advisory Service, Ref. Book No. 433, or the series of publications, *Nutrient Requirements of Domestic Animals,* by the USA National Research Council or hybrid livestock producers' published guidelines for their specific strains.

---

### Note 13 a  Digestibility and Energy Units

---

#### a. Digestibility and energy value of feeds

Different units are used to measure the energy content of feedstuffs in the various countries. The diversity of units often causes confusion. The most common units are:

**(i)**   Starch equivalent (SE), which measures the amount of pure starch that, when fed, produces as much fat as 100 kg of the feed.

**(ii)**  Russian oat unit (OU), which is similar to the unit above but based on oats.

**(iii)** Scandinavian feed unit (FE, from *Foderenhet*), which is based on barley and refers to milking rather than fattening.

**(iv)**  Modified feed unit ($FE_C$, UF or French feed unit).

**(v)**   Metabolizable energy (ME), which is usually expressed in megacalories per kilogram of feed, but can also be expressed in joules or megajoules per kilogram of feed. To convert calories to joules the value should be multiplied by 4.18.

**(vi)**  Total digestible nutrients (TDN), a unit which is almost identical to the German *Gesamtnährstoff* (GNS).

**Table 237** will enable the reader to calculate the desired energy unit from the proximate composition and the digestibility data.

The 'availability' *(Wertigkeit)* factor needed for the calculation of SE, FE and OU can be obtained from references.

If digestibility data are not available, the energy content may be approximated with the following regression equations.

Fresh grasses: TDN = 54.6 + 3.66 $Log_e$ CP - 0.26 CF + 6.85 $Log_e$ EE

Hay: TDN = 51.78 + 6.44 $Log_e$ CP

The following is an example of how to calculate the metabolizable energy in barley for pigs:

| | Dry Matter (%) | (g/kg) | Digestibility (coefficient) | Digestible Nutrient (g/kg) | (kcal/g) | Total (kcal) |
|---|---|---|---|---|---|---|
| CP | 10.0 | 100 | 79 | 79 | 5.01 | 395.8 |
| CF | 3.9 | 39 | 20 | 7.8 | 3.44 | 26.8 |
| EE | 1.7 | 17 | 50 | 8.5 | 8.93 | 75.9 |
| NFE | 68.7 | 687 | 90 | 618.3 | 4.08 | 2523.5 |
| | | | | | Sum | 3022.0 |

Therefore, the barley contains 3022 kilocalories (3.02 megacalories) of energy per kg of dry matter.

---

**Note 13 b  Digestibility and Energy Units**

---

## b. Conversion of energy units

SE, FE and OU. These three units may be reciprocally converted by multiplication, as $FE_C$ and OU are fractions (0.7 and 0.6) of SE.

TDN and ME. The relationship between TDN and ME depends on the amount of crude protein and crude fibre in the feed. If the dry matter in a feed consists of 25% crude fibre containing 15% digestible crude protein, a kilogram of TDN corresponds to 0.274 megacalories of ME.

SE and TDN. The relationship between SE and TDN depends on the digestible CP content of the feed. The following approximation can be used:

SE (kg) = 0.995 TDN (kg) - 0.051 digestible CP (kg/kg feed).

*Source:* FAO (1981) *Tropical Feeds. Animal Production and Health Series. No. 12.*
Reproduced with kind permission of the Food and Agriculture Organization of the United Nations.

**TABLE 237   Determination of the Nutritive Value of Feedstuffs in Different Feed Units and Energy Units**

| Feed unit | Unit of the digestible nutrient[1] | Multiply digestible nutrient by | | | | Further procedure |
|---|---|---|---|---|---|---|
| | | Crude protein | Ether extract[1] | Crude fibre | N-free extract | |
| TDS or GNS (weight units or %) | Weight units or % | 1 | 2.25 (TDN) 2.30 (GNS) | 1 | 1 | Sum up |
| SE (weight units or %) | Weight units or % | 0.94 | 2.41 2.12 or 1.91 | 1 | 1 | Sum up and multiply by the availability *(Wertigkeit)* |
| FEc (per kg) | g/kg | 0.94 | 2.41 2.12 or 1.91 | 1 | 1 | Sum up, multiply by the availability *(Wertigkeit)* and divide by 700 |
| OU (per kg) | g/kg | 0.94 | 2.41 2.12 or 1.91 | 1 | 1 | Sum up, multiply by the availability *(Wertigkeit)* and divide by 600 |
| FE (per kg) | g/kg | 1.43 | 2.41 2.12 or 1.91 | 1 | 1 | Sum up, multiply by the availability *(Wertigkeit)* and divide by 750 |
| DE ruminants (k/cal/kg) | g/kg | 5.79 | 8.15 | 4.42 | 4.06 | Sum up |
| DE pigs (kcal/kg) | g/kg | 5.78 | 9.42 | 4.40 | 4.07 | Sum up |
| ME ruminants (kcal/kg) | g/kg | 4.32 | 7.73 | 3.59 | 3.63 | Sum up |
| ME pigs (kcal/kg) | g/kg | 5.01 | 8.93 | 3.44 | 4.08 | Sum up |

[1]The digestible ether extract of oily seeds, cakes and feeds of animal origin should be multiplied by 2.41; that of leguminous seeds, cereal grains and their by-products by 2.12; that of hays, straws, chaffs, green fodders, silages, roots and tubers by 1.19.

*Source:* FAO (1981) *Tropical Feeds. Animal Production & Health Series. No. 12.*
Reproduced with kind permission of the Food and Agriculture Organization of the United Nations.

**TABLE 238  Common Mineral Deficiency Symptoms in Livestock**

| Deficient element | Symptoms | Aggravating factors |
|---|---|---|
| Ca | Rickets, osteomalacia, milk fever, weak egg shells | Ca:P ratio, Vitamin D implicated in metabolism of Ca |
| Co | Emaciation and anaemia - 'pine' - mainly in ruminants | |
| Cu | Spontaneous bone fractures, lack of pigmentation in black hair or wool, lack of crimp in wool, Myocardial fibrosis, e.g. falling disease in cattle, demyelination of central nervous system causing severe ataxias, e.g. swayback, diarrhoea in association with excess Mo, aortic rupture in chickens | Excess Mo and sulphate, also F |
| Fe | Anaemia particularly in pigs, poor growth | |
| I | Goiter | Goitrogens in feed |
| K | Deficiency rate - concerned with osmotic regulation together with Na, $HCO_3$ and Cl ions | |
| Mg | Seizures, e.g. Grass tetany 'Staggers' | Excess Ca, excess protein, breed susceptibility, excess K fertilizer |
| Mn | Reduced growth and reproduction, shortened and thickened legs, parrot beak in chickens, abortion, lameness in pigs | High levels of Ca and P |
| Mo | Not observed under natural conditions | |
| Na | Dehydration, poor growth, reduced production, depraved appetite | |
| P | Rickets, osteomalacia, 'Pica' or depraved appetite, poor fertility, reduced growth | Ca:P ratio, Vitamin D implicated in metabolism of P |
| Se | Necrotic liver in pigs, muscular dystrophies in lambs, calves, horses, turkeys, infertility in sheep, unthriftiness in cattle and sheep, poor hair and feather in pigs, horses and chickens | Linked with Vitamin E metabolism |
| Zn | Retarded growth; skin lesions, hair and fur disorders, reduced fertility: Parakeratosis in pigs | Excess Ca, genetic factors |

*NB*  1. This list is intended only as a simplified summary of symptoms and some causative factors.
   2. When considering the source of interacting mineral factors the mineral contents of both feed and the water supply need to be considered.

*Sources:* Mortvedt, J.J. *et al.* (1982) *Micronutrients in Agriculture.* ARC (1965) *The Nutrient Requirements of Farm Livestock. No. 2.* McDonald, P., Edwards, R.A. & Greenhalgh, J.F.D. (1990) *Animal Nutrition.* Spedding C.R.W., ed. (1983) *Fream's Elements of Agriculture.*

**TABLE 239** Animal Tissue Tests for Detecting Mineral Deficiency and Toxicity in Cattle

| Element | Tissue | Critical level indicating deficiency | Critical level indicating toxicity | Corresponding dietary level |
|---|---|---|---|---|
| Ca | Plasma | 8 mg/100 ml | | 0.18 - 0.6% |
| Mg | Serum | 1 - 2 mg/100 ml | | |
| | Urine | 2 - 10mg/100 ml | | |
| K | | | | 0.6 - 0.80% |
| P | Plasma | 4.5 mg/100 ml | | 0.18 - 0.43% |
| Na | Saliva | 100 - 200 mg/100 ml | | 0.1% |
| S | | | | 0.1% (deficient) |
| Co | Liver | 0.05 mg/kg | | 0.05 - 0.1 mg/kg |
| Cu | Liver | 25 mg/kg | | |
| | Liver | | 700 mg/kg | |
| I | Milk | 10 - 20 µg/l | | |
| | Serum | 40 µg/l | | |
| Fe | Haemoglobin | 10 g/100 ml | | |
| | Transferrin | 13 - 15% saturation | | |
| Mn | Liver | 6 - 10 mg/kg | | 20 - 40 mg/kg |
| | Hair | | 1000 - 2000 mg/kg | 70 mg/kg |
| Se | Liver | 0.25 - 0.50 mg/kg | | 0.05 - 0.1 mg/kg |
| | Liver | | 5 - 15 mg/kg | 5 mg/kg |
| Zn | Plasma | 0.04 mg/100 ml | | 10 - 50 mg/kg |
| F | Bone | | 4500 - 5000 mg/kg | |
| Mo | | | | 6 - 20 mg/kg (toxicity) |

*Source:* Chamberlain, A. (1989) *Milk Production in the Tropics.*
Reproduced with kind permission of Longman Group UK Ltd.

TABLE 240   Suggested Mineral Requirements and Toxicities for Selected
Livestock and Poultry (dry basis) – ppm

| | Growing and finishing steers and heifers | Lactating dairy cows | Sheep | Pigs | Starting chicks (0-8 weeks) |
|---|---|---|---|---|---|
| **Required elements** | | | | | |
| Co | 0.05 - 0.1 | 0.1 | 0.1 | – | – |
| Cu | 4 | 10 | 5 | 6 | 4 |
| I | 0.05 - 0.1 | 0.6 | 0.10 - 0.8 | 0.2 | 0.35 |
| Fe | 10 | 100 | 30 - 50 | 80 | 80 |
| Mn | 1 - 10 | 20 | 20 - 40 | 20 | 55 |
| Mo | – | – | >0.5 | – | – |
| Se | 0.05 - 0.1 | 0.1 | 0.1 | 0.1 | 0.1 |
| Zn | 10 - 30 | 40 | 35 - 50 | 50 | 50 |
| **Toxic elements** | | | | | |
| Cu | – | 100 | 8 - 25 | 300 - 500 | 1270 |
| F | 30 - 100 | 40 | 60 - 200 | – | 500 - 1000 |
| Mo | – | 6 | 5 - 20 | – | 200 - 500 |
| Se | 5 | 5 | >2 | 5 - 8 | 10 |

*Source:* McDowell, L.R. & Conrad, J.H. (1977) *Trace Element Nutrition in Latin America.* FAO
World Animal Reviews. Reproduced with kind permission of the Food and Agriculture
Organization of the United Nations.

# Ruminant Nutrition – General Notes, Equations etc.

### Note 14  Ruminant Nutrition

A range of tables is presented in this section illustrating feed requirements and allowances for ruminants. These are in the main of UK origin but others are included from different sources to illustrate the use of different parameters.

The UK table makes extensive use of the ME system (Metabolizable Energy) and also Net Energy. The use of both parameters is illustrated where relevant. Tables from other sources also use the TDN system (Total Digestible Nutrients) sometimes along with the ME system thereby enabling a degree of dual usage. The use of Net Energy values with animal production levels is also illustrated.

A table of general equations and parameter relationships for ruminant nutrition is also given at the beginning of the section to assist in the conversion of analytical data to energy values etc.

### TABLE 241  General Equations and Parameter Relationships for Ruminant Nutrition

The following formulae are useful for general planning purposes when considering the energy metabolism of ruminants. They are mainly sourced from many to be found in *Energy Allowances and Feeding Systems for Ruminants* prepared by the UK Ministry of Agriculture Fisheries and Food Agricultural Development and Advisory Service (RB 433. 1987); also the *Zimbabwe Cattle Breeders Handbook* (1988), *Milk Production in the Tropics* (Chamberlain, A. 1989) and *Atlas of Nutritional Data on United States and Canadian Feeds* (1971). Since the relationships are mainly derived from British temperate experience it follows that they must be applied with caution to tropical situations. Abbreviations are presented at the end of the list (page 353).

**1.**  ME of a feed = 0.81 DE

**2.**  ME(MJ/kg) = 0.0152 DCP + 0.342 DEE + 0.0128 DCF + 0.0159 DNFE (all in g/kg) This formula is conveniently used for concentrate feeds.

Other formulae for compound feeds are:
ME = 12 + 0.008 CP + 0.023 EE - 0.018 CF - 0.012 TA

$$ME = \left[ \frac{DOMD\%}{100 - TA} \right] (0.0152CP + 0.342EE + 0.0128\ CF + 0.0159\ NFE)$$
$$\text{(all in g/kg DM)}$$

**3.**   Gross energy of feed (MJ/kg DM)
= 0.0226 CP + 0.0407 EE + 0.0192 CF + 0.0177 NFE
(all in g/kg DM = 10 x % in DM)

**4.**   ME (MJ/kg DM) = 16.37 - 0.0205 MADF(g/kg DM)
This formula is conveniently used for roughage feeds.

More specific formulae are:

| | |
|---|---|
| **For fresh grass:** | ME = 15.3 - 0.0153 MADF |
| | ME = 0.165 IVD - 0.31 |
| **For legumes:** | ME = 12.3 - 0.012 MADF |
| | ME = 0.130 IVD + 0.01 CP + 1.09 |
| **For grass hays:** | ME = 16.53 - 0.0213 MADF |
| | ME = 0.185 IVD - 1.88 |
| **For dried grass:** | ME = 15.0 - 0.0166 MADF |
| | ME = 0.143 IVD + 1.0 |
| **For dried alfalfa:** | ME = 14.7 - 0.0167 MADF |
| | ME = 0.163 IVD - 0.23 |
| **For grass silage:** | ME = 14.6 - 0.013 MADF |
| | ME = 0.11 IVD + 3.2 |
| | (all in g/kg DM) |

**5.**   ME of a forage = 0.15 DOMD% (MJ/kg DM)
(assumes average value of digested organic matter of 19 MJ/kg DM.
Use coeffcents 0.14 for straw, 0.155 for hay and dried grass, and 0.16 for roots,
green feeds, grasses, legumes, miscellaneous, cereals and bi-products for greater
precision).

**6.**   DOMD% = $\dfrac{OMD\%(100 - ash\%)}{100}$

**7.**   Energy value of DOM = 0.0124 CP + 17.3 (M J/kg DM) (proposed for grass and
fresh maize; CP in g/kg DM)

**8.**   DOMD% = 0.98 DMD - 4.8 (for forages)

**9.**   DOMD% = 0.92 OMD% - 1.2 (for forages)

**10.**   ME concentration – $\dfrac{\text{Total ME of ration}}{\text{Total DMI}}$ (MJ/kg DM)

**11.**   1 kg liveweight loss in a dairy cow = 28 MJ of dietary energy contribution to
milk production.

**12.**   1 kg of liveweight gain in a lactating dairy cow = 34 MJ of dietary energy
requirement.

*(continued over)*

**TABLE 241** *(continued)*    **General Equations and Parameter Relationships for Ruminant Nutrition**

**13.**  DMI kg/day of dairy cow = 0.025 W(kg) + 0.1Y (kg/day)
In first 10 weeks of lactation DMI will be 2-3 kg/day below value given by equation.

**14.**  ME for maintenance of sheep:
$$\text{indoors} = 1.4 + 0.09 \text{ W}$$
$$\text{outdoors} = 1.8 + 0.1 \text{ W}$$

**15.**  ME for maintenance of growing and fattening sheep:
$$\text{indoors} = 1.2 + 0.13 \text{ W}$$
$$\text{outdoors} = 1.4 + 0.15 \text{ W}$$

**16.**  ME for maintenance of housed beef cattle = 8.3 + 0.091 W

**17.**  TDN/100 kg feed = % apparently digested protein
+ % apparently digested NFE
+ % apparently digested CF
+ 2.25 x % apparently digested fat

**18.**  18.4 MJ of DE is approximately equal to 1 kg of TDN

**19.**  15 MJ of ME is approximately equal to 1 kg TDN

**20.**  Net Energy (NE) = ME less the heat increments of fermentation and metabolism

**21.**  Digestible crude protein is approximately equal to (CP x 0.9) - 3.0

**22.**  CP in diet is approximately equal to 6.25 x % nitrogen in the diet

**23.**  DE kcal/kg = $\dfrac{\text{TDN\%} \times 4409}{100}$

**TABLE 241** *(continued)*   **General Equations and Parameter Relationships for Ruminant Nutrition**

**Abbreviations:**

| | |
|---|---|
| APL | Animal Production Level (a factor taking into account the size of the animal and the live weight gain) |
| BF | Butter Fat |
| CF | Crude Fibre |
| CP | Crude Protein |
| DCF | Digestible Crude Fibre |
| DCP | Digestible Crude Protein |
| DE | Digestible Energy |
| DEE | Digestible Ether Extract |
| DM | Dry Matter |
| DMD | % Digestibility of the Dry Matter |
| DMI | Dry Matter Intake |
| DNFE | Digestible Nitrogen Free Extractive |
| DOMD | % Digestible Organic Matter in the Dry Matter |
| DP | Digestible Protein |
| EE | Ether Extract |
| $E_{mp}$ | Net Energy for Maintenance and Production |
| IVD | In Vitro DOMD |
| LWG | Live Weight gain |
| M | Maintenance |
| MADF | Modified Acid Detergent Fibre |
| ME | Metabolizable Energy |
| MEF | Metabolizable Energy of Fibre |
| MJ | Megajoule |
| $M_m$ | Metabolizable Energy for Maintenance |
| $M_{mp}$ | Metabolizable Energy for Maintenance and Production |
| NE | Net Energy |
| NFE | Nitrogen Free Extractive |
| OMD | % Digestibility of the Organic Matter |
| RDP | Rumen Degradable Protein |
| SNF | Solids Not Fat |
| TA | Total Ash |
| TDN | Total Digestible Nutrients |
| UDP | Undegradable Protein in the rumen |
| W & LW | Liveweight (kg) |
| Y | Yield |

$$\frac{M}{D} = \frac{\text{Metabolizable energy of ration}}{\text{Dry Matter Intake}}$$

$$\text{Metabolizability 'q'} = \frac{\text{Metabolizable energy}}{\text{Gross Energy}}$$

qm          Metabolizability at Maintenance

These abbreviations apply to the above equations and in the following tables where used.

# Feed and Nutrient Requirements – Milking Cows

**TABLE 242  Daily Maintenance Allowance of ME for Dairy Cows**

| Body weight (kg) | MJ/ head | Body weight (kg) | MJ/ head |
|---|---|---|---|
| 100 | 17 | 400 | 45 |
| 150 | 22 | 450 | 49 |
| 200 | 27 | 500 | 54 |
| 250 | 31 | 550 | 59 |
| 300 | 36 | 600 | 63 |
| 350 | 40 | | |

(including safety margin) Based on $M_m = 8.3 + 0.091\ W$

*Source:* MAFF/ADAS (1987) *Energy allowances and feeding systems for ruminants.* RB 433.
Reproduced with kind permission of MAFF ADAS (UK)

**TABLE 243  ME Allowances for 1 kg Milk**

| Type of milk | BF (g/kg) | SNF (g/kg) | Energy value $EV_1$. (MJ/kg) | ME allowance $M_1$. (MJ/kg) |
|---|---|---|---|---|
| Channel Island | 48 | 91 | 3.482 | 5.90 |
| Shorthorn | 36 | 87 | 2.937 | 4.98 |
| Ayrshire | 37 | 88 | 2.996 | 5.08 |
| Friesian | 36 | 86 | 2.917 | 4.94 |
| Average | 38 | 87 | 3.014 | 5.10 |
| Solids corrected | 40 | 89 | 3.133 | 5.31 |

(including safety margin) Based on $M_1 = 1.694\ EV_1$    ($M_1$ = ME required for milk production, $EV_1$ = energy value of milk)

*Source:* MAFF/ADAS (1987) *Energy allowances and feeding systems for ruminants.* RB 433.
Reproduced with kind permission of MAFF ADAS (UK).

**TABLE 244    Metabolizable Energy Allowance (MJ) to Produce 1 kg Milk of Varying Composition**

| SNF content (g/kg) | Fat content of milk (g/kg) | | | | | | | | | | | |
|---|---|---|---|---|---|---|---|---|---|---|---|---|
| | 30 | 32 | 34 | 36 | 38 | 40 | 42 | 44 | 46 | 48 | 50 | 52 |
| 84 | 4.48 | 4.61 | 4.74 | 4.87 | 5.00 | 5.13 | 5.26 | 5.39 | 5.52 | 5.62 | 5.79 | 5.92 |
| 85 | 4.51 | 4.64 | 4.77 | 4.90 | 5.04 | 5.17 | 5.30 | 5.43 | 5.56 | 5.69 | 5.82 | 5.95 |
| 86 | 4.55 | 4.68 | 4.81 | 4.94 | 5.07 | 5.20 | 5.33 | 5.46 | 5.59 | 5.72 | 5.85 | 5.99 |
| 87 | 4.58 | 4.71 | 4.84 | 4.98 | 5.10 | 5.24 | 5.37 | 5.50 | 5.63 | 5.76 | 5.89 | 6.02 |
| 88 | 4.62 | 4.75 | 4.88 | 5.01 | 5.14 | 5.27 | 5.40 | 5.53 | 5.66 | 5.79 | 5.92 | 6.05 |
| 89 | 4.65 | 4.78 | 4.91 | 5.04 | 5.17 | 5.31 | 5.44 | 5.57 | 5.70 | 5.83 | 5.96 | 6.09 |
| 90 | 4.69 | 4.82 | 4.95 | 5.08 | 5.21 | 5.34 | 5.47 | 5.60 | 5.73 | 5.86 | 5.99 | 6.12 |
| 91 | 4.72 | 4.85 | 4.98 | 5.11 | 5.24 | 5.37 | 5.51 | 5.64 | 5.77 | 5.90 | 6.03 | 6.16 |
| 92 | 4.76 | 4.89 | 5.02 | 5.15 | 5.28 | 5.41 | 5.54 | 5.67 | 5.80 | 5.93 | 6.06 | 6.19 |
| 93 | 4.79 | 4.92 | 5.05 | 5.18 | 5.31 | 5.44 | 5.57 | 5.71 | 5.84 | 5.97 | 6.10 | 6.23 |
| 94 | 4.82 | 4.96 | 5.09 | 5.22 | 5.35 | 5.48 | 5.61 | 5.74 | 5.87 | 6.00 | 6.13 | 6.26 |
| 95 | 4.86 | 4.99 | 5.12 | 5.25 | 5.38 | 5.51 | 5.64 | 5.77 | 5.91 | 6.04 | 6.17 | 6.30 |

(Including safety margin)

[  ] Milk of average composition          [  ] Solids corrected milk (SCM)

*Source:* MAFF/ADAS (1987) *Energy allowances and feeding systems for ruminants.* RB 433.
Reproduced with kind permission of MAFF ADAS (UK).

**TABLE 245   Daily ME Allowances for Three Breeds of Dairy Cattle (MJ/head)**

| Breed | Liveweight change | Main-tenance | Milk yield kg/day | | | | | | |
|---|---|---|---|---|---|---|---|---|---|
| | | | 5 | 10 | 15 | 20 | 25 | 30 | 35 |
| **JERSEY** | | | | | | | | | |
| **400 kg** | Losing 0.5 kg/day | – | 61 | 91 | 121 | 152 | 183 | | |
| **49 g/kg BF** | No weight change | 45 | 75 | 105 | 135 | 166 | 197 | | |
| **95 g/kg SNF** | Gaining 0.5 kg/day | – | 92 | 122 | 152 | 183 | 214 | | |
| **AYRSHIRE** | | | | | | | | | |
| **500 kg** | Losing 0.5 kg/day | – | 66 | 92 | 118 | 144 | 169 | 195 | |
| **38 g/kg BF** | No weight change | 54 | 80 | 106 | 132 | 158 | 183 | 209 | |
| **89 g/kg SNF** | Gaining 0.5 kg/day | – | 97 | 123 | 149 | 175 | 200 | 226 | |
| **FRIESIAN** | | | | | | | | | |
| **590 kg** | Losing 0.5 kg/day | – | 73 | 97 | 122 | 147 | 172 | 196 | 221 |
| **36 g/kg BF** | No weight change | 62 | 87 | 111 | 136 | 161 | 186 | 210 | 235 |
| **86 g/kg SNF** | Gaining 0.5 kg/day | – | 104 | 128 | 153 | 178 | 203 | 227 | 252 |

(including safety margin)

*Source:* MAFF/ADAS (1987) *Energy allowances and feeding systems for ruminants.* RB 433. Reproduced with kind permission of MAFF ADAS (UK)

**TABLE 246   Daily ME Allowances for Cattle During Pregnancy (MJ/head)**

| Liveweight (kg) | Month of pregnancy | | | |
|---|---|---|---|---|
| | 6 | 7 | 8 | 9 |
| 350 | 48 | 51 | 55 | 60 |
| 400 | 52 | 55 | 59 | 65 |
| 450 | 57 | 60 | 64 | 69 |
| 500 | 61 | 64 | 68 | 74 |
| 550 | 66 | 69 | 73 | 78 |
| 600 | 71 | 73 | 77 | 83 |
| 650 | 75 | 78 | 82 | 87 |
| 700 | 80 | 83 | 86 | 92 |

(including safety margin) Based on $M_{mp} = M_m + 1.13e^{0.0106t}$

where t = number of days pregnant

*Source:* MAFF/ADAS (1987) *Energy allowances and feeding systems for ruminants.* RB 433. Reproduced with kind permission of MAFF ADAS (UK).

TABLE 247  **Probable Dry Matter Intake of Cows in Mid and Late Lactation (kg/day)**

| Liveweight W (kg) | Milk yield (Y kg/day) | | | | | | | |
|---|---|---|---|---|---|---|---|---|
| | 5 | 10 | 15 | 20 | 25 | 30 | 35 | 40 |
| 350 | 9.3 | 9.8 | 10.3 | 10.8 | 11.3 | 11.8 | | |
| 400 | 10.5 | 11.0 | 11.5 | 12.0 | 12.5 | 13.0 | | |
| 450 | 11.8 | 12.3 | 12.8 | 13.3 | 13.8 | 14.3 | 14.8 | |
| 500 | 13.0 | 13.5 | 14.0 | 14.5 | 15.0 | 15.5 | 16.0 | |
| 550 | 14.3 | 14.8 | 15.3 | 15.8 | 16.3 | 16.8 | 17.3 | 17.8 |
| 600 | 15.5 | 16.0 | 16.5 | 17.0 | 17.5 | 18.0 | 18.5 | 19.0 |
| 650 | 16.8 | 17.3 | 17.8 | 18.3 | 18.8 | 19.3 | 19.8 | 20.3 |
| 700 | 18.0 | 18.5 | 19.0 | 19.5 | 20.0 | 20.5 | 21.0 | 21.5 |

Based on DMI (kg/day) = 0.025 W + 0.1 Y

Note: In the first 6 weeks of lactation, reduce these values by 2 - 3 kg DMI per day

Source: MAFF/ADAS (1987) *Energy allowances and feeding systems for ruminants.* RB 433.
Reproduced with kind permission of MAFF ADAS (UK).

TABLE 248  **Minimum Metabolizable Energy Concentrations of Diets for Cows – M/D (MJ/kg DM)**

| Breed and rate of liveweight change | Milk yield (kg/day) | | | | | | | |
|---|---|---|---|---|---|---|---|---|
| | 0 | 5 | 10 | 15 | 20 | 25 | 30 | 35 |
| **JERSEY** | | | | | | | | |
| - 0.5 kg/day | – | (5.9) | (8.6) | 11.1 | 13.4 | | | |
| No change | (4.6) | (7.5) | 10.2 | 12.6 | | | | |
| + 0.5 kg/day | – | (9.3) | 11.8 | 14.1 | | | | |
| **AYRSHIRE** | | | | | | | | |
| - 0.5 kg.day | – | (5.0) | (6.7) | (8.3) | 9.8 | 11.2 | 12.5 | |
| No change | (4.3) | (6.1) | (7.8) | (9.4) | 10.8 | 12.2 | 13.5 | |
| + 0.5 kg/day | – | (7.4) | (9.1) | 10.6 | 12.0 | 13.3 | | |
| **FRIESIAN** | | | | | | | | |
| - 0.5 kg/day | – | (4.7) | (6.1) | (7.4) | (8.7) | 9.8 | 10.9 | 12.0 |
| No change | (4.2) | (5.7) | (7.1) | (8.4) | (9.6) | 10.7 | 11.8 | 12.8 |
| + 0.5 kg/day | – | (6.8) | (8.1) | (9.4) | 10.6 | 11.7 | 12.7 | |

( ) indicate theoretical values only as appetite limits on poor quality forage make rations unfeasible

Source: MAFF/ADAS (1987) *Energy allowances and feeding systems for ruminants.* RB 433.
Reproduced with kind permission of MAFF ADAS (UK).

TABLE 249  Desirable Pattern of Liveweight Change During Lactation

| Week number | Liveweight change (kg/day) | Change during 10 weeks (kg) | Net effect on liveweight (kg) |
|---|---|---|---|
| 1 - 10 | – 0.5 | – 35 | – 35 |
| 11 - 20 | 0 | 0 | – 35 |
| 21 - 30 | + 0.5 | + 35 | 0 |
| 31 - 40 | + 0.5 | + 35 | + 35 |
| 41 - 52 | + 0.75 | + 63 | + 98 |

*Source:* MAFF/ADAS (1987) *Energy allowances and feeding systems for ruminants.* RB 433.
Reproduced with kind permission of MAFF ADAS (UK).

TABLE 250  Probable Dry Matter Intakes (DMI) in kg/day of Milking Cows

**EARLY LACTATION (Weeks 1 - 10)**

**Milk yield (kg/day)**

| kg LW | (Dry Cow) | 5 | 10 | 15 | 20 | 25 | 30 | 35 | 40 | 45 | 50 | kg LW |
|---|---|---|---|---|---|---|---|---|---|---|---|---|
| 350 | 4.8 | 6.5 | 8.3 | 10.0 | 11.8 | 13.5 | 15.3 | 17.1 | 18.8 | 20.6 | 22.3 | 350 |
| 400 | 5.4 | 7.1 | 8.9 | 10.7 | 12.4 | 14.2 | 15.9 | 17.7 | 19.4 | 21.2 | 22.9 | 400 |
| 450 | 6.0 | 7.8 | 9.5 | 11.3 | 13.0 | 14.8 | 16.5 | 18.3 | 20.1 | 21.8 | 23.6 | 450 |
| 475 | 6.3 | 8.1 | 9.8 | 11.6 | 13.3 | 15.1 | 16.9 | 18.6 | 20.4 | 22.1 | 23.9 | 475 |
| 500 | 6.6 | 8.4 | 10.2 | 11.9 | 13.7 | 15.4 | 17.2 | 18.9 | 20.7 | 22.4 | 24.2 | 500 |
| 525 | 7.0 | 8.7 | 10.5 | 12.2 | 14.0 | 15.7 | 17.5 | 19.2 | 21.0 | 22.7 | 24.5 | 525 |
| 550 | 7.3 | 9.0 | 10.8 | 12.5 | 14.3 | 16.0 | 17.8 | 19.6 | 21.3 | 23.1 | 24.8 | 550 |
| 575 | 7.6 | 9.3 | 11.1 | 12.8 | 14.6 | 16.4 | 18.1 | 19.9 | 21.6 | 23.4 | 25.1 | 575 |
| 600 | 7.9 | 9.6 | 11.4 | 13.2 | 14.9 | 16.7 | 18.4 | 20.2 | 21.9 | 23.7 | 25.4 | 600 |
| 625 | 8.2 | 10.0 | 11.7 | 13.5 | 15.2 | 17.0 | 18.7 | 20.5 | 22.2 | 24.0 | 25.8 | 625 |
| 650 | 8.5 | 10.3 | 12.0 | 13.8 | 15.5 | 17.3 | 19.0 | 20.8 | 22.6 | 24.3 | 26.1 | 650 |
| 675 | 8.8 | 10.6 | 12.3 | 14.1 | 15.8 | 17.6 | 19.4 | 21.2 | 22.9 | 24.6 | 26.4 | 675 |

Mid lactation (weeks 11 - 29) increase above values by 1.7 kg/day
Late lactation (weeks 30 on) increase above values by 4.1 kg/day

*Source:* Wilson, P.N. & Brigstocke, T.D.A. *Improved Feeding of Cattle and Sheep,*
in *Collins Farmers Diary* (1992). Reproduced with kind permission of Blackwell Scientific Publications.

**TABLE 251 a & b   Feeding Standards for Lactating and Pregnant Cattle**

**a. Daily allowances for cows producing milk of 38 g fat and 34 g protein/kg and weighing 500 kg ($q_m$ = 0.55)**

| Milk yield (kg/day) | 0 | 5 | 10 | 15 | 20 | 25 | Month of pregnancy | |
|---|---|---|---|---|---|---|---|---|
| Liveweight change (kg/day) | 0 | + 0.6 | + 0.4 | + 0.25 | 0 | – 0.2 | 8 | 9 |
| DMI (kg) | 9 | 11 | 13 | 15 | 16 | 16 | 10 | 10 |
| ME (MJ) | 56 | 112 | 130 | 150 | 166 | 186 | 72 | 85 |
| RDP (g) | 465 | 934 | 1082 | 1253 | 1383 | 1553 | 600 | 709 |
| UDP (g) | 2 | 15 | 95 | 174 | 256 | 335 | 0 | 0 |
| $E_{mp}$ (MJ) | 39 | 71 | 82 | 94 | 103 | 114 | 41 | 43 |
| APL | 1.00 | 1.84 | 2.11 | 2.42 | 2.65 | 2.95 | 1.06 | 1.10 |
| Ca (g) | 19 | 32 | 45 | 58 | 71 | 85 | 30 | 30 |
| P (g) | 16 | 26 | 36 | 45 | 55 | 65 | 23 | 23 |
| Mg (g) | 9 | 13 | 17 | 21 | 25 | 29 | 12 | 12 |
| Na (g) | 4 | 8 | 11 | 14 | 18 | 21 | 6 | 6 |
| Vit.A (i.u.) | | | | ←— 50 000 —→ | | | | |
| Vit D (i.u.) | | | | ←— 5000 —→ | | | | |
| Vit E (i.u.) | | | | ←— 300 —→ | | | | |

**b. Daily allowances for cows producing milk of 38 g fat and 34 g protein/kg and weighing 500 kg ($q_m$ = 0.60)**

| Milk yield (kg/day) | 0 | 5 | 10 | 15 | 20 | 25 | Month of pregnancy | |
|---|---|---|---|---|---|---|---|---|
| Liveweight change (kg/day) | 0 | +0.6 | +0.4 | +0.25 | 0 | -0.2 | 8 | 9 |
| DMI (kg) | 9 | 11 | 13 | 15 | 16 | 16 | 10 | 10 |
| ME (MJ) | 54 | 109 | 126 | 146 | 161 | 181 | 70 | 83 |
| RDP (g) | 450 | 909 | 1051 | 1218 | 1343 | 1510 | 584 | 692 |
| UDP (g) | 16 | 35 | 122 | 201 | 289 | 369 | 0 | 0 |
| $E_{mp}$ (MJ) | 37 | 72 | 82 | 94 | 103 | 114 | 41 | 43 |
| APL | 1.00 | 1.86 | 2.12 | 2.42 | 2.65 | 2.95 | 1.06 | 1.10 |
| Ca (g) | 19 | 32 | 45 | 58 | 71 | 85 | 30 | 30 |
| P (g) | 16 | 26 | 36 | 45 | 55 | 65 | 23 | 23 |
| Mg (g) | 9 | 13 | 17 | 21 | 25 | 29 | 12 | 12 |
| Na (g) | 4 | 8 | 11 | 14 | 18 | 21 | 6 | 6 |
| Vit.A (i.u.) | | | | ←— 50 000 —→ | | | | |
| Vit.D (i.u.) | | | | ←— 5000 —→ | | | | |
| Vit. E (i.u.) | | | | ←— 300 —→ | | | | |

*(continued over)*

**TABLE 251 c & d  Feeding Standards for Lactating and Pregnant Cattle**

**c. Daily allowances for cows producing milk of 38g fat and 32 g protein/kg and weighing 500 kg ($q_m = 0.65$)**

| Milk yield (kg/day) | 0 | 5 | 10 | 15 | 20 | 25 | 30 | Month of pregnancy | |
|---|---|---|---|---|---|---|---|---|---|
| Liveweight change (kg/day) 0 | | + 0.6 | + 0.4 | + 0.25 | 0 | − 0.2 | − 0.4 | 8 | 9 |
| DMI (kg) | 9 | 11 | 13 | 15 | 16 | 16 | 16 | 10 | 10 |
| ME (MJ) | 53 | 106 | 123 | 142 | 157 | 176 | 196 | 69 | 82 |
| RDP (g) | 443 | 886 | 1026 | 1188 | 1311 | 1471 | 1634 | 575 | 684 |
| UDP (g) | 22 | 55 | 142 | 230 | 316 | 400 | 483 | 0 | 0 |
| $E_{mp}$ (MJ) | 39 | 71 | 82 | 94 | 103 | 114 | 126 | 41 | 43 |
| APL | 1.00 | 1.84 | 2.11 | 2.42 | 2.65 | 2.95 | 3.24 | 1.06 | 1.10 |
| Ca (g) | 19 | 32 | 45 | 58 | 71 | 85 | 98 | 30 | 30 |
| P (g) | 16 | 26 | 36 | 45 | 55 | 65 | 75 | 23 | 23 |
| Mg (g) | 9 | 13 | 17 | 21 | 25 | 29 | 33 | 12 | 12 |
| Na (g) | 4 | 8 | 11 | 14 | 18 | 21 | 24 | 6 | 6 |
| Vit.A (i.u.) | ← | | | | 50 000 | | | | → |
| Vit D (i.u.) | ← | | | | 5000 | | | | → |
| Vit. E (i.u.) | ← | | | | 300 | | | | → |

**d. Daily allowances for cows producing milk of 38 g fat and 34 g protein/kg and weighing 650 kg ($q_m = 0.55$)**

| Milk yield (kg/day) | 0 | 5 | 10 | 15 | 20 | Month of pregnancy | |
|---|---|---|---|---|---|---|---|
| Liveweight change (kg/day)  0 | | + 0.6 | + 0.4 | + 0.25 | 0 | 8 | 9 |
| DMI (kg) | 10.5 | 13 | 15 | 17 | 17.5 | 12 | 12 |
| ME (MJ) | 67 | 123 | 141 | 161 | 177 | 88 | 105 |
| RDP (g) | 560 | 1028 | 1175 | 1344 | 1473 | 734 | 876 |
| UDP (g) | 11 | 23 | 104 | 182 | 265 | 0 | 0 |
| $E_{mp}$ (MJ) | 47 | 79 | 90 | 102 | 111 | 50 | 52 |
| APL | 1.00 | 1.69 | 1.92 | 2.18 | 2.37 | 1.06 | 1.11 |
| Ca (g) | 24 | 37 | 51 | 64 | 77 | 39 | 39 |
| P (g) | 19 | 29 | 39 | 49 | 59 | 27 | 27 |
| Mg (g) | 12 | 16 | 20 | 24 | 28 | 15 | 15 |
| Na (g) | 5 | 9 | 12 | 16 | 19 | 8 | 8 |
| Vit. A (i.u.) | ← | | | 65 000 | | | → |
| Vit. D (i.u.) | ← | | | 6500 | | | → |
| Vit.E (i.u.) | ← | | | 385 | | | → |

**TABLE 251 e & f   Feeding Standards for Lactating and Pregnant Cattle**

**e. Daily allowances for cows producing milk of 38 g fat and 34 g protein/kg and weighing 650 kg ($q_m = 0.60$)**

| Milk yield (kg/day) | 0 | 5 | 10 | 15 | 20 | 25 | Month of pregnancy | |
|---|---|---|---|---|---|---|---|---|
| Liveweight change (kg/day) | 0 | + 0.6 | + 0.4 | + 0.25 | 0 | − 0.2 | 8 | 9 |
| DMI (kg) | 10.5 | 13 | 15 | 17 | 17.5 | 17.5 | 12 | 12 |
| ME (MJ) | 66 | 120 | 137 | 157 | 172 | 191 | 87 | 104 |
| RDP (g) | 550 | 1000 | 1143 | 1309 | 1434 | 1597 | 726 | 867 |
| UDP (g) | 17 | 43 | 130 | 209 | 298 | 384 | 0 | 0 |
| $E_{mp}$ (MJ) | 47 | 80 | 90 | 102 | 111 | 122 | 50 | 52 |
| APL | 1.00 | 1.71 | 1.93 | 2.19 | 2.37 | 2.62 | 1.06 | 1.11 |
| Ca (g) | 24 | 37 | 51 | 64 | 77 | 90 | 39 | 39 |
| P (g) | 19 | 29 | 39 | 49 | 59 | 69 | 27 | 27 |
| Mg (g) | 12 | 16 | 20 | 24 | 28 | 31 | 15 | 15 |
| Na (g) | 5 | 9 | 12 | 16 | 19 | 22 | 8 | 8 |
| Vit.A (i.u.) | | | | ← 65 000 → | | | | |
| Vit.D (i.u.) | | | | ← 6500 → | | | | |
| Vit.E (i.U.) | | | | ← 385 → | | | | |

**f. Daily allowances for cows producing milk of 38 g fat and 34 g protein/kg, and weighing 650 kg ($q_m = 0.65$)**

| Milk yield(kg/day) | 0 | 5 | 10 | 15 | 20 | 25 | 30 | Month of pregnancy | |
|---|---|---|---|---|---|---|---|---|---|
| Liveweight change (kg/day) | 0 | + 0.6 | + 0.4 | + 0.25 | 0 | − 0.2 | − 0.4 | 8 | 9 |
| DMI (kg) | 10.5 | 13 | 15 | 17 | 17.5 | 17.5 | 18 | 14 | 14 |
| ME (MJ) | 64 | 117 | 134 | 153 | 167 | 187 | 206 | 85 | 102 |
| RDP (g) | 534 | 976 | 1115 | 1275 | 1397 | 1556 | 1716 | 709 | 851 |
| UDP (g) | 31 | 63 | 150 | 236 | 331 | 410 | 496 | 0 | 0 |
| $E_{mp}$ (MJ) | 47 | 79 | 90 | 102 | 111 | 122 | 134 | 50 | 52 |
| APL | 1.00 | 1.69 | 1.92 | 2.18 | 2.37 | 2.62 | 2.86 | 1.06 | 1 11 |
| Ca (g) | 24 | 37 | 51 | 64 | 77 | 90 | 103 | 39 | 39 |
| P (g) | 19 | 29 | 39 | 49 | 59 | 69 | 79 | 27 | 27 |
| Mg (g) | 12 | 16 | 20 | 24 | 28 | 31 | 35 | 15 | 15 |
| Na (g) | 5 | 9 | 12 | 16 | 19 | 22 | 26 | 8 | 8 |
| Vit.A (i.u.) | | | | ← 65 000 → | | | | | |
| Vit.D (i.u.) | | | | ← 6500 → | | | | | |
| Vit.E (i.u.) | | | | ← 385 → | | | | | |

*Source:* McDonald, P., Edwards, R.A., & Greenhalgh, J.F.D. (1988) *Animal Nutrition.*
Reproduced with kind permission of Longman Group UK Ltd.

---

**TABLE 252    Utilizable Metabolizable Energy Calculation (UME)**

---

UME is an estimate of how efficiently grassland is used to supply the energy requirements of cows. In the analysis the UME value is calculated as follows.

|  |  |
|---|---|
|  | Total energy required by cows for milk, maintenance and pregnancy |
| **less:** | Total energy provided by concentrates and purchased bulks |
| **Multiplied by:** | Stocking rate (LSU per ha) |
| **equals:** | UME per hectare (GJ) |

**Assumptions:**

| | |
|---|---|
| Milk production | 5.5 MJ per litre |
| Maintenance and pregnancy | 25 000 MJ per cow |
| Concentrates | 11.2 MJ per kg fresh weight |
| Bulk feeds | 2.8 MJ per kg fresh weight |

**Example:** **Yield 5689 litres, concentrates 1455 kgs, bulk feeds 640 kgs, stocking rate 2.30**

|  |  | MJ | GJ |
|---|---|---|---|
|  | Energy for production | 31 290 | |
|  | Energy for maintenance and pregnancy | 25 000 | |
| **Less** | Energy supplied by concentrates | 16 296 | |
|  | Energy supplied by bulk feed | 1792 | |
| ∴ | Energy supplied by forage | 38 202 | |
| **Multiplied by** | | | |
|  | Stocking rate | 87 865 | |
|  | UME per hectare (GJ) | | **88** |

*Source:* Genus Management (1991) *Genus Milkminder. Dairy Herd Costing Service.*
Reproduced with kind permission of the Milk Marketing Board.

# Feed and Nutrient Requirements – Beef Cattle

**TABLE 253  Daily Maintenance Allowance of ME for Housed Beef Cattle**

| Body weight (kg) | MJ/ head | Body weight (kg) | MJ/ head |
|---|---|---|---|
| 100 | 17 | 400 | 45 |
| 150 | 22 | 450 | 49 |
| 200 | 27 | 500 | 54 |
| 250 | 31 | 550 | 59 |
| 300 | 36 | 600 | 63 |
| 350 | 40 | | |

(including safety margin) Based on $M_m = 8.3 + 0.091\ W$

(*Note:* The values are the same as those for diary cattle but do not include an activity allowance)

*Source:* MAFF/ADAS (1987) *Energy allowance and feeding systems for Ruminants.* RB. 433. Reproduced with kind permission of MAFF ADAS (UK).

**TABLE 254  Daily ME Allowances for Growing and Fattening Cattle (MJ/day)**

| Liveweight (kg) | Ration M/D (MJ/kg DM) | Rate of gain (kg/day) | | | | | | |
|---|---|---|---|---|---|---|---|---|
| | | 0 | 0.25 | 0.50 | 0.75 | 1.00 | 1.25 | 1.50 |
| **100** | 8 | 17 | 24 | | | | | |
| | 10 | 17 | 22 | 29 | | | | |
| | 12 | 17 | 21 | 27 | 33 | | | |
| | 14 | 17 | 21 | 25 | 31 | 37 | | |
| **150** | 8 | 22 | 29 | | | | | |
| | 10 | 22 | 28 | 35 | | | | |
| | 12 | 22 | 27 | 33 | 40 | 48 | | |
| | 14 | 22 | 26 | 31 | 37 | 44 | 53 | |
| **200** | 8 | 27 | 35 | | | | | |
| | 10 | 27 | 34 | 41 | 51 | | | |
| | 12 | 27 | 33 | 39 | 47 | 56 | | |
| | 14 | 27 | 32 | 37 | 45 | 52 | 62 | 74 |

*(continued over)*

**TABLE 254** *(continued)*  **Daily ME Allowances for Growing and Fattening Cattle (MJ/day)**

| Liveweight (kg) | Ration M/D (MJ/kg DM) | Rate of gain (kg/day) | | | | | | |
|---|---|---|---|---|---|---|---|---|
| | | 0 | 0.25 | 0.50 | 0.75 | 1.00 | 1.25 | 1.50 |
| **250** | 8 | 31 | 40 | 51 | | | | |
| | 10 | 31 | 38 | 47 | 57 | | | |
| | 12 | 31 | 37 | 44 | 52 | 63 | 75 | |
| | 14 | 31 | 36 | 42 | 49 | 58 | 69 | 83 |
| **300** | 8 | 36 | 46 | 57 | | | | |
| | 10 | 36 | 44 | 53 | 64 | | | |
| | 12 | 36 | 43 | 50 | 59 | 70 | 84 | |
| | 14 | 36 | 42 | 48 | 56 | 65 | 77 | 92 |
| **350** | 8 | 40 | 51 | 63 | | | | |
| | 10 | 40 | 48 | 58 | 70 | 84 | | |
| | 12 | 40 | 47 | 55 | 65 | 77 | 92 | |
| | 14 | 40 | 46 | 53 | 62 | 72 | 84 | 101 |
| **400** | 8 | 45 | 56 | 70 | | | | |
| | 10 | 45 | 54 | 65 | 77 | 93 | | |
| | 12 | 45 | 53 | 61 | 72 | 85 | 101 | |
| | 14 | 45 | 51 | 59 | 68 | 79 | 93 | 110 |
| **450** | 8 | 49 | 61 | 75 | | | | |
| | 10 | 49 | 59 | 70 | 83 | | | |
| | 12 | 49 | 57 | 67 | 78 | 91 | 108 | |
| | 14 | 49 | 56 | 64 | 74 | 85 | 100 | 118 |
| **500** | 8 | 54 | 67 | 82 | | | | |
| | 10 | 54 | 64 | 76 | 91 | | | |
| | 12 | 54 | 63 | 73 | 85 | 99 | 117 | |
| | 14 | 54 | 61 | 70 | 80 | 93 | 108 | 128 |
| **550** | 8 | 59 | 73 | 89 | | | | |
| | 10 | 59 | 70 | 83 | 98 | | | |
| | 12 | 59 | 68 | 79 | 91 | 107 | 126 | |
| | 14 | 59 | 67 | 76 | 87 | 100 | 116 | 137 |
| **600** | 8 | 63 | 77 | 94 | | | | |
| | 10 | 63 | 75 | 88 | 104 | | | |
| | 12 | 63 | 73 | 84 | 97 | 114 | 134 | |
| | 14 | 63 | 71 | 81 | 92 | 106 | 124 | 146 |

*Source:* MAFF/ADAS (1987) *Energy allowance and feeding systems for Ruminants.* RB 433.
Reproduced with kind permission of MAFF ADAS (UK).

**TABLE 255   Minimum Metabolizable Energy Concentration of Beef Cattle Diets. M/D(MJ/kg DM) at Stated Levels of Dry Matter Intake DMI(kg/day)**

| Gain (kg/day) | Liveweight W(kg) | | | | | | | | | | |
|---|---|---|---|---|---|---|---|---|---|---|---|
| | 100 | 150 | 200 | 250 | 300 | 350 | 400 | 450 | 500 | 550 | 600 |
| 0 | (5.8) | (5.2) | (4.8) | (4.7) | (4.6) | (4.7) | (4.8) | (4.9) | (5.1) | (5.3) | (5.4) |
| 0.1 | (6.8) | (6.0) | (5.5) | (5.4) | (5.3) | (5.4) | (5.5) | (5.5) | (5.7) | (5.9) | (6.0) |
| 0.2 | 7.6 | (6.8) | (6.2) | (6.1) | (5.9) | (6.0) | (6.0) | (6.1) | (6.2) | (6.4) | (6.6) |
| 0.3 | 8.4 | 7.5 | (6.9) | (6.7) | (6.5) | (6.5) | (6.6) | (6.6) | (6.8) | 7.0 | 7.1 |
| 0.4 | 9.1 | 8.1 | 7.5 | 7.2 | 7.0 | 7.0 | 7.1 | 7.1 | 7.3 | 7.5 | 7.6 |
| 0.5 | 9.8 | 8.7 | 8.0 | 7.8 | 7.5 | 7.5 | 7.6 | 7.6 | 7.7 | 7.9 | 8.1 |
| 0.6 | 10.5 | 9.3 | 8.5 | 8.3 | 8.1 | 8.0 | 8.0 | 8.1 | 8.2 | 8.4 | 8.5 |
| 0.7 | 11.1 | 9.8 | 9.1 | 8.8 | 8.5 | 8.5 | 8.5 | 8.5 | 8.7 | 8.8 | 9.0 |
| 0.8 | 11.8 | 10.4 | 9.6 | 9.3 | 9.0 | 9.0 | 9.0 | 9.0 | 9.1 | 9.3 | 9.5 |
| 0.9 | 12.5 | 11.0 | 10.1 | 9.8 | 9.5 | 9.4 | 9.5 | 9.5 | 9.6 | 9.8 | 9.9 |
| 1.0 | 13.1 | 11.6 | 10.7 | 10.3 | 10.0 | 9.9 | 9.9 | 9.9 | 10.1 | 10.2 | 10.4 |
| 1.1 | 13.8 | 12.1 | 11.2 | 10.8 | 10.5 | 10.4 | 10.4 | 10.4 | 10.6 | 10.7 | 10.9 |
| 1.2 | 14.5 | 12.8 | 11.8 | 11.3 | 11.0 | 10.9 | 10.9 | 10.9 | 11.1 | 11.2 | 11.4 |
| 1.3 | | 13.4 | 12.4 | 11.9 | 11.6 | 11.5 | 11.4 | 11.4 | 11.6 | 11.7 | 11.9 |
| 1.4 | | 14.0 | 13.0 | 12.5 | 12.1 | 12.0 | 12.0 | 12.0 | 12.1 | 12.3 | 12.5 |
| 1.5 | | 13.6 | 13.1 | 12.7 | 12.6 | 12.5 | 12.5 | 12.7 | 12.9 | 13.0 | |
| DMI (kg/day) | 2.94 | 4.26 | 5.48 | 6.60 | 7.62 | 8.54 | 9.36 | 10.08 | 10.70 | 11.22 | 11.65 |

( ) indicates values are theoretical only, appetite limits on poor quality forage make M/D unfeasible.

$$\text{Based on M/D} = \frac{M_m + \sqrt{M_m^2 + (96.4 \text{ DMI} \times E_g)}}{2 \text{ DMI}}$$ where $E_g$ = net energy required for body gain.

*Source:* MAFF/ADAS (1987) *Energy Allowance and Feeding Systems for Ruminants.* RB 433. Reproduced with kind permission of MAFF ADAS (UK).

**TABLE 256 a & b Ration Formulation for Growing and Fattening Cattle Using Net Energy System**

**a. Animal Production Level (APL)**

| Liveweight W (kg) | Liveweight gain LWG (kg/day) | | | | | |
|---|---|---|---|---|---|---|
| | 0.25 | 0.50 | 0.75 | 1.00 | 1.25 | 1.50 |
| 100 | 1.19 | 1.40 | 1.66 | 1.98 | | |
| 150 | 1.16 | 1.36 | 1.59 | 1.87 | | |
| 200 | 1.15 | 1.33 | 1.54 | 1.79 | 2.11 | |
| 250 | 1.14 | 1.30 | 1.50 | 1.74 | 2.03 | |
| 300 | 1.13 | 1.29 | 1.47 | 1.70 | 1.97 | 2.33 |
| 350 | 1.13 | 1.27 | 1.45 | 1.67 | 1.93 | 2.27 |
| 400 | 1.12 | 1.26 | 1.43 | 1.64 | 1.90 | 2.22 |
| 450 | 1.12 | 1.26 | 1.42 | 1.62 | 1.87 | 2.18 |
| 500 | 1.11 | 1.25 | 1.41 | 1.60 | 1.84 | 2.15 |
| 550 | 1.11 | 1.24 | 1.40 | 1.59 | 1.83 | 2.13 |
| 600 | 1.11 | 1.24 | 1.39 | 1.58 | 1.81 | 2.13 |

$$\text{Based on APL} = 1 + \left[ \frac{\text{LWG}\ (6.28 + 0.0188W)}{(1 - 0.3\ \text{LWG})\ (5.67 + 0.061\ W)} \right]$$

**b. Net Energies of foods for maintenance and production, $NE_{mp}$ (MJ/kg DM)**

| APL | ME of food, MEF(MJ/kg DM) | | | | | | | | |
|---|---|---|---|---|---|---|---|---|---|
| | 6 | 7 | 8 | 9 | 10 | 11 | 12 | 13 | 14 |
| 1.00 | 4.3 | 5.0 | 5.8 | 6.5 | 7.2 | 7.9 | 8.6 | 9.4 | 10.1 |
| 1.10 | 3.7 | 4.5 | 5.2 | 6.0 | 6.8 | 7.6 | 8.3 | 9.1 | 9.9 |
| 1.15 | 3.5 | 4.3 | 5.1 | 5.8 | 6.6 | 7.4 | 8.2 | 9.0 | 9.8 |
| 1.20 | 3.3 | 4.1 | 4.9 | 5.7 | 6.5 | 7.3 | 8.1 | 8.9 | 9.8 |
| 1.25 | 3.2 | 4.0 | 4.7 | 5.5 | 6.4 | 7.2 | 8.0 | 8.9 | 9.7 |
| 1.30 | 3.1 | 3.8 | 4.6 | 5.4 | 6.3 | 7.1 | 7.9 | 8.8 | 9.7 |
| 1.35 | 3.0 | 3.7 | 4.5 | 5.3 | 6.2 | 7.0 | 7.8 | 8.7 | 9.6 |
| 1.40 | 2.9 | 3.6 | 4.4 | 5.2 | 6.1 | 6.9 | 7.8 | 8.7 | 9.6 |
| 1.45 | 2.8 | 3.5 | 4.3 | 5.1 | 6.0 | 6.8 | 7.7 | 8.6 | 9.5 |
| 1.50 | 2.7 | 3.5 | 4.2 | 5.1 | 5.9 | 6.8 | 7.7 | 8.6 | 9.5 |
| 1.55 | 2.7 | 3.4 | 4.2 | 5.0 | 5.8 | 6.7 | 7.6 | 8.5 | 9.5 |
| 1.65 | 2.6 | 3.3 | 4.1 | 4.9 | 5.7 | 6.6 | 7.5 | 8.4 | 9.4 |
| 1.75 | 2.5 | 3.2 | 3.9 | 4.8 | 5.6 | 6.5 | 7.4 | 8.4 | 9.3 |
| 2.00 | 2.3 | 3.0 | 3.8 | 4.6 | 5.4 | 6.3 | 7.3 | 8.2 | 9.2 |
| 2.25 | 2.2 | 2.9 | 3.6 | 4.4 | 5.3 | 6.2 | 7.1 | 8.1 | 9.1 |

$$\text{Based on NE}_{mp} = \frac{(\text{MEF})^2 \times \text{APL}}{1.39\ \text{MEF} + 23\ (\text{APL} - 1)}$$

**TABLE 256 c   Ration Formulation for Growing and Fattening Cattle Using Net Energy System**

**c. Net Energy allowances (MJ/day) for maintenance and liveweight gain in growing and fattening animals**

| Gain (kg) | Liveweight, W(kg) | | | | | | | | | | |
|---|---|---|---|---|---|---|---|---|---|---|---|
| | 100 | 150 | 200 | 250 | 300 | 350 | 400 | 450 | 500 | 550 | 600 |
| 0 | 12.4 | 15.6 | 18.8 | 22.0 | 25.2 | 28.4 | 31.6 | 34.8 | 38.0 | 41.2 | 44.4 |
| 0.1 | 13.3 | 16.6 | 19.9 | 23.2 | 26.5 | 29.8 | 33.1 | 36.4 | 39.7 | 43.0 | 46.3 |
| 0.2 | 14.2 | 17.6 | 21.0 | 24.5 | 27.9 | 31.3 | 34.7 | 38.1 | 41.5 | 44.9 | 48.3 |
| 0.3 | 15.2 | 18.8 | 22.3 | 25.8 | 29.3 | 32.9 | 36.4 | 39.9 | 43.4 | 47.0 | 50.5 |
| 0.4 | 16.3 | 19.9 | 23.6 | 27.2 | 30.9 | 34.5 | 38.2 | 41.8 | 45.5 | 49.1 | 52.8 |
| 0.5 | 17.4 | 21.2 | 25.0 | 28.8 | 32.6 | 36.3 | 40.1 | 43.9 | 47.7 | 51.5 | 55.2 |
| 0.6 | 18.7 | 22.6 | 26.5 | 30.4 | 34.4 | 38.3 | 42.2 | 46.1 | 50.2 | 54.0 | 57.9 |
| 0.7 | 20.0 | 24.2 | 28.1 | 32.2 | 36.3 | 40.4 | 44.4 | 48.5 | 52.6 | 56.7 | 60.7 |
| 0.8 | 21.4 | 25.7 | 29.9 | 34.1 | 38.4 | 42.6 | 46.9 | 51.1 | 55.3 | 59.6 | 63.8 |
| 0.9 | 23.0 | 27.4 | 31.8 | 36.2 | 40.6 | 45.0 | 49.5 | 53.9 | 58.3 | 62.7 | 67.1 |
| 1.0 | 24.6 | 29.3 | 33.9 | 38.5 | 43.1 | 47.7 | 54.3 | 56.9 | 61.5 | 66.1 | 70.7 |
| 1.1 | | 31.3 | 36.1 | 40.9 | 45.7 | 50.6 | 55.4 | 60.2 | 65.0 | 69.9 | 74.7 |
| 1.2 | | | 38.6 | 43.6 | 48.7 | 53.7 | 58.8 | 63.8 | 68.9 | 73.9 | 79.0 |
| 1.3 | | | | 46.6 | 51.9 | 57.2 | 62.5 | 67.8 | 73.1 | 78.4 | 83.7 |
| 1.4 | | | | | 55.4 | 61.0 | 66.6 | 72.2 | 77.7 | 83.3 | 88.9 |
| 1.5 | | | | | | 65.2 | 71.1 | 77.0 | 82.9 | 88.8 | 94.7 |

(including safety margin)

Based on NE allowance = $E_m + E_g$, where

$$E_m = 1.05 \, [5.67 + 0.061 \, W]$$

and $E_g = 1.05 \left[ \dfrac{LWG(6.28 + 0.0188 \, W)}{(1 - 0.3 \, LWG)} \right]$

$\left( E_m = \text{Net energy required for maintenance} \right.$

$\left. E_g = \text{Net energy required for body gain} \right)$

*Source:* MAFF/ADAS (1987) *Energy Allowances and Feeding Systems for Ruminants.* RB 433.
Reproduced with kind permission of MAFF ADAS (UK).

**TABLE 257  RDP and UDP Requirements (g/day) for Beef Cattle**

| ME/GE (q) | Live weight (kg) | Form of protein | 0 | 0.25 | 0.50 | 0.75 | 1.0 | 1.25 | 1.50 |
|---|---|---|---|---|---|---|---|---|---|
| | | | | | **Weight gain(kg/day)** | | | | |
| 0.4 | 100 | RDP | 160 | 185 | 215 | 250 | – | – | – |
| | | UDP | – | 15 | 80 | 135 | – | – | – |
| | 200 | RDP | 255 | 290 | 330 | 385 | – | – | – |
| | | UDP | – | – | – | 35 | – | – | – |
| | 300 | RDP | 335 | 380 | 435 | 500 | – | – | – |
| | 400 | RDP | 410 | 465 | 525 | 605 | – | – | – |
| | 500 | RDP | 480 | 540 | 6l5 | 705 | – | – | – |
| | 600 | RDP | 540 | 610 | 695 | 795 | – | – | – |
| 0.5 | 100 | RDP | 150 | 170 | 195 | 225 | 260 | – | – |
| | | UDP | – | 25 | 95 | 155 | 205 | – | – |
| | 200 | RDP | 240 | 270 | 305 | 345 | 395 | – | – |
| | | UDP | – | – | 20 | 65 | 100 | – | – |
| | 300 | RDP | 320 | 355 | 400 | 450 | 515 | – | – |
| | 400 | RDP | 390 | 435 | 485 | 550 | 620 | – | – |
| | 500 | RDP | 455 | 505 | 565 | 640 | 725 | – | – |
| | 600 | RDP | 515 | 575 | 640 | 720 | 815 | – | – |
| 0.6 | 100 | RDP | 145 | 160 | 185 | 205 | 235 | 270 | 310 |
| | | UDP | – | 35 | 105 | 170 | 230 | 275 | 310 |
| | 200 | RDP | 230 | 255 | 285 | 320 | 360 | 405 | 460 |
| | | UDP | – | – | 35 | 85 | 130 | 160 | 180 |
| | 300 | RDP | 305 | 335 | 375 | 420 | 465 | 525 | 595 |
| | | UDP | – | – | – | 10 | 35 | 55 | 60 |
| | 400 | RDP | 370 | 410 | 455 | 505 | 565 | 635 | 720 |
| | 500 | RDP | 430 | 475 | 530 | 590 | 660 | 740 | 835 |
| | 600 | RDP | 490 | 540 | 600 | 665 | 745 | 835 | 945 |
| 0.7 | 100 | RDP | 135 | 155 | 170 | 195 | 215 | 245 | 275 |
| | | UDP | – | 40 | 115 | 185 | 245 | 295 | 340 |
| | 200 | RDP | 220 | 245 | 270 | 300 | 330 | 370 | 415 |
| | | UDP | – | – | 50 | 105 | 150 | 190 | 215 |
| | 300 | RDP | 290 | 320 | 350 | 390 | 430 | 480 | 535 |
| | | UDP | – | – | – | 30 | 65 | 90 | 105 |
| | 400 | RDP | 355 | 390 | 430 | 475 | 525 | 580 | 650 |
| | | UDP | – | – | – | – | – | 5 | 10 |
| | 500 | RDP | 410 | 450 | 500 | 550 | 610 | 675 | 755 |
| | 600 | RDP | 465 | 515 | 565 | 625 | 690 | 765 | 850 |

*Source: Collins Farmers Diary 1992*
© Collins 1960, 1991, a division of Harper Collins.

**TABLE 258 a   Feeding Standards for Growing Cattle**

**a. Daily allowances for early maturing heifers**

| W (kg) | $q_m$ | Component | | Liveweight gain (kg/day) | | | | | DMI (kg/day) |
|---|---|---|---|---|---|---|---|---|---|
| | | | | 0 | 0.5 | 0.75 | 1.00 | 1.25 | |
| **200** | 0.55 | ME | (MJ) | 28 | 44 | 55 | | | 5.0 |
| | | RDP | (g) | 231 | 365 | 460 | | | |
| | | UDP | (g) | 2 | 0 | 0 | | | |
| | 0.65 | ME | (MJ) | 26 | 40 | 49 | 60 | 75 | 6.5 |
| | | RDP | (g) | 220 | 335 | 410 | 503 | 624 | |
| | | UDP | (g) | 16 | 15 | 0 | 0 | 0 | |
| | | $E_{mp}$ | (MJ) | 19 | 27 | 31 | 36 | 41 | |
| | | APL | | | 1.00 | 1.40 | 1.62 | 1.86 | 2.13 |
| | | Ca | (g) | 6.4 | 19.6 | 26.2 | 32.8 | 39.4 | |
| | | P | (g) | 5.0 | 13.2 | 17.4 | 21.5 | 25.6 | |
| | | Mg | (g) | 3.5 | 4.9 | 5.5 | 6.2 | 6.9 | |
| | | Na | (g) | 1.6 | 2.5 | 2.9 | 3.3 | 3.7 | |
| | | Vit. A | (i.u.) | ← | | 14 000 | | → | |
| | | Vit. D | (i.u.) | ← | | 1200 | | → | |
| | | Vit. E | (i.u.) | ← | | 115 | | → | |
| **400** | 0.55 | ME | (MJ) | 45 | 69 | 86 | | | 8.5 |
| | | RDP | (g) | 373 | 576 | 719 | | | |
| | | UDP | (g) | 18 | 0 | 0 | | | |
| | 0.65 | ME | (MJ) | 43 | 64 | 77 | 94 | 116 | 11.0 |
| | | RDP | (g) | 355 | 530 | 644 | 784 | 964 | |
| | | UDP | (g) | 31 | 0 | 0 | 0 | 0 | |
| | | $E_{mp}$ | (MJ) | 31 | 43 | 49 | 57 | 64 | |
| | | APL | | | 1.00 | 1.38 | 1.59 | 1.82 | 2.07 |
| | | Ca | (g) | 14.2 | 26.6 | 32.7 | 38.9 | 45.1 | |
| | | P | (g) | 9.9 | 18.2 | 22.3 | 26.5 | 30.6 | |
| | | Mg | (g) | 7.0 | 8.4 | 9.1 | 9.7 | 10.4 | |
| | | Na | (g) | 3.2 | 4.1 | 4.5 | 4.9 | 5.3 | |
| | | Vit. A | (i.u.) | ← | | 28 000 | | → | |
| | | Vit. D | (i.u.) | ← | | 2400 | | → | |
| | | Vit. E | (i.u.) | ← | | 195 | | → | |

*(continued over)*

**TABLE 258 b   Feeding Standards for Growing Cattle**

**b. Daily allowances for steers of medium maturity**

| W (kg) | $q_m$ | Component | | Liveweight gain (kg/day) | | | | | DMI (kg/day) |
|---|---|---|---|---|---|---|---|---|---|
| | | | | 0 | 0.5 | 0.75 | 1.00 | 1.25 | |
| **200** | 0.55 | ME | (MJ) | 28 | 40 | 48 | | | 5.0 |
| | | RDP | (g) | 231 | 331 | 397 | | | |
| | | UDP | (g) | 2 | 37 | 37 | | | |
| | 0.65 | ME | (MJ) | 26 | 37 | 43 | 51 | 61 | 6.5 |
| | | RDP | (g) | 220 | 307 | 361 | 426 | 505 | |
| | | UDP | (g) | 16 | 57 | 70 | 66 | 45 | |
| | | $E_{mp}$ | (MJ) | 19 | 25 | 28 | 32 | 36 | |
| | | APL | | | 1.00 | 1.31 | 1.48 | 1.66 | 1.87 |
| | | Ca | (g) | 6.4 | 19.6 | 26.2 | 32.8 | 39.4 | |
| | | P | (g) | 5.0 | 13.2 | 17.4 | 21.5 | 25.6 | |
| | | Mg | (g) | 3.5 | 4.9 | 5.5 | 6.2 | 6.9 | |
| | | Na | (g) | 1.6 | 2.5 | 2.9 | 3.3 | 3.7 | |
| | | Vit. A | (i.u.) | ← | | | 14 000 | | → |
| | | Vit. D | (i.u.) | ← | | | 1200 | | → |
| | | Vit. E | (i.u.) | ← | | | 115 | | → |
| **400** | 0.55 | ME | (MJ) | 45 | 63 | 75 | 90 | | 8.5 |
| | | RDP | (g) | 373 | 525 | 625 | 750 | | |
| | | UDP | (g) | 18 | 1 | 0 | 0 | | |
| | 0.65 | ME | (MJ) | 43 | 58 | 68 | 80 | 94 | 11.0 |
| | | RDP | (g) | 355 | 487 | 570 | 668 | 787 | |
| | | UDP | (g) | 31 | 35 | 15 | 0 | 0 | |
| | | $E_{mp}$ | (MJ) | 31 | 40 | 45 | 51 | 57 | |
| | | APL | | | 1.00 | 1.29 | 1.45 | 1.63 | 1.83 |
| | | Ca | (g) | 14.2 | 26.6 | 32.7 | 38.9 | 45.1 | |
| | | P | (g) | 9.9 | 18.2 | 22.3 | 26.5 | 30.6 | |
| | | Mg | (g) | 7.0 | 8.4 | 9.1 | 9.7 | 10.4 | |
| | | Na | (g) | 3.2 | 4.1 | 4.5 | 4.9 | 5.3 | |
| | | Vit. A | (i.u.) | ← | | | 28 000 | | → |
| | | Vit. D | (i.u.) | ← | | | 2400 | | → |
| | | Vit. E | (i.u.) | ← | | | 195 | | → |

**TABLE 258 c   Feeding Standards for Growing Cattle**

**c. Daily allowances for bulls of late maturity**

| W (kg) | $q_m$ | Component | | Liveweight gain (kg/day) | | | | | DMI (kg/day) |
|---|---|---|---|---|---|---|---|---|---|
| | | | | 0 | 0.5 | 0.75 | 1.00 | 1.25 | |
| **200** | 0.55 | ME | (MJ) | 32 | 40 | 45 | 51 | | 5.0 |
| | | RDP | (g) | 266 | 333 | 374 | 423 | | |
| | | UDP | (g) | 0 | 62 | 92 | 111 | | |
| | 0.65 | ME | (MJ) | 30 | 37 | 42 | 47 | 52 | 6.5 |
| | | RDP | (g) | 253 | 312 | 347 | 388 | 435 | |
| | | UDP | (g) | 0 | 82 | 112 | 138 | 160 | |
| | | $E_{mp}$ | (MJ) | 22 | 26 | 29 | 31 | 34 | |
| | | APL | | | 1.00 | 1.19 | 1.29 | 1.40 | 1.53 |
| | | Ca | (g) | 6.4 | 19.6 | 26.2 | 32.8 | 39.4 | |
| | | P | (g) | 5.0 | 13.2 | 17.4 | 21.5 | 25.6 | |
| | | Mg | (g) | 3.5 | 4.9 | 5.5 | 6.2 | 6.9 | |
| | | Na | (g) | 1.6 | 2.5 | 2.9 | 3.3 | 3.7 | |
| | | Vit. A | (i.u.) | | | ← 14 000 → | | | |
| | | Vit. D | (i.u.) | | | ← 1200 → | | | |
| | | Vit. E | (i.u.) | | | ← 115 → | | | |
| **400** | 0.55 | ME | (MJ) | 51 | 64 | 71 | 80 | 91 | 8.5 |
| | | RDP | (g) | 429 | 531 | 594 | 668 | 755 | |
| | | UDP | (g) | 0 | 16 | 27 | 20 | 0 | |
| | 0.65 | ME | (MJ) | 49 | 60 | 66 | 74 | 82 | 11.0 |
| | | RDP | (g) | 408 | 498 | 552 | 613 | 685 | |
| | | UDP | (g) | 0 | 43 | 60 | 60 | 57 | |
| | | $E_{mp}$ | (MJ) | 36 | 42 | 46 | 49 | 54 | |
| | | APL | | | 1.00 | 1.18 | 1.28 | 1.39 | 1.50 |
| | | Ca | (g) | 14.2 | 26.6 | 32.7 | 38.9 | 45.1 | |
| | | P | (g) | 9.9 | 18.2 | 22.3 | 26.5 | 30.6 | |
| | | Mg | (g) | 7.0 | 8.4 | 9.1 | 9.7 | 10.4 | |
| | | Na | (g) | 3.2 | 4.1 | 4.5 | 4.9 | 5.3 | |
| | | Vit. A | (i.u.) | | | ← 28 000 → | | | |
| | | Vit. D | (i.u.) | | | ← 2400 → | | | |
| | | Vit. E | (i.u.) | | | ← 195 → | | | |

*Source:* McDonald, P., Edwards, R.A., & Greenhalgh, J.F.D. (1988) *Animal Nutrition.*
Reproduced with kind permission of Longman Group UK Ltd.

**TABLE 259  Daily Nutrient Requirements of Breeding Cows**

## DRY MATTER BASIS

| Body Mass kg) | Daily gain (kg) | DM intake (kg) | CP (kg) | DP (kg) | Energy ME MJ | TDN (kg) | Ca (g) | P (g) | Vit. A IU x 1000 |
|---|---|---|---|---|---|---|---|---|---|
| **Dry pregnant cows** | | | | | | | | | |
| **350** | – | 5.8 | 0.34 | 0.16 | 43.1 | 2.8 | 9 | 9 | 14.0 |
| **400** | – | 6.4 | 0.38 | 0.18 | 48.1 | 3.2 | 10 | 10 | 15.5 |
| **450** | – | 6.8 | 0.40 | 0.19 | 51.9 | 3.4 | 12 | 12 | 16.8 |
| **500** | – | 7.6 | 0.44 | 0.21 | 56.9 | 3.8 | 12 | 12 | 18.2 |
| **550** | – | 8.0 | 0.47 | 0.22 | 60.3 | 4.0 | 12 | 12 | 19.5 |
| **Lactating cows, first 3 – 4 months** | | | | | | | | | |
| **350** | – | 8.6 | 0.79 | 0.46 | 74.1 | 4.9 | 25 | 20 | 33.2 |
| **400** | – | 9.3 | 0.86 | 0.50 | 80.4 | 5.3 | 26 | 21 | 36.0 |
| **450** | – | 9.9 | 0.91 | 0.53 | 85.4 | 5.6 | 28 | 22 | 38.5 |
| **500** | – | 10.5 | 0.97 | 0.57 | 90.4 | 6.0 | 28 | 23 | 41.0 |

*Source:* Zimbabwe Cattle Producers Association (1988) *Beef Production Manual.*
Reproduced with kind permission of Zimbabwe Cattle Producers Association.

**TABLE 260  Daily Nutrient Requirements of Growing Steers and Heifers**

| | | | | | DRY MATTER BASIS | | | | |
|---|---|---|---|---|---|---|---|---|---|
| | | | | | Energy | | | | Vit. A |
| Body Mass (kg) | Daily gain (kg) | DM intake (kg) | CP (kg) | DP (kg) | ME MJ | TDN (kg) | Ca (g) | P (g) | IU x 1000 |
| **Growing steers** | | | | | | | | | |
| | 0.00 | 2.7 | 0.21 | 0.11 | 23.4 | 1.5 | 5 | 5 | 6.0 |
| 150 | 0.25 | 3.1 | 0.34 | 0.22 | 29.7 | 2.0 | 8 | 7 | 6.8 |
| | 0.50 | 3.2 | 0.39 | 0.26 | 35.2 | 2.3 | 12 | 10 | 7.0 |
| | 0.75 | 3.2 | 0.43 | 0.29 | 37.7 | 2.5 | 17 | 13 | 7.0 |
| | 0.00 | 3.3 | 0.26 | 0.14 | 28.5 | 1.9 | 6 | 6 | 7.4 |
| 200 | 0.25 | 4.5 | 0.45 | 0.27 | 38.9 | 2.6 | 8 | 8 | 10.0 |
| | 0.50 | 4.9 | 0.54 | 0.35 | 46.9 | 3.1 | 13 | 10 | 10.8 |
| | 0.75 | 5.0 | 0.56 | 0.36 | 52.3 | 3.5 | 18 | 14 | 11.2 |
| | 0.00 | 4.5 | 0.35 | 0.19 | 38.9 | 2.6 | 8 | 8 | 10.0 |
| 300 | 0.25 | 6.1 | 0.54 | 0.32 | 52.7 | 3.5 | 11 | 11 | 13.6 |
| | 0.50 | 7.7 | 0.77 | 0.47 | 66.6 | 4.4 | 14 | 14 | 17.4 |
| | 0.75 | 8.0 | 0.89 | 0.57 | 76.2 | 5.0 | 17 | 15 | 17.8 |
| | 0.00 | 5.6 | 0.44 | 0.24 | 48.1 | 3.2 | 10 | 10 | 12.4 |
| 400 | 0.25 | 7.7 | 0.64 | 0.35 | 66.6 | 4.4 | 14 | 14 | 17.2 |
| | 0.50 | 9.7 | 0.86 | 0.50 | 83.7 | 5.5 | 17 | 17 | 21.6 |
| | 0.75 | 9.9 | 0.88 | 0.51 | 94.6 | 6.3 | 18 | 18 | 22.0 |
| **Growing heifers** | | | | | | | | | |
| | 0.00 | 2.7 | 0.21 | 0.11 | 23.4 | 1.5 | 5 | 5 | 6.0 |
| 150 | 0.25 | 3.2 | 0.36 | 0.23 | 30.6 | 2.0 | 8 | 7 | 7.0 |
| | 0.50 | 3.2 | 0.39 | 0.26 | 35.2 | 2.3 | 12 | 10 | 7.2 |
| | 0.75 | 3.3 | 0.44 | 0.30 | 38.9 | 2.6 | 17 | 13 | 7.4 |
| | 0.00 | 3.3 | 0.26 | 0.14 | 28.5 | 1.9 | 6 | 6 | 7.4 |
| 200 | 0.25 | 4.6 | 0.46 | 0.28 | 39.8 | 2.6 | 8 | 8 | 10.2 |
| | 0.50 | 5.0 | 0.56 | 0.36 | 47.7 | 3.2 | 13 | 10 | 11.2 |
| | 0.75 | 5.4 | 0.60 | 0.38 | 56.5 | 3.7 | 18 | 14 | 12.0 |
| | 0.00 | 4.5 | 0.35 | 0.19 | 38.9 | 2.6 | 8 | 8 | 10.0 |
| 300 | 0.25 | 6.2 | 0.55 | 0.32 | 53.6 | 3.5 | 11 | 11 | 13.8 |
| | 0.50 | 8.2 | 0.82 | 0.50 | 70.7 | 4.7 | 15 | 15 | 18.2 |
| | 0.75 | 8.6 | 0.95 | 0.61 | 82.0 | 5.4 | 17 | 15 | 19.0 |
| | 0.00 | 5.6 | 0.44 | 0.24 | 48.1 | 3.2 | 10 | 10 | 12.4 |
| 400 | 0.25 | 7.7 | 0.64 | 0.35 | 66.6 | 4.4 | 14 | 14 | 17.2 |
| | 0.50 | 10.2 | 0.91 | 0.53 | 87.9 | 5.8 | 18 | 18 | 22.6 |
| | 0.75 | 10.6 | 0.94 | 0.55 | 101.3 | 6.7 | 19 | 19 | 23.6 |

*Source:* Zimbabwe Cattle Producers Association (1988) *Beef Production Manual.*
Reproduced with kind permission of Zimbabwe Cattle Producers Association.

**TABLE 261 TDN Content of *ad lib* Dry Fattening Diets and Predicted Daily Dry Matter Intake and Carcass Gain in Beef Steers**

| | | | | | % TDN in dry matter | | | | | |
|---|---|---|---|---|---|---|---|---|---|---|
| Live mass kg | 66 | 68 | 70 | 72 | 74 | 76 | 78 | 80 | 82 | 84 |
| | | | | | DM consumption kg/day | | | | | |
| 250 | 5.7 | 6.0 | 6.3 | 6.5 | 6.8 | 7.1 | 7.3 | 7.6 | 7.8 | 8.1 |
| 300 | 6.6 | 6.9 | 7.2 | 7.5 | 7.8 | 8.1 | 8.4 | 8.7 | 9.0 | 9.3 |
| 350 | 7.4 | 7.7 | 8.1 | 8.4 | 8.7 | 9.1 | 9.4 | 9.8 | 10.1 | 10.4 |
| 400 | 8.2 | 8.6 | 9.0 | 9.3 | 9.7 | 10.1 | 10.4 | 10.9 | 11.2 | 11.5 |
| 450 | 8.9 | 9.3 | 9.8 | 10.2 | 10.5 | 11.0 | 11.4 | 11.9 | 12.2 | 12.6 |
| | | | | | Carcass gain per day (kg) | | | | | |
| 250 | 0.20 | 0.26 | 0.32 | 0.38 | 0.44 | 0.51 | 0.58 | 0.66 | 0.74 | 0.83 |
| 300 | 0.24 | 0.30 | 0.37 | 0.43 | 0.50 | 0.58 | 0.67 | 0.76 | 0.85 | 0.97 |
| 350 | 0.27 | 0.33 | 0.41 | 0.48 | 0.56 | 0.65 | 0.75 | 0.86 | 0.96 | 1.06 |
| 400 | 0.30 | 0.37 | 0.46 | 0.54 | 0.63 | 0.73 | 0.83 | 0.96 | 1.06 | 1.18 |
| 450 | 0.32 | 0.40 | 0.50 | 0.59 | 0.68 | 0.79 | 0.91 | 1.04 | 1.16 | 1.29 |

*Source:* Zimbabwe Cattle Producers Association (1988) *Beef Production Manual.*
Reproduced with kind permission of Zimbabwe Cattle Producers Association.

**TABLE 262 Daily Nutrient Requirements of Bulls. Growth and Maintenance with Moderate Activity**

| Body Mass (kg) | Daily gain (kg) | DM intake (kg) | CP (kg) | DP (kg) | Energy | | CA (g) | P (g) | Vit. A. IU x 1000 |
|---|---|---|---|---|---|---|---|---|---|
| | | | | | ME MJ | TDN (kg) | | | |
| 300 | 1.00 | 8.7 | 1.21 | 0.84 | 85.4 | 5.6 | 23 | 18 | 34.0 |
| 400 | 0.90 | 10.0 | 1.33 | 0.90 | 98.4 | 6.5 | 19 | 18 | 38.8 |
| 500 | 0.70 | 12.0 | 1.60 | 1.08 | 108.0 | 7.1 | 21 | 21 | 46.6 |
| 600 | 0.50 | 11.6 | 1.42 | 0.94 | 104.2. | 6.9 | 21 | 21 | 45.2 |
| 700 | 0.30 | 12.7 | 1.41 | 0.90 | 109.7 | 7.2 | 23 | 23 | 49.4 |
| 800 | 0.00 | 9.9 | 0.99 | 0.60 | 85.4 | 5.6 | 18 | 18 | 38.5 |
| 900 | 0.00 | 10.7 | 1.07 | 0.65 | 92.1 | 6.1 | 19 | 19 | 41.6 |

*Source:* Zimbabwe Cattle Producers Association (1988) *Beef Production Manual.*
Reproduced with kind permission of Zimbabwe Cattle Producers Association.

**TABLE 263   Mineral and Vitamin Requirements of Beef Cattle (per unit dry matter in diet)**

| Nutrient | | Growing and finishing steers and heifers | Dry pregnant cows | Breeding bulls and lactating cows |
|---|---|---|---|---|
| Vitamin A | 1000 IU/kg | 2.2 | 2.8 | 3.9 |
| Vitamin D | IU/kg | 275 | 275 | 275 |
| Vitamin E | IU/kg | 15 to 60 | – | 15 to 60 |
| Minerals | | | | |
| Sodium | % | 0.1 | 0.1 | 0.1 |
| Calcium | % | 0.18 to 0.6 | 0.18 | 0.18 to 0.29 |
| Phosphorus | % | 0.18 to 0.43 | 0.18 | 0.18 to 0.23 |
| Magnesium | mg/kg | 400 to 1000 | – | – |
| Potassium | % | 0.6 to 0.8 | – | – |
| Sulphur | % | 0.1 | | |
| Iodine | microgram/kg | – | 50 to 100 | 50 to 100 |
| Iron | mg/kg | 10 | – | – |
| Copper | mg/kg | 4 | – | – |
| Cobalt | mg/kg | 0.05 to 0.10 | 0.05 to 0.10 | 0.05 to 0.10 |
| Manganese | mg/kg | 1.0 to 10.0 | – | – |
| Zinc | mg/kg | 10 to 30 | – | – |
| Selenium | mg/kg | 0.05 to 0.10 | 0.05 to 0.10 | 0.05 to 0.10 |

*Source:* Zimbabwe Cattle Producers Association (1988) *Beef Production Manual.*
Reproduced with kind permission of Zimbabwe Cattle Producers Association.

**TABLE 264   Estimates of Energy Requirements of Cattle for Survival**

| Normal body mass kg | Energy required MJ ME/day | | | | kg feed dry matter minimum/day |
|---|---|---|---|---|---|
| | Minimum (kg TDN)* | | Recommended (kg TDN)* | | |
| 150 | 13.9 | (0.93) | 15.0 | (1.00) | 1.2 |
| 175 | 15.6 | (1.04) | 16.8 | (1.12) | 1.4 |
| 200 | 17.3 | (1.15) | 18.6 | (1.24) | 1.6 |
| 225 | 18.9 | (1.26) | 20.3 | (1.36) | 1.8 |
| 250 | 20.4 | (1.36) | 22.0 | (1.47) | 2.0 |
| 275 | 22.0 | (1.47) | 23.6 | (1.58) | 2.2 |
| 300 | 23.4 | (1.57) | 25.2 | (1.69) | 2.4 |
| 325 | 24.9 | (1.66) | 26.8 | (1.79) | 2.6 |
| 350 | 26.3 | (1.76) | 28.3 | (1.89) | 2.8 |
| 375 | 27.7 | (1.85) | 29.8 | (1.99) | 3.0 |
| 400 | 29.1 | (1.95) | 31.3 | (2.09) | 3.2 |
| 425 | 30.4 | (2.03) | 32.8 | (2.19) | 3.6 |
| 450 | 31.8 | (2.13) | 34.2 | (2.28) | 3.8 |
| 475 | 33.1 | (2.21) | 35.6 | (2.38) | 4.0 |
| 500 | 34.4 | (2.30) | 37.0 | (2.47) | 4.2 |

* TDN was calculated from ME by assuming 14.95 MJ of ME per kg TDN

*Source:* Zimbabwe Cattle Producers Association (1988) *Beef Production Manual.*
Reproduced with kind permission of Zimbabwe Cattle Producers Association.

## Feed & Nutrient Requirements – Sheep

**TABLE 265    Probable Dry Matter Intakes, in kg/day/kg Liveweight for Pregnant and Lactating Ewes**

| Stage | Single Lamb | Twin Lamb |
|---|---|---|
| 6 weeks before lambing | 0.026 | 0.026 |
| 4 weeks before lambing | 0.025 | 0.0245 |
| 2 weeks before lambing | 0.023 | 0.022 |
| 0 weeks before lambing | 0.021 | 0.019 |
| 1st month of lactation | 0.035 | 0.0375 |
| 2nd month of lactation | 0.045 | 0.0475 |
| 3rd month of lactation | 0.045 | 0.0475 |

*Source:* MAFF/ADAS (1987) *Energy allowances and feeding systems for ruminants.* RB 433. Reproduced with kind permission of MAFF ADAS UK.

**TABLE 266    ME Allowances (MJ/day) of Pregnant Ewes Outdoors**

| Liveweight W(kg) | | Maintenance | Weeks before lambing | | | | |
|---|---|---|---|---|---|---|---|
| | | | 8 | 6 | 4 | 2 | Birth |
| 30 | S | 4.8 | 5.1 | 5.7 | 6.3 | 6.9 | 7.7 |
| | T | * (− 0.7) | 5.1 | 5.9 | 6.8 | 7.9 | 9.2 |
| 40 | S | 5.8 | 6.1 | 6.7 | 7.4 | 8.2 | 9.1 |
| | T | * (− 0.8) | 6.1 | 7.1 | 8.2 | 9.5 | 11.0 |
| 50 | S | 6.8 | 7.0 | 7.8 | 8.6 | 9.5 | 10.5 |
| | T | * (− 0.9) | 7.1 | 8.3 | 9.6 | 11.1 | 12.8 |
| 60 | S | 7.8 | 8.0 | 8.8 | 9.8 | 10.8 | 11.9 |
| | T | * (− 1.0) | 8.1 | 9.4 | 10.9 | 12.7 | 14.7 |
| 70 | S | 8.8 | 8.9 | 9.9 | 10.9 | 12.1 | 13.4 |
| | T | * (− 1.1) | 9.2 | 10.6 | 12.3 | 14.2 | 16.5 |
| 80 | S | 9.8 | 9.9 | 10.9 | 12.1 | 13.4 | 14.8 |
| | T | * (− 1.2) | 10.2 | 11.8 | 13.7 | 15.8 | 18.3 |

(including safety margin)

* For ewes indoors decrease by allowance shown thus (− 1.2)

Based on $M_m = 1.8 + 0.1$ (outdoors), $M_m = 1.4 + 0.09$ W (indoors)

$$S = \text{singles:} \quad M_m = (1.2 + 0.05\ W)\ e^{0.0072t}$$
$$T = \text{twins:} \quad M_m = (0.8 + 0.04\ W)\ e^{0.0105t}$$

(where t = number of days pregnant, e = 2.718, the base of natural logarithms)

*Source:* MAFF/ADAS (1987) *Energy allowances and feeding systems for ruminants.* RB 433. Reproduced with kind permission of MAFF ADAS UK.

**TABLE 267   ME Allowances (MJ/day) for Lactating Hill Ewes**

| Liveweight, W (kg) | Lambs | Stage of lactation | | |
|---|---|---|---|---|
| | | Month 1 | Month 2 | Month 3 |
| **30** | Single | 15.3 | 14.9 | 12.2 |
| | Twins | 22.3 | 19.2 | 14.6 |
| **40** | Single | 16.3 | 15.9 | 13.2 |
| | Twins | 23.3 | 20.2 | 15.6 |
| **50** | Single | 17.3 | 16.9 | 14.2 |
| | Twins | 24.3 | 21.2 | 16.6 |
| **60** | Single | 18.3 | 17.9 | 15.2 |
| | Twins | 25.3 | 22.2 | 17.6 |

(including safety margin)

*Source:* MAFF/ADAS (1987) *Energy allowances and feeding systems for ruminants.* RB 433. Reproduced with kind permission of MAFF ADAS UK.

**TABLE 268   ME Allowances (MJ/day) for Lactating Lowland Ewes**

| Liveweight, W(kg) | Lambs | Stage of lactation | | |
|---|---|---|---|---|
| | | Month 1 | Month 2 | Month 3 |
| **50** | Single | 19.3 | 18.9 | 15.7 |
| | Twins | 26.6 | 23.2 | 18.0 |
| **60** | Single | 20.3 | 19.9 | 16.7 |
| | Twins | 27.6 | 24.2 | 19.0 |
| **70** | Single | 21.3 | 20.9 | 17.7 |
| | Twins | 28.6 | 25.2 | 20.0 |
| **80** | Single | 22.3 | 21.9 | 18.7 |
| | Twins | 29.6 | 26.2 | 21.0 |

(Including safety margin)

*Source:* MAFF/ADAS (1987) *Energy allowances and feeding systems for ruminants.* RB 433. Reproduced with kind permission of MAFF ADAS UK.

**TABLE 269   Daily ME Allowances (MJ/day) for Indoor Fed Growing Sheep**

| Liveweight (kg) | Ration M/D (MJ/kg DM) | Rate of gain (g/day) | | | | | | | |
|---|---|---|---|---|---|---|---|---|---|
| | | 50 | 100 | 150 | 200 | 250 | 300 | 350 | 400 |
| **10** (+ 0.4) * | 8 | 4.4 | | | | | | | |
| | 10 | 4.0 | 5.8 | | | | | | |
| | 12 | 3.8 | 5.3 | 6.9 | | | | | |
| | 14 | 3.6 | 4.9 | 6.2 | | | | | |
| **15** (+ 0.5) * | 8 | 5.2 | 7.5 | | | | | | |
| | 10 | 4.8 | 6.6 | 8.6 | | | | | |
| | 12 | 4.5 | 6.1 | 7.7 | 9.4 | | | | |
| | 14 | 4.3 | 5.6 | 7.1 | 8.5 | | | | |
| **20** (+ 0.6) * | 8 | 5.9 | 8.4 | 11.0 | | | | | |
| | 10 | 5.5 | 7.5 | 9.5 | 11.7 | | | | |
| | 12 | 5.2 | 6.8 | 8.6 | 10.4 | 12.2 | 14.1 | | |
| | 14 | 5.0 | 6.4 | 7.9 | 9.4 | 11.0 | 12.6n | | |
| **25** (+ 0.7) * | 8 | 6.7 | 9.2 | 12.0 | | | | | |
| | 10 | 6.2 | 8.3 | 10.5 | 12.7 | 15.0 | | | |
| | 12 | 5.9 | 7.6 | 9.5 | 11.3 | 13.3 | 15.3 | 17.3 | |
| | 14 | 5.7 | 7.2 | 8.7 | 10.4 | 12.0 | 13.7 | 15.4 | |
| **30** (+ 0.8) * | 8 | 7.4 | 10.1 | 13.0 | | | | | |
| | 10 | 7.0 | 9.1 | 11.4 | 13.8 | 16.2 | | | |
| | 12 | 6.6 | 8.4 | 10.3 | 12.3 | 14.3 | 16.4 | 18.5 | |
| | 14 | 6.4 | 8.0 | 9.6 | 11.3 | 13.0 | 14.8 | 16.6 | |
| **35** (+ 0.9) * | 8 | 8.2 | 11.0 | 14.0 | 17.1 | | | | |
| | 10 | 7.7 | 9.9 | 12.3 | 14.8 | 17.4 | 20.0 | | |
| | 12 | 7.4 | 9.2 | 11.2 | 13.3 | 15.4 | 17.6 | 19.8 | 22.1 |
| | 14 | 7.1 | 8.7 | 10.5 | 12.2 | 14.0 | 15.9 | 17.8 | 19.7 |
| **40** (+ 1.0) * | 8 | 8.9 | 11.9 | 15.0 | 18.3 | | | | |
| | 10 | 8.4 | 10.8 | 13.3 | 15.9 | 18.6 | 21.3 | | |
| | 12 | 8.1 | 10.1 | 12.1 | 14.3 | 16.5 | 18.8 | 21.1 | 23.5 |
| | 14 | 7.9 | 9.5 | 11.3 | 13.2 | 15.1 | 17.0 | 19.0 | 21.0 |
| **45** (+ 1.1) * | 8 | 9.7 | 12.8 | 16.1 | 19.5 | | | | |
| | 10 | 9.2 | 11.7 | 14.3 | 17.0 | 19.8 | 22.6 | | |
| | 12 | 8.8 | 10.9 | 13.1 | 15.3 | 17.7 | 20.0 | 22.5 | 24.9 |
| | 14 | 8.6 | 10.3 | 12.2 | 14.1 | 16.1 | 18.2 | 20.2 | 22.4 |
| **50** (+ 1.2) * | 8 | 10.5 | 13.7 | 17.2 | 20.7 | | | | |
| | 10 | 9.9 | 12.5 | 15.3 | 18.1 | 21.0 | 24.0 | | |
| | 12 | 9.6 | 11.7 | 14.0 | 16.4 | 18.8 | 21.3 | 23.8 | 26.4 |
| | 14 | 9.3 | 11.1 | 13.1 | 15.1 | 17.2 | 19.4 | 21.5 | 23.7 |

(including safety margin)        * Outdoor-fed growing sheep, increase in maintenance allowance of 0.15 indicated as MJ/head daily thus (+ 0.8)

*Source:* MAFF/ADAS (1987) *Energy allowance and feeding systems for ruminants.* RB 433.
Reproduced with kind permission of MAFF ADAS UK.

**TABLE 270   Maintenance Allowances for Growing Sheep**

| Liveweight, W (kg) | ME allowance (MJ/day) | |
|---|---|---|
| | Indoors [1] | Outdoors [2] |
| 10 | 2.5 | 2.9 |
| 15 | 3.2 | 3.7 |
| 20 | 3.8 | 4.4 |
| 25 | 4.5 | 5.2 |
| 30 | 5.1 | 5.9 |
| 35 | 5.8 | 6.7 |
| 40 | 6.4 | 7.4 |
| 45 | 7.1 | 8.2 |
| 50 | 7.7 | 8.9 |

[1] Based on $M_m = 1.2 + 0.13\ W$ 
[2] Based on $M_m = 1.4 + 0.15\ W$

} including safety margin

*Source:* MAFF/ADAS (1987) *Energy allowance and feeding systems for ruminants*. RB 433. Reproduced with kind permission of MAFF ADAS UK.

**TABLE 271 a   Ration Formulation for Growing Sheep using Net Energy System**

**a. Animal Production Levels (APL) for growing sheep**

| LGW (g/day) | Liveweight, W (kg) | | | | | | | | |
|---|---|---|---|---|---|---|---|---|---|
| | 10 | 15 | 20 | 25 | 30 | 35 | 40 | 45 | 50 |
| 50 | 1.38 | 1.32 | 1.28 | 1.25 | 1.23 | 1.21 | 1.20 | 1.19 | 1.18 |
| 100 | 1.83 | 1.69 | 1.60 | 1.53 | 1.49 | 1.45 | 1.43 | 1.41 | 1.39 |
| 150 | 2.30 | 2.08 | 1.94 | 1.84 | 1.77 | 1.71 | 1.67 | 1.64 | 1.61 |
| 200 | 2.79 | 2.49 | 2.29 | 2.15 | 2.05 | 1.98 | 1.92 | 1.88 | 1.84 |
| 250 | 3.29 | 2.90 | 2.65 | 2.48 | 2.35 | 2.25 | 2.18 | 2.12 | 2.08 |
| 300 | | 3.33 | 3.02 | 2.81 | 2.65 | 2.54 | 2.44 | 2.37 | 2.32 |
| 350 | | | 3.40 | 3.15 | 2.96 | 2.82 | 2.71 | 2.63 | 2.56 |
| 400 | | | | 3.49 | 3.27 | 3.11 | 2.99 | 2.89 | 2.81 |

$$\text{Based on APL} = \frac{E_m + E_p}{E_m}\ ^*$$

\* $E_m$ and $E_p$ = Net energy allowance for maintenance and production

*(continued over)*

**TABLE 271  b & c  Ration Formulation for Growing Sheep using Net Energy System**

**b. Net Energies of foods for maintenance and production, NE$_{mp}$ (MJ/kg DM)**

| APL | Metabolizable energy food, MEF (MJ/kg DM) | | | | | | | | |
| | 6 | 7 | 8 | 9 | 10 | 11 | 12 | 13 | 14 |
| --- | --- | --- | --- | --- | --- | --- | --- | --- | --- |
| 1.0 | 4.2 | 4.9 | 5.6 | 6.3 | 7.0 | 7.7 | 8.4 | 9.1 | 9.8 |
| 1.1 | 3.6 | 4.4 | 5.1 | 5.9 | 6.6 | 7.4 | 8.1 | 8.9 | 9.7 |
| 1.2 | 3.3 | 4.0 | 4.8 | 5.6 | 6.3 | 7.1 | 7.9 | 8.7 | 9.6 |
| 1.3 | 3.0 | 3.8 | 4.5 | 5.3 | 6.1 | 7.0 | 7.8 | 8.6 | 9.5 |
| 1.4 | 2.8 | 3.6 | 4.3 | 5.1 | 6.0 | 6.8 | 7.6 | 8.5 | 9.4 |
| 1.5 | 2.7 | 3.4 | 4.2 | 5.0 | 5.8 | 6.7 | 7.5 | 8.4 | 9.3 |
| 1.6 | 2.6 | 3.3 | 4.1 | 4.9 | 5.7 | 6.6 | 7.4 | 8.3 | 9.3 |
| 1.7 | 2.5 | 3.2 | 4.0 | 4.8 | 5.6 | 6.5 | 7.4 | 8.3 | 9.2 |
| 1.8 | 2.4 | 3.1 | 3.9 | 4.7 | 5.5 | 6.4 | 7.3 | 8.2 | 9.2 |
| 1.9 | 2.3 | 3.0 | 3.8 | 4.6 | 5.4 | 6.3 | 7.2 | 8.2 | 9.1 |
| 2.0 | 2.3 | 3.0 | 3.7 | 4.5 | 5.4 | 6.2 | 7.2 | 8.1 | 9.1 |
| 2.2 | 2.2 | 2.9 | 3.6 | 4.4 | 5.3 | 6.1 | 7.1 | 8.0 | 9.1 |
| 2.4 | 2.1 | 2.8 | 3.5 | 4.3 | 5.2 | 6.1 | 7.0 | 8.0 | 9.0 |
| 2.6 | 2.1 | 2.7 | 3.4 | 4.2 | 5.1 | 6.0 | 6.9 | 7.9 | 9.0 |
| 2.8 | 2.0 | 2.7 | 3.4 | 4.2 | 5.0 | 5.9 | 6.9 | 7.9 | 8.9 |
| 3.0 | 2.0 | 2.6 | 3.3 | 4.1 | 5.0 | 5.9 | 6.8 | 7.8 | 8.9 |

$$\text{Based on } NE_{mp} = \frac{(MEF)^2 \times APL}{1.43\,MEF + 23\,(APL - 1)}$$

**c. Net Energy allowances (MJ) for growing lambs (indoors)**

| LGW (g/day) | Liveweight, W (kg) | | | | | | | | |
| | 10 | 15 | 20 | 25 | 30 | 35 | 40 | 45 | 50 |
| --- | --- | --- | --- | --- | --- | --- | --- | --- | --- |
| 50 | 2.4 | 2.9 | 3.4 | 3.9 | 4.4 | 4.9 | 5.4 | 5.9 | 6.4 |
| 100 | 3.2 | 3.7 | 4.2 | 4.8 | 5.3 | 5.8 | 6.4 | 6.9 | 7.5 |
| 150 | 4.0 | 4.6 | 5.2 | 5.7 | 6.3 | 6.9 | 7.5 | 8.1 | 8.7 |
| 200 | 4.9 | 5.5 | 6.1 | 6.7 | 7.3 | 8.0 | 8.6 | 9.3 | 9.9 |
| 250 | 5.8 | 6.4 | 7.1 | 7.7 | 8.4 | 9.1 | 9.8 | 10.5 | 11.2 |
| 300 | | 7.3 | 8.0 | 8.8 | 9.5 | 10.2 | 11.0 | 11.7 | 12.5 |
| 350 | | | 9.0 | 9.8 | 10.6 | 11.4 | 12.2 | 13.0 | 13.8 |
| 400 | | | | 10.9 | 11.7 | 12.5 | 13.4 | 14.3 | 15.2 |
| **Outdoors** | + 0.3 | + 0.4 | + 0.4 | + 0.5 | + 0.5 | + 0.6 | + 0.6 | + 0.7 | + 0.7 |

(including safety margin)                    Based on NE allowance = $E_m + E_p$

Requirements for lambs *outdoors* should be increased by amounts shown in final row at base of table.

**TABLE 272 a    Feeding Standards for Pregnant Ewes**

**a. daily allowances for ewes, assuming zero weight change**

| Ewe weight (kg) | Component | | Weeks before lambing | | | | | | | |
|---|---|---|---|---|---|---|---|---|---|---|
| | | | Single lamb | | | | Twin lambs | | | |
| | | | 8 - 7 | 6 - 5 | 4 - 3 | 2 - 1 | 8 - 7 | 6 - 5 | 4 - 3 | 2 - 1 |
| 55 | DMI | (kg) | 1.2 | 1.2 | 1.2 | 1.1 | 1.3 | 1.3 | 1.3 | 1.2 |
| | ME | (MJ) | 8.9 | 9.9 | 11.1 | 12.8 | 10.0 | 11.5 | 13.6 | 16.4* |
| | RDP | (g) | 74 | 82 | 93 | 107 | 83 | 96 | 114 | 137 |
| | UDP | (g) | 22 | 23 | 27 | 36 | 21 | 23 | 26 | 33 |
| | $E_{mp}$ | (MJ) | 5.4 | 5.5 | 5.6 | 5.9 | 5.5 | 5.7 | 6.0 | 6.3 |
| | APL | | 1.04 | 1.07 | 1.10 | 1.14 | 1.07 | 1.11 | 1.16 | 1.23 |
| | Ca | (g) | 5.0 | 5.0 | 7.4 | 7.4 | 5.0 | 5.0 | 8.9 | 8.9 |
| | P | (g) | 4.0 | 4.0 | 5.1 | 5.1 | 4.0 | 4.0 | 5.7 | 5.7 |
| | Mg | (g) | 1.0 | 1.0 | 1.3 | 1.3 | 1.0 | 1.0 | 1.5 | 1.5 |
| | Na | (g) | ← | | 1.8 | → | ← | | 1.8 | → |
| | Vit. A | (i.u.) | ← | | 3500 | → | ← | | 3500 | → |
| | Vit. D | (i.u.) | ← | | 550 | → | ← | | 550 | → |
| | Vit. E | (i.u.) | ← | | 35 | → | ← | | 35 | → |
| 75 | DMI | (kg) | 1.5 | 1.5 | 1.5 | 1.4 | 1.6 | 1.6 | 1.6 | 1.5 |
| | ME | (MJ) | 11.3 | 12.5 | 14.1 | 16.2 | 12.7 | 14.6 | 17.3 | 20.7* |
| | RDP | (g) | 94 | 104 | 118 | 135 | 106 | 122 | 144 | 173 |
| | UDP | (g) | 26 | 28 | 31 | 40 | 26 | 27 | 30 | 38 |
| | $E_{mp}$ | (MJ) | 6.8 | 7.0 | 7.2 | 7.4 | 7.0 | 7.2 | 7.6 | 8.0 |
| | APL | | 1.04 | 1.07 | 1.10 | 1.14 | 1.07 | 1.11 | 1.16 | 1.23 |
| | Ca | (g) | 6.7 | 6.7 | 9.8 | 9.8 | 6.7 | 6.7 | 11.7 | 11.7 |
| | P | (g) | 5.4 | 5.4 | 6.8 | 6.8 | 5.4 | 5.4 | 7.6 | 7.6 |
| | Mg | (g) | 1.3 | 1.3 | 1.7 | 1.7 | 1.3 | 1.3 | 2.0 | 2.0 |
| | Na | (g) | ← | | 2.5 | → | ← | | 2.5 | → |
| | Vit. A | (i.u.) | ← | | 5000 | → | ← | | 5000 | → |
| | Vit. D | (i.u.) | ← | | 750 | → | ← | | 750 | → |
| | Vit. E | (i.u.) | ← | | 45 | → | ← | | 45 | → |

\* These energy levels are not attainable because of inadequate DMI

*Source:* McDonald, P., Edwards, R.A., & Greenhalgh, J.F.D. (1988) *Animal Nutrition.*
Reproduced with kind permission of Longman Group UK Ltd.

*(continued over)*

**TABLE 272 b   Feeding Standards for Pregnant Ewes**

**b. Daily allowances for 55 kg pregnant ewes losing 50 g liveweight/day, and 75 kg pregnant ewes losing 70 g liveweight/day**

| Ewe weight (kg) | Component | | **Weeks before lambing** | | | | | | | |
| | | | **Single lamb** | | | | **Twin lambs** | | | |
| | | | 8 - 7 | 6 - 5 | 4 - 3 | 2 - 1 | 8 - 7 | 6 - 5 | 4 - 3 | 2 - 1 |
| **55** | DMI | (kg) | 1.2 | 1.2 | 1.2 | 1.1 | 1.3 | 1.3 | 1.3 | 1.2 |
| | ME | (MJ) | 6.8 | 7.8 | 9.0 | 10.7 | 7.9 | 9.4 | 11.5 | 14.2 |
| | RDP | (g) | 57 | 65 | 75 | 89 | 66 | 79 | 96 | 119 |
| | UDP | (g) | 29 | 30 | 34 | 43 | 28 | 30 | 33 | 41 |
| | $E_{mp}$ | (MJ) | 5.1 | 5.2 | 5.4 | 5.6 | 5.2 | 5.4 | 5.7 | 6.0 |
| | APL | | 1.00 | 1.01 | 1.05 | 1.09 | 1.02 | 1.06 | 1.11 | 1.18 |
| | Ca | (g) | 5.0 | 5.0 | 7.4 | 7.4 | 5.0 | 5.0 | 8.9 | 8.9 |
| | P | (g) | 4.0 | 4.0 | 5.1 | 5.1 | 4.0 | 4.0 | 5.7 | 5.7 |
| | Mg | (g) | 1.0 | 1.0 | 1.3 | 1.3 | 1.0 | 1.0 | 1.5 | 1.5 |
| | Na | (g) | ← | | 1.8 | → | ← | | 1.8 | → |
| | Vit. A | (i.u.) | ← | | 3500 | → | ← | | 3500 | → |
| | Vit. D | (i.u.) | ← | | 550 | → | ← | | 550 | → |
| | Vit. E | (i.u.) | ← | | 35 | → | ← | | 35 | → |
| **75** | DMI | (kg) | 1.5 | 1.5 | 1.5 | 1.4 | 1.6 | 1.6 | 1.6 | 1.5 |
| | ME | (MJ) | 8.4 | 9.6 | 11.1 | 13.2 | 9.7 | 11.7 | 14.3 | 17.7 |
| | RDP | (g) | 70 | 80 | 93 | 110 | 81 | 97 | 119 | 148 |
| | UDP | (g) | 36 | 38 | 42 | 50 | 36 | 37 | 40 | 48 |
| | $E_{mp}$ | (MJ) | 6.4 | 6.6 | 6.8 | 7.1 | 6.6 | 6.9 | 7.2 | 7.6 |
| | APL | | 1.00 | 1.01 | 1.04 | 1.08 | 1.01 | 1.05 | 1.10 | 1.17 |
| | Ca | (g) | 6.7 | 6.7 | 9.8 | 9.8 | 6.7 | 6.7 | 11.7 | 11.7 |
| | P | (g) | 5.4 | 5.4 | 6.8 | 6.8 | 5.4 | 5.4 | 7.6 | 7.6 |
| | Mg | (g) | 1.3 | 1.3 | 1.7 | 1.7 | 1.3 | 1.3 | 2.0 | 2.0 |
| | Na | (g) | ← | | 2.5 | → | ← | | 2.5 | → |
| | Vit. A | (i.u.) | ← | | 5000 | → | ← | | 5000 | → |
| | Vit. D | (i.u.) | ← | | 750 | → | ← | | 750 | → |
| | Vit. E | (i.u.) | ← | | 45 | → | ← | | 45 | → |

*Source:* McDonald, P., Edwards, R.A., & Greenhalgh, J.F.D. (1988) *Animal Nutrition.*
Reproduced with kind permission of Longman Group UK Ltd.

**TABLE 273 a Feeding Standards for Lactating Ewes**

**a. Daily allowances for lactating ewes, assuming zero weight change ($q_m$ = 0.625)**

| Ewe weight (kg) | Component | | Week of lactation: Single lamb 1 - 4 | 5 - 8 | 9 - 12 | Twin lambs 1 - 4 | 5 - 8 | 9 - 12 |
|---|---|---|---|---|---|---|---|---|
| **55** | DMI | (kg) | 1.5 | 1.7 | 1.6 | 1.6 | 1.8 | 1.7 |
| | ME | (MJ) | 16.6 | 15.7 | 13.0 | 22.3* | 20.0 | 15.8 |
| | RDP | (g) | 139 | 131 | 108 | 186 | 167 | 132 |
| | UDP | (g) | 57 | 53 | 43 | 75 | 68 | 54 |
| | $E_{mp}$ | (MJ) | 11.0 | 10.4 | 8.8 | 14.3 | 13.1 | 10.5 |
| | APL | | 2.14 | 2.03 | 1.71 | 2.79 | 2.53 | 2.04 |
| | Ca | (g) | 8.1 | 7.7 | 6.7 | 10.3 | 9.4 | 7.8 |
| | P | (g) | 6.8 | 6.6 | 5.8 | 8.4 | 7.8 | 6.6 |
| | Mg | (g) | 2.2 | 2.1 | 1.7 | 2.9 | 2.6 | 2.1 |
| | Na | (g) | 2.1 | 2.0 | 1.9 | 2.4 | 2.3 | 2.0 |
| | Vit. A | (i.u.) | ← | 5500 | → | ← | 5500 | → |
| | Vit. D | (i.u.) | ← | 550 | → | ← | 550 | → |
| | Vit. E | (i.u.) | ← | 40 | → | ← | 45 | → |
| **75** | DMI | (kg) | 1.9 | 2.2 | 2.0 | 2.0 | 2.3 | 2.1 |
| | ME | (MJ) | 24.9 | 23.3 | 18.4 | 32.3* | 27.4 | 21.3 |
| | RDP | (g) | 208 | 194 | 154 | 270 | 229 | 178 |
| | UDP | (g) | 84 | 78 | 62 | 108 | 92 | 72 |
| | $E_{mp}$ | (MJ) | 16.2 | 15.2 | 12.3 | 20.6 | 17.7 | 14.1 |
| | APL | | 2.48 | 2.33 | 1.89 | 3.14 | 2.71 | 2.15 |
| | Ca | (g) | 12.2 | 11.5 | 9.6 | 15.0 | 13.1 | 10.8 |
| | P | (g) | 10.1 | 9.7 | 8.3 | 12.1 | 10.8 | 9.1 |
| | Mg | (g) | 3.3 | 3.1 | 2.5 | 4.2 | 3.6 | 2.9 |
| | Na | (g) | 3.0 | 2.9 | 2.6 | 3.4 | 3.1 | 2.8 |
| | Vit. A | (i.u.) | ← | 7500 | → | ← | 7500 | → |
| | Vit. D | (i.u.) | ← | 750 | → | ← | 750 | → |
| | Vit. E | (i.u.) | ← | 55 | → | ← | 60 | → |

* These energy levels are not attainable because of inadequate DMI

*Source:* McDonald, P., Edwards, R.A., & Greenhalgh, J.F.D. (1988) *Animal Nutrition*. Reproduced with kind permission of Longman Group UK Ltd.

*Feed & Nutrient Requirements – Sheep*

---

**TABLE 273 b   Feeding Standards for Lactating Ewes**

---

**b. Daily allowances for 55 kg lactating ewes losing 50 g liveweight/day, and 75 kg lactating ewes losing 75 g liveweight/day**

$$(q_m = 0.625)$$

| Ewe weight (kg) | Component | | Single lamb | | | Twin lambs | | |
|---|---|---|---|---|---|---|---|---|
| | | | Week of lactation | | | | | |
| | | | 1 - 4 | 5 - 8 | 9 - 12 | 1 - 4 | 5 - 8 | 9 - 12 |
| 55 | DMI | (kg) | 1.5 | 1.7 | 1.6 | 1.6 | 1.8 | 1.7 |
| | ME | (MJ) | 14.7 | 13.8 | 11.1 | 22.3* | 18.1 | 13.9 |
| | RDP | (g) | 123 | 115 | 93 | 169 | 151 | 116 |
| | UDP | (g) | 62 | 58 | 49 | 81 | 74 | 59 |
| | $E_{mp}$ | (MJ) | 9.8 | 9.3 | 7.6 | 13.2 | 11.9 | 9.4 |
| | APL | | 1.92 | 1.80 | 1.48 | 2.56 | 2.31 | 1.82 |
| | Ca | (g) | 8.1 | 7.7 | 6.7 | 10.3 | 9.4 | 7.8 |
| | P | (g) | 6.8 | 6.6 | 5.8 | 8.4 | 7.8 | 6.6 |
| | Mg | (g) | 2.2 | 2.1 | 1.7 | 2.9 | 2.6 | 2.1 |
| | Na | (g) | 2.1 | 2.0 | 1.9 | 2.4 | 2.3 | 2.0 |
| | Vit. A | (i.u.) | ← 5500 → | | | ← 5500 → | | |
| | Vit. D | (i.u.) | ← 550 → | | | ← 550 → | | |
| | Vit. E | (i.u.) | ← 40 → | | | ← 45 → | | |
| 75 | DMI | (kg) | 1.9 | 2.2 | 2.0 | 2.0 | 2.3 | 2.1 |
| | ME | (MJ) | 22.0 | 20.4 | 15.6 | 29.4* | 24.5 | 18.5 |
| | RDP | (g) | 184 | 170 | 130 | 245 | 205 | 154 |
| | UDP | (g) | 93 | 87 | 70 | 117 | 101 | 80 |
| | $E_{mp}$ | (MJ) | 14.5 | 13.5 | 10.6 | 18.8 | 16.0 | 12.4 |
| | APL | | 2.21 | 2.07 | 1.62 | 2.88 | 2.44 | 1.89 |
| | Ca | (g) | 12.2 | 11.5 | 9.6 | 15.0 | 13.1 | 10.8 |
| | P | (g) | 10.1 | 9.7 | 8.3 | 12.1 | 10.8 | 9.1 |
| | Mg | (g) | 3.3 | 3.1 | 2.5 | 4.2 | 3.6 | 2.9 |
| | Na | (g) | 3.0 | 2.9 | 2.6 | 3.4 | 3.1 | 2.8 |
| | Vit. A | (i.u.) | ← 7500 → | | | ← 7500 → | | |
| | Vit. D | (i.u.) | ← 750 → | | | ← 750 → | | |
| | Vit. E | (i.u.) | ← 55 → | | | ← 60 → | | |

\* These energy levels are not attainable because of inadequate DMI

*Source:* McDonald, P., Edwards, R.A., & Greenhalgh, J.F.D. (1988) *Animal Nutrition.* Reproduced with kind permission of Longman Group UK Ltd.

**TABLE 274  a  Feeding Standards for Growing Lambs**

**a. Daily allowances for castrated male lambs**

| Weight (kg) | $q_m$ | Component | | Liveweight gain (g/day) | | | | DMI (kg/day) |
|---|---|---|---|---|---|---|---|---|
| | | | | 0 | 50 | 100 | 150 | |
| 20 | 0.55 | ME | (MJ) | 3.7 | 4.8 | | | 0.46 |
| | | RDP | (g) | 31 | 40 | | | |
| | | UDP | (g) | 14 | 18 | | | |
| | 0.65 | ME | (MJ) | 3.5 | 4.5 | 5.6 | 6.8 | 0.56 |
| | | RDP | (g) | 29 | 38 | 47 | 57 | |
| | | UDP | (g) | 16 | 20 | 23 | 26 | |
| | | $E_{mp}$ | (MJ) | 2.6 | 3.1 | 3.7 | 4.3 | |
| | | APL | | 1.00 | 1.22 | 1.44 | 1.66 | |
| | | Ca | (g) | 1.5 | 2.3 | 3.1 | 3.9 | |
| | | P | (g) | 1.0 | 1.2 | 1.5 | 1.8 | |
| | | Mg | (g) | 0.35 | 0.47 | 0.59 | 0.71 | |
| | | Na | (g) | 0.56 | 0.62 | 0.68 | 0.74 | |
| | | Vit. A | (i.u.) | ← | | 660 | → | |
| | | Vit. D | (i.u.) | ← | | 120 | → | |
| | | Vit. E | (i.u.) | ← | | 21 | → | |
| 35 | 0.55 | ME | (MJ) | 5.7 | 7.3 | | | 0.77 |
| | | RDP | (g) | 48 | 61 | | | |
| | | UDP | (g) | 18 | 17 | | | |
| | 0.65 | ME | (MJ) | 5.4 | 6.8 | 8.4 | 10.1 | 0.92 |
| | | RDP | (g) | 45 | 57 | 70 | 84 | |
| | | UDP | (g) | 20 | 20 | 20 | 18 | |
| | | $E_{mp}$ | (MJ) | 4.0 | 4.8 | 5.6 | 6.4 | |
| | | APL | | 1.00 | 1.21 | 1.41 | 1.62 | |
| | | Ca | (g) | 2.5 | 3.4 | 4.2 | 5.0 | |
| | | P | (g) | 1.7 | 2.0 | 2.2 | 2.5 | |
| | | Mg | (g) | 0.62 | 0.74 | 0.86 | 0.98 | |
| | | Na | (g) | 0.98 | 1.0 | 1.1 | 1.2 | |
| | | Vit. A | (i.u.) | ← | | 1200 | → | |
| | | Vit. D | (i.u.) | ← | | 210 | → | |
| | | Vit. E | (i.u.) | ← | | 25 | → | |

*Source:* McDonald, P., Edwards, R.A., & Greenhalgh, J.F.D. (1988) *Animal Nutrition.*
Reproduced with kind permission of Longman Group UK Ltd.

**TABLE 274 b   Feeding Standards for Growing Lambs**

**b. Daily allowances for entire male lambs**

| Weight (kg) | $q_m$ | Component | | Liveweight gain (g/day) 0 | 50 | 100 | 150 | DMI (kg/day) |
|---|---|---|---|---|---|---|---|---|
| 20 | 0.55 | ME | (MJ) | 4.3 | | | | 0.46 |
| | | RDP | (g) | 36 | | | | |
| | | UDP | (g) | 10 | | | | |
| | 0.65 | ME | (MJ) | 4.1 | 4.9 | 5.8 | 6.8 | 0.56 |
| | | RDP | (g) | 34 | 41 | 49 | 57 | |
| | | UDP | (g) | 12 | 17 | 22 | 26 | |
| | | $E_{mp}$ | (MJ) | 3.0 | 3.5 | 4.0 | 4.5 | |
| | | APL | | 1.00 | 1.17 | 1.34 | 1.51 | |
| | | Ca | (g) | 1.5 | 2.3 | 3.1 | 3.9 | |
| | | P | (g) | 1.0 | 1.2 | 1.5 | 1.8 | |
| | | Mg | (g) | 0.35 | 0.47 | 0.59 | 0.71 | |
| | | Na | (g) | 0.56 | 0.62 | 0.68 | 0.74 | |
| | | Vit. A | (i.u.) | ← | | 660 | → | |
| | | Vit. D | (i.u.) | ← | | 120 | → | |
| | | Vit. E | (i.u.) | ← | | 21 | → | |
| 35 | 0.55 | ME | (MJ) | 6.6 | | | | 0.77 |
| | | RDP | (g) | 55 | | | | |
| | | UDP | (g) | 12 | | | | |
| | 0.65 | ME | (MJ) | 6.2 | 7.6 | 9.0 | 10.5 | 0.92 |
| | | RDP | (g) | 52 | 63 | 75 | 88 | |
| | | UDP | (g) | 14 | 15 | 16 | 16 | |
| | | $E_{mp}$ | (MJ) | 4.6 | 5.3 | 6.1 | 6.9 | |
| | | APL | | 1.00 | 1.17 | 1.34 | 1.51 | |
| | | Ca | (g) | 2.5 | 3.4 | 4.2 | 5.0 | |
| | | P | (g) | 1.7 | 2.0 | 2.3 | 2.5 | |
| | | Mg | (g) | 0.62 | 0.74 | 0.86 | 0.98 | |
| | | Na | (g) | 0.98 | 1.0 | 1.1 | 1.2 | |
| | | Vit. A | (i.u.) | ← | | 1200 | → | |
| | | Vit. D | (i.u.) | ← | | 210 | → | |
| | | Vit. E | (i.u.) | ← | | 25 | → | |

*Source:* McDonald, P., Edwards, R.A., & Greenhalgh, J.F.D. (1988) *Animal Nutrition.*
Reproduced with kind permission of Longman Group UK Ltd.

**TABLE 274 c   Feeding Standards for Growing Lambs**

## c. Daily allowances for female lambs

| Weight (kg) | $q_m$ | Component | | Liveweight gain (g/day) 0 | 50 | 100 | 150 | DMI (kg/day) |
|---|---|---|---|---|---|---|---|---|
| 20 | 0.55 | ME | (MJ) | 3.7 | 4.9 | | | |
|    |      | RDP | (g) | 31 | 40 | | | 0.46 |
|    |      | UDP | (g) | 14 | 16 | | | |
|    | 0.65 | ME | (MJ) | 3.5 | 4.5 | 5.7 | 6.9 | |
|    |      | RDP | (g) | 29 | 38 | 47 | 58 | 0.56 |
|    |      | UDP | (g) | 16 | 18 | 20 | 22 | |
|    |      | $E_{mp}$ | (MJ) | 2.6 | 3.2 | 3.7 | 4.3 | |
|    |      | APL | | 1.00 | 1.23 | 1.45 | 1.68 | |
|    |      | Ca | (g) | 1.5 | 2.3 | 3.1 | 3.9 | |
|    |      | P | (g) | 1.0 | 1.2 | 1.5 | 1.8 | |
|    |      | Mg | (g) | 0.35 | 0.47 | 0.59 | 0.71 | |
|    |      | Na | (g) | 0.56 | 0.62 | 0.68 | 0.74 | |
|    |      | Vit. A | (i.u.) | ←——— | 660 | ———→ | | |
|    |      | Vit. D | (i.u.) | ←——— | 120 | ———→ | | |
|    |      | Vit. E | (i.u.) | ←——— | 21 | ———→ | | |
| 35 | 0.55 | ME | (MJ) | 5.7 | 7.6 | | | |
|    |      | RDP | (g) | 48 | 63 | | | 0.77 |
|    |      | UDP | (g) | 18 | 13 | | | |
|    | 0.65 | ME | (MJ) | 5.4 | 7.1 | 8.9 | 10.9 | |
|    |      | RDP | (g) | 45 | 59 | 74 | 91 | 0.92 |
|    |      | UDP | (g) | 20 | 17 | 13 | 8 | |
|    |      | $E_{mp}$ | (MJ) | 4.0 | 4.9 | 5.8 | 6.8 | |
|    |      | APL | | 1.00 | 1.24 | 1.47 | 1.71 | |
|    |      | Ca | (g) | 2.5 | 3.4 | 4.2 | 5.0 | |
|    |      | P | (g) | 1.7 | 2.0 | 2.3 | 2.5 | |
|    |      | Mg | (g) | 0.62 | 0.74 | 0.85 | 1.0 | |
|    |      | Na | (g) | 0.98 | 1.0 | 1.1 | 1.2 | |
|    |      | Vit. A | (i.u.) | ←——— | 1200 | ———→ | | |
|    |      | Vit. D | (i.u.) | ←——— | 210 | ———→ | | |
|    |      | Vlt. E | (i.u.) | ←——— | 25 | ———→ | | |

*Source:* McDonald, P., Edwards, R.A., & Greenhalgh, J.F.D. (1988) *Animal Nutrition.*
Reproduced with kind permission of Longman Group UK Ltd.

*Feed & Nutrient Requirements – Sheep*

**TABLE 275   Daily Nutrient Requirements of Sheep (100% dry matter basis)**

| Body weight (lb) | Gain or loss (lb) | Dry matter [a] Per animal (lb) | Dry matter [a] % live wt. | Energy TDN (kg) | Energy DE [b] (Mcal) | Total protein (g) | Grams DP per Mcal DE | Ca (g) | P (g) | Vita-min A (IU) | Vita-min D (IU) |
|---|---|---|---|---|---|---|---|---|---|---|---|
| **EWES [c]** | | | | | | | | | | | |
| **Maintenance** | | | | | | | | | | | |
| 110 | 0.02 | 2.2 | 2.0 | 0.55 | 2.42 | 95 | 20 | 2.0 | 1.8 | 2350 | 278 |
| 132 | 0.02 | 2.4 | 1.8 | 0.61 | 2.68 | 104 | 20 | 2.3 | 2.1 | 2820 | 333 |
| 154 | 0.02 | 2.6 | 1.7 | 0.66 | 2.90 | 113 | 20 | 2.5 | 2.4 | 3290 | 388 |
| 176 | 0.02 | 2.9 | 1.6 | 0.72 | 3.17 | 122 | 20 | 2.7 | 2.8 | 3760 | 444 |
| **Nonlactating and first 15 weeks of gestation** | | | | | | | | | | | |
| 110 | 0.07 | 2.6 | 2.4 | 0.67 | 3.0 | 112 | 20 | 2.9 | 2.1 | 2350 | 278 |
| 132 | 0.07 | 2.9 | 2.2 | 0.72 | 3.17 | 121 | 20 | 3.2 | 2.5 | 2820 | 333 |
| 154 | 0.07 | 3.1 | 2.0 | 0.77 | 3.39 | 130 | 20 | 3.5 | 2.9 | 3290 | 388 |
| 176 | 0.07 | 3.3 | 1.9 | 0.82 | 3.61 | 139 | 20 | 3.8 | 3.3 | 3760 | 444 |
| **Last 4 weeks of gestation or last 4-6 weeks of lactation suckling singles [d] (130 -150% lambing expected)** | | | | | | | | | | | |
| 110 | 0.10 | 3.5 | 3.2 | 0.94 | 4.1 | 175 | 20 | 5.9 | 4.8 | 4250 | 278 |
| 132 | 0.10 | 3.7 | 2.8 | 1.00 | 4.4 | 184 | 20 | 6.0 | 5.2 | 5100 | 333 |
| 154 | 0.10 | 4.0 | 2.6 | 1.06 | 4.7 | 193 | 20 | 6.2 | 5.6 | 5950 | 388 |
| 176 | 0.10 | 4.2 | 2.4 | 1.12 | 4.9 | 202 | 20 | 6.3 | 6.1 | 6800 | 444 |
| **First 6 - 8 weeks of lactation suckling singles or last 4-6 weeks of lactation suckling twins [e]** | | | | | | | | | | | |
| 110 | - 0.20 | 4.6 | 4.2 | 1.36 | 5.98 | 304 | 22 | 8.9 | 6.1 | 4250 | 278 |
| 132 | - 0.20 | 5.1 | 3.9 | 1.50 | 6.60 | 319 | 22 | 9.1 | 6.6 | 5100 | 333 |
| 154 | - 0.20 | 5.5 | 3.6 | 1.63 | 7.17 | 334 | 22 | 9.3 | 7.0 | 5950 | 388 |
| 176 | - 0.20 | 5.7 | 3.2 | 1.69 | 7.44 | 344 | 22 | 9.5 | 7.4 | 6800 | 444 |
| **First 6 - 8 weeks of lactation suckling twins** | | | | | | | | | | | |
| 110 | - 0.13 | 5.3 | 4.8 | 1.56 | 6.86 | 389 | 25 | 10.5 | 7.3 | 5000 | 278 |
| 132 | - 0.13 | 5.7 | 4.3 | 1.69 | 7.44 | 405 | 25 | 10.7 | 7.7 | 6000 | 333 |
| 154 | - 0.13 | 6.2 | 4.0 | 1.82 | 8.01 | 420 | 25 | 11.0 | 8.1 | 7000 | 388 |
| 176 | - 0.13 | 6.6 | 3.7 | 1.95 | 8.58 | 435 | 25 | 11.2 | 8.6 | 8000 | 444 |
| **Replacement lambs and yearlings [f]** | | | | | | | | | | | |
| 66 | 0.50 | 2.6 | 4.0 | 0.78 | 3.40 | 185 | 21 | 6.4 | 2.6 | 1410 | 166 |
| 88 | 0.40 | 3.1 | 3.5 | 0.91 | 4.00 | 176 | 20 | 5.9 | 2.6 | 1880 | 222 |
| 110 | 0.26 | 3.3 | 3.0 | 0.88 | 3.90 | 136 | 20 | 4.8 | 2.4 | 2350 | 278 |
| 132 | 0.22 | 3.3 | 2.5 | 0.88 | 3.90 | 134 | 20 | 4.5 | 2.5 | 2820 | 333 |

**TABLE 275** *(continued)*  **Daily Nutrient Requirements of Sheep (100% dry matter basis)**

| | | | | Nutrients per animal | | | | | | | |
|---|---|---|---|---|---|---|---|---|---|---|---|
| | | Dry matter [a] | | Energy | | | | | | | |
| Body weight (lb) | Gain or loss (lb) | Per animal (lb) | % live wt. | TDN (kg) | DE [b] (Mcal) | Total protein (g) | Grams DP per Mcal DE | Ca (g) | P (g) | Vita-min A (IU) | Vita-min D (IU) |
| **RAMS** | | | | | | | | | | | |
| **Replacement lambs and yearlings** | | | | | | | | | | | |
| 88 | 0.73 | 4.0 | 4.5 | 1.17 | 5.50 | 243 | 21 | 7.8 | 3.7 | 1880 | 222 |
| 132 | 0.70 | 5.3 | 4.0 | 1.50 | 6.70 | 263 | 20 | 8.4 | 4.2 | 2820 | 333 |
| 176 | 0.64 | 6.2 | 3.5 | 1.80 | 7.80 | 268 | 20 | 8.5 | 4.6 | 3760 | 444 |
| 220 | 0.55 | 6.6 | 3.0 | 1.90 | 8.40 | 264 | 20 | 8.2 | 4.8 | 4700 | 555 |
| 265 | 0.50 | 6.6 | 2.5 | 1.90 | 8.50 | 264 | 20 | 8.2 | 4.8 | 5500 | 666 |
| **LAMBS 4 - 7 months old** | | | | | | | | | | | |
| **Finishing [g]** | | | | | | | | | | | |
| 66 | 0.65 | 2.9 | 4.3 | 0.94 | 4.10 | 191 | 24 | 6.6 | 3.2 | 1410 | 166 |
| 77 | 0.65 | 2.9 | 3.8 | 0.94 | 4.14 | 191 | 23 | 6.6 | 3.2 | 1410 | 194 |
| 88 | 0.60 | 3.5 | 4.0 | 1.22 | 5.40 | 185 | 22 | 6.6 | 3.3 | 1880 | 222 |
| 99 | 0.60 | 3.5 | 3.5 | 1.22 | 5.40 | 185 | 22 | 6.6 | 3.3 | 1880 | 250 |
| 110 | 0.45 | 4.0 | 3.6 | 1.26 | 5.54 | 160 | 22 | 5.6 | 3.0 | 2350 | 278 |
| 121 | 0.45 | 4.0 | 3.3 | 1.33 | 5.85 | 160 | 22 | 5.6 | 3.0 | 2350 | 305 |
| **Early weaned [h] - rapid growth potential** | | | | | | | | | | | |
| 22 | 0.55 | 1.3 | 6.0 | 0.48 | 2.10 | 157 | 36 | 4.9 | 2.2 | 470 | 67 |
| 44 | 0.66 | 2.6 | 6.0 | 0.92 | 4.00 | 205 | 36 | 6.5 | 2.9 | 940 | 133 |
| 66 | 0.72 | 3.1 | 4.7 | 1.10 | 4.80 | 216 | 30 | 7.2 | 3.4 | 1440 | 200 |

a.  To convert dry matter to be an as-fed basis, divide dry matter by percentage of dry matter.

b.  1 kg TDN = 4.4 Mcal DE (digestible energy). DE may be converted to ME (metabolizable energy) by multiplying by 82%.

c.  Values are for ewes in moderate condition, not excessively fat or thin. Fat ewes should be fed at the next lower weight, thin ewes at the next higher weight. Once maintenance weight is established such weight would follow through all production phases.

d.  Values in parentheses are for ewes suckling singles last 4 - 6 weeks of lactation.

e  Values in parentheses are for ewes suckling twins last 4 - 6 weeks of lactation.

f.  Requirements for replacement lambs (ewe and ram) start when the lambs are weaned.

g.  Maximum gains expected. If lambs are held for later market, they should be fed as replacement ewe lambs are fed. Lambs capable of gaining faster than indicated should be fed at a higher level. Lambs finish at the maximum rate if they are self-fed.

h.  A 40- kg early weaned lamb should be fed the same as a finishing lamb of the same weight.

*Source:* Haenlein, A.F.W. (1986) *Dietary Nutrient Allowances for Sheep and Goats. Feedstuffs 58.*
Reproduced with kind permission of A.F.W. Haenlein, University of Delaware.

# Feed and Nutrient Requirements – Goats

TABLE 276  Suggested Maximum Dry Matter Intakes for Housed Goats

| Growing goats | Liveweight (kg) | Dry matter intake (kg/day) |
|---|---|---|
| | 10 | 0.45 |
| | 20 | 1.1 |
| | 30 | 1.3 |
| | 40 | 1.4 |

**Adult goats**

| | Milk yield (kg/day at 3.5% fat) | | | | | | |
|---|---|---|---|---|---|---|---|
| | 0 | 1 | 2 | 3 | 4 | 5 | 6 |
| Liveweight (kg) | Dry matter intake (kg/day) | | | | | | |
| 50 | 1.5 | 1.7 | 1.9 | 2.1 | 2.3 | 2.4 | 2.5 |
| 60 | 1.8 | 2.0 | 2.2 | 2.4 | 2.6 | 2.8 | 3.0 |
| 70 | 2.1 | 2.3 | 2.5 | 2.7 | 2.9 | 3.1 | 3.3 |
| 80 | 2.4 | 2.6 | 2.8 | 3.1 | 3.4 | 3.7 | 4.0 |

*Notes:* Dry matter intake is assumed to increase with milk yield from 3% of body weight for growing kids and dry animals to 5% of body weight for animals giving 6 kg milk/day. In early lactation appetite is likely to be reduced.

*Source:* Wilkinson, J.M. & Stark, B.A. (1987) *Commercial Goat Production.* Reproduced with kind permission of Blackwell Scientific Publications Ltd.

TABLE 277  Suggested Energy Requirements of Housed Adult Dairy Goats

| | Milk yield (kg/day at 3.5% fat) | | | | | | |
|---|---|---|---|---|---|---|---|
| | 0 | 1 | 2 | 3 | 4 | 5 | 6 |
| Liveweight (kg) | Energy requirement (MJ ME/day) | | | | | | |
| 50 | 8.0 | 13.1 | 18.2 | 23.3 | 28.4 | 33.5 | 38.6 |
| 60 | 9.2 | 14.3 | 19.4 | 24.5 | 29.6 | 34.7 | 39.8 |
| 70 | 10.3 | 15.4 | 20.5 | 25.6 | 30.7 | 35.8 | 41.1 |
| 80 | 11.3 | 16.5 | 21.6 | 26.8 | 31.9 | 37.1 | 42.2 |

*Notes:* For grazing animals, increase daily energy requirements by 2.0, 2.3, 2.6 and 2.8 MJ ME for goats weighing 50, 60, 70 and 80 kg respectively.
For milk of higher fat content, increase daily energy requirement by 0.6 MJ ME per percentage unit of fat.

*Source:* Wilkinson, J.M. & Stark, B.A. (1987) *Commercial Goat Production.* Reproduced with kind permission of Blackwell Scientific Publications Ltd.

TABLE 278    Suggested Energy Requirements of Housed Young Goats

| Liveweight (kg) | Liveweight gain (g/day) | | | | |
|---|---|---|---|---|---|
| | 0 | 50 | 100 | 150 | 200 |
| | Energy requirement (MJ ME/day) | | | | |
| 10 | 3.0 | 4.5 | 6.0 | 7.5 | 9.0 |
| 20 | 5.0 | 6.5 | 8.0 | 9.5 | 11.0 |
| 30 | 6.8 | 8.3 | 9.8 | 11.3 | 12.8 |
| 40 | 8.5 | 10.0 | 11.5 | 13.0 | 14.5 |
| 50 | 10.0 | 11.5 | 13.0 | 14.5 | 16.0 |
| 60 | 11.4 | 12.9 | 14.4 | 15.9 | 17.4 |

*Source:* Wilkinson, J.M. & Stark, B.A. (1987) *Commercial Goat Production.*
Reproduced with kind permission of Blackwell Scientific Publications Ltd.

TABLE 279    Suggested Digestible Crude Protein (DCP) Requirements of Housed Goats[1]

| Growing goats Liveweight (kg) | Liveweight gain (g/day) | | | | |
|---|---|---|---|---|---|
| | 0 | 50 | 100 | 150 | 200 |
| | DCP (g/day) | | | | |
| 10 | 35 | 45 | 55 | 65 | 75 |
| 20 | 46 | 56 | 66 | 76 | 86 |
| 30 | 50 | 60 | 70 | 80 | 90 |
| 40 | 53 | 63 | 73 | 83 | 93 |
| 50 | 61 | 71 | 81 | 91 | 101 |
| 60 | 69 | 79 | 89 | 99 | 109 |

| Adult goats[2] Liveweight (kg) | Milk yield (kg/day) | | | | | | |
|---|---|---|---|---|---|---|---|
| | 0 | 1 | 2 | 3 | 4 | 5 | 6 |
| 50 | 51 | 106 | 161 | 216 | 271 | 326 | 381 |
| 60 | 59 | 114 | 169 | 224 | 279 | 334 | 389 |
| 70 | 66 | 121 | 176 | 231 | 200 | 341 | 396 |
| 80 | 73 | 128 | 183 | 238 | 293 | 348 | 403 |

1. For grazing, increase daily DCP requirements by 25%.
2. Increase DCP requirements by 57 g/day in the last 2 months of pregnancy.

*Source:* Wilkinson, J.M. & Stark, B.A. (1987) *Commercial Goat Production.*
Reproduced with kind permission of Blackwell Scientific Publications Ltd.

**TABLE 280  Suggested Requirements for Calcium, Phosphorus and Magnesium of Goats**

| Maintenance: | Calcium (g/day) | Phosphorous (g/day) | Magnesium (g/day) |
|---|---|---|---|
| liveweight (kg) | | | |
| 10 | 1 | 0.70 | 0.18 |
| 20 | 1 | 0.70 | 0.35 |
| 30 | 2 | 1.4 | 0.53 |
| 40 | 2 | 1.4 | 0.70 |
| 50 | 3 | 2.1 | 0.88 |
| 60 | 3 | 2.1 | 1.06 |
| 70 | 4 | 2.8 | 1.23 |
| 80 | 4 | 2.8 | 1.41 |
| Plus additional requirements for: | | | |
| liveweight gain(g/day) | | | |
| 50 | 1 | 0.7 | 0.14 |
| 100 | 1 | 0.7 | 0.27 |
| 150 | 2 | 1.4 | 0.41 |
| 200 | 2 | 1.4 | 0.55 |
| **Late pregnancy** | 2 | 1.4 | 0.60 |
| **Milk (per kg)** | | | |
| 2.5 to 3.5% fat | 2 | 1.4 | 1.0 |
| 3.5 to 5.0% fat | 3 | 2.1 | 1.0 |

*Source:* Wilkinson, J.M. & Stark, B.A. (1987) *Commercial Goat Production.*
Reproduced with kind permission of Blackwell Scientific Publications Ltd.

**TABLE 281  Suggested Dietary Allowances for Trace Elements of Goats**

| Element | Concentration in diet (mg/kg dry matter) |
|---|---|
| Iron | 40 |
| Zinc | 40 |
| Manganese | 40 |
| Copper | 10 |
| Cobalt | 0.11 |
| Selenium | 0.10 |
| Iodine: | |
| housed | 0.5 |
| at pasture | 0.15 |
| with kale, cabbage or clover in diet, i.e. | |
| goitrogenic feeds | 2.0 |

*Source:* Wilkinson, J.M. & Stark, B.A. (1987) *Commercial Goat Production.*
Reproduced with kind permission of Blackwell Scientific Publications Ltd.

**TABLE 282   Suggested Requirements for Vitamins A and D of Goats**

| | Vitamin A | Vitamin D |
|---|---|---|
| **Maintenance:** | (i.u. per day) | |
| liveweight (kg) | | |
| 10 | 400 | 84 |
| 20 | 700 | 144 |
| 30 | 900 | 195 |
| 40 | 1200 | 243 |
| 50 | 1400 | 285 |
| 60 | 1600 | 327 |
| 70 | 1800 | 369 |
| 80 | 2000 | 411 |
| Plus additional requirements for: | | |
| liveweight gain (g/day) | | |
| 50 | 300 | 54 |
| 100 | 500 | 108 |
| 150 | 800 | 162 |
| 200 | 1100 | 216 |
| **Late pregnancy** | 1100 | 213 |
| **Milk (per kg)** | 3800 | 760 |

*Source:* Wilkinson, J.M. & Stark, B.A. (1987) *Commercial Goat Production.*
Reproduced with kind permission of Blackwell Scientific Publications Ltd.

**TABLE 283   Daily Nutrient Requirements of Goats (100% dry matter basis)**

| Body [a] weight (kg) | Feed energy [b] TDN (g) | DE (Mcal) | ME (Mcal) | NE (Mcal) | Crude protein TP (g) | DP (g) | Ca (g) | P (g) | Vita-min A 1000 IU | Vita-min D IU | Dry matter per animal 1kg = 2.0 Mcal ME Total (kg) | % of kg BW | 1kg = 2.4 Mcal ME Total (kg) | % of kg BW |
|---|---|---|---|---|---|---|---|---|---|---|---|---|---|---|
| **Maintenance only** (includes stable feeding conditions, minimal activity, and early pregnancy) | | | | | | | | | | | | | | |
| 10 | 159 | 0.70 | 0.57 | 0.32 | 22 | 15 | 1 | 0.7 | 0.4 | 84 | 0.28 | 2.8 | 0.24 | 2.4 |
| 20 | 267 | 1.18 | 0.96 | 0.54 | 38 | 26 | 1 | 0.7 | 0.7 | 144 | 0.48 | 2.4 | 0.40 | 2.0 |
| 30 | 362 | 1.59 | 1.30 | 0.73 | 51 | 35 | 2 | 1.4 | 0.9 | 195 | 0.65 | 2.2 | 0.54 | 1.8 |
| 40 | 448 | 1.98 | 1.61 | 0.91 | 63 | 43 | 2 | 1.4 | 1.2 | 243 | 0.81 | 2.0 | 0.67 | 1.7 |
| 50 | 530 | 2.34 | 1.91 | 1.08 | 75 | 51 | 3 | 2.1 | 1.4 | 285 | 0.95 | 1.9 | 0.79 | 1.6 |
| 60 | 608 | 2.68 | 2.19 | 1.23 | 86 | 59 | 3 | 2.1 | 1.6 | 327 | 1.09 | 1.8 | 0.91 | 1.5 |
| 70 | 682 | 3.01 | 2.45 | 1.38 | 96 | 66 | 4 | 2.8 | 1.8 | 369 | 1.23 | 1.8 | 1.02 | 1.5 |
| 80 | 754 | 3.32 | 2.71 | 1.53 | 106 | 73 | 4 | 2.8 | 2.0 | 408 | 1.36 | 1.7 | 1.13 | 1.4 |
| 90 | 824 | 3.63 | 2.96 | 1.67 | 116 | 80 | 4 | 2.8 | 2.2 | 444 | 1.48 | 1.6 | 1.23 | 1.4 |
| 100 | 891 | 3.93 | 3.21 | 1.81 | 126 | 86 | 5 | 3.5 | 2.4 | 480 | 1.60 | 1.6 | 1.34 | 1.3 |

**Maintenance plus low activity**
(= 25% increment, intensive management, tropical range and early pregnancy)

**Maintenance plus medium activity**
(= 50% increment, semi-arid rangeland, slightly hilly pastures and early pregnancy)

**Maintenance plus high activity**
(= 75% increment, arid rangeland, sparse vegetation, mountainous pastures and early pregnancy)

**Additional requirements for late pregnancy (for all goat sizes)**

| | 397 | 1.74 | 1.42 | 0.80 | 82 | 57 | 2 | 1.4 | 1.1 | 213 | 0.71 | | 0.59 | |
|---|---|---|---|---|---|---|---|---|---|---|---|---|---|---|

**Additional requirements for growth – weight gain at 50 g per day (for all goat sizes)**

| | 100 | 0.44 | 0.36 | 0.20 | 14 | 10 | 1 | 0.7 | 0.3 | 54 | 0.18 | | 0.15 | |
|---|---|---|---|---|---|---|---|---|---|---|---|---|---|---|

**Additional requirements for growth – weight gain at 100 g per day (for all goat sizes)**

| | 200 | 0.88 | 0.72 | 0.40 | 28 | 20 | 1 | 0.7 | 0.5 | 108 | 0.36 | | 0.30 | |
|---|---|---|---|---|---|---|---|---|---|---|---|---|---|---|

**Additional requirements for growth – weight gain at 150 g per day (for all goat sizes)**

| | 300 | 1.32 | 1.08 | 0.60 | 42 | 30 | 2 | 1.4 | 0.8 | 162 | 0.54 | | 0.45 | |
|---|---|---|---|---|---|---|---|---|---|---|---|---|---|---|

**TABLE 283** *(continued)*  **Daily Nutrient Requirements of Goats (100% dry matter basis)**

| Body [a] weight (kg) | Feed energy [b] TDN (g) | DE (Mcal) | ME (Mcal) | NE (Mcal) | Crude protein TP (g) | DP (g) | Ca (g) | P (g) | Vita- min A (1000 IU) | Vita- min D IU | Dry matter per animal 1kg = 2.0 Mcal ME Total (kg) | % of kg BW | 1kg = 2.4 Mcal ME Total (kg) | % of kg BW |
|---|---|---|---|---|---|---|---|---|---|---|---|---|---|---|

**Additional requirements for milk production per kg at different fat percentages**
(including requirements for nursing single, twin or triplet kids at the respective milk production level)

**(% Fat)**

| | | | | | | | | | | | | | | |
|---|---|---|---|---|---|---|---|---|---|---|---|---|---|---|
| 2.5 | 333 | 1.47 | 1.20 | 0.68 | 59 | 42 | 2 | 1.4 | 3.8 | 760 | | | | |
| 3.0 | 337 | 1.49 | 1.21 | 0.68 | 64 | 45 | 2 | 1.4 | 3.8 | 760 | | | | |
| 3.5 | 342 | 1.51 | 1.23 | 0.69 | 68 | 48 | 2 | 1.4 | 3.8 | 760 | | | | |
| 4.0 | 346 | 1.53 | 1.25 | 0.70 | 72 | 51 | 3 | 2.1 | 3.8 | 760 | | | | |
| 4.5 | 351 | 1.55 | 1.26 | 0.71 | 77 | 54 | 3 | 2.1 | 3.8 | 760 | | | | |
| 5.0 | 356 | 1.57 | 1.28 | 0.72 | 82 | 57 | 3 | 2.1 | 3.8 | 760 | | | | |
| 5.5 | 360 | 1.59 | 1.29 | 0.73 | 86 | 60 | 3 | 2.1 | 3.8 | 760 | | | | |
| 6.0 | 365 | 1.61 | 1.31 | 0.74 | 90 | 63 | 3 | 2.1 | 3.8 | 760 | | | | |

**Additional requirements for mohair production by Angora at different production levels**

**Annual fleece yield (kg)**

| | | | | | | |
|---|---|---|---|---|---|---|
| 2 | 16 | 0.07 | 0.06 | 0.03 | 9 | 6 |
| 4 | 34 | 0.15 | 0.12 | 0.07 | 17 | 12 |
| 6 | 50 | 0.22 | 0.18 | 0.10 | 26 | 18 |
| 8 | 66 | 0.29 | 0.24 | 0.14 | 34 | 24 |

a.  1 kg = 2.2 lb.

b.  100 g TDN = 0.44 Mcal DE.
    1 Mcal = 1000 kcal.
    1 Kcal = 4.184 kg.
    1 kg European Starch Equivalent = 5.082 Mcal DE.

*Source:* Haenlein, G.F.W. (1986) *Dietary Nutrient Allowances for Sheep and Goats. Feedstuffs 58.*
Reproduced with kind permission of G.F.W. Haenlein, University of Delaware

# Feed and Nutrient Requirements – Non-Ruminants

### TABLE 284 a   Feeding Standards for Pigs

**a. Typical dietary nutrient levels for growing pigs (fresh basis)** *

| | Liveweight (kg) | | |
|---|---|---|---|
| **Component** | **20** | **50** | **90** |
| Feed (kg/day | 1.2 | 2.2 | 2.4 |
| Digestible energy (MJ/kg) | 14.0 | 13.5 | 13.0 |
| Crude protein (g/kg) | 220 | 180 | 140 |
| Ideal protein (g/kg) | 194 | 149 | 94 |
| Lysine (g/kg) | 13.6 | 10.4 | 6.6 |
| Methionine + cystine (g/kg) | 6.8 | 5.2 | 3.3 |
| Threonine (g/kg) | 8.2 | 6.3 | 4.0 |
| Tryptophan (g/kg) | 1.9 | 1.5 | 1.0 |
| Calcium (g/kg) | 9.8 | 8.1 | 7.8 |
| Phosphorus (g/kg) | 7.0 | 6.1 | 5.9 |
| Salt (g/kg) | 3.2 | 3.1 | 3.0 |
| Iron (mg/kg) | 62 | 59 | 57 |
| Magnesium (mg/kg) | 308 | 230 | 220 |
| Zinc (mg/kg) | 56 | 49 | 47 |
| Copper (mg/kg) | 5.6 | 5.4 | 5.2 |
| Manganese (mg/kg) | 11 | 11 | 11 |
| Iodine (mg/kg) | 0.15 | 0.15 | 0.15 |
| Selenium (mg/kg) | 0.15 | 0.15 | 0.15 |
| Vitamin A.(i.u./kg) | 8000 | 6000 | 6000 |
| Vitamin D.(i.u./kg) | 1000 | 750 | 750 |
| Vitamin E.(i.u./kg) | 15 | 15 | 15 |
| Thiamin (mg/kg) | 2.0 | 1.5 | 1.5 |
| Riboflavin (mg/kg) | 3.0 | 3.0 | 3.0 |
| Nicotinic acid (mg/kg) | 15 | 15 | 15 |
| Pantothenic acid (mg/kg) | 10 | 10 | 10 |
| Pyridoxine (mg/kg) | 2.5 | 2.5 | 2.5 |
| Choline (mg/kg) | 1000 | 1000 | 1000 |
| Biotin (mg/kg) | 0.2 | 0.2 | 0.2 |
| Vitamin $B_{12}$(mg/kg) | 0.01 | 0.01 | 0.01 |

* Growth rate 0.7 kg/day

**TABLE 284 b   Feeding Standards for Pigs**

**b. Typical dietary nutrient levels for breeding sows (fresh basis)**

| Component | Pregnancy | Lactation 5 - 8 weeks weaning 11 piglets | 9 piglets | 7 piglets | 3 weeks weaning All piglets |
|---|---|---|---|---|---|
| Feed (kg/day) | 2.0 | 5.9 | 5.2 | 4.4 | 5.2 |
| Digestible energy (MJ/kg) | 13.0 | 13.0 | 13.0 | 13.0 | 13.0 |
| Crude protein (g/kg) | 130 | 160 | 160 | 160 | 170 |
| Lysine (g/kg) | 4.5 | 7.0 | 7.0 | 7.0 | 8.0 |
| Methionine + cystine (g/kg) | 3.1 | 3.8 | 3.8 | 3.8 | 4.4 |
| Threonine (g/kg) | 3.8 | 4.9 | 4.9 | 4.9 | 5.6 |
| Isoleucine (g/kg) | 3.9 | 4.9 | 4.9 | 4.9 | 5.6 |
| Leucine (g/kg) | 3.4 | 8.1 | 8.1 | 8.1 | 9.2 |
| Tryptophan (g/kg) | 0.7 | 1.3 | 1.3 | 1.3 | 1.5 |
| Calcium (g/kg) | 8.5 | 8.5 | 8.5 | 8.5 | 8.5 |
| Phosphorus (g/kg) | 6.5 | 6.5 | 6.5 | 6.5 | 6.5 |
| Salt (g/kg) | 3.0 | 3.0 | 3.0 | 3.0 | 3.0 |
| Iron (mg/kg) | 60 | 60 | 60 | 60 | 60 |
| Zinc (mg/kg) | 50 | 50 | 50 | 50 | 50 |
| Copper (mg/kg) | 6 | 6 | 6 | 6 | 6 |
| Manganese (mg/kg) | 16 | 16 | 16 | 16 | 16 |
| Iodine (mg/kg) | 0.5 | 0.5 | 0.5 | 0.5 | 0.5 |
| Selenium (mg/kg) | 0.15 | 0.15 | 0.15 | 0.15 | 0.15 |
| Vitamin A.(i.u./kg) | 8000 | 8000 | 8000 | 8000 | 8000 |
| Vitamin D.(i.u./kg) | 1000 | 1000 | 1000 | 1000 | 1000 |
| Vitamin E (i.u./kg) | 15 | 15 | 15 | 15 | 15 |
| Thiamin (mg/kg) | 1.5 | 1.5 | 1.5 | 1.5 | 1.5 |
| Riboflavin (mg/kg) | 3.0 | 3.0 | 3.0 | 3.0 | 3.0 |
| Nicotinic acid (mg/kg) | 15 | 15 | 15 | 15 | 15 |
| Pantothenic acid (mg/kg) | 10 | 10 | 10 | 10 | 10 |
| Pyridoxine (mg/kg) | 1.5 | 1.5 | 1.5 | 1.5 | 1,5 |
| Choline (mg/kg) | 1500 | 1500 | 1500 | 1500 | 1500 |
| Biotin (mg/kg) | 0.30 | 0.30 | 0.30 | 0.30 | 0.30 |
| Vitamin $B_{12}$ (mg/kg) | 0.015 | 0.015 | 0.015 | 0.015 | 0.015 |

*Source:* McDonald, P., Edwards R.A. & Greehalgh, J.F.D. (1988) *Animal Nutrition.*
Reproduced with kind permission of Longman Group UK Ltd.

**TABLE 285 a & c   Nutrient Guide for 'Cobb 500' Broiler Breeding Stock**

**a. Recommended Nutrient Levels**

| Nutrient | | Chick 0 - 42 days | Grower 43 - 119 days | Pre-Breeder 120 - 147 days | 1 Breeder 148 - 280 days | 2 Breeder 280 days - depletion |
|---|---|---|---|---|---|---|
| Protein | % | 18 | 15 | 15.5 | 16 | 16 |
| Lysine | % | 0.90 | 0.60 | 0.70 | 0.75 | 0.75 |
| Methionine | % | 0.40 | 0.24 | 0.30 | 0.36 | 0.34 |
| Methionine + Cystine | % | 0.72 | 0.50 | 0.56 | 0.65 | 0.60 |
| Tryptophan | % | 0.20 | 0.16 | 0.17 | 0.19 | 0.19 |
| Calcium | % | 1.0 | 1.2 | 1.5 | 2.8 | 3.2 |
| Available Phosphorus | % | 0.45 | 0.33 | 0.40 | 0.33 | 0.30 |
| Salt | % | 0.36 | 0.36 | 0.36 | 0.36 | 0.36 |
| Sodium | % | 0.15 | 0.15 | 0.15 | 0.15 | 0.15 |
| MEQ/100g* | | 20 | 20 | 20 | 20 | 20 |
| Linoleic Acid | | 1.25 | 1.0 | 1.35 | 1.50 | 1.00 |
| ME: MJ/kg | | 11.70 | 11.20 | 11.50 | 11.6 - 11.70 | 11.6 - 11.70 |
| kcal/kg | | 2796 | 2677 | 2750 | 2772 - 2800 | 2772 - 2800 |
| kcal/lb | | 1268 | 1214 | 1250 | 1258 - 1270 | 1258 - 1270 |
| Vitamin premix | | Chick | Grower | Breeder | Breeder | Breeder |

\* An expression of acid-base balance where sodium + potassium minus chloride is greater than 20 MEQ/100g.

**c. Feed Allowances relative to production and energy levels**

Daily feed allowance (grams/bird) based on changes in breeder ration energy level and rate of egg production

| Breeder Ration Energy ME:kcal/kg | 85% | 80% | 75% | 70% | 65% | 60% | 55% | 50% | Breeder Ration Energy ME:MJ/kg |
|---|---|---|---|---|---|---|---|---|---|
| 2970 | 159 | 157 | 152 | 150 | 148 | 146 | 144 | 142 | 12.43 |
| 2860 | 166 | 163 | 159 | 157 | 154 | 152 | 150 | 149 | 11.96 |
| 2800 | 169 | 166 | 163 | 160 | 157 | 155 | 153 | 151 | 11.71 |
| 2640 | 179 | 177 | 172 | 170 | 168 | 166 | 163 | 160 | 11.04 |
| 2530 | 186 | 184 | 179 | 177 | 174 | 172 | 169 | 167 | 10.58 |
| Calorie intake at each ration ME level | 1.98 / 473 | 1.95 / 467 | 1.91 / 456 | 1.87 / 448 | 1.83 / 443 | 1.82 / 437 | 1.79 / 428 | 1.77 / 423 | MJ/bird/day kcal/bird/day |

**TABLE 285 b & d   Nutrient Guide for 'Cobb 500' Broiler Breeding Stock**

### b. Recommended Vitamin and Trace Element Levels

(divided into those for feeds composed mainly of maize or wheat since the base levels provided by these grains may vary significantly)

| Micro-Nutrient | Unit | Chick | | Grower | | Pre-Breeder Breeder 1 - 2 | |
|---|---|---|---|---|---|---|---|
| | | maize | wheat | maize | wheat | maize | wheat |
| Vitamin A | m.iu | 7 | 8 | 7 | 8 | 12 | 13 |
| $D_3$ | m.iu | 3 | 3 | 3 | 3 | 3 | 3 |
| E | k.iu | 30 | 30 | 20 | 20 | 80 | 80 |
| K | g | 2 | 2 | 2 | 2 | 4 | 4 |
| $B_1$ | g | – | – | – | – | 1 | 2 |
| $B_2$ | g | 3 | 3 | 3 | 3 | 12 | 12 |
| Pantothenic Acid | g | 7 | 5 | 5 | 3 | 25 | 20 |
| Nicotinic Acid | g | 25 | 20 | 15 | 10 | 40 | 30 |
| $B_2$ | g | 1 | 3 | – | 2 | 3 | 6 |
| $B_{12}$ | mg | 10 | 10 | 10 | 10 | 15 | 15 |
| Folic Acid | g | 0.5 | 0.5 | – | – | 3 | 3 |
| Biotin | mg | 30 | 80 | 20 | 50 | 150 | 300 |
| Choline | g | 200 | 100 | – | – | 200 | 100 |
| Cobalt | g | 0.4 | 0.4 | 0.4 | 0.4 | 0.6 | 0.6 |
| Copper | g | 10 | 10 | 10 | 10 | 10 | 10 |
| Iodine | g | 0.5 | 0.5 | 0.5 | 0.5 | 2.5 | 2.5 |
| Iron | g | 15 | 15 | 15 | 15 | 70 | 70 |
| Manganese | g | 100 | 100 | 100 | 100 | 100 | 100 |
| Molybdenum | g | – | – | – | – | 0.5 | 0.5 |
| Selenium | g | 0.20 | 0.20 | 0.15 | 0.15 | 0.25 | 0.25 |
| Zinc | g | 80 | 80 | 80 | 80 | 100 | 100 |

(Levels per tonne of finished feed)

### d. Typical UK male ration

| | | | | | |
|---|---|---|---|---|---|
| Protein | % | 11.0 - 12.0 | Salt | % | 0.36 - 0.40 |
| Lysine | % | 0.50 | Sodium | % | 0.15 - 0.17 |
| Methionine | % | 0.26 | Linoleic Acid | % | 1.26 - 1.60 |
| Methionine + Cystine | % | 0.44 | ME: MJ/kg | % | 11.5 |
| Tryptophan | % | 0.13 | kcal/kg | | 2750 |
| Calcium | % | 0.80 - 1.0 | kcal/lb | | 1250 |
| Available Phosphorus | % | 0.33 - 0.35 | | | |

*Note:* Males are fed a different ration to females by use of differently designed troughs.
Separate sex feeding gives improved fertility and feed consumption.

*Source:* Data kindly supplied and reproduced with kind permission of The Cobb Breeding Co., East Hanningfield, Chelmsford, Essex, UK ('Cobb 500' is a strain of broilers produced by this company).

A G R I   I N F O

**TABLE 286 a   Feeding Standards for Poultry. Typical Dietary Nutrient Levels (Fresh Basis)**

| a. Chickens | Growing chicks | | Pullets 12 - 18 wks | Laying hens | Breeding hens | Broiler starter | Broiler finisher |
|---|---|---|---|---|---|---|---|
| | 0 - 6 wks | 6 - 12 wks | | | | | |
| ME (MJ/kg) | 11.5 | 10.9 | 10.9 | 11.1 | 11.1 | 12.6 | 12.6 |
| Crude protein (g/kg) | 210 | 145 | 120 | 160 | 160 | 230 | 190 |
| Amino acids | | | | | | | |
|   Arginine | 11 | 7.1 | 6.7 | 4.9 | 4.9 | 12.6 | 9.5 |
|   Glycine + serine | 13.2 | 9.4 | 8.0 | – | – | 12.0 | 11.0 |
|   Histidine | 5.1 | 3.3 | 2.4 | 1.6 | 1.6 | 5.0 | 5.0 |
|   Isoleucine | 9 | 5.9 | 4.5 | 5.3 | 5.3 | 9.0 | 8.0 |
|   Leucine | 14.7 | 9.9 | 8.4 | 6.6 | 6.6 | 16.0 | 13.0 |
|   Lysine | 11 | 7.4 | 6.6 | 7.3 | 7.3 | 12.5 | 10.0 |
|   Methionine + Cystine | 9.2 | 6.2 | 4.5 | 5.5 | 4.6 | 9.2 | 8.0 |
|   Phenylalanine + Tyrosine | 15.8 | 10.8 | 8.0 | 7.0 | 7.0 | 15.8 | 14.0 |
|   Threonine | 7.4 | 4.9 | 4.2 | 3.5 | 3.5 | 8.0 | 6.5 |
|   Tryptophan | 2.0 | 1.4 | 1.2 | 1.4 | 1.4 | 2.3 | 1.9 |
|   Valine | 10.4 | 6.6 | 5.3 | 5.3 | 5.3 | 10 | 9.0 |
| Major minerals (g/kg) | | | | | | | |
|   Calcium | 12 | 10 | 8 | 35 | 33 | 12 | 10 |
|   Phosphorus (av.) | 5 | 5 | 5 | 5 | 5 | 5 | 5 |
|   Magnesium | 0.3* | 0.3* | 0.3* | 0.3* | 0.3* | 0.3* | 0.3* |
|   Sodium | 1.5 | 1.5 | 1.5 | 1.5 | 1.5 | 1.5 | 1.5 |
|   Potassium | 3.0 | – | – | – | – | 3.0 | 3.0 |
| Trace minerals (mg/kg) | | | | | | | |
|   Copper | 3.5* | 3.5* | 3.5* | 3.5* | 3.5* | 3.5* | 3.5* |
|   Iodine | 0.4* | 0.4* | 0.4* | 0.4* | 0.4* | 0.4* | 0.4* |
|   Iron | 80* | 80* | 80* | 80* | 80* | 80* | 45* |
|   Manganese | 100* | 100* | 100* | 100* | 100* | 100* | 100* |
|   Zinc | 50* | 50* | 50* | 50* | 50* | 50* | 50* |
|   Selenium | 0.15 | – | – | – | – | 0.15 | 0.15 |
| Vitamins (i.u/kg) | | | | | | | |
|   A | 2000* | 2000* | 2000* | 6000* | 6000* | 2000* | 2000* |
|   $D_3$ | 600* | 600* | 600* | 800* | 800* | 600* | 600* |
|   E | 25* | 25* | 25* | 25* | 25* | 25* | 25* |
| Vitamins (mg\kg) | | | | | | | |
|   K | 1.3* | 1.3* | 1.3* | 1.3* | 1.3* | 1.3* | 1.3* |
|   Thiamin | 3 | – | – | – | 2 | 3 | – |
|   Riboflavin | 4* | 4* | 4* | 4* | 4* | 4* | 4* |
|   Nicotinic acid | 28* | 28* | 28* | 28* | 28* | 28* | 28* |
|   Pantothenic acid | 10* | 10* | 10* | 10* | 10* | 10* | 10* |
|   Choline | 1300 | – | – | – | 1100 | 1300 | 1300 |
|   Vitamin $B_{12}$ | – | – | – | – | 0.01 | – | – |

* Added as supplement

**TABLE 286 b   Feeding Standards for Poultry. Typical Dietary Nutrient Levels (Fresh Basis)**

| b.Turkeys | Growing turkeys | | | Breeding turkeys |
|---|---|---|---|---|
| | 0 - 6 wks | 6 - 12 wks | 12 + wks | |
| ME (MJ/kg) | 12.6 | 11.9 | 11.9 | 11.3 |
| Crude protein (g/kg) | 300 | 260 | 180 | 160 |
| Amino acids | | | | |
| Arginine | 16 | 13 | 8 | 5 |
| Glycine + serine | 9 | – | 7 | – |
| Histidine | 6 | 5 | 3.5 | 2 |
| Isoleucine | 11 | 10 | 5.5 | 5.5 |
| Leucine | 19 | 15 | 8 | 7 |
| Lysine | 17 | 13 | 8 | 7.5 |
| Methionine + Cystine | 10 | 8 | 6 | 5.5 |
| Phenylalanine + Tyrosine | 16 | 15 | 10 | 8 |
| Threonine | 10 | 9 | 5.5 | 4 |
| Tryptophan | 2.6 | 2.3 | 1.5 | 1.7 |
| Valine | 12 | 10 | 6 | 5 |
| Major minerals (g/kg) | | | | |
| Calcium | 9 | 10 | 8 | 30 |
| Phosphorus (av.) | 4.5 | 5 | 4 | 5 |
| Magnesium | 0.36* | 0.36* | 0.36* | 0.30* |
| Sodium | 1.75 | 1.75 | 1.75 | 1.75 |
| Potassium | – | – | – | – |
| Trace minerals (mg/kg) | | | | |
| Copper | 4.2* | 4.2* | 4.2* | 3.5* |
| Iodine | 0.48* | 0.48* | 0.48* | 0.4* |
| Iron | 96* | 96* | 96* | 80* |
| Manganese | 120* | 120* | 120* | 100* |
| Zinc | 60* | 60* | 60* | 50* |
| Selenium | 0.2 | 0.15 | 0.15 | – |
| Vitamins (i.u./kg) | | | | |
| A | 12 000* | 12 000* | 12 000* | 10 000* |
| D$_3$ | 1800* | 1800* | 1800* | 1500* |
| E | 36* | 36* | 36* | 30* |
| Vitamins (mg/kg) | | | | |
| K | 4.8* | 4.8* | 4.8* | 4.0* |
| Thiamin | 4.8* | 4.8* | 4.8* | 4.0* |
| Riboflavin | 12* | 12* | 12* | 10* |
| Nicotinic acid | 60* | 60* | 60* | 50* |
| Pantothenic acid | 19.2* | 19.2* | 19.2* | 16* |
| Pyridoxine | 6* | 6* | 6* | 5* |
| Biotin | 0.12* | 0.12* | 0.12* | 0.1* |
| Folic acid | 2.4* | 2.4* | 2.4* | 2.0* |
| Vitamin B$_{12}$ | 0.024* | 0.024* | 0.024* | 0.02* |
| Choline | 1760 | – | – | 1350 |

* Added as supplement

*Source:* McDonald, P., Edwards R.A., & Greehalgh, J.F.D. (1988) *Animal Nutrition.*
Reproduced with kind permission of Longman Group UK Ltd.

**TABLE 287 a & b   Feed Requirements of Hybrid 'Isabrown' Hen Layers**

**a. Controlled Feed Programme 0 - 18 Weeks of Age**

| | | Energy Kcal | Daily Intake Protein | Daily Feed | Cumulative Consumption | Average Bodyweight |
|---|---|---|---|---|---|---|
| **CHICK FEED – 2850 Kcal/kg (11.9 MJ)** | | | | | | |
| **19% Crude Protein** | | | g | g | g | g |
| Wk | Days | | | | | |
| 1 | 1 - 7 ad - lib | 34 | 2.3 | 12 | 84 | 65 |
| 2 | 8 - 14 ad - lib | 57 | 3.8 | 20 | 224 | 125 |
| 3 | 15 - 21 ad - lib | 74 | 4.9 | 26 | 406 | 210 |
| 4 | 22 - 28 ad - lib | 88 | 5.9 | 31 | 623 | 300 |
| 5 | 29 - 35 ad - lib | 103 | 6.8 | 36 | 875 | 390 |
| 6 | 36 - 42 ad - lib | 117 | 7.8 | 41 | 1162 | 480 |
| 7 | 43 - 49 ad - lib | 131 | 8.7 | 46 | 1484 | 570 |
| 8 | 50 - 56 ad - lib | 145 | 9.7 | 51 | 1841 | 660 |
| 9 | 57 - 63 controlled | 154 | 10.8 | 57 | 2240 | 745 |
| 10 | 64 - 70 controlled | 165 | 11.6 | 61 | 2667 | 835 |
| **GROWER FEED - 2700 Kcal/kg (11.3 MJ)** | | | | | | |
| **15.5% Crude Protein** | | | | | | |
| 11 | 71 - 77 controlled | 173 | 9.9 | 64 | 3115 | 925 |
| 12 | 78 - 84 controlled | 181 | 10.4 | 67 | 3584 | 1015 |
| 13 | 85 - 91 controlled | 189 | 10.9 | 70 | 4074 | 1100 |
| 14 | 92 - 98 controlled | 197 | 11.3 | 73 | 4585 | 1180 |
| 15 | 99 - 105 controlled | 205 | 11.8 | 76 | 5117 | 1260 |
| 16 | 106 - 112 controlled | 213 | 12.2 | 79 | 5670 | 1340 |
| 17 | 113 - 119 ad - lib | 221 | 12.7 | 82 | 6244 | 1420 |
| 18 | 120 - 126 ad - lib | 232 | 13.3 | 86 | 6848 | 1500 |

Bodyweights are intended as a guide; the average may vary from the figures above without adversely affecting performance. If bodyweight is below target at 10 weeks, delay change to grower feed.

**b. Daily Nutrient requirements per bird in cages (house temp 20°C)**

| | | **1.** 18 - 40 wks | **2.** 40 - 55 wks | **3.** After 55 wks |
|---|---|---|---|---|
| Met Energy | Kcal | 325 | 325 | 325 |
| Crude Protein | g | 19 | 18.5 | 18 |
| Methionine | mg | 410 | 395 | 380 |
| Meth & Cystine | mg | 740 | 710 | 680 |
| Lysine | mg | 860 | 830 | 800 |
| Threonine | mg | 570 | 550 | 530 |
| Tryptophan | mg | 190 | 180 | 170 |

The recommended layer feed will provide the necessary daily nutrient requirements (Layer 1 to 40 weeks of age, Layer 2 from 40 weeks to 55 and Layer 3 from 55 weeks to end of lay) at feed intakes of 116g, 117g and 118g respectively.

Feeding less than 18g of protein per day towards the end of the laying period is not recommended. The effect would be detrimental to those birds that are maintaining a high rate of egg production.

*Source:* 1991 data supplied and reproduced with kind permission of ISA Poultry Services Ltd., Green Road, Eye, Peterborough, UK.

**TABLE 288   Food Consumption and Body Weight Guide to New Zealand White Rabbits**

| Age (weeks) | Weekly Food Intake (kg) | | | Body Weight | |
|---|---|---|---|---|---|
| | Doe and Litter (kg) | Litter (kg) | Per Young Rabbit (g) | Total Litter (kg) | Per Young Rabbit (g) |
| Kindling to: | | | | | |
| 1st week | 1.91 | – | – | 0.45 | 56.7 |
| 2nd week | 2.29 | – | – | 1.09 | 136.2 |
| 3rd week | 2.31 | – | – | 1.91 | 239.0 |
| 4th week | 3.10 | 0.91 | 313 | 4.40 | 550.0 |
| 5th week | 5.24 | 2.32 | 417 | 7.08 | 885.0 |
| 6th week | 7.00 | 4.35 | 378 | 9.80 | 1226.0 |
| 7th week | 8.02 | 5.02 | 428 | 12.35 | 1544.0 |
| 8th week | 9.31 | 6.31 | 428 | 15.61 | 1952.2 |

*Source:* Portsmouth, J. (1979) *Commercial Meat Production.*
Reproduced with kind permission of Nimrod Press.

**TABLE 289   Suggested Nutrient Levels for Rabbits**

| Nutrient | Classification | | |
|---|---|---|---|
| | Adults, non pregnant does, early pregnancy | Late pregnant does, Lactating does 7 Litter | Growing Rabbits |
| Protein % | 12 - 16 | 17 - 18 | 17 - 18 |
| Energy T.D.N. | 65 | 70 - 80 | 80 |
| Energy MJ/kg | 11.42 | 12.30 - 14.06 | 14.06 |
| Fat % | 2 - 4 | 2 - 6 | 2 - 6 |
| Fibre % | 12 - 14 | 10 - 12 | 10 - 12 |
| Calcium % | 1.0 | 1.0 - 1.2 | 1.0 - 1.2 |
| Phosphorus % | 0.40 | 0.40 - 0.80 | 0.40 - 0.80 |
| Salt % | 0.50 | 0.65 | 0.65 |
| Magnesium % | 0.25 | 0.25 | 0.25 |
| Potassium % | 1.0 | 1.5 | 1.5 |
| Manganese mg/kg | 30 | 50 | 50 |
| Zinc mg/kg | 20 | 30 | 30 |
| Iron mg/kg | 100 | 100 | 100 |
| Copper mg/kg | 10 | 10 | 10 |
| Metrionive % | 0.50 | 0.56 | 0.56 |
| Lysine % | 0.60 | 0.80 | 0.80 |
| Vitamin A in/kg | 8000 | 9000 | 9000 |
| Vitamin D$_3$ in/kg | 2000 | 2000 | 2000 |
| Vitamin E in/kg | 20 | 40 | 40 |
| Vitamin K mg/kg | 1.0 | 1.0 | 1.0 |
| Nicotinic Acid mg/kg | 30 | 50 | 50 |
| Chlorine mg/kg | 1300 | 1300 | 1300 |
| B12 mcg/kg | 10 | 10 | 10 |
| B6 mg/kg | 1.0 | 1.0 | 1.0 |

*Source:* Portsmouth, J. (1979) *Commercial Meat Production.*
Reproduced with kind permission of Nimrod Press.

**TABLE 290   Digestible Energy (DE) and TDN Values for Some Rabbit Feed Ingredients**

|  | DE Kcals/kg | TDN % |  | DE Kcals/kg | TDN % |
|---|---|---|---|---|---|
| Grass meal | 2610 | 59 | Ground Nut Meal | 4120 | 90 |
| Dried Bakery Waste | 4190 | 101 | Rye | 3590 | 77 |
| Barley | 3330 | 75 | Sorghum | 3330 | 82 |
| Dried Sugar Beet | 3080 | 70 | Soya Bean Meal | 3770 | 82 |
| Maize | 3790 | 83 | Wheat | 3680 | 79 |
| Cotton Seed Meal | 3090 | 67 |  |  |  |

*Source:* Portsmouth, J. (1979) *Commercial Meat Production.*
Reproduced with kind permission of Nimrod Press.

**TABLE 291   Daily Energy Requirements of Horses for Various Functions and Amounts of Hay and**

|  | Mature bodyweight | | | | | |
|---|---|---|---|---|---|---|
|  | 200 kg | | | 400 kg | | |
|  | Digestible energy (MJ) | Hay * (kg) | Concentrate mixture † (kg) | Digestible energy (MJ) | Hay * (kg) | Concentrate mixture † (kg) |
| **Mature Horse ‡**, maintenance | 34.4 | 4.2 | – | 58.2 | 7.1 | – |
| **Mares,** last 90 days of gestation | 38.8 | 3.2 | 1.1 | 67.4 | 5.3 | 2.1 |
| **Lactating mare,** first 3months | 69.7 (12) § | 2.8 | 4.1 | 103.2 (18) § | 4.8 | 5.6 |
| **Lactating mare,** 3 months to weaning | 55.4 | 3.7 | 2.2 | 86.2 | 6.2 | 3.1 |
| **Stallion:** |  |  |  |  |  |  |
| Breeding | 54.2 | 3.0 | 2.6 | 79.9 | 3.9 | 4.2 |
| Non-breeding | 45.5 | 3.6 | 1.4 | 64.1 | 4.9 | 2.1 |
| **Weanling** (6 months old) | 35.6 | 1.0 | 2.4 | 55.8 | 1.8 | 3.6 |
| **Yearling** (12 months old) | 35.8 | 2.0 | 1.7 | 58.5 | 3.1 | 2.9 |
| **Long yearling** (18 months old) | 34.4 | 2.8 | 1.0 | 61.2 | 4.4 | 2.2 |
| **Two-year old** |  |  |  |  |  |  |
| (24 months old excluding work) | 34.4 | 2.8 | 1.0 | 61.2 | 4.4 | 2.2 |
| **Maintenance plus 1 h** |  |  |  |  |  |  |
| moderately hard work | 54.0 | 2.8 | 2.7 | 98.0 | 5.0 | 5.0 |

\*    Hay containing 8.2 MJ/kg and 88% dry matter.

†    Concentrate mixture containing 11.4 MJ/kg and 88% dry matter. Quantities of concentrates up to
     1.5% bodyweight daily may be fed if a minimum roughage allowance of 1kg/100 bodyweight daily
     may be fed if a minimum roughage allowance of 1kg/100 kg bodyweight is given.

‡    2.3 kg extra feed daily should produce 0.5 - 0.6 kg. gain.

§    Assumed peak daily milk yield (kg).

## Concentrates needed to Provide the Energy

| | | | | | | |
|---|---|---|---|---|---|---|
| **Mature bodyweight** | | | | | | |
| **500 kg** | | | **600 kg** | | | |
| Digestible energy (MJ) | Hay * (kg) | Concentrate mixture † (kg) | Digestible energy (MJ) | Hay * (kg) | Concentrate mixture † (kg) | |
| 69.7 | 8.5 | – | 79.5 | 9.7 | – | **Mature Horse ‡,** maintenance |
| 80.5 | 6.2 | 2.6 | 89.8 | 7.2 | 2.7 | **Mares,** last 90 days of gestation |
| 122.2 (21) § | 6.0 | 6.4 | 141.1 (23) § | 7.2 | 7.2 | **Lactating mare,** first 3months |
| 103.4 | 7.6 | 3.6 | 120.4 | 8.7 | 4.3 | **Lactating mare,** 3 months to weaning |
| | | | | | | **Stallion:** |
| 97.2 | 4.9 | 5.0 | 113.7 | 5.8 | 5.8 | Breeding |
| 77.7 | 6.0 | 2.5 | 90.5 | 7.0 | 2.9 | Non-breeding |
| 67.1 | 2.2 | 4.3 | 74.7 | 2.3 | 4.9 | **Weanling** (6 months old) |
| 71.4 | 3.7 | 3.6 | 80.4 | 4.1 | 4.1 | **Yearling** (12 months old) |
| 73.4 | 5.2 | 2.7 | 80.6 | 5.8 | 2.9 | **Long yearling** (18 months old) |
| | | | | | | **Two-year old** |
| 71.1 | 5.2 | 2.5 | 82.9 | 5.8 | 3.1 | (24 months old excluding work) |
| | | | | | | **Maintenance plus 1 h** |
| 119.0 | 5.2 | 6.7 | 147.0 | 5.8 | 8.7 | moderately hard work |

(Based on National Research Council, 1978)

*Source:* Frape, D. (1986) *Equine Nutrition and Feeding.*
Reproduced with kind permission of Longman Group UK Ltd.

TABLE 292   Digestible Energy (DE) Demands of Maintenance and Work* for Horses on the Flat

| | 200 | 400 | 600 |
|---|---|---|---|
| Bodyweight (kg) | 200 | 400 | 600 |
| Approx. feed capacity per day (MJ of DE) | 60 | 100 | 150 |
| Maintenance requirement per day (MJ of DE) | 35 | 58 | 79 |

| | Energy requirements for work above maintenance (MJ of DE) [†] | | |
|---|---|---|---|
| Walking (1 hour) | 0.4 | 0.8 | 1.3 |
| Slow trotting, some cantering (1 hour) | 4.2 | 8.4 | 12.5 |
| Fast trotting, cantering, some jumping (1 hour) | 10.5 | 20.9 | 31.4 |
| Cantering, galloping, jumping (1 hour) | 25.0 | 50.0 | 75.0 |
| Strenuous effort, racing, polo (1 hour) | 42.0 | 85.0 | 127.0 |
| Slow trotting, some cantering (10.4 h,100 km ) | | | |
| calculated from above | 43.5 | 87.0 | 130.5 |

\*   Based on National Research Council (1978) and more recent evidence

†   1 kg concentrate provides about 12 MJ of DE.

*Source:* Frape, D. (1986) *Equine Nutrition and Feeding.*
Reproduced with kind permission of Longman Group UK Ltd.

TABLE 293   Effect of a Range of Required Energy Densities (MJ of DE per kg air-dry feed) on the Cereal Content of the Daily Ration of Horses when Hays of Two Energy Contents are Available

| Energy density of ration required | Oats (%) | | Barley (%) | |
|---|---|---|---|---|
| | 7.2 * | 7.8 * [†] | 7.2 * | 7.8 * [†] |
| 7.5 | 7 | 0 | 5 | 0 |
| 8.0 | 19 | 5 | 14 | 4 |
| 8.5 | 30 | 19 | 23 | 14 |
| 9.0 | 42 | 32 | 32 | 24 |
| 9.5 | 54 | 46 | 41 | 34 |
| 10.0 | 65 | 60 | 50 | 44 |
| 10.5 | 77 | 73 | 59 | 54 |
| 11.0 | 88 | 86 | 68 | 64 |

\*   Energy content of hay (MJ of DE per kg) 7.2 MJ/kg, medium quality; 7.8 MJ/kg, good quality.

†   Hay can be assumed to contain 86% dry matter and where haylage of 45% dry matter is to be used it may be substituted for the hay of 7.8 MJ of DE in the proportions 1.8 - 1.9 kg haylage per 1 kg hay. Similarly, 1.6 - 1.7 kg haylage of 50% dry matter could be used.

*Source:* Frape, D. (1986) *Equine Nutrition and Feeding.*
Reproduced with kind permission of Longman Group UK Ltd.

**TABLE 294    Nutrient Concentration in Diets for Horses and Ponies Expressed on the Basis of 90 per cent Dry Matter**

|  | Crude protein (g/kg) | Calcium (g/kg) | Phosphorus (g/kg) |
|---|---|---|---|
| Mature horses and ponies at maintenance | 80 | 3.2 | 2.0 |
| Mare, last 90 days of gestation | 100 | 4.5 | 3.0 |
| Lactating mare, first 3 months | 125 | 4.5 | 3.0 |
| Lactating mare, 3 months to weaning | 110 | 4.0 | 2.5 |
| Creep feed | 160 | 8.0 | 5.5 |
| Foal (3 months old) | 160 | 8.0 | 5.5 |
| Weanling (6 months old) | 145 | 6.0 | 4.5 |
| Yearling (12 months old) | 120 | 5.0 | 3.5 |
| Long yearling (18 months old) | 100 | 4.0 | 3.0 |
| Two-year-old (light training) | 90 | 4.0 | 3.0 |
| Mature working horse, light to intense work | 80 | 3.2 | 2.0 |

(Based on National Research Council 1978)

*Source:* Frape, D. (1986) *Equine Nutrition and Feeding.*
Reproduced with kind permission of Longman Group UK Ltd.

**TABLE 295   Minerals and Vitamins per kg Diet Adequate for Horses**

| | Adequate levels | |
|---|---|---|
| | Maintenance of mature horses | Mare last 90 days of gestation and lactation and growing horses |
| Sodium (g) | 3.5 | 3.5 |
| Potassium (g) | 4.0 | 5.0 |
| Magnesium (g) | 0.9 | 1.0 |
| Sulphur (g) | 1.5 | 1.5 |
| Iron (mg) | 40 | 50 |
| Zinc (mg) | 60 | 60 |
| Manganese (mg) | 40 | 40 |
| Copper (mg) | 15 | 15 |
| Iodine (mg) | 0.1 | 0.2 |
| Cobalt (mg) | 0.1 | 0.1 |
| Selenium (mg) | 0.2 | 0.2 |
| Cholecalciferol ($\mu$g*) | 10 (400 iu) | 10 (400 iu) |
| Retinol (mg[†]) | 1.5 (5000 iu) | 2.0 (6666 iu) |
| D-$\alpha$-tocopherol (mg[‡]) | 15 | 15 |
| Thiamin (mg) | 3.0 | 3.0 |
| Riboflavin (mg) | 2.2 | 2.2 |
| Pantothenic acid (mg) | 12 | 12 |
| Available biotin (mg) | 0.2 | 0.2 |
| Folic acid (mg) | 1.0 | 1.0 |

(Based on National Research Council 1978)

\*   1 iu (international unit) is equal to biopotency of 0.025 $\mu$g of cholecalciferol (vitamin $D_3$) or ergocalciferol (vitamin $D_2$).

†   1 iu is equal to the biopotency of 0.3 $\mu$g of retinol (vitamin A alcohol). Grass carotene has 0.025 of value of vitamin A on a weight basis.

‡   1 iu vitamin E is the biopotency of 1 mg of DL-$\alpha$-tocopheryl. Where 50 g of supplementary fat of average composition are added per kg feed, the requirement rises to 25 - 30 mg of $\alpha$-tocopherol per kg, equivalent to 44 - 53 mg of DL-$\alpha$-tocopheryl acetate.

*Source:* Frape, D. (1986) *Equine Nutrition and Feeding.*
Reproduced with kind permission of Longman Group UK Ltd.

# Livestock Feed Analyses

## General Standards

---

### Note 15  General Information

---

The following section contains lists of feedstuff analyses from several sources as assessed for different species. They are far from comprehensive as this requires a large text in its own right. The reader seeking a broader range of analyses should initially consult the KEY texts on the subject and if these do not suffice then seek professional guidance.

Values of individual nutrient parameters may vary according to species of livestock to which the feed is to be fed. It is therefore critical that species appropriate tables are consulted. Analyses of specific feeds may vary between batches from different sources. A nutrient analysis is therefore specific to that batch. The tables contained herein should be considered as indicating average levels.

It is important to appreciate that many feeds have to be fed with caution and only up to a limited level in the diet. This is because they may have properties which cause undesirable side-effects and which are not apparent from a simple consideration of nutrient analysis of the feed. For example individual feeds may:

- **have undesirable effects on carcass composition**
- **contain toxins which limit the amount that can be included in the diet**
- **depress intakes of the whole diet when included above a certain level in the diet**

---

**TABLE 296  Classification of Metabolizability of Ruminant Diets**

| *Metabolizability 'q' | ME content MJ/kg DM | Type of Diet |
|---|---|---|
| 0.4 | 7 | Poor roughage |
| 0.5 | 9 | Average roughage |
| 0.6 | 11 | Roughage/concentrate mix |
| 0.7 | 13 | Concentrates |

$$* \text{ Metabolizability} = \frac{\text{Metabolizable Energy}}{\text{Gross Energy}}$$

---

**TABLE 297  Classification of Different Feed Types According to Level of Metabolizable Energy**

| Food type | ME (MJ/kg DM) |
|---|---|
| Straws | 4 - 7.2 |
| Hay and Dried grass | 6.3 - 10.8 |
| Silages | 7.4 - 12 |
| Concentrates | 10 - 14 |

**TABLE 298   Estimated Daily Intakes of a Range of Pasture Types**

| Pasture description | Example | Maximum possible daily intake as % liveweight |
| --- | --- | --- |
| Good digestibility >70% | Lush, temperate | 2.5 - 3 |
| Medium digestibility 65% | Tropical grass / legume | 2.5 |
| Average digestibility <65% | Tropical grass (pangola, kikuyu) | 2 - 2.5 |
| Poor digestibility <50% | Standover feed and native pasture | 1.5 - 2 |

*Source:* Chamberlain, A. (1989) *Milk Production in the Tropics.*
Reproduced with kind permission of Longman Group UK Ltd.

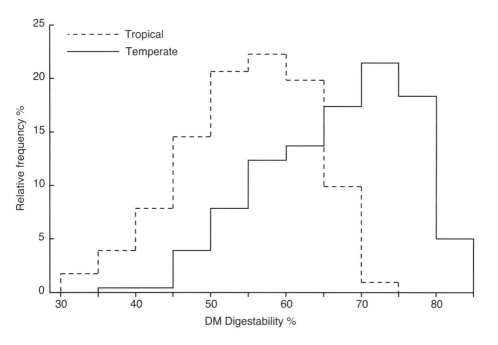

**Fig. 52** Frequency distribution of the dry matter digestibility of 543 cuts of tropical grasses and 592 temperate grasses

*Source:* Smith, A.J., ed. (1974) *Beef Cattle Production in Developing Countries* (after Minson & McLeod, 1970). Reproduced with kind permission of A.J. Smith.

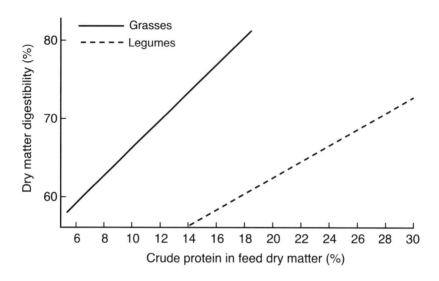

**Fig. 53** Relation between dry matter digestibility of grasses and legumes and the crude protein content of the feed (Equations of Minson and Brown, 1959)

*Source:* Minson, D.J. (1982) *Effect of Composition on Feed Digestibility and Metabolizable Energy. Nutrition Abstracts & Reviews.* 52.10. © CAB International, Wallingford, Oxon.

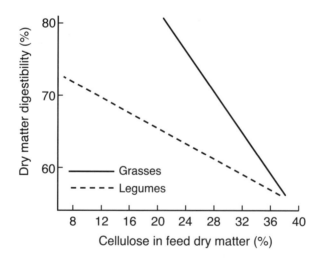

**Fig. 54** Relation between the dry matter digestibility of grasses and legumes and the cellulose content of the feed (Equations of Sullivan, 1964)

*Source:* Minson, D.J. (1982) *Effect of Composition on Feed Digestibility and Metabolizable Energy. Nutrition Abstracts & Reviews.* 52.10. © CAB International, Wallingford, Oxon.

**TABLE 299  Mean Values for Mineral Content in the Dry Matter of Tropical and Temperate Grasses and Legumes. Values in parentheses are the number of samples**

| | | Temperate | | | Tropical | | |
| | Grass | | Legume | | Grass | | Legume | |
|---|---|---|---|---|---|---|---|---|
| Phosphorus (%) | 0.33 | (400) | 0.36 | (320) | 0.22 | (586) | 0.26 | (165) |
| Calcium (%) | 0.59 | (428) | 1.86 | (291) | 0.40 | (390) | 1.21 | (154) |
| Magnesium (%) | 0.18 | (335) | 0.29 | (193) | 0.36 | (280) | 0.40 | (48) |
| Sodium (%) | 0.23 | (318) | 0.19 | (121) | 0.26 | (192) | 0.07 | (40) |
| Copper (ppm) | 6 | (127) | 12 | (93) | 15 | (94) | 10 | (17) |
| Zinc (ppm) | 32 | (31) | 55 | (34) | 36 | (119) | 42 | (7) |
| Cobalt (ppm) | 0.20 | (111) | 0.42 | (21) | 0.16 | (45) | 0.07 | (3) |

*Source:* Hacker, J.B. ed. (1981) *Nutritional Limits to Animal Production from Pastures.*
© CAB International, Wallingford, Oxon.

**TABLE 300  The Range of Concentrations in Plant Material of Various Minerals (taken from Whitehead, 1966; Minson, 1977, 1982)**

| | % in DM | | ppm in DM |
|---|---|---|---|
| Ca | 0.04 - 6.0 | Zn | 1 - 120 |
| P | 0.02 - 0.71 | Cu | 1.1 - 100 |
| Mg | 0.03 - 1.00 | Mn | 9 - 2400 |
| Na | 0.001 - 2.12 | Co | 0.016 - 4.7 |
| | | I | 0.09 - 5 |
| | | Se | <0.01 - 4000 |
| | | Mo | 0.01 - 156 |

*Source:* Hacker J.B. ed. (1981) *Nutritional Limits to Animal Production from Pastures.*
© CAB International, Wallingford, Oxon.

---

**TABLE 301   General Quality Parameters and Values for Grass in UK**

---

**a. Target Dry matter for Silage**

**Clamp**   18 - 28%          **Big Bale**   30 - 40%          **Tower Silo**   35 - 45%

**b. Field Hay**

| Quality | High | Avg | Poor |
|---|---|---|---|
| Dry Matter % | 90 | 80 | 70 |
| Crude Protein in Dry Matter | 16 | 11 | 6 |
| Digestible Crude Protein | 11.2 | 6.6 | 3 |
| 'D' value | 65 | 55 | 40 |
| Metabolizable Energy (MJ/kg DM) | 10.4 | 8.8 | 6.4 |

**c. Dried Grass**

| Quality Rating | 5* | 4* | 3* | 2* |
|---|---|---|---|---|
| Max. crude fibre in grass of 90% DM | 17 | 21 | 25 | 29 |
| 'D' value | 70 | 65 | 60 | 55 |
| ME (MJ/kg DM) | 11.2 | 10.4 | 9.6 | 8.8 |

**d. Critical Levels of Trace Minerals in Herbage for Dietary Purposes**

| | Cattle | Sheep |
|---|---|---|
| Mn ppm | 70 | 70 |
| Cu ppm | 11 | 7 |
| Mo ppm | max 2 | max 2 |
| I ppm | 1.8 | 1.8 |
| Zn ppm | 40 | 40 |
| Mg % | 0.2 | 0.2 |

*Source:* Phosyn Chemical Co., Pocklington, York, UK. Reproduced with kind permission.

**TABLE 301** *(continued)*   **General Quality Parameters and Values for Grass in UK**

### e. Grass Silage

| | | |
|---|---|---|
| **Dry Matter %** | >25 | Good |
| | 20 - 25 | Medium |
| | <20 | Poor |
| **pH** | 3.8 - 4.1 is ideal when DM% 18 - 22, else not ideal | |
| | 4.2 - 4.4 is ideal when DM% 22 - 26, else not ideal | |
| | 4.4 - 4.6 is ideal when DM% 26 - 30, else not ideal | |
| **MAD Fibre %** | <32 | Good |
| | 32 - 38 | Medium |
| | >38 | Poor |
| **Crude Protein %** | >15 | Good |
| | 10 - 15 | Medium |
| | <10 | Poor |
| **Digestible Crude Protein g/kg** | >100 | Good |
| | 60 - 100 | Medium |
| | <60 | Poor |
| **Ammonia Nitrogen as a % of total N** | <10 | Well fermented |
| | 10 - 15 | Moderate fermentation |
| | 15 - 20 | Poor fermentation |
| | >20 | Very poor fermentation |
| **D value %** | >65 | Good |
| | 60 - 65 | Medium |
| | <60 | Poor |
| **Metabolizable energy MJ/kg DM** | >10.4 | Good |
| | 9.6 - 10.4 | Medium |
| | <9.6 | Poor |
| **Ca : P** | <1:1 | Low - possible problem |
| | >6:1 | High - possible problem |

*Source:* Phosyn Chemicals Ltd., Pocklington, York, UK. Reproduced with kind permission.

# Tables of Feed Analyses

**TABLE 302    Analyses of Ruminant Feedstuffs**

| Food name | Dry Matter Content g/kg | Metabolizable Energy MJ/kg DM | Digestible Crude Protein g/kg DM | Analysis of Dry Matter g/kg — Crude Protein | Ether Extract | Crude Fibre | N free Extract | Total Ash | Gross Energy MJ/kg DM | Q {ME/GE} | Digestible Organic Matter in Dry Matter DOMD% | Digestibility Coefficients (decimal) — Crude Protein | Ether Extract | Crude Fibre | N free Extract |
|---|---|---|---|---|---|---|---|---|---|---|---|---|---|---|---|
| **1 Roots** | | | | | | | | | | | | | | | |
| Artichoke, Jerusalem | 200 | 13.2 | 50 | 75 | 10 | 35 | 825 | 55 | 17.4 | 0.76 | 84 | 0.67 | 0.00 | 0.29 | 0.94 |
| Carrots | 130 | 12.8 | 62 | 92 | 15 | 108 | 715 | 69 | 17.4 | 0.73 | 81 | 0.67 | 0.50 | 0.50 | 0.96 |
| Kohlrabi | 130 | 10.8 | 54 | 154 | 8 | 108 | 654 | 77 | 17.4 | 0.62 | 69 | 0.35 | 0.00 | 0.43 | 0.90 |
| Mangels, white-fleshed globe | 110 | 12.5 | 64 | 91 | 9 | 64 | 773 | 64 | 17.3 | 0.72 | 79 | 0.70 | 0.00 | 0.43 | 0.91 |
| Mangels, intermediate | 120 | 12.4 | 58 | 83 | 8 | 58 | 783 | 67 | 17.2 | 0.72 | 79 | 0.70 | 0.00 | 0.43 | 0.90 |
| Mangels, yellow-fleshed globe | 130 | 12.4 | 54 | 92 | 8 | 62 | 769 | 69 | 17.2 | 0.72 | 78 | 0.58 | 0.00 | 0.38 | 0.92 |
| Mangels, long red | 130 | 12.6 | 54 | 77 | 8 | 62 | 785 | 69 | 17.1 | 0.74 | 80 | 0.70 | 0.00 | 0.38 | 0.92 |
| Parsnips | 150 | 13.3 | 67 | 87 | 20 | 80 | 753 | 60 | 17.6 | 0.76 | 84 | 0.77 | 0.33 | 0.58 | 0.96 |
| Potatoes | 210 | 12.5 | 47 | 90 | 5 | 38 | 824 | 43 | 17.6 | 0.71 | 79 | 0.52 | 0.00 | 0.00 | 0.90 |
| Sugar beet | 230 | 13.7 | 35 | 48 | 4 | 48 | 870 | 30 | 17.6 | 0.78 | 87 | 0.73 | 0.00 | 0.36 | 0.94 |
| Swede turnip | 120 | 12.8 | 91 | 108 | 17 | 100 | 717 | 58 | 17.7 | 0.72 | 82 | 0.84 | 0.00 | 0.66 | 0.93 |
| Turnip | 90 | 11.2 | 73 | 122 | 22 | 111 | 667 | 78 | 17.6 | 0.64 | 72 | 0.60 | 0.00 | 0.33 | 0.91 |
| **2 Leaves of roots** | | | | | | | | | | | | | | | |
| Artichoke tops | 320 | 8.8 | 65 | 106 | 34 | 169 | 534 | 156 | 16.5 | 0.53 | 55 | 0.61 | 0.45 | 0.41 | 0.75 |
| Carrot leaves | 180 | 7.9 | 123 | 189 | 50 | 133 | 389 | 239 | 15.7 | 0.50 | 48 | 0.65 | 0.56 | 0.56 | 0.66 |
| Kohlrabi leaves | 140 | 10.2 | 141 | 207 | 29 | 121 | 529 | 114 | 17.5 | 0.58 | 65 | 0.68 | 0.50 | 0.56 | 0.80 |
| Mangel leaves | 110 | 9.0 | 146 | 218 | 36 | 145 | 418 | 182 | 16.6 | 0.54 | 57 | 0.67 | 0.50 | 0.57 | 0.76 |
| Potato haulm | 230 | 6.5 | 48 | 109 | 43 | 270 | 443 | 135 | 17.3 | 0.38 | 42 | 0.44 | 0.20 | 0.36 | 0.60 |
| Sugar beet tops | 160 | 9.9 | 88 | 125 | 31 | 100 | 531 | 212 | 15.4 | 0.64 | 62 | 0.70 | 0.60 | 0.69 | 0.83 |
| Turnip leaves | 120 | 9.2 | 130 | 192 | 42 | 125 | 458 | 183 | 16.5 | 0.55 | 58 | 0.68 | 0.40 | 0.53 | 0.79 |
| **3 Other green foods** | | | | | | | | | | | | | | | |
| Cabbage, drumhead | 110 | 10.4 | 100 | 136 | 36 | 182 | 536 | 109 | 17.5 | 0.59 | 66 | 0.73 | 0.50 | 0.70 | 0.78 |
| Cabbage, open leaved | 150 | 10.8 | 115 | 160 | 47 | 160 | 527 | 107 | 17.9 | 0.60 | 68 | 0.72 | 0.57 | 0.71 | 0.80 |
| Comfrey | 120 | 8.5 | 130 | 217 | 25 | 150 | 433 | 175 | 16.5 | 0.52 | 54 | 0.60 | 0.67 | 0.47 | 0.74 |
| Kale, thousandhead | 160 | 11.1 | 106 | 137 | 25 | 200 | 531 | 106 | 17.4 | 0.64 | 71 | 0.77 | 0.50 | 0.56 | 0.90 |
| Kale, marrow stem (unthinned) | 140 | 11.0 | 123 | 157 | 36 | 179 | 493 | 136 | 17.2 | 0.64 | 69 | 0.78 | 0.60 | 0.64 | 0.88 |
| Kale, marrow stem (singled) | 140 | 11.0 | 114 | 150 | 21 | 179 | 521 | 129 | 16.9 | 0.65 | 70 | 0.76 | 0.67 | 0.60 | 0.89 |
| Broccoli, purple sprouting | 120 | 11.1 | 117 | 158 | 25 | 125 | 592 | 100 | 17.5 | 0.63 | 70 | 0.74 | 0.67 | 0.74 | 0.80 |
| Mustard | 150 | 9.0 | 126 | 193 | 27 | 193 | 493 | 93 | 17.9 | 0.50 | 57 | 0.65 | 0.50 | 0.52 | 0.68 |
| Rape | 140 | 9.5 | 144 | 200 | 57 | 250 | 400 | 93 | 18.7 | 0.51 | 59 | 0.72 | 0.63 | 0.54 | 0.68 |

| Food name | Dry Matter Content g/kg | Metabolizable Energy MJ/kg DM | Digestible Crude Protein g/kg DM | Analysis of Dry Matter g/kg | | | | | Gross Energy MJ/kg DM | Q {ME/GE} | Digestible Organic Matter in Dry Matter DOMD% | Digestibility Coefficients (decimal) | | | |
|---|---|---|---|---|---|---|---|---|---|---|---|---|---|---|---|
| | | | | Crude Protein | Ether Extract | Crude Fibre | N free Extract | Total Ash | | | | Crude Protein | Ether Extract | Crude Fibre | N free Extract |
| **4 Cereals** | | | | | | | | | | | | | | | |
| Barley in flower | 250 | 10.0 | 146 | 68 | 16 | 316 | 536 | 64 | 17.7 | 0.56 | 66 | 0.68 | 0.60 | 0.64 | 0.75 |
| Maize | 190 | 8.8 | 53 | 89 | 26 | 289 | 532 | 63 | 18.1 | 0.49 | 57 | 0.59 | 0.60 | 0.55 | 0.64 |
| Millet | 130 | 7.9 | 54 | 100 | 15 | 315 | 477 | 92 | 17.4 | 0.45 | 52 | 0.54 | 0.50 | 0.54 | 0.61 |
| Oats in flower | 230 | 8.6 | 61 | 83 | 26 | 365 | 448 | 78 | 17.9 | 0.48 | 57 | 0.74 | 0.66 | 0.58 | 0.62 |
| Rye in flower | 230 | 9.5 | 88 | 126 | 39 | 322 | 439 | 74 | 18.4 | 0.52 | 62 | 0.70 | 0.56 | 0.65 | 0.68 |
| **5 Grasses** | | | | | | | | | | | | | | | |
| Pasture grass, close grazing: Non-rotational | 200 | 12.1 | 225 | 265 | 55 | 130 | 445 | 105 | 18.6 | 0.65 | 75 | 0.85 | 0.64 | 0.81 | 0.87 |
| Rotational, with 3 weekly intervals | 200 | 12.1 | 185 | 225 | 65 | 155 | 465 | 90 | 18.9 | 0.64 | 75 | 0.82 | 0.61 | 0.81 | 0.86 |
| Rotational, with monthly intervals | 200 | 11.2 | 130 | 175 | 50 | 225 | 460 | 90 | 18.5 | 0.61 | 72 | 0.74 | 0.50 | 0.82 | 0.82 |
| Pasture grass, extensive grazing Spring value, running off during summer | 200 | 10.0 | 124 | 175 | 40 | 200 | 485 | 100 | 18.0 | 0.56 | 64 | 0.71 | 0.50 | 0.65 | 0.75 |
| Winter pasturage (after close grazing allowing free growth from end July to December) | 200 | 9.7 | 101 | 155 | 30 | 220 | 515 | 80 | 18.1 | 0.53 | 63 | 0.65 | 0.16 | 0.59 | 0.77 |
| Rice grass | 220 | 7.0 | 59 | 132 | 27 | 227 | 505 | 109 | 17.4 | 0.40 | 46 | 0.45 | 0.50 | 0.66 | 0.46 |
| Ryegrass perennial post flowering | 250 | 8.4 | 72 | 116 | 28 | 288 | 464 | 104 | 17.5 | 0.48 | 55 | 0.62 | 0.43 | 0.56 | 0.65 |
| Ryegrass, Italian post flowering | 250 | 8.7 | 84 | 136 | 40 | 248 | 464 | 112 | 17.7 | 0.49 | 55 | 0.62 | 0.50 | 0.58 | 0.66 |
| Sorghum | 200 | 8.0 | 60 | 105 | 30 | 310 | 485 | 70 | 18.1 | 0.44 | 53 | 0.57 | 0.33 | 0.53 | 0.60 |
| Timothy, in flower | 250 | 8.5 | 52 | 96 | 32 | 280 | 524 | 68 | 18.1 | 0.47 | 55 | 0.54 | 0.50 | 0.53 | 0.63 |
| **6 Green legumes** | | | | | | | | | | | | | | | |
| Alsike | 150 | 8.8 | 141 | 220 | 40 | 300 | 340 | 100 | 18.4 | 0.48 | 56 | 0.64 | 0.67 | 0.49 | 0.71 |
| Crimson clover | 190 | 9.5 | 114 | 153 | 37 | 337 | 374 | 100 | 18.0 | 0.53 | 61 | 0.75 | 0.71 | 0.57 | 0.75 |
| Red clover, beginning to flower | 190 | 10.2 | 132 | 179 | 37 | 274 | 426 | 84 | 18.3 | 0.56 | 65 | 0.74 | 0.71 | 0.58 | 0.78 |
| White clover, beginning to flower | 190 | 9.0 | 152 | 237 | 42 | 232 | 374 | 116 | 18.1 | 0.50 | 57 | 0.64 | 0.63 | 0.60 | 0.68 |
| Beans, beginning to flower | 150 | 9.2 | 154 | 213 | 53 | 220 | 380 | 133 | 17.9 | 0.51 | 57 | 0.72 | 0.62 | 0.49 | 0.72 |
| Kidney vetch | 180 | 8.7 | 77 | 133 | 33 | 283 | 478 | 72 | 18.3 | 0.48 | 56 | 0.58 | 0.50 | 0.53 | 0.66 |
| Lucerne, early flower | 240 | 8.2 | 130 | 171 | 17 | 300 | 413 | 100 | 17.6 | 0.47 | 54 | 0.76 | 0.25 | 0.44 | 0.67 |
| Lucerne, in bud | 220 | 9.4 | 164 | 205 | 23 | 282 | 409 | 82 | 18.2 | 0.52 | 62 | 0.80 | 0.20 | 0.50 | 0.76 |
| Lucerne, before bud | 150 | 10.2 | 213 | 253 | 27 | 220 | 380 | 120 | 17.8 | 0.57 | 67 | 0.84 | 0.25 | 0.64 | 0.81 |
| Peas, beginning to flower | 170 | 8.5 | 140 | 206 | 35 | 353 | 335 | 71 | 18.8 | 0.45 | 56 | 0.68 | 0.50 | 0.50 | 0.66 |
| Sainfoin, early flower | 230 | 10.3 | 143 | 196 | 26 | 209 | 509 | 61 | 18.5 | 0.56 | 65 | 0.73 | 0.67 | 0.45 | 0.78 |
| Sainfoin, full flower | 250 | 8.4 | 116 | 176 | 24 | 236 | 500 | 64 | 18.3 | 0.46 | 54 | 0.66 | 0.50 | 0.46 | 0.61 |

*(continued over)*

| Food name | Dry Matter Content g/kg | Metabolizable Energy MJ/kg DM | Digestible Crude Protein g/kg DM | Analysis of Dry Matter g/kg | | | | | Gross Energy MJ/kg DM | Q {ME/GE} | Digestible Organic Matter in Dry Matter DOMD% | Digestibility Coefficients (decimal) | | | |
|---|---|---|---|---|---|---|---|---|---|---|---|---|---|---|---|
| | | | | Crude Protein | Ether Extract | Crude Fibre | N free Extract | Total Ash | | | | Crude Protein | Ether Extract | Crude Fibre | N free Extract |
| Trefoil | 200 | 9.0 | 121 | 175 | 40 | 285 | 420 | 80 | 18.5 | 0.49 | 57 | 0.69 | 0.50 | 0.49 | 0.70 |
| Vetches, in flower | 180 | 8.6 | 123 | 178 | 28 | 294 | 417 | 83 | 18.2 | 0.48 | 56 | 0.69 | 0.60 | 0.45 | 0.68 |
| **7 Miscellaneous** | | | | | | | | | | | | | | | |
| Brushwood | 750 | 6.3 | 28 | 61 | 25 | 356 | 537 | 20 | 18.8 | 0.34 | 41 | 0.46 | 0.42 | 0.28 | 0.50 |
| Buckwheat | 160 | 9.1 | 100 | 156 | 38 | 262 | 475 | 69 | 18.5 | 0.49 | 59 | 0.64 | 0.50 | 0.57 | 0.67 |
| Gorse | 500 | 6.8 | 44 | 104 | 22 | 468 | 352 | 54 | 18.5 | 0.37 | 45 | 0.42 | 0.45 | 0.40 | 0.60 |
| Heather | 500 | 6.0 | 28 | 70 | 86 | 454 | 332 | 58 | 19.7 | 0.31 | 37 | 0.40 | 0.35 | 0.31 | 0.52 |
| Artichoke tops (dried) | 870 | 8.5 | 87 | 145 | 25 | 162 | 551 | 117 | 17.2 | 0.49 | 53 | 0.60 | 0.50 | 0.29 | 0.70 |
| Elm leaves (dried) | 880 | 10.3 | 132 | 181 | 33 | 98 | 567 | 122 | 17.3 | 0.59 | 65 | 0.73 | 0.24 | 0.57 | 0.81 |
| Hop leaves and bine (dried) | 890 | 8.2 | 90 | 140 | 39 | 273 | 426 | 121 | 17.6 | 0.47 | 51 | 0.64 | 0.72 | 0.31 | 0.71 |
| Leaves of trees in July (dried) | 840 | 9.1 | 74 | 125 | 36 | 169 | 587 | 83 | 17.9 | 0.51 | 55 | 0.59 | 0.80 | 0.37 | 0.66 |
| Nettles (dried) | 890 | 10.4 | 145 | 207 | 87 | 119 | 429 | 158 | 18.1 | 0.57 | 61 | 0.70 | 0.64 | 0.57 | 0.79 |
| Poplar leaves in October (dried) | 840 | 9.7 | 72 | 129 | 104 | 207 | 471 | 89 | 19.4 | 0.50 | 53 | 0.56 | 0.79 | 0.32 | 0.66 |
| **8a Silage - Clamp** | | | | | | | | | | | | | | | |
| Alsike | 250 | 8.6 | 80 | 136 | 72 | 272 | 436 | 84 | 13.9 | 0.45 | 51 | 0.59 | 0.67 | 0.50 | 0.57 |
| Clover (red) | 220 | 8.8 | 135 | 205 | 55 | 300 | 327 | 114 | 18.4 | 0.48 | 56 | 0.66 | 0.54 | 0.53 | 0.72 |
| Grass(very high digestibility) | 200 | 10.2 | 116 | 170 | 40 | 300 | 390 | 100 | 18.1 | 0.57 | 67 | 0.68 | 0.67 | 0.81 | 0.72 |
| Grass (high digestibility) | 200 | 9.3 | 107 | 170 | 40 | 305 | 390 | 95 | 18.1 | 0.51 | 61 | 0.63 | 0.62 | 0.76 | 0.63 |
| Grass (moderate digestibility) | 200 | 8.8 | 102 | 160 | 35 | 340 | 375 | 90 | 18.2 | 0.48 | 58 | 0.64 | 0.57 | 0.73 | 0.56 |
| Grass (low digestibility) | 200 | 7.6 | 98 | 160 | 35 | 380 | 345 | 80 | 18.4 | 0.41 | 52 | 0.61 | 0.35 | 0.69 | 0.42 |
| Lucerne | 250 | 8.5 | 113 | 168 | 84 | 296 | 352 | 100 | 19.1 | 0.45 | 52 | 0.67 | 0.48 | 0.42 | 0.69 |
| Maize | 210 | 10.8 | 70 | 110 | 57 | 233 | 538 | 62 | 18.8 | 0.57 | 65 | 0.64 | 0.90 | 0.68 | 0.69 |
| Mangel leaves | 220 | 6.9 | 88 | 132 | 50 | 145 | 450 | 223 | 15.8 | 0.44 | 43 | 0.67 | 0.40 | 0.55 | 0.54 |
| Marrowstem kale | 160 | 9.8 | 95 | 125 | 31 | 231 | 456 | 156 | 16.6 | 0.59 | 65 | 0.76 | 0.00 | 0.74 | 0.85 |
| Mustard | 150 | 9.6 | 107 | 167 | 27 | 253 | 400 | 153 | 16.8 | 0.57 | 60 | 0.64 | 1.00 | 0.50 | 0.85 |
| Oats (green) | 240 | 8.0 | 47 | 79 | 33 | 358 | 454 | 75 | 18.1 | 0.44 | 53 | 0.60 | 0.50 | 0.60 | 0.55 |
| Overheated ryegrass and clover | 320 | 7.1 | 16 | 134 | 34 | 319 | 422 | 91 | 18.0 | 0.39 | 45 | 0.12 | 0.73 | 0.55 | 0.56 |
| Pea haulm & pods (canning) | 210 | 8.7 | 95 | 167 | 67 | 290 | 276 | 200 | 16.9 | 0.51 | 51 | 0.57 | 0.93 | 0.56 | 0.69 |
| Pea pods (canning) | 280 | 10.6 | 85 | 129 | 36 | 307 | 464 | 64 | 18.5 | 0.58 | 67 | 0.66 | 0.90 | 0.65 | 0.77 |
| Potatoes | 270 | 11.8 | 39 | 81 | 19 | 26 | 822 | 52 | 17.6 | 0.67 | 74 | 0.48 | 0.20 | 0.00 | 0.85 |
| Potato haulm | 250 | 6.4 | 49 | 128 | 108 | 176 | 364 | 224 | 17.1 | 0.38 | 36 | 0.38 | 0.44 | 0.39 | 0.55 |
| Rye | 130 | 8.3 | 71 | 123 | 38 | 338 | 431 | 69 | 18.5 | 0.45 | 55 | 0.58 | 0.38 | 0.60 | 0.60 |
| Sainfoin | 240 | 8.4 | 124 | 179 | 63 | 333 | 342 | 83 | 19.0 | 0.44 | 52 | 0.69 | 0.50 | 0.42 | 0.67 |
| Sugar beet pulp (wet) | 120 | 9.7 | 42 | 83 | 17 | 200 | 625 | 75 | 17.5 | 0.55 | 62 | 0.50 | 0.50 | 0.50 | 0.75 |
| Sugar beet tops | 230 | 7.9 | 65 | 104 | 30 | 148 | 396 | 322 | 13.4 | 0.59 | 50 | 0.62 | 0.50 | 0.73 | 0.80 |
| Sugar beet tops and pulp | 160 | 11.3 | 100 | 150 | 38 | 131 | 556 | 125 | 17.3 | 0.65 | 70 | 0.67 | 0.67 | 0.81 | 0.85 |
| Sunflowers | 220 | 8.4 | 51 | 95 | 45 | 305 | 450 | 105 | 17.8 | 0.47 | 53 | 0.53 | 0.67 | 0.49 | 0.66 |

| Food name | Dry Matter Content g/kg | Metabolizable Energy MJ/kg DM | Digestible Crude Protein g/kg DM | Crude Protein | Ether Extract | Crude Fibre | N free Extract | Total Ash | Gross Energy MJ/kg DM | Q {ME}{GE} | Digestible Organic Matter in Dry Matter DOMD% | Crude Protein | Ether Extract | Crude Fibre | N free Extract |
|---|---|---|---|---|---|---|---|---|---|---|---|---|---|---|---|
| | | | | Analysis of Dry Matter g/kg | | | | | | | | Digestibility Coefficients (decimal) | | | |
| Turnip tops | 170 | 8.4 | 88 | 124 | 35 | 159 | 353 | 329 | 13.5 | 0.62 | 52 | 0.71 | 0.83 | 0.68 | 0.83 |
| Vetch and oats | 270 | 9.6 | 82 | 126 | 44 | 293 | 456 | 81 | 18.3 | 0.52 | 60 | 0.65 | 0.73 | 0.58 | 0.70 |
| **8b Silage - Tower** | | | | | | | | | | | | | | | |
| Barley (whole crop) | 400 | 9.6 | 50 | 95 | 22 | 250 | 570 | 63 | 18.0 | 0.54 | 62 | 0.53 | 0.61 | 0.53 | 0.74 |
| Grass(very high digestibility) | 400 | 10.4 | 121 | 170 | 38 | 313 | 383 | 97 | 18.1 | 0.57 | 68 | 0.71 | 0.65 | 0.80 | 0.74 |
| Grass (high digestibility) | 400 | 9.3 | 87 | 142 | 28 | 313 | 433 | 85 | 18.0 | 0.52 | 61 | 0.61 | 0.67 | 0.73 | 0.64 |
| Wheat (whole crop) | 400 | 8.4 | 36 | 78 | 17 | 300 | 563 | 42 | 18.2 | 0.46 | 55 | 0.47 | 0.56 | 0.43 | 0.66 |
| **9 Hay** | | | | | | | | | | | | | | | |
| Barley (just past milk stage) | 850 | 8.8 | 54 | 81 | 22 | 289 | 533 | 74 | 17.7 | 0.50 | 58 | 0.67 | 0.42 | 0.62 | 0.63 |
| Clover, crimson | 850 | 8.2 | 100 | 144 | 29 | 314 | 426 | 87 | 18.0 | 0.46 | 54 | 0.70 | 0.40 | 0.47 | 0.65 |
| Clover, red very good | 850 | 9.6 | 128 | 184 | 39 | 266 | 428 | 84 | 18.4 | 0.52 | 61 | 0.70 | 0.64 | 0.50 | 0.75 |
| Clover, red good | 850 | 8.9 | 103 | 161 | 35 | 287 | 445 | 72 | 18.5 | 0.48 | 57 | 0.64 | 0.57 | 0.47 | 0.70 |
| Clover, red poor | 850 | 7.8 | 67 | 131 | 25 | 340 | 445 | 60 | 18.4 | 0.42 | 50 | 0.51 | 0.48 | 0.40 | 0.65 |
| Clover, red damaged | 850 | 6.9 | 73 | 141 | 18 | 394 | 364 | 84 | 17.9 | 0.38 | 46 | 0.52 | 0.47 | 0.40 | 0.60 |
| Grass (very high digestibility) | 850 | 10.1 | 90 | 132 | 20 | 291 | 473 | 85 | 17.7 | 0.57 | 67 | 0.68 | 0.37 | 0.76 | 0.75 |
| Grass (high digestibility) | 850 | 9.0 | 58 | 101 | 16 | 320 | 480 | 82 | 17.6 | 0.51 | 61 | 0.57 | 0.30 | 0.70 | 0.67 |
| Grass (moderate digestibility) | 850 | 8.4 | 39 | 85 | 16 | 328 | 496 | 74 | 17.7 | 0.48 | 57 | 0.46 | 0.27 | 0.61 | 0.65 |
| Grass (low digestibility) | 850 | 7.5 | 45 | 92 | 16 | 366 | 456 | 69 | 17.8 | 0.42 | 51 | 0.49 | 0.27 | 0.56 | 0.56 |
| Grass (very low digestibility) | 850 | 7.0 | 38 | 88 | 16 | 340 | 478 | 78 | 17.6 | 0.40 | 47 | 0.43 | 0.35 | 0.54 | 0.51 |
| Lucerne, before flowering | 850 | 8.3 | 143 | 193 | 28 | 321 | 371 | 87 | 18.2 | 0.46 | 54 | 0.74 | 0.46 | 0.42 | 0.68 |
| Lucerne, half flower | 850 | 8.2 | 166 | 225 | 13 | 302 | 365 | 95 | 17.9 | 0.46 | 55 | 0.74 | 0.00 | 0.48 | 0.66 |
| Lucerne, full flower | 850 | 7.7 | 116 | 171 | 31 | 353 | 349 | 96 | 18.1 | 0.43 | 51 | 0.68 | 0.46 | 0.45 | 0.62 |
| Millet hay | 850 | 8.4 | 71 | 125 | 26 | 339 | 445 | 66 | 18.2 | 0.46 | 56 | 0.57 | 0.41 | 0.60 | 0.61 |
| Mineral deficient hay (mainly purple molinia and brown bent) | 850 | 6.4 | 63 | 129 | 22 | 344 | 469 | 35 | 18.7 | 0.34 | 44 | 0.49 | 0.11 | 0.55 | 0.39 |
| Oats, milk stage | 850 | 7.8 | 52 | 94 | 31 | 324 | 473 | 79 | 18.0 | 0.43 | 50 | 0.55 | 0.62 | 0.52 | 0.56 |
| Rice grass, poor | 850 | 7.0 | 31 | 79 | 16 | 306 | 499 | 100 | 17.2 | 0.41 | 47 | 0.39 | 0.36 | 0.63 | 0.49 |
| Rye, before flowering | 850 | 9.5 | 85 | 121 | 29 | 332 | 455 | 62 | 18.4 | 0.52 | 62 | 0.70 | 0.60 | 0.60 | 0.70 |
| Sainfoin, before flowering | 850 | 9.2 | 129 | 182 | 38 | 296 | 404 | 80 | 18.5 | 0.50 | 58 | 0.71 | 0.66 | 0.43 | 0.74 |
| Sainfoin, in flower | 850 | 9.0 | 115 | 158 | 31 | 335 | 389 | 87 | 18.1 | 0.50 | 58 | 0.73 | 0.62 | 0.42 | 0.78 |
| Trefoil | 850 | 8.8 | 139 | 184 | 40 | 292 | 395 | 89 | 18.4 | 0.48 | 56 | 0.76 | 0.47 | 0.44 | 0.70 |
| Vetches, beginning to flower | 850 | 8.8 | 181 | 238 | 27 | 281 | 342 | 112 | 17.9 | 0.49 | 57 | 0.76 | 0.61 | 0.54 | 0.65 |
| Vetches, full flower | 850 | 8.0 | 113 | 171 | 29 | 306 | 394 | 100 | 17.9 | 0.45 | 52 | 0.66 | 0.60 | 0.50 | 0.60 |
| Vetches, and oats (vetches in flower) | 850 | 8.1 | 77 | 138 | 39 | 288 | 433 | 102 | 17.9 | 0.46 | 52 | 0.56 | 0.52 | 0.51 | 0.64 |
| Wheat, milk stage | 850 | 8.5 | 36 | 66 | 19 | 293 | 551 | 72 | 17.6 | 0.48 | 56 | 0.55 | 0.56 | 0.58 | 0.62 |

*(continued over)*

| Food name | Dry Matter Content g/kg | Metabolizable Energy MJ/kg DM | Digestible Crude Protein g/kg DM | Crude Protein | Ether Extract | Crude Fibre | N free Extract | Total Ash | Gross Energy MJ/kg DM | Q {ME/GE} | Digestible Organic Matter in Dry Matter DOMD% | Crude Protein | Ether Extract | Crude Fibre | N free Extract |
|---|---|---|---|---|---|---|---|---|---|---|---|---|---|---|---|
| **10 Dried grasses and legumes** | | | | | | | | | | | | | | | |
| Grass, very leafy | 900 | 10.8 | 113 | 161 | 28 | 217 | 471 | 123 | 17.3 | 0.62 | 70 | 0.70 | 0.58 | 0.83 | 0.83 |
| Grass, leafy | 900 | 10.6 | 136 | 187 | 38 | 213 | 460 | 102 | 18.0 | 0.59 | 68 | 0.73 | 0.61 | 0.79 | 0.77 |
| Grass, early flower | 900 | 9.7 | 97 | 154 | 28 | 258 | 453 | 107 | 17.6 | 0.55 | 64 | 0.63 | 0.49 | 0.75 | 0.73 |
| Lucerne, just in bud | 900 | 9.4 | 174 | 244 | 32 | 198 | 400 | 126 | 17.7 | 0.53 | 60 | 0.71 | 0.45 | 0.53 | 0.78 |
| Lucerne, early flower | 900 | 8.7 | 128 | 178 | 27 | 269 | 414 | 112 | 17.6 | 0.49 | 57 | 0.72 | 0.29 | 0.46 | 0.74 |
| Lucerne leaf meal (Amer.) | 900 | 9.3 | 179 | 236 | 21 | 176 | 446 | 122 | 17.4 | 0.54 | 61 | 0.76 | 0.00 | 0.49 | 0.78 |
| **11 Straws and chaff** | | | | | | | | | | | | | | | |
| Barley straw, spring | 860 | 7.3 | 9 | 38 | 21 | 394 | 493 | 53 | 13.0 | 0.40 | 49 | 0.24 | 0.33 | 0.54 | 0.53 |
| Barley straw, winter | 860 | 5.8 | 8 | 37 | 16 | 488 | 392 | 66 | 17.8 | 0.32 | 39 | 0.22 | 0.29 | 0.38 | 0.50 |
| Bean straw (including pods) | 860 | 7.4 | 26 | 52 | 9 | 501 | 384 | 53 | 13.0 | 0.41 | 50 | 0.49 | 0.63 | 0.43 | 0.67 |
| Buckwheat straw | 860 | 6.6 | 27 | 57 | 14 | 455 | 413 | 62 | 17.9 | 0.37 | 45 | 0.47 | 0.42 | 0.45 | 0.52 |
| Clover, straw, red | 840 | 5.6 | 48 | 108 | 21 | 531 | 271 | 68 | 18.3 | 0.31 | 38 | 0.44 | 0.33 | 0.37 | 0.49 |
| Maize straw | 850 | 7.3 | 20 | 59 | 18 | 461 | 406 | 56 | 18.1 | 0.40 | 51 | 0.34 | 0.33 | 0.60 | 0.50 |
| Oat straw, spring | 860 | 6.7 | 11 | 34 | 22 | 394 | 493 | 57 | 18.0 | 0.38 | 46 | 0.34 | 0.32 | 0.54 | 0.46 |
| Oat straw, winter | 860 | 6.8 | 9 | 22 | 17 | 402 | 501 | 57 | 17.8 | 0.38 | 46 | 0.40 | 0.33 | 0.57 | 0.44 |
| Pea straw | 860 | 6.5 | 50 | 105 | 19 | 410 | 390 | 77 | 17.9 | 0.36 | 43 | 0.48 | 0.44 | 0.39 | 0.55 |
| Rape straw | 840 | 6.5 | 21 | 30 | 14 | 450 | 461 | 45 | 18.0 | 0.36 | 44 | 0.72 | 0.42 | 0.37 | 0.53 |
| Rye straw, spring | 860 | 6.2 | 7 | 37 | 19 | 429 | 485 | 30 | 18.4 | 0.33 | 42 | 0.19 | 0.50 | 0.47 | 0.41 |
| Rye straw, winter | 860 | 6.3 | 7 | 36 | 16 | 465 | 437 | 45 | 18.1 | 0.35 | 43 | 0.19 | 0.50 | 0.51 | 0.41 |
| Soya bean straw | 840 | 7.5 | 44 | 88 | 24 | 311 | 456 | 121 | 17.0 | 0.44 | 48 | 0.50 | 0.60 | 0.38 | 0.66 |
| Tare or vetch straw | 860 | 6.3 | 48 | 105 | 20 | 472 | 342 | 62 | 18.3 | 0.34 | 42 | 0.46 | 0.47 | 0.40 | 0.52 |
| Wheat straw, spring | 860 | 5.6 | 1 | 34 | 15 | 417 | 463 | 71 | 17.6 | 0.32 | 39 | 0.03 | 0.31 | 0.50 | 0.37 |
| Wheat straw, winter | 860 | 5.7 | 1 | 24 | 15 | 426 | 473 | 62 | 17.7 | 0.32 | 39 | 0.03 | 0.31 | 0.50 | 0.37 |
| Linseed chaff | 880 | 5.0 | 16 | 40 | 39 | 460 | 395 | 66 | 18.3 | 0.27 | 32 | 0.40 | 0.50 | 0.30 | 0.37 |
| Lupin pods | 870 | 7.4 | 39 | 103 | 8 | 334 | 485 | 69 | 17.7 | 0.42 | 50 | 0.38 | 0.29 | 0.48 | 0.61 |
| Millet chaff and husks | 880 | 5.2 | 19 | 55 | 25 | 464 | 330 | 127 | 17.0 | 0.31 | 35 | 0.35 | 0.32 | 0.37 | 0.47 |
| Oat chaff, spring | 860 | 6.4 | 26 | 70 | 24 | 265 | 521 | 120 | 16.9 | 0.38 | 41 | 0.37 | 0.48 | 0.45 | 0.49 |
| Rice husks | 900 | 2.5 | 5 | 42 | 16 | 421 | 364 | 157 | 16.1 | 0.15 | 15 | 0.11 | 0.64 | 0.01 | 0.35 |
| Rye chaff | 860 | 5.8 | 13 | 41 | 15 | 515 | 340 | 90 | 17.4 | 0.33 | 41 | 0.31 | 0.31 | 0.50 | 0.39 |
| Soya bean pods | 890 | 8.6 | 30 | 67 | 17 | 340 | 482 | 93 | 17.3 | 0.50 | 56 | 0.44 | 0.53 | 0.51 | 0.73 |
| Wheat chaff | 860 | 5.9 | 13 | 43 | 14 | 322 | 495 | 126 | 16.5 | 0.36 | 39 | 0.30 | 0.29 | 0.48 | 0.45 |
| **12a Grains and seeds – cereals** | | | | | | | | | | | | | | | |
| Barley | 860 | 13.7 | 82 | 108 | 17 | 53 | 795 | 26 | 18.3 | 0.75 | 86 | 0.76 | 0.80 | 0.56 | 0.92 |
| Sorghum | 860 | 13.4 | 87 | 108 | 43 | 21 | 801 | 27 | 18.8 | 0.72 | 81 | 0.80 | 0.79 | 0.53 | 8592 |
| Maize | 860 | 14.2 | 78 | 98 | 42 | 24 | 823 | 13 | 19.0 | 0.75 | 87 | 0.80 | 0.61 | 0.36 | 0.92 |

| Food name | Dry Matter Content g/kg | Metabolizable Energy MJ/kg DM | Digestible Crude Protein g/kg DM | Analysis of Dry Matter g, kg | | | | | Gross Energy MJ/kg DM | Q {ME/GE} | Digestible Organic Matter in Dry Matter DOMD% | Digestibility Coefficients (decimal) | | | |
|---|---|---|---|---|---|---|---|---|---|---|---|---|---|---|---|
| | | | | Crude Protein | Ether Extract | Crude Fibre | N free Extract | Total Ash | | | | Crude Protein | Ether Extract | Crude Fibre | N free Extract |
| Millet | 860 | 11.3 | 92 | 121 | 44 | 93 | 698 | 44 | 18.7 | 0.61 | 68 | 0.76 | 0.80 | 0.33 | 0.75 |
| Oats | 860 | 11.5 | 84 | 109 | 49 | 121 | 688 | 33 | 19.0 | 0.61 | 68 | 0.77 | 0.83 | 0.25 | 0.77 |
| Rice (polished) | 860 | 15.0 | 67 | 77 | 5 | 17 | 892 | 9 | 18.0 | 0.83 | 94 | 0.87 | 0.50 | 0.47 | 0.97 |
| Rye | 860 | 14.0 | 110 | 133 | 20 | 22 | 802 | 23 | 18.4 | 0.76 | 87 | 0.83 | 0.65 | 0.53 | 0.92 |
| Wheat | 860 | 14.0 | 105 | 124 | 19 | 26 | 810 | 21 | 18.4 | 0.76 | 87 | 0.84 | 0.63 | 0.47 | 0.92 |
| **12b Legumes** | | | | | | | | | | | | | | | |
| Beans, field spring | 860 | 12.8 | 248 | 314 | 15 | 80 | 551 | 40 | 19.0 | 0.67 | 81 | 0.79 | 0.80 | 0.58 | 0.91 |
| Beans, field winter | 860 | 12.8 | 209 | 265 | 15 | 90 | 591 | 40 | 18.8 | 0.68 | 81 | 0.79 | 0.80 | 0.58 | 0.91 |
| Beans, butter | 860 | 12.6 | 175 | 265 | 13 | 42 | 631 | 49 | 18.5 | 0.68 | 80 | 0.66 | 0.64 | 0.62 | 0.93 |
| Gram | 860 | 12.4 | 173 | 263 | 13 | 57 | 610 | 57 | 18.4 | 0.67 | 78 | 0.66 | 0.64 | 0.57 | 0.93 |
| Lentils | 860 | 13.6 | 255 | 297 | 22 | 40 | 607 | 35 | 19.1 | 0.71 | 85 | 0.86 | 0.63 | 0.53 | 0.93 |
| Lupins, sweet (yellow) | 860 | 13.2 | 432 | 480 | 63 | 120 | 285 | 52 | 20.8 | 0.64 | 81 | 0.90 | 0.84 | 0.91 | 0.76 |
| Lupins, sweet (blue) | 860 | 13.3 | 346 | 388 | 67 | 83 | 423 | 38 | 20.6 | 0.65 | 81 | 0.89 | 0.81 | 0.97 | 0.77 |
| Peas | 860 | 13.4 | 225 | 262 | 19 | 63 | 624 | 33 | 18.9 | 0.71 | 85 | 0.86 | 0.63 | 0.46 | 0.93 |
| Vetches | 860 | 13.6 | 264 | 300 | 20 | 69 | 574 | 37 | 19.1 | 0.71 | 85 | 0.88 | 0.88 | 0.65 | 0.92 |
| **12c Oil seeds** | | | | | | | | | | | | | | | |
| Beech mast | 900 | 15.2 | 121 | 149 | 308 | 208 | 288 | 48 | 25.0 | 0.61 | 66 | 0.81 | 0.88 | 0.40 | 0.66 |
| Cottonseed, Egyptian | 900 | 14.1 | 147 | 216 | 261 | 233 | 236 | 54 | 24.1 | 0.59 | 67 | 0.68 | 0.87 | 0.76 | 0.50 |
| Cottonseed, Bombay | 900 | 13.1 | 135 | 196 | 212 | 219 | 327 | 47 | 23.0 | 0.57 | 65 | 0.69 | 0.87 | 0.76 | 0.50 |
| Cottonseed, Brazilian | 900 | 14.1 | 159 | 233 | 256 | 188 | 276 | 48 | 24.2 | 0.58 | 66 | 0.68 | 0.88 | 0.76 | 0.50 |
| Groundnuts or peanuts | 900 | 21.1 | 256 | 284 | 478 | 28 | 187 | 23 | 29.7 | 0.71 | 85 | 0.90 | 0.90 | 0.08 | 0.84 |
| Hemp seed | 900 | 17.5 | 150 | 200 | 359 | 164 | 231 | 46 | 26.4 | 0.66 | 76 | 0.75 | 0.90 | 0.60 | 0.80 |
| Linseed | 900 | 19.3 | 208 | 260 | 392 | 59 | 248 | 41 | 27.4 | 0.71 | 80 | 0.80 | 0.95 | 0.33 | 0.80 |
| Palm nut kernels | 900 | 23.0 | 87 | 92 | 532 | 63 | 292 | 20 | 30.4 | 0.76 | 88 | 0.94 | 0.95 | 0.60 | 0.84 |
| Rape seed | 900 | 21.0 | 172 | 212 | 484 | 63 | 194 | 46 | 29.2 | 0.72 | 80 | 0.81 | 0.95 | 0.25 | 0.80 |
| Sesame seed | 900 | 20.8 | 195 | 217 | 499 | 67 | 159 | 59 | 29.3 | 0.71 | 77 | 0.90 | 0.95 | 0.22 | 0.56 |
| Soya bean | 900 | 14.9 | 328 | 369 | 194 | 46 | 339 | 52 | 23.1 | 0.64 | 75 | 0.89 | 0.90 | 0.42 | 0.68 |
| Sunflower seed | 900 | 16.6 | 138 | 153 | 350 | 303 | 157 | 37 | 26.3 | 0.63 | 68 | 0.90 | 0.95 | 0.34 | 0.71 |
| **12d Miscellaneous seeds** | | | | | | | | | | | | | | | |
| Acorns, fresh | 500 | 13.6 | 54 | 66 | 48 | 136 | 726 | 24 | 18.9 | 0.72 | 83 | 0.82 | 0.79 | 0.60 | 0.90 |
| Acorns, dried | 860 | 13.6 | 55 | 67 | 49 | 136 | 724 | 23 | 18.9 | 0.72 | 83 | 0.81 | 0.80 | 0.60 | 0.90 |
| Buckwheat | 860 | 10.6 | 99 | 131 | 30 | 167 | 638 | 33 | 18.7 | 0.57 | 65 | 0.75 | 0.73 | 0.24 | 0.77 |
| Corozo nut (vegetable ivory) | 900 | 13.6 | 21 | 51 | 10 | 77 | 850 | 12 | 18.1 | 0.75 | 86 | 0.41 | 0.33 | 0.65 | 0.93 |
| Horse chestnut, fresh | 500 | 12.1 | 50 | 84 | 30 | 50 | 804 | 32 | 18.3 | 0.66 | 74 | 0.60 | 0.80 | 0.32 | 0.81 |

*(continued over)*

| Food name | Dry Matter Content g/kg | Metabolizable Energy MJ/kg DM | Digestible Crude Protein g/kg DM | Analysis of Dry Matter g/kg | | | | | Gross Energy MJ/kg DM | Q {ME/GE} | Digestible Organic Matter in Dry Matter DOMD% | Digestibility Coefficients (decimal) | | | |
|---|---|---|---|---|---|---|---|---|---|---|---|---|---|---|---|
| | | | | Crude Protein | Ether Extract | Crude Fibre | N free Extract | Total Ash | | | | Crude Protein | Ether Extract | Crude Fibre | N free Extract |
| Horse chestnut, dry | 860 | 11.2 | 50 | 85 | 29 | 51 | 803 | 31 | 18.3 | 0.61 | 68 | 0.59 | 0.83 | 0.29 | 0.74 |
| Locust beans (pods plus seeds) | 860 | 13.8 | 47 | 69 | 15 | 76 | 812 | 29 | 18.0 | 0.77 | 87 | 0.69 | 0.54 | 0.58 | 0.95 |
| Lucerne seed meal | 880 | 14.1 | 316 | 376 | 119 | 92 | 363 | 50 | 21.5 | 0.65 | 79 | 0.84 | 0.86 | 0.62 | 0.87 |
| Mangel seed | 880 | 7.5 | 83 | 139 | 61 | 386 | 334 | 80 | 19.0 | 0.40 | 46 | 0.60 | 0.60 | 0.35 | 0.62 |
| Red clover seed meal | 880 | 13.4 | 313 | 373 | 89 | 106 | 357 | 76 | 20.4 | 0.66 | 78 | 0.84 | 0.87 | 0.82 | 0.86 |
| Rye grass seed meal (perennial and Italian) | 880 | 11.5 | 72 | 108 | 22 | 105 | 719 | 47 | 16.1 | 0.64 | 71 | 0.67 | 0.84 | 0.21 | 0.83 |
| Sainfoin seed meal (unmilled seed) | 880 | 11.6 | 258 | 300 | 68 | 203 | 383 | 45 | 20.2 | 0.57 | 69 | 0.86 | 0.88 | 0.41 | 0.75 |
| Sugar beet seed | 880 | 7.1 | 82 | 136 | 61 | 451 | 277 | 74 | 19.1 | 0.37 | 44 | 0.60 | 0.60 | 0.34 | 0.60 |
| **13 Oil cakes and meals** | | | | | | | | | | | | | | | |
| Beech mast cake, shelled | 900 | 12.6 | 357 | 406 | 94 | 76 | 337 | 88 | 20.4 | 0.62 | 72 | 0.88 | 0.90 | 0.24 | 0.76 |
| Beech mast cake, unshelled | 900 | 8.9 | 162 | 217 | 101 | 299 | 328 | 56 | 20.6 | 0.43 | 47 | 0.75 | 0.91 | 0.16 | 0.51 |
| Castor bean meal (de-toxicated) | 900 | 6.2 | 263 | 324 | 16 | 412 | 177 | 71 | 19.0 | 0.32 | 39 | 0.81 | 0.93 | 0.09 | 0.43 |
| Coconut cake | 900 | 13.0 | 184 | 236 | 81 | 127 | 491 | 66 | 19.7 | 0.66 | 75 | 0.78 | 0.97 | 0.63 | 0.83 |
| Coconut cake meal | 900 | 12.7 | 174 | 220 | 76 | 153 | 479 | 72 | 19.5 | 0.65 | 74 | 0.79 | 0.97 | 0.63 | 0.83 |
| Cotton cake, Bombay | 900 | 8.5 | 178 | 231 | 54 | 248 | 401 | 66 | 19.3 | 0.44 | 50 | 0.77 | 0.94 | 0.20 | 0.54 |
| Cotton cake, Brazilian | 900 | 8.9 | 234 | 304 | 61 | 280 | 304 | 50 | 20.1 | 0.44 | 51 | 0.77 | 0.93 | 0.21 | 0.54 |
| Cotton cake, Egyptian | 900 | 8.7 | 203 | 263 | 57 | 242 | 372 | 66 | 19.5 | 0.45 | 51 | 0.77 | 0.92 | 0.21 | 0.54 |
| Cotton cake, decorticated | 900 | 12.3 | 393 | 457 | 89 | 87 | 293 | 74 | 20.8 | 0.59 | 70 | 0.86 | 0.94 | 0.28 | 0.67 |
| Cotton cake, semi-decorticated | 900 | 11.4 | 366 | 426 | 69 | 143 | 297 | 66 | 20.4 | 0.56 | 66 | 0.86 | 0.93 | 0.27 | 0.66 |
| Ground nut cake, decorticated | 900 | 12.9 | 449 | 504 | 67 | 72 | 293 | 63 | 20.7 | 0.62 | 76 | 0.89 | 0.90 | 0.08 | 0.85 |
| Ground nut cake, undecorticated | 900 | 11.4 | 310 | 337 | 101 | 256 | 243 | 63 | 20.9 | 0.55 | 63 | 0.92 | 0.90 | 0.11 | 0.84 |
| Ground nut meal, decorticated extracted | 900 | 11.7 | 491 | 552 | 8 | 88 | 289 | 63 | 19.6 | 0.60 | 75 | 0.89 | 0.86 | 0.08 | 0.85 |
| Ground nut meal, undecorticated extracted | 900 | 9.2 | 316 | 343 | 21 | 273 | 316 | 47 | 19.5 | 0.47 | 58 | 0.92 | 0.79 | 0.11 | 0.69 |
| Hemp seed cake | 900 | 9.0 | 255 | 344 | 97 | 268 | 204 | 87 | 20.5 | 0.44 | 48 | 0.74 | 0.90 | 0.08 | 0.58 |
| Hemp seed meal | 900 | 6.9 | 296 | 394 | 19 | 291 | 190 | 106 | 13.6 | 0.37 | 43 | 0.75 | 0.77 | 0.08 | 0.53 |
| Kapok cake | 900 | 8.7 | 232 | 313 | 81 | 299 | 233 | 73 | 20.3 | 0.43 | 48 | 0.74 | 0.91 | 0.20 | 0.50 |
| Linseed cake, English made | 900 | 13.4 | 286 | 332 | 107 | 102 | 400 | 59 | 23.9 | 0.64 | 75 | 0.86 | 0.92 | 0.49 | 0.80 |
| Linseed cake, foreign | 900 | 12.9 | 305 | 354 | 77 | 104 | 402 | 62 | 23.3 | 0.63 | 75 | 0.86 | 0.93 | 0.50 | 0.80 |
| Linseed meal, extracted | 900 | 11.9 | 348 | 404 | 36 | 102 | 384 | 73 | 23.4 | 0.62 | 74 | 0.86 | 0.90 | 0.50 | 0.80 |
| Media cake | 900 | 9.6 | 250 | 358 | 117 | 233 | 207 | 86 | 21.0 | 0.46 | 51 | 0.70 | 0.80 | 0.20 | 0.60 |
| Niger cake | 900 | 10.5 | 292 | 364 | 66 | 203 | 262 | 104 | 19.4 | 0.54 | 62 | 0.80 | 0.81 | 0.27 | 0.84 |
| Olive cake | 900 | 12.7 | 69 | 71 | 201 | 338 | 329 | 61 | 22.1 | 0.57 | 60 | 0.97 | 0.95 | 0.33 | 0.70 |

| Food name | Dry Matter Content g/kg | Metabolizable Energy MJ/kg DM | Digestible Crude Protein g/kg DM | Analysis of Dry Matter g/kg — Crude Protein | Ether Extract | Crude Fibre | N free Extract | Total Ash | Gross Energy MJ/kg DM | Q {ME/GE} | Digestible Organic Matter in Dry Matter DOMD% | Digestibility Coefficients (decimal) — Crude Protein | Ether Extract | Crude Fibre | N free Extract |
|---|---|---|---|---|---|---|---|---|---|---|---|---|---|---|---|
| Palm kernel cake | 900 | 12.8 | 196 | 216 | 68 | 150 | 522 | 44 | 19.8 | 0.65 | 76 | 0.91 | 0.88 | 0.38 | 0.85 |
| Palm kernel meal, extracted | 900 | 12.2 | 204 | 227 | 10 | 167 | 552 | 44 | 18.5 | 0.66 | 78 | 0.90 | 0.89 | 0.50 | 0.88 |
| Poppy seed cake | 900 | 11.3 | 322 | 408 | 108 | 92 | 240 | 152 | 19.6 | 0.58 | 62 | 0.79 | 0.93 | 0.49 | 0.64 |
| Rape cake | 900 | 11.4 | 322 | 388 | 106 | 91 | 280 | 136 | 19.8 | 0.58 | 64 | 0.83 | 0.79 | 0.08 | 0.80 |
| Rape meal, extracted | 900 | 10.9 | 343 | 413 | 34 | 104 | 366 | 82 | 19.2 | 0.57 | 67 | 0.83 | 0.77 | 0.11 | 0.80 |
| Sesame cake, English | 900 | 13.0 | 442 | 491 | 131 | 49 | 231 | 98 | 21.5 | 0.61 | 70 | 0.90 | 0.90 | 0.31 | 0.56 |
| Sesame cake, French | 900 | 11.7 | 371 | 412 | 121 | 187 | 183 | 97 | 21.1 | 0.56 | 64 | 0.90 | 0.92 | 0.31 | 0.56 |
| Sesame meal, extracted | 900 | 10.4 | 444 | 493 | 26 | 82 | 284 | 114 | 18.8 | 0.55 | 65 | 0.90 | 0.91 | 0.31 | 0.56 |
| Soya bean cake | 900 | 13.3 | 454 | 504 | 66 | 60 | 308 | 62 | 20.7 | 0.64 | 79 | 0.90 | 0.93 | 0.72 | 0.77 |
| Soya bean meal, extracted | 900 | 12.3 | 453 | 503 | 17 | 58 | 360 | 62 | 19.5 | 0.63 | 79 | 0.90 | 0.88 | 0.71 | 0.77 |
| Sunflower cake, decorticated | 900 | 13.3 | 372 | 413 | 152 | 134 | 226 | 74 | 22.1 | 0.60 | 71 | 0.90 | 0.88 | 0.30 | 0.71 |
| Sunflower cake, undecorticated | 900 | 9.5 | 185 | 206 | 80 | 323 | 311 | 80 | 19.6 | 0.48 | 53 | 0.90 | 0.90 | 0.18 | 0.71 |
| Sunflower meal, extracted | 900 | 10.4 | 381 | 423 | 11 | 181 | 312 | 72 | 19.0 | 0.54 | 67 | 0.90 | 0.90 | 0.30 | 0.71 |
| Walnut cake | 900 | 14.7 | 364 | 404 | 141 | 77 | 319 | 59 | 22.0 | 0.67 | 79 | 0.90 | 0.95 | 0.25 | 0.85 |
| **14 Feedingstuffs of animal crgin** | | | | | | | | | | | | | | | |
| Blood meal | 900 | 13.2 | 848 | 942 | 9 | 0 | 18 | 31 | 22.0 | 0.60 | 86 | 0.90 | 1.00 | 0.00 | 0.00 |
| Fish meal, white | 900 | 11.1 | 631 | 701 | 40 | 0 | 18 | 241 | 17.8 | 0.62 | 68 | 0.90 | 0.94 | 0.00 | 0.80 |
| Greaves | 900 | 18.2 | 615 | 648 | 281 | 0 | 0 | 71 | 26.1 | 0.70 | 87 | 0.95 | 0.92 | 0.00 | 0.00 |
| Pure meat meal | 900 | 16.3 | 753 | 810 | 148 | 0 | 0 | 42 | 24.3 | 0.67 | 89 | 0.93 | 0.95 | 0.00 | 0.00 |
| Feeding meat meal (high fat) | 900 | 13.3 | 624 | 663 | 121 | 0 | 6 | 210 | 20.0 | 0.66 | 74 | 0.94 | 0.89 | 0.00 | 1.00 |
| Feeding meat meal (low fat) | 900 | 11.1 | 631 | 717 | 31 | 0 | 43 | 209 | 18.2 | 0.61 | 70 | 0.88 | 0.83 | 0.00 | 0.98 |
| Meat and bone meal (high protein) | 900 | 9.7 | 465 | 597 | 50 | 0 | 62 | 291 | 16.6 | 0.58 | 57 | 0.78 | 0.95 | 0.00 | 0.98 |
| Meat and bone meal (medium protein) | 900 | 7.9 | 411 | 527 | 44 | 0 | 17 | 412 | 14.0 | 0.57 | 47 | 0.78 | 0.95 | 0.00 | 0.98 |
| Milk, cows' whole | 128 | 20.2 | 250 | 266 | 305 | 0 | 375 | 55 | 25.0 | 0.81 | 93 | 0.94 | 1.00 | 0.00 | 1.00 |
| Milk, buttermilk | 92 | 15.7 | 368 | 391 | 87 | 0 | 446 | 76 | 20.3 | 0.77 | 90 | 0.94 | 1.00 | 0.00 | 1.00 |
| Milk, separated | 94 | 14.1 | 350 | 372 | 11 | 0 | 532 | 85 | 18.3 | 0.77 | 89 | 0.94 | 1.00 | 0.00 | 1.00 |
| Milk, skimmed, deep set | 97 | 14.8 | 339 | 361 | 41 | 0 | 515 | 82 | 19.0 | 0.78 | 90 | 0.94 | 1.00 | 0.00 | 1.00 |
| Milk, skimmed, shallow set | 100 | 15.3 | 329 | 350 | 70 | 0 | 500 | 80 | 19.6 | 0.78 | 90 | 0.94 | 1.00 | 0.00 | 1.00 |
| Milk, whey | 66 | 14.5 | 91 | 106 | 30 | 0 | 758 | 106 | 17.0 | 0.85 | 88 | 0.86 | 1.00 | 0.00 | 1.00 |
| **15 By-products** | | | | | | | | | | | | | | | |
| Apple pomace, fresh | 250 | 8.4 | 28 | 60 | 44 | 184 | 656 | 56 | 18.3 | 0.46 | 51 | 0.47 | 0.45 | 0.00 | 0.70 |
| Apple pomace, dried | 900 | 7.7 | 18 | 46 | 40 | 308 | 587 | 20 | 19.0 | 0.41 | 47 | 0.40 | 0.49 | 0.06 | 0.70 |
| Fine barley dust | 860 | 13.5 | 101 | 136 | 26 | 52 | 750 | 36 | 18.4 | 0.73 | 83 | 0.74 | 0.91 | 0.24 | 0.92 |

*(continued over)*

| Food name | Dry Matter Content g/kg | Metabolizable Energy MJ/kg DM | Digestible Crude Protein g/kg DM | Analysis of Dry Matter g/kg: Crude Protein | Ether Extract | Crude Fibre | N free Extract | Total Ash | Gross Energy MJ/kg DM | Q {ME/GE} | Digestible Organic Matter in Dry Matter DOMD% | Digestibility Coefficients (decimal): Crude Protein | Ether Extract | Crude Fibre | N free Extract |
|---|---|---|---|---|---|---|---|---|---|---|---|---|---|---|---|
| Barley, brewers' grains, fresh | 220 | 10.0 | 149 | 205 | 64 | 186 | 500 | 45 | 19.6 | 0.51 | 59 | 0.73 | 0.86 | 0.39 | 0.62 |
| Barley, brewers' grains, ensiled | 280 | 10.0 | 149 | 204 | 64 | 189 | 500 | 43 | 19.7 | 0.51 | 59 | 0.73 | 0.86 | 0.39 | 0.62 |
| Barley, brewers' grains, dried | 900 | 10.3 | 145 | 204 | 71 | 169 | 512 | 43 | 19.8 | 0.52 | 60 | 0.71 | 0.88 | 0.48 | 0.60 |
| Barley, distillers' grains, fresh | 250 | 11.8 | 237 | 320 | 116 | 136 | 396 | 32 | 21.6 | 0.55 | 65 | 0.74 | 0.87 | 0.47 | 0.62 |
| Barley, distillers' grains, dried | 900 | 12.1 | 214 | 301 | 126 | 110 | 443 | 20 | 21.9 | 0.55 | 65 | 0.71 | 0.88 | 0.48 | 0.62 |
| Barley, ale and porter grains, fresh | 250 | 10.2 | 178 | 240 | 76 | 212 | 428 | 44 | 20.2 | 0.51 | 59 | 0.74 | 0.86 | 0.39 | 0.62 |
| Barley, ale and porter grains, dried | 900 | 10.3 | 153 | 219 | 74 | 194 | 477 | 36 | 20.1 | 0.51 | 60 | 0.70 | 0.88 | 0.48 | 0.60 |
| Barley malt culms | 900 | 11.2 | 222 | 271 | 22 | 156 | 471 | 80 | 18.4 | 0.61 | 72 | 0.82 | 0.75 | 0.91 | 0.73 |
| Bean husks (chaff or hulls) | 900 | 9.4 | 0 | 40 | 2 | 488 | 421 | 49 | 17.8 | 0.53 | 67 | 0.00 | 1.00 | 0.88 | 0.57 |
| Broad bean pod meal | 900 | 10.4 | 112 | 167 | 11 | 178 | 571 | 73 | 17.7 | 0.59 | 67 | 0.67 | 0.60 | 0.58 | 0.79 |
| Fodder-cellulose (from wheat straw by paper process) | 900 | 10.3 | 0 | 3 | 6 | 798 | 162 | 31 | 18.5 | 0.56 | 79 | 0.00 | 0.00 | 0.91 | 0.38 |
| Flax chaff (containing about 10 per cent seed) | 860 | 6.5 | 55 | 92 | 57 | 369 | 402 | 80 | 18.6 | 0.35 | 40 | 0.60 | 0.74 | 0.46 | 0.32 |
| Hominy chop, high grade | 900 | 14.7 | 78 | 118 | 89 | 49 | 716 | 29 | 19.9 | 0.74 | 84 | 0.66 | 0.91 | 0.75 | 0.90 |
| Hominy chop, low grade | 900 | 14.1 | 70 | 106 | 69 | 94 | 701 | 30 | 19.4 | 0.73 | 83 | 0.66 | 0.90 | 0.75 | 0.90 |
| Hops, spent, fresh | 250 | 6.3 | 52 | 172 | 76 | 236 | 456 | 60 | 19.6 | 0.32 | 35 | 0.30 | 0.63 | 0.17 | 0.47 |
| Hops, spent, dried | 900 | 6.4 | 53 | 172 | 77 | 236 | 443 | 72 | 19.4 | 0.33 | 36 | 0.31 | 0.65 | 0.17 | 0.48 |
| Horse-chestnut meal (alcohol-extracted) | 900 | 10.1 | 0 | 73 | 74 | 80 | 746 | 27 | 19.4 | 0.52 | 58 | 0.00 | 0.60 | 0.00 | 0.72 |
| Horse-chestnut meal (water extracted) | 900 | 9.9 | 0 | 78 | 78 | 84 | 736 | 24 | 19.6 | 0.51 | 57 | 0.00 | 0.66 | 0.00 | 0.70 |
| Lentil husks (chaff or hulls) | 900 | 9.0 | 15 | 127 | 8 | 291 | 539 | 36 | 18.3 | 0.49 | 60 | 0.12 | 1.00 | 0.67 | 0.70 |
| Maize, flaked | 900 | 15.0 | 106 | 110 | 49 | 17 | 814 | 10 | 19.2 | 0.78 | 92 | 0.96 | 0.47 | 0.33 | 0.97 |
| Maize germ meal, high fat | 900 | 14.9 | 116 | 146 | 140 | 46 | 628 | 41 | 21.0 | 0.71 | 80 | 0.80 | 0.92 | 0.61 | 0.84 |
| Maize germ meal, low fat | 900 | 13.2 | 90 | 112 | 36 | 34 | 779 | 39 | 18.4 | 0.71 | 80 | 0.80 | 0.92 | 0.61 | 0.84 |
| Maize meal, degermed, cooked | 900 | 15.6 | 99 | 104 | 17 | 14 | 856 | 9 | 18.5 | 0.85 | 97 | 0.95 | 0.87 | 0.92 | 0.99 |
| Maize bran | 900 | 12.5 | 62 | 96 | 47 | 132 | 704 | 21 | 19.1 | 0.66 | 75 | 0.65 | 0.86 | 0.33 | 0.86 |
| Maize, gluten feed | 900 | 13.5 | 223 | 262 | 38 | 39 | 633 | 28 | 19.4 | 0.70 | 83 | 0.85 | 0.79 | 0.71 | 0.87 |
| Maize, gluten meal | 900 | 14.2 | 339 | 394 | 52 | 23 | 518 | 12 | 23.7 | 0.69 | 85 | 0.86 | 0.94 | 0.00 | 0.90 |
| Maize, malt culms | 860 | 15.1 | 201 | 240 | 167 | 67 | 457 | 69 | 21.6 | 0.70 | 80 | 0.84 | 0.86 | 0.78 | 0.88 |
| Maize, feeding meal from corn flour | 900 | 13.8 | 188 | 227 | 49 | 56 | 658 | 11 | 19.8 | 0.69 | 84 | 0.83 | 0.80 | 0.66 | 0.87 |

| Food name | Dry Matter Content g/kg | Metabolizable Energy MJ/kg DM | Digestible Crude Protein g/kg DM | Analysis of Dry Matter g/kg | | | | | Gross Energy MJ/kg DM | Q {ME/GE} | Digestible Organic Matter in Dry Matter DOMD% | Digestibility Coefficients (decimal) | | | |
|---|---|---|---|---|---|---|---|---|---|---|---|---|---|---|---|
| | | | | Crude Protein | Ether Extract | Crude Fibre | N free Extract | Total Ash | | | | Crude Protein | Ether Extract | Crude Fibre | N free Extract |
| Maize, starch feed | 900 | 14.1 | 211 | 251 | 76 | 83 | 580 | 10 | 20.6 | 0.69 | 83 | 0.84 | 0.90 | 0.72 | 0.85 |
| Malt, dry | 900 | 12.9 | 118 | 148 | 33 | 97 | 694 | 28 | 18.8 | 0.68 | 80 | 0.80 | 0.77 | 0.50 | 0.87 |
| Oat bran } from preparation | 900 | 8.8 | 44 | 89 | 40 | 242 | 562 | 67 | 18.2 | 0.49 | 55 | 0.50 | 0.56 | 0.37 | 0.70 |
| Oat-meal } of oat-meal | 900 | 12.4 | 131 | 174 | 73 | 18 | 712 | 22 | 19.9 | 0.62 | 72 | 0.75 | 0.81 | 0.50 | 0.73 |
| Oat husks | 900 | 4.9 | 0 | 21 | 11 | 351 | 574 | 42 | 17.8 | 0.28 | 33 | 0.00 | 0.40 | 0.33 | 0.36 |
| Pea husks (chaff or hulls) | 860 | 12.5 | 41 | 60 | 8 | 545 | 355 | 31 | 18.4 | 0.68 | 88 | 0.68 | 0.71 | 0.94 | 0.90 |
| Pea pod meal (from canning industry) | 900 | 10.7 | 108 | 150 | 13 | 169 | 604 | 63 | 17.9 | 0.60 | 69 | 0.72 | 0.67 | 0.63 | 0.77 |
| Potato sludge | 860 | 9.9 | 0 | 40 | 1 | 102 | 793 | 64 | 16.9 | 0.58 | 62 | 0.00 | 0.00 | 0.13 | 0.77 |
| Potato slump | 900 | 6.6 | 135 | 270 | 41 | 106 | 453 | 130 | 17.8 | 0.37 | 40 | 0.50 | 0.49 | 0.21 | 0.50 |
| Potato pulp (dry) | 860 | 10.8 | 0 | 40 | 1 | 102 | 793 | 64 | 16.9 | 0.64 | 68 | 0.00 | 0.00 | 0.24 | 0.83 |
| Potato cossettes (meal) | 900 | 12.4 | 54 | 98 | 6 | 22 | 834 | 40 | 17.6 | 0.70 | 78 | 0.55 | 0.00 | 0.50 | 0.86 |
| Potato flakes | 900 | 13.3 | 42 | 91 | 3 | 23 | 840 | 42 | 17.5 | 0.76 | 84 | 0.46 | 0.00 | 0.48 | 0.94 |
| Potato slices | 900 | 13.1 | 45 | 104 | 2 | 18 | 832 | 43 | 17.5 | 0.75 | 83 | 0.43 | 0.00 | 0.50 | 0.93 |
| Rice meal | 900 | 12.7 | 82 | 141 | 150 | 70 | 544 | 94 | 20.3 | 0.62 | 66 | 0.58 | 0.85 | 0.25 | 0.79 |
| Rice sludge, dried | 860 | 13.6 | 250 | 305 | 24 | 13 | 642 | 16 | 19.5 | 0.70 | 85 | 0.82 | 0.48 | 0.64 | 0.91 |
| Rye bran | 880 | 11.2 | 143 | 191 | 35 | 59 | 664 | 51 | 18.6 | 0.60 | 68 | 0.75 | 0.77 | 0.33 | 0.74 |
| Seaweed meal(dried):*Laminaria* | 860 | 8.8 | 73 | 136 | 13 | 102 | 548 | 201 | 15.3 | 0.58 | 56 | 0.54 | 0.82 | 0.73 | 0.73 |
| Seaweed meal(dried):*Fucus* | 860 | 8.8 | 0 | 58 | 48 | 106 | 605 | 184 | 16.0 | 0.55 | 51 | 0.00 | 0.95 | 0.66 | 0.66 |
| Sugar beet pulp, pressed | 180 | 12.7 | 66 | 106 | 6 | 206 | 644 | 39 | 18.0 | 0.71 | 84 | 0.63 | 0.00 | 0.90 | 0.91 |
| Sugar beet pulp, dried | 900 | 12.7 | 59 | 99 | 7 | 203 | 657 | 34 | 18.0 | 0.71 | 84 | 0.60 | 0.00 | 0.89 | 0.91 |
| Sugar beet pulp, molassed | 900 | 12.2 | 61 | 106 | 6 | 144 | 662 | 82 | 17.1 | 0.71 | 79 | 0.58 | 0.00 | 0.89 | 0.91 |
| Sugar beet molasses | 750 | 12.9 | 16 | 47 | 0 | 0 | 884 | 69 | 16.7 | 0.77 | 81 | 0.34 | 0.00 | 0.00 | 0.90 |
| Sugar cane molasses | 750 | 12.7 | 14 | 41 | 0 | 0 | 872 | 87 | 16.4 | 0.78 | 80 | 0.35 | 0.00 | 0.00 | 0.90 |
| Tapioca flour | 900 | 15.0 | 13 | 20 | 6 | 29 | 922 | 23 | 17.6 | 0.86 | 95 | 0.67 | 0.20 | 0.76 | 0.99 |
| Wheat feeds, middlings | 880 | 11.9 | 129 | 176 | 41 | 86 | 650 | 47 | 18.8 | 0.63 | 72 | 0.73 | 0.87 | 0.23 | 0.82 |
| Wheat feeds, bran | 880 | 10.1 | 126 | 170 | 45 | 114 | 603 | 67 | 18.6 | 0.55 | 61 | 0.74 | 0.69 | 0.22 | 0.71 |
| Yeast, dried | 900 | 11.7 | 381 | 443 | 11 | 2 | 441 | 102 | 18.3 | 0.64 | 75 | 0.86 | 0.40 | 0.00 | 0.82 |
| Yeast, wood sugar(dried) | 900 | 12.6 | 471 | 523 | 14 | 0 | 381 | 81 | 19.2 | 0.66 | 81 | 0.90 | 0.23 | 0.00 | 0.88 |

*Source:* MAFF/ADAS (1987) *Energy Allowances and Feed Systems for Ruminants.* RB433 Reproduced with kind permission of MAFF ADAS (UK).

**TABLE 303   Typical Composition of Feeds for Cattle (USA) (All Values Except Dry Matter Are Shown on a Dry Matter Basis)**

| FEEDSTUFFS | DM % | CP % | EE % | CF % | ADF % | NDF % | ASH % | Ca % | P % | K % | S % | Zn PPM | TDN % | DE Mcal/Lb | NE_M Mcal/Lb | NE_G Mcal/Lb |
|---|---|---|---|---|---|---|---|---|---|---|---|---|---|---|---|---|
| Alfalfa Cubes | 91 | 18 | 2.0 | 29 | 34 | 45 | 11 | 1.3 | 0.23 | 1.9 | 0.35 | 18 | 57 | 1.14 | 0.56 | 0.25 |
| Alfalfa Dehydrated 17% Protein | 92 | 19 | 3.3 | 26 | 32 | 45 | 10 | 1.4 | 0.25 | 2.7 | 0.29 | 17 | 62 | 1.24 | 0.62 | 0.33 |
| Alfalfa Fresh | 26 | 21 | 2.2 | 27 | 32 | 44 | 8 | 1.6 | 0.32 | 2.3 | 0.34 | 21 | 57 | 1.14 | 0.56 | 0.25 |
| Alfalfa Hay Early Bloom | 90 | 18 | 2.2 | 29 | 35 | 47 | 8 | 1.4 | 0.25 | 2.3 | 0.30 | 18 | 57 | 1.14 | 0.56 | 0.25 |
| Alfalfa Hay Mid-Bloom | 89 | 17 | 2.0 | 30 | 38 | 50 | 10 | 1.4 | 0.23 | 1.8 | 0.30 | 17 | 56 | 1.12 | 0.55 | 0.23 |
| Alfalfa Hay full Boom | 88 | 16 | 1.8 | 34 | 41 | 56 | 8 | 1.3 | 0.20 | 1.7 | 0.29 | 17 | 53 | 1.06 | 0.52 | 0.18 |
| Alfalfa Hay Mature | 90 | 14 | 1.7 | 38 | 45 | 59 | 8 | 1.3 | 0.19 | 1.4 | 0.25 | 17 | 50 | 1.00 | 0.49 | 0.12 |
| Alfalfa Silage Wilted | 36 | 18 | 2.8 | 30 | 35 | 46 | 10 | 1.5 | 0.28 | 2.4 | 0.30 | 17 | 56 | 1.12 | 0.55 | 0.23 |
| Alfalfa Seed Screenings | 90 | 28 | 7.0 | 16 | – | – | 7 | 0.03 | 0.67 | – | – | – | 76 | 1.52 | 0.80 | 0.52 |
| Alfalfa Stems | 89 | 11 | 1.3 | 44 | 51 | 68 | 6 | 0.9 | 0.18 | 2.5 | – | – | 47 | 0.94 | 0.46 | 0.07 |
| Almond Hulls | 90 | 3 | 3.0 | 11 | 28 | 30 | 10 | 0.4 | 0.06 | 0.5 | 0.11 | – | 67 | 1.34 | 0.68 | 0.40 |
| Apple Pomace Wet | 20 | 6 | 5.9 | 17 | – | – | 4 | 0.1 | 0.10 | 0.5 | 0.02 | – | 73 | 1.46 | 0.76 | 0.48 |
| Apple Pomace Dried | 89 | 4 | 3.4 | 15 | – | – | 2 | 0.1 | 0.12 | 0.5 | 0.02 | – | 69 | 1.38 | 0.70 | 0.43 |
| Bagasse Sugar Cane | 91 | 1 | 0.7 | 49 | 59 | 86 | 3 | 0.9 | 0.29 | 0.5 | 0.10 | – | 36 | 0.72 | 0.37 | 0.0 |
| Bakery Product Dried | 91 | 11 | 13.0 | 1 | – | – | 3 | 0.1 | 0.35 | 0.6 | 0.02 | 15 | 89 | 1.78 | 0.98 | 0.67 |
| Barley Silage | 32 | 10 | 4.0 | 34 | – | – | 10 | 0.3 | 0.30 | 1.6 | 0.17 | – | 50 | 1.00 | 0.49 | 0.12 |
| Barley Silage Mature | 40 | 9 | 4.0 | 34 | – | – | 10 | 0.2 | 0.30 | 1.5 | 0.15 | – | 67 | 1.34 | 0.68 | 0.40 |
| Barley Straw | 88 | 4 | 1.9 | 42 | 57 | 82 | 7 | 0.3 | 0.05 | 2.0 | 0.15 | 7 | 44 | 0.88 | 0.43 | 0.01 |
| Barley Grain | 89 | 12 | 2.0 | 6 | 7 | 19 | 3 | 0.1 | 0.40 | 0.5 | 0.16 | 20 | 82 | 1.64 | 0.88 | 0.59 |
| Barley Grain Light Weight | 88 | 13 | 2.3 | 9 | – | – | 4 | 0.0 | – | – | – | – | 79 | 1.58 | 0.84 | 0.55 |
| Barley Grain Screenings | 89 | 13 | 2.6 | 9 | – | – | 4 | 0.0 | – | – | 0.15 | – | 81 | 1.62 | 0.87 | 0.58 |
| Beans Navy Cull | 90 | 24 | 1.4 | 5 | – | – | 6 | 0.1 | 0.40 | 0.1 | 0.26 | – | 84 | 1.68 | 0.91 | 0.61 |
| Beet Pulp Wet | 11 | 10 | 2.0 | 20 | 34 | 59 | 5 | 0.9 | 0.50 | 1.4 | 0.22 | 1 | 68 | 1.36 | 0.69 | 0.41 |
| Beet Pulp Dried | 91 | 9 | 0.8 | 21 | 34 | 59 | 5 | 0.8 | 0.10 | 0.2 | 0.22 | 1 | 72 | 1.44 | 0.74 | 0.47 |
| Beet Pulp Wet With Molasses | 24 | 12 | 0.5 | 16 | 27 | 47 | 9 | 0.6 | 0.08 | 0.2 | 0.36 | 11 | 76 | 1.52 | 0.80 | 0.52 |
| Beet Pulp Dry With Molasses | 92 | 12 | 0.5 | 16 | 27 | 47 | 9 | 0.6 | 0.10 | 1.8 | 0.36 | 11 | 76 | 1.52 | 0.80 | 0.52 |
| Beet Tops Sugar | 20 | 13 | 1.4 | 9 | – | – | 25 | 0.7 | 0.24 | 1.8 | 0.45 | 20 | 58 | 1.16 | 0.57 | 0.26 |
| Beet Top Silage | 25 | 10 | 2.0 | 10 | – | – | 38 | 1.2 | 0.22 | 4.8 | – | – | 52 | 1.04 | 0.51 | 0.16 |
| Bermuda Grass Coastal Dehydrated | 90 | 16 | 3.8 | 27 | 24 | 40 | 7 | 0.3 | 0.25 | 5.7 | 0.22 | – | 62 | 1.24 | 0.62 | 0.33 |
| Bermuda Grass Hay Coastal | 91 | 9 | 2.0 | 30 | 33 | 75 | 5 | 0.4 | 0.18 | 1.5 | 0.14 | – | 49 | 0.98 | 0.48 | 0.11 |
| Bermuda Grass Hay | 91 | 9 | 1.9 | 30 | 35 | 77 | 8 | 0.5 | 0.21 | 1.7 | 0.21 | – | 49 | 0.98 | 0.48 | 0.11 |
| Birdsfoot Trefoil Fresh | 22 | 21 | 4.7 | 21 | – | – | 7 | 1.8 | 0.25 | 2.3 | 0.25 | – | 70 | 1.40 | 0.72 | 0.44 |
| Birdsfoot Trefoil Hay | 89 | 16 | 2.2 | 30 | 34 | 44 | 8 | 1.8 | 0.22 | 1.8 | 0.25 | – | 57 | 1.4 | 0.56 | 0.25 |
| Bluegrass Kentucky Fresh Early Bloom | 36 | 14 | 3.9 | 27 | – | – | 7 | 0.5 | 0.39 | 2.3 | 0.12 | 25 | 72 | 1.44 | 0.74 | 0.47 |
| Blue Grass Straw | 93 | 6 | 1.1 | 40 | 50 | 78 | 6 | 0.2 | 0.10 | – | – | – | 45 | 0.90 | 0.44 | 0.03 |

| FEEDSTUFFS | DM % | CP % | EE % | Fibre |  |  |  | Minerals |  |  |  |  | TDN % | Energy |  |  |
|---|---|---|---|---|---|---|---|---|---|---|---|---|---|---|---|---|
|  |  |  |  | CF % | ADF % | NDF % | ASH % | Ca % | P % | K % | S % | Zn PPM |  | DE Mcal/Lb | NE_M Mcal/Lb | NE_G Mcal/Lb |
| Bluestem Fresh Mature | 61 | 6 | 2.5 | 34 | – | – | 4 | 0.3 | 0.14 | 1.0 | 0.05 | 28 | 46 | 0.92 | 0.45 | 0.05 |
| Bone Meal Steamed | 95 | 13 | 11.6 | 1 | – | – | 79 | 30.7 | 12.86 | 0.2 | 0.25 | 130 | 16 | 0.32 | 0.34 | 0.0 |
| Brewers Grains Wet | 24 | 26 | 6.5 | 16 | 21 | 34 | 5 | 0.3 | 0.57 | 0.1 | 0.36 | 100 | 81 | 1.62 | 0.87 | 0.58 |
| Brewers Dried Grains | 92 | 28 | 7.5 | 15 | 19 | 32 | 4 | 0.3 | 0.57 | 0.1 | 0.36 | 100 | 81 | 1.62 | 0.87 | 0.58 |
| Brewers Yeast Dried | 94 | 48 | 1.0 | 3 | – | – | 7 | 0.1 | 1.56 | 1.8 | 0.41 | 41 | 79 | 1.58 | 0.84 | 0.55 |
| Brome Grass Fresh Immature | 32 | 15 | 4.1 | 28 | 33 | 54 | 10 | 0.4 | 0.39 | 2.7 | 0.20 | – | 64 | 1.28 | 0.64 | 0.36 |
| Brome Grass Hay | 89 | 10 | 2.5 | 32 | 39 | 69 | 9 | 0.5 | 0.23 | 2.5 | 0.16 | – | 52 | 1.04 | 0.51 | 0.16 |
| Buckwheat Grain | 88 | 12 | 2.8 | 11 | – | 0 | 2 | 0.1 | 0.36 | 0.5 | 0.16 | 10 | 79 | 1.58 | 0.84 | 0.55 |
| Buttermilk Dried | 94 | 34 | 5.6 | 0 | 0 | 0 | 10 | 1.5 | 1.03 | 1.1 | 0.09 | 44 | 88 | 1.76 | 0.97 | 0.65 |
| Cactus | 32 | 6 | 2.1 | 27 | 23 | 29 | 16 | 2.9 | – | – | – | – | 65 | 1.30 | 0.65 | 0.37 |
| Calcium Carbonate | 99 | 0 | 0.0 | 0 | 0 | 0 | 99 | 39.0 | 0.04 | – | 0.09 | – | 0 | 0.0 | 0.0 | 0.0 |
| Cattle Manure Dried | 92 | 15 | 2.6 | 34 | 37 | 57 | 10 | 1.3 | 1.00 | 0.5 | – | 240 | 41 | 0.82 | 0.41 | 0.41 |
| Cheatgrass Fresh immature | 21 | 16 | 2.7 | 23 | – | – | 10 | 0.6 | 0.28 | – | – | – | 68 | 1.36 | 0.69 | 0.0 |
| Citrus Pulp Dried | 90 | 7 | 4.1 | 13 | 22 | 23 | 6 | 1.8 | 0.12 | 1.0 | 0.20 | 14 | 77 | 1.54 | 0.81 | 0.53 |
| Clover Ladino Fresh | 19 | 25 | 4.8 | 14 | – | – | 11 | 1.3 | 0.42 | 2.2 | 0.20 | 39 | 69 | 1.38 | 0.70 | 0.43 |
| Clover Ladino Hay | 90 | 21 | 2.0 | 24 | – | – | 9 | 1.7 | 0.32 | 2.4 | 0.22 | 17 | 61 | 1.22 | 0.61 | 0.31 |
| Clover Red Fresh | 24 | 18 | 4.0 | 24 | 33 | 44 | 9 | 1.7 | 0.26 | 2.0 | 0.17 | 23 | 64 | 1.28 | 0.64 | 0.36 |
| Clover Red Hay | 88 | 15 | 2.9 | 30 | 41 | 56 | 8 | 1.4 | 0.22 | 1.9 | 0.17 | 17 | 59 | 1.18 | 0.58 | 0.28 |
| Coffee Grounds | 88 | 13 | 22.3 | 41 | 68 | 80 | 1 | 0.1 | 0.08 | – | – | – | 13 | 0.26 | 0.20 | 0.0 |
| Corn Whole Plant Pellets | 91 | 9 | 2.4 | 21 | 29 | 48 | 6 | 0.5 | 0.24 | 1.0 | 0.14 | – | 63 | 1.26 | 0.63 | 0.34 |
| Corn Fodder | 80 | 9 | 2.4 | 27 | 41 | 71 | 7 | 0.3 | 0.18 | 1.0 | 0.14 | – | 65 | 1.30 | 0.65 | 0.37 |
| Corn Stover Mature | 80 | 5 | 1.3 | 35 | 41 | – | 7 | 0.5 | 0.09 | 1.6 | 0.17 | – | 59 | 1.18 | 0.58 | 0.28 |
| Corn Silage Milk Stage | 26 | 8 | 2.8 | 26 | 31 | – | 6 | 0.3 | 0.24 | 1.6 | 0.12 | 25 | 67 | 1.34 | 0.68 | 0.40 |
| Corn Silage Mature Well Eared | 36 | 8 | 2.7 | 23 | – | – | 7 | 0.3 | 0.20 | 1.0 | 0.11 | 24 | 69 | 1.38 | 0.70 | 0.43 |
| Corn Grain Dent Yellow | 88 | 10 | 4.0 | 3 | 3 | 10 | 2 | 0.0 | 0.30 | 0.4 | 0.14 | 14 | 90 | 1.80 | 0.99 | 0.68 |
| Corn Grain High Lysine | 92 | 12 | 4.4 | 4 | – | – | 2 | 0.0 | 0.24 | 0.3 | 0.11 | – | 84 | 1.68 | 0.91 | 0.61 |
| Corn and Cob Meal | 87 | 9 | 3.7 | 9 | 10 | 25 | 2 | 0.1 | 0.26 | 0.5 | 0.21 | 10 | 82 | 1.64 | 0.88 | 0.59 |
| Corn Cobs | 90 | 3 | 0.5 | 36 | 43 | 88 | 2 | 0.1 | 0.04 | 0.8 | 0.47 | 5 | 48 | 0.96 | 0.47 | 0.09 |
| Corn Gluten Feed | 90 | 27 | 2.9 | 8 | – | – | 7 | 0.5 | 0.86 | 0.6 | 0.24 | 100 | 82 | 1.64 | 0.88 | 0.59 |
| Corn Gluten Meal | 91 | 43 | 2.5 | 5 | – | – | 4 | 0.2 | 0.51 | 0.0 | 0.85 | 45 | 84 | 1.68 | 0.91 | 0.61 |
| Corn Cannery Waste | 29 | 8 | 3.0 | 28 | 36 | 59 | 5 | 0.1 | 0.29 | 1.0 | 0.13 | 25 | 68 | 1.36 | 0.69 | 0.41 |
| Cotton Gin Trash | 90 | 7 | 1.7 | 37 | – | – | 9 | 0.3 | 0.16 | – | – | – | 44 | 0.88 | 0.43 | 0.01 |
| Cottonseed Hulls | 90 | 4 | 1.5 | 48 | 67 | 86 | 3 | 0.1 | 0.07 | 1.0 | 0.09 | 22 | 44 | 0.88 | 0.43 | 0.01 |
| Cottonseed Meal Screw Press 41% Protein | 93 | 45 | 6.0 | 13 | 22 | 30 | 7 | 0.2 | 1.18 | 1.4 | 0.33 | 63 | 78 | 1.56 | 0.82 | 0.54 |
| Cottonseed Meal Solvent 41% Protein | 92 | 46 | 2.2 | 13 | 22 | 30 | 7 | 0.2 | 1.16 | 1.4 | 0.36 | 66 | 75 | 1.50 | 0.78 | 0.50 |
| Crambe Meal Solvent | 92 | 34 | 1.1 | 25 | – | – | 6 | 1.2 | 1.30 | – | – | – | 70 | 1.40 | 0.72 | 0.44 |

*(continued over)*

| FEEDSTUFFS | Fibre | | | | | | | Minerals | | | | | | Energy | | |
|---|---|---|---|---|---|---|---|---|---|---|---|---|---|---|---|---|
| | DM % | CP % | EE % | CF % | ADF % | NDF % | ASH % | Ca % | P % | K % | S % | Zn PPM | TDN % | DE Mcal/Lb. | NE_M Mcal/Lb. | NE_G Mcal/Lb. |
| Cranberry Pulp Dried | 88 | 7 | 15.7 | 26 | 47 | 54 | 2 | – | – | – | – | – | 49 | 0.98 | 0.48 | 0.11 |
| Curaco Phosphate | 99 | 0 | 0.0 | 0 | 0 | 0 | 95 | 34.0 | 15.00 | – | – | – | 0 | 0.0 | 0.0 | 0.0 |
| Defluorinated Phosphate | 99 | 0 | 0.0 | 0 | 0 | 0 | 95 | 32.6 | 18.07 | 1.0 | – | 100 | 0 | 0.0 | 0.0 | 0.0 |
| Diammonium Phosphate | 98 | 115 | 0.0 | 0 | 0 | 0 | 35 | 0.5 | 20.41 | – | – | – | 0 | 0.0 | 0.0 | 0.0 |
| Dicalcium Phosphate | 96 | 0 | 0.0 | 0 | 0 | 0 | 94 | 21.0 | 18.65 | 0.1 | – | 70 | 0 | 0.0 | 0.0 | 0.0 |
| Distillers Grains Corn | 92 | 30 | 8.2 | 14 | 16 | 39 | 2 | 0.1 | 0.43 | 0.2 | 0.46 | 35 | 84 | 1.68 | 0.91 | 0.61 |
| Distillers Grains Corn with Solubles | 92 | 29 | 9.8 | 10 | – | – | 5 | 0.2 | 0.85 | 0.7 | 0.32 | 90 | 87 | 1.74 | 0.95 | 0.64 |
| Distillers Dried Solubles | 92 | 30 | 9.5 | 4 | – | – | 8 | 0.4 | 1.48 | 1.9 | 0.40 | 91 | 88 | 1.76 | 0.97 | 0.65 |
| Fat Animal-Poultry | 99 | 0 | 99.0 | 0 | 0 | 0 | 0 | 0.0 | 0.0 | 0.0 | 0.0 | – | 198 | 3.96 | 2.43 | 1.84 |
| Feather Meal Hydrolyzed | 94 | 91 | 3.3 | 2 | 20 | 20 | 4 | 0.2 | 0.78 | 0.3 | 2.00 | 53 | 68 | 1.36 | 0.69 | 0.41 |
| Fescue Kentucky 31 Fresh | 29 | 15 | 5.5 | 25 | – | – | 10 | 0.4 | 0.35 | 2.6 | – | 22 | 66 | 1.32 | 0.67 | 0.38 |
| Fescue Kentucky 31 Hay Early Bloom | 88 | 18 | 6.6 | 25 | – | – | 9 | 0.5 | 0.37 | – | – | – | 64 | 1.28 | 0.64 | 0.36 |
| Fescue Kentucky 31 Hay Manure | 89 | 14 | 5.2 | 28 | 38 | 65 | 7 | 0.4 | 0.25 | – | – | – | 60 | 1.20 | 0.59 | 0.30 |
| Fescue Straw (Red) | 94 | 4 | 1.1 | 41 | – | – | 6 | 0.0 | 0.06 | – | – | – | 43 | 0.86 | 0.43 | 0.0 |
| Garbage Municpal Cooked | 23 | 16 | 23.3 | 8 | 50 | 59 | 11 | 1.6 | 0.45 | – | – | – | 75 | 1.50 | 0.78 | 0.50 |
| Grain Screenings | 90 | 15 | 5.5 | 14 | – | – | 9 | 0.5 | 0.43 | – | – | 17 | 63 | 1.26 | 0.63 | 0.34 |
| Grain Dust | 91 | 11 | 3.2 | 15 | – | – | 12 | 0.3 | 0.18 | – | – | 42 | 72 | 1.44 | 0.74 | 0.47 |
| Grape Pomac Stemless | 91 | 12 | 7.5 | 32 | 50 | 53 | 9 | 0.6 | 0.06 | – | – | 24 | 30 | 0.60 | 0.37 | 0.0 |
| Grass Silage | 26 | 12 | 4.6 | 34 | 38 | 66 | 9 | 0.8 | 0.22 | 0.6 | – | 29 | 61 | 1.22 | 0.61 | 0.31 |
| Hominy Feed | 90 | 12 | 7.7 | 6 | 12 | 56 | 3 | 0.1 | 0.58 | 2.0 | 0.06 | 3 | 95 | 1.90 | 1.06 | 0.73 |
| Hop Leaves | 37 | 15 | 3.6 | 15 | – | – | 35 | 2.8 | 0.64 | 0.7 | – | – | 49 | 0.98 | 0.48 | 0.11 |
| Hop Vine Silage | 30 | 15 | 3.1 | 21 | – | – | 20 | 3.3 | 0.37 | – | 0.22 | 44 | 53 | 1.06 | 0.52 | 0.18 |
| Hops Spent | 89 | 22 | 4.0 | 28 | – | – | 7 | 1.6 | 0.60 | 1.8 | – | – | 39 | 0.78 | 0.40 | 0.0 |
| Lespedeza Fresh Early Bloom | 25 | 16 | 2.0 | 32 | – | – | 10 | 1.4 | 0.21 | – | 0.21 | – | 67 | 1.34 | 0.68 | 0.40 |
| Linestone Ground | 98 | 0 | 0.0 | 0 | 0 | 0 | 96 | 33.8 | 0.02 | 1.1 | – | – | 0 | 0.0 | 0.0 | 0.0 |
| Linseed Meal Solvent | 91 | 40 | 1.9 | 10 | 17 | 25 | 6 | 0.4 | 0.91 | – | 0.40 | 35 | 76 | 1.52 | 0.80 | 0.52 |
| Meadow Hay | 92 | 8 | 2.5 | 33 | – | – | 9 | 0.6 | 0.17 | 1.5 | – | – | 46 | 0.92 | 0.45 | 0.05 |
| Milo Grain | 89 | 11 | 3.2 | 3 | 8 | 17 | 2 | 0.0 | 0.32 | – | 0.18 | 15 | 80 | 1.60 | 0.85 | 0.56 |
| Molasses Beet | 77 | 9 | 0.2 | 0 | 0 | 0 | 11 | 0.2 | 0.04 | 0.4 | 0.60 | 18 | 79 | 1.58 | 0.84 | 0.55 |
| Molasses Cane | 76 | 5 | 0.0 | 0 | 0 | 0 | 10 | 1.0 | 0.10 | 6.1 | 0.46 | 30 | 81 | 1.62 | 0.87 | 0.58 |
| Molasses Cane Dried | 94 | 11 | 0.6 | 3 | 0 | 0 | 14 | 1.2 | 0.15 | 4.0 | 0.46 | 30 | 81 | 1.62 | 0.87 | 0.58 |
| Molasses Citrus | 65 | 11 | 0.3 | 0 | 0 | 0 | 9 | 2.0 | 0.25 | 4.0 | 0.23 | 137 | 77 | 1.54 | 0.81 | 0.53 |
| Molasses Wood | 59 | 1 | 0.5 | 1 | – | – | 7 | 1.9 | 0.04 | 0.2 | 0.05 | – | 77 | 1.54 | 0.81 | 0.53 |
| Monoammonium Phosphate | 98 | 74 | 0.0 | 0 | 0 | 0 | 24 | 0.5 | 24.77 | 0.1 | 0.71 | 81 | 0 | 0.0 | 0.0 | 0.0 |
| Mono-Dicalcium Phosphate | 97 | 0 | 0.0 | 0 | 0 | 0 | 94 | 16.7 | 21.10 | – | – | 70 | 0 | 0.0 | 0.0 | 0.0 |
| Oat Hay | 87 | 9 | 2.1 | 30 | 38 | 63 | 9 | 0.2 | 0.22 | 1.0 | 0.30 | 39 | 59 | 1.18 | 0.58 | 0.28 |
| Oat Silage | 34 | 11 | 3.8 | 30 | – | – | 10 | 0.4 | 0.25 | 3.4 | 0.32 | 35 | 58 | 1.16 | 0.57 | 0.26 |
| Oat Straw | 90 | 4 | 2.3 | 41 | 46 | 70 | 8 | 0.3 | 0.10 | 2.2 | 0.22 | 6 | 52 | 1.04 | 0.51 | 0.16 |

| FEEDSTUFFS | Fibre DM % | CP % | EE % | CF % | ADF % | NDF % | ASH % | Minerals Ca % | P % | K % | S % | Zn PPM | Energy TDN % | DE Mcal/Lb | NE_M Mcal/Lb | NE_G Mcal/Lb |
|---|---|---|---|---|---|---|---|---|---|---|---|---|---|---|---|---|
| Oats Grain | 89 | 14 | 4.0 | 12 | 17 | 31 | 4 | 0.1 | 0.38 | 0.5 | 0.19 | 18 | 74 | 1.48 | 0.77 | 0.49 |
| Oat Groats | 91 | 18 | 6.6 | 3 | – | – | 2 | 0.1 | 0.47 | 0.4 | 0.22 | – | 93 | 1.86 | 1.03 | 0.71 |
| Oat Meal Feeding | 90 | 17 | 6.0 | 4 | – | – | 3 | 0.1 | 0.51 | 0.5 | 0.29 | – | 94 | 1.88 | 1.05 | 0.72 |
| Oat Mill Byproduct | 90 | 7 | 2.5 | 23 | – | – | 6 | 0.1 | 0.24 | 0.6 | – | – | 33 | 0.66 | 0.37 | 0.0 |
| Oat Hulls | 93 | 5 | 1.5 | 32 | 44 | 81 | 7 | 0.2 | 0.15 | 0.6 | 0.15 | – | 37 | 0.74 | 0.39 | 0.0 |
| Orange Pulp Dried | 89 | 9 | 1.8 | 9 | – | – | 4 | 0.7 | 0.11 | – | – | – | 87 | 1.74 | 0.95 | 0.64 |
| Orchard Grass Fresh Immature | 24 | 18 | 5.0 | 24 | 27 | 45 | 11 | 0.4 | 0.40 | 2.7 | 0.22 | 20 | 65 | 1.30 | 0.65 | 0.37 |
| Orchard Grass Hay | 88 | 11 | 3.3 | 34 | 40 | 70 | 7 | 0.3 | 0.28 | 2.8 | 0.26 | 18 | 59 | 1.18 | 0.58 | 0.28 |
| Pea Vine Hay | 89 | 10 | 1.8 | 32 | – | – | 7 | 1.2 | 0.21 | 1.8 | 0.17 | 15 | 60 | 1.20 | 0.59 | 0.30 |
| Pea Vine Silage | 24 | 13 | 3.3 | 31 | – | – | 8 | 1.3 | 0.24 | – | 0.29 | – | 57 | 1.14 | 0.56 | 0.25 |
| Pea Straw | 89 | 7 | 1.3 | 45 | – | – | 7 | – | 0.11 | 1.1 | – | – | 56 | 1.12 | 0.55 | 0.23 |
| Peas Cull | 89 | 25 | 1.5 | 8 | – | – | 5 | 0.2 | 0.43 | 1.1 | 0.20 | 30 | 79 | 1.58 | 0.84 | 0.55 |
| Peanut Hulls | 92 | 7 | 1.3 | 65 | 65 | 74 | 5 | 0.2 | 0.07 | 0.9 | 0.26 | – | 22 | 0.44 | 0.35 | 0.0 |
| Peanut Meal Solvent | 92 | 52 | 1.3 | 14 | – | – | 8 | 0.2 | 0.71 | 1.2 | 0.30 | 22 | 77 | 1.54 | 0.81 | 0.53 |
| Pineapple Green Chop | 17 | 9 | 2.6 | 23 | 35 | 64 | 3 | – | – | – | – | – | 45 | 0.90 | 0.44 | 0.03 |
| Pineapple Bran | 89 | 5 | 1.2 | 20 | 28 | 59 | 3 | 0.3 | 0.10 | – | – | – | 72 | 1.44 | 0.74 | 0.47 |
| Pineapple Presscake | 21 | 5 | 1.1 | 20 | 36 | 69 | 5 | 0.2 | 0.12 | – | – | – | 72 | 1.44 | 0.74 | 0.47 |
| Potatoes Cull | 21 | 10 | 0.4 | 2 | – | – | 3 | 0.0 | 0.24 | 2.2 | 0.09 | – | 80 | 1.60 | 0.85 | 0.56 |
| Potato Waste Wet | 14 | 7 | 1.5 | 9 | – | – | 5 | 0.2 | 0.26 | 1.3 | 0.11 | 12 | 82 | 1.64 | 0.88 | 0.59 |
| Potato Waste Dried | 89 | 8 | 0.5 | 7 | – | – | 9 | 0.1 | 0.13 | 1.2 | – | – | 85 | 1.70 | 0.92 | 0.62 |
| Potato Waste Wet With Lime | 17 | 5 | 0.3 | 10 | – | – | 16 | 4.2 | 0.18 | – | – | – | 80 | 1.60 | 0.85 | 0.56 |
| Poultry Litter Dried | 86 | 30 | 2.8 | 20 | 16 | 33 | 28 | 2.7 | 1.85 | 1.8 | – | 235 | 64 | 1.28 | 0.64 | 0.36 |
| Poultry Manure Dried | 89 | 30 | 2.1 | 15 | – | – | 8 | 8.6 | 2.30 | 1.7 | – | 400 | 53 | 1.06 | 0.52 | 0.18 |
| Prairie Hay | 91 | 7 | 2.0 | 35 | – | – | 8 | 0.4 | 0.13 | 1.1 | – | 34 | 50 | 1.00 | 0.49 | 0.12 |
| Rapeseed Meal Solvent | 91 | 41 | 2.2 | 14 | 53 | 68 | 6 | 0.7 | 1.10 | 1.4 | 0.28 | 66 | 70 | 1.40 | 0.72 | 0.44 |
| Rice Straw Ammoniated | 87 | 9 | 1.3 | 39 | – | – | 8 | 0.3 | 0.10 | 1.0 | 0.11 | – | 45 | 0.90 | 0.44 | 0.03 |
| Rice Polishings | 90 | 14 | 13.1 | 4 | 20 | 26 | 8 | 0.1 | 1.37 | 1.1 | 0.19 | 28 | 89 | 1.78 | 0.98 | 0.67 |
| Rice Bran | 91 | 13 | 15.2 | 13 | – | – | 13 | 0.1 | 1.68 | 1.9 | 0.20 | 33 | 66 | 1.32 | 0.67 | 0.38 |
| Rice Hulls | 92 | 3 | 0.9 | 44 | 70 | 80 | 20 | 0.1 | 0.08 | 0.3 | 0.09 | – | 13 | 0.26 | 0.33 | 0.0 |
| Rice Mill Feed | 91 | 7 | 5.4 | 33 | – | – | – | 0.4 | 0.62 | – | – | – | 42 | 0.84 | 0.42 | 0.01 |
| Rye Straw | 89 | 4 | 1.5 | 44 | 55 | 71 | 6 | 0.3 | 0.10 | 1.0 | 0.11 | – | 44 | 0.88 | 0.43 | 0.01 |
| Rye Grain | 89 | 13 | 1.7 | 2 | – | – | 2 | 0.1 | 0.38 | 0.5 | 0.17 | – | 81 | 1.62 | 0.87 | 0.58 |
| Safflower Meal Solvent | 91 | 22 | 1.0 | 33 | 41 | 59 | 6 | 0.3 | 0.80 | 0.8 | 0.20 | 34 | 55 | 1.10 | 0.54 | 0.21 |
| Safflower Meal Dehulled Solvent | 91 | 49 | 0.6 | 9 | – | – | 7 | 0.3 | 1.83 | 1.3 | 0.22 | 44 | 76 | 1.52 | 0.80 | 0.52 |
| Sagebrush Fresh | 50 | 13 | 9.2 | 25 | 0 | 0 | 10 | 1.0 | 0.25 | – | – | 36 | 50 | 1.00 | 0.49 | 0.12 |
| Sodium Tripolyphosphate | 96 | 0 | 0.0 | 0 | 0 | 0 | 96 | 0.0 | 25.98 | – | – | – | 0 | 0.0 | 0.0 | 0.0 |
| Sorghum Stover | 85 | 5 | 2.1 | 33 | – | – | 10 | 0.4 | 0.11 | 1.5 | – | – | 57 | 1.14 | 0.56 | 0.25 |

*(continued over)*

| FEEDSTUFFS | DM % | CP % | Fibre EE % | CF % | ADF % | NDF % | ASH % | Minerals Ca % | P % | K % | S % | Zn PPM | Energy TDN % | DE Mcal/Lb | NE$_M$ Mcal/Lb | NE$_G$ Mcal/Lb |
|---|---|---|---|---|---|---|---|---|---|---|---|---|---|---|---|---|
| Sorghum Silage | 31 | 8 | 2.8 | 24 | 21 | 39 | 7 | 0.4 | 0.11 | 1.5 | 0.11 | 32 | 57 | 1.14 | 0.56 | 0.25 |
| Soybean Hay | 89 | 15 | 2.2 | 37 | – | – | 8 | 1.3 | 0.32 | 1.0 | 0.24 | 24 | 52 | 1.04 | 0.51 | 0.16 |
| Soybean Straw | 88 | 5 | 1.4 | 44 | 54 | 70 | 6 | 1.6 | 0.06 | 0.6 | 0.26 | – | 42 | 0.84 | 0.42 | 0.0 |
| Soybeans Whole | 91 | 42 | 19.2 | 6 | – | – | 5 | 0.3 | 0.63 | 1.8 | 0.24 | 60 | 92 | 1.84 | 1.02 | 0.70 |
| Soybean Meal Solvent 44% Protein | 89 | 52 | 1.3 | 6 | 10 | 12 | 7 | 0.3 | 0.73 | 2.1 | 0.48 | 48 | 82 | 1.64 | 0.88 | 0.59 |
| Soybean Meal Solvent 49% Protein | 90 | 56 | 1.2 | 3 | – | – | 6 | 0.3 | 0.71 | 2.2 | 0.48 | 61 | 84 | 1.68 | 0.91 | 0.61 |
| Soybran Flakes (Hulls) | 91 | 12 | 2.8 | 39 | 44 | 60 | 4 | 0.6 | 0.17 | 1.0 | 0.09 | 24 | 65 | 1.30 | 0.65 | 0.37 |
| Sudangrass Fresh Immature | 18 | 17 | 3.9 | 31 | – | – | 9 | 0.5 | 0.31 | 2.0 | 0.04 | – | 70 | 1.40 | 0.72 | 0.44 |
| Sudangrass Hay | 89 | 10 | 1.8 | 31 | 43 | 68 | 10 | 0.4 | 0.30 | 2.1 | 0.06 | – | 59 | 1.18 | 0.58 | 0.28 |
| Sudangrass Silage | 23 | 10 | 3.1 | 34 | – | – | 10 | 0.4 | 0.25 | 3.5 | 0.05 | – | 57 | 1.14 | 0.56 | 0.25 |
| Sunflower Meal Solvent | 93 | 50 | 3.1 | 12 | – | – | 8 | 0.6 | 0.54 | 1.1 | – | – | 65 | 1.30 | 0.65 | 0.37 |
| Sunflower Meal With Hulls | 90 | 32 | 1.4 | 27 | – | – | 7 | 0.4 | 1.04 | 0.9 | 0.33 | – | 57 | 1.14 | 0.56 | 0.25 |
| Sunflower Hulls | 90 | 5 | 2.2 | 25 | 63 | – | 3 | 0.0 | 0.11 | – | – | – | 40 | 0.80 | 0.41 | 0.0 |
| Timothy Fresh Pre-Bloom | 26 | 11 | 3.8 | 32 | – | – | 7 | 0.4 | 0.28 | 2.1 | 0.21 | 24 | 61 | 1.22 | 0.61 | 0.31 |
| Timothy Hay Early Bloom | 88 | 8 | 2.6 | 33 | 43 | 68 | 6 | 0.5 | 0.25 | 0.9 | 0.21 | – | 59 | 1.18 | 0.58 | 0.28 |
| Timothy Hay Full Bloom | 88 | 7 | 2.5 | 34 | 45 | 70 | 5 | 0.4 | 0.20 | 1.6 | 0.13 | 17 | 57 | 1.14 | 0.56 | 0.25 |
| Tomato Pomace Dried | 92 | 23 | 10.6 | 26 | 50 | 55 | 6 | 0.4 | 0.59 | 3.6 | – | – | 67 | 1.34 | 0.68 | 0.40 |
| Triticale Silage | 38 | 12 | – | – | – | – | – | – | – | – | – | – | – | – | – | – |
| Triticale | 90 | 16 | 4.6 | 4 | – | – | 2 | 0.1 | 0.34 | 0.4 | 0.17 | – | 86 | 1.72 | 0.94 | 0.63 |
| Turnip Tops (Purple) | 17 | 16 | 2.6 | 10 | – | – | 14 | 3.2 | 0.34 | 3.0 | 0.27 | – | 69 | 1.38 | 0.70 | 0.43 |
| Turnip Roots | 9 | 12 | 1.5 | 11 | – | – | 8 | 0.8 | 0.40 | 3.4 | 0.43 | 40 | 86 | 1.72 | 0.94 | 0.63 |
| Urea 45% N | 98 | 287 | 0.0 | 0 | 0 | 0 | 0 | 0.0 | 0.0 | 0.0 | 0.0 | 0 | 0 | 0.0 | 0.0 | 0.0 |
| Wheat Fresh (Pasture) | 21 | 28 | 4.0 | 18 | – | – | 14 | 0.4 | 0.40 | 3.5 | – | – | 69 | 1.38 | 0.70 | 0.43 |
| Wheat Silage | 28 | 10 | 3.2 | 28 | – | – | 8 | 0.3 | 0.27 | 1.2 | 0.23 | – | 63 | 1.26 | 0.63 | 0.34 |
| Wheat Straw | 88 | 3 | 1.5 | 42 | 56 | 85 | 7 | 0.2 | 0.08 | 1.2 | 0.14 | 7 | 44 | 0.88 | 0.43 | 0.01 |
| Wheat Grain Hard | 89 | 14 | 2.0 | 3 | – | – | 2 | 0.1 | 0.45 | 0.5 | 0.17 | 16 | 89 | 1.78 | 0.98 | 0.67 |
| Wheat Grain Soft | 89 | 12 | 2.0 | 3 | – | – | 2 | 0.1 | 0.35 | 0.4 | 0.17 | 16 | 89 | 1.78 | 0.98 | 0.67 |
| Wheat Bran | 89 | 18 | 4.8 | 11 | 12 | 44 | 7 | 0.2 | 1.32 | 1.4 | 0.25 | 105 | 70 | 1.40 | 0.72 | 0.44 |
| Wheat Middlings | 88 | 18 | 3.9 | 3 | – | – | 3 | 0.1 | 0.57 | 0.6 | 0.22 | 70 | 90 | 1.80 | 0.99 | 0.68 |
| Wheat Mill Run | 90 | 17 | 4.7 | 9 | – | – | 6 | 0.1 | 1.15 | 1.4 | 0.28 | – | 75 | 1.50 | 0.78 | 0.50 |
| Wheat Shorts | 89 | 20 | 5.4 | 7 | – | – | 5 | 0.1 | 0.99 | 1.1 | 0.19 | 118 | 80 | 1.60 | 0.85 | 0.56 |
| Wheatgrass Crested Fresh Early Bloom | 37 | 11 | 1.6 | 30 | – | – | 7 | 0.3 | 0.30 | – | – | – | 58 | 1.16 | 0.57 | 0.26 |
| Wheatgrass Crested Fresh Full Bloom | 50 | 10 | 1.6 | 33 | – | – | 7 | 0.4 | 0.28 | – | – | – | 55 | 1.10 | 0.54 | 0.21 |
| Wheatgrass Crested Hay | 92 | 10 | 2.4 | 33 | – | – | 7 | 0.3 | 0.15 | – | – | 32 | 54 | 1.08 | 0.53 | 0.20 |
| Whey Dried | 94 | 16 | 0.9 | 0 | 0 | 0 | 10 | 1.0 | 0.81 | 1.6 | 1.10 | 3 | 84 | 1.68 | 0.91 | 0.61 |

*Source:* Perry, J.W. (1980) *Beef Cattle Feeding and Nutrition.* Reproduced with kind permission of Academic Press.

**TABLE 304** **Nutrient Composition of Cattle Feeds in Zimbabwe**

The analyses of feeds given in these tables are compiled from a wide range of sources, mostly local but some from the USA. The final figures are 'preferred' analyses which give the most commonly expected composition of local feeds. A dash indicates that figures are not available.

| Feed | DM % | on Dry Matter basis | | | | | | |
|---|---|---|---|---|---|---|---|---|
| | | ME MJ/kg | TDN % | CP % | DP % | Ca % | P % | Vit.A IU/g |
| **(a) Cereals and other energy concentrates** | | | | | | | | |
| Butu - munga | 91.0 | 9.0 | 60.0 | 13.8 | 8.9 | 0.11 | 0.57 | – |
| Corn and cob meal | 88.3 | 12.8 | 85.7 | 8.7 | 4.6 | 0.05 | 0.35 | – |
| Maize bran | 90.1 | 11.5 | 76.8 | 8.2 | 4.4 | 0.03 | 0.15 | – |
| Maize germ | 90.0 | 12.2 | 81.3 | 8.4 | 4.6 | 0.04 | 0.50 | – |
| Maize grain - white | 90.0 | 13.9 | 93.1 | 9.4 | 5.2 | – | 0.40 | – |
| Maize grain - yellow | 88.0 | 13.6 | 91.0 | 9.7 | 5.7 | – | 0.40 | 4.4 |
| Maize mill screenings | 90.0 | 11.7 | 78.0 | 10.0 | 6.3 | – | 0.32 | – |
| * Masese - rapoko | 90.5 | 10.3 | 68.6 | 20.5 | 14.6 | 0.04 | 0.21 | – |
|    - sorghum | 32.2 | 10.7 | 71.7 | 29.7 | 23.7 | 0.04 | 0.31 | – |
| Molasses | 75.0 | 10.8 | 72.0 | 3.0 | 0.0 | 0.70 | 0.10 | – |
| Molasses (US analyses) | 75.0 | 13.8 | 91.0 | 4.3 | 2.4 | 1.19 | 0.11 | – |
| HE Molasses finishing concentrate | 84.3 | 9.2 | 61.5 | 14.5 | 10.9 | – | – | – |
| Mhunga/nyouti | 90.8 | 13.2 | 88.6 | 11.6 | 7.4 | 0.05 | 0.37 | – |
| Oats grain | 90.0 | 11.8 | 78.6 | 12.6 | 8.3 | 0.11 | 0.39 | – |
| Rapoko | 88.6 | 12.9 | 86..0 | 9.0 | 5.1 | – | – | – |
| Snapcorn (Corn, cob and husk) | 91.1 | 12.3 | 82.0 | 8.2 | 4.3 | – | – | – |
| Snapcorn and high protein concentrate (complete HE mix) | 93.3 | 13.3 | 88.9 | 13.9 | 10.5 | – | – | – |
| Sorghum grain | 90.0 | 12.6 | 84.0 | 11.8 | 7.6 | 0.04 | 0.33 | – |
| Sorghum head meal and HP concentrate (complete HE mix) | 92.7 | 11.6 | 77.8 | 12.4 | 6.8 | – | – | – |
| Sugar (Canex) | 90.0 | 14.5 | 97.0 | – | – | – | – | – |
| Wheat grain | 89.6 | 13.1 | 87.5 | 14.0 | 10.9 | 0.06 | 0.34 | – |
| Wheat, bran | 89.0 | 9.7 | 65.0 | 14.0 | 10.9 | 0.10 | 1.10 | 1.0 |
| Wheat, pollard | 90.0 | 11.4 | 76.0 | 16.0 | 11.4 | 0.10 | 1.00 | 1.0 |
| Wheat, hominy chop | 88.0 | 11.5 | 77.0 | 10.5 | 5.5 | 0.02 | 0.60 | – |

* Extremely variable. Specific analysis always preferred

*(continued over)*

A G R I  I N F O

**TABLE 304** *(continued)* **Nutrient Composition of Cattle Feeds in Zimbabwe**

| Feed | DM % | on Dry Matter basis | | | | | | |
|---|---|---|---|---|---|---|---|---|
| | | ME MJ/kg | TDN % | CP % | DP % | Ca % | P % | Vit.A IU/g |
| **(b) Legume seeds and oil seeds** | | | | | | | | |
| Cowpeas | 86.2 | 13.2 | 88.6 | 26.2 | 19.5 | 0.11 | 0.52 | 0.3 |
| Cottonseed (whole) | 92.1 | 13.5 | 90.6 | 24.0 | 17.7 | 0.15 | 0.73 | – |
| Groundnuts (kernels) | 95.0 | 21.8 | 145.8 | 31.0 | 23.6 | 0.06 | 0.47 | 0.4 |
| Groundnuts (in pods) | 94.1 | 16.4 | 110.0 | 26.5 | 19.8 | – | 0.35 | – |
| Jackbeans | 89.3 | 13.7 | 91.4 | 25.8 | 19.2 | – | – | – |
| Jackbeans and pods | 90.4 | 10.8 | 72.0 | 13.6 | 8.8 | – | – | – |
| Soyabean seed | 90.7 | 14.4 | 96.4 | 39.7 | 31.0 | 0.26 | 0.61 | 0.9 |
| Sunflower seed (with hulls) | 92.4 | 12.2 | 81.5 | 19.1 | 13.5 | 0.18 | 0.56 | – |
| Sunflower head entire | 90.3 | 9.7 | 65.0 | 10.3 | 5.9 | – | – | – |
| Sunhemp seed | 92.0 | 12.2 | 84.8 | 32.0 | 24.5 | – | – | – |
| Velvet beans | 90.0 | 13.6 | 90.7 | 26.0 | 19.4 | – | – | – |
| Velvet beans in pods | 89.0 | 12.1 | 81.0 | 17.2 | 11.8 | 0.27 | 0.42 | – |
| **(c) Oil-seed cakes and meals** | | | | | | | | |
| Cottonseed Cake I | 94.1 | 11.9 | 79.7 | 44.4 | 36.2 | 0.20 | 1.20 | 0.30 |
| Cottonseed Cake II | 94.9 | 11.0 | 73.5 | 36.5 | 29.8 | 0.20 | 1.20 | 0.30 |
| Cottonseed Cake III | 92.4 | 10.3 | 68.7 | 29.8 | 23.2 | 0.20 | 1.20 | 0.30 |
| Groundnut Cake I | 93.3 | 13.2 | 88.9 | 48.0 | 43.7 | 0.16 | 0.80 | 0.30 |
| Groundnut Cake II | 93.0 | 13.0 | 87.0 | 52.0 | 47.3 | 0.16 | 0.80 | 0.33 |
| Soyabean meal (exp) | 92.0 | 12.2 | 82.0 | 44.0 | 34.8 | 0.25 | 0.60 | 0.34 |
| Soyabean meal (solv) | 91.2 | 12.1 | 81.0 | 45.1 | 35.7 | 0.30 | 0.70 | – |
| Sunflower cake | 93.9 | 10.5 | 70.0 | 38.0 | 29.6 | 0.41 | 0.95 | – |
| **(d) Animal protein feeds and by-products** | | | | | | | | |
| Blood meal | 89.6 | 9.0 | 60.0 | 93.7 | 66.5 | 0.28 | 0.22 | – |
| Feather meal (hydrolized) | 90.7 | 9.9 | 66.0 | 93.4 | 65.4 | – | – | – |
| Fish meal (white) | 94.4 | 11.6 | 77.4 | 67.2 | 64.5 | 3.00 | 2.50 | – |
| Meat and bone meal | 96.0 | 9.7 | 65.0 | 59.0 | 47.2 | 11.20 | 5.40 | – |
| Poultry manure (battery)* | 88.6 | 6.3 | 42.0 | 22.5 | *16.4 | 6.60 | 1.20 | – |
| Poultry manure (litter)* | 90.0 | – | – | 17.1 | *11.8 | – | – | – |

* Very variable. Actual analysis preferred.

**TABLE 304** *(continued)*  **Nutrient Composition of Cattle Feeds in Zimbabwe**

| Feed | DM % | on Dry Matter basis | | | | | | |
|---|---|---|---|---|---|---|---|---|
| | | ME MJ/kg | TDN % | CP % | DP % | Ca % | P % | Vit.A IU/g |
| **(e) Dry roughages** | | | | | | | | |
| Cottonseed hulls | 93.2 | 6.1 | 41.0 | 5.7 | 0.3 | 0.15 | 0.9 | – |
| Cowpea hay | 91.0 | 8.2 | 55.0 | 14.0 | 9.1 | 1.34 | 0.32 | – |
| Cowpea straw (seed harvested) | 91.5 | 6.2 | 41.8 | 8.5 | 2.2 | – | – | – |
| Dolichos hay | 90.0 | 8.5 | 56.8 | 13.9 | 9.1 | – | – | – |
| **Grasses:** | | | | | | | | |
| Rhodes grass hay early cut | 91.0 | 8.1 | 54.1 | 7.2 | 2.3 | – | – | – |
| Rhodes grass hay late cut | 91.0 | 6.9 | 46.5 | 3.4 | 0.04 | – | – | – |
| Veld hay | 93.4 | 7.2 | 48.5 | 4.1 | 0.40 | 0.27 | 0.15 | – |
| Veld (Henderson) | | | | | | | | |
| end December | 30.0 | 9.3 | 62.5 | 10.6 | 6.2 | – | – | – |
| February | 32.0 | 7.4 | 49.2 | 7.5 | 2.9 | – | – | – |
| April | 40.0 | 7.4 | 49.2 | 3.8 | – | – | – | – |
| July | 55.0 | 3.8 | 25.3 | 1.4 | – | – | – | – |
| Veld (Shangani) | | | | | | | | |
| November | 30.0 | 7.5 | 50.4 | 8.1 | 3.1 | 0.27 | 0.08 | – |
| February | 32.0 | 7.0 | 47.1 | 5.4 | 1.6 | 0.24 | 0.14 | – |
| April | 45.0 | 6.7 | 45.1 | 3.3 | 0.3 | 0.25 | 0.09 | – |
| July | 58.0 | 7.4 | 49.7 | 2.3 | 0.0 | 0.32 | 0.06 | – |
| Groundnut hay (tops) | 91.9 | 7.7 | 51.3 | 10.3 | 5.5 | 1.24 | 0.14 | – |
| Groundnut shells | 93.0 | 2.7 | 18.6 | 7.6 | 1.8 | 0.31 | 0.24 | – |
| Jackbean hay | 90.0 | 8.6 | 57.6 | 14.9 | 8.4 | – | – | – |
| Lucerne (early cut) | 90.0 | 8.6 | 57.8 | 21.8 | 15.2 | 1.59 | 0.35 | 74.4 |
| Lucerne, good average | 90.5 | 7.8 | 52.1 | 15.9 | 10.9 | 1.35 | 0.22 | 31.2 |
| Lucerne, stemmy mature | 90.5 | 6.8 | 45.3 | 14.4 | 8.6 | 1.28 | 0.20 | 12.1 |
| Maize cobs | 94.0 | 7.0 | 47.0 | 2.1 | 0.0 | 0.12 | 0.04 | – |
| Maize husk (cob-sheath) | 94.8 | 6.8 | 45.8 | 3.2 | 0.4 | 0.18 | 0.25 | – |
| Maize leaves | 93.8 | 9.0 | 60.1 | 5.9 | 2.5 | 0.70 | 0.25 | – |
| Maize stalk | 94.8 | 7.3 | 49.2 | 3.2 | 0.5 | 0.39 | 0.28 | – |
| Maize stover | 90.8 | 8.5 | 55.6 | 4.0 | 1.4 | 0.55 | 0.10 | – |
| Maize hay | 92.2 | 9.8 | 65.7 | 6.6 | 3.2 | 0.4 | 0.21 | 6.6 |
| Russian Comfrey | 93.1 | 7.8 | 51.9 | 15.8 | 10.9 | 1.30 | 0.80 | – |
| Sorghum whole plant | 93.4 | 8.5 | 56.6 | 6.0 | 3.5 | 0.38 | 0.16 | – |
| Sorghum plant (poor) | 81.1 | 8.2 | 54.6 | 3.4 | 0.1 | 0.23 | 0.14 | – |
| Soyabean hay (good) | 93.5 | 9.2 | 61.9 | 14.9 | 10.9 | 1.25 | 0.17 | – |
| Soyabean hay (mature) | 92.4 | 8.6 | 57.2 | 8.9 | 5.4 | 1.25 | 0.17 | 11.0 |
| Soyabean pod husks | 86.9 | 6.0 | 40.3 | 4.1 | 1.1 | – | – | – |
| Soyabean straw | 87.6 | 5.7 | 38.0 | 5.5 | 1.7 | 1.59 | 0.06 | – |
| Sunhemp hay | 91.4 | 9.1 | 60.7 | 15.1 | 10.6 | – | – | – |
| Sugar cane tops | 92.3 | 7.3 | 48.7 | 3.4 | 0.1 | 0.29 | 0.35 | – |
| Sweet potato vines | 87.0 | 8.5 | 57.0 | 14.2 | 9.3 | – | – | – |
| Velvet bean hay | 90.4 | 7.9 | 52.7 | 14.5 | 8.6 | – | 0.26 | – |

*(continued over)*

TABLE 304 *(continued)*   Nutrient Composition of Cattle Feeds in Zimbabwe

| Feed | DM % | ME MJ/kg | TDN % | CP % | DP % | Ca % | P % | Vit.A IU/g |
|---|---|---|---|---|---|---|---|---|
| **(f) Silages** | | | | | | | | |
| Citrus pulp | 18.8 | 11.5 | 76.9 | 6.3 | 3.9 | – | – | – |
| Maize:    good,high DM | 40.0 | 10.5 | 70.0 | 8.1 | 4.7 | 0.27 | 0.20 | – |
| average | 22.9 | 9.5 | 63.6 | 7.9 | 4.2 | 0.34 | 0.23 | 21.0 |
| poor | 27.0 | 8.3 | 55.5 | 7.4 | 2.6 | 0.37 | 0.19 | – |
| Maize and velvet bean | 31.7 | 9.9 | 66.1 | 7.8 | 3.2 | – | – | – |
| Millet | 19.8 | 8.3 | 55.5 | 9.5 | 4.1 | 0.65 | 0.19 | – |
| Sorghum | 26.2 | 8.4 | 56.0 | 7.8 | 3.9 | 0.30 | 0.22 | – |
| Sugardrip sorghum (soft dough) | 23.5 | 8.9 | 59.8 | 5.4 | 1.9 | 0.22 | 0.17 | – |
| Sugardrip sorghum (medium) | 31.9 | 10.0 | 67.0 | 5.9 | 1.5 | 0.21 | 0.14 | – |
| Sugardrip sorghum (hard dough) | 33.2 | 7.5 | 50.3 | 5.4 | 0.3 | 0.22 | 0.15 | – |
| **(g) Green roughage and succulents** | | | | | | | | |
| Chicory tops | 11.4 | 10.8 | 72.0 | 13.5 | 8.7 | – | – | – |
| Citrus pulp (fresh) | 14.6 | 10.9 | 73.2 | 5.3 | 4.1 | – | – | – |
| Clover | 13.9 | 9.6 | 64.4 | 20.9 | 15.2 | 1.40 | 0.50 | 76.0 |
| Lucerne (green) | 18.2 | 8.8 | 58.7 | 25.3 | 18.8 | 1.60 | 0.25 | 104.0 |
| Maize fodder | 30.2 | 9.6 | 64.0 | 5.5 | 3.2 | 0.29 | 0.21 | 32.0 |
| Napier fodder | 14.5 | 8.0 | 53.6 | 10.3 | 6.4 | 0.36 | 0.32 | – |
| Pumpkins | 12.2 | 12.9 | 86.5 | 11.5 | 8.6 | – | 0.38 | – |
| Rhodes grass, Giant | 18.7 | 9.3 | 62.0 | 13.7 | 8.1 | 0.51 | 0.32 | – |
| Rhodes grass, Katambora | 14.4 | – | – | 15.3 | 9.0 | – | – | – |
| Sugar cane tops | 35.2 | 7.6 | 50.9 | 5.2 | 2.2 | 0.35 | 0.27 | – |
| Sweet potato vines | 14.0 | 9.6 | 64.0 | 11.9 | 7.3 | – | – | – |

*Source:* Zimbabwe Cattle Producers Association. (1988) *Beef Production Manual.*
Reproduced with kind permission of Zimbabwe Cattle Producers Association.

**TABLE 305** **Nutritive Value of Food for Pigs**

| Food | Fresh basis | | | | | | | DM basis | |
|---|---|---|---|---|---|---|---|---|---|
| | DM (g/kg) | CP (g/kg) | CF (g/kg) | EE (g/kg) | Ash (g/kg) | DCP (g/kg) | DE (MJ/kg) | DCP (g/kg) | DE (MJ/kg |
| **Green crops** | | | | | | | | | |
| Pasture grass, closely grazed | 200 | 52 | 34 | 8 | 17 | 35 | 2.2 | 175 | 11.0 |
| Pasture grass, rotational grazing | 200 | 34 | 39 | 6 | 16 | 19 | 2.1 | 95 | 10.3 |
| Kale, marrow-stem, minced | 140 | 22 | 25 | 5 | 19 | 14 | 1.6 | 100 | 11.4 |
| Lucerne meal, dried | 900 | 197 | 223 | 46 | 91 | 116 | 8.5 | 129 | 9.4 |
| **Roots and tubers** | | | | | | | | | |
| Cassava, dried | 900 | 25 | 78 | 3 | 31 | 17 | 13.4 | 19 | 14.9 |
| Mangels | 130 | 10 | 8 | 1 | 9 | 8 | 2.0 | 62 | 15.4 |
| Potato meal | 900 | 88 | 21 | 5 | 36 | 35 | 14.2 | 39 | 15.8 |
| Swedes | 120 | 15 | 12 | 2 | 13 | 9 | 1.7 | 75 | 14.2 |
| Turnips | 90 | 11 | 10 | 2 | 7 | 9 | 1.4 | 100 | 15.6 |
| **Cereals and by-products** | | | | | | | | | |
| Barley | 860 | 93 | 46 | 15 | 22 | 76 | 12.9 | 88 | 15.0 |
| Barley, brewers' yeast, dried | 940 | 502 | 11 | 6 | 88 | 487 | 17.6 | 518 | 18.7 |
| Maize | 860 | 84 | 21 | 36 | 11 | 67 | 14.5 | 78 | 16.9 |
| Maize, flaked | 860 | 95 | 14 | 42 | 9 | 90 | 15.6 | 105 | 18.1 |
| Oats | 860 | 94 | 104 | 42 | 28 | 73 | 11.4 | 85 | 13.3 |
| Rice | 860 | 66 | 15 | 4 | 8 | 57 | 15.3 | 66 | 17.8 |
| Rye | 860 | 114 | 19 | 17 | 20 | 92 | 13.9 | 107 | 16.2 |
| Sorghum | 860 | 93 | 18 | 37 | 23 | 72 | 14.3 | 84 | 16.6 |
| Wheat | 860 | 107 | 22 | 16 | 18 | 98 | 14.0 | 114 | 16.3 |
| **Oilseed by-products** | | | | | | | | | |
| Coconut meal | 870 | 232 | 119 | 75 | 53 | 169 | 13.4 | 194 | 15.4 |
| Cottonseed meal, dec. | 930 | 408 | 112 | 68 | 67 | 335 | 12.9 | 360 | 13.9 |
| Groundnut meal, dec. | 900 | 497 | 79 | 7 | 57 | 462 | 13.7 | 513 | 15.2 |
| Palm kernel meal | 900 | 204 | 150 | 9 | 40 | 122 | 10.4 | 136 | 11.6 |
| Rapeseed meal | 900 | 360 | 137 | 26 | 72 | 310 | 11.8 | 344 | 13.1 |
| Soyabean meal | 900 | 453 | 52 | 15 | 56 | 390 | 13.9 | 433 | 15.4 |
| Sunflower meal | 900 | 381 | 120 | 23 | 65 | 310 | 9.0 | 344 | 10.0 |
| **Animal by-products** | | | | | | | | | |
| Fish meal | 900 | 662 | 8 | 58 | 175 | 586 | 15.1 | 662 | 16.8 |
| Herring meal | 900 | 686 | 0 | 82 | 92 | 644 | 18.0 | 715 | 20.0 |
| Meat meal, low fat | 900 | 645 | 0 | 28 | 188 | 568 | 12.1 | 631 | 13.4 |
| Meat and bone meal | 900 | 474 | 0 | 40 | 371 | 422 | 9.6 | 469 | 10.7 |
| Milk, cows, whole | 124 | 32 | 0 | 38 | 8 | 30 | 3.0 | 242 | 24.2 |
| Milk, separated fresh | 90 | 34 | 0 | 1 | 8 | 33 | 1.5 | 367 | 16.7 |
| Whey, fresh | 66 | 9 | 0 | 3 | 6 | 9 | 1.1 | 136 | 16.7 |

*Source:* McDonald, P., Edwards, R.A., & Greenhalgh, F.F.D. (1988) *Animal Nutrition.*
Reproduced with kind permission of Longman Group UK Ltd.

**TABLE 306  Nutritive Value of Foods for Poultry**

| Food | Fresh basis | | | | | | DM basis | |
|---|---|---|---|---|---|---|---|---|
| | DM (g/kg) | CP (g/kg) | EE (g/kg) | Ash (g/kg) | DCP (g/kg) | DE (MJ/kg) | DCP (g/kg) | DE (MJ/kg |
| **Green crops and tubers** | | | | | | | | |
| Dried grass | 921 | 178 | 37 | 77 | 156 | 5.82 | 169 | 6.32 |
| Dried lucerne | 887 | 145 | 27 | 73 | 123 | 4.60 | 139 | 9.19 |
| Potato meal | 913 | 87 | 2 | 32 | 63 | 12.1 | 69 | 13.3 |
| **Cereals and by-products** | | | | | | | | |
| Barley | 891 | 113 | 15 | 27 | 90 | 11.1 | 101 | 12.5 |
| Malt distillers' dried solubles | 949 | 268 | 2 | 172 | - | 6.82 | - | 7.19 |
| Brewers' yeast, dried | 867 | 425 | 21 | 89 | 374 | 11.0 | 431 | 12.7 |
| Maize | 882 | 82 | 32 | 12 | 67 | 13.2 | 76 | 15.0 |
| Maize gluten feed | 897 | 250 | 19 | 53 | 223 | 9.75 | 249 | 10.9 |
| Millet | 856 | 119 | 39 | 29 | 82 | 12.0 | 96 | 14.0 |
| Oats | 876 | 100 | 49 | 27 | 85 | 11.1 | 97 | 12.7 |
| Rice, brown | 907 | 101 | 21 | 8 | 84 | 15.0 | 93 | 16.5 |
| Rye | 846 | 85 | 11 | 19 | 67 | 12.1 | 79 | 14.3 |
| Sorghum (milo) | 867 | 107 | 29 | 18 | 84 | 13.0 | 97 | 15.0 |
| Wheat | 891 | 104 | 14 | 18 | 88 | 12.2 | 99 | 13.7 |
| Wheat germ meal | 889 | 248 | 73 | 43 | 198 | 11.1 | 223 | 12.5 |
| Wheat middlings, coarse | 874 | 149 | 39 | 42 | 127 | 9.75 | 145 | 11.2 |
| Wheat middlings, fine | 875 | 177 | 52 | 32 | 150 | 11.8 | 171 | 13.5 |
| **Oilseed by-products** | | | | | | | | |
| Coconut meal | 887 | 195 | 67 | 64 | 109 | 6.90 | 123 | 7.78 |
| Cottonseed meal, dec. | 901 | 378 | 61 | 67 | 280 | 10.91 | 311 | 12.1 |
| Groundnut meal, dec. | 912 | 454 | 51 | 64 | 408 | 13.2 | 447 | 14.5 |
| Linseed meal | 888 | 341 | 63 | 53 | 300 | 8.66 | 338 | 9.75 |
| Palm kernel meal | 900 | 190 | 20 | 40 | 171 | 6.74 | 190 | 7.49 |
| Soyabean meal | 873 | 499 | 15 | 47 | 428 | 10.7 | 490 | 12.3 |
| Sunflower seed meal, dec. | 916 | 321 | 27 | 64 | 248 | 8.83 | 270 | 9.6 |
| **Leguminous seeds** | | | | | | | | |
| Bean meal | 866 | 250 | 13 | 39 | 211 | 10.4 | 244 | 12.0 |
| Pea meal | 871 | 271 | 17 | 28 | 206 | 11.1 | 237 | 12.7 |
| **Animal by-products** | | | | | | | | |
| Blood meal | 868 | 800 | 8 | 35 | 720 | 13.0 | 829 | 15.0 |
| Fish meal | 910 | 655 | 42 | 215 | 590 | 11.5 | 648 | 12.6 |
| Herring meal | 905 | 740 | 70 | 95 | 666 | 13.4 | 736 | 14.8 |
| Meat meal | 902 | 722 | 132 | 38 | 650 | 15.7 | 721 | 17.4 |
| Meat and bone meal | 935 | 515 | 112 | 275 | 412 | 11.0 | 441 | 11.8 |
| Milk, dried skim | 934 | 340 | 9 | 80 | 275 | 12.3 | 294 | 13.2 |
| Milk, dried whey | 937 | 125 | 7 | 85 | 101 | 12.0 | 108 | 12.8 |

*Source:* McDonald, P., Edwards, R.A., & Greenhalgh, F.F.D. (1988) *Animal Nutrition.*
Reproduced with kind permission of Longman Group UK Ltd.

**TABLE 307  Amino Acid Composition of Foods (g/kg) (Fresh Basis)**

| Food | DM (g/kg) | Nitrogen (g/kg) | Arginine | Cystine | Glycine | Histidine | Isoleucine | Leucine | Lysine | Methionine | Phenylalanine | Serine | Threonine | Tryptophan | Tyrosine | Valine |
|---|---|---|---|---|---|---|---|---|---|---|---|---|---|---|---|---|
| **Green crops** | | | | | | | | | | | | | | | | |
| Dried grass | 897 | 23.5 | 7.6 | 3.4 | 7.5 | 2.9 | 5.8 | 10.9 | 7.1 | 3.0 | 7.1 | 6.1 | 6.5 | 1.2 | 4.8 | 4.9 |
| Dried lucerne | – | 35.7 | 10.9 | 5.0 | 10.2 | 4.7 | 9.3 | 16.1 | 11.7 | 2.8 | 10.5 | 9.0 | 9.2 | 1.6 | 8.2 | 11.3 |
| **Cereals and by-products** | | | | | | | | | | | | | | | | |
| Barley | 856 | 15.6 | 5.4 | 4.3 | 4.1 | 4.1 | 3.5 | 6.9 | 3.8 | 2.1 | 5.0 | 4.3 | 3.4 | 1.0 | 3.4 | 5.1 |
| Brewers' yeast, dried | 930 | 71.0 | 21.9 | 5.0 | 21.9 | 10.7 | 21.4 | 31.9 | 32.3 | 7.0 | 18.1 | – | 20.6 | 4.9 | 14.9 | 23.2 |
| Distillers' dark grains | 900 | 39.4 | 10.1 | 8.8 | 10.6 | 4.5 | 8.5 | 15.5 | 9.6 | 4.3 | 8.9 | 9.3 | 8.8 | 2.1 | 7.5 | 11.7 |
| Distillers' solubles | – | 42.9 | 3.8 | 4.8 | 12.9 | 4.0 | 8.0 | 13.0 | 6.8 | 3.4 | 7.7 | 6.4 | 6.0 | 3.6 | 8.5 | 12.8 |
| Maize | 852 | 13.5 | 4.3 | 3.8 | 3.3 | 2.6 | 3.0 | 11.1 | 2.5 | 2.3 | 4.5 | 4.3 | 3.2 | 0.4 | 3.9 | 4.3 |
| Maize gluten meal | – | 106.2 | 24.1 | 25.2 | 17.4 | 14.0 | 28.4 | 117.7 | 10.8 | 24.5 | 41.0 | 37.7 | 24.0 | 2.6 | 34.7 | 33.0 |
| Oats | 869 | 16.8 | 7.0 | 8.0 | 5.7 | 2.3 | 3.7 | 7.3 | 4.5 | 2.6 | 5.1 | 5.7 | 3.7 | 0.7 | 4.1 | 5.1 |
| Rice, polished, broken (Brewers' rice) | 890 | 13.9 | 6.2 | 0.8 | 6.3 | 1.7 | 3.5 | 5.2 | 2.4 | 1.5 | 3.6 | 13.6 | 2.9 | 1.3 | 4.1 | 5.0 |
| Sorghum | 870 | 14.1 | 3.4 | 1.6 | 3.5 | 1.9 | 4.2 | 11.8 | 2.1 | 1.6 | 4.2 | 3.9 | 2.9 | 1.0 | 3.8 | 5.3 |
| Wheat | 858 | 16.2 | 5.2 | 4.5 | 4.1 | 2.5 | 3.5 | 7.1 | 3.1 | 2.1 | 4.8 | 4.8 | 3.1 | 1.2 | 3.3 | 4.5 |
| Wheat feed | 858 | 22.6 | 10.2 | 7.8 | 7.8 | 4.0 | 4.7 | 9.5 | 6.4 | 3.2 | 6.1 | 6.5 | 4.0 | 2.2 | 4.6 | 7.1 |
| **Oilseed by-products** | | | | | | | | | | | | | | | | |
| Cottonseed meal | 900 | 66.2 | 45.9 | 6.4 | 17.0 | 11.0 | 13.3 | 24.1 | 17.1 | 5.2 | 22.2 | – | 13.2 | 4.7 | 10.2 | 18.8 |
| Groundnut Meal | 897 | 75.5 | 57.0 | 11.2 | 26.4 | 11.3 | 15.7 | 29.9 | 16.4 | 5.6 | 25.1 | 23.1 | 13.5 | 3.0 | 19.9 | 20.7 |
| Lupinseed meal | – | 60.8 | 42.7 | 12.4 | 13.6 | 8.5 | 16.5 | 26.8 | 17.0 | 3.0 | 13.1 | 17.4 | 12.1 | 1.8 | 18.5 | 14.6 |
| Rapeseed meal | 899 | 50.0 | 23.2 | 15.1 | 18.5 | 9.9 | 14.2 | 25.9 | 21.5 | 7.9 | 14.3 | 16.3 | 16.8 | 1.7 | 11.5 | 18.6 |
| Soyabean meal | 861 | 70.9 | 35.3 | 12.0 | 19.5 | 12.6 | 20.3 | 35.0 | 28.5 | 7.9 | 23.0 | 23.5 | 17.9 | 5.5 | 17.7 | 22.2 |
| Sunflower seed meal | – | 44.5 | 23.1 | 9.2 | 15.6 | 7.2 | 11.6 | 18.5 | 10.1 | 7.6 | 13.4 | 11.9 | 10.4 | 1.4 | 8.1 | 14.3 |
| **Leguminous seeds** | | | | | | | | | | | | | | | | |
| Beans (*Vica faba*) | – | 39.8 | 22.2 | 7.8 | 10.5 | 6.1 | 9.7 | 18.3 | 15.8 | 1.8 | 10.1 | 11.7 | 9.1 | 1.6 | 8.9 | 11.2 |
| Peas (*Pisum sativum*) | – | 31.4 | 17.1 | 6.0 | 8.7 | 5.3 | 8.2 | 14.4 | 15.2 | 2.5 | 8.9 | 9.5 | 8.0 | 0.9 | 7.2 | 9.2 |
| **Animal by-products** | | | | | | | | | | | | | | | | |
| Fishmeal | 918 | 100.0 | 40.5 | 6.4 | 50.6 | 14.1 | 26.1 | 44.6 | 48.2 | 15.2 | 28.8 | 28.3 | 24.9 | 6.9 | 21.4 | 30.7 |
| Meat and bone meal | 957 | 73.3 | 32.4 | 5.4 | 70.6 | 7.7 | 11.6 | 26.1 | 22.0 | 6.5 | 14.6 | 16.0 | 14.3 | 2.1 | 9.6 | 19.6 |
| Whey dried | 930 | 19.2 | 3.4 | 3.0 | 3.0 | 1.8 | 8.2 | 11.9 | 9.7 | 1.9 | 3.3 | 3.2 | 8.9 | 1.9 | 2.5 | 6.8 |

*Source:* McDonald, P., Edwards, R.A., & Greenhalgh, J.F.D., (1988) *Animal Nutrition.* Reproduced with kind permission of Longman Group UK Ltd.

**TABLE 308  Vitamin Potency of Foods ( Fresh Basis)**

| Food | Vitamin A potency* i.u./g | Vitamin E i.u./kg | Thiamin mg/kg | Riboflavin mg/kg | Nicotinic acid mg/kg | Pantothenic acid mg/kg | Vitamin B$_6$ mg/kg | Vitamin B$_{12}$ mg/kg | Choline mg/kg |
|---|---|---|---|---|---|---|---|---|---|
| **Green crops** | | | | | | | | | |
| Grass, dried | 328 | 150 | – | 15.5 | 74 | – | – | – | 890 |
| Lucerne, dried | 267 | 200 | – | 16.6 | 43 | – | – | 0.003 | 1110 |
| **Cereals and by-products** | | | | | | | | | |
| Barley | 0.7 | 20 | 1.9 | 1.8 | 55 | 8 | 3.0 | – | 990 |
| Brewers' yeast, dried | – | – | 91.8 | 37.0 | 448 | 109 | 42.8 | – | 3984 |
| Maize | 5.0 | 22 | 3.5 | 1.0 | 24 | 4 | 7.0 | – | 620 |
| Oats | 0.6 | 20 | 6.0 | 1.1 | 12 | – | 1.0 | – | 946 |
| Rice | – | 12 | – | 0.4 | 15 | – | – | – | 780 |
| Rye | 0.2 | 17 | 3.6 | 1.6 | 19 | 8 | 2.6 | – | 419 |
| Sorghum | 0.7 | 12 | 4.0 | 1.1 | 41 | 12 | 3.2 | – | 450 |
| Wheat | 0.4 | 13 | 4.5 | 1.4 | 48 | 10 | 3.4 | – | 1090 |
| Wheat, fine middlings | 0.5 | 20 | – | 2.2 | 100 | – | – | – | 1110 |
| Wheat, coarse middlings | 0.4 | 57 | – | 2.4 | 95 | – | – | – | 1170 |
| **Oilseed by-products** | | | | | | | | | |
| Coconut meal | – | 16 | – | 3.3 | 27 | – | – | – | 1110 |
| Cottonseed meal (dec. exp) | 0.3 | 39 | 6.4 | 5.1 | 38 | 10 | 5.3 | – | 2753 |
| Groundnut meal (dec. extr) | – | 3 | 5.7 | 11.0 | 170 | 53 | 10.0 | – | 2396 |
| Groundnut meal (dec. exp) | 0.3 | 3 | 7.1 | 5.2 | 166 | 47 | 10.0 | – | 1655 |
| Linseed meal (extr) | 0.4 | – | – | 3.5 | 40 | – | – | – | 1660 |
| Soya bean meal (extr) | – | 2 | 4.5 | 2.9 | 29 | 16 | 6.0 | – | 2794 |
| **Animal by-products** | | | | | | | | | |
| Fish meal | – | 8 | 2.1 | 6.0 | 49 | 10 | 4.1 | 0.081 | 5180 |
| Meat meal | – | 1 | 0.2 | 5.5 | 57 | 5 | 3.0 | 0.068 | 2077 |
| Milk, dried skim | 0.3 | 1 | – | 21.0 | 12 | – | – | 0.055 | 1060 |

* For chicks. The values for pigs and ruminants are about half those quoted for plant products.

*Source:* McDonald, P., Edwards, R.A., & Greenhalgh F.F.D. (1988) *Animal Nutrition.*
Reproduced with kind permission of Longman Group UK Ltd.

## TABLE 309  Mineral Contents of Foods (DM Basis)

| Food | Calcium (g/kg) | Phosphorus (g/kg) | Magnesium (g/kg) | Sodium (g/kg) | Copper (mg/kg) | Manganese (mg/kg) | Zinc (mg/kg) | Cobalt (mg/kg) | Selenium (mg/kg) |
|---|---|---|---|---|---|---|---|---|---|
| **Green crops** | | | | | | | | | |
| Grass, close grazing | 5.0 | 3.5 | 1.7 | 1.9 | 8.0 | – | – | 0.10 | 0.05 |
| Grass, extensive grazing | 4.8 | 2.8 | 1.7 | 1.7 | 7.0 | 16 | 5.0 | 0.08 | 0.04 |
| Kale | 21.0 | 3.2 | 2.5 | 2.0 | 4.5 | 38 | – | 0.10 | 0.05 |
| Lucerne, late vegetative | 21.9 | 3.3 | 2.7 | 2.1 | 11.0 | 41 | – | 0.17 | – |
| Turnip tops | 24.2 | 3.1 | 2.8 | 3.1 | 8.0 | – | – | 0.08 | 0.06 |
| **Silages** | | | | | | | | | |
| Cereal, vegetative | 4.0 | 2.7 | 1.0 | 1.8 | 6.0 | 80 | 25 | 0.07 | 0.06 |
| Grass, early | 8.0 | 4.0 | 3.0 | 3.0 | 11.0 | 90 | 25 | – | 0.10 |
| Grass, mature | 3.0 | 2.0 | 0.9 | 1.0 | 3.0 | 94 | 30 | 0.05 | 0.02 |
| **Hays** | | | | | | | | | |
| Clover | 15.3 | 2.5 | 4.3 | 1.9 | 11.0 | 73 | 17 | 0.16 | – |
| Grass, poor quality | 2.5 | 1.5 | 0.8 | 1.0 | 2.0 | 70 | 17 | 0.05 | 0.01 |
| Grass, good quality | 7.0 | 3.5 | 2.5 | 2.5 | 9.0 | 100 | 21 | 0.20 | 0.07 |
| Lucerne, mature | 11.3 | 1.8 | 2.7 | 0.8 | 14.0 | 44 | 24 | 0.09 | – |
| **Straws** | | | | | | | | | |
| Barley | 4.5 | 0.7 | 0.8 | 1.1 | 3.2 | 84 | 16 | 0.04 | 0.04 |
| Oat | 4.0 | 0.7 | 1.3 | 3.7 | 4.0 | 69 | 29 | 0.04 | 0.02 |
| **Roots and tubers** | | | | | | | | | |
| Cassava, dried | 2.0 | 1.0 | – | 0.2 | – | 20 | – | – | – |
| Mangels | 2.9 | 2.1 | 5.3 | 9.9 | 9.4 | – | – | 0.09 | 0.03 |
| Potatoes | 1.0 | 2.1 | 1.0 | 0.5 | 4.5 | 42 | 28 | 0.06 | 0.03 |
| Sugar beet | | | | | | | | | |
| pulp, molassed, dried | 5.7 | 0.8 | 2.4 | 2.5 | 11.0 | 51 | 32 | 0.10 | 0.02 |
| Swedes | 3.6 | 3.2 | 1.2 | 2.6 | 3.8 | 21 | 19 | 0.07 | 0.03 |
| Turnips | 5.0 | 3.6 | 1.4 | 2.2 | 2.7 | 35 | 36 | 0.04 | 0.03 |

*(continued over)*

A G R I   I N F O

TABLE 309 *(continued)* Mineral Contents of Foods (DM Basis)

| Food | Calcium (g/kg) | Phosphorus (g/kg) | Magnesium (g/kg) | Sodium (g/kg) | Copper (mg/kg) | Manganese (mg/kg) | Zinc (mg/kg) | Cobalt (mg/kg) | Selenium (mg/kg) |
|---|---|---|---|---|---|---|---|---|---|
| **Cereals and by-products** | | | | | | | | | |
| Barley | 0.5 | 4.0 | 1.3 | 0.2 | 4.8 | 18 | 19 | 0.04 | 0.02 |
| Brewers' grains, dried | 3.2 | 7.8 | 1.8 | 0.4 | 25.0 | 50 | – | 0.03 | – |
| Brewers' yeast | 1.3 | 15.1 | 2.5 | 0.8 | 35.3 | 6 | 42 | — | – |
| Distillers' grains, malt | 1.7 | 3.7 | 1.4 | 0.9 | 10.0 | – | – | 0.02 | 0.02 |
| Maize | 0.3 | 2.7 | 1.1 | 0.2 | 2.5 | 6 | 16 | 0.02 | 0.02 |
| Maize gluten meal | 1.6 | 5.0 | 0.6 | 1.0 | 30.0 | 8 | 190 | 0.08 | – |
| Millet | 0.6 | 3.1 | 1.8 | 0.4 | 24.4 | 32 | 16 | 0.04 | – |
| Oats | 0.8 | 3.7 | 1.3 | 0.2 | 3.6 | 42 | 41 | 0.04 | 0.03 |
| Oat feed | 1.5 | 2.9 | 1.0 | 0.2 | 3.9 | – | – | 0.04 | 0.03 |
| Rice | 0.7 | 3.2 | 1.5 | 0.6 | 3.0 | 20 | 17 | 0.05 | – |
| Rye | 0.7 | 3.7 | 1.4 | 0.3 | 8.0 | 66 | 36 | – | – |
| Sorghum | 0.5 | 3.5 | 1.9 | 0.4 | 10.8 | 16 | 15 | 0.14 | – |
| Wheat | 0.5 | 3.5 | 1.2 | 0.1 | 5.0 | 42 | 50 | 0.05 | 0.02 |
| Wheat bran | 1.6 | 13.6 | 5.0 | 0.4 | 12.9 | 143 | 189 | 0.03 | 0.40 |
| Wheat feed | 1.1 | 8.0 | 3.3 | 0.4 | 17.5 | – | – | 0.03 | 0.04 |
| **Oilseeds and by-products** | | | | | | | | | |
| Coconut meal | 2.3 | 6.6 | 2.8 | 0.4 | 20.4 | 59 | – | 0.14 | – |
| Cottonseed meal, dec. | 1.9 | 12.4 | 5.0 | 0.6 | 16.0 | 25 | 79 | 0.05 | – |
| Groundnut meal, dec. | 2.9 | 6.8 | 1.7 | 0.8 | 17.0 | 29 | 22 | 0.12 | – |
| Linseed meal | 4.1 | 8.6 | 5.8 | 0.7 | 25.0 | 42 | – | 0.55 | 0.91 |
| Soyabean meal | 3.5 | 6.8 | 3.0 | 0.4 | 25.0 | 32 | 61 | 0.20 | 0.55 |
| **Leguminous seeds** | | | | | | | | | |
| Beans | 1.0 | 5.5 | 2.0 | 0.1 | 14.0 | 16 | 46 | 0.20 | – |
| Peas | 1.5 | 4.4 | 1.4 | 0.5 | – | – | 33 | – | – |
| **Animal by-products** | | | | | | | | | |
| Fishmeal | 79.0 | 44.0 | 3.6 | 4.5 | 9.0 | 21 | 119 | 0.14 | 2.00 |
| Meat and bone meal | 120 | 58.0 | 2.5 | 7.2 | 24.0 | – | – | 0.20 | 0.20 |
| Whey, dried | 9.2 | 8.2 | 1.4 | 7.0 | 50.0 | 6 | 3 | 0.13 | – |

*Source:* McDonald, P., Edwards, R.A., & Greenhalgh, F.F.D. (1988) *Animal Nutrition.*
Reproduced with kind permission of Longman Group UK Ltd.

# Livestock
# Performance

---

**Note 16  Introductory Note**

---

This livestock performance section has been compiled to provide mostly indicative and generalized data from a range of sources and environments. Some performance guidelines are given for specific hybrid strains as examples of stock types but these must not be regarded as typical of hybrids from all breeders. It is noted that hybrid livestock strains are being bred increasingly to suit specific markets requiring critical carcass characteristics for butchering purposes. Individual strains also increasingly benefit from specifically formulated diets according to age.

The section is divided into two sections as far as has been practicably possible, given the nature of available summarized data. The two sections are 'between' and 'within' specie comparisons. There is however some overlap of these two broad sections, particularly in the 'within' specie section.

# Between Specie Comparisons

---

TABLE 310   Average Performance of Different Animal Species and Energy Cost of Proteins they Produce

| | No.of young per breeding female per year | Live-weight of breeding female | Live-weight at slaughter | Slaughter yield | Daily weight increase | | Fat content of carcass | Food kcal per g of usable protein |
|---|---|---|---|---|---|---|---|---|
| | | Kg | | % | g | g/kgW $^{0.75}$ | % | |
| **Broilers** | 100 | 3.0 | 2.0 | 63 | 31 | 30.5 | 13.0 | 80 |
| **Turkeys** | 60 | 10 | 10.1 | 79 | 65 | 19.2 | 13.0 | 87 |
| **Rabbits** | 40 | 4.5 | 3.2 | 60 | 32 | 22.3 | 6.8 | 105 |
| **Pigs** | 12 | 170 | 100 | 73 | 540 | 28.4 | 32.0 | 151 |
| **Sheep** | 1.4 | 70 | 50 | 50 | 220 | 18.2 | 36.0 | 427 |
| **Beef cattle** | 0.8 | 500 | 475 | 61 | 950 | 14.8 | 32.0 | 442 |
| **Dairy cattle[1]** | 0.8 | 500 | 475 | 61 | 950 | 14.8 | 32.0 | 184 |

*Note:* Performance levels noted for each species are not peaks, but fall within the easy range of most breeders. For instance 18 - 22 piglets would be expected in high performance herds.

[1] Theoretical calculation for beef production of a dairy cattle breed, arbitrarily assigning total reproduction and maintenance costs of adult animals to milk production, retaining only feed portions consumed by slaughter animals, i.e. 43.6 per cent of total energy expenditure.

*Source:* FAO (1986) *The Rabbit; Husbandry, health and Production.* Animal Production and Health Series. No. 21 (after Dickerson 1978). Reproduced with kind permission of the Food and Agricultural Organization of the United Nations.

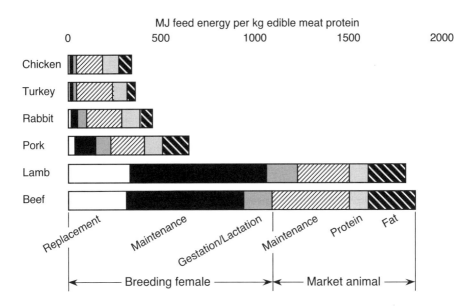

**Fig. 55** Efficiency of transformation of feed to meat. (From EAAP long-range study on the future of pig production in Europe)

*Source:* Whitemore, P.T. (1987) *Elements of Pig Science* .
Reproduced with kind permission of Longman Group UK Ltd.

---

**TABLE 311   Efficiency (E) of Individual Animals**

---

where $E = \dfrac{\text{carcass (kg)}}{\text{G.E. (MJ)}^*}$

| | |
|---|---|
| **Domestic fowl** | 18.9 |
| **Pig (pork)** | 18.6 |
| **Rabbit** | 13.3 |
| **Sheep** | 5.2 |
| **Cattle** | 6.6 |

\* of food fed to individual animals – excludes overhead of maintaining breeding stock.

*Source:* Lister, D., Rhodes, P.N., Fowler, V.R., Fuller, M.F., eds. (1974) *Meat Animals. Growth & Productivity* . Reproduced with kind permission of Plenum Publishing Corporation.

**TABLE 312  Reproductive Rate of Domestic Animals**

| Species | Gestation length | No. of parities per annum | Mean litter size | Total no. of progeny per annum |
|---|---|---|---|---|
| Domestic fowl | – | – | – | 120  (240) |
| Pigs | 112  days | 2 | 8  (15) | 16  (30) |
| Rabbits | 32  days | 6  (10) | 8  (10) | 48  (100) |
| Sheep | 5  months | 1  (2) | 1.5  (3) | 1.5  (6) |
| Cattle | 9  months | 1 | 1  (2) | 1  (2) |

(Average values derived from the literature. Figures in parentheses are suggested potentials).

*Source:* Lister, D., Rhodes, P.N., Fowler, V.R., Fuller, M.F., eds (1974) *Meat Animals. Growth & Productivity* . Reproduced with kind permission of Plenum Publishing Corporation.

**TABLE 313  Indicative Livestock Birth Weights and Number of Live Births per Parturition**

| Livestock type / breed | Birth wt. (kg) | Live births / parturition (does NOT reflect overall population fertility) |
|---|---|---|
| *Bos taurus* cattle: | | |
| (bull calves; for heifers assume 7% lower birth wt.) | | |
| Jersey | 26 | 1 |
| Angus | 27 | 1 |
| Guernsey | 33 | 1 |
| Beef Shorthorn | 33 | 1 |
| Ayrshire | 35 | 1 |
| Devon | 35 | 1 |
| Hereford | 36 | 1 |
| Lincoln Red | 37 | 1 |
| Sussex | 37 | 1 |
| Limousin | 39 | 1 |
| Friesian | 39 | 1 |
| South Devon | 43 | 1 |
| Simmental | 44 | 1 |
| Charolais | 44 | 1 |
| Holstein | 45 | 1 |
| *Bos indicus* cattle: | | |
| Indian cattle (general) | 20 - 24.5 | 1 |
| Sahiwal | 25 | 1 |
| Arabian Baladi | 16 - 20 | 1 |
| Buffalo: | | |
| Nili-Ravi | 37 - 39 | 1 |

**TABLE 313** *(continued)*  Indicative Livestock Birth Weights and Number of Live Births
per Parturition

| Livestock Type / breed | Birth wt. (kg) | | | Live births / parturition (does NOT reflect overall population fertility) |
|---|---|---|---|---|
| **UK Sheep Breeds:** | **Total lamb wt.** | | | |
| Ewe wt. (kg) | single | twin | triplet | |
| 40 | 3.3 | 5.4 | 6.3 | 1 - 3  Smaller breeds |
| 50 | 3.9 | 6.4 | 7.5 | tend to produce |
| 60 | 4.5 | 7.3 | 8.7 | mostly singles |
| 70 | 5.0 | 8.2 | 9.7 | whilst heavier |
| 80 | 5.5 | 9.0 | 10.8 | breeds produce |
| 90 | 6.0 | 9.8 | 11.8 | an increasing % |
| | | | | of twins or singles |
| **Goats:** | | | | |
| Israeli dairy breeds | | | | 1.75 - 1.9 |
| Cyprus dairy breeds | | | | 1.85 |
| Lower Congo small breeds | 1.2 - 1.7 | | | 2  (and may breed 2 times/yr) |
| European breeds | 2 - 7 | | | 1 - 3 |
| **Rabbits:** | | | | N.B. |
| Small/local | | | | 4  weaned litter |
| Large improved | | | | 8.7  size often less |
| **Horses:** | | | | 1 |
| **Donkeys:** | | | | 1 |
| **Camels** | | | | 1 |
| **Llamas, Alpacas:** | | | | 1 |

*Sources:* AFRC Technical Committee on Responses to Nutrients (1990) Report No. 5. *Nutrient Requirements of Ruminant Animals. Energy.* Nutrition Abstracts and Reviews. Series B 60 10.
*The Water Buffalo.* FAO Animal Production and Health Paper No. 4. Chamberlain, A. (1989) *Milk Production in the Tropics.* ILACO (1985) *Agricultural Compendium.* Pritchard, C.J.R. (1975) *A Note on the Characteristics and Performance of Native Zebu Cattle at the Hofuf Agricultural Research Centre.* Publn. No. 60. Ministry of Agriculture and Water, Saudi Arabia.
Maule, J.P. (1966) *A Note on Dairy Goats in the Tropics.* Animal Breeding Abstracts. 34.2.

**TABLE 314   Indicative Adult Weights and Rate of Weight Gain of Different Livestock Types**

| Livestock type | Adult wt. (kg) female | male | Rate of gain kg/day (good conditions, favourable growth stage) |
|---|---|---|---|
| **A. CATTLE *B. Taurus*** | | | |
| Friesian/heavy beef types | >600 | >1000 | 1.4 |
| Ayrshire/Guernsey/Shorthorn/ | | | |
| Angus/beef types | 430 - 600 | 900 | |
| Jersey type | 325 - 430 | 650 | |
| Very light Dexter type | <325 | 400 | |
| **B. CATTLE *B. Indicus*** | | | |
| Pure-bred and non-descript zebus | 350 | 500 | 0.3 - 0.5 |
| Milk breed zebus | 350 | 600 | 0.5 - 1.0 |
| Zebu in feedlot conditions | | | 0.85 |
| **C. CATTLE Zebu x European** | | | |
| New beef breeds | 400 - 600 | 700 - 1000 | 0.8 - 1.3 |
| Zebu crossbreds in feedlot | | | 1.1 |
| **D. SHEEP** | | | |
| Very light | <30 | | 0.1 |
| Light | 30 - 50 | | |
| Medium | 50 - 70 | | |
| Heavy | >70 | 200 - 250 | 0.3 - 0.4 |
| **E. GOATS** | | | |
| Dwarf, e.g. West African | 20 (m + f) | | 0.05 |
| Light, e.g. Indian Barbari | 25 - 45 (m + f) | | 0.05 |
| Medium, e.g. Angora | 50 - 60 (m + f) | | 0.07 - 0.17 |
| Heavy, e.g. Saanen,Damascus | 70 - 100 (m + f) | | 0.19 - 0.24 |
| V.heavy, e.g. Boer | 75 | 120 | 0.18 - 0.29 |
| Angora | 50 | | |
| Cashmere | 40 | | |

**TABLE 314** *(continued)* **Indicative Adult Weights and Rate of Weight Gain of Different Livestock Types**

| Livestock type | Adult wt. (kg) female | male | Rate of gain kg/day (good conditions, favourable growth stage) |
|---|---|---|---|
| **F. CAMELS** | | | |
| Bactrian | 500 - 800 (m + f) | | |
| Dromedary | 450 - 700 (m + f) | | |
| **G. BUFFALO** | 400 | 700 | 0.4 - 1.5 |
| **H. POULTRY** | | | |
| Indigenous fowl | 0.6 - 0.9 | 0.7 - 1.2 | |
| Hybrid layers | 2 - 3 | 2.5 - 4.0 | |
| Broilers | 1 - 2 kg at 9 week slaughter | | |
| Ducks | 1 - 3 | 1.25 - 4.5 | 0.04 |
| Geese | 4.5 - 12 (m+f) | | 0.04 |
| Turkeys: | | | |
| Broad Breasted Bronze types | 5 - 15 (m + f) | | |
| **I. RABBITS** (mean 4 breeds) | 4 | | 0.03 - 0.04 (78 days at slaughter) |

*NB* Classification of some specie types and rounding of numbers is by the author.

*Sources:* ADAS/MAFF (1983) *Livestock Units Handbook. Booklet 2267.*
 ILACO (1985) *Agricultural Compendium.* Casey, N. (1985) *Meat production and Meat quality from Boer Goats,* University of Pretoria. Campbell, Q.P. (1977) *The Enobled Boer Goat of South Africa.* Glen Agric. 6.2. Copland, J.N., ed. (1984) *Goat Production and Research in the Tropics.* University of Queensland. FAO (1982) *Camels and Camel Milk.* Animal Production and Health Paper No.26. US Feeds Grain Council, *New Methods of Sheep Management for Mediterranean Countries.* FAO (1986) *The Rabbit; Husbandry, Health and Production.* Animal Production and Health Series No. 21. FAO (1977) *The Water Buffalo.* Animal Production and Health Series No. 4. Sainsbury, D. (1984) *Poultry Health and Management.* Wilson, P.N. (1980) *Agriculture in the Tropics.* Wilkinson, J.M., & Stark, B.A. (1987) *Commercial Goat Production.*

**TABLE 315   Indicative Lactation Yields and Durations**

|  | Yield kg | Duration months |
|---|---|---|
| **DAIRY CATTLE – *B. taurus* type** | | |
| Friesian, UK average | 5300 | 10 |
| Jersey,UK average | 3950 | 10 |
| Holsteins, USA 1981 | 6600 | 10 |
| **DAIRY CATTLE – *B. indicus* type** | | |
| (select herds) | | |
| Hariana | 1112 | 8 |
| Kenana | 1712 | 8 |
| Red Sindhi | 1445 | 9 |
| Mean of 7 Indian breeds | 1598 | 10 |
| (*N.B.* Indicative dry periods of 163 days, 426 day calving interval and 6.3 year generation interval. Individual animals up to 4000 kg) | | |
| Sahiwal | 1826 | 10 |
| Non-descript Zebu breeds | 250 - 700 | 5 - 7 |
| Beef Zebu breeds, e.g. Brahman | 200 - 400 | 5 |
| **DAIRY CATTLE – *B. taurus* x *indicus*** | | |
| Friesian x Sahiwal or Sindhi | 772 - 1027 | 10 |
| **BUFFALO** | | |
| Milking | 2000 - 4000 | 10 |
| Working | 300 - 800 | 6 |
| **CAMEL** | | |
| Dromedary | 3200 | 14 |
| Bactrian | 1800 | |
| **LLAMA** | up to 1000 | |
| **YAK/YAK x CATTLE** | 103 - 3 181 | 5 - 10 |
| **GOATS- improved, recorded** | | |
| UK Crossbred | 1220 | |
| UK Saanens | 1243 | |
| UK Toggenburgs | 1169 | |
| UK Anglo Nubian | 1040 | |
| UK Alpine | 1094 | |
| UK Golden Guernsey | 992 | |
| USA Saanen | Avg. 977   High 1909 | 8 - 10 |

**TABLE 315** *(continued)*   **Indicative Lactation Yields and Durations**

| | Yield kg | Duration months |
|---|---|---|
| **GOATS – improved, recorded** | | |
| USA Toggenburg | Avg. 973   High 2000 | 7 - 10 |
| USA Nubian | Avg. 768   High 1909 | 7 - 10 |
| USA Alpine | Avg. 968   High 2090 | 8 - 10 |
| Swiss Alpine | 634 | 9 |
| Swiss Graubundner | 480 | 8 |
| Chinese Developing breeds | 400 - 800 | 7 - 10 |
| Damascus (Cyprus) | 560 - 700 (after weaning) | |
| | | |
| **SHEEP** | | |
| Meat Producers, suckled : | | |
| UK Hill type, single lamb | 100 | 3 month yield |
| UK Hill type, twins | 150 | 3 month yield |
| UK Lowland type, single lamb | 120 | 3 month yield |
| UK Lowland type, twins | 170 | 3 month yield |
| Important milk breeds : | | |
| Awassi (Israel) | 114+ | |
| Chios (Cyprus) | 240 - 280 | 7 |
| Langhe (Italy) | 119 | |
| Mancha (Spain) | 124 | |
| Lacaune (France) | 73 | |
| E. Friesian (Germany) | 215 | |
| Valachian (Czechoslovakia) | 34 | |
| | | |
| **HORSE (Russia, milked)** | 856 | 5 - 7 |

*Sources:* FAO (1982) *Camels and Camel Milk.* Animal Production and Health Paper No. 26. Wilkinson, J.M. & Stark, B.A. (1987) *Commercial Goat Production.* ILACO (1985) *Agricultural Compendium.* MAFF/ADAS (1987) *Energy Allowances and Feeding Systems for Ruminants.* Mahadevan, P. *Dairy Cattle Breeding in the Tropics.* CAB Tech. comm. No. 11. (1940-50s data). Khurody, D.N. (1941 ICAR records) *Dairying in India.* Cole, H. & Ronning, M.R. (1974) *Animal Agriculture.* American Dairy Science Annual Meeting (1978) Dairy Goat Symposium. *J. Dairy Sci.* **61**, 7. Ying, J. *China. Some Goat Breeds.* World Animal Review. No. 58. Constantinou, A. *Damascus Goats in Cyprus.* Dept of Agriculture Cyprus. Lysandrides, P. *The Chios Sheep in Cyprus.* Dept of Agriculture Cyprus. ARS-USDA (1981) *Dairy Herd Improvement Letter* 57.4.

**TABLE 316   Indicative Milk Qualities**

**a. CATTLE (UK-1987 recorded herds)**

|  | Fat% | Protein% | SNF% |
|---|---|---|---|
| Friesian | 3.91 | 3.22 | |
| Holstein | 3.87 | 3.16 | |
| Ayrshire | 3.97 | 3.33 | 8.5 - 9 |
| Jersey | 5.35 | 3.81 | |
| Guernsey | 4.7 | 3.54 | |

**b. CATTLE – Zebu types**

|  | Fat% | Protein% | SNF% |
|---|---|---|---|
| Non-descript herds | 5.8 | 2.8 | 8.5 |
| Milk Zebu, e.g. Sindhi | 5.7 | 2.8 | |

**c. GOATS (UK)**

|  | Fat% | Protein% | SNF% |
|---|---|---|---|
| Crossbred | 3.7 | 2.8 | |
| British Saanen/Saanen | 3.7 | 2.8 | |
| British Toggenburg/Toggenburg | 3.7 | 2.7 | 9.3 |
| Anglo-Nubian | 5.0 | 3.5 | |
| British Alpine | 4.1 | 3.0 | |
| Golden/English Guernsey | 4.1 | 2.9 | |

**d. PIG**

|  | Fat% | Protein% | SNF% |
|---|---|---|---|
| | 8.5 | 5.8 | 11.6 |

**e. SHEEP**

|  | Fat% | Protein% | SNF% |
|---|---|---|---|
| Meat breeds | 7.4 | 6.1 | 11.9 |
| Milk breeds | 6.5 - 7.5 | 5.5 | |

**f. CAMEL**

|  | Fat% | Protein% | SNF% |
|---|---|---|---|
| Dromedary | 3.8 | 3.7 | 9.6 |
| Bactrian | 4.5 | 2.7 | |

**g. BUFFALO**

|  | Fat% | Protein% | SNF% |
|---|---|---|---|
| | 7 - 9 | 3.9 - 5.5 | 9.6 |

**h. HORSE**

|  | Fat% | Protein% | SNF% |
|---|---|---|---|
| h. HORSE | 1.9 | 2.5 | 8.5 |
| i. ASS | 1.5 | 2 | 8.6 |
| j.DONKEY | 1.4 | 2 | |
| k. RABBIT | 17.9 | 13.5 | |

**Hygiene Quality**

The following are recognized standards in UK:

| Avg. monthly total bacterial cell count/ml of milk | Payment Quality Assessment |
|---|---|
| 20 000 or less | Premium |
| 20 000 - 100 000 | Standard |
| >100 000 | Penalty |

*Sources:* Milk Marketing Board (1986-87) *Recorded Herds.* Wilkinson, J.M. & Stark, B.A. (1987) *Commercial Goat Production.* McDonald, P., Edwards, R.A., & Greenhalgh, J.F.D. (1978) *Animal Nutrition.* FAO (1982) *Camels and Camel Milk. Animal Production and Health Paper No. 26.* ILACO (1985) *Agricultural Compendium.* van den Berg, J.C.T. (1990) *Strategy for Dairy Development in the Tropics and Sub Tropics.* Collins Farmers Diary (1992). Wilson, P.N. (1980) *Agriculture in the Tropics.* Cole, H.H. & Ronning, M.R. (1974) *Animal Agriculture.* American Dairy Science Annual Meeting (1978) Dairy Goat Symposium. *J. Dairy Sci.* **61**, 7. Gatenby, R.M. (1986) *Sheep Production in the Tropics and Sub-Tropics.*

**TABLE 317   Indicative Killing Out (KO) % and Slaughter Liveweights**

| Livestock Type | KO% | Slaughter Live wt. kg |
|---|---|---|
| **CATTLE** | | |
| **UK typical types** | | |
| Small bobby calves | 50 | |
| Veal calves | 60 | |
| Lightweight bullocks and heifers | 53 | 350 - 450 |
| Medium-weight bullocks and heifers | 55 | 450 - 550 |
| Heavy-weight bullocks and heifers | 60 | >550 |
| Over-fat exhibition animals | 65 | |
| Fat cull | 52 | |
| Grazing cows | 48 | |
| (*NB* For cereal fed animals add 5%; for roughage/bulky fed subtract 5%) | | |
| **SHEEP** | | |
| **Tropical range** | 35 - 45* | 20 - 90 |
| **UK typical types** | | |
| Early suckled lambs | 52 | |
| Lightweight lambs | 50 | up to 30 |
| Standard lambs | 48 | 30 - 35 |
| Medium  lambs | 46 | 35 - 40 |
| Heavy lambs | 44 | 40 - 45 |
| Hoggets | 42 | >45 |
| Cull ewes and rams | 40 | |
| (*NB* If newly shorn add 3%; if heavily wooled subtract 3%) | | |
| * Occasionally up to 52% e.g. Desert Sudanese, Indian Malpara | | |
| | | |
| **Goats** | | |
| **Tropical types** | 35 - 45 | 15 - 70 |
| **UK Saanen males** | 64 | 8.1 |
| | 48.6 | 24.5 |
| | 52.3 | 36.5 |
| | | |
| **PIGS** | | |
| Unimproved breeds, traditional management | 60 - 65 | 30 - 40 (180 days) |
| Improved breeds, high management standard | 70 - 75 | 85 - 95 (180 days) |
| **UK typical types** | | |
| Porkers | 70 | 40 - 60 |
| Cutters | 72 | 60 - 80 |
| Baconers | 74 | 80 - 100 |
| 'Heavies' | 76 | >100 |
| Cull sows and boars | 78 | |
| | | |
| **Water Buffalo** | | |
| Male, milk type | 35 - 48 | 400 - 700 |
| Female, milk type | 35 - 48 | 350 - 650 |
| Male, work type | 35 - 48 | 400 - 800 |
| Female, work type | 35 - 48 | 300 - 500 |

*(continued over)*

**TABLE 317** *(continued)*  Indicative Killing Out (KO) % and Slaughter Liveweights

| Livestock Type | | KO% | Slaughter Live wt. kg |
|---|---|---|---|
| **CAMELS** | | | |
| Bactrian | | 40 | 500 - 800 |
| Dromedary | | 40 | 450 - 700 |
| Llamas | | 40 | up to 150 |
| Alpacas | | 40 | up to 70 |
| **POULTRY** | | | |
| Layers, indigenous breeds, | male | 75 | 0.7 - 1.2 |
| traditional management | female | 75 | 0.6 - 0.9 |
| Layers, hybrids, | male | 80 | 2.5 - 4.0 |
| good management | female | 80 | 2.0 - 3.0 |
| Broilers, modern production | (6 - 9 wks) | 80 | 1.05 - 1.88 |
| Ducks, indigenous breeds, | male | 75 | 1.25 - 4.5 |
| traditional management | female | 75 | 1.0 - 3.0 |
| Geese | m and f | 70 | 5.4 - 6.4 |
| **RABBIT** | | | |
| New Zealand White | | 70 | 2.7 (at 15 wks) |
| 'Hylyne Carolina' | | 61 (skinned and eviscerated) | 2.3 |
| | | 75 (eviscerated only) | |

*Sources:* ILACO (1985) *Agricultural Compendium.* Collins Farmers Diary (1992). Mtenga, L.A. (1979) *Meat Production from Saanen Goats.* Ph.D. thesis University of Reading. FAO (1986) *The Rabbit; husbandry, health and production.* Animal Production and Health Series No. 21. ADAS/MAFF (UK) *Ducks and Geese.* Ref book No. 70. Hylyne Rabbits Ltd. Lymm, Cheshire, UK (1992). Gatenby, R.M. (1986) *Sheep Production in the Tropics and Sub-Tropics.*

**TABLE 318  Indicative Wool Qualities and Fleece Weights**

### a. SHEEP

| Major Type | Example breeds | Fleece wt. kg | Staple length cm | Wool Class* | Quality Rating** |
|---|---|---|---|---|---|
| Fine Wool ('Merino') | Merino | 6 - 7 | 8 | | 60s - 90s |
| Quality >60s | Merino* Cheviot | up to 8 | | | 26 - 28 micron*** |
| Medium Wool ('Crossbred') | Border Leicester | 2.75 - 4.5 | 15 - 20 | M | 48s - 52s |
| Quality 46s - 58s | Cheviot | 2 - 2.5 | 8 - 10 | M and H | 48s - 56s |
| | Dorset Down | 2.25 - 3 | 5 - 8 | S and D | 56s - 58s |
| | Dorset Horn | 2.25 - 3 | 8 - 10 | S and D | 54s - 58s |
| | Hampshire Down | 2.25 - 3 | 5 - 10 | S and D | 56s - 58s |
| | Romney | 3 - 4.5 | 10 - 20 | M | 48s - 54s |
| | Southdown | 1.5 - 2.25 | 4 - 6 | S and D | 58s - 60s |
| | Suffolk | 2.5 - 3 | 5 - 10 | S and D | 54s - 58s |
| | Welsh Mountain | 1.25 - 2 | 5 - 15 | M and H | 36s - 48s |
| | Leicester Longwool | 5 - 7 | 20 - 25 | L and L | 40s - 46s |
| | Lincoln Longwool | 7 - 10 | 15 - 35 | L and L | 36s - 40s |
| | Oxford Down | 3 - 4 | 10 - 15 | S and D | 50s - 54s |
| | Ryeland | 2.25 - 3 | 8 - 10 | S and D | 56s - 58s |
| | Shropshire | 2 - 3 | 8 - 12 | S and D | 54s - 56s |
| | Charollais | 2 - 2.5 | 4 - 6 | S and D | 56s - 60s |
| | Friesland | 4 - 6 | 10 - 15 | M | 50s - 54s |
| | Texel | 3.5 - 5.5 | 8 - 15 | M | 46s - 56s |
| | Polwarth | 6 | | | 56s - 58s |
| | Corriedale | 7.4 | | | 50s - 56s |
| Carpet Quality 28s - 44s | Blackface | 1.75 - 3 | 15 - 30 | M and H | Coarse |
| Desert Sheep a) Saudi arabia | Awassi | 2 | 44% coarse fibres, 47 micron avg. diameter | | |
| | | | 43% fine fibres, 24 micron avg. diameter | | |
| | Najdi | 2.2 | 88% coarse fibres, 68 micron avg. diameter | | |
| | | | 12% coarse fibres, 24 micron avg. diameter | | |
| b) Pakistan | 21 breed survey | 0.8 - 5.6 | Staple length 4.1 - 8.3 cm (6 month growth) | | |
| c) Libya | Barbary | 2.7 | fibre length 12.7 cms, 32.5 micron diameter | | |
| Fur Sheep | Karakul | | | | |
| 'No' Fleece | Wiltshire Horn | No significant fleece | | | |

\*     Wool Classes: M = Medium, and H = Mountain and Hill, S and D = Shortwool and Down, L and L = Longwool and Lustre

\*\*    UK quality rating relates to the fineness of individual fibres in terms of theoretical spinning capacity of the wool. One pound of wool of 56 quality would theoretically spin out to 56 hanks each 560 yards long. Similarly one pound of 64 quality would spin out to a length of 64 x 560 yards etc etc. Hence finer wool has a higher number. This rating is also known as the Bradford count. Other ratings are based on the fibre diameter in microns.

\*\*\*   (micron = 0.001mm)

*(continued over)*

---

**TABLE 318 *(continued)*   Indicative Wool Qualities and Fleece Weights**

---

## b. GOATS

Mohair yield from Angora goats can be around 6 kg/yr from two shearings. Cashmere yields are much lower at 0.5kg/yr from combing. During cleaning similar fleece weight losses are experienced as in sheep. Fibres of adult Angoras are around 36 - 46 microns diameter and those of Cashmere about 15.5. A general figure for typical goat fleece yields is considered to be 1.5 kg depending on breed and size. Mohair is typically 14 cm long and Cashmere 4. Many goats with a hair fleece have a fibre length of over 14 cm.

.

## c. CAMELS, etc.

Llamas yield about 2 kg/yr of long, coarse fleece. Camels yield between 1 - 5 kg/yr with Bactrian camels generally giving more than Dromedaries.

*Major source:* British Wool Marketing Board (1990) *British Sheep and Wool.* Reproduced with kind permission
Also: Wilkinson, J.M. & Stark, B.A. (1987) *Commercial Goat Production.* ILACO (1985) *Agricultural Compendium.* Pritchard, C.J.R. (1975) *A Note on the Wool Characteristics of Sheep at the Hofuf Agricultural Research Centre.* Publication No. 61. Ministry of Agriculture and Water, Saudi Arabia. FAO (1982) *Camels and Camel Milk.* Animal Production and Health Paper No. 26. FAO (1985) *Sheep and Goats in Pakistan.* Animal Production and Health Paper No 56. National Sheep Association (1987) *British Sheep.* Hunt, J. (1992) Merino Cross: Shear Value. *Farmers Weekly,* Jan. 31. Barnard, A., ed. *The Simple Fleece.* Australian National University. Hugo, W.T. (1968) *The Small Stock Industry of South Africa.* Gatenby, R.M. (1986) *Sheep Production in the Tropics and Sub - Tropics.*

---

**TABLE 319   Comparison of Layer Duck Efficiency with that of Laying Hens**

|  | Brown Egg* Mean | White Egg* Mean | Layer Duck (CV 2000) |
|---|---|---|---|
| H.H.A. (52 weeks) | 269 | 277 | 285 |
| Egg Weight (g) | 59.7 | 56.9 | 75 |
| Feed: D.O. - P.O.L. | 8.0 | 7.1 | 11.2 |
| P.O.L. + 52 weeks | 42.2 | 41.5 | 43.7 |
| Total food/kg eggs | 3.13 | 3.1 | 2.56 |
| Point of lay weight (kg) | 1.61 | 1.30 | 1.50 |

*1980 Data

*Source:* 1991 data Kindly provided by Cherry Valley Farms Livestock Division, Rothwell, Lincs, UK.

# Within Specie Comparisons – Cattle

---

**TABLE 320    Basic Performance of UK Cattle Milking Herds**

---

**a. Analysis by yield from forage**

| Litres from Forage | <1500 | 1500-<br>2000 | 2001-<br>2500 | 2501<br>3000 | >3000 |
|---|---|---|---|---|---|
| Milk yield/cow (1) | 5662 | 5636 | 5602 | 5681 | 5823 |
| Yield from forage/cow (1) | 853 | 1773 | 2270 | 2744 | 3377 |
| Concentrate use/1 (kg) | 0.33 | 0.30 | 0.27 | 0.24 | 0.20 |
| Concentrate use/cow (kg) | 1859 | 1679 | 1496 | 1361 | 1158 |

**b.Analysis by Breed**

| Breed | Black and White | Channel Island | Others |
|---|---|---|---|
| Milk yield/cow (1) | 5689 | 4145 | 4801 |
| Yield from forage/cow (1) | 2393 | 1430 | 2248 |
| Concentrate use/1 (kg) | 0.26 | 0.32 | 0.25 |
| Concentrate use/cow (kg) | 1455 | 1356 | 1200 |

*Note:* UK Dairy herds are predominantly Holstein/Friesian type grazed on pasture for about 7 months per year.

*Source:* Genus Milkminder Dairy Herd Costing Service (1991) Report for 4200 costed farms. (Genus is a Milk Marketing Board Business). Figures reproduced with kind permission of the Milk Marketing Board.

---

**TABLE 321    UK Dairy Cattle: Lactation Trends, Yields, Fat and Protein (1989/90 and 1979/80)**

---

**Lactation Trends.  National.  All Breeds**

| All Breeds | 1979/80 | 1989/90 | 10 years | Last year |
|---|---|---|---|---|
| Yield | 5428 | 5887 | +459 | +129 |
| Fat | 3.81 | 3.97 | +0.16 | +0.09 |
| Protein | 3.28 | 3.25 | -0.03 | +0.04 |
| Wt. Fat + Protein | 3.85 | 4.24 | +39 | +15 |

**Lactation Trends by breed. Milk yield (kg)**

| | | | Change | |
|---|---|---|---|---|
| | 1979/80 | 1989/90 | 10 years | Last year |
| Ayrshire | 4938 | 5352 | +322 | +92 |
| British Holstein | 6233 | 6703 | +250 | +220 |
| Holstein Friesian | 5523 | 5925 | +277 | +125 |
| Dairy Shorthorn | 4834 | 5195 | +341 | +20 |
| Guernsey | 3974 | 4280 | +301 | +5 |
| Jersey | 3814 | 4163 | +233 | +116 |

*(continued over)*

**TABLE 321** *(continued)*   **UK Dairy Cattle: Lactation Trends, Yields, Fat and Protein (1989/90 and 1979/80)**

**Lactation Trends by breed. Butterfat Percentage**

|  | 1979/80 | 1989/90 | Change 10 years | Last year |
|---|---|---|---|---|
| Ayrshire | 3.90 | 3.98 | +0.08 | +0.06 |
| British Holstein | 3.73 | 3.91 | +0.18 | +0.09 |
| Holstein/Friesian | 3.77 | 3.94 | +0.17 | +0.08 |
| Dairy Shorthorn | 3.58 | 3.75 | +0.17 | +0.08 |
| Guernsey | 4.62 | 4.74 | +0.12 | +0.08 |
| Jersey | 5.13 | 5.39 | 0.26 | 0.08 |

**Lactation Trends by Breed. Protein Percentage**

|  | 1979/80 | 1989/90 | Change 10 years | Last year |
|---|---|---|---|---|
| Ayrshire | 3.37 | 3.33 | -0.04 | +0.04 |
| British Holstein | 3.21 | 3.18 | -0.03 | +0.04 |
| Holstein/Friesian | 3.26 | 3.24 | -0.02 | +0.04 |
| Dairy Shorthorn | 3.31 | 3.28 | -0.03 | +0.02 |
| Guernsey | 3.60 | 3.58 | -0.02 | +0.04 |
| Jersey | 3.84 | 3.84 | No change | +0.04 |

**Lactation Trends by Breed. Measured solids – fat and protein (kg)**

|  | 1979/80 | 1989/90 | Change 10 years | Last year |
|---|---|---|---|---|
| Ayrshire | 360 | 391 | +30 | +12 |
| British Holstein | 433 | 475 | +42 | +23 |
| Holstein/Friesian | 389 | 425 | +36 | +16 |
| Dairy Shorthorn | 333 | 365 | +32 | +6 |
| Guernsey | 327 | 356 | +61 | +6 |
| Jersey | 342 | 384 | +42 | +15 |

*Source:* Milk Marketing Board (Year ended Sept 1990) *National Milk Records.*
Reproduced with kind permission of the Milk Marketing Board.

TABLE 322   UK Dairy Cattle: Yield Averages Lactational and Annual; Percentages and Weights of Butterfat and Protein 1989/90

| Breed | Average Yield (kg) | | Butterfat % Average (Weight kg) Lactation | Protein % Average (Weight kg) Lactation |
|---|---|---|---|---|
| | Lactation | Annual | | |
| Ayrshire | 5352 | 5020 | 3.98 (213) | 3.33 (178) |
| Blue Albion | 4355 | 3530 | 3.62 (158) | 3.18 (139) |
| British Holstein | 6703 | 6336 | 3.91 (262) | 3.18 (213) |
| Dairy Shorthorn | 5195 | 4861 | 3.75 (195) | 3.28 (170) |
| Dexter | 2362 | 1748 | 4.16 (98) | 3.47 (82) |
| Guernsey | 4280 | 3985 | 4.74 (203) | 3.58 (153) |
| Holstein/Friesian | 5925 | 5624 | 3.94 (233) | 3.24 (192) |
| Jersey | 4163 | 3935 | 5.39 (224) | 3.84 (160) |
| Red Poll/British Dane | 4213 | 3905 | 3.91 (165) | 3.32 (140) |
| Simmental | 4377 | 4214 | 4.02 (176) | 3.33 (146) |
| South Devon | 3276 | 2783 | 4.03 (132) | 3.36 (110) |
| Mixed | 5903 | 5420 | 3.93 (232) | 3.24 (191) |
| All Breeds | 5887 | 5586 | 3.97 (233) | 3.25 (191) |

*Source:* Milk Marketing Board (Year ended Sept 1990) *National Milk Records.*
Reproduced with kind permission of the Milk Marketing Board.

TABLE 323   Numbers of Females, and Daily Requirements of Mcal ME for Herds of Various Breed Groups that will Provide 700 kg Milk per day Throughout the Year

| Groups | Holstein/Friesian | | Blanco Orejinegro ** | Horro † | Harianna †† | Egyptian Buffalo |
|---|---|---|---|---|---|---|
| | New York* | Puerto Rico * | | | | |
| **Cows** | | | | | | |
| Lactating | 40 | 47 | 350 | 226 | 241 | 110 |
| Dry (4 yr+) | 8 | 19 | 927 | 264 | 234 | 77 |
| **Young stock** | | | | | | |
| 37 - 51 mo. | 0 | 6 | 100 | 146 | 94 | 56 |
| 24 - 36 mo. | 0 | 10 | 168 | 156 | 95 | 63 |
| 13 - 24 mo. | 10 | 18 | 315 | 164 | 100 | 70 |
| 4 - 12 mo. | 12 | 22 | 394 | 197 | 120 | 84 |
| Total no. | 70 | 122 | 2254 | 1153 | 884 | 460 |
| **Mcal ME** | | | | | | |
| Req./days§ | 1 080 | 1 385 | 13 268 | 6 685 | 5 593 | 3 400 |
| Req./kg milk | 1.54 | 1.98 | 18.95 | 9.55 | 7.99 | 4.86 |
| % to production | 45.4 | 33.9 | 4.0 | 8.2 | 9.8 | 22.2 |

\*   Estimates based on averages for DHIA herds.
\*\*  *Bos taurus* type native to north-central Colombia.
†   *Bos indicus* type native to northern Ethiopia.
††  *Bos indicus* type native to north-central India.
§   Fat content of milk was considered in proportion of Mcal required for milk production.

*Source:* Proceedings of 3rd World Conference on Animal Production (1975)
Reproduced with kind permission of the University of Sydney.

**TABLE 324  Average Performance, Animal Numbers and Daily Feed used to Produce 700 kg Milk from Hariana and F1 Crosses of Brown Swiss, Holstein and Jersey with Hariana at Haringhata, India**

|  | Hariana | Crosses | Diff. from Hariana |
|---|---|---|---|
|  |  | **Average performance** |  |
| Daily yield/cow (kg) | 1.8 | 7.2 | +5.4 |
| Calving interval (days) | 515 | 395 | -120 |
| Lactation length (days) | 290 | 290 | 0 |
| Age 1st calving (mo.) | 35 | 26 | -9 |
|  |  | **Number animals** |  |
| **Cows** |  |  |  |
| Lactating | 389 | 97 | -292 |
| Dry | 302 | 35 | -267 |
| **Young stock** |  |  |  |
| >36 mo. | 48 | 0 | -48 |
| 25 - 36 mo. | 104 | 29 | -75 |
| 12 - 24 mo. | 167 | 32 | -135 |
| 12 mo. | 200 | 39 | -161 |
| **Total herd** | 1210 | 232 | -978 |
|  |  | **Feed/day for herds (kg)** |  |
| Concentrates | 1500 | 800 | -700 |
| Green fodder | 15 000 | 3500 | -11 500 |
| Dry fodder | 4000 | 900 | -3100 |

*Source: Proceedings of 3rd World Conference on Animal Production* (1975).
Reproduced with kind permission of the University of Sydney.

**TABLE 325  Milk Yields of Crossbred Cattle F₁ Stock**

| Country | Breeding | Lactation milk yield of F₁ | Milk yield of parents | | % increase over Zebu |
|---|---|---|---|---|---|
|  |  |  | Zebu | European |  |
| India | Friesian x Zebu | 6977 | 4272 * | 8023 | 63 |
| India | Friesian x Hariana | 6000 | 3000 | – | 100 |
| India | Friesian x Zebu | 6881 | 3798 | 9406 | 81 |
| Tanganyika | Friesian x Zebu | 3570 | – | – | – |
| Trinidad | Friesian x Zebu (Ongole) | 4316 | – | 2979 | – |
| India | Ayrshire x Sahiwal | 7651 | 3024 | – | 53 |
| India | Ayrshire x Hariana | 4500 | 3000 | – | 50 |
| Philippines | Ayrshire x Nellore | 3214 | – | 2925 | – |
|  | Jersey x Red Sindhi | 4551 | 3084 | – | 47 |
| India | Jersey x Red Sindhi | 4323 | 3000 | – | – |
| Jamaica | Jersey x Zebu (1911-1923) | 4294 | 3288 | 4137 | 30 |
| Jamaica | Jersey x Zebu † | 4835 | – | 4894 | – |
| USA | Jersey x Red Sindhi ‡ | 8697 | – | 9588 | – |
| India | Brown Swiss x Red Sindhi § | 5790 | 3000 | – | 75 |
| Egypt | Shorthorn x Damietta | 4764 | 2832 | 5982 | 68 |

\* calculated from average of several breeds   ‡ mainly 365-day yields
† 305-day yield, corrected for years   § converted from 1lb butterfat at 3.8%

*Source:* Maule, J.P. (1953) *Crossbreeding Experiments with Dairy Cattle. Animal Breeding Abstracts* 21.2..
© CAB International, Wallingford, Oxon, UK.

TABLE 326  Comparison of Milk Yield of Tropical Milch Animals on the Basis of Yield
per Unit Liveweight

| Type of stock | Mean lactation yield (kg) | Mean liveweight (kg) | Yield (kg)/10kg liveweight |
|---|---|---|---|
| **Goat** | | | |
| Tropical type | 200 | 40 | 50 |
| European crossbred | 400 | 50 | 80 |
| European type | 600 | 60 | 100 |
| **Cow** | | | |
| Unimproved zebu | 400 | 300 | 13 |
| Improved zebu | 1000 | 350 | 29 |
| European crossbred | 2000 | 400 | 50 |
| **Water buffalo** | | | |
| Unimproved breed | 1000 | 500 | 20 |
| Improved breed | 1500 | 500 | 30 |

*Source:* Wilson, P.N. (1980) *Agriculture in the Tropics.*
Reproduced with kind permission of Longman Group UK Ltd.

TABLE 327  Data for Mean Lactation Yield of Various Breeds of Water Buffalo
and Zebu Cattle in India

| Item | Water buffalo (*Bos bubalis*) | Zebu cattle (*Bos indicus*) |
|---|---|---|
| Number of records | 6160 | 4310 |
| Mean daily yield (kg) | 3.6 | 1.7 |
| Estimated lactation yield (kg) | 982 | 429 |
| Average length of lactation (days) | 300 | 264 |
| Calving interval (months) | 18 | 18.2 |

*Source:* Wilson, P.N. (1980) *Agriculture in the Tropics* (after Maule, 1953-54).
Reproduced with kind permission of Longman Group UK Ltd.

**TABLE 328  Representative Breeds of Cattle Milked in the Tropical World**

| Cattle breed | Type | Usual environment | Milk production for lactation (litres) Normal | Max. | Lactation length (days) | Butterfat % | Age at 1st calving (mths) | Calving interval (mths) | Usual purpose |
|---|---|---|---|---|---|---|---|---|---|
| **Asian** | | | | | | | | | |
| Chinese yellow | Bt | Alpine subtropic | 14 – 285 | | 50 – 200 | 5.9 | 23 – 30 | 23 – 29 | Milk (butter) |
| Bhagnari | Bi | Semi-arid | 455 –1590 | 2270 | 260 – 340 | 5.2 | 29 – 50 | 13 – 15 | Draught/milk/beef |
| Damascus | Bt/Bi | Dry subtropic | 1500 – 3000 | 5000 | 190 – 300 | 4 – 5 | | | Milk |
| Gir | Bi | Monsoonnal | 1225 – 2270 | 3175 | 240 – 380 | 4.5 – 4.6 | 31 – 51 | 14 – 16 | Milk/Draught |
| Hariana | Bi | Semi-arid | 640 –1500 | 4540 | 260 – 320 | 4 – 4.8 | 32 – 72 | 19 – 21 | Milk/Draught |
| Kangayam | Bi | Humid tropic | 230 –1135 | 2270 | 220 – 310 | 5.7 | 39 – 47 | | Draught/Milk |
| Kankrej | Bi | Dry tropic | 680 – 2495 | 3310 | 250 – 370 | 4.6 – 4.7 | 33 – 78 | | Milk/Draught |
| Malaysian LID | Bi | Humid tropic | 455 – 680 | | 180 – 210 | | 48 | 13 – 14 | Milk |
| Nagori | Bi | Semi-arid | 680 –1360 | 1590 | 220 – 320 | | 40 | 15 – 16 | Draught/Milk |
| Oksh | Bt | Semi-arid | 600 – 910 | | 150 – 250 | 4 – 5 | | | Milk/Beef/Draught |
| Ongole | Bi | Dry tropic | 1180 –1635 | 3265 | 300 – 330 | 5.1 | 36 – 51 | 16 – 18 | Draught/Milk |
| Red Sindhi | Bi | Subtropic to semi-arid | 680 – 2270 | 5445 | 270 – 490 | 4 – 5 | 30 – 40 | 13 – 18 | Milk/Draught |
| Sahiwal | Bi | Arid subtropic | 1135 – 3.175 | 4535 | 290 – 490 | 4 – 6 | 30 – 43 | 13 – 18 | Milk/Draught |
| Sinhala | Bi | Humid | 230 – 545 | | 230 – 280 | 5 – 6 | 27 – 43 | 12 | Draught/Milk |
| Siri | Bt/Bi | Alpine subtropic | 455 –1360 | 1905 | 280 | 6 – 10 | 60 – 63 | 12 – 20 | Draught/Milk |
| Tharparkar | Bi | Arid subtropic | 680 – 2270 | 4765 | 280 – 440 | 4.7 | 24 – 47 | 14 – 18 | Milk (ghee)/Draught |
| **African** | | | | | | | | | |
| Angoni | Bi | Wet tropic | 630 – 800 | | 245 – 270 | 5.7 | | | Draught/Milk/Beef/Ceremonial |
| Ankole | Bt/Bi | Hot dry | 320 – 820 | 900 | 212 – 239 | 3 – 7 | 42 – 60 | 16 – 24 | Milk/Ceremonial |
| Azaouak | Bi | Dry tropic | 360 – 670 | 1500 | 180 – 300 | 3 – 6 | 27 – 46 | | Milk |
| Boran | Bi | Semi-arid | 455 –1815 | 2640 | 139 – 303 | 4 – 6.8 | 36 – 52 | 11 – 14 | Milk/Beef |
| Brown Atlas | Bt | Arid | 450 – 600 | 1200 | 150 – 180 | 4 | 22 – 24 | | Milk/Beef/Draught |
| Bukedi | Bi | Wet tropic | 230 –1000 | 1940 | 223 – 280 | 4.7 – 7 | 25 – 61 | 11 – 14 | Beef/Blood/Milk/Draught |
| Drakensberger | Bt/Bi Recent | Subtropic | 820 –1635 | | 180 – 250 | | | | Beef/Milk/Draught |
| Dwarf Shorthorn | Bt | Humid | 120 – 360 | | 120 – 180 | | | | Beef/Draught/Ceremonial |
| Kuri | Bt | Humid to semi-arid | 360 – 800 | 1800 | 180 – 300 | | 30 – 48 | 12 – 24 | Milk/Beef/Pack |
| Libyan | Bt | Arid | 455 –1360 | 2000 | 200 – 320 | 3.2 | 36 – 43 | | Milk/Beef/Draught |
| Madagascar Zebu | Bi | Wet tropic | 150 – 250 | 800 | 120 – 180 | | 27 – 48 | 18 – 24 | Beef/Draught/Milk |
| Maure | Bi | Semi-arid tropic | 600 – 700 | 2000 | 210 – 240 | | 36 – 48 | | Milk/Draught/Beef |
| Nandi | Bi | Mild tropic | 340 –1135 | 2265 | 175 – 300 | 5.5 – 6 | 43 | 11 – 13 | Draught/Milk/Beef/Ceremonial |

| Breed | Type | Environment | | | | | | | | Use |
|---|---|---|---|---|---|---|---|---|---|---|
| N'dama | Bt | Humid tropic | 150 – 270 | 450 | 150 – 300 | 6.5 – 7 | 27 – 72 | 14 – 42 | | Beef/Milk/Draught |
| Nganda | Bt/Bi | Humid tropic | 135 – 1135 | 2790 | 247 – 380 | 4.7 – 7 | 24 – 55 | 14 | | Beef/Milk |
| Nguni | Bt/Bi | Subtropic varied | 570 – 1400 | 3600 | 140 – 300 | 4 – 7 | 27 – 48 | 18 – 24 | | Beef/Milk/Draught |
| Senegal Fulani | Bi | Dry tropic | 450 – 500 | | 180 – 200 | 5.5 | 24 – 60 | 14 – 24 | | Beef/Pack/Milk |
| Shuwa | Bi | Arid tropic | 455 – 1815 | 3420 | 240 – 396 | | 45 | 18 | | Milk/Draught |
| Sokoto | Bi | Varied tropic | 455 – 1360 | 1940 | 230 – 280 | 5.8 | 36 | 15 | | Milk/Draught |
| Sudanese (Kenana type) | Bi | Dry tropic | 454 – 2720 | 4660 | 168 – 339 | 4.7 – 5.5 | 24 – 54 | 12 – 24 | | Milk (Semni) |
| Rana | Bt/Bi Recent | Wet mountain tropic | 900 – 1800 | 2800 | 180 – 300 | 4.5 | | | | Milk/Beef |
| Tanzania Zebu | Bi | Dry to humid tropic | 230 – 545 | 1205 | 247 – 301 | 4.9 | 36 – 42 | 11 – 13 | | Blood/Milk/Beef |
| Tuli | Bt/Bi | Dry subtropic | 1635 – 3630 | 4050 | 180 – 270 | | | | | Beef/Milk/Draught |
| Tuni | Bt/Bi | Humid to semi-arid | 451 – 360 | 1980 | 252 – 305 | 5.3 | | 12 – 14 | | Draught/Milk/Beef |
| White Fulani | Bi | Semi-arid tropic | 635 – 1225 | 2300 | 190 – 360 | 5 – 7.5 | 36 – 48 | 12 – 15 | | Milk/Draught/Beef |
| **Tropical American** | | | | | | | | | | |
| Jamaica Hope | Bt/Bi Recent | Humid tropic | 2000 – 3000 | 9075 | 250 – 305 | 5 – 5.2 | 27 – 33 | 12 – 14 | | Milk |
| Milking Criollo | Bt | Humid tropic | 500 – 1800 | 3600 | 250 – 300 | 4.6 – 5.1 | 30 | 12 – 14 | | Milk |
| Mocha Nacional | Bt | Subtropic | | 1600 | | 4.5 | | | | Milk/Beef |
| **Tropical Australian** | | | | | | | | | | |
| Australian Friesian | Bt/Bi Recent | Varied, subtropic to tropic | | | | | | | | |
| Sahiwal | Bt | to tropic | 2405 | 5500 | 265 | 4.2 | 28 | | | Milk |
| Australian Illawarra Shorthorn | Bt | Varied | 3030 | 11 855 | 270 | 3.8 | 24 – 36 | 12 – 20 | | Milk/Beef/Draught |
| Australian Milking Zebu | Bt/Bi | Subtropic | 2280 | 4850 | 275 | 4.5 | 28 | | | Milk |
| **Temperate breeds used in the tropics** | | | | | | | | | | |
| Brown Swiss | Bt | Highlands | 2585 – 3400 | | 270 | 3.8 | | 14 | | Milk (Tropic Aust.*) / Milk/Beef (S.Am.*) |
| Friesian | Bt | Varied | 3455 | 12 350 | 280 | 3.7 | 24 – 30 | | | Milk (Tropic Aust.*) |
| Ayrshire | Bt | Varied | 2710 | 8180 | 272 | 3.8 | | | | Milk (Tropic Aust.*) |
| Guernsey | Bt | Subtropic | 2445 | 6310 | 275 | 4.2 | | | | Milk (Tropic Aust.*) |
| Jersey | Bt | Varied | 2460 | 6890 | 277 | 4.5 | 24 | | | Milk (Tropic Aust.*) |
| Red danish | Bt | Humid | 2500 – 3800 | | 390 | 4.2 | 28 | | | Milk/Beef (India*) |

*Source:* Chamberlain, A. (1989) *Milk Production in the Tropics.* Reproduced with kind permission of Longman Group UK Ltd.

Bi = *Bos indicus* (humped). Bt = *Bos taurus* (humpless). Bi/Bt = *Bos indicus* x *Bos taurus* crosses *source of production figures

**TABLE 329   Reproductive Aims in Well-managed Dairy Cattle Herds in the Tropics**

| | |
|---|---|
| Calving interval | 365 days |
| Calving to first heat | 40 days or less |
| Calving to first service | 60 days or less |
| Calving to conception | 85 days |
| Average interval between heats | 18 to 24 days |
| First service pregnancy rate | 60% or better |
| 60 - to 90-day non-return date | 70% |
| Services per conception | 1.5 or less |
| Abortion rate | 5% or less |
| Lactation Length | 300 days |
| Dry period | 60 days |

*Source:* Chamberlain, A. (1989) *Milk Production in the Tropics.*
Reproduced with kind permission of Longman Group UK Ltd.

**TABLE 330   Average First-calving Age for Cattle Breeds in the Tropics**

| Breed and country | Average first-calving age (months) | Breed and country | Average first-calving age (months) |
|---|---|---|---|
| **India** | | **West Africa** | |
| Jersey | 30 | Shorthorn types | 30 - 48 |
| Jersey/Red Sindhi | 29 | Zebu x Holstein | 32 |
| European/Red Sindhi | 42 | Sokoto | 36 |
| Red Sindhi | 41 - 44 | Fulani | 30 - 60 |
| Sahiwal | 38 - 42 | Zebu | 50 |
| Tharparkar | 35 - 44 | | |
| Gir | 31 - 51 | **Egypt** | |
| Ongole | 36 - 51 | Egypt Native Cattle | 34 |
| Mewati | 48 | | |
| Hariana | 53 - 59 | **Sudan** | |
| Siri | 60 - 63 | Kenana | 38 - 44 |
| | | Butana | 44 |
| **Sri Lanka** | | Gezira | 41 |
| Sinhala | 42 | | |
| Sinhala x Jersey | 37 | **S. America/Caribbean** | |
| | | Jamaica Hope | 27 - 33 |
| **East Africa** | | Criollo | 30 |
| Ankole | 42 - 60 | | |
| Boran | 40 | **Queensland, Australia** | |
| Nandi (Zebu) | 43 | Friesian | 35 |
| Nganda (Uganda) | 42 | Aust. Friesian Sahiwal | 33 |
| Sudanese | 44 | Aust. Illawarra Shorthorn | 33 |
| Zebu (Uganda) | 52 | Aust. Milking Zebu | 28 |
| | | Jersey | 24 |
| **Central/S. Africa** | | | |
| Nguni | 27 - 48 | | |

*Source:* Chamberlain, A. (1989) *Milk Production in the Tropics.* Reproduced with kind permission of Longman Group UK Ltd.

**TABLE 331** **Average Age at First Calving in Months in Different Cattle Breeds under Tropical Conditions**

| Breeds | Number of studies | Mean | Standard deviation | Standard error | Minimum | Maximum | 95% confidence interval for the mean |
|---|---|---|---|---|---|---|---|
| Holstein | 14 | 32.8[a] | 7.0 | 1.8 | 25.3 | 54.7 | 28.7 - 36.8 |
| Sahiwal | 10 | 40.8[c] | 7.2 | 2.2 | 27.1 | 52.6 | 35.6 - 46.0 |
| Gir | 7 | 48.0[c] | 5.9 | 2.2 | 41.1 | 58.2 | 42.5 - 53.5 |
| Milking Zebu | 62 | 33.1[a] | 4.3 | 0.5 | 25.6 | 50.7 | 32.0 - 34.2 |
| Jersey | 5 | 27.7[a] | 3.2 | 1.4 | 24.7 | 31.6 | 23.7 - 31.7 |
| Nelore | 5 | 43.9[b] | 3.0 | 1.3 | 39.4 | 47.3 | 40.1 - 47.7 |
| Guzerat | 3 | 44.2[b] | 3.9 | 2.2 | 39.7 | 47.0 | 34.4 - 54.0 |
| Brown Swiss | 5 | 35.6[a] | 9.4 | 4.2 | 28.5 | 51.9 | 23.8 - 47.3 |
| Hariana | 13 | 48.0[c] | 7.6 | 2.1 | 31.6 | 58.4 | 43.4 - 52.7 |
| Tharparkar | 9 | 43.2[b] | 8.1 | 2.7 | 29.2 | 53.4 | 37.0 - 49.5 |
| Beef Zebu | 18 | 42.9[b] | 9.1 | 2.1 | 27.2 | 67.1 | 38.3 - 47.4 |
| Total | 153 | 37.9 | 8.5 | 0.6 | 24.7 | 67.1 | 36.5 - 39.2 |

Different letters in the mean column indicate that values differ significantly $P < 0.05$

*Source:* Galina, L.S. and Arthur, G.H. (1989) *Review of Cattle Reproduction in the Tropics. Pt. 1 Puberty and Age at First Calving.* Animal Breeding Abstracts 57. 8. © CAB International, Wallingford, Oxon, UK.

---

**TABLE 332** **Average Calving Interval in Days in Different Cattle Breeds under Tropical Conditions**

| Breeds | Number of studies | Mean | Standard deviation | Standard error | Minimum | Maximum | 95% confidence interval for the mean |
|---|---|---|---|---|---|---|---|
| Santa Gertrudis | 3 | 17.9[a] | 7.3 | 4.2 | 13.5 | 26.4 | 0.3 to 36.1 |
| Jersey | 4 | 14.5[a] | 0.9 | 0.4 | 13.4 | 15.6 | 13.1 to 15.9 |
| Hariana | 8 | 15.7[a] | 1.6 | 0.5 | 13.7 | 18.2 | 14.3 to 17.1 |
| Holstein | 20 | 15.0[a] | 1.9 | 0.4 | 11.8 | 19.9 | 14.1 to 15.9 |
| Tharparkar | 5 | 14.8[a] | 2.6 | 1.1 | 12.0 | 19.0 | 11.5 to 18.0 |
| Australian Milking Zebu | 3 | 15.7[a] | 2.1 | 1.2 | 14.0 | 18.1 | 10.3 to 21.0 |
| Brown Swiss | 5 | 15.8[a] | 1.5 | 0.6 | 13.5 | 17.7 | 13.8 to 17.7 |
| Milking Zebu * | 17 | 14.3[a] | 2.8 | 0.7 | 11.5 | 24.9 | 12.8 to 15.8 |
| Criollo | 3 | 15.1[a] | 1.1 | 0.6 | 14.4 | 16.5 | 12.1 to 18.0 |
| Sahiwal | 3 | 14.7[a] | 1.4 | 0.8 | 13.1 | 15.8 | 11.1 to 18.3 |
| Charolais | 3 | 17.9[b] | 3.5 | 2.0 | 14.1 | 21.0 | 9.2 to 26.7 |
| Brahman | 6 | 15.0[a] | 1.7 | 0.7 | 12.8 | 17.9 | 13.1 to 16.9 |
| Nellore | 3 | 15.7[a] | 1.0 | 0.6 | 14.7 | 16.8 | 13.0 to 18.3 |
| Beef Zebu * | 13 | 17.6[b] | 3.2 | 0.8 | 12.7 | 22.8 | 15.7 to 19.6 |
| Total | 96 | 15.5 | 2.6 | 0.8 | 11.5 | 26.4 | 15.0 to 16.1 |

* Unspecified breed types. Different letters in mean column differ significantly ($P < 0.05$)

*Source:* Galina, L.S. and Arthur, G.H. *Review of Cattle Reproduction in the Tropics. Parturition and Calving Intervals. Animal Breeding Abstracts* (57.8. 1989) © CAB International, Wallingford, Oxon, UK.

**TABLE 333** **Comparison of Age at First Calving and Calving Interval in Cattle with Different Genetic Make-up**

| Genetic type | Number of studies | Age at first calving (months) | Number of studies | Calving interval (months) |
|---|---|---|---|---|
| *Bos taurus* | 24 | 32.3 [a] | 33 | 15.4 [a] |
| *Bos taurus* x *Bos indicus* | 61 | 33.1 [a] | 20 | 15.1 [a] |
| *Bos indicus* | 50 | 44.8 [b] | 25 | 15.2 [a] |
| *Bos indicus* x *Bos indicus* | 7 | 41.3 [b] | 4 | 16.2 [a] |
| Unspecified crossbred cattle | 10 | 43.7 [b] | 11 | 17.2 [a] |
| *Bos taurus* x *Bos taurus* | | − | 3 | 15.1 [a] |

Different letters within columns differ significantly ($P <0.05$)

*Source:* Galina, L.S. & Arthur, G.H. *(1989) Review of Cattle Reproduction in the Tropics. Parturition and Calving Intervals :* Animal Breeding Abstracts 57.8. © CAB International, Wallingford, Oxon, UK.

**TABLE 334** **Summary of Comparative Losses and Herd Life of European Dairy Cattle Breeds in the Tropics at Different Stages of Life, Compared with Zebus and their Crosses**

| Trait | n * | | | Losses (%) and herd life, according to fraction of European inheritance | | | |
|---|---|---|---|---|---|---|---|
| | | | | **0 - 1/4** | **1/2 - 5/8** | **3/4 - 7/8** | **1** |
| **Abortions (%)** | 7 | x̄: | ** | 1.8 | 3.5 | 5.8 | 11.0 |
| | | range: | ** | 0 - 3.5 | 0 - 4.6 | 1.5 - 9.6 | 5.3 - 21.5 |
| **Stillbirths (%)** | 6 | x̄: | | 1.8 | 2.0 | 4.2 | 6.7 |
| | | range: | | 0 - 4.4 | 0 - 4.2 | − | 5.1 - 10.6 |
| **Death and culling (%):** | | | | | | | |
| Young stock | 20 | x̄: | | 15.4 | 11.6 | 18.5 | 25.9 |
| | | range: | | 4.4 - 57.9 | 2.1 - 28.7 | 8.7 - 62.0 | 12.5 - 44.6 |
| Adult cows | 3 | x̄: | | 14.8 | 12.6 | 15.4 | 18.0 |
| | | range: | | 12.0 - 20.4 | 4.0 - 27.4 | 9.0 - 22.2 | 15.9 - 21.2 |
| **Herd life:** | | | | | | | |
| Lactations | 1 | x̄: | | 3.4 | 6.6 | 3.9 | 2.9 |
| | | range: | | − | − | − | − |
| Years | 3 | x̄: | | 6.3 | 6.3 | 6.4 | − |
| | | range: | | 5.3 - 7.0 | 6.1 - 6.6 | 6.0 - 6.6 | − |

\* Number of studies reviewed
\*\* Simple means and ranges of published values

*Source: Survival of European Dairy Breeds and their crosses with Zebus in the Tropics.* Animal Breeding Abstracts 58.6. (1990) © CAB International, Wallingford, Oxon.

**TABLE 335  Superiority Percentage of F1 Criollo x Zebu for Growth Traits in Tropical Latin America [a]**

| | | Location | | | | | | | |
| Item | Turrialba Costa Rica | Calabozo Venezuela A[b] | Calabozo Venezuela B[b] | Turipaná Colombia | La Libertad Colombia | El Nus Colombia | Beni Bolivia A[b] | Beni Bolivia B[b] | Unweighted mean |
|---|---|---|---|---|---|---|---|---|---|
| Number of observations at birth | 129 | 215 | 496 | 953 | 381 | 349 | 5 109[c] | 3 881[c] | |
| Years | 1961-66 | 1965 | 1966-71 | 1970-74 | 1971-74 | 1972-76 | 1968-72 | 1973-78 | |
| Birth weight | 11 | 4 | 19[d] | 6 | 13 | 16 | | | 12 |
| Pre-weaning average daily gain | | 6 | 2[d] | 3 | 12 | 5 | | | 6 |
| Weaning Weight | 14 | 4 | 5[d] | 4 | 12 | 6 | 6[d] | 3[d] | 7 |
| Postweaning average daily gain | 19 | | | 10 | 37 | 20 | | | 22 |
| Postweaning weight | 16 | | 11[d] | 7 | 18 | 9 | | 8[d] | 12 |

a  Reciprocals, except where indicated
b  Each letter refers to a different data set of the same population
c  Number of observations at weaning
d  Crossbred calves out of Criollo cows. No reciprocals available

*Source:* Plasse, D. (1983) *Cross breeding results from beef cattle in the Latin American Tropics.* Animal Breeding Abstracts 51.11.
© CAB International, Wallingford, Oxon, UK.

**TABLE 336    Potential Performance of Some Indigenous Dairy Cattle Breeds**

| Breed and Country | | Present average performance [1] | | Present elite performance [2] (Milk, kg) | | Potential average yield |
|---|---|---|---|---|---|---|
| | | MY (kg) | CI (days) | Elite group | Max/cow [3] | MY (kg) |
| 1  **Sahiwal** | India Pakistan | 1600 - 1800 | 450 | 3000+ | 5900 in 306 days | 2500 - 2800 |
| 2  **Sahiwal** | Kenya | 1500 - 1600 | 400 | ? 2000 | – | 1800 - 2000 |
| 3  **Kenya** | Sudan | 1250 - 1800 | 390 | 3000 - 3500 | 4670 in 339 days | 2000 - 2500 |
| 4  **Dairy Criollo** | Costa Rica | 1500 - 2000 | | | | 2500 - 2800 |
| 5  **Gir** | India | 1200 - 1500 | 466 | ? 1800 | | 1800 - 2000 |
| 6  **Gir** | Brazil | 1800 - 2000 | 520 | ? 2000 - 2250 | | 2500 - 2700 |
| 7  **Rath** | India | 1600 + | 600 | ? | | 1800 - 2000 |
| 8  **Shuwa** | Nigeria | 900 - 1200 | 420 | 1500 - 1800 | 3430 in 330 days | 1500 - 1800 |
| 9  **Nguni** | Natal Swaziland | 1000 - 1200 | ? | ? | | 1500 - 1600 |
| 10  **White Fulani** | Nigeria | 1000 - 1200 | | 2000 - 2200 | 2337 | 1500 - 1800 |

*Notes:*
1.  These are typical yields of the best recorded herds          MY  =  Milk Yield
2.  Based on data for elite groups (e.g. top 5 - 10%)          CI  =  Calving Interval
3.  Actual yields recorded in the 1940s

*Source:*  Maule, J.P. (1990) *The Cattle of the Tropics.* CTVM, University of Edinburgh.
Reproduced with kind permission of J.P. Maule.

**TABLE 337 Milk Production of Five African Cattle Breeds**

| | Kenana | Butana | Sahiwal (Naivasha) | Boran (Tanga) | Shuwa (Nigeria) |
|---|---|---|---|---|---|
| Age at first calving (months) | 41 ± 8 | 47±11 | 37 | 40(12) | 50 ± 7.7 |
| Lactation yield (kg) | 1204 - 1423 | 1095 - 1213 | 1455 - 1600 | 1050 | 1212 ± 665 |
| Lactation length (days) | 287 - 251 | 242 - 220 | 274 | – | 259 ± 74 |
| Calving interval (days) | 398 ± 90 | 400 ± 75 | 400 - 420 | 382 | 413 ± 110 |
| Daily yield in lactation (kg) | 4.2 - 5.6 | 4.5 - 5.5 | 5.3 - 6.0 | – | 4.9 |
| Daily yield of calving interval (kg) | 3.6 | 3.0 | 3.5 - 4.0 | 2.9 | 3.9 |

*Source:* Maule, J.P. (1990) *The Cattle of the Tropics.* CTVM, University of Edinburgh.
Reproduced with kind permission of J.P. Maule.

**TABLE 338   The Performance of Some 'New' Dairy Cattle Breeds**

| Breed | Country | Lactation yield kg | Fat (yield or %) kg | Days in milk [1] or calving interval [2] |
|---|---|---|---|---|
| AFS | Australia | 2200 - 2400 | 94 - 100 | – |
| | Malaysia | 1670 | – | – |
| AMZ | Australia | 1763 | 82 | 262 [1] |
| | | (651 1st lacts) | | |
| | Overall | 1900 - 2100 | – | – |
| | Malaysia | 1250 | – | – |
| Jamaica Hope | Jamaica (Bodles herd, 1972) | 3218 | 4.88% | 439 [2] |
| | Commercial herds (1981) | 2678 | – | 397 [2] |
| Karan Swiss | N. India | 3249 | – | 396 [2] |
| | | (955 lacts) | | |
| | (1966 - 83) | 3088 ± 5.3 | | 319 ± 3.8 [1] |
| | | (1593 lacts) | | |
| Pitangueiras | Brazil | 2865 | 4.2% | 283 [1] |
| Siboney | Cuba | 3283 ± 6.7 | – | 288 [1] |
| Sunandini | S. India | 2359 | – | 432 [2] |
| Mpwapwa | Tanzania | 1500 - 1800 | – | 390 - 437 [2] |

*Source:* Maule, J.P. (1990) *The Cattle of the Tropics.* CTVM, University of Edinburgh.
Reproduced with kind permission of J.P. Maule.

**TABLE 339 a   Production Standards for Beef from Pure or Cross Calves
(UK Friesian / Holstein dairy herds)**

**a. Calf rearing standards for Friesian / Holstein, Hereford x Friesan, Limousin x Friesian, Belgian Blue x Friesian ***

| | Bucket rearing | | Teat rearing |
|---|---|---|---|
| | Twice - daily | Once - daily | |
| Feeds (kg): milk replacer | 16 | 13 | 25 - 30 |
| concentrates | 175 | 180 | 170 |
| Daily gain (kg). | | | |
| pre-weaning | 0.5 | 0.4 | 0.6 |
| post weaning | 0.9 | 1.0 | 0.9 |
| Reared calf weight after 3 months (kg) * | 110 | 110 | 115 |

* Add 10 kg for Charolais and Simmental x Friesian

*(continued over)*

**TABLE 339 b & c   Production Standards for Beef from Pure or Cross Calves**
**(UK Friesian / Holstein dairy herds)**

### b. For cereal beef *

|  | Friesian /<br>Holstein | Limousin<br>x Friesian | Charolais<br>x Friesian |
|---|---|---|---|
| Slaughter age (months) | 11.5 | 12.5 | 12 |
| Concentrate (t) | 1.9 | 1.9 | 2.0 |
| Reared calf (kg) | 110 | 110 | 120 |
| Daily gain (kg) | 1.3 | 1.3 | 1.4 |
| Slaughter weight (kg) | 460 | 500 | 525 |
| Carcass weight (kg) | 245 | 280 | 290 |

* Usually housed for whole life

### c. For 18-month steers

|  | Friesian /<br>Holstein | Hereford<br>x Friesian | Limousin<br>x Friesian | Charolais<br>x Friesian |
|---|---|---|---|---|
| **Slaughter age (months)** | **18** | **17** | **19** | **19** |
| **Feeds** | | | | |
| 25% DM silage (t) | 5.5 | 4.5 | 5.5 | 6.0 |
| Cereal (t) | 1.0 | 0.8 | 1.0 | 1.1 |
| Stocking (cattle / ha) | 3.5 | 4.0 | 3.5 | 3.2 |
| Reared calf (kg) | 110 | 110 | 110 | 120 |
| Daily gain (kg) | | | | |
| first winter | 0.8 | 0.8 | 0.8 | 0.8 |
| grazing | 0.8 | 0.8 | 0.8 | 0.9 |
| finishing | 0.9 | 0.8 | 0.9 | 1.0 |
| overall | 0.85 | 0.80 | 0.85 | 0.90 |
| Slaughter wt. (kg) | 520 | 470 | 540 | 570 |
| Carcass wt. (kg) | 275 | 265 | 295 | 310 |

*Source:* MLC (1991) *Dairy Beef.* Reproduced with kind permission of the Meat and Livestock Commission (UK).

**TABLE 340 a - c   Production Standards for Suckler Beef in UK**

### a. For autumn and spring-calving lowland herds

| Sire breed | Charolais, Simmental, Limousin | | Aberdeen-Angus, Hereford | |
|---|---|---|---|---|
| Calving season | Autumn | Spring | Autumn | Spring |
| **Feeds** | | | | |
| Silage (t) | 5.0 | 5.5 | 5.0 | 5.5 |
| Cow concentrate (kg) | 250 | 100 | 250 | 100 |
| Calf concentrate (kg) | 150 | 100 | 100 | 75 |
| Calving period (max) (weeks) | 12 | 12 | 12 | 12 |
| Calves per 100 cows bulled | 92 | 92 | 95 | 95 |
| Calf gain (kg) | 1.0 | 1.1 | 0.9 | 1.1 |
| Rearing period (months) | 10 | 7 | 10 | 7 |
| Calf weaning weight (kg) | 350 | 280 | 315 | 250 |

### b. For winter finishing Charolais-cross suckled calves

| | Steers | Heifers | Bulls (rapid finishing) |
|---|---|---|---|
| **Feeding period (months)** | **8** | **7** | **6** |
| Feeds: silage (t) | 5.1 | 3.8 | – |
| concentrates (t) | 0.6 | 0.4 | 1.4 |
| Start weight (kg) | 300 | 250 | 250 |
| Daily gain (kg) | 1.1 | 0.8 | 1.4 |
| Slaughter weight (kg) | 540 | 420 | 500 |
| Carcass weight (kg) | 300 | 220 | 280 |

### c. For overwintering and grass finishing Charolais-cross spring-born suckled calves

| | Steers | Heifers |
|---|---|---|
| Feeding period (months) | 6 winter & 5 grazing | |
| Feeds: silage (t) | 4.5 | 3.5 |
| concentrates (t) | 0.3 | 0.2 |
| Start weight (kg) | 250 | 230 |
| Daily gain (kg) | | |
| overwinter | 0.8 | 0.6 |
| at grass | 0.9 | 0.8 |
| Slaughter weight (kg) | 540 | 460 |
| Carcass weight (kg) | 300 | 245 |

*Source: Source:* MLC (1991) *Suckler Beef.* Reproduced with kind permission of the Meat and Livestock Commission (UK).

**TABLE 341  Effect of Sire Breed on Calf Weaning Weights**

| Sire breed | Lowland herd | Upland herd | Hill herd |
|---|---|---|---|
| | | 200-day weight (kg) | |
| Hereford | 208 | 194 | 184 |
| | Difference from Hereford cross (kg) | | |
| Charolais | +32 | +33 | +21 |
| Simmental | +24 | +28 | +14 |
| South Devon | +23 | +27 | +16 |
| Devon | +17 | +21 | +7 |
| Lincoln Red | +14 | +20 | +5 |
| Sussex | +7 | +13 | +2 |
| Limousin | +7 | +10 | +2 |
| Aberdeen-Angus | -14 | -12 | -8 |

*Source: Source:* MLC (1991) *Suckler Beef.* Reproduced with kind
permission of the Meat and Livestock Commission (UK).

**TABLE 342  Assisted Calvings and Calf Mortality for Different Sire Breeds**

| Sire breed | Assisted calving (%) | Calf mortality (%) |
|---|---|---|
| Charolais | 9.0 | 4.8 |
| Simmental | 8.9 | 4.2 |
| South Devon | 8.7 | 4.0 |
| Devon | 6.4 | 2.6 |
| Limousin | 7.4 | 3.8 |
| Lincoln Red | 6.7 | 2.0 |
| Sussex | 4.5 | 1.5 |
| Hereford | 4.0 | 1.6 |
| Aberdeen-Angus | 2.4 | 1.3 |

*Source: Source:* MLC (1991) *Suckler Beef.* Reproduced with kind
permission of the Meat and Livestock Commission (UK).

**TABLE 343  Liveweights of Some African Cattle Breeds**

| | Breed | Birth weights | Adult Liveweights | | |
|---|---|---|---|---|---|
| | | | Bulls kg | Cows kg | Steers kg |
| **LARGE** | Africander | | 817 | 590 | 450 - 650 |
| | Barotse | 25 - 27 | 580 - 710 | 400 - 485 | |
| | Tuli | 32 - 33 | 770 - 820 | 500 - 550 | |
| | Nkone | | 727 - 820 | 455 - 500 | |
| **MEDIUM** | Boran | 23.5 - 27 | 550 - 675 | 350 - 450 | Up to 660 |
| | Angoni | 22 - 24 | 545 - 725 | 325 - 400 | 600 |
| | Mashona | | 545 - 660 | 320 - 410 | 500 + |
| | Nguni | | 430 - 680 | 225 - 450 | 590 - 725 |
| | White Fulani | 20 - 25 | 350 - 600 (?) | 250 - 400 (?) | 265 + |
| | Ankole | 15 - 22 | 400 - 500 | 250 - 400 | – |
| | Shuwa | – | 360 - 450 | 300 - 340 | 360 |
| **SMALL** | Small East African | 17 - 19 | 315 - 415 | 230 - 330 | |
| | Maasai | | 350 - 450 | | |
| | Nandi | | 315 - 415 | 200 - 300 | |
| | N'Dama | | 260 - 420 | 210 - 350 | |
| | West African Shorthorn | | 200 - 250 | 150 - 200 | |

*Source:* Maule, J.P. (1990) *The Cattle of the Tropics.* CTVM, University of Edinburgh.
Reproduced with kind permission of J.P. Maule.

**TABLE 344  Cold Dressed Mass (CDM) Performance of Beef Cattle of Various Ages Fed a Standard High Energy Diet According to Total Feed Consumed**

| | Total feed consumed kg | | | | | | | | |
|---|---|---|---|---|---|---|---|---|---|
| | 200 | 400 | 600 | 800 | 1000 | 1200 | 1400 | 1600 | 1800 |
| **Light weaner steers, 175 kg** | | | | | | | | | |
| Marginal conversion rate | 8.6 | 9.3 | 10.2 | 11.2 | 12.7 | 14.4 | 16.8 | 19.8 | 24.5 |
| Total CDM gain | 23.3 | 44.8 | 64.4 | 82.2 | 98.0 | 111.9 | 123.9 | 133.9 | 142.1 |
| Days in pens | 33 | 57 | 81 | 105 | 128 | 152 | 176 | 199 | 223 |
| Average conversion rate | 8.6 | 8.9 | 9.3 | 9.7 | 10.2 | 10.7 | 11.3 | 12.0 | 12.7 |
| **Good weaner steers 210 kg** | | | | | | | | | |
| Marginal conversion rate | 10.1 | 10.7 | 11.3 | 12.3 | 13.2 | 14.3 | 15.7 | 17.0 | 19.1 |
| Total CDM gain | 19.8 | 38.5 | 56.2 | 72.5 | 87.7 | 101.7 | 114.4 | 126.2 | 136.7 |
| Days in pens | 31 | 53 | 75 | 96 | 118 | 140 | 161 | 183 | 205 |
| Average conversion rate | 10.1 | 10.4 | 10.7 | 11.0 | 11.4 | 11.8 | 12.2 | 12.7 | 13.2 |
| **Light 1½ year old steers 250 kg** | | | | | | | | | |
| Marginal conversion rate | 9.3 | 10.0 | 10.8 | 11.9 | 13.1 | 14.6 | 16.5 | 19.0 | 22.4 |
| Total CDM gain | 21.5 | 41.5 | 60.0 | 76.8 | 92.1 | 105.8 | 117.9 | 128.4 | 137.3 |
| Days in pens | 20 | 38 | 56 | 73 | 91 | 109 | 126 | 144 | 162 |
| Average conversion rate | 9.3 | 9.6 | 10.0 | 10.4 | 10.9 | 11.3 | 11.9 | 12.5 | 13.1 |
| **Average 1½ year old steers 300 kg** | | | | | | | | | |
| Marginal conversion rate | 9.4 | 10.0 | 10.6 | 11.3 | 12.2 | 13.2 | 14.3 | 15.7 | 17.3 |
| Total CDM gain | 21.3 | 41.3 | 60.2 | 77.9 | 94.3 | 109.5 | 123.5 | 136.2 | 147.8 |
| Days in pens | 24 | 41 | 58 | 75 | 91 | 108 | 125 | 142 | 159 |
| Average conversion rate | 9.4 | 9.7 | 10.0 | 10.3 | 10.6 | 11.0 | 11.3 | 11.7 | 12.2 |
| **2½ year old steers 385 kg** | | | | | | | | | |
| Marginal conversion rate | 11.2 | 11.8 | 12.2 | 12.8 | 13.3 | 14.0 | 14.7 | 15.7 | 16.4 |
| Total CDM gain | 17.9 | 34.9 | 51.3 | 66.9 | 81.9 | 96.2 | 109.8 | 122.5 | 134.5 |
| Days in pens | 18 | 33 | 47 | 62 | 77 | 92 | 107 | 121 | 136 |
| Average conversion rate | 11.2 | 11.5 | 11.7 | 12.0 | 12.2 | 12.5 | 12.8 | 13.1 | 13.4 |

(CDM is taken as 97% of hot dressed mass)

In this table conversion rate is indicated as a single value which represents the number of kg of feed required to produce 1 kg of cold dressed mass, e.g. light weaners will have a marginal conversion rate of 8.6 kg feed per kg cold dressed mass gain for the first 200 kg feed they consume. For the next margin of 200 kg of feed, conversion drops to 9.3 kg feed per kg CDM gain, hence the term marginal. However average conversion rate for a total of 400 kg of feed consumed is 8.9 kg of feed per kg CDM gain.

*Source:* Zimbabwe Cattle Producers Association (1988) *Beef Production Manual*. Reproduced with kind permission of Zimbabwe Cattle Producers Association.

**TABLE 345   Feed Conversion Efficiency as Affected by Calving Percentage
and Rate of Gain of Beef Animals**

| | Growth rate | |
| Calving % | 1 kg / day | 2 kg / day |
| --- | --- | --- |
| 25 | 5.6 | 6.2 |
| 50 | 8.3 | 9.8 |
| 75 | 10.0 | 12.2 |
| 100 | 11.1 | 13.9 |

Conversion efficiency = Body weight gain (kg for each 100 kg TDN fed to herd)

*Source:* Smith, A.J., ed. (1974) *Beef Cattle Production in Developing Countries.*
Reproduced with kind permission of A.J. Smith.

**TABLE 346   Estimated Average Production Levels of Cattle in the
Latin American Tropics**

| Trait | Production level [1] |
| --- | --- |
| Calving percentage | 35 - 60% |
| Death loss to first service | 10 - 25% |
| Weaning weight (7 months) [2] | 120 - 150 kg |
| Weaning weight of calf/cow | 35 - 80 kg |
| Age at first calving | 3 - 4 years |
| Slaughter age (males) | 3.5 - 5 years |
| Slaughter weight | 350 - 450 kg |
| Extraction | 8 - 15% |

1.  Does not express total range but the extreme values of the averages corresponding
    to the main areas.

2.  In most herds systematic weaning is not practiced but natural weaning occurs at
    around 7 months of age.

*Source:* Smith, A.J., ed. (1974) *Beef Cattle Production in Developing Countries.*
Reproduced with kind permission of A.J. Smith.

---

**TABLE 347   Cattle Production Systems; Technical Coefficients, Biological Efficiency and Breed and Management Requirements**

| System | Technical coefficients | Biological * (feed) efficiency | | Management level |
|---|---|---|---|---|
| **Specialist milk** | 3000 litres milk per cow per year | 60 | Friesian | High |
| **Intensive fattening** | 800 g daily LW gain | 50 | Crosses and indigenous | High |
| **Dual purpose (beef / milk)** | 720 litres/cow/year; 180 kg weaned calf; 90% calf crop | 44 | Crosses [1] | Medium |
| **Specialist beef rearing** | 180 kg weaned calf; 80% calf crop | 30 | Crosses and indigenous | Low |

1   Should contain from 30 to 60% genes from recognized milk breeds

*   For simplicity biological efficiency is defined as the proportion of the total feed consumed by the animal which subsequently is used for productive purposes

*Source:* Smith, A.J., ed. (1974) *Beef Cattle Production in Developing Countries.*
Reproduced with kind permission of A.J. Smith.

---

**TABLE 348   The Eight-year Mean and Annual Range of Liveweight Gains and Losses by Steers Grazing High Altitude Sand-Veld in Rhodesia Using a Three Herd-Four Camp System of Management (after Barnes, 1965)**

| Changes in liveweight over 8 years | Stocking rate (ha grazed / steer) | | |
|---|---|---|---|
| | 2.5 | 4 | 5.7 |
| **Wt. gains (rain season) kg / head** | | | |
| Mean | 132 | 132 | 143 |
| Range | 40 - 161 | 47 - 155 | 56 - 164 |
| **Wt. losses (dry season) kg / head** | | | |
| Mean | 43 | 39 | 26 |
| Range | 28 - 76 | 21 - 76 | 1 - 57 |
| **Net annual wt. gain kg / head** | | | |
| Mean | 84 | 93 | 117 |
| Range | -2 - 133 | 10 - 117 | 23 - 165 |
| **Mean liveweight gain / hectare** | | | |
| Grazed | 80.3 | 50.7 | 45.7 |
| Used | 60.0 | 38.1 | 34.3 |

*Source:* Smith, A.J., ed. (1974) *Beef Cattle Production in Developing Countries.*
Reproduced with kind permission of A.J. Smith.

**TABLE 349 Estimated Annual Beef Production (kg LWG / ha) from Nitrogen Fertilized Pastures (200 - 300 kg N / ha / yr) in Monsoonal and Humid Tropical Environments**

| Rainfall (mm) | Monsoonal (6 months dry) | Humid tropics (long growing season) |
|---|---|---|
| 750 - 1000 | 150 - 250 | 300 - 400 |
| 1000 - 1500 | 250 - 400 | 400 - 700 |
| 1500 - 2000 | 350 - 550 | 1000 - 1500 |
| >2001 | 450 - 600 | 1300 - 1600 |
| Irrigated | 1600 - 2000 | |

*Source:* Smith, A.J., ed. (1974) *Beef Cattle Production in Developing Countries.*
Reproduced with kind permission of A.J. Smith.

**TABLE 350 Estimated Beef Production from Natural Grasslands and Sown Grasslands in the Tropics (kg LWG / ha / yr)**

| | Monsoonal (5 - 6 months dry) | Humid tropics (long growing season) |
|---|---|---|
| **Natural grasslands** | | |
| Improved grazing | 10 - 80 | 60 - 100 |
| Oversown legumes and fertilizer | 120 - 170 | 250 - 450 |
| **Cultivated grasslands** | | |
| Grass / legume mixtures with fertilizer | 200 - 300 | 300 - 600 |
| Nitrogen fertilized grass | 300 - 500 | 800 - 1500 |

*Source:* Smith, A.J., ed. (1974) *Beef Cattle Production in Developing Countries.*
Reproduced with kind permission of A.J. Smith..

**TABLE 351 Herd Structure and Daily Energy Requirements of a Representative Herd of *Bos indicus* Cattle**

| Age group (years) | No. | Required daily TDN[1] (kg) | Herd total TDN (kg) | Average body weight (kg) |
|---|---|---|---|---|
| 0 - 1 | 20 | 1.8 | 36 | 125 |
| 1 - 2 | 18 | 2.3 | 41 | 200 |
| 2 - 3 | 17 | 2.7 | 48 | 250 |
| 3 - 4 | 15 | 3.7 | 56 | 300 |
| Aged | 50 | 4.5 | 225 | 350 |
| TOTAL | 120 | | 406 | |

1 Total Digestible Nutrients

*Source:* Smith, A.J., ed. (1974) *Beef Cattle Production in Developing Countries.*
Reproduced with kind permission of A.J. Smith.

**TABLE 352  Herd Structure and Daily Energy Requirements of a Representative Herd of *Bos taurus* and *Bos indicus* Cattle**

| Age group (years) | No. | Required daily TDN (kg) | Herd total TDN (kg) | Average body weight (kg) |
|---|---|---|---|---|
| 0 - 1 | 22 | 2.3 | 51 | 200 |
| 1 - 2 | 20 | 2.7 | 54 | 275 |
| 2 - 3 | 18 | 3.7 | 67 | 350 |
| Aged | 50 | 5.5 | 275 | 425 |
| TOTAL | 110 | | 447 | |

*Source:* Smith, A.J., ed. (1974) *Beef Cattle Production in Developing Countries.*
Reproduced with kind permission of A.J. Smith.

**TABLE 353  Herd Structure and Daily Energy Requirements of a Representative Herd of *Bos taurus* Cattle**

| Age group (years) | No. | Required daily TDN (kg) | Herd total TDN (kg) | Average body weight (kg) |
|---|---|---|---|---|
| 0 - 1 | 20 | 2.5 | 50 | 275 |
| 1 - 2 | 18 | 4.0 | 72 | 425 |
| Aged | 50 | 6.0 | 300 | 500 |
| TOTAL | 88 | | 422 | |

*Source:* Smith, A.J., ed. (1974) *Beef Cattle Production in Developing Countries.*
Reproduced with kind permission of A.J. Smith.

**TABLE 354** **Comparative Performance of Pakistani Buffalo and Indian Cattle (averages from several sources)**

|  | Nili-Ravi Buffalo (Pakistan) | | Indian cattle | |
| --- | --- | --- | --- | --- |
| Birth weight | 37 - 39 | kg | 20 - 24.5 | kg |
| Weight at puberty | 469 | kg | 220 - 281 | kg |
| Weight at first calving | 624 - 628 | kg | 302 - 331 | kg |
| Weight at maturity | 760 | kg | 313 - 367 | kg |
| Age at puberty | 955 - 960 | days | 812 - 930 | days |
| Age at first calving | 1366 | days | 1166 - 1323 | days |
| Calving interval | 524 | days | 413 - 539 | days |
| Gestation period | 308 | days | 285 | days |
| Daily growth rate | 0.5 | kg | 0.2 - 0.4 | kg |
| Milk, lactation | 2444 | kg | 1355 - 1760 | kg |
| Daily milk yield | 4.58 | kg | 4.35 | kg |
| Butterfat | 6.7 | % | 4.3 | % |
| Solids nonfat | 9.1 | % | 8.5 | % |

*Source:* FAO, *The Water Buffalo. Animal Production and Health Series No. 4.* Reproduced with kind permission of the Food and Agriculture Organization of the United Nations.

**TABLE 355** **Carcass Weights of Grade 1 Italian Buffaloes, and Dressing-out Percentages**

|  | Average carcass weight Kg | Dressing-out % |
| --- | --- | --- |
| **Males** | | |
| Adult male | 263.19 | 48.50 |
| Adult castrate | 252.79 | 48.23 |
| Yearling | 127.46 | 45.60 |
| Calf under one year | 74.14 | 45.29 |
| Suckling calf, grade 2 | 27.14 | 55.99 |
| **Females** | | |
| Adult female | 241.18 | 47.23 |
| Heifer | 200.19 | 49.75 |
| Yearling | 122.75 | 45.26 |
| Calf under one year | 83.35 | 46.68 |
| Suckling calf, grade 2 | 28.38 | 54.92 |

*Source:* FAO, *The Water Buffalo. Animal Production and Health Series No. 4.* Reproduced with kind permission of the Food and Agriculture Organization of the United Nations

# Within Specie Comparisons – Sheep and Goats

**TABLE 356  a & b  Physical Results for Sheep Flocks in UK (1990)**

### a. Physical results early lambing flocks *

|  | Bottom third | Average | Top third |
|---|---|---|---|
| Average flock size | 315 | 289 | 287 |
| Ewe to ram ratio | 33 | 33 | 30 |
| **Per 100 ewes to ram** | | | |
| Ewe lambs | 21 | 20 | 24 |
| Empty ewes | 7 | 5 | 4 |
| Ewe deaths | 5 | 4 | 3 |
| Productive ewes | 90 | 94 | 96 |
| Total lambs born | 154 | 162 | 170 |
| Lambs born dead | 7 | 8 | 9 |
| Lambs born alive | 147 | 154 | 161 |
| Lamb deaths after birth | 8 | 8 | 8 |
| Lambs reared | 139 | 146 | 153 |
| Breeding | 4 | 3 | 1 |
| Slaughter | 110 | 123 | 127 |
| Sold or retained for feeding | 14 | 9 | 10 |
| kg concentrates / ewe | | 63 | |
| kg lamb concentrates / ewe | | 77 | |

* Avg. of 50 recorded flocks

### b. Physical results for lowland spring lambing flocks selling most of their lambs off grass in summer and autumn *

|  | Bottom third | Average | Top third |
|---|---|---|---|
| Average flock size | 643 | 616 | 536 |
| Ewe to ram ratio | 39 | 39 | 39 |
| **Per 100 ewes to ram** | | | |
| Ewe lambs | 16 | 16 | 17 |
| Empty ewes | 5 | 5 | 4 |
| Ewe deaths | 4 | 4 | 4 |
| Productive ewes | 93 | 93 | 94 |
| Total lambs born | 161 | 168 | 176 |
| Lambs born dead | 8 | 8 | 8 |
| Lambs born alive | 153 | 160 | 168 |
| Lamb deaths after birth | 8 | 7 | 6 |
| Lambs reared | 145 | 153 | 162 |
| Breeding | 9 | 9 | 9 |
| Slaughter | 65 | 87 | 105 |
| Sold or retained for feeding | 70 | 55 | 43 |
| kg concentrates / ewe | | 54 | |
| kg lamb concentrates / ewe | | 14 | |

* Avg. of 308 recorded flocks

*(continued over)*

A G R I   I N F O

**TABLE 356 c   Physical Results for Sheep Flocks in UK (1990)**

**c. Physical results for upland spring lambing flocks selling most of their lambs off grass in summer and autumn ***

|  | Bottom third | Average | Top third |
|---|---|---|---|
| Average flock size | 767 | 679 | 617 |
| Ewe to ram ratio | 40 | 42 | 43 |
| **Per 100 ewes to ram** | | | |
| Ewe lambs | 16 | 19 | 22 |
| Empty ewes | 6 | 5 | 5 |
| Ewe deaths | 5 | 4 | 4 |
| Productive ewes | 91 | 92 | 93 |
| Total lambs born | 141 | 150 | 154 |
| Lambs born dead | 6 | 7 | 7 |
| Lambs born alive | 135 | 143 | 147 |
| Lamb deaths after birth | 6 | 6 | 5 |
| Lambs reared | 129 | 137 | 142 |
| Breeding | 27 | 23 | 21 |
| Slaughter | 48 | 67 | 79 |
| Sold or retained for feeding | 53 | 47 | 39 |
| kg concentrates / ewe | | 49 | |

\* Avg. of 134 recorded flocks

*Source:* MLC (1991) *Sheep Yearbook.* Reproduced with kind permission of the Meat and Livestock Commission (UK).

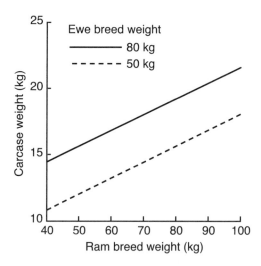

**Fig. 56** General relationship between UK sheep breed weights and carcass weights at fat class 3

*Source:* Croston, D. & Pollott, G.E. (1985) *Planned Sheep Production.* (after Meat and Livestock Commission, UK) Reproduced with kind permission of G.E. Pollott.

**TABLE 357   Growth-rates of Lambs before Weaning**

| Breed | Location | LWG (g / day) | Age at weaning (days) |
|---|---|---|---|
| Awassi | Cyprus | 240 | 35 |
| Awassi | Iraq | 165 | 120 |
| Baggara | Sudan | 140 | 91 |
| Barbados Blackbelly | Trinidad | 102 | 21 |
| Barbados Blackbelly | Trinidad | 129 | 84 |
| Blackhead Persian | Brazil | 117 | 112 |
| Blackhead Persian | Trinidad | 103 | 21 |
| Blackhead Persian | Trinidad | 188 | 84 |
| Blackhead Persian | Trinidad | 125 | 21 |
| Blackhead Persian | Trinidad | 136 | 84 |
| Chios | Cyprus | 230 | 35 |
| Chios | Cyprus | 183 | 42 |
| Chokla x Rambouillet | India | 156 | 21 |
| Chokla x Rambouillet | India | 118 | 84 |
| Corriedale x Ile-de-France | Brazil | 240 | 30 |
| Cyprus Fat-tailed | Cyprus | 230 | 35 |
| Deccani | India | 116 | 28 |
| Deccani | India | 98 | 84 |
| Deccani x Merino | India | 105 | 28 |
| Deccani x Merino | India | 102 | 84 |
| Deccani x Rambouillet | India | 118 | 28 |
| Deccani x Rambouillet | India | 99 | 84 |
| Desert Sudanese | Sudan | 157 | 42 |
| Desert Sudanese | Sudan | 96 | 112 |
| Jaffina | Sri Lanka | 45 | 91 |
| Jaisalmeri x Rambouillet | India | 160 | 21 |
| Jaisalmeri x Rambouillet | India | 108 | 85 |
| Javanese Thin-tailed | Indonesia | 151 | 91 |
| Javanese Thin-tailed x Suffolk | Indonesia | 173 | 91 |
| Karakul | Iran | 21 | – |
| Madras Red | India | 100 | 119 |
| Malpura | India | 89 | 28 |
| Malpura | India | 77 | 84 |
| Malpura x Rambouillet | India | 153 | 21 |
| Malpura x Rambouillet | India | 117 | 84 |
| Mandya | India | 81 | 119 |
| Masai | Kenya | 73 | 153 |
| Morada Nova | Brazil | 119 | 112 |
| Muzaffarnagari | India | 122 | 80 |
| Nali | India | 90 | 84 |
| Polwarth | India | 89 | 61 |
| Rambouillet | India | 162 | 21 |
| Romney Marsh | Mexico | 193 | 30 |
| Romney Marsh | Mexico | 156 | 60 |
| Santa Inês | Brazil | 152 | 112 |
| Uda | Nigeria | 118 | 91 |
| West African | Trinidad | 116 | 21 |
| West African | Trinidad | 141 | 84 |
| West African | Venezuela | 122 | – |

*Source:* Gatenby, R.M. (1986) *Sheep Production in the Tropics and Sub Tropics.*
Reproduced with kind permission of Longman Group UK Ltd.

**TABLE 358   Recorded Milk Production for Various Sheep Breeds**

| Breed | Milk yield (l) | Lactation length (days) | Notes |
|---|---|---|---|
| **Tropical sheep** | | | |
| Awassi (Western Asia) | 130 - 270 | 260 | Ordinary flocks |
| | 360 - 890 | | Good flocks |
| | | | Max. yld at 4th lactation |
| | | | Fat 6 to 8% |
| D'Man (Morroco) | 90 - 140 | | Very prolific |
| Dwarf West African | 40 - 50 | 120 - 135 | Range conditions |
| | 75 - 85 | | Well fed |
| Iran fat-tailed breeds | 169 | 75 | Under good management |
| Kuka (India) | 1.8 - 3.6 daily | | |
| Lohi (India) | 3.6 daily max. | | Prolific, good meat producers |
| Priangan (Indonesia) | 21 - 53 | | 5% milk fat |
| Sonedi (India) | 0.9 to 1.4 daily | | |
| Sudan Desert | 2.3 to 2.7 daily | | |
| Somali Blackhead (Africa) | 50 to 60 | | Arid range conditions |
| **Temperate sheep** | | | |
| Chios | 100 - 250 | 170 - 260 | Prolific |
| East Friesian | 600 - 700 | 260 | Most productive milk breed |
| Lacaune | 135 | 100 - 210 | Large sheep, low milk |
| Sardinian | 100 - 230 (average 150) | 170 - 250 | Easy milkers, hardy |

*Source:* Chamberlain, A. (1989) *Milk Production in the Tropics.*
Reproduced with kind permission of Longman Group UK Ltd.

**TABLE 359 Tropical Goat Breeds Used for Milking, Their Origin and Typical Milking Performance (if known)**

| Breed | Country of origin | Typical milk yield per Lactation (l) | Lactation length (days) | Average daily milk yield (l) |
|---|---|---|---|---|
| Barbari | India, Pakistan (dry tropical, subtropical) | 118 | 183 | 0.6 |
| Beetal | India, Pakistan (dry tropical) | 200 | 208 | 1.0 |
| Black Bedouin | Israel, Egypt (very dry tropical) | | | |
| Criollo | Venezuela | 60 | 150 | 0.4 |
| Damani | Pakistan (dry tropical) | | | |
| Damascus | Syria, Cyprus, Lebanon (dry subtropical) | 563 | 416 | 1.4 |
| Dera Din Panah | Pakistan (dry tropical) | | | |
| Gaddi | India, Pakistan | 45 | 56 | 0.8 |
| Jamnapari | India (dry tropical, subtropical) | 182 | 168 | 1.1 |
| Kamori | Pakistan (dry subtropical) | 227 | 120 | 1.9 |
| Kilis | Turkey (dry subtropical) | 280 | 260 | 1.1 |
| Maiabar | India (humid tropical) | 181 | 180 | 1.0 |
| Mamber | Israel (dry tropical) | 400 | 267 | 1.5 |
| Maradi | Niger (dry tropical) | 70 | 100 | 0.7 |
| Marwari | India (dry tropical) | | | |
| Nubian | Egypt, Sudan (dry tropical) | 70 | 65 | 1.1 |
| Zairabi | Egypt (dry tropical) | | | |

*Source:* Chamberlain, A. (1989) *Milk Production in the Tropics.*
Reproduced with kind permission of Longman Group UK Ltd.

**TABLE 360 Milking Performance of European Goat Breeds in the Tropics**

| Breed | Type of environment | Location | Lactation milk yield (l) | Lactation length (days) | Average daily yield (l) |
|---|---|---|---|---|---|
| Anglo-Nubian | Dry tropics, subtropics | Mauritius | 222 | 247 | 0.75 |
| | | Malaysia | 250 - 300 | 300 | 0.8 to 1.0 |
| | | Philippines | 167 | 294 | 0.6 |
| | | Trinidad | 143 | 124 | 1.2 |
| British Alpine | Humid tropics, subtropics | Malaysia | 309 | 106 - 253 | 2.0 |
| | | Trinidad | 274 | 209 | 1.3 |
| Saanen | Humid tropics, subtropics | Australia | 886 | 240 - 270 | 3.5 |
| | | Cyprus | 536 | 240 - 300 | 2.0 - 2.25 |
| | | Israel | 500 | | 2.0 - 3.4 |
| | | Puerto Rico | 704 | 344 | 1.0 - 2.0 |

*Source:* Wilkinson, G., Payne, W.J.A., eds. (1978) *Introduction to Animal Husbandry in the Tropics.*
Reproduced with kind permission of Longman Group UK Ltd.

---

**TABLE 361  UK Goat Performance Data**

---

**a. Average lactation yields**
(From recorded herds. A 365-day lactation is the basis of calculation and the data below are for animals recorded for more than 270 days. The data are for 1983 and little change in yields is considered to have occurred since.)

| Breed | Highest | Average Lactation yield |
|---|---|---|
| British Saanen | 2298 | 1261 |
| British | 2022 | 1235 |
| Grading up, English etc | 1787 | 1197 |
| British Toggenburg | 2443 | 1180 |
| Saanen | 1864 | 1153 |
| British Alpine | 1912 | 1099 |
| Toggenburg | 1254 | 1047 |
| Anglo Nubian | 1632 | 1040 |
| Golden Guernsey | 1273 | 992 |

**b. Milk Quality (from the same animals as above)**

| Breed | Butterfat % Highest | Average | Lowest | Protein % Highest | Average | Lowest |
|---|---|---|---|---|---|---|
| Anglo Nubian | 6.77 | 5.04 | 3.66 | 4.22 | 3.55 | 2.7 |
| British Alpine | 6.09 | 4.12 | 2.94 | 3.96 | 2.97 | 2.5 |
| Golden Guernsey | 4.92 | 4.09 | 3.23 | 3.27 | 2.95 | 2.54 |
| Saanen | 4.87 | 3.81 | 2.71 | 3.30 | 2.78 | 2.37 |
| British | 5.23 | 3.71 | 2.19 | 3.95 | 2.79 | 2.15 |
| British Saanen | 4.87 | 3.68 | 2.32 | 3.61 | 2.80 | 2.10 |
| British Toggenburg | 5.13 | 3.68 | 2.20 | 3.38 | 2.71 | 2.13 |
| Grades, English | 5.90 | 3.66 | 2.40 | 3.48 | 2.76 | 2.22 |
| Toggenburg | 4.35 | 3.55 | 2.21 | 3.24 | 2.84 | 1.82 |

**c. Other Parameters**
**Age at first kidding :** 1 - 2 years       **Kidding Interval :** Every year or every other year

**Food Conversion Efficiency :** Extremely variable and little researched. Experimentally castrates and entires have given a range between 3.53 and 6.45 kg feed DM/kg liveweight gain. Good FCE's are only obtained when growth rates are high at about 200g / day.

**Kid Weights :** Up to 7 kg for singles and 2 - 3 kg in multiple births.

**Adult weights :** Dairy - female 55 - 105 kg, male 75 - 100 kg
Angora - female 35 - 50 kg, male 50 - 65 kg

**Meat potential :** Dairy breeds are all good meat animals, particularly Anglo Nubians with their high prolificacy (frequent triplets and quads), e.g. Seale Hayne College trial on British Saanens:

| | Entires | Castrates |
|---|---|---|
| Age at slaughter (days) | 159 | 159 |
| Liveweight (kg) | 36.5 | 30.9 |
| Carcass weight (kg) | 17.3 | 14.9 |

*Source:* Data collected and supplied by the British Goat Society, 34-36 Fore Street, Bovey Tracey, Newton Abbot, Devon, UK.

TABLE 362  Mature Size of Some Common Breeds of Goats (from Devendra & Burns, 1973, and others) and Selected Growth Rates Reported in Literature

| Breed & location | Mature Weight kg | sex | Weight range kg | Growth rate g/d | Feed |
|---|---|---|---|---|---|
| **Improved Boer** | | | | | |
| South Africa | 100 - 110 | M | 3 - 30 | 291 | Concentrate and excellent grazing |
| | | | 3 - 70 | 250 | |
| | | F | 3 - 29 | 272 | Highly selected meat goats |
| | | | 3 - 52 | 186 | |
| **Damascus** | | | | | |
| Cyprus | 80 - 90 | M | 20 - 50 | 240 | Barley & soybean 20% CP |
| | | C | 18 - 40 | 210 | |
| | | F | 18 - 37 | 190 | |
| **Alpine** | | | | | |
| France | 80 - 90 | M | 3 - 22 | 219 | Milk replacer |
| | | | 3 - 22 | 241 | Milk replacer & concentrates |
| **Saanen** | | | | | |
| Britain | 90 - 100 | M | 12 - 24 | 222 | Concentrates |
| Australia | 90 - 100 | C | 15 - 25 | 210 | Milk fed |
| **Anglo-Nubian** | | | | | |
| World wide | 80 - 90 | | | | |
| **Jamnapari** | | | | | |
| India | 70 - 85 | | | | |
| **Angora** | | | | | |
| Texas | 50 - 60 | M | ? | 165 | 15% CP pellets |
| Australia | 50 - 60 | C | 27 - 31 | 154 | Spring pasture |
| | | C | 16 - 27 | 122 | 16.5% CP pellets |
| | | C | 20 - 27 | 47 | Pellets & hay as above + zeranol |
| | | | | 73 | |
| **Feral** | | | | | |
| Australia | 45 - 55(?) | M+F | 2 - 10 | 175 | Does fed oats & lucerne |
| | | M+F | 3 - 17 | 157 | Spring pasture |
| | | M+F | 17 - 24 | 122 | |
| | | M | 15 - 24 | 88 | Pigeon pea grazing as above |
| | | | | 119 | with sorghum |
| **Beetal** | | | | | |
| India | 45 - 60 | | | | |
| **Barbari** | | | | | |
| India | 35 - 45 | M+F | 6 - 12 | 53 | Berseem clover + barley |
| | | | | 29 | Berseem only |
| **Kambing/Kajang** | | | | | |
| Malaysia & Indonesia | 25 - 30 | M+F | 2 - 22 | 57 | Various |

*Source:* Copland, J.W., ed. In. *Goat Production and Research in the Tropics.* McGregor, B.A. *Growth Development and Carcass Composition of Goats: A Review.* Reproduced with kind permission of B.A. McGregor.

**TABLE 363    Some Reproductive Parameters and Pre-weaning Mortality in Small Ruminants in African Countries**

| Country | Breed / Species | Age 1st parturition (mths) | Parturition interval (days) | Avg. litter size | % multiple births | Pre-weaning mortality % |
|---|---|---|---|---|---|---|
| Sudan | Native goats | 10 | 238 | 1.34 | 52 | 16 |
| Shana | W. Afr. dwarf goats | 12 | 266 | 1.87 | 68 | 21 |
| Mali | Native goats | 16 | 250 | 1.20 | 23 | 35 |
| Mali | Native sheep | 16 | 257 | 1.05 | 5 | 30 |
| Kenya | Native goats | 15 | 289 | NR | 18 | NR |
| Kenya | Masai sheep | 18 | 344 | 1.02 | NR | NR |
| Uganda | Mubende goats | 19 | 297 | NR | 32 | 46 |
| Nigeria | Sokota red goats | 19 | NR | 1.34 | 32 | 35 |
| Nigeria | W. Afr. dwarf sheep | 18 | 279 | 1.30 | NR | NR |
| Nigeria | Yankasa sheep | 19 | 227 | 1.04 | 3 | 39 |
| Nigeria | W. Afr. dwarf goats | 18 | 278 | 1.60 | 70 | 15 |
| Senegal | Djallonke sheep | 19 | 307 | 1.12 | 12 | 33 |
| Tchad | Fellata sheep | 15 | NR | 1.07 | NR | NR |

NR = Not reported

*Source:* FAO (1986) *Small Ruminant Production in the Developing Countries. Animal Production and Health Paper No. 58.* Reproduced with kind permission of the Food and Agriculture Organization of the United Nations.

**TABLE 364    Litter Size, Parturition Interval and Annual Reproductive Rate (ARR) for Sheep in Some African Livestock Systems**

| Country – System | | Litter size | Parturition interval (days) | ARR |
|---|---|---|---|---|
| Sudan: | - pastoral | 1.14 | 275 | 1.51 |
| Ethiopia: | - pastoral | 1.03 | 315 | 1.20 |
| | - agro-pastoral | 1.05 | 365 | 1.05 |
| Kenya: | - pastoral | 1.05 | 312 | 1.23 |
| Rwanda: | - station | 1.54 | 365 | 1.54 |
| Mali: | - agro-pastoral | 1.04 | 261 | 1.45 |

ARR = Litter Size x 365 / Parturition interval

*Source:* FAO *Small Ruminant Production in the Developing Countries. Animal Production and Health Paper No. 58.* Reproduced with kind permission of the Food and Agriculture Organization of the United Nations.

**TABLE 365  Pre-weaning Lamb Death Rate (%) for Different Variables at Six Sites in Africa**

| Variable | | Sudan | Mali | Kenya 1[1] | Kenya 11[1] | Nigeria | Ethiopia |
|---|---|---|---|---|---|---|---|
| Overall | LS Mean | 30.2 | 28.0 | 19.1 | 16.3 | 16.0 | 12.6 |
| System [2] : | A | 35.3 $^a$ | 32.1 $^a$ | – | 16.6 | – | – |
| | B | 25.1 $^b$ | 23.9 $^b$ | – | 14.6 | – | – |
| Sex : | male | 30.9 | 27.9 | 18.2 | 16.7 | 15.0 | 10.5 |
| | female | 29.5 | 28.1 | 20.0 | 15.8 | 17.0 | 14.6 |
| Birth type : | single | 23.1 $^a$ | 22.9 $^a$ | 6.7 $^a$ | 9.6 $^a$ | 16.0 | 12.1 |
| | twin | 36.7 $^b$ | 33.0 $^b$ | 31.4 $^b$ | 22.9 $^b$ | 14.0 | 13.1 |
| Birth season[3]: | A | 26.9 $^a$ | 32.8 $^a$ | 17.2 $^a$ | 12.2 $^a$ | – | 4.6 $^a$ |
| | B | 43.3 $^b$ | 28.1 $^a$ | 13.9 $^b$ | – | – | – |
| | C | 27.9 $^a$ | 23.0 $^b$ | 22.4 $^c$ | 23.3 $^b$ | – | – |
| | D | 22.7 $^a$ | 27.1 $^a$ | 22.7 $^c$ | 13.2 $^a$ | – | 21.4 $^b$ |
| Parity : | 1 | 27.6 $^a$ | 38.6 $^a$ | 28.6 $^a$ | – | 19.0 | 19.5 $^a$ |
| | 2 | 26.8 $^a$ | 22.7 $^b$ | 24.4 $^b$ | – | – | 12.7 $^b$ |
| | 3 | 14.8 $^b$ | 24.1 $^b$ | 20.7 | – | – | – |
| | 9 [4] | 34.4 $^c$ | 26.2 $^b$ | 12.2 $^c$ | – | 13.0 | 5.4 $^c$ |
| Flock : | best | 21.1 $^a$ | 9.4 $^a$ | 12.1 $^a$ | – | – | 2.1 $^a$ |
| | worst | 42.2 $^b$ | 55.9 $^b$ | 26.8 $^b$ | – | – | 29.3 $^b$ |

*Notes:*  1.  Kenya 1 = Elangata Wuas Group Ranch 1978-81;
Kenya 11 = Three Kaputeosection Group Ranches 1981-83.

2.  Sudan A = sedentary, B = migratory; Mali A = rainfed millet B = irrigated rice;
Kenya 11A = small flocks, B = large flocks

3.  Sudan, Mali, A = cold dry, B = hot dry, C = rains, D = post rains;
Kenya, A = short dry, B = long rains, C = long dry, D = short rains;
Ethiopia, A = best month, D = worst month

4.  Parity 9 = parities $\geq 4$ except Nigeria $\geq 2$ and Ethiopia $\geq 3$

Within variables means in same column with different superscript differ significantly.

*Source:* FAO, *Small Ruminant Production in the Developing Countries, Animal Production and Health Paper No. 58*. Reproduced with kind permission of the Food and Agriculture Organization of the United Nations.

# Within Specie Comparisons – Non-Ruminants

**TABLE 366   Physical Results for Pig Rearing Herds in UK \*\***

|  | Bottom third * | Average | Top third * | Top 10% |
|---|---|---|---|---|
| No. of herds | 105 | 314 | 105 | 31 |
| Av. no. of pigs | 588 | 753 | 848 | 809 |
| **Pig performance** | | | | |
| Weight of pigs at start (kg) | 6.5 | 6.4 | 6.4 | 6.5 |
| Weight of pigs produced (kg) | 30.9 | 32.2 | 31.2 | 30.8 |
| Mortality (%) | 3.2 | 2.7 | 2.3 | 2.8 |
| Feed conversion ratio | 1.90 | 1.77 | 1.61 | 1.60 |
| Daily gain (g) | 399 | 425 | 434 | 427 |
| Feed per pig produced (kg) | 46 | 46 | 40 | 39 |
| Sale weight (kg) | 30.9 | 30.6 | 29.8 | 29.1 |

\*     Selected on feed cost per kg liveweight gain
\*\*    Year ending Sept 1990

*Source:* MLC (1991) *Pig Yearbook.* Reproduced with kind permission of the Meat and Livestock Commission (UK).

**TABLE 367   Comparison of Pig Rearing Herd Results for Outdoor Breeding Herds and Indoor Breeding Herds in UK \***

|  | From outdoor breeding herds | From indoor breeding herds |
|---|---|---|
| No. of herds | 37 | 277 |
| Av. no. of pigs | 1222 | 690 |
| **Pig performance** | | |
| Weight of pigs at start (kg) | 6.2 | 6.5 |
| Weight of pigs produced (kg) | 32.2 | 32.2 |
| Mortality (%) | 2.0 | 3.0 |
| Feed conversion ratio | 1.76 | 1.77 |
| Daily gain (g) | 448 | 414 |
| Feed per pig produced (kg) | 45.8 | 45.5 |

\* Year ending Sept 1990

*Source:* MLC (1991) *Pig Yearbook.* Reproduced with kind permission of the Meat and Livestock Commission (UK).

**TABLE 368  Physical Results for Pig Breeding Herds * in UK**

|  | Bottom third ** | Average | Top third ** | Top 10% ** |
|---|---|---|---|---|
| No. of herds | 169 | 508 | 169 | 51 |
| **Herd structure** | | | | |
| Av. no. of sows & gilts | 197 | 227 | 266 | 192 |
| Av. no. of unserved gilts | 14 | 16 | 20 | 14 |
| Sow replacements (%) | 44.1 | 43.1 | 42.6 | 42.3 |
| Sow sales and deaths (%) | 40.6 | 40.0 | 38.9 | 39.4 |
| Sow mortality (%) | 5.3 | 4.4 | 3.6 | 3.1 |
| **Sow performance** | | | | |
| Successful services (%) | 82.4 | 85.1 | 87.5 | 89.8 |
| Litters per sow per year *** | 2.08 | 2.23 | 2.35 | 2.40 |
| Pigs born per litter: | | | | |
| alive | 10.35 | 10.72 | 11.11 | 11.41 |
| dead | 0.84 | 0.85 | 0.87 | 0.86 |
| Total | 11.19 | 11.57 | 11.98 | 12.27 |
| Mortality of pigs born alive (%) | 13.3 | 12.0 | 10.5 | 9.8 |
| Pigs reared per litter | 8.96 | 9.43 | 9.94 | 10.29 |
| Pigs reared per sow per year *** | 18.7 | 21.1 | 23.4 | 24.7 |
| Weight of pigs produced (kg) | 6.9 | 6.6 | 6.4 | 6.3 |
| Av weaning age (days) | 27 | 25 | 24 | 24 |
| **Feed usage** (incl. unserved gilts) | | | | |
| Sow feed per sow per year (t) | 1.200 | 1.206 | 1.218 | 1.228 |
| Feed per pig reared (kg): | | | | |
| sow feed | 70 | 62 | 57 | 54 |
| piglet feed | 0.9 | 0.7 | 0.4 | 0.4 |
| Total | 71 | 63 | 57 | 54 |

\*    Pigs sold or transferred out at weaning
\*\*   Selected on pigs reared per sow per year
\*\*\*  Per sow figures exclude unserved gilts
†    508 Recorded herds, 1990

*Source:* MLC (1991) *Pig Yearbook.* Reproduced with kind permission of the Meat and Livestock Commission (UK).

**TABLE 369   Comparison ** of Outdoor Pig Breeding Herds and Indoor Breeding Herds in UK**

|  | Outdoor breeding herds | Indoor breeding herds |
|---|---|---|
| Av. no. of sows & gilts | 410 | 205 |
| Litters per sow per year* | 2.21 | 2.23 |
| Pigs born alive per litter | 10.56 | 10.77 |
| Mortality of pigs born alive (%) | 11.4 | 12.1 |
| Pigs reared per sow per year* | 20.7 | 21.8 |
| Av. weaning age (days) | 24 | 25 |
| Sow feed per sow per year (t) | 1.347 | 1.170 |

\*     Per sow figures exclude unserved gilts
\*\*    Year ending Sept 1990

*Source:* MLC (1991) *Pig Yearbook.* Reproduced with kind permission of the Meat and Livestock Commission (UK).

**TABLE 370   Pig Feeding Herd Results in UK ****

|  | Bottom third * | Average | Top third * | Top 10% * |
|---|---|---|---|---|
| No. of herds | 136 | 407 | 136 | 41 |
| Av. no. of pigs | 925 | 1111 | 1256 | 1109 |
| **Pig performance** | | | | |
| Weight of pigs at start (kg) | 23.2 | 20.8 | 20.4 | 22.8 |
| Weight of pigs produced (kg) | 80.3 | 80.5 | 80.6 | 82.1 |
| Mortality (%) | 3.3 | 3.1 | 2.9 | 2.6 |
| Feed conversion ratio | 2.85 | 2.66 | 2.48 | 2.40 |
| Daily gain (g) | 594 | 615 | 644 | 675 |
| Feed per pig reared (kg) | 163 | 159 | 149 | 142 |
| Carcass weight (kg) | 61.5 | 61.3 | 61.2 | 62.2 |

\*     Selected on feed cost per kg liveweight gain
\*\*    Year ending Sept 1990

*Source:* MLC (1991) *Pig Yearbook.* Reproduced with kind permission of the Meat and Livestock Commission (UK).

**TABLE 371  Prolificacy of UK Pig Breeds**

| Breed | Average born | Average reared |
|-------|--------------|----------------|
| Berkshire | 9.07 | 8.29 |
| British Saddleback | 10.45 | 9.47 |
| Chester White | 10.8 | 9.38 |
| Duroc | 9.24 | 7.82 |
| Gloucester Old Spot | 9.56 | 8.46 |
| Large Black | 9.5 | 8.52 |
| Hampshire | 10.28 | 9.17 |
| Large White | 10.17 | 8.98 |
| Landrace | 10.31 | 9.4 |
| Middle White | 8.96 | 7.85 |
| Welsh | 10.77 | 9.61 |
| Tamworth | 7.92 | 7.2 |

*Source:* 1990 full year data kindly supplied and compiled by the British Pig Association, 7 Rickmansworth Rd, Watford, Herts, UK. Reproduced with kind permission of the British Pig Association.

**TABLE 372  Body Weights of fowl (g)**

| Age (weeks) | Light (egg laying) strain | | Heavy strain (including Broilers, roasters and breeders) | |
|-------------|--------|------|--------|------|
| | Female | Male | Female | Male |
| 0 | 35 | 35 | 40 | 40 |
| 1 | 70 | 70 | 110 | 120 |
| 2 | 135 | 140 | 230 | 250 |
| 3 | 190 | 280 | 400 | 440 |
| 4 | 265 | 560 | 600 | 800 |
| 5 | 350 | 600 | 800 | 1000 |
| 6 | 450 | 650 | 1000 | 1200 |
| 7 | 550 | 750 | 1000 | 1350 |
| 8 | 590 | 800 | 1250 | 1480-3000 * |
| 9 | 630 | 900 | 1330 | 1620 |
| 10 | 650 | 1000 | 1420 | 1750 |
| 11 | 700 | 1050 | 1500 | 1900 |
| 12 | 800 | 1100 | 1600 | 2000 |
| 13 | 850 | 1200 | 1700 | 2150 |
| 14 | 900 | 1050 | 1750 | 2300 |
| 15 | 950 | 1300 | 1850 | 2400 |
| 16 | 1000 | 1350 | 1950 | 2550 |
| 17 | 1050 | 1400 | 2000 | 2700 |
| 18 | 1100 | 1450 | 2100 | 2800 |
| 19 | 1150 | 1500 | 2200 | 2960 |
| 20 | 1250 | 1550 | 2250 | 3000 |
| 21 | 1300 | 1560 | 2300 | 3200 |
| 22 | 1350 | 1600 | 2450 | 2400 |

*Notes:*  Medium strains have respective intermediate body weights. Restricted feeding and tropical conditions, unless feed is well adjusted, may cause lower body weights than stated for light strains. The heavier weights refer to broilers.

\*      Some broiler Strains

*Source:* Oluyemi, J.A. & Roberts, F. (1979) *Poultry Production in Wet Climates.* Reproduced with kind permission of Macmillan Press.

**TABLE 373  Feed and Water Consumption of the Fowl**

| Age (weeks) | Feed consumption per bird | | | Water consumption | | Cumulative feed conversion ratio (feed efficiency) |
|---|---|---|---|---|---|---|
| | per week | | per day | At 10 - 24 °C | Tropics | |
| | Light (g) | Heavy (broiler and roaster) (g) | (g) | l / 100 birds per day | l / 100 birds per day (estimated) | |
| 1 | 68 | 88 - 130 | 7 - 11 | 1 - 5 | 4 - 10 | |
| 2 | 113 | 127 - 190 | | 4 - 6.5 | 8 - 12 | 1.31 - 1.8 |
| 3 | 136 | 191 - 340 | | 4.5 - 8.5 | 9 - 15 | 1.50 - 2.8 |
| 4 | 200 | 286 - 440 | 25 - 40 | 5.5 - 10.0 | 11 - 18 | 1.64 - 2.0 |
| 5 | 260 | 317 - 560 | | 6.5 - 11.5 | 12 - 20 | 1.76 - 2.8 |
| 6 | 290 | 354 - 650 | 35 - 45 | 7 - 12.5 | 13 - 22 | 1.88 |
| 7 | 315 | 368 - 820 | | 8 - 13.0 | 14 - 25 | 1.99 |
| 8 | 360 | 409 - 960 | 40 - 60 | 8 - 13.5 | 15 - 29 | 2.10 |
| 9 | 405 | 418 - 1040 | | 9 - 16 | 17 - 32 | 2.21 |
| 10 | 450 | 429 - 1050 | | 9.5 - 18 | 18 - 36 | 2.32 |
| 11 | 467 | 437 - 1060 | | 10 - 18.5 | 19 - 37 | 2.44 |
| 12 | 489 | 445 - 1070 | 45 - 70 | 10.5 - 19 | 20 - 38 | 2.57 |
| 13 | 511 | 446 - 1080 | | 12 - 20 | 21 - 40 | 2.71 |
| 14 | 551 | 447 - 1110 | | 12 - 20.5 | 22 - 41 | 3.1 |
| 15 | 586 | 474 - 1190 | | 17 - 21 | 23 - 42 | 3.6 |
| 16 | 626 | 508 - 1300 | 60 - 90 | 18 - 22 | 24 - 44 | 3.6 |
| 17 | 626 | 524 - 1300 | | 19 - 22.5 | 25 - 45 | |
| 18 | 693 | 540 - 1300 | 54 - 77 | 19.5 - 23 | 26 - 46 | |
| 19 | 733 | 556 - 1300 | | 19.5 - 23.5 | 27 - 47 | |
| 20 | 778 | 570 - 1300 | 68 - 110 | 20 | 28 - 48 | |
| 21 | | | | | 29 - 48 | |
| 22 | | | | | 29 - 48 | |

*Notes:* * Higher values for the male broiler or roaster.
Cumulative feed consumption from one day old to

       (a)  4 weeks  =  0.440 - 1.10 kg
       (b)  8 weeks  =  1.453 - 4.13 kg
       (c)  12 weeks  =  3.00 - 8.34 kg
       (d)  20 weeks  =  from about 8 kg to 11 or 12 kg

The low level of consumption refers to the tropics. Feed consumption from one day old to point of lay is about 10 - 11 kg. Water consumption (in litres) per 100 birds per day in cold weather is at least about the age of the birds in weeks. The lower values for water consumption are for the light strains like the White Leghorn. They are claimed to be about equal to twice the age of the fowl in hot weather.

*Source:* Oluyemi, J.A. & Roberts, F. (1979) *Poultry Production in Wet Climates.*
Reproduced with kind permission of Macmillan Press.

**TABLE 374    Hybrid 'Isabrown' Laying Hen Performance Targets Under Specified
Controlled Environmental Conditions**

**Growing period (0 - 18 weeks of age)**

| | |
|---|---|
| Liveability | 97% |
| Feed consumption (controlled feeding) | 6.85 kg |
| Bodyweight at 18 weeks of age (controlled feeding) | 1.47 kg |

**Laying period (18 - 76 weeks of age)**

| | |
|---|---|
| Liveability | 93% |
| Hen housed average egg production | 307 |
| Hen day average egg production | 318 |
| Peak production | 93% |
| Average egg weight | 63 g |
| Egg mass (HHA egg numbers x egg weight) | 19.3 kg |
| Egg grading (minimum cumulative) | |
|        Size 1 (70 g +) | 10% |
|        Size 2 (65 - 70 g) | 26% |

**Feed**

| | |
|---|---|
|        Daily consumption (2800 Kcal / kg feed) | 116 g |
|        Conversion | 2.3 - 2.4 : 1 |
| End of lay bodyweight | 2.1 kg |

*NB* Where extensive production systems are in use productive output will vary from published standards due to the variation in environmental factors to which the birds are subjected.

*Source:* 1991 data supplied and reproduced with kind permission of ISA Poultry Services Ltd, Green Rd, Eye, Peterborough, UK.

**TABLE 375   Hybrid 'Isabrown' Laying Hen Weekly Production Targets**

| Age Weeks | Live- ability | HDA Prod | Cumul. Eggs / Bird | | Average Egg Wt. | Hen Housed Egg Mass | |
|---|---|---|---|---|---|---|---|
| | | | Hen Housed | Hen Day | | Weekly | Cumul. |
| | % | % | | | g | g | kg |
| 20 | 100.0 | 2.0 | 0.1 | 0.1 | 48.0 | 6.7 | – |
| 21 | 99.8 | 15.0 | 1.1 | 1.1 | 51.5 | 53.9 | 0.06 |
| 22 | 99.7 | 50.0 | 4.6 | 4.6 | 53.5 | 186.6 | 0.24 |
| 23 | 99.6 | 80.0 | 10.2 | 10.2 | 55.0 | 306.7 | 0.55 |
| 24 | 99.5 | 88.0 | 16.3 | 16.4 | 56.0 | 343.2 | 0.89 |
| 25 | 99.4 | 90.0 | 22.6 | 22.7 | 57.0 | 356.9 | 1.25 |
| 26 | 99.3 | 92.0 | 29.0 | 29.1 | 57.7 | 368.9 | 1.62 |
| 27 | 99.2 | 93.0 | 35.4 | 35.6 | 58.3 | 376.4 | 1.99 |
| 28 | 99.0 | 93.0 | 41.5 | 42.1 | 59.0 | 380.2 | 2.38 |
| 29 | 98.8 | 93.0 | 48.3 | 48.6 | 59.5 | 382.6 | 2.76 |
| 30 | 98.7 | 93.0 | 54.8 | 55.1 | 60.0 | 385.5 | 3.14 |
| 31 | 98.6 | 93.0 | 61.2 | 61.6 | 60.5 | 388.3 | 3.53 |
| 32 | 98.5 | 93.0 | 67.6 | 68.1 | 60.9 | 390.5 | 3.92 |
| 33 | 98.4 | 92.5 | 74.0 | 74.6 | 61.2 | 389.9 | 4.31 |
| 34 | 98.3 | 92.0 | 80.3 | 81.0 | 61.6 | 389.9 | 4.70 |
| 35 | 98.2 | 91.5 | 86.6 | 87.4 | 61.8 | 388.7 | 5.09 |
| 36 | 98.0 | 91.0 | 92.8 | 93.8 | 62.0 | 387.0 | 5.48 |
| 37 | 97.8 | 90.5 | 99.0 | 100.1 | 62.2 | 385.3 | 5.86 |
| 38 | 97.7 | 90.0 | 105.3 | 106.4 | 62.4 | 384.0 | 6.25 |
| 39 | 97.6 | 89.0 | 111.2 | 112.6 | 62.6 | 380.6 | 6.63 |
| 40 | 97.5 | 88.5 | 117.3 | 118.8 | 62.8 | 379.3 | 7.01 |
| 41 | 97.4 | 88.0 | 123.3 | 125.0 | 63.0 | 377.9 | 7.39 |
| 42 | 97.3 | 87.5 | 129.3 | 131.1 | 63.2 | 376.6 | 7.76 |
| 43 | 97.2 | 87.0 | 135.2 | 137.2 | 63.3 | 374.4 | 8.14 |
| 44 | 97.0 | 86.5 | 141.0 | 143.3 | 63.4 | 372.3 | 8.51 |
| 45 | 96.8 | 86.0 | 146.9 | 149.3 | 63.6 | 370.6 | 8.88 |
| 46 | 96.7 | 85.5 | 152.7 | 155.3 | 63.7 | 368.6 | 9.25 |
| 47 | 96.6 | 85.0 | 158.4 | 161.3 | 63.8 | 366.7 | 9.61 |
| 48 | 96.5 | 84.5 | 164.1 | 167.2 | 64.0 | 365.3 | 9.98 |
| 49 | 96.4 | 84.0 | 169.8 | 173.1 | 64.0 | 362.7 | 10.34 |
| 50 | 96.3 | 83.5 | 175.4 | 178.9 | 64.2 | 361.3 | 10.70 |
| 51 | 96.2 | 83.0 | 181.0 | 184.7 | 64.2 | 358.8 | 11.06 |
| 52 | 96.0 | 85.2 | 186.5 | 190.5 | 64.4 | 357.0 | 11.42 |
| 53 | 95.8 | 82.0 | 192.0 | 196.2 | 64.4 | 354.1 | 11.77 |
| 54 | 95.7 | 81.5 | 197.5 | 201.9 | 64.6 | 352.6 | 12.13 |
| 55 | 95.6 | 81.0 | 202.9 | 207.6 | 64.6 | 350.1 | 12.48 |
| 56 | 95.5 | 80.5 | 208.3 | 213.2 | 64.8 | 348.7 | 12.83 |

TABLE 375 *(continued)*  Hybrid 'Isabrown' Laying Hen Weekly Production Targets

| Age Weeks | Live-ability | HDA Prod | Cumul. Eggs / Bird | | Average Egg Wt. | Hen Housed Egg Mass | |
|---|---|---|---|---|---|---|---|
| | | | Hen Housed | Hen Day | | Weekly | Cumul. |
| | % | % | | | g | g | kg |
| 57 | 95.4 | 80.0 | 213.6 | 218.8 | 64.8 | 346.1 | 13.17 |
| 58 | 95.3 | 79.5 | 218.9 | 224.4 | 64.9 | 344.1 | 13.52 |
| 59 | 95.2 | 79.0 | 224.2 | 229.9 | 64.9 | 341.6 | 13.86 |
| 60 | 95.0 | 78.5 | 229.4 | 235.4 | 65.0 | 339.3 | 14.20 |
| 61 | 94.8 | 78.0 | 234.6 | 240.9 | 65.0 | 336.4 | 14.53 |
| 62 | 94.7 | 77.5 | 239.7 | 246.3 | 65.0 | 333.9 | 14.87 |
| 63 | 94.6 | 77.0 | 244.8 | 251.7 | 65.1 | 331.9 | 15.20 |
| 64 | 94.5 | 76.5 | 249.9 | 257.1 | 65.1 | 329.4 | 15.53 |
| 65 | 94.4 | 76.0 | 254.9 | 262.4 | 65.2 | 327.4 | 15.86 |
| 66 | 94.3 | 75.5 | 259.9 | 267.8 | 65.2 | 324.9 | 16.18 |
| 67 | 94.2 | 75.0 | 264.9 | 273.1 | 65.3 | 322.9 | 16.50 |
| 68 | 94.0 | 74.5 | 269.8 | 278.3 | 65.3 | 320.1 | 16.82 |
| 69 | 93.8 | 74.0 | 274.6 | 283.5 | 65.4 | 317.7 | 17.14 |
| 70 | 93.7 | 73.5 | 279.4 | 288.6 | 65.4 | 315.2 | 17.46 |
| 71 | 93.6 | 73.0 | 284.2 | 293.7 | 65.5 | 313.2 | 17.77 |
| 72 | 93.5 | 72.5 | 289.0 | 298.8 | 65.5 | 310.8 | 18.08 |
| 73 | 93.4 | 72.0 | 293.7 | 303.8 | 65.6 | 308.8 | 18.39 |
| 74 | 93.3 | 71.5 | 298.3 | 308.8 | 65.6 | 306.3 | 18.70 |
| 75 | 93.2 | 71.0 | 303.0 | 313.8 | 65.7 | 304.3 | 19.00 |
| 76 | 93.0 | 70.5 | 307.6 | 318.7 | 65.7 | 301.5 | 19.30 |

The data, based on selected field and test results, are reproduced here only to show the performance capability under carefully controlled conditions and in no way constitute a warranty or guarantee that equal or similar performance will be achieved.

*Source:* 1991 data supplied and reproduced with kind permission of ISA Poultry Services Ltd, Green Rd, Eye, Peterborough, UK.

---

**TABLE 376    Adjustments in Feed Allowances in g / bird / day at Different Ambient Temperatures (Hybrid 'Isabrown' Hen Layers)**

---

| Age Wks | 14°C g | 16°C g | 18°C g | 20°C g | 22°C g | 24°C g |
|---|---|---|---|---|---|---|
| 8 | + 2.7 | + 1.8 | + 0.9 | 0 | - 0.9 | - 1.8 |
| 10 | + 3.6 | + 2.4 | + 1.2 | 0 | - 1.2 | - 2.4 |
| 12 | + 4.4 | + 3.0 | + 1.5 | 0 | - 1.5 | - 3.0 |
| 14 | + 5.1 | + 3.4 | + 1.7 | 0 | - 1.7 | - 3.4 |
| 16 | + 6.0 | + 4.0 | + 2.0 | 0 | - 2.0 | - 4.0 |
| 18 | + 6.6 | + 4.4 | + 2.2 | 0 | - 2.2 | - 4.4 |
| 20 | + 7.1 | + 4.9 | + 2.5 | 0 | - 2.5 | - 4.9 |

*Source:* 1991 data supplied and reproduced with kind permission of ISA Poultry Services Ltd, Green Rd, Eye, Peterborough, UK.

---

**TABLE 377    Relationship Between Egg Production and Bodyweight over Time for Hybrid 'Isabrown' Laying Hens**

---

| Production | Bodyweight |
|---|---|
| First Egg | 1650 g |
| 2% | 1700 g |
| 10% | 1750 g |
| 50% | 1825 g |
| 80% | 1875 g |
| 90% | 1900 g |
| 30 weeks of age | 1950 g |
| 35 weeks of age | 2000 g |

These weights are for birds fed ad-lib and weighed in the afternoon. This performance will only be achieved if:

a) Daily feed intake at onset of lay meets the requirements for growth, production and maintenance;

b) Bodyweight increases in relation to sexual maturity as measured by level of egg production.

*Source:* 1991 data supplied and reproduced with kind permission of ISA Poultry Services Ltd, Green Rd, Eye, Peterborough, UK.

**TABLE 378  Hybrid Layer Efficiency Guide**

| | | |
|---|---|---|
| **52 week average** | white breeds | 285 eggs / bird |
| | brown breeds | 280 eggs / bird |
| **Feed consumption** | white breeds | 104 g / day |
| | brown breeds | 117 g / day |
| **Feed conversion** | 133 - 153 g / egg | |
| **Layer mortality** | 5.5% | |

**TABLE 379  Hen Egg Size Grades (UK)**

| Grade | Wt. / egg gms |
|---|---|
| 1 | 70 + |
| 2 | 65 - 70 |
| 3 | 60 - 65 |
| 4 | 55 - 60 |
| 5 | 50 - 55 |
| 6 | 45 - 50 |
| 7 | <45 |

**TABLE 380  Hen Egg Production under Different Housing Systems**

| | Eggs / bird (dozen) | Food / bird (kg) | Food conversion |
|---|---|---|---|
| Free range | 14 | 54 | 11.4 |
| Intensive (poor control) | 17 | 50 | 8.6 |
| Intensive (fully controlled environment) | 20 | 43 | 6.3 |

*Source:* Sámsburg, D. (1984) *Poulltry Health & Management.*
Reproduced with kind permission of Blackwell Scientific Publishers Ltd.

**TABLE 381  Broiler Performance: Liveweight Feed Consumption and Conversion**

| | AS HATCHED | | | | MALES | | | | FEMALES | | | |
|---|---|---|---|---|---|---|---|---|---|---|---|---|
| Age (days) | Weight g | Feed g/bird/day | Total Feed g | FCR (Cumulative) | Weight g | Feed g/bird/day | Total Feed g | FCR (Cumulative) | Weight g | Feed g/bird/day | Total Feed g | FCR (Cumulative) |
| 0 | 42 | 00 | 00 | 0.00 | 42 | 00 | 00 | 0.00 | 42 | 00 | 00 | 0.00 |
| 1 | 53 | 12 | 12 | 0.22 | 53 | 12 | 12 | 0.22 | 53 | 12 | 12 | 0.22 |
| 2 | 66 | 16 | 28 | 0.42 | 67 | 16 | 28 | 0.42 | 65 | 16 | 27 | 0.42 |
| 3 | 83 | 20 | 48 | 0.58 | 84 | 21 | 48 | 0.58 | 81 | 20 | 47 | 0.58 |
| 4 | 102 | 23 | 71 | 0.70 | 103 | 24 | 72 | 0.70 | 100 | 23 | 70 | 0.70 |
| 5 | 124 | 27 | 99 | 0.79 | 127 | 28 | 100 | 0.79 | 122 | 27 | 97 | 0.79 |
| 6 | 148 | 30 | 129 | 0.87 | 151 | 31 | 132 | 0.87 | 144 | 30 | 126 | 0.88 |
| 7 | 173 | 33 | 162 | 0.94 | 177 | 34 | 166 | 0.94 | 168 | 32 | 159 | 0.94 |
| 8 | 199 | 36 | 198 | 1.00 | 204 | 37 | 203 | 0.99 | 194 | 35 | 194 | 1.00 |
| 9 | 228 | 40 | 238 | 1.04 | 234 | 41 | 244 | 1.04 | 221 | 38 | 232 | 1.05 |
| 10 | 260 | 43 | 281 | 1.08 | 267 | 45 | 288 | 1.08 | 252 | 42 | 273 | 1.09 |
| 11 | 294 | 48 | 328 | 1.12 | 304 | 49 | 338 | 1.11 | 285 | 46 | 319 | 1.12 |
| 12 | 330 | 52 | 381 | 1.15 | 342 | 54 | 392 | 1.15 | 319 | 50 | 369 | 1.16 |
| 13 | 368 | 56 | 436 | 1.19 | 382 | 59 | 451 | 1.18 | 354 | 53 | 422 | 1.19 |
| 14 | 407 | 59 | 496 | 1.22 | 423 | 62 | 513 | 1.21 | 391 | 57 | 479 | 1.23 |
| 15 | 448 | 64 | 560 | 1.25 | 466 | 67 | 579 | 1.24 | 429 | 61 | 540 | 1.26 |
| 16 | 491 | 69 | 628 | 1.28 | 512 | 72 | 652 | 1.27 | 469 | 65 | 605 | 1.29 |
| 17 | 536 | 72 | 700 | 1.31 | 561 | 75 | 727 | 1.30 | 511 | 68 | 673 | 1.32 |
| 18 | 584 | 76 | 776 | 1.33 | 612 | 80 | 807 | 1.32 | 556 | 72 | 745 | 1.34 |
| 19 | 633 | 79 | 855 | 1.35 | 664 | 84 | 891 | 1.34 | 601 | 75 | 820 | 1.36 |
| 20 | 683 | 83 | 938 | 1.37 | 718 | 87 | 979 | 1.36 | 647 | 78 | 898 | 1.39 |
| 21 | 733 | 86 | 1024 | 1.40 | 772 | 90 | 1069 | 1.38 | 695 | 80 | 978 | 1.41 |

| | | | | | | | | | | | | | | |
|---|---|---|---|---|---|---|---|---|---|---|---|---|---|---|
| 22 | 785 | 88 | 1112 | 1.42 | 22 | 828 | 94 | 1163 | 1.40 | 22 | 743 | 84 | 1061 | 1.43 |
| 23 | 838 | 93 | 1205 | 1.44 | 23 | 885 | 98 | 1261 | 1.43 | 23 | 792 | 88 | 1149 | 1.45 |
| 24 | 893 | 97 | 1302 | 1.46 | 24 | 943 | 103 | 1364 | 1.45 | 24 | 843 | 91 | 1240 | 1.47 |
| 25 | 949 | 101 | 1403 | 1.48 | 25 | 1004 | 107 | 1470 | 1.47 | 25 | 895 | 95 | 1334 | 1.49 |
| 26 | 1006 | 104 | 1507 | 1.50 | 26 | 1064 | 112 | 1582 | 1.49 | 26 | 947 | 97 | 1432 | 1.51 |
| 27 | 1063 | 108 | 1615 | 1.52 | 27 | 1126 | 114 | 1696 | 1.51 | 27 | 999 | 100 | 1531 | 1.53 |
| 28 | 1120 | 110 | 1725 | 1.54 | 28 | 1188 | 119 | 1815 | 1.53 | 28 | 1052 | 103 | 1635 | 1.55 |
| 29 | 1178 | 114 | 1839 | 1.56 | 29 | 1250 | 122 | 1937 | 1.55 | 29 | 1105 | 106 | 1741 | 1.58 |
| 30 | 1236 | 117 | 1956 | 1.58 | 30 | 1313 | 125 | 2062 | 1.57 | 30 | 1159 | 108 | 1849 | 1.60 |
| 31 | 1294 | 117 | 2073 | 1.60 | 31 | 1377 | 126 | 2188 | 1.59 | 31 | 1212 | 109 | 1958 | 1.62 |
| 32 | 1354 | 120 | 2193 | 1.62 | 32 | 1441 | 128 | 2316 | 1.61 | 32 | 1266 | 111 | 2069 | 1.63 |
| 33 | 1413 | 123 | 2316 | 1.64 | 33 | 1506 | 133 | 2449 | 1.63 | 33 | 1319 | 116 | 2185 | 1.66 |
| 34 | 1472 | 127 | 2443 | 1.66 | 34 | 1571 | 136 | 2585 | 1.65 | 34 | 1373 | 118 | 2303 | 1.68 |
| 35 | 1531 | 131 | 2575 | 1.68 | 35 | 1636 | 139 | 2724 | 1.66 | 35 | 1426 | 121 | 2424 | 1.70 |
| 36 | 1590 | 132 | 2707 | 1.70 | 36 | 1701 | 142 | 2866 | 1.68 | 36 | 1479 | 124 | 2548 | 1.72 |
| 37 | 1650 | 136 | 2843 | 1.72 | 37 | 1767 | 145 | 3011 | 1.70 | 37 | 1532 | 127 | 2675 | 1.75 |
| 38 | 1709 | 141 | 2983 | 1.75 | 38 | 1832 | 149 | 3160 | 1.72 | 38 | 1585 | 133 | 2808 | 1.77 |
| 39 | 1768 | 145 | 3128 | 1.77 | 39 | 1898 | 152 | 3312 | 1.74 | 39 | 1638 | 135 | 2943 | 1.80 |
| 40 | 1827 | 146 | 3274 | 1.79 | 40 | 1964 | 157 | 3469 | 1.77 | 40 | 1691 | 137 | 3080 | 1.82 |
| 41 | 1887 | 150 | 3424 | 1.82 | 41 | 2030 | 158 | 3628 | 1.79 | 41 | 1743 | 143 | 3223 | 1.85 |
| 42 | 1946 | 153 | 3577 | 1.84 | 42 | 2096 | 161 | 3789 | 1.81 | 42 | 1795 | 142 | 3365 | 1.87 |
| 43 | 2005 | 156 | 3733 | 1.86 | 43 | 2163 | 164 | 3953 | 1.83 | 43 | 1847 | 146 | 3512 | 1.90 |
| 44 | 2064 | 157 | 3890 | 1.88 | 44 | 2229 | 167 | 4120 | 1.85 | 44 | 1899 | 147 | 3659 | 1.93 |
| 45 | 2123 | 159 | 4049 | 1.91 | 45 | 2295 | 167 | 4287 | 1.87 | 45 | 1951 | 151 | 3810 | 1.95 |
| 46 | 2183 | 162 | 4211 | 1.93 | 46 | 2362 | 172 | 4460 | 1.89 | 46 | 2003 | 152 | 3963 | 1.98 |
| 47 | 2242 | 164 | 4375 | 1.95 | 47 | 2428 | 173 | 4633 | 1.91 | 47 | 2055 | 154 | 4117 | 2.00 |
| 48 | 2301 | 167 | 4542 | 1.97 | 48 | 2495 | 178 | 4811 | 1.93 | 48 | 2107 | 159 | 4276 | 2.03 |
| 49 | 2360 | 170 | 4712 | 2.00 | 49 | 2562 | 176 | 4987 | 1.95 | 49 | 2158 | 157 | 4433 | 2.05 |

*(continued over)*

**TABLE 381** *(continued)* **Broiler Performance: Liveweight Feed Consumption and Conversion**

### AS HATCHED

| Age (days) | Weight g | Feed g / bird / day | Total Feed g | FCR (Cumulative) |
|---|---|---|---|---|
| 50 | 2419 | 170 | 4882 | 2.02 |
| 51 | 2478 | 172 | 5054 | 2.04 |
| 52 | 2537 | 175 | 5229 | 2.06 |
| 53 | 2596 | 177 | 5406 | 2.08 |
| 54 | 2654 | 179 | 5585 | 2.10 |
| 55 | 2713 | 183 | 5768 | 2.13 |
| 56 | 2770 | 182 | 5950 | 2.15 |

### MALES

| Age (days) | Weight g | Feed g / bird / day | Total Feed g | FCR (Cumulative) |
|---|---|---|---|---|
| 50 | 2628 | 181 | 5168 | 1.97 |
| 51 | 2695 | 184 | 5352 | 1.99 |
| 52 | 2762 | 184 | 5536 | 2.00 |
| 53 | 2829 | 189 | 5725 | 2.02 |
| 54 | 2895 | 188 | 5913 | 2.04 |
| 55 | 2962 | 191 | 6104 | 2.06 |
| 56 | 3028 | 196 | 6300 | 2.08 |
| 57 | 3094 | 196 | 6495 | 2.10 |
| 58 | 3161 | 198 | 6693 | 2.12 |
| 59 | 3227 | 200 | 6894 | 2.14 |
| 60 | 3293 | 202 | 7096 | 2.15 |
| 61 | 3360 | 205 | 7301 | 2.17 |
| 62 | 3426 | 207 | 7508 | 2.19 |
| 63 | 3492 | 206 | 7714 | 2.21 |
| 64 | 3558 | 212 | 7926 | 2.23 |
| 65 | 3624 | 214 | 8140 | 2.25 |
| 66 | 3690 | 213 | 8353 | 2.26 |
| 67 | 3756 | 219 | 8572 | 2.28 |
| 68 | 3822 | 217 | 8790 | 2.30 |
| 69 | 3887 | 220 | 9009 | 2.32 |
| 70 | 3953 | 222 | 9231 | 2.34 |

### FEMALES

| Age (days) | Weight g | Feed g / bird / day | Total Feed g | FCR (Cumulative) |
|---|---|---|---|---|
| 50 | 2210 | 162 | 4595 | 2.08 |
| 51 | 2261 | 162 | 4757 | 2.10 |
| 52 | 2312 | 164 | 4922 | 2.13 |
| 53 | 2363 | 168 | 5090 | 2.15 |
| 54 | 2414 | 167 | 5257 | 2.18 |
| 55 | 2463 | 173 | 5430 | 2.20 |
| 56 | 2512 | 171 | 5602 | 2.23 |

*Note:* Data relate to 'Cobb 500' broilers recorded in commercial trials and commercial flocks worldwide.

*Source:* 1991 data supplied and reproduced with kind permission of The Cobb Breeding Company, East Hanningfield, Chelmsford, Essex, UK.

## TABLE 382 a & b   'Big 6' Turkey Performance Goals

### a. Egg production and hatchability

| | EGG PRODUCTION | | | HATCHABILITY | | | |
|---|---|---|---|---|---|---|---|
| Weeks in lay | % Hen housed prodn. | Settable eggs per hen per week | Settable eggs per hen to date | % Hatch of eggs set | Number of poults per hen week | Total poults per hen to date | Weeks in lay |
| 1 | 39.7 | 2.78 | 2.78 | 74.6 | 2.07 | 2.07 | 1 |
| 2 | 63.8 | 4.47 | 7.25 | 79.0 | 3.53 | 5.60 | 2 |
| 3 | 69.0 | 4.83 | 12.08 | 81.8 | 3.95 | 9.55 | 3 |
| 4 | 68.1 | 4.77 | 16.85 | 82.7 | 3.94 | 13.49 | 4 |
| 5 | 66.8 | 4.68 | 21.53 | 83.1 | 3.88 | 17.37 | 5 |
| 6 | 67.1 | 4.70 | 26.23 | 82.6 | 3.88 | 21.25 | 6 |
| 7 | 67.7 | 4.74 | 30.97 | 82.2 | 3.89 | 25.14 | 7 |
| 8 | 67.3 | 4.71 | 35.68 | 81.5 | 3.84 | 28.98 | 8 |
| 9 | 67.1 | 4.70 | 40.38 | 80.6 | 3.79 | 32.77 | 9 |
| 10 | 65.4 | 4.58 | 44.96 | 79.8 | 3.65 | 36.42 | 10 |
| 11 | 64.6 | 4.52 | 49.48 | 78.7 | 3.56 | 39.98 | 11 |
| 12 | 63.6 | 4.45 | 53.93 | 77.8 | 3.46 | 43.44 | 12 |
| 13 | 62.2 | 4.35 | 58.28 | 76.8 | 3.35 | 46.79 | 13 |
| 14 | 60.6 | 4.24 | 62.52 | 76.0 | 3.22 | 50.01 | 14 |
| 15 | 59.5 | 4.17 | 66.69 | 75.0 | 3.12 | 53.13 | 15 |
| 16 | 58.0 | 4.06 | 70.75 | 73.9 | 3.00 | 56.13 | 16 |
| 17 | 56.6 | 3.96 | 74.71 | 73.0 | 2.89 | 59.02 | 17 |
| 18 | 55.1 | 3.86 | 78.57 | 72.1 | 2.78 | 61.80 | 18 |
| 19 | 53.6 | 3.75 | 82.32 | 71.0 | 2.67 | 64.47 | 19 |
| 20 | 51.8 | 3.63 | 85.95 | 70.0 | 2.54 | 67.01 | 20 |
| 21 | 49.5 | 3.47 | 89.42 | 69.1 | 2.39 | 69.40 | 21 |
| 22 | 47.5 | 3.33 | 92.75 | 67.9 | 2.26 | 71.66 | 22 |
| 23 | 45.7 | 3.20 | 95.95 | 66.8 | 2.14 | 73.80 | 23 |
| 24 | 44.1 | 3.09 | 99.04 | 65.2 | 2.01 | 75.81 | 24 |

### b. Breeding stock feed consumption according to weight and age

The following Table shows the amount of feed required for the number of males and females needed to bring 1000 females into lay

#### PARENT MALE

| Age (days) | Feed used per live bird (kg) | Average no. of birds in period | Total feed use in period (kg) | Selection rejects | L.Wt at end of period (kg) |
|---|---|---|---|---|---|
| 0 to 28 | 1.73 | 168 | 291 | | 1.18 |
| 28 to 56 | 6.10 | 165 | 1007 | | 4.43 |
| 56 to 84 | 10.51 | 163 | 1713 | | 9.05 |
| 84 to 112 | 14.33 | 160 | 2293 | 70 at 16 weeks | 14.29 |
| 112 to 168 | 40.04 | 88 | 3524 | 10 at 24 weeks | 23.59 |
| 168 to 203 | 27.64 | 76 | 2101 | | 26.41 |
| | 100.35 | 75 | 10 929 | | |

Feed used per male alive at 203 days 145.72 kg
Feed used in semen production 5.85 kg per male per week

*(continued over)*

A G R I   I N F O

**TABLE 382** *(continued)* **b & c 'Big 6' Turkey Performance Goals**

### PARENT FEMALE

| Age (days) | Feed used per live bird (kg) | Average no. of birds in period | Total feed use in period (kg) | Selection rejects | L.Wt at end of period (kg) |
|---|---|---|---|---|---|
| 0 to 28 | 1.30 | 1163 | 1512 | | 0.86 |
| 28 to 56 | 3.97 | 1143 | 4538 | | 2.67 |
| 56 to 84 | 6.78 | 1126 | 7634 | | 4.84 |
| 84 to 112 | 9.07 | 1115 | 10 113 | 90 at 16 weeks | 7.01 |
| 112 to 203 | 33.83 | 1010 | 34 168 | | 11.34 |
| | 54.95 | 1000 | 57 965 | | |

Feed used per female alive at 203 days 57.97 kg.
Feed use in lay 2.17 kg per female per week.
End of lay weights: Male 31.57 kg. Female 11.02 kg.

### c. Commercial stock weight, liveability and feed intake

| | MALE | | | | | FEMALE | | | |
|---|---|---|---|---|---|---|---|---|---|
| Age (days) | Weight (kg) | Live-ability % | Weekly Feed Intake (kg) | Cumulative FCR | | Weight (kg) | Live-ability | Weekly Feed Intake (kg) | Cumulative FCR |
| 7 | 0.16 | 98.5 | 0.14 | 0.87 | | 0.16 | 98.5 | 0.14 | 0.87 |
| 14 | 0.39 | 98.2 | 0.32 | 1.18 | | 0.36 | 98.2 | 0.29 | 1.18 |
| 21 | 0.74 | 98.0 | 0.53 | 1.34 | | 0.65 | 98.0 | 0.46 | 1.37 |
| 28 | 1.22 | 97.8 | 0.78 | 1.46 | | 1.03 | 97.8 | 0.66 | 1.50 |
| 35 | 1.80 | 97.6 | 1.02 | 1.56 | | 1.50 | 97.6 | 0.87 | 1.61 |
| 42 | 2.49 | 97.4 | 1.29 | 1.65 | | 2.05 | 97.5 | 1.09 | 1.71 |
| 49 | 3.26 | 97.2 | 1.54 | 1.73 | | 2.65 | 97.3 | 1.30 | 1.82 |
| 56 | 4.10 | 97.0 | 1.78 | 1.82 | | 3.30 | 97.2 | 1.50 | 1.92 |
| 63 | 4.99 | 96.8 | 2.00 | 1.90 | | 3.98 | 97.0 | 1.67 | 2.01 |
| 70 | 5.94 | 96.6 | 2.21 | 1.97 | | 4.69 | 96.9 | 1.84 | 2.10 |
| 77 | 6.93 | 96.4 | 2.43 | 2.04 | | 5.40 | 96.7 | 1.98 | 2.20 |
| 84 | 7.94 | 96.2 | 2.62 | 2.12 | | 6.11 | 96.6 | 2.14 | 2.29 |
| 91 | 8.96 | 96.0 | 2.80 | 2.19 | | 6.81 | 96.4 | 2.28 | 2.40 |
| 98 | 9.98 | 95.7 | 2.96 | 2.27 | | 7.49 | 96.3 | 2.45 | 2.51 |
| 105 | 10.99 | 95.4 | 3.16 | 2.36 | | 8.14 | 96.1 | 2.57 | 2.63 |
| 112 | 11.99 | 95.1 | 3.35 | 2.45 | | 8.76 | 96.0 | 2.65 | 2.75 |
| 119 | 12.97 | 94.8 | 3.50 | 2.54 | | 9.34 | 95.8 | 2.69 | 2.87 |
| 126 | 13.94 | 94.4 | 3.65 | 2.64 | | 9.88 | 95.6 | 2.73 | 3.00 |
| 133 | 14.91 | 93.9 | 3.83 | 2.73 | | 10.38 | 95.4 | 2.79 | 3.13 |
| 140 | 15.88 | 93.2 | 4.00 | 2.84 | | 10.84 | 95.2 | 2.87 | 3.27 |
| 147 | 16.84 | 92.4 | 4.13 | 2.94 | | | | | |
| 154 | 17.79 | 91.6 | 4.26 | 3.05 | | | | | |
| 161 | 18.74 | 90.70 | 4.42 | 3.16 | | | | | |
| 168 | 19.68 | 89.7 | 4.57 | 3.27 | | | | | |

*NB* 1. Under the best conditions 10% better liveweights than above should be achieved.
    2. The feed programme used will have a big influence on both liveweights and feed conversions.
    3. Weekly feed intake is per live bird. Cumulative FCR includes an allowance for feed consumed by dead birds.

**TABLE 382** *(continued)*  **d  'Big 6' Turkey Performance Goals**

### d. Commercial stock live and eviscerated weight by age

| COMMERCIAL MALES | | | COMMERCIAL FEMALES | | |
|---|---|---|---|---|---|
| Age (days) | Live-weight (kg) | Eviscerated Weight (kg) | Age (days) | Live-weight (kg) | Eviscerated Weight (kg) |
| 84 | 7.9 | 5.8 | 84 | 6.1 | 4.5 |
| 91 | 9.0 | 6.6 | 91 | 6.8 | 5.0 |
| 98 | 10.0 | 7.4 | 98 | 7.5 | 5.6 |
| 105 | 11.0 | 8.2 | 105 | 8.1 | 6.2 |
| 112 | 12.0 | 9.1 | 112 | 8.8 | 6.7 |
| 119 | 13.0 | 9.9 | 119 | 9.3 | 7.2 |
| 126 | 13.9 | 10.7 | 126 | 9.9 | 7.6 |
| 133 | 14.9 | 11.5 | 133 | 10.4 | 8.0 |
| 140 | 15.9 | 12.2 | 140 | 10.8 | 8.4 |
| 147 | 16.8 | 13.0 | | | |
| 154 | 17.8 | 13.8 | | | |
| 161 | 18.7 | 14.6 | | | |
| 168 | 19.7 | 15.3 | | | |

*Source:* 1991 data kindly supplied and reproduced with kind permission of British United Turkeys Ltd. Warren Hall, Broughton, Chester, UK. ("Big 6 " is one of the strains of turkeys produced by this company.)

---

**TABLE 383   Liveweight (kg) of Medium Sized Geese at Various Ages**

| | |
|---|---|
| At hatching | 0.09 - 0.11 |
| 4  weeks | 0.9 - 1.6 |
| 8  weeks | 2.5 - 3.1 |
| 12  weeks | 3.4 - 4.3 |
| 16  weeks | 3.8 - 5.0 |
| 20  weeks | 4.3 - 5.4 |
| 24  weeks | 4.8 - 5.9 |
| 28  weeks | 5.9 - 7.2 |

*Source:* MAFF/ADAS (1980) *Ducks and Geese. Ref. Book No. 70.* Reproduced with kind permission of MAFF ADAS (UK).

**TABLE 384   Growth Rate, Feed Efficiency and Feed Consumption of White Pekin ducklings \* (mixed sexes)**

| Age in weeks | Average weight | Feed conversion | | Feed per bird | |
| | | Weight of feed per unit of liveweight | | | |
| | | For period | To date | For period | To date |
| | kg | kg | kg | kg | kg |
|---|---|---|---|---|---|
| 1 | 0.19 | 1.14 | 1.14 | 0.21 | 0.21 |
| 2 | 0.60 | 1.82 | 1.61 | 0.75 | 0.97 |
| 3 | 1.11 | 2.50 | 2.02 | 1.28 | 2.25 |
| 4 | 1.68 | 2.73 | 2.26 | 1.55 | 3.79 |
| 5 | 2.18 | 3.22 | 2.48 | 1.62 | 5.42 |
| 6 | 2.58 | 4.44 | 2.78 | 1.75 | 7.17 |
| 7 | 2.95 | 5.16 | 3.08 | 1.92 | 9.09 |
| 8 | 3.29 | 6.01 | 3.38 | 2.02 | 11.11 |
| 9 | 3.57 | 7.82 | 3.73 | 2.20 | 13.31 |

(\* Source: New York State Poultry Industry Co-ordinated Effort)

*NB* Modern strains can achieve a liveweight of 3.2 kg at 49 days with an FCE of well under 3:1

*Source:* MAFF/ADAS (1980) *Ducks and Geese. Ref. Book No. 70.*
Reproduced with kind permission of MAFF ADAS (UK).

---

**TABLE 385   Performance of Two Modern Meat and a Layer Strain \* of Duck**

**MEAT STRAIN CV 'BARBARY'**

**Parent Stock**

| | |
|---|---|
| Age at sexual maturity | 28 weeks |
| Female bodyweight at 28 weeks | 2.1 kg |
| Production period | 24 + 22 weeks |
| Egg production per female housed (in 58 weeks allowing for 12 weeks moult and for standard mortality of 1% per month) | 182 eggs |
| Hatchability | 75% |
| Day-old ducks per female parent | 136 day-olds |

| Commercial Stock | Females @ 70Days of Age | Males @ 84Days of Age |
|---|---|---|
| Liveweight | 2.59 kg | 4.35 kg |
| Eviscerated Weight \* | 1.66 kg | 2.78 kg |
| Feed Conversion on Liveweight | 2.70:1 | 2.90:1 |
| Feed Conversion on Eviscerated Weight \* | 4.29:1 | 4.60:1 |

**MEAT HYBRID CV 'SUPER M'**

**Parent Stock**

| | |
|---|---|
| Age at sexual maturity | 26 weeks |
| Female bodyweight at 26 weeks | 3.1 kg |
| Production period | 40 weeks |
| Egg production per female housed (in 40 weeks of lay allowing for standard mortality of 1% per month) | 220 eggs |
| Day-old ducks per female parent (in 40 weeks of lay) | 172 |
| Yield of day-olds from eggs | 78% |

**Commercial Stock**

| | |
|---|---|
| Liveweight at 47 and 52 days | 3.07 - 3.24 kg |
| Oven ready weight at 47 and 52 days | 2.23 - 2.42 kg |
| Feed conversion liveweight at 47 days | 2.81:1 |
| Feed conversion oven-ready weight at 47 days | 3.86:1 |

\* Not including any offal

**TABLE 385    Performance of Two Modern Meat and a Layer Strain of Duck**

**LAYER STRAIN CV 2000**

| Parent Stock | | Commercial Egg Layers | |
|---|---|---|---|
| Age at sexual maturity | 20 weeks | Age at sexual maturity | 20 weeks |
| Egg production to 72 weeks of age | 275 | Egg production to 72 weeks of age | 285 |
| Mean egg weight | 75 g | Mean egg weight | 75 g |
| Total food per egg (day-old to death) | 280 g | Total food per egg (day-old to death) | 206 g |
| Percentage hatch of settable eggs | 78 % | Body weight at point of lay | 1.50 kg |
| Total food per egg (day-old to death) | 359 g | | |

*NB* Performance of all strains achieved under normal conditions at Cherry Valley Farms

*Source:* 1991 data kindly provided by Cherry Valley Farms Livestock Division, Rothwell, Lincoln LN7 6BR, UK.

**TABLE 386    'Hylyne Carolina' Rabbit Indicative Technical Performance Data**

| Mating | Age at first mating | 150 days |
|---|---|---|
| | Doe wt. at first mating | 3 - 3.4 kg |
| | Mating to birth of litter | 31 days |
| | Remating after parturition | 18 days |
| Feed consumption | Mating to birth per doe | 4.55 kg |
| | Birth to 6 weeks doe + litter | 29.0 kg |
| | Litter to slaughter weight | 2.27 kg |
| Weaning age | | 42 - 47 days |
| Slaughter age | | 8 - 10 weeks |
| Slaughter Liveweight | | 2.27 kg |
| Killing out % | (skinned, eviscerated) | 61 % |
| | (eviscerated only) | 75 % |
| Mortality | Annual doe mortality | 11 - 17 % |
| | Rearing to 10 weeks | 10 % |
| Feed conversion | overall litters + doe + buck | 3 : 1 - 3.15 |
| Litter size | reared / parturition | 8 |
| Litters / year / doe | | 6 - 7 |

*Source:* 1992 data supplied and reproduced with kind permission of Hylyne Rabbits Ltd., Lymm, Cheshire, UK.

**TABLE 387   Production Trends in France from 1950 to 1980 in the Most Productive Rabbitries**

| Criteria | 1950 | 1960 | 1970 | 1980 |
|---|---|---|---|---|
| Rabbits produced (sold) per breeding doe | 20 - 25 | 30 | 45 | 60 |
| Average interval between litters (days) | 90 - 100 | 70 | 54 | 42 |
| Concentrate fed necessary to produce 1 kg live rabbit (kg) | (*) | 6 | 4.5 | 3.6 |
| Type of rabbit | Common, of no specific breed | Pure breeds | Purebred Does crossed with improver buck | Specialized hybrid strains |
| Man-hours per doe per year | 16 | 16 | 10 | 7.5 |
| Labour used to produce 1 kg carcass (minutes) | 27 | 22 | 9.5 | 6.2 |
| No. of breeding does in breeding units | 80 - 100 | 100 - 150 | 200 - 250 | 350 - 400 |
| Percentage of investment in retail price of rabbit (%) | < 3 | 5 - 8 | 12 - 15 | 18 - 20 |

* Rabbits were not fed concentrates at this date

*Source:* FAO (1986) *The Rabbit; Husbandry, Health and Production.* Animal Production and Health Series No. 21 (after Lebas). Reproduced with kind permission of the Food and Agriculture Organization of the United Nations.

# Livestock
# Environment

## Note 17  Livestock Environment

Tolerance of livestock to varying climates is difficult to define but general recommendations have become increasingly available and are continuously being refined. Factors influencing tolerance to climate are:

- **specie**
- **breed**
- **origin of animal (if re-located outside its environment of origin)**
- **nutritional status of the animal**
- **disease status of the animal**
- **premature animals are less tolerant of heat and cold**
- **size of the group of animals**
- **quality of housing including bedding**
- **general management standards**
- **level of environmental amelioration, e.g. shade, cooling, ventilation**

Economic as well as technical considerations need to be assessed in defining any 'optimum' environment since livestock, like crops, can perform at sub-optimal technical levels over a range of climates. Optimal climatic requirements are most precisely known for hybrid poultry and pigs kept in controlled environments.

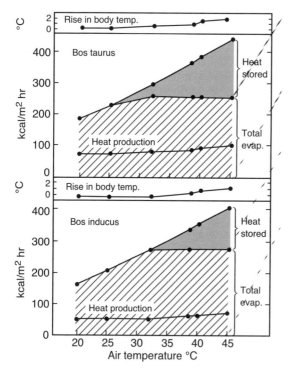

*NB* The graph indicates the effect of different air temperatures on heat storage and thereby gives the critical air temperature below which a fasting animal can dissipate both the metabolic and solar heat loads.

The heat production values used in the calculation only relate to the fasting animal. Increments in heat production must be added for the energy cost of maintenance and production together with the energy cost of grazing. This will reduce the air temperature at which heat storage commences.

**Fig. 57** The predicted heat balance of *Bos indicus* and *Bos taurus* subjected to solar radiation at noon

*Source:* Smith, A.J., ed. (1974) *Beef Cattle Production in Developing Countries.* Reproduced with kind permission of A.J. Smith.

**TABLE 388  Surface to Mass Ratios of Mammals of Different Sizes**

| Animal | Weight kg | Surface area m$^2$ | kg m$^{-2}$ | cm$^2$ kg$^{-1}$ |
|--------|-----------|--------------------|-------------|------------------|
| Kid    | 5    | 0.22 | 22.7  | 440 |
| Lamb   | 10   | 0.34 | 29.4  | 340 |
| Sheep  | 50   | 1.04 | 48.1  | 208 |
| Pig    | 100  | 1.59 | 62.8  | 159 |
| Horse  | 500  | 4.66 | 107.2 | 93  |
| Ox     | 1000 | 7.46 | 134.0 | 75  |

Animals such as camels with long neck and legs would have greater areas for a given mass than these approximations. The ratio of surface to mass influences the gain or loss of radiation, the convective exchange of heat, the ratio of heat storage to heat exchange and the metabolic rate, and the rate of water turnover.

*Source:* Clarke, J.A., ed. (1981) *Environmental Aspects of Housing for Animal Production.* Reproduced with permission.

**TABLE 389  Approximate Outputs of Heat and Water of Adult Livestock in Warm Environments (c. 30° C)**

| Species | Weight kg | Daily heat output | | | Daily water turnover * | | |
|---------|-----------|------|-------------|----------------|-------------|-----|----------------|
|         |           | kJ   | kJ kg$^{-1}$ | kJ kg$^{-0.75}$ | ml kg$^{-1}$ | (l) | ml kg$^{-0.82}$ |
| Elephant      | 3 672 | 206 000 | 56  | 436 | 68  | 250 | 300 |
| Horse         | 703   | 49 900  | 71  | 360 | 78  | 61  | 280 |
| *B. taurus*   | 600   | 50 600  | 84  | 418 | 145 | 87  | 460 |
| *B. indicus*  | 530   | 41 000  | 77  | 372 | 113 | 60  | 350 |
| Camel         | 500   | 23 000  | 46  | 219 | 36  | 18  | 110 |
| Donkey        | 300   | 20 000  | 67  | 278 | 70  | 21  | 200 |
| Pig           | 120   | 10 000  | 83  | 277 | 108 | 13  | 250 |
| Sheep (wet)   | 50    | 6200    | 124 | 326 | 88  | 4.4 | 180 |
| Sheep (arid)  | 45    | 4850    | 108 | 280 | 64  | 2.9 | 130 |
| Goat (milk)   | 45    | 5210    | 115 | 300 | 91  | 4.1 | 180 |
| Goat (arid)   | 30    | 3360    | 112 | 262 | 47  | 1.4 | 90  |

* Metabolic, food and drinking water combined, the amount released daily to the environment

Measurements have been made on different breeds of animal, in different thermal conditions on diverse foods and by different methods so that it is difficult to produce a unified table. There is a 20-30 per cent range on either side of the values given, because of individual and breed variation and differences in nutrition. The estimates given (based on Brody, 1945; Macfarlane & Howard, 1972; Macfarlane, 1976) lie at the upper range of water turnovers–for evaporative cooling–and the lower range of metabolism, for hot climates. Metabolic rates are standard basal values.

*Source:* Clarke, J.A., ed. (1981) *Environmental Aspects of Housing for Animal Production.* Reproduced with permission.

**TABLE 390   Lower Critical Temperature of Ruminants Housed in Conditions of Very Low Air Movement (0.2 m s -1) and in a Draught (2 m S $^{-1}$) (Dry Enclosure)**

| | Body weight kg | Coat depth cm | Heat production W m $^{-2}$ | Thermal insulation m $^2$ KW $^{-1}$ | | Lower critical temperature °C | |
|---|---|---|---|---|---|---|---|
| | | | | $I_t$ | $I_e$† | u = 0.2 ms $^{-1}$ | u = 2 ms $^{-1}$ |
| **Cattle** | | | | | | | |
| newborn calf | 35 | 1.2 | 100 | 0.09 | 0.25 | + 9 | + 17 |
| 1 month old | 50 | 1.4 | 120 | 0.10 | 0.26 | 0 | + 9 |
| veal | 100 | 1.2 | 154 | 0.25 | 0.25 | - 14 | - 1 |
| Store cattle, maintenance | 250 | 2.0 | 157 | 0.17 | 0.32 | - 32 | - 20 |
| Beef cow, maintenance | 450 | 2.9 | 107 | 0.22 | 0.36 | - 17 | - 9 |
| Dairy cow, 22 kg milk per day | 500 | 1.2 | 154 | 0.20 | 0.25 | - 26 | - 13 |
| **Sheep** | | | | | | | |
| ewe, maintenance | 50 | 6.0 | 75 | 0.15 | 0.68 | - 11 | - 4 |
| ewe, shorn | 50 | 1.0 | 75 | 0.15 | 0.19 | + 17 | + 20 |
| newborn lamb, coat dry | 4 | 0.8 | 80 | 0.10 | 0.19 | + 19 | + 24 |
| growing lamb, 0.2 kg gain per day | 35 | 4.0 | 100 | 0.13 | 0.48 | - 13 | - 3 |
| **Red deer** | | | | | | | |
| calf, 0.2 kg gain per day | 45 | 1.6 | 135 | 0.10 | 0.28 | - 7 | + 3 |

† measured in still air (0.2 m s$^{-1}$)

$I_t$, tissue insulation; $I_e$, external insulation; u, wind speed

*Source:* Clarke, J.A., ed. (1981) *Environmental Aspects of Housing for Animal Production.* Reproduced with permission.

TABLE 391    Estimated Heat Output of Farm Livestock at Different Weights under Summer Conditions

| Animal | Approximate age at weight given | Liveweight range kg | Sensible heat output per animal at 21°C (W) |
|---|---|---|---|
| Calf | birth | 45 | 110 |
|  | 12 wk | 90 | 170 |
|  |  | 135 | 210 |
| Growing cattle |  | 180 | 250 |
|  |  | 226 | 270 |
|  |  | 272 | 280 |
|  |  | 317 | 300 |
|  |  | 362 | 320 |
| Cow |  | per 450 | 480 |
| Fattening pig | 7 wk | 13.5 | 35 |
|  | 8 wk | 18 | 40 |
|  | 9 wk | 22.6 | 50 |
|  | 10 wk | 27.2 | 55 |
|  | 13 wk | 45.3 | 75 |
|  | 15 wk | 56.7 | 85 |
|  | 17 wk | 68 | 90 |
|  | 21 wk | 90 | 110 |
|  | 26 wk | 113 | 145 |
| Farrowing sow |  | 147 | 200 |
| and litter of 10 |  | 193 | 230 |
| Meat chicken | day-old | 0.036 | 0.35 |
| (broiler) | 3 wk | 0.27 | 3.1 |
|  | 4 wk | 0.453 | 3.9 |
|  | 5 wk | 0.60 | 4.6 |
|  | 6 wk | 0.9 | 5.4 |
|  | 7 wk | 1.225 | 6.5 |
|  | 8 wk | 1.495 | 7.2 |
|  | 9 wk | 1.723 | 8.3 |
| Chicken reared | day-old | 0.036 | 0.35 |
| for eggs | 3 wk | 0.113 | 1.6 |
|  | 8 wk | 0.680 | 4.6 |
|  | 12 wk | 1.135 | 6.2 |
|  | 16 wk | 1.580 | 7.5 |
|  | 18 wk | 1.815 | 8.3 |
| Laying poultry | adult | 1.815 | 8.3 |
|  |  | 2.268 | 9.9 |
|  |  | 2.720 | 10.9 |
|  |  | 3.175 | 12.4 |

*Source:* Sainsbury, D. & Sainsbury, P. (1988) *Livestock Health and Housing.*
Reproduced with kind permission of Bailliere Tindall.

TABLE 392  Estimated Heat Output and Moisture Respiration from Farm Animals When Housed under Approximately Optimum Conditions in Winter

| Animal | Approx. age at weight given | Practical critical temperature | Liveweight range | Sensible heat output per animal | Moisture respired per animal by evaporation at critical temperature |
|---|---|---|---|---|---|
| | | °C | kg | W | g / h |
| Calf | birth | | 45 | 120 | 95 |
| | 12 wk | 16 | 90 | 200 | 150 |
| | | | 135 | 250 | 192 |
| Growing cattle | | 2 | 181 | 375 (290) | 155 (221) |
| | | | 227 | 390 (310) | 163 (237) |
| | | | 272 | 425 (330) | 176 (251) |
| | | | 318 | 435 (350) | 180 (269) |
| | | | 363 | 445 (370) | 185 (279) |
| Cow | | 2 | per 450 kg | 810 (570) | 385 (521) |
| Piglet | birth | 27 | 1.1 | 4 | 12.6 |
| | 2-3 wk | 21 | 4.5 | 10 | 330 |
| Fattening pig | 7 wk | | 14 | 37 | 43 |
| | 8 wk | | 18 | 45 | 52 |
| | 9 wk | | 23 | 55 | 60 |
| | 10 wk | | 27 | 60 | 68 |
| | 13 wk | 18 | 45 | 82 | 94 |
| | 15 wk | | 57 | 95 | 109 |
| | 17 wk | | 68 | 102 | 117 |
| | 21 wk | | 90 | 125 | 144 |
| | 26 wk | | 113 | 162 | 185 |
| Farrowing sow | | 13 | 135 | 200 | 176 |
| | | | 181 | 245 | 196 |
| Meat chicken (broiler) | day-old | 21 | 0.036 | 0.35 | 0.20 |
| | 3 wk | | 0.27 | 3.1 | 1.7 |
| | 4 wk | | 0.45 | 4.3 | 1.5 |
| | 5 wk | | 0.68 | 5.2 | 1.8 |
| | 6 wk | 16 | 0.90 | 6.1 | 2.1 |
| | 7 wk | | 1.22 | 7.1 | 2.5 |
| Chicken reared for eggs | day-old | 21 | 0.036 | 0.35 | 0.20 |
| | 3 wk | | 0.11 | 1.6 | 0.91 |
| | 8 wk | | 0.7 | 5.3 | 1.56 |
| | 12 wk | 13 | 1.1 | 7.0 | 2.06 |
| | 16 wk | | 1.6 | 8.7 | 2.52 |
| | 18 wk | | 1.8 | 9.7 | 2.78 |
| Laying poultry | adult | 4 | 1.8 | 10.3 (9.2) | 1.58 (3.25) |
| | | | 2.2 | 12.3 (11.0) | 1.91 (3.92) |
| | | | 2.7 | 13.6 (12.1) | 2.12 (4.34) |
| | | | 3.1 | 15.5 (13.8) | 2.42 (4.92) |

Notes:    1. The figures in parentheses in columns 5 and 6 for growing cattle, cows and laying poultry give sensible heat outputs and moisture respiration for these animals at 16°C.

2. Figures for moisture respiration are quantities respired through skin and lungs and exclude moisture voided in dung and urine, for which an extra allowance must be made.

Source: Sainsbury, D. & Sainsbury, P. (1988) Livestock Health and Housing.
Reproduced with kind permission of Bailliere Tindall.

**TABLE 393   Minimum Area for the Outlet of Stale Air in Farm Buildings**

| Type of stock | Minimum outlet area m$^2$ per animal |
|---|---|
| Cows | 0.09 |
| Farrowing pigs | 0.01 |
| Fattening pigs | 0.008 |
| Calves | 0.006 |
| Laying birds | 0.003 |
| Broiler poultry | 0.0015 |

*Source:* Sainsbury, D. (1986) *Animal Health*. Reproduced with kind permission of Blackwell Scientific Publications Ltd.

**TABLE 394   Ventilation Rates for Varying Conditions in Temperate Climates for Farm Livestock**

| Animal | (m$^3$ / hour per kg body weight) | |
|---|---|---|
| | Maximum summer rate | Minimum winter rate |
| Adult cattle | 0.75 - 1.4 | 0.19 |
| Young calf | 0.94 - 1.9 | 0.38 |
| Sow and litter | 0.94 - 1.9 | 0.38 |
| Fattening pig | 0.94 - 1.9 | 0.38 |
| Broiler chicken | 2.8 - 4.7 | 0.75 |
| Laying poultry | 5.6 - 9.4 | 1.50 |

*Source:* Sainsbury, D. (1986) *Animal Health*. Reproduced with kind permission of Blackwell Scientific Publications Ltd.

**TABLE 395   Lower Critical Temperatures for Housed Livestock**

| Type of livestock | °C * | Type of livestock | °C * |
|---|---|---|---|
| Adult cattle | 0 | Fatteners | 18 |
| Young calves | 15 | Adult poultry | 7 |
| Farrowing sows | 15 | Broilers, 3 - 7 weeks | 16 |
| Newly born piglets | 27 | Day-old chicks to 3 weeks | 21 |

* Air temperatures below which intensively housed livestock should not be allowed to go

*Source:* Sainsbury, D. & Sainsbury, P. (1988) *Livestock Health and Housing*. Reproduced with kind permission of Bailliere Tindall.

**Table 396   Ambient Temperature Ranges and Housing Systems Suitable for Housed Livestock**

| Type of animal | Ambient temperature range | Housing system |
|---|---|---|
| **Adult milking cattle** | Milk production optimum 10 - 15° C (20) but little effect on yield from - 7 to + 21° C (25) | Climatic housing usual and generally satisfactory |
| **Beef cattle** (from 3 months of age) | - 7 to + 15° C (25) the optimum range | Climatic housing appropriate but thermal insulation may be needed if bedding is absent |
| **Calves** | 10 - 15° C at birth, which may fall gradually thereafter. Higher temperatures,15 - 21° C, are used in veal houses | Controlled environment housing or kennels required |
| **Lambs** | 4 - 21° C | Controlled environment may be used if housing intensive; otherwise kennels satisfactory |
| **Adult sheep** | - 7 to + 30° C | Climatic, or kennels |
| **Adult pigs** | 4 - 30° C | Climatic housing suitable where bedding is used, but not with individually housed adults without bedding when minimum temperature should be at least 10° C higher |
| **Fattening pigs** | 15 - 27° C | Controlled environment or kennel housing required |
| **Young piglets** | 21 - 27° C | Artificial heating needed to supplement controlled environment |
| **Brooding poultry** | 30 - 35° C | Artificial heating essential in controlled environment housing |
| **Broiler chickens** | 15 - 30° C | Artificial heating essential in controlled environment housing |
| **Laying poultry** | 15 - 21° C (20 - 25) | Controlled environment housing required with high-standard insulation or occasionally climatic  housing with deep straw or other litter |

*Definitions:*
| | |
|---|---|
| 'Climatic housing' | – only a cover and protection from the elements |
| 'Controlled environment' | – regulation of microclimate as completely as is required by the housed stock |
| 'Kennel' | – half way between 'climatic' and 'controlled environment giving two environments in one building and allowing the animal choice |

*Principal Source:* Clark, J.A., ed. (1981) *Environmental Aspects of Housing for Animal Production.* Reproduced with permission. Figures in parentheses from Sainsbury, D. & Sainsbury, P. (1988) *Livestock Health and Housing.*

**TABLE 397  Critical Temperatures for Pigs on a Solid Concrete Floor**

| Live-weight (kg) | Number in group | Lower critical temp. (°C) Feed level (multiple of maintenance) : | | | | Upper critical temp. (dry) (°C) Feed level (multiple of maintenance) : | | | | Upper critical temp. (wet) (°C) Feed level (multiple of maintenance) : | | | |
|---|---|---|---|---|---|---|---|---|---|---|---|---|---|
| | | 1 | 2 | 3 | 4 | 1 | 2 | 3 | 4 | 1 | 2 | 3 | 4 |
| 1 | 1 | 32 | 31 | 29 | 27 | 37 | 36 | 36 | 35 | 38 | 38 | 37 | 37 |
| | 10 | 31 | 29 | 26 | 24 | 35 | 34 | 32 | 30 | 37 | 37 | 36 | 35 |
| 5 | 1 | 30 | 28 | 26 | 23 | 37 | 35 | 34 | 33 | 38 | 37 | 36 | 36 |
| | 10 | 28 | 25 | 22 | 20 | 34 | 32 | 30 | 28 | 37 | 36 | 35 | 34 |
| 10 | 1 | 29 | 27 | 24 | 21 | 36 | 35 | 34 | 32 | 38 | 37 | 36 | 36 |
| | 10 | 27 | 24 | 21 | 17 | 34 | 32 | 29 | 27 | 37 | 36 | 35 | 34 |
| 20 | 1 | 29 | 26 | 22 | 18 | 36 | 35 | 33 | 31 | 38 | 37 | 36 | 35 |
| | 15 | 26 | 21 | 16 | 11 | 34 | 31 | 28 | 24 | 37 | 36 | 34 | 33 |
| 40 | 1 | 27 | 23 | 19 | 15 | 36 | 34 | 32 | 30 | 37 | 36 | 35 | 34 |
| | 15 | 24 | 18 | 13 | 7 | 33 | 30 | 26 | 23 | 37 | 35 | 34 | 32 |
| 60 | 1 | 26 | 22 | 18 | 14 | 35 | 33 | 31 | 29 | 37 | 36 | 35 | 34 |
| | 15 | 22 | 16 | 11 | 5 | 33 | 29 | 26 | 22 | 36 | 35 | 34 | 32 |
| 80 | 1 | 25 | 21 | 17 | 13 | 35 | 33 | 31 | 29 | 37 | 36 | 35 | 34 |
| | 15 | 21 | 15 | 10 | 4 | 32 | 29 | 26 | 23 | 36 | 35 | 34 | 32 |
| 100 | 1 | 24 | 21 | 17 | 13 | 35 | 33 | 31 | 29 | 37 | 36 | 35 | 34 |
| | 15 | 19 | 14 | 9 | 4 | 32 | 29 | 26 | 23 | 36 | 35 | 34 | 32 |
| 140 | 1 | 23 | 19 | 14 | 10 | 34 | 32 | 30 | 28 | 37 | 36 | 35 | 34 |
| | 5 | 20 | 12 | 9 | 3 | 32 | 30 | 26 | 23 | 36 | 34 | 33 | 32 |
| 180 | 1 | 22 | 18 | 13 | 8 | 34 | 32 | 29 | 27 | 36 | 35 | 34 | 33 |
| | 5 | 19 | 10 | 7 | 1 | 32 | 27 | 25 | 22 | 36 | 34 | 33 | 32 |

*Source:* Clark, J.A., ed. (1981) *Environmental Aspects of Housing for Animal Production.* Reproduced with permission.

**TABLE 398   Critical Temperatures for Pigs on a Perforated Metal Floor**

| Live-weight (kg) | Number in group | Lower critical temp. (°C) Feed level (multiple of maintenance) : | | | | Upper critical temp. (dry) (°C) Feed level (multiple of maintenance) : | | | | Upper critical temp. (wet) (°C) Feed level (multiple of maintenance) : | | | |
|---|---|---|---|---|---|---|---|---|---|---|---|---|---|
| | | 1 | 2 | 3 | 4 | 1 | 2 | 3 | 4 | 1 | 2 | 3 | 4 |
| 1 | 1 | 30 | 28 | 25 | 23 | 35 | 33 | 31 | 29 | 37 | 37 | 36 | 35 |
| | 10 | 29 | 26 | 23 | 20 | 34 | 32 | 29 | 27 | 37 | 36 | 36 | 35 |
| 5 | 1 | 28 | 26 | 23 | 20 | 34 | 32 | 30 | 28 | 37 | 36 | 36 | 35 |
| | 10 | 27 | 24 | 20 | 17 | 34 | 31 | 29 | 26 | 37 | 36 | 35 | 34 |
| 10 | 1 | 28 | 25 | 21 | 18 | 34 | 32 | 30 | 27 | 37 | 36 | 35 | 35 |
| | 10 | 26 | 22 | 19 | 15 | 33 | 31 | 28 | 26 | 37 | 36 | 35 | 34 |
| 20 | 1 | 28 | 24 | 20 | 15 | 35 | 32 | 29 | 26 | 37 | 36 | 35 | 34 |
| | 15 | 26 | 21 | 17 | 12 | 34 | 31 | 28 | 24 | 37 | 36 | 34 | 33 |
| 40 | 1 | 26 | 22 | 17 | 13 | 34 | 32 | 29 | 26 | 37 | 36 | 35 | 34 |
| | 15 | 24 | 19 | 14 | 8 | 34 | 30 | 27 | 24 | 37 | 35 | 34 | 32 |
| 60 | 1 | 25 | 20 | 16 | 11 | 34 | 31 | 28 | 26 | 37 | 36 | 35 | 33 |
| | 15 | 23 | 17 | 12 | 7 | 33 | 30 | 27 | 24 | 36 | 35 | 34 | 32 |
| 80 | 1 | 24 | 20 | 15 | 11 | 34 | 31 | 29 | 26 | 37 | 36 | 35 | 33 |
| | 15 | 22 | 16 | 11 | 6 | 33 | 30 | 27 | 24 | 36 | 35 | 34 | 32 |
| 100 | 1 | 23 | 19 | 15 | 11 | 33 | 31 | 29 | 26 | 37 | 36 | 35 | 34 |
| | 15 | 21 | 16 | 11 | 6 | 33 | 30 | 27 | 24 | 36 | 35 | 34 | 33 |
| 140 | 1 | 22 | 17 | 13 | 8 | 33 | 30 | 28 | 25 | 36 | 35 | 35 | 33 |
| | 5 | 20 | 14 | 9 | 3 | 32 | 29 | 26 | 23 | 36 | 35 | 34 | 32 |
| 180 | 1 | 21 | 16 | 11 | 6 | 33 | 30 | 27 | 25 | 36 | 35 | 34 | 33 |
| | 5 | 19 | 13 | 7 | 1 | 32 | 29 | 26 | 23 | 36 | 34 | 33 | 32 |

*Source:* Clark, J.A., ed. (1981) *Environmental Aspects of Housing for Animal Production.* Reproduced with permission.

**TABLE 399  Critical Temperatures for Pigs on a Straw-bedded Floor**

| Live-weight (kg) | Number in group | Lower critical temp. (°C) Feed level (multiple of maintenance): | | | | Upper critical temp. (dry) (°C) Feed level (multiple of maintenance): | | | | Upper critical temp. (wet) (°C) Feed level (multiple of maintenance): | | | |
|---|---|---|---|---|---|---|---|---|---|---|---|---|---|
| | | 1 | 2 | 3 | 4 | 1 | 2 | 3 | 4 | 1 | 2 | 3 | 4 |
| 1 | 1 | 30 | 27 | 25 | 22 | 34 | 32 | 30 | 28 | 37 | 37 | 36 | 35 |
| | 10 | 27 | 23 | 20 | 17 | 33 | 30 | 27 | 24 | 37 | 36 | 35 | 34 |
| 5 | 1 | 27 | 24 | 21 | 18 | 34 | 31 | 29 | 26 | 37 | 36 | 35 | 35 |
| | 10 | 24 | 20 | 16 | 12 | 32 | 29 | 26 | 23 | 37 | 36 | 35 | 34 |
| 10 | 1 | 26 | 23 | 19 | 16 | 33 | 31 | 28 | 26 | 37 | 36 | 35 | 34 |
| | 10 | 23 | 18 | 14 | 9 | 32 | 29 | 26 | 22 | 37 | 35 | 34 | 33 |
| 20 | 1 | 26 | 22 | 17 | 12 | 34 | 31 | 28 | 24 | 37 | 36 | 35 | 34 |
| | 15 | 23 | 17 | 11 | 4 | 33 | 29 | 25 | 20 | 37 | 35 | 34 | 32 |
| 40 | 1 | 24 | 20 | 14 | 9 | 34 | 30 | 27 | 23 | 37 | 36 | 35 | 33 |
| | 15 | 20 | 13 | 7 | 0 | 32 | 28 | 24 | 19 | 36 | 35 | 33 | 32 |
| 60 | 1 | 23 | 17 | 12 | 7 | 33 | 30 | 27 | 23 | 37 | 36 | 34 | 33 |
| | 15 | 18 | 12 | 5 | -2 | 32 | 27 | 23 | 19 | 36 | 35 | 33 | 32 |
| 80 | 1 | 22 | 16 | 11 | 6 | 33 | 30 | 27 | 23 | 37 | 35 | 34 | 33 |
| | 15 | 17 | 10 | 4 | -2 | 31 | 27 | 23 | 20 | 36 | 35 | 33 | 32 |
| 100 | 1 | 21 | 16 | 11 | 6 | 32 | 30 | 27 | 24 | 36 | 35 | 34 | 33 |
| | 15 | 16 | 10 | 4 | -2 | 31 | 27 | 24 | 20 | 36 | 35 | 33 | 32 |
| 140 | 1 | 19 | 14 | 8 | 2 | 32 | 29 | 26 | 22 | 36 | 35 | 34 | 32 |
| | 5 | 15 | 8 | 2 | -5 | 31 | 27 | 23 | 19 | 36 | 34 | 33 | 31 |
| 130 | 1 | 18 | 12 | 6 | 0 | 32 | 28 | 25 | 22 | 36 | 35 | 33 | 32 |
| | 5 | 14 | 5 | -1 | -8 | 30 | 26 | 22 | 18 | 35 | 34 | 32 | 31 |

*Source:* Clark, J.A., ed. (1981) *Environmental Aspects of Housing for Animal Production.* Reproduced with permission.

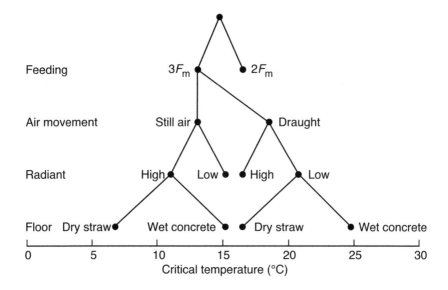

(Feed intakes, 2 $F_m$ and 3 $F_m$ (where $F_m$ = 440 kJ kg $^{-0.75}$ d $^{-1}$ of ME);
air movement, still air (0.1 m s $^{-1}$) and draughts (0.8 m s $^{-1}$);
radiant temperature, high (summer) and low (winter); floor, dry straw and wet concrete)

**Fig. 58** A diagrammatic representation of the effects of nutritional and environmental factors influencing the critical temperature of a group of nine 60 kg pigs

*Source:* Clark, J.A., ed. (1981) *Environmental Aspects of Housing for Animal Production.* Reproduced with permission.

---

**TABLE 400   a & b   Temperature Requirements for Pigs**

---

**a.** The relation between body-weight, food intake and critical temperature for groups of pigs at levels of intake normally applied in practice.

| | | | Food intake (kg / day) | | | | |
|---|---|---|---|---|---|---|---|
| Body | 0.5 | 1.0 | 1.5 | 2.0 | 2.5 | 3.0 | 3.5 |
| wt. (kg) | | | Critical temperature (°C) | | | | |
| 20 | 21 | 14 | | | | | |
| 40 | | 20 | 14 | 8 | | | |
| 60 | | | 18 | 13 | 8 | | |
| 80 | | | | 16 | 11 | 7 | |
| 100 | | | | 18 | 13 | 9 | |
| 120 | | | | | 15 | 11 | 8 |

## TABLE 400   b  Temperature Requirements for Pigs

**b.** The additional food requirements necessary to maintain optimum growth rates at environmental temperatures below the optimum.

| Body wt. (kg) | Range of optimum temps (°C) | Range of food intakes equivalent to ranges of critical temperature (kg / day) | Decrease in temperature below optimum (°C) | | | | |
| --- | --- | --- | --- | --- | --- | --- | --- |
| | | | -2 | -4 | -6 | -8 | -10 |
| | | | Additional food requirements (g / day) | | | | |
| 20 | 20 - 14 | 0.5 - 1.0 | 25 | 50 | 75 | 100 | 125 |
| 40 | 20 - 8 | 1.0 - 2.0 | 31 | 62 | 92 | 123 | 154 |
| 60 | 18 - 8 | 1.5 - 2.5 | 36 | 72 | 108 | 145 | 181 |
| 80 | 16 - 7 | 2.0 - 3.0 | 42 | 83 | 125 | 166 | 208 |
| 100 | 18 - 9 | 2.0 - 3.0 | 47 | 94 | 141 | 188 | 235 |
| 120 | 15 - 8 | 2.5 - 3.5 | 53 | 105 | 158 | 210 | 263 |

*Source:* Sainsbury, D. & Sainsbury, P. (1988) *Livestock Health and Housing.* Reproduced with kind permission of Bailliere Tindall.

## TABLE 401   The Combined Effect of Ambient Temperature and Air Movement on the Comfort of Pigs

| Temperature | Air movement Below 0.15 | Air movement from 0.15 m - 0.25 m / sec | Air movement from 0.25 m - 0.38 m / sec |
| --- | --- | --- | --- |
| 21° C | Pigs of all ages comfortable | Pigs of all ages comfortable | Young piglets uncomfortable (1-8 weeks) |
| 18° C | Pigs below 1 week uncomfortable | Pigs below 5 weeks uncomfortable | Pigs below 12 weeks uncomfortable |
| 15° C | Pigs below 10 days uncomfortable | Young piglets (c.1-3 weeks old) uncomfortable | Pigs below 12 weeks uncomfortable |
| 13° C | Pigs below 8 weeks uncomfortable | Pigs below 12 weeks uncomfortable | Pigs below 14 weeks uncomfortable |
| 10° C | Pigs below 15 weeks uncomfortable | Pigs below c.16 weeks uncomfortable | Pigs below 16 weeks uncomfortable |
| 7° C | Pigs below 20 weeks uncomfortable | Pigs below 14 weeks uncomfortable | Pigs below 20 week uncomfortable |
| 4° C | Pigs below 20 weeks uncomfortable | Pigs below 20 weeks uncomfortable | Pigs below 20 weeks uncomfortable |
| 2° C | All fattening pigs uncomfortable | | |

*Source:*  Sainsbury, D. & Sainsbury, P. (1988) *Livestock Health and Housing.*
Reproduced with kind permission of Bailliere Tindall.

---

**TABLE 402   Temperature Requirements for Housed Pigs in UK**

|  | °C |  | °C |
|---|---|---|---|
| Dry Sows | 16 - 20 | Porkers | 18 - 21 |
| Farrowing sows | 15 - 18 | Baconers | 15 - 18 |
| Creep area | 24 - 27 | Heavies | 10 - 15 |
| Weaners | 21 - 24 |  |  |

Ventilation: general requirement for growing pigs is 2 m$^3$ / hr / kg liveweight summer maximum and 20% of this in winter.

*Source: Collins Farmers Diary* (1992). © Collins 1990, 1991, A division of Harper Collins Publishers.

---

**TABLE 403   Optimal Environment for Cage Laying Flocks**

| | |
|---|---|
| Temperature | 21° C |
| Ventilation | max 25 m$^3$ / s / t feed / day |
| | min 2 m$^3$ / s / t feed / day |
| Relative humidity | 50 - 70 % |
| Light | 10 - 12 lux |

*Source: Collins Farmers Diary* (1992). © Collins 1990, 1991, A division of Harper Collins Publishers.

*NB* Temperatures above 26° C depress egg yield and production. At temperatures above 32° C feed intake, egg weight and shell thickness are depressed. Most serious thermal stress comes with high temperature and high humidity. Below 21° C there is a reduction of 1 egg / hen / year for each deg C. Below 5° C there is a steep rise in appetite. See Sainsbury, D. (1988) *Livestock health and Housing* and Oluyemi J.A. (1979) *Poultry Production in Warm Climates* .

---

**TABLE 404   Suggested Lighting Programme for Commercial Hybrid Layers and Broilers**

| a. Layers | 0 - 1 week | 18 hrs light, 6 hrs darkness |
|---|---|---|
| | 2 - 18 weeks | 6 hrs light, 18 hrs darkness |
| | 19 - 22 weeks | Increase light by 45 mins / week |
| | 23 - 49 weeks | Increase light by 20 mins / week |
| | 49 weeks onwards | Hold light steady at 18 hrs light / day |
| b. Broilers | 0 - 3 weeks | Lighting 23 hours out of 24 |
| | 3 - 5 weeks | 3 hours on, 1 hour off |
| | 5 - 7 weeks | 2 hours on, 2 hours off |
| | 7 weeks onwards | 1 hour on, 3 hours off |

(This broiler schedule is a relatively new development; previously most producers gave light for 23 out of 24 hours)

*Source:* Sainsbury, D. & Sainsbury, P. (1988) *Livestock Health and Housing.*
Reproduced with kind permission of Bailliere Tindall.

---

**TABLE 405   Hybrid 'Isabrown' Laying Hens Lighting Programme
(Rearing and Laying in Windowless Houses)**

---

All systems should follow these basic rules:

a) Never increase the lighting period during rearing;
b) Never reduce the lighting period after commencement of lay;
c) Use a light meter to check intensity at cage level.

Birds can be stimulated by lighting to mature early, but this will reduce average egg weight. This reduced egg weight can persist through to the end of the laying period.

|  |  | Daily Lighting Hours | Intensity W / m$^2$ | Lux |
|---|---|---|---|---|
| 1 - 2 | days | 22 | 3 - 4 | 20 - 40 |
| 3 - 4 | days | 20 | 3 | 20 - 30 |
| 5 - 6 | days | 18 | 3 | 20 - 30 |
| 1 - 2 | weeks | 16 | 2 | 10 - 20 |
| 2 - 3 | weeks | 14 | 2 | 10 - 20 |
| 3 - 4 | weeks | 13 | 2 | 10 - 20 |
| 4 - 5 | weeks | 12 | 1 | 5 - 10 |
| 5 - 6 | weeks | 11 | 1 | 5 - 10 |
| 6 - 7 | weeks | 10 | 1 | 5 - 10 |
| 7 - 8 | weeks | 9 | 1 | 5 - 10 |
| 8 - 16 | weeks | 8 | 1 | 5 - 10 |
| 16 - 17 | weeks | 8 | 2 | 10 - 15 |
| 17 - 18 | weeks | 8 | 2 | 10 - 15 |
| 18 - 19 | weeks | 9 | 2 | 10 - 15 |
| 19 - 20 | weeks | 10 | 2 | 10 - 15 |
| 20 - 21 | weeks | 11 | 2 | 10 - 15 |
| 21 - 22 | weeks | 12 | 2 | 10 - 15 |

Then: Increase by ½ hour each week until 15 hours is attained and then maintain at 15 hours to end of lay.

*NB*  An effective light control programme requires that daylight entering the house should be less than 0.4 lux.

The light intensity in the laying house should never be less than in the rearing house, otherwise sexual development will be delayed. A large increase in light intensity on housing in the laying quarters is not recommended as it may lead to vent and feather pecking.

*Source:* 1991 data supplied and reproduced with kind permission of ISA Poultry Services Ltd., Green Road, Eye, Peterborough, UK.

**TABLE 406    Brooding Temperatures for Hybrid 'Isabrown' Laying Hens**

Radiant Heaters: Ensure that there are adequate brooders. All chicks must be able to settle under the canopy.

### Brooder Temperatures

| | | |
|---|---|---|
| 1 day old | 35° C | under brooder (32.5 at edge) |
| 1 week old | 33° C | |
| 2 weeks old | 30° C | |
| 3 weeks old | 27° C | |
| 4 weeks old | 25° C | |

Whole house heaters: Ensure the temperature is even throughout the house, especially at chick level. Temperature should be 32° C at day old reducing by 2.5° C / week.

*NB* Temperatures given are only a guide; bird behaviour should be observed to check that optimum conditions are being provided.

**Post Brooding:** The ideal ambient temperature if 20° C.

*Source:* 1991 data supplied and reproduced with kind permission of ISA Poultry Services Ltd., Green Road, Eye, Peterborough, UK.

**TABLE 407 a    Outline Housing Equipment Requirements for Hybrid Hen Layers**

**a. Floor systems - rearing**

**Brooding Equipment**

For 1000 pullets the following are required:

| | |
|---|---|
| **Heating** | 2 gas or radiant heat brooders 10 000 BTUs each |
| **Feeding** | 20 trays |
| **Drinkers** | 10 small fountains or supplementary drinkers |
| **Light** | 1 bulb per brooder |

**Growing Equipment**

For 1000 pullets the following are required:

| | |
|---|---|
| **Area** | 100 m $^2$ (1000 ft $^2$) |
| **Feeding** | 50 m Chain (speed : 1 circuit of the house in 15 minutes) |
| | or 50 tube feeders or pans |
| **Drinkers** | 10 large circular drinkers or 10 m trough |
| **Ventilation** | 6000 cfm (maximum capacity of fans) |
| **Lighting** | 450 w (45 lux) windowed houses |
| | 100 w (lux) windowless houses |
| **Time clock** | Yes |
| **Dimmer** | Yes |

---

**TABLE 407 b & c   Outline Housing Equipment Requirements for Hybrid Hen Layers**

---

### b. Cage systems - rearing

The advantage of cages are: savings in feed used (provided a good ambient temperature is maintained); more effective parasite control; higher stocking density per sq. metre of building.

**Brooding Equipment**

| | |
|---|---|
| Feeding | On paper or trays in each cage for the first 5 days |
| Drinkers | 1 supplementary fount / cage (minimum) |
| Light | 4 w / m $^2$ (40 lux) |

**Growing Equipment**

| | |
|---|---|
| Feeding | 8 cm trough / bird |
| Drinkers | Minimum 1 nipple / 10 birds |
| Ventilation | 6 cfm / bird |
| Light | 1 w / m $^2$ (10 lux) |
| Time clock | Yes |

### c. Stocking density - layers

Following EEC Directive 86/113, regulations have been placed before the UK Parliament to cover stocking densities for all new cages:

| | |
|---|---|
| 4 birds / cage | 450 cm $^2$ / bird |
| 3 birds / cage | 550 cm $^2$ / bird |
| 2 birds / cage | 750 cm $^2$ / bird |
| 1 birds / cage | 1000 cm $^2$ / bird |

The results of over-stocking will be:

a) increased percentage mortality

b) reduced egg numbers per bird

c) reduced average egg weight

d) increased percentage broken eggs

*Source:* ISA Poultry Services (1991) *Management Guide for Isabrown Layers.* Data supplied and reproduced with kind permission of ISA Poultry Services Ltd., Green Road, Eye, Peterborough, UK.

**TABLE 408  Broiler Environment Checklist**

| ITEM | GUIDELINES | NOTES |
|---|---|---|
| **Stocking Density** | | |
| Controlled environment | 34.22 kg / m$^2$ (7 lb / ft$^2$) * | May need to reduce at higher temperatures |
| Open housing | 9 - 12 birds / m$^2$ (1.15 - 0.86 ft$^2$ / bird) | |
| **Ventilation** | | |
| Maximum | 20 m$^3$ / sec / tonne feed / day (19.2 cfm / lb feed / day), or 6 m$^3$ / hour / kg liveweight (1.5 cfm / lb liveweight) | May need to increase ventilation and combine with cooling at higher temperatures. Different building designs for temperate and hot climates |
| Minimum | 2 m$^3$ / sec / tonne feed / day (1.9 cfm / lb feed / day), or 1.5 m$^2$ / hour / kg liveweight (0.4 cfm / lb liveweight) | |
| **Inlet Area** | 0.5 m$^2$ of inlet per 100 m$^3$ extracted (5 ft$^2$ per 1000 ft$^3$) | |
| **Lighting** | Dimmer controlled, providing between 25 lux (2.5 ft candles) and 1 lux (0.1 ft candles) at bird level | Can have 23 hours / day light or intermittent lighting programme. Reduce from 25 to 1 lux during period 7 - 21 days of age |
| **Drinkers** | 8 x 400 mm auto drinkers / 1000 birds. When brooding add 6 extra drinker points / 1000 birds | Provide additional drinkers above 30° C |
| **Feeders** | Trough space 2.5 cm / bird (1 inch / bird), 18 - 20 pan or tube type feeders / 1000 birds | |
| **Litter** | Wile wood shavings to depth of 10 cm (4 in), equivalent to 550 kg / 100 m$^2$ (0.5 tonne / 1000 ft$^2$) | |

| **Ammonia in Air :** | Concentration (ppm) | Effect |
|---|---|---|
| | 10 to 20 | Can be detected by smell |
| | 20 to 25 | Detrimental effect on poultry begins and staff will complain |
| | 25 | UK Factory Inspectorate limit for standard 8-hour working day |
| | 30 to 35 | Increased risk of respiratory disease in the stock |
| | 35 to 40 | Effect on appetite starts |
| | Over 50 | Birds' eyes become watery and inflamed Depression in growth rate. High risk of respiratory disease |

| **Brooder Temperature Requirement** | Guide: 32.5° C at day old decreasing to 21° C at 35 days and 18° C at 70 days |
|---|---|

* UK Animal Welfare Code Standard

*Source: Broiler Management Guide* (1991). Data reproduced with kind permission of the Cobb Breeding Co., East Hanningfield, Chelmsford, Essex, UK.

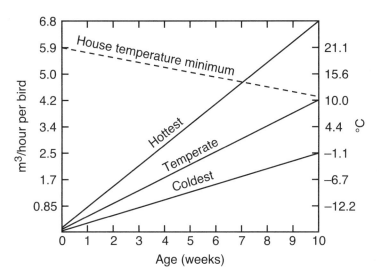

**Fig. 59** Suggested ventilation scheme for broilers

*Source:* Sainsbury, D. & Sainsbury, P. (1988) *Livestock Health and Housing.*
Reproduced with kind permission of Bailliere Tindall.

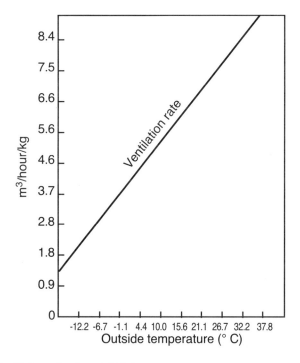

**Fig. 60** Ventilation rates for adult birds

*Source:* Sainsbury, D. & Sainsbury, P. (1988) *Livestock Health and Housing.*
Reproduced with kind permission of Bailliere Tindall.

**TABLE 409  Environmental Conditions Affecting the Performance of Layers**

| Condition | Age in weeks from point of lay of 25 to 50 hen-day production | | | | | | | |
|---|---|---|---|---|---|---|---|---|
| | 1 | 2 | 3 | 4 | 5 | 6 | 7 | 11 |
| **Light period** | | | | | | | | |
| (1)  Tropics | 12 | 12.5 | 13 | 13.5 | 14 | 14.5 | 15 | 17 |
| (2)  Temperate regions | 6.0 | 6.5 | 7.0 | 7.5 | 8.0 | 8.5 | | 14 or 17 |

| **Light intensity** | 0.5 lux (5 foot candles)<br>not less than 0.05 lux (0.5 foot candle) |
|---|---|

| | Rate ($m^3$ / h per bird) at | | | | | |
|---|---|---|---|---|---|---|
| | 1 wk | 3 wk | 6 wk | 12 wk | 20 wk | Beyond 20 wk |
| **Ventilation** | 1.8 – 2 | 2.5 – 3 | 3.5 – 4 | 5 – 6 | 7 – 8 | 10 – 14 |

| **Atmospheric oxygen** | At least 11 % |
|---|---|
| **Atmospheric CO$_2$** | Not more than 5 % of its air |
| **Ammonia** | 20 – 40 ppm |
| **Temperature** | 12.8 – 24° C |
| **Relative Humidity** | 50 – 80 % |

Notes:
1.  Temperatures in the tropics are higher than optimum.
2.  For adequate light intensity, light bulbs of 40, 60, 75 and 100 watts should be placed at heights of 1.5, 1.8 - 2.1, 2.4 and 2.7 m respectively above the floor. The distances between light bulbs should be 2.4, 2.7 - 3.2, 3.6 and 4.1 m respectively.

*Source:* Oluyemi, J.A. & Roberts, F. (1979) *Poultry Production in Wet Climates.*
Reproduced with kind permission of Macmillan Press.

**TABLE 410  The Influence of Ambient Temperature on Layer Performance**

| Mean temp * | Relative production | Relative egg size | Relative food required per egg | Relative food required per unit of egg mass |
|---|---|---|---|---|
| °C | % | % | % | % |
| 16 | 100 | 100 | 100 | 100 |
| 18 | 100 | 100 | 95 | 95 |
| 21 | 100 | 100 | 91 | 91 |
| 24 | 100 | 99 | 88 | 89 |
| 27 | 99 - 100 | 96 | 86 | 89 |
| 29 | 97 - 100 | 93 | 85 | 91 |
| 32 | 94 - 100 | 86 | 84 | 98 |

* The daily temperature range was 4.7 - 7.1° C

*Source:* Clark, J.A., ed. (1981) *Environmental Aspects of Housing for Animal Production* (after Zimmerman and Snetsinger, 1975). Reproduced with permission.

**TABLE 411 a - c  Turkey Environment Checklist**

### a. General

| | |
|---|---|
| **Feeder Space** | 40 birds / 1 tube feeder or 120 cm linear trough |
| **Drinker Space** | 100 birds / 1 bell type drinker or 100 cm linear trough |
| **Ventilation** | The minimum ventilation rate is normally defined as the smallest rate of air flow capable of removing dust, carbon dioxide, ammonia and other pollutants, as well as providing adequate oxygen requirements. This is equivalent to: $3 \text{ m}^3$ / second / tonne of feed consumed / day. The maximum ventilation rate should be capable of moving $25 \text{ m}^3$ / second / tonne of feed consumed / day |
| **Water Consumption** | Water consumption is directly correlated to several environmental factors and is often doubled in temperatures above 35 °C. Feed and health also significantly affect consumption, which monitored on a daily basis, can provide an early warning of impending problems. Turkeys normally consume 2.2 times the amount of water (litres) as they do feed (kg). |

### b. Commercial Stock

| | |
|---|---|
| **Lighting Programme** | Commercial turkeys can be grown successfully under a number of different lighting programmes. The following is most commonly adopted: |
| Day old - 36 hours | Continuous light, 100 Lux with 1 hour conditioning darkness after 24 hours |
| 36 hours - Kill | 14 hours light with reduced intensity to control pecking. It is advantageous to maintain a very low light level during the dark period to prevent panic problems |

| **Stocking Density** | **Males** | **Females** |
|---|---|---|
| Day old - 8 weeks | 8 | 10 poults / $m^2$ |
| 8 weeks - kill | A maximum liveweight of 38 kg / $m^2$ depending on environment | |

### c. Parent Stock
#### 1. Lighting Programme

| Age | PARENT MALE: Hours / Day | Minimum (Lux) |
|---|---|---|
| Day old - 36 hours | Continuous with 1 hour conditioning darkness after 24 hrs | 100 |
| 36 hours - 14 weeks | 14 | 50 |
| 14 weeks - 25 weeks | 10 or 14 | 25 |
| 25 weeks - end of production | 14 | 25 |

| Age | PARENT FEMALE: Hours / Day | Minimum (Lux) |
|---|---|---|
| Day old - 36 hours | Continuous with 1 hour conditioning darkness after 24 hrs | 100 |
| 36 hours - 18 weeks | 14 | 75 |
| 18 weeks - 29.5 weeks | 7 | 75 |
| 29.5 weeks - end of lay | 14 | 100 |

Never decrease the light period or intensity during production and always ensure that the intensity during lay is greater than during the pre-lay period.

*(continued over)*

**TABLE 411 *(continued)*   Turkey Environment Checklist**

### 2. Stocking Density (birds m $^2$)

|  | Parent Females | Parent Males |
|---|---|---|
| Day old - 8 weeks | 7.0 | 5.0 |
| 8 weeks - 14 weeks | 3.5 | 2.0 |
| 14 weeks - 29 weeks | 3.0 | 1.5 |
| 29 weeks - | 1.5 | 1.0 BIG 6 and |
| end of production | 2.0 | 1.0 BUT 8 |

Provision of adequate space allows parent stock to develop properly, promoting uniformity and optimum production.

*Source:* 1991 data supplied and reproduced with kind permission of British United Turkeys Ltd., Warren Hall, Broughton, Chester, UK (Data is that recommended for their 'Big 6' strain of turkey.)

**TABLE 412   Range of Climatic Environmental Requirements of Stabled Horses**

| | |
|---|---|
| Ambient temperature | 0 - 30° C |
| Relative humidity | 30 - 70 % |
| Air movement | 0.15 - 0.5 m/s |
| Ventilation rate | 0.2 - 2.0 m $^3$ / h per kg bodyweight |
| Outlet ventilation area | 0.1 m $^2$ per horse |
| Inlet ventilation area | 0.3 m $^2$ per horse |

*Source:* Frape, D. (1986) *Equine Nutrition and Feeding* (after Sainsbury, 1981).
Reproduced with kind permission of Longman Group UK Ltd.

**TABLE 413   Ventilation Standards Used in France for Rabbitries Enclosed in Buildings**

| Temperature | Humidity | Air speed | Air flow m $^3$ / h / kg liveweight |
|---|---|---|---|
| °C | % | m / sec | |
| 12 - 15 | 60 - 65 | 0.10 - 0.15 | 1 - 1.5 |
| 16 - 18 | 70 - 75 | 0.15 - 0.20 | 2 - 2.5 |
| 19 - 22 | 75 - 80 | 0.20 - 0.30 | 3 - 3.5 |
| 22 - 25 | 80 | 0.30 - 0.40 | 3.5 - 4 |

*NB* Rabbits are sensitive to low humidity, and high humidity at high temperature.

*Source:* FAO, *The Rabbit. Husbandry, Health and Production.* Animal Production and Health Series No. 21 (after Morrisse 1981). Reproduced with kind permission of the Food and Agriculture Oganization of the United Nations.

# Farm Buildings

# Crop Storage

**TABLE 414   Typical Crop and Feedstuff Bulk Densities**

| Crop | Density M$^3$/t | Crop | Density M$^3$/t |
|---|---|---|---|
| Alfalfa meal | 3.3 | Maize, flaked | 4.8 |
| Apples | 1.2 - 1.9 | Maize silage | 1.3 |
| Barley, grain | 1.45 | Maize cobs, dehusked | 2.1 |
| Barley, milled | 2 | Malt culms | 4.5 |
| Barley straw, conventional bales | 11.5 | Mangolds | 1.8 |
| Barley straw, large round bales | 18 | Meat meal | 1.6 |
| Beans, whole | 1.2 | Meat and bone meal | 1.6 |
| Beans, milled | 2.8 | Millet grain | 1.5 |
| Beet pulp dried | 4.8 | Molasses | 0.75 |
| Beetroot | 1.4 - 1.7 | Oats, grain | 1.95 |
| Blood meal | 2.3 | Oats, milled | 2.4 |
| Bone meal | 1.2 | Onions | 1.8 - 2.8 |
| Brewers grains | 0.9 | Onions - Salad | 1.8 - 2.8 |
| Brewers grains, dried | 4 | Palm kernel oilcake | 2.3 |
| Cabbage, hard white | 2.55 | Parsnips | 1.2 - 2 |
| Carrots | 1.7 - 1.8 | Peas, whole | 1.3 |
| Cauliflower | 2.3 - 4.5 | Peas, milled | 1.7 |
| Cherries | 2.5 | Pears | 1.7 |
| Coconut oilcake | 2.2 | Potatoes | 1.4 - 1.7 |
| Coffee, green | 1.8 | Rice grain | 1.4 |
| Coffee, washed | 3.3 | Rice bran | 3.3 |
| Coffee, unwashed | 4.7 | Rye | 1.4 |
| Compost from produce | 1.8 - 6.3 | Ryegrass seed | 3.4 - 4 |
| Concentrate feeds, meal | 2 | Rapeseed | 1.5 - 1.8 |
| Concentrate feeds, cubes | 1.6 | Sesame oilcake | 2.2 |
| Cotton, lightly pressed | 7.7 - 12.5 | Silage, grass, clamp, 18% DM | 1.3 |
| Cotton in woolpacks | 4.7 | Silage, grass, clamp, 22% DM | 1.4 |
| Dried grass, 10% MC, 7.5 mm chop | 8.5 | Silage, grass, clamp, 30% DM | 1.6 |
| Fish meal | 1.8 | Silage, grass, tower, 40% DM | 1.7 |
| Fodder beet | 1.4 | Sorghum | 1.3 |
| Gooseberries | 2.1 | Sorghum, ground | 1.8 |
| Grain, rolled | 1.9 - 2.8 | Soybean oilcake | 1.9 |
| Gram seeds | 1.4 | Sprouts - Brussels | 2.3 |
| Groundnuts, shelled | 1.6 | Strawberries | 2.2 |
| Groundnut oilcake | 1.7 | Swedes | 1.8 |
| Hay, temperate, conventional bales | 6 | Timothy seed | 1.8 |
| Hay, temperate, large round bales | 8 | Tobacco, baled for sale | 2.8 |
| Hay, temperate, long, in stack | 9 | Tobacco, loose bales | 4.4 - 5.3 |
| Hay, temperate, chopped | 12 | Tobacco, loose in bulk | 6.3 |
| Hay, temperate, barn dried baled | 7.8 | Tomatoes | 2.4 |
| Hay, tropical, in stack | 7.7 | Turnips | 1.8 |
| Hemp oilcake | 2.3 | Wheat, grain | 1.35 |
| Herbage seed | 4.25 | Wheat, milled | 1.9 |
| Hominy | 4.8 | Wheat bran | 4.2 |
| Lentil seeds | 1.3 | Wheat shorts | 2.4 |
| Lettuce | 4-5.7 | Wheat straw, conventional bales | 13 |
| Linseed | 1.4 | Wheat straw, large round bales | 20 |
| Maize, grain | 1.3 - 1.4 | Whey, dried | 1.7 |
| Maize, milled | 1.7 | | |

*Sources;* Nix, J. (1990) *Farm Management Pocketbook.* Wye College. University of London. Scottish Agricultural Colleges (1986-87) *Farm Management Handbook.* MAFF (1978) *Farm Buildings Pocketbook in Metric.* FAO (1981) *Tropical Feeds.* Animal Production and Health Series.

**TABLE 415   Drying Capacities of Ambient and Warmed Air**

| Condition of ambient air | | | Drying capacity in kg per 24 h of 1.0 m³ / s | | |
|---|---|---|---|---|---|
| Temperature °C | | RH % | | | |
| Dry bulb | Wet bulb | | Ambient air | + 5.5 °C | + 11 °C |
| 7.2 | 3.9 | 60 | 118.8 | 237.0 | 361.7 |
| 7.2 | 5.0 | 70 | 77.9 | 203.1 | 332.4 |
| 7.2 | 5.6 | 80 | 47.9 | 167.2 | 298.7 |
| 7.2 | 6.7 | 90 | 12.9 | 139.2 | 274.6 |
| 10 | 6.7 | 60 | 135.6 | 255.4 | 386.4 |
| 10 | 7.2 | 70 | 88.7 | 218.8 | 352.6 |
| 10 | 8.3 | 80 | 52.3 | 184.8 | 320.2 |
| 10 | 9.4 | 90 | 15.2 | 146.8 | 286.5 |
| 15.6 | 11.1 | 60 | 158.2 | 295.3 | 435.3 |
| 15.6 | 12.2 | 70 | 108.9 | 249.9 | 398.3 |
| 15.6 | 13.3 | 80 | 62.4 | 205.3 | 358.7 |
| 15.6 | 14.4 | 90 | 17.7 | 165.7 | 316.5 |
| 21.1 | 16.1 | 60 | 182.7 | 327.9 | 486.4 |
| 21.1 | 17.8 | 70 | 125.8 | 275.4 | 438.7 |
| 21.1 | 18.9 | 80 | 72.0 | 229.0 | 382.7 |
| 21.1 | 20.0 | 90 | 21.5 | 178.3 | 336.9 |

*Source:* Electricity Association (1984) *Grain Drying and Storage.* Farm Electric Handbook. Reproduced with kind permission of Electricity Association Technology Ltd.

**Fig. 61** Air and grain, moisture and relative humidity relationship

*NB* Raising the temperature, or dehumidifying the drying air, will dehumidify the air and increase its drying potential.

*Source:* Electricity Association (1984) *Grain Drying and Storage.* Farm Electric Handbook. Reproduced with kind permission of Electricity Association Technology Ltd.

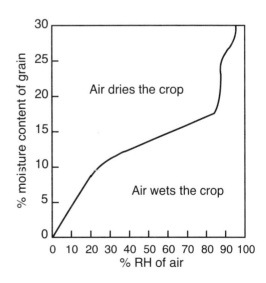

**TABLE 416   Hay and Grain Air Humidity and Equilibrium Moisture Storage Levels**

| RH of air % | Hay Moisture Content % | RH of air % | Grain Moisture Content % |
|---|---|---|---|
| 95 | 35 | 88 | 20 |
| 90 | 30 | 87 | 19 |
| 85 | 25 | 86 | 18 |
| 80 | 21.5 | 82 | 17 |
| 77 | 20 | 77 | 16 |
| 70 | 16 | 72 | 15 |
| 60 | 12.5 | 65 | 14 |
| 50 | 10 | 50 | 12 |
| | | 36 | 10 |

*Main sources:* Electricity Association (1985) *Grain Drying and Storage* and *Hay Drying.* Farm Electric Handbooks. Data reproduced with kind permission of Electricity Association Technology Ltd. Also: *Collins Farmers Diary* (1992).

**TABLE 417   Safe Drying Temperatures**

| Crop | Max Temp °C |
|---|---|
| Barley, wheat and oats for stock feed | 82 |
| Cowpea - seed | 38 |
| Grain for stock feed | 82 - 104 |
| Groundnut - seed | 37 |
| Malting Barley and seed (up to 24 % moisture) | 49 |
| Malting Barley and Seed (over to 24 % moisture) | 43 |
| Maize - seed | 44 |
| Maize for starch | 55 |
| Maize for feed | 82 |
| Millet - seed | 44 |
| Millet - feed | 65 |
| Oily seeds | 46 |
| Rice up to 20% moisture | 44 |
| Rice over 20 % moisture | 40 |
| Sorghum - seed | 44 |
| Sorghum for starch | 60 |
| Sorghum for feed | 82 |
| Soyabean seed | 38 |
| Soyabean for manufacturing | 48 |
| Wheat - seed up to 24% moisture | 49 |
| Wheat - seed over 24% moisture | 44 |
| Wheat for milling | 45-66 |

*Sources:* Electricity Association (1985) *Grain Drying and Storage.* Farm Electric Handbook. ILACO (1985) *Agricultural Compendium.* Reproduced with kind permission of Electricity Association, Technology Ltd. and ILACO BV.

**TABLE 418    Maximum Moisture Content for Safe Storage at 27° C**

| Product | Moisture content (%) | Product | Moisture content (%) |
|---|---|---|---|
| Beans | 15.0 | Maize flour | 11.5 |
| Bulgur wheat | 13.5 | Milled rice | 12.0 |
| Cassava chips | 12 - 13.0 | Millet | 15.0 |
| Cassava flour, viz: | | Paddy | 14.0 |
|    Gari, white | 21.1 | Palm kernels | 5.0 |
|    yellow | 13.0 | Peas | 14.0 |
| Cocoa beans | 7.0 | Sorghum | 13.5 |
| Copra | 7.0 | Soyabean | 11.0 |
| Cottonseed | 10.0 | Sunflower | 9.5 |
| Cowpeas | 15.0 | Safflower | 9.5 |
| Groundnuts: shelled | 7.0 | Tapioca | 12.0 |
|    unshelled | 9.0 | Wheat | 13.5 |
| Maize (yellow) | 13.0 | Wheat flour | 12.0 |
| Maize (white) | 13.5 | | |

*Source:* Muckle, T.B. & Sterling, H.G. (1971) *Review of the drying of cereals in the tropics.* Tropical Stored Product Information 22, pp. 11-30.

**TABLE 419    Indicative Storage Temperatures, Relative Humidities and Expected Storage Life of Selected Crops**

*NB* Expected storage life varies with variety and husbandry

| Crop | °C | RH % | Storage Days |
|---|---|---|---|
| Almond in shell | 0 to 7 | 60 - 75 | 300 - 365 |
| Apples | - 2 to 5 | 90 | 60 - 285 |
| Apricot | - 1 to 1 | 85 - 90 | 7 - 14 |
| Artichoke, Globe | 0 | 95 | 14 - 28 |
| Artichoke, Jerusalem | - 0.5 to 0 | 90 - 95 | 14 - 35 |
| Asparagus | 0 to 0.5 | 85 - 95 | 14 - 28 |
| Avocado | 5 to 13 | 85 - 90 | 14 - 28 |
| Banana, green | 11.5 to 14.5 | 90 - 95 | 10 - 20 |
| Banana, coloured | 13 to 16 | 85 - 90 | 5 - 10 |
| Beans, French and Runner | 5 to 7 | 95 | 8 |
| Bean, Lima | 0 to 4.4 | 90 | 7 - 14 |
| Beans, Broad | 0 to 1 | 85 - 95 | 14 - 21 |
| Beetroot, bunched | 0 | 90 - 95 | 10 - 14 |
| Beetroot, topped | 0 to 1 | 90 - 96 | 30 - 180 |
| Blackberry | - 1 to 1 | 90 | 5 - 7 |
| Blackcurrant | - 1 to 1 | 90 | 7 - 14 |
| Broccoli | 0 | 90 - 95 | 7 - 21 |
| Brussels sprouts | - 1 to 1 | 90 - 95 | 14 - 42 |
| Cabbage, Chinese | 0 | 90 - 95 | 30 - 60 |

*(continued over)*                                    A G R I   I N F O

**TABLE 419 *(continued)*   Indicative Storage Temperatures, Relative Humidities and Expected Storage Life of Selected Crops**

*NB* Expected storage life varies with variety and husbandry

| Crop | °C | RH % | Storage Days |
|---|---|---|---|
| Cabbage, spring | 0 | 95 | 28 |
| Cabbage, white winter | 0 | 95 | 180 |
| Carrots, bunched | 0 to 1 | 90 | 14 |
| Carrots, topped | - 1 to 1 | 90 - 95 | 120 - 180 |
| Cashew apples | 0 to 1.5 | 85 - 90 | 21 - 49 |
| Cauliflower | 0 to 1.5 | 85 - 90 | 21 - 49 |
| Celeriac | 0 to 1 | 85 - 95 | 60 - 150 |
| Celery | 0 to 1 | 95+ | 14 - 75 |
| Cherry | - 1 to 1 | 85 - 90 | 7 - 28 |
| Chestnut | 0 | 70 | 240 - 365 |
| Chicory | 0 to 1 | 85 - 95 | 14 - 21 |
| Citrus      - Grapefruit | 0 to 15.5 | 85 - 90 | 21 - 90 |
|             - Lemon, green | 11 to 14.5 | 85 - 90 | 30 - 120 |
|             - Lemon, coloured | 0 to 10 | 85 - 90 | 21 - 42 |
|             - Lime | 8 to 10 | 95 - 90 | 21 - 56 |
|             - Mandarin | 4 to 7 | 85 - 90 | 21 - 84 |
|             - Orange | - 1 to 7 | 85 - 90 | 30 - 180 |
| Coconut | 0 | 80 - 85 | 30 - 60 |
| Collards | 0 | 90 - 95 | 10 - 14 |
| Cucumber | 7 to 12 | 85 - 95 | 7 - 14 |
| Dates, cured | - 2 to 0 | 70 | 120 - 140 |
| Eggplant | 7 to 10 | 85 - 90 | 10 |
| Endive | 0 to 1 | 90 - 95 | 14 - 21 |
| Fig, fresh | - 1 to 0 | 90 | 7 - 14 |
| Garlic, dry | - 1.5 to 0 | 70 - 75 | 180 - 240 |
| Ginger | 12.8 | 65 | 180 |
| Ginger (alt. recommendation) | 1.5 to 3.5 | 85 - 90 | 105 |
| Gooseberry, Indian | 0 to 1.5 | 85 - 90 | 56 |
| Grape | - 1 to 0 | 85 - 90 | 21 - 140 |
| Guava | 7 to 10 | 85 - 90 | 21 - 28 |
| Horseradish | - 1 to 0 | 90 - 95 | 300 - 365 |
| Kale | 0 | 90 - 95 | 10 - 14 |
| Kohlrabi | 0 | 90 - 95 | 14 - 28 |
| Leek | 0 | 95 | 45 |
| Lettuce | 0 to 1 | 95 | 7 - 42 |
| Lima bean | 0 to 2 | 85 - 90 | 7 |
| Litchi | 0 to 1.5 | 85 - 90 | 35 - 77 |
| Loganberry | 0 | 90 | 7 |
| Mango | 7 to 10 | 85 - 90 | 28 - 50 |
| Mangosteen | 4 to 5.5 | 85 - 90 | 49 |
| Melon    - Casaba | 7.2 to 10 | 85 - 90 | 28 - 42 |
|          - Crenshaw | 7.2 to 10 | 85 - 90 | 14 |
|          - Honeydew | 7.2 to 10 | 85 - 90 | 21 - 28 |
|          - Musk, 3/4 slip | 2.2 to 4.4 | 85 - 90 | 15 |
|          - Musk, full slip | 0 to 1.7 | 85 - 90 | 5 - 14 |
|          - Persian | 7.2 to 10 | 85 - 90 | 14 |
|          - Watermelon | 7.2 to 10 | 80 - 85 | 14 - 21 |

**TABLE 419 *(continued)*   Indicative Storage Temperatures, Relative Humidities and Expected Storage Life of Selected Crops**

*NB* Expected storage life varies with variety and husbandry

| Crop | °C | RH % | Storage Days |
|---|---|---|---|
| Mushroom | 0 to 1 | 85 - 90 | 3 - 7 |
| Nectarine | - 1 to 0 | 85 - 90 | 21 - 49 |
| Nuts: Brazil and pecan | 0 | 70 | 240 - 365 |
| Nuts, other | 7 | 70 | 365 |
| Okra | 7.2 to 10 | 90 - 95 | 7 - 10 |
| Onion, bulb | 0 | 70 | 30 - 250 |
| Onion, spring | 0 | 95 | 3 |
| Parsley | 0 to 1 | 85 - 90 | 30 - 60 |
| Parsnip | 0 | 95 | 60 - 180 |
| Passion Fruit | 5.5 to 7 | 80 - 85 | 28 - 35 |
| Pawpaw | 4 to 10 | 85 - 90 | 14 - 35 |
| Peach | - 1 to 1 | 85 - 90 | 7 - 56 |
| Pear | - 1.5 to 4 | 90 | 30 - 120 |
| Peas in pod | 0 | 95 | 15 |
| Pepper, green | 7 to 10 | 90 - 95 | 14 - 21 |
| Peppers, ripe | 4.4 to 7.2 | 90 - 95 | 7 |
| Persimmon, Chinese | - 0.5 to 0 | 85 - 90 | 21 |
| Pimento | 0 | 95 - 90 | 28 - 35 |
| Pineapple, green | 10 | 90 | 14 - 28 |
| Pineapple, ripe | 4.5 to 10 | 85 - 90 | 14 - 42 |
| Plum | 0.5 to 1 | 85 - 90 | 14 - 56 |
| Potato, early | 10 to 15 | 90 | 14 - 21 |
| Potato, late | 4 to 5 | 90 | 120 - 270 |
| Pomegranate | 1 to 2 | 90 | 60 - 120 |
| Pumpkin | 10 to 13 | 70 - 75 | 60 - 180 |
| Quince | 0 to 4 | 90 | 60 - 90 |
| Radish, bunched | 0 | 95 | 10 |
| Radish, leafless | 0 to 1 | 85 - 95 | 20 |
| Radish, winter | 0 | 90 - 95 | 60 - 120 |
| Raspberry | 0 | 85 - 90 | 3 - 5 |
| Red Currant | 0 | 90 | 14 - 21 |
| Rhubarb | 0 | 95 | 14 - 28 |
| Rutabaga | 0 | 90 - 95 | 60 - 120 |
| Salsify | 0 to 1 | 90 - 95 | 60 - 120 |
| Spinach | - 0.5 to 0 | 90 - 95 | 7 - 14 |
| Strawberry | 0 to 4.5 | 85 to 90 | 1 - 5 |
| Summer Squash | 0 to 4.5 | 85 - 95 | 60 - 180 |
| Sweet Corn | 0 | 90 - 95 | 4 - 8 |
| Sweet Potato | 11 to 15 | 80 - 90 | 90 - 180 |
| Tomato, ripe | 0 to 7 | 85 - 00 | 7 - 28 |
| Tomato, green | 1.5 to 13 | 85 - 90 | 7 - 42 |
| Turnip | 0 | 95 | 150 |
| Watercress | 1 | 95 | 4 |
| Winter Squash | 10 to 13 | 70 - 75 | 120 - 180 |

*Sources:* Adapted from ILACO (1985) *Agricultural Compendium.* Electricity Association, (1990 and 1981) *Vegetable Storage.* Farm Electric Handbooks. Samson, J.D. (1986) *Tropical Fruits.* Lorenz, O.A. & Maynard, D.N. (1980) *Knott's Handbook for Vegetable Growers.*

---

### TABLE 420  Storage of Plant Parts for Vegetable Propagation

| Plant Part | Temp °F | RH % |
|---|---|---|
| Asparagus crown | 35-40 | 80 - 85 |
| Garlic bulbs | 50 | 50 - 65 |
| Horseradish roots | 32 | 85 - 90 |
| Onion sets | 32 | 70 - 75 |
| Potato tubers | 36 - 40 | 90 |
| Sweet Potato | 55 - 60 | 85 - 90 |
| Rhubarb crowns | 32 - 35 | 80 - 85 |
| Chicory roots | 32 | 90 - 95 |

*Source:* Lorenz, O.A. & Maynard, D.N. (1987) *Knott's Handbook for Vegetable Growers.*
© 1980 and reprinted with kind permission of John Wiley and Sons Inc.

---

### TABLE 421  Target Dry Matter Values for Gross Silage

| System | % DM |
|---|---|
| Clamp | 18 - 28 |
| Big Bale | 30 - 40 |
| Tower | 35 - 45 |

---

### TABLE 422  'Humid-Cold'* Storage Period Data for Some Vegetables, Fruits and Flowers

| Specie | Storage period in good conditions (days) | Specie | Storage period in good conditions (days) |
|---|---|---|---|
| **VEGETABLES** | | **FRUITS** | |
| Cauliflower | 21 - 35 | Strawberries | 6 - 9 |
| Carrots | 365 | Raspberries | 6 - 9 |
| Potatoes | 365 | Blackberries | 8 - 10 |
| Celery | 42 - 49 | Gooseberries | 21 - 28 |
| Calabrese | 21 - 42 | Damson | 12 - 14 |
| Spinach | 7 | Grapefruit | 180 - 210 |
| Salad Onions | 10 - 14 | Lemon | 180 - 210 |
| Asparagus | 21 - 28 | **FLOWERS** | |
| Green Beans | 10 - 14 | Tulip | 14 - 28 |
| Lettuce (all types) | 14 - 28 | Rose | 7 - 14 |
| Red / Green Peppers | 14 - 21 | Orchid | 14 - 21 |
| Cucumber | 14 - 28 | Lily | 14 - 28 |
| Salad Cress | 35 - 42 | Hyacinth | 14 - 28 |
| Dutch W. Cabbage | 42 - 49 | Gerbera | 14 - 28 |
| Spring Cabbage | 42 - 56 | Freesia | 14 - 28 |
| Radish | 21 - 28 | Narcissus | 21 - 28 |
| Rhubarb | 21 - 28 | Chrysanthemum | 10 - 21 |
| Sprouts | 21 - 35 | Carnation | 28 - 56 |

*NB* Most flower species are adapted to Humid-Cold Storage.
\* High humidity, low temperature air ventilated through the produce.

*Source:* 1991 data based on experience of installations around the world installed by Grove Coldstore Construction Group, Grove Refrigeration Ltd., Unit 23, Mereview Industrial Estate, Yaxley, Peterborough, UK. Reproduced with kind permission of Grove Coldstore Construction Group.

# Machinery and Materials Storage

**TABLE 423   Typical Agricultural Material Bulk Densities**

| Material | Density $m^3$ / t |
| --- | --- |
| Anthracite | 0.6 |
| Aviation gasoline 100 | 1.39 |
| Bricks | 0.5 - 0.6 |
| Cement | 0.7 |
| Clay | 0.5 |
| Coal | 0.8 |
| Coke | 1 |
| Concrete | 0.4 |
| Concrete work | 0.5 - 0.6 |
| Cordwood (gums) | 2.5 incl. bark at 12% MC |
| Cordwood (indigenous) | 2 incl. bark at 12% MC |
| Diesel | 1.2 |
| Earth | 0.6 |
| Farm Yard Manure | 1.1 |
| Fertilizer | 2.5 In 50 kg bags |
| | 1 Bulk - big bags |
| | 0.9 Loose |
| Fresh Poultry Droppings | 0.9 |
| Gear oil | 1.1 |
| Glass | 0.4 |
| Gravel | 0.6 Coarse |
| Ground Limestone | 0.9 |
| Illuminating paraffin | 1.26 |
| Molasses | 0.6 |
| Motor oil | 1.1 |
| Petrol, super | 1.37 |
| Petrol, regular | 1.4 |
| Power paraffin | 1.23 |
| Sand | 0.6 |
| Sawdust | 4.2 |
| Stone, crushed | 0.0 |
| Stone, granite block | 0.4 |
| Stone, granite chips | 0.7 |
| Timber, softwood | 1.8 - 2.5 Sawn, dry |
| Timber, hardwood | 1.25 - 1.8 Sawn, dry |
| Water | 1 |

**TABLE 424   Indicative Machinery Dimensions for Storage**

| Machine | Height (m) | Width (m) | Length (m) |
|---|---|---|---|
| 200 HP 4WD Tractor (Non-artic) | 3.4 | 3.7 (dual wheels) | 5 |
| 100 HP Tractor | 2.7 | 2.2 | 4.3 |
| 60-70 HP Tractor | 2.4 - 2.7 | 1.82 | 3.5 |
| Tractor + loader, 80 HP, with counterweight, cab | 2.44 | 2.24 | 4.65 |
| Beet harvester, single row trailed | 2.4 | 2.43 | 5.18 |
| Potato harvester, 2 row | 2.4 | 3.65 | 7.31 |
| Pneumatic seed drill. 5t cap | 3.05 | 3.3 | 7.1 |
| Pneumatic fertilizer spreader 6t  cap | 3.12 | 2.6 | 5.45 |
| Grain and fertilizer end wheel drill 4m width | 1.45 | 5.15 | 1.7 |
| Planter, 6 row | 1.38 | 1.5 | 1.8 |
| Crop sprayer. Trailed. 4 100 litre | 4.3 | 3 | 8.75 |
| Combine harvester. 180 HP tractor unit | 4 | 3.74 | 7.74 |
| (on transporter) Header unit | 3 | 2.5 | 5.45 |
| Tractor + header | 4 | 4.6 | 7.74 |
| Combine harvester, 310 HP,with header attached | 4.05 | 7.3 | 8.72 |
| Baler, big square | 3.6 | 3 | 7.2 |
| Baler, small rectangular | 1.8 | 2.96 | 5.2 |
| Baler, round | 2.6 | 2.42 | 4.35 |
| Forage harvester, trailed up to 60t/hr capacity | 3.6 | 2.72 | 5.15 |
| Self-propelled forage harvester | | | |
| 340 HP. Pickup attached | 3.64 | 2.8 | 5.39 |
| Hay tedder | | 3.2 | 3.81 |
| Trailers. 6t tipping | 3.0 | 2.13 | 5 |
| 10t tipping | 3.5 | 2.43 | 5.48 |
| Liquid manure tanker, 3000 l | 2.5 | 1.47 | 4.34 |
| Plough, 5 furrow, mounted | | 2.74 | 3.04 |
| Plough, 3 furrow, mounted, reversible | | 1.44 | 2.43 |
| Roller, single | | 1.52 | 1.82 - 3.04 |
| Tool bar/cultivators, mounted | | 1.2 - 2.4 | 2 - 5.7 |
| Land rover LWB | 2.0 | 1.7 | 4.5 |
| Truck, LWB | 2.75 - 3.25 | 2.45 | 9.2 |
| Articulated truck | | 2.45 | 10.7 - 12.2 |

*Sources:* Assorted commercial literature. MAFF (1978) *Farm Buildings.* Pocketbook in metric.
Massey Ferguson.

# Livestock Housing

**TABLE 425   Floor Area Housing Requirements for Livestock (UK)**

| Livestock type | | Floor area m $^2$ / animal |
|---|---|---|
| **Dairy cows** | Kennels or cubicle house excluding feeding and slurry areas, but including a proportion of the passage area | 3.3 - 3.8 |
| | Kennels or cubicle house, including a proportion of feeding areas and passages | 6.3 - 6.8 |
| | Collecting yard | 1.0 - 1.3 |
| **Suckler cows** | Bedded house with calf creeps, including a proportion of feeding passages | 7.5 - 8.0 |
| | + for calf | 1.1 - 1.9 |
| **Other cattle** | Calves up to 6 months old | 1.2 - 2.2 |
| | 1 - 2 year old / finishing: | |
| | Bedded house including a proportion of feeding passages | 3.2 - 4.8 |
| | Slatted floor house including a proportion of feeding passages | 2.2 - 3.3 |
| **Deer** | Red deer: overwintering calves | 2.0 - 2.2 |
| | yearlings | 2.5 - 2.7 |
| | adult hinds & de-antlered yearling stags | 3.0 - 3.2 |
| **Goats** | | |
| | Adults: individually penned | 2.0 - 2.3 |
| | communal with outside access | 1.5 |
| **Pigs** | Sows in yards | 3.0 - 3.5 |
| | Sows in stalls, excluding passage areas | 1.2 - 1.3 |
| | Pigs 6 - 12 weeks old, excluding passage areas | 0.2 - 0.3 |
| | Porkers, cutters and baconers, including passages | 0.6 - 0.9 |
| **Poultry** | Laying hens: deep litter | 0.28 - 0.29 |
| | deep litter + slats / wire | 0.23 - 0.24 |
| | Rearing pullets 1 - 16 weeks old, deep litter | 0.03 - 0.28 |
| | Broilers | 0.05 - 0.08 |
| | Capons / roasters | 0.18 - 0.20 |
| | Ducks 3 - 7 weeks old: wire floor | 0.04 - 0.14 |
| | solid floor | 0.06 - 0.19 |
| | Turkeys: breeding | 0.50 - 0.90 |
| | finishing | 0.09 - 0.46 |
| **Rabbits** | Doe and litter: wire cages | 0.55 - 0.70 |
| | hutches | 0.75 - 0.85 |
| **Sheep** | Buildings with no internal feeding passages: | |
| | small ewe and lamb | 1.3 - 1.8 |
| | large ewe and lamb | 1.4 - 1.9 |
| | fattening lambs | 0.6 - 0.8 |

*Source:* Agro Business Consultants (Nov. 1990) *The Agricultural Budgeting and Costing Book. No. 31.* Reproduced with kind permission of Agro Business Consultants, Church Lane, Twyford, Melton Mowbray, Leicestershire, UK.

**TABLE 426   Generalized Housing Space Allowances for Pigs and Cattle in UK**

**A. Pigs**

| Animal type | Sleeping / feeding $m^2$ | Dunging $m^2$ | Trough cm |
|---|---|---|---|
| Farrowing / rearing | 3.7 - 5.5 | 2.7 - 3.7 | 45 |
| Creep area | 1.4 - 1.8 | | |
| Early weaned pigs | 0.16 | 0.07 | 10 |
| Weaners | 0.23 | 0.09 | 15 |
| Porkers | 0.32 | 0.14 | 25 |
| Baconers | 0.46 | 0.18 | 30 |
| Heavies | 0.5 | 0.18 | 35 |

*per pig*

**B. Cattle**

| Age of Animal | Area $m^2$ * | Trough cm |
|---|---|---|
| 6 months | 1.8 - 2.8 (4) | 30 - 45 |
| 1 year | 2.8 - 3.7 (6) | 45 - 55 |
| Adult cows | 3.7 - 4.6 | 60 - 70 |
| Bullocks - 9 months | 1.8 - 2.8 | 30 - 45 |
| 18 months | 3.2 - 3.7 | 45 - 60 |
| Large, horned bullocks | (12) | 85 |

\* smaller area is required with slatted floor accommodation

*Sources: Collins Farmers Diary* (1992). © Collins 1960, 1991, A division of Harper Collins
Publishers. (Figures in parentheses from Sainsbury, D. & Sainsbury, P. (1988) *Livestock Health
and Housing* - the figures include area for trough.)

**TABLE 427   Recommended Hen Housing Facilities for Hot Weather Growing**

| | | White egg pullets | Brown egg pullets |
|---|---|---|---|
| Cage growing | **Floorspace** | 340 sq cm | 425 sq cm |
| | **Waterer space** | | |
| | Birds / cup | 8 | 8 |
| | Birds / nipple | 8 | 8 |
| | Trough / bird | 3 cm | 4 cm |
| | **Feeder space** | | |
| | Trough / bird | 6 cm | 8 cm |
| Floor growing | **Floor space** | 8.6 birds / $m^2$ | 7.2 birds / $m^2$ |
| | **Waterer space** | | |
| | Bird / cup | 20 | 20 |
| | Birds / nipple | 10 | 10 |
| | Birds / fountain | 75 | 50 |
| | Trough / bird | 3 cm | 4 cm |
| | **Feeder space** | | |
| | Trough / bird | 8 cm | 8 cm |
| | Birds / pan | 25 | 20 |

**TABLE 428 Environmental Conditions for Rearing Chicks: Floor, Feeding and Drinking Space**

| Age of fowl (weeks) | Floor space (cm$^2$/bird) | | | Linear feeding space (cm) | | Linear Drinking space (cm) |
|---|---|---|---|---|---|---|
| | All litter or litter and wire | | All wire | Dry mash | Wet mash | |
| | Pullets | Broilers | | | | |
| 0 | 65 - 120 | | 30 | 2.5 | 3 | 1 - 2.5 |
| 1 | 120 - 210 | 210 | 60 | 2.5 | 3 | 1 - 2.5 |
| 2 | 180 - 300 | 240 | 90 | 2.5 | 3 | 1 - 2.5 |
| 3 | 240 - 390 | 280 | 120 | 2.5 - 3.8 | 4.5 | 1 - 2.5 |
| 4 | 300 - 460 | 350 | 150 - 230 | 3.8 - 7.5 | 4.5 | 1 - 2.5 |
| 5 | 360 - 550 | 420 | 360 - 350 | 6 - 7.5 | 7.5 | 1 - 2.5 |
| 6 | 450 - 640 | 450 | 450 - 470 | 6 - 7.5 | 7.5 | 1 - 2.5 |
| 7 | 680 - 730 | 500 | 680 - 590 | 6 - 7.5 | 7.5 | 1 - 2.5 |
| 8 | 820 - 900 | 550 | 900 - 710 | 6 - 7.5 | 7.5 | 2.5 - 5 |
| 9 | 910 - 940 | 600 | 940 - 830 | 7.5 - 10 | 10 | 2.5 - 5 |
| 10 | 960 - 1010 | 700 | 960 - 950 | 7.5 - 10 | 10 | 2.5 - 5 |
| 11 | 1020 - 1100 | | 1020 - 1070 | 7.5 - 10 | 10 | 2.5 - 5 |
| 12 | 1060 - 1200 | | 1020 - 1200 | 7.5 - 10 | 10 | 2.5 - 5 |
| 13 | 1080 - 1270 | | 1080 - 1210 | 9 - 10 | 12 | 2.5 - 5 |
| 14 | 1160 - 1340 | | 1160 - 1220 | 9 - 10 | 12 | 2.5 - 5 |
| 15 | 1200 - 1420 | | 1240 - 1230 | 9 - 10 | 12 | 2.5 - 5 |
| 16 | 1240 - 1500 | | 1240 | 9 - 10 | 12 | 2.5 - 5 |
| 17 | 1390 - 1580 | | 1390 | 9 - 10 | 12 | 2.5 - 5 |
| 18 | 1540 - 1660 | | 1540 | 9 - 10 | 12 | 2.5 - 5 |
| 19 | 1690 - 1740 | | 1690 | 12 | 15 | 2.5 - 5 |
| 20 | 1860 | | 1860 | 12 | 15 | 2.5 - 5 |

*Notes:*    Number of chicks or growers per m$^2$ floor space at

    (a)  one day old:    50 - 80 (depending on the strain)

    (b)  6 weeks old:    20

    (c)  8 weeks old:    10 - 12 pullet chicks, 15 - 18 broilers

    (d)  20 weeks old:    10 pullets and 6 broiler strains

| Cage | Age (weeks) | Floor space per bird (cm$^2$) | Feeding space per bird (cm) | Drinking space per bird (cm) |
|---|---|---|---|---|
| | 0 - 6 | 161 | 2.5 | 2.5 |
| | 7 - 18 / 20 | 277 | 5.0 | 2.5 |

*Source:* Oluyemi, J.A. & Roberts, F. (1979) *Poultry Production in Warm Wet Climates.*
Reproduced with kind permission of Macmillan Press.

**TABLE 429   Environmental Conditions for Rearing Chicks: Feeders and Drinkers**

| Type of feeder or drinker | | Length or capacity of feeder or drinker | Age of the fowl (weeks) | No. of birds per feeder or drinker |
|---|---|---|---|---|
| (a)  Keyes egg tray | | 30 egg size | 0 - 4 | 100 |
| (b)  Pan of tube feeder | | 30 egg size | 0 - 4 | 100 |
| Trough feeders | (a) | 0.3 - 0.5 m | 3 or 4 | 50 |
| | (b) | 0.6 m | 3 or 4 | 100 |
| | (c) | 1.0 - 1.2 m | 4 - 8 | 30 - 40 |
| | (d) | 1.0 - 1.2 m | 9 - 20 | 20 - 25 |
| | (e) | 1.5 m | 4 - 8 | 40 - 50 |
| | (f) | 1.5 m | 9 - 20 | 30 |
| Hanging feeders | (a) | 20 kg | 1 - 3 | 100 |
| | (b) | 25 kg | 4 - 9 | 35 - 50 |
| | | | 10 - 18 | 30 - 35 |
| Drinkers | | 2 - 4 litre | 0.4 | 100 |
| | | 2 - 4 litre | 0 - 4 | 50 |
| | | 6 - 8 litre | 0 - 4 | 50 |
| | | 10 litre | 4 - 20 | 50 |

*Notes:*
The number and size of feeders and drinkers may be changed at 4 and 9 weeks of age of the birds. Feeders for chicks should be placed about 3 m apart and 1.5 m from wall. Drinkers should not be more than 2.4 m apart.

*Source:* Oluyemi, J.A. & Roberts, F. (1979) *Poultry Production in Warm Wet Climates.*
Reproduced with kind permission of Macmillan Press.

**TABLE 430  Space Conditions for Adult Fowls (Layers and Breeders)**

| Condition | | Specification | | | |
|---|---|---|---|---|---|
| | | Layers | | Breeders | |
| Definition | Units | Light | Heavy | Pullets | Broilers |
| **Floor Space** | | | | | |
| (a) Litter | per bird (m²) | 0.1 - 0.20 | 0.20 - 0.30 | 0.23 - 0.30 | 0.23 - 0.30 |
| (b) Wire and litter | per bird (m²) | 0.14 | 0.16 | 0.14 | 0.16 |
| (c) Wire slats | per bird (m²) | 0.09 | 0.11 | 0.09 | 0.11 |
| **Feeding space** | per bird (cm) | 6.5 - 12.5 | 10 - 15 | 12.5 | 15 |
| | **Trough feeders** | | | | |
| | (1.5 - 1.8 m long) per 100 birds | 4 | 5 | 4 | 5 |
| | **Tube feeders** | | | | |
| | (20 - 25 kg) for 100 layers | 4 | 5 | 4 | 5 |
| **Drinking space** | per bird (cm) | 2 - 10 | 5 - 10 | 5 - 10 | 5 - 10 |
| | **Trough drinkers** | | | | |
| | (1.8 m) per 100 layers | 4 | 5 | 5 | 5 |
| | 10 litre water fountains | | | | |
| | per 100 layers | 2 - 4 | 2 - 3 | 2 - 3 | 2 - 3 |

*Source:* Oluyemi, J.A. & Roberts, F. (1979) *Poultry Production in Warm Wet Climates.*
Reproduced with kind permission of Macmillan Press.

**Table 431  Stocking Densities for Hens**

| | | Floor space / bird (m²) |
|---|---|---|
| **Layers** | Deep-litter | 0.27 |
| | Deep-litter and slats in ratio 2:1 | 0.18 |
| | Deep-litter and slats in ratio 1:2 | 0.14 |
| | All slats or welded mesh | 0.09 |
| | Multi-bird batteries and house | 0.18 - 0.27 |
| | Totally covered straw-yard | 0.27 |
| **Rearing** | Tier brooders | 0.02 at 3 weeks |
| | Floor rearing | 0.09 - 0.14 |
| | Hay-box brooder | 0.03 to 8 weeks |
| | Sussex night ark | 0.04 to point of lay |
| | Range shelter    plus pasture | 0.06 to point of lay |
| | Fold units | 0.06 to point of lay |

*Source:* Sainsbury, D. (1984) *Poultry Health and Management.*
Reproduced with kind permission of Blackwell Scientific Publishers Ltd.

**TABLE 432   Housing Data for Sheep**

| Type of sheep | | Requirements | |
|---|---|---|---|
| | | Slat | Straw |
| **COVERED AREA** | | (m $^2$) | (m $^2$) |
| Ewe: | large | 1.0 | 1.2 |
| | small | 0.8 | 1.2 |
| With lambs, add | | 0.3 | |
| Hoggs up to 30 kg | | 0.6 | 0.7 |
| Lamb creeps to 5 weeks | | | 0.3 |

**VENTILATION**

| | |
|---|---|
| Mechanical | Up to 3 m $^3$ / h/kg body weight |
| Natural | Open ridge up to 0.6 m |
| | Side walls at least 1 m all round building |

| | | Troughs | Hay rack |
|---|---|---|---|
| **TROUGHS AND RACKS** | | (mm) | (mm) |
| Ewe: | large | 450 | 200 |
| | small | 400 | 150 |
| Lambs: | Up to 50 kg | 350 | 150 |
| | Up to 40 kg | 300 | 125 |

*Source:* Sainsbury, D. & Sainsbury, P. (1988) *Livestock Health and Housing.*
Reproduced with kind permission of Bailliere Tindall.

**TABLE 433   Trough Space Allowances for Pigs**

150 mm at weaning

230 mm for small porkers

250 mm for heavy porkers

300 mm for baconers

350 mm for heavy hogs

*Source:* Sainsbury, D. & Sainsbury, P. (1988) *Livestock Health & Housing.*
Reproduced with kind permission of Bailliere Tindall.

**TABLE 434  Guidelines for Yard, Shade and Trough Requirements for Cattle and Sheep in Hot, Arid Climates**

|  | Yard (sq m) | Shade (sq m) | Trough (sq m) |
|---|---|---|---|
| Mature cattle | 50 | 5 | 75 |
| Mated heifers | 35 | 4 | 60 |
| Growing heifers | 25 | 3 | 50 |
| Weaned calves | 10 | 1.5 | 35 |
| Ewes | 10 | 1.5 | 35 |
| Suckling lambs (12 weeks) | 7 | 1 | 30 |

*NB* These are guidelines only. Breed, rainfall, soil type and feeding regime must be taken into consideration.

*Source:* Farnworth, J. *Accommodating cattle and sheep in hot dry climates.* Middle East Agribusiness.

**TABLE 435  Housing and Feeding Requirements for Suckler Cows, Suckled Calves and Rearing Stock**

|  | Housing ($m^2$ per head) | | Feeding (mm trough space) | |
|---|---|---|---|---|
|  | Straw bedding | Slats * | Restricted feeding | Ad lib feeding |
| Cows | 5.0 - 6.0 | 3.0 - 3.5 | 600 - 700 | 140 - 175 |
| Creep area | 1.0 - 2.0 | 1.0 | na | 75 |
| Suckled calves | 3.5 - 4.0 | 1.5 - 1.8 | 450 - 550 | 110 - 140 |
| Stores / finishing cattle | 3.8 - 4.3 | 1.8 - 2.1 | 500 - 650 | 140 - 170 |

\*   Excludes feed troughs and feeding passages
na  Not applicable

*Source:* MLC (1991) *Suckler Beef.* Reproduced with kind permission of the Meat and Livestock Commission (UK).

**TABLE 436   General Space Requirements of Different Classes of Dairy Cattle in the Tropics**

| Class | Type of construction | | m | m$^2$ |
|---|---|---|---|---|
| **Milking cows** | Milking stalls (barn) | Width single stall | 1.03 | |
| | | Width double stall | 1.98 | |
| | | Length small breed | 1.37 | |
| | | Length large breed | 1.52 | |
| | Length of feeding manger per cow | | 0.76 | |
| | Height of floor of stall above dunging passage | | 0.15 - 0.18 | |
| | Width of feeding passage | | 0.91 - 1.22 | |
| | Milking stalls (bail) used with | Width of stall | 0.76 | |
| | line machine | Width of bail | 0.60 | |
| | Collection and dispersal yards. | Polled cattle / cow | | 2.3 |
| | | Horned cattle / cow | | 3.7 |
| | Loafing yards. | Total area / cow | | 6.5 - 7.4 |
| | | Shade area / cow | | 3.7 - 5.6 |
| | Length of feeding manger per cow | Access at all times | 0.30 - 0.46 | |
| | | Limited access | 0.61 - 0.76 | |
| | Loose boxes for calving | Area / cow | | 14.9 - 15.8 |
| **Bulls** | Bull houses. Area within house per bull | | | 16.7 |
| | Area outside house per bull | | | 33.5 - 37.2 |
| **Calves** | Individual pens for small calves | Dimensions | 1.83 x 1.22 | |
| | Collective pens: area required. | | | |
| | Calves up to 3 months of age | | | 1.9 - 2.8 |
| | Calves 3 to 6 months of age | | | 2.8 |

*Source:* Wilkinson, G. & Payne, W.J.A., eds. (1978) *Introduction to Animal Husbandry in the Tropics* 3rd Ed. Reproduced by kind permission of Longman Group UK Ltd.

# Building Materials

**TABLE 437   Indicative Volume Mixes for Different Types of Concrete**

| | Volumes | | | |
|---|---|---|---|---|
| Cement | Damp sand | Coarse aggregate | Use | |
| 2 | 2 | 3 (10) | Floor screed | |
| 3 | 5 | 9 (20) | Bases of screed floors, yards, steps, floors | |
| 2 | 5 | 8 (40) | Foundations | |
| 3 | 4 | 7 (20) | Floors subject to chemical attack | |
| 1 | 3 | 2 (10) | Infilling of hollow blocks | |

( ) Maximum aggregate size in mm.

*NB* Take specialist advice where concrete is to be used for special and critical purposes including exposure to chemicals, sulphate in water or soil, exposure to salt; also critical standards for aggregate and sand may be required and appropriate quality cement must always be used.

**TABLE 438  Concrete Mixes for Farm Use**

| Concrete mix | C7.5P (replaces 1 : 3 : 6) | C20P (replaces 1 : 2 : 4) | C25P (replaces 1 : 1.5 : 3) |
|---|---|---|---|
| Application | Strip footings; trench-fill foundations; oversite concrete and blinding under slabs; stanchion bases; mass concrete | Yards and slabs not exposed to de-icing salts [1], slabs for farmyard manure, floors for dairy and beef cattle, calves, pigs (not insulated) and poultry grains and potato stores; hay barns; machinery sheds | All concrete in silos; brewer's grain stores; parlours and dairies; floors used by small-wheeled fork lift trucks or other handling equipment; walls, infill cavity walls and retaining walls; suspended floors, beams and lintels |
| **Ready mixed concrete** | C7.5P to BS 5328, Ordinary Portland Cement 20 mm max aggregate [2] Medium workability [4] | C20P to BS 5328, Ordinary Portland Cement 20 mm max aggregate Medium workability | C25P to BS 5328, Ordinary Portland Cement 20 mm max aggregate [3] Medium workability |
| **Site-mixed by weight** | | | |
| Cement | 50 kg (1 whole bag) | 50 kg (1 whole bag) | 50 kg (1 whole bag) |
| Damp sand | 190 kg | 115 kg | 90 kg |
| Coarse aggregate (20 mm maximum) | 270 kg [2] | 195 kg | 170 kg [3] |
| Approximate yield | C.24 cu. m | 0.17 cu. m | 0.14 cu. m |
| **Site-mixed by proportion (volume)** | | | |
| Cement | 2 parts | 3 parts | 3 parts |
| Damp sand | 5 parts | 5 parts | 4 parts |
| Coarse aggregate (20 mm maximum) | 8 parts [2] | 9 parts | 7 parts [3] |

1. For roads and pavings exposed to de-icing salts specify an SP300 (a mix with 300 kg of cement in each cubic metre of concrete) using ordinary Portland cement, 20 mm maximum-sized aggregate, a 50 mm slump (75 mm for hand tamping) and an open air-entraining agent to give a target air content of 5 per cent.
2. 40 mm maximum-size aggregate may be used for large blocks of mass concrete.
3. Use 10 mm maximum-size aggregate for cavity wall infill and for thin sections (less than 60 mm).
4. Specify high workability for concrete used in trench fill foundations.
NB  Concrete exposed to sulphate in soil or water requires special specifications.

*Source:* Barnes, M. & Mander, C. (1986) *Farm Building Construction: the farmer's guide.*
Reproduced with kind permission of Farming Press Books, Wharfedale Road, Ipswich, Suffolk, UK.

**TABLE 439   Guideline Mix Proportions for Mortars and Plasters**

| USE | WHOLE SACK MIXES Recommended standard practice | | | LOOSE VOLUME MIXES for small batches; proportions given by number of containers for all constituents | | |
|---|---|---|---|---|---|---|
| | Cement bags | Lime kg | Sand m³ | Cement | Lime | Sand |
| Brickwork mortar for lightly loaded internal walls | 1 | 75 | 0.40 | 1 | 3 | 8·5 |
| Brickwork mortar for solid walls greater than 110 mm thick under mild conditions of exposure; internal first coat plaster | 1 | 50 | 0.30 | 1 | 2 | 6.5 |
| Brickwork mortar for lightly loaded walls under average conditions of exposure | 1 | 0 - 40 | 0.25 | 1 | 0 - 1.5 | 5 |
| Brickwork mortar for external and load-bearing, walls, and walls below damp-proof coursing. Also for external rendering and strong internal plasters | 1 | 0 - 25 | 0.20 | 1 | 0 - 1 | 4 |
| Brickwork mortar for load-bearing walls carrying heavy loads, and for walls in damp conditions. Also for mortar where a high early strength is desired. Also for durable external rendering | 1 | 0 - 6 | 0.13 | 1 | 0 - 0.5 | 3 |
| Waterproof plaster and rendering (when fully covered and protected from early cracking) | 1 | – | 0.10 | 1 | – | 3 |
| Lime finishing coat plaster (using fine clean sand) | – | 50 | 0.04 | – | 1 | 1 |

*Source: Rhodesian Farm Management Handbook* (1970).
Reproduced with kind permission of the Ministry of Lands, Agriculture and Resettlement, Zimbabwe.

**TABLE 440    Typical Thermal Transmission Co-efficients (U values) *
of Various Materials**

| Material | U value (W / m $^2$ °C) |
|---|---|
| Horticultural glass | 5.7 |
| 500 gauge polythene film | 5.7 |
| Corrugated asbestos | 6.5 |
| Sheet asbestos | 5.0 |
| Wood, 25 mm | 2.6 |
| Wood, 38 mm | 2.4 |
| Brickwork (unplastered) 114 mm | 3.4 |
| Brickwork (unplastered) 228 mm | 2.7 |
| Brickwork (unplastered) 342 mm | 2.1 |
| Concrete, 100 mm (solid) | 4.0 |
| Concrete 152 mm (solid) | 3.6 |
| Earth floor | 1.9 |
| Concrete floor | 0.7 |

\* U value = W / m $^2$ °C   or   Btu / h. ft $^2$ °F
(note 'U' value differs according to units used)

*Source:* Electricity Association (1985) *Vegetable Storage.* A Farm Electric Handbook.
Reproduced with kind permission of Electricity Association Technology Ltd.

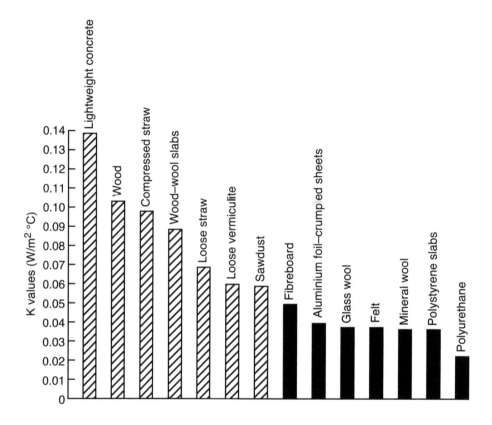

(K = amount of heat in watts flowing through a square metre when a temperature difference of 1° C is maintained between opposite surfaces of a metre thickness)

**Fig. 62** K values of a series of materials used for thermal insulation

*Source:* Sainsbury, D. (1984) *Poultry Health and Management.*
Reproduced with kind permission of Blackwell Scientific Publishers Ltd.

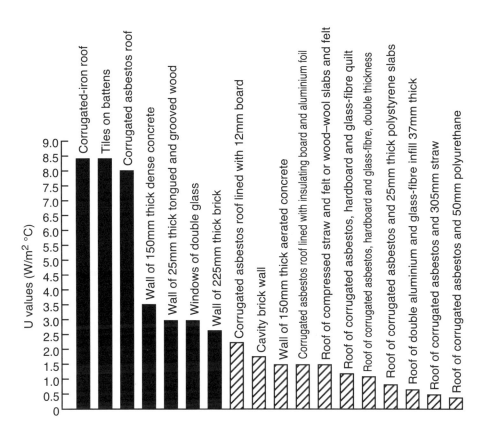

(U = amount of heat in watts transmitted through one square metre of construction when there is a temperature difference of 1° C between inside and outside)

**Fig. 63** U values of some roof and wall constructions

*Source:* Sainsbury, D. (1984) *Poultry Health and Management.*
Reproduced with kind permission of Blackwell Scientific Publishers Ltd.

**TABLE 441  Solar Absorptivity and Long-wave Emissivity over the Temperature Range 0 - 100° C for Some Common Building Materials**

| Material | | Solar absorptivity, $\alpha$ | Long-wave emissivity, $\varepsilon$ |
|---|---|---|---|
| **Aluminium** | metal | 0.15 - 0.26 | 0.08 - 0.14 |
| | foil | – | 0.03 - 0.09 |
| | paint | 0.54 | 0.29 - 0.55 |
| **Asbestos** | cement new / aged | 0.61 - 0.75 | 0.95 |
| | insulating board | – | 0.93 - 0.96 |
| **Asphalt** | new | 0.91 | – |
| | weathered | 0.85 | 0.96 |
| **Bitumen felt** | | 0.88 | 0.91 |
| **Bricks** | colour - dependent | 0.40 - 0.89 | 0.94 |
| **Concrete** | rough | 0.65 | 0.94 |
| **Glass** | smooth | 0.83 | 0.92 - 0.95 |
| **Iron** | galvanized, new | 0.64 - 0.66 | 0.22 - 0.28 |
| | galvanized, aged | 0.90 | – |
| **Slates** | | 0.86 - 0.93 | – |
| **Tiles** | colour - dependent | 0.43 - 0.91 | – |

*Source:* Clark, J.A., ed. (1981) *Environmental aspects of housing for animal production.* (after Holden & Greenland 1951, and Diamant 1977). Reproduced with permission.

# Appendices

# Appendix I Accounting

---

**TABLE 442  Indicative Components of an Annual Farm Trading Account**

---

(Components to be included will depend on the purpose of the account and statutory requirements)

## A. * 'Asset' valuation

|  | Opening Valuation | Closing Valuation |
|---|---|---|
|  | (financial | valuations) |
| Productive Livestock | | |
| Replacement Livestock | | |
| Fatstock | | |
| Fleeces in Store | | |
| Livestock Feedstuffs | | |
| Seeds | | |
| Crop Protection Chemicals | | |
| Fertilizers and Lime | | |
| Veterinary Pharmaceuticals | | |
| Fuel and Oil | | |
| Crops in Store | | |
| Crops in Field | | |
| Land and Buildings | | |
| Moving and Fixed Machinery | | |
| Sundry unconsumed supplies | | |
| Cultivations and Residual Manures | | |

## B. * 'Actual' Transacted products, supplies, services and charges
### (financial expenditure and income)

| Expenditure | Income |
|---|---|
| Livestock | Livestock |
| Livestock Feeds | Crops |
| Livestock Medicines | Milk / Milk Products |
| Veterinary Services | Fleeces |
| Artificial Inseminations etc. | Eggs |
| Dairy Expenses | Crop Grazings |
| Sundry Livestock Expenses | Rent of Land |
| Seeds | Rent of Buildings |
| Fertilizers, Lime, Manures | Contracting |
| Crop Protection Chemicals | Compensations |
| Sundry Crop Expenses | Refunds |
| Crop and Livestock Processing and Packaging Costs | Surplus Machines / Equipment |
| Machinery including Transport | Processed Products |
| Advisory Charges | Manure |
| Contractors Charges | Services |
| Property Repairs | Use of Land for Non-Farming Activities |
| Machinery Repairs | Wayleave Income |
| Wages and Statutory Associated Charges | Sundry Income |
| Fuel and Oil | Sale of Land / Buildings |
| Electricity and Gas | Subsidies, Supports and Grants |
| Water Charges | |
| Office Expenses, Telephone etc. | |
| Insurance (staff, vehicles, general liabilities etc.) | |
| Local Taxes | |
| Staff Recruitment Costs | |
| Drainage | |
| Accounting Expenses, Professional Fees | |
| Interest Charges | |
| Leasing and Finance Charges | |
| Pension Provisions | |
| Advertising | |
| Sundry Overheads, e.g. Memberships | |
| Litigation Charges | |
| Commission Charges | |
| Sundry Expenses | |

## C. * 'Notional' expenditures and income (assessed values)

| Expenditure | Income |
|---|---|
| • Depreciation of Machinery and Equipment | • Products Retained for Direct on Farm Consumption by Management e.g. milk |
| • Tenants Improvements | • Rental Value of Farmhouse |
| | • Use of Car |

**\*** In broad terms the left-hand columns of **A, B** and **C** are inputs to the year's trading and the right-hand columns are the outputs. The difference between the summation of the left-hand columns and right-hand columns is the annual farm profit or loss. Net Farm Income can be deduced by subtracting from this a notional rent figure for any land owned.

---

**TABLE 443   Indicative List of Variable and Fixed Cost Items**

---

### A. Variable Costs (normally considered as enterprise specific items and vary in proportion to the size of the enterprise)

| **Crops** | **Livestock** |
|---|---|
| Seeds / Plants | Purchased Feedstuffs |
| Fertilizers and Lime | Forage Produced on Farm |
| Manure and Composts | Veterinary Products |
| Crop Protection Chemicals | Veterinary Services |
| Crop Supports / Special Housing | Dairy Supplies |
| String | Enterprise Specific Hand Tools |
| Packing Materials | Casual Labour |
| Crop Specific Hand Tools | Specialist Labour, e.g. Shearers |
| Crop Specific Contract Work | Levy or Quota Charges |
| Casual Labour | Inseminations / Transplants |
| Levy or Quota Charges | Advisory Charges |
| Advisory Charges | Marketing Costs |
| Crop Dessicants | Stock Purchases |
| Storage Costs | Bedding |
| Drying Costs | Sundries |
| Marketing Costs | |
| Sundries | |

### B. Fixed Costs or Overheads (normally considered to be shared between enterprises and not readily allocated to specific enterprises; where any of the costs below are enterprise specific they may be considered as variable costs)

| | | |
|---|---|---|
| Permanent Labour | Insurance | Water Charges |
| Machinery | Advertising | Rent |
| Transport | Fuel and Oil | Building Maintenance |
| Office Costs | Electric and Gas | Personnel Vehicles |
| Processing Equipment | Drainage | Irrigation |
| Storage Facilities | Depreciation | Memberships |
| Local Taxes | Contractors Charges | Advisory Charges |
| Interest / Financing Charges | Litigation Charges | Pensions |
| Infrastructure Improvements | Shelter Belts | Conservation Areas |
| Accounting Expenses | | |

*NB* The gross margin of an enterprise is the value of its output less the variable costs usually expressed on a per-hectare or per-livestock unit basis. When comparisons of gross margins are made they should contain the same component costs and returns.

# Appendix II

# Units, Physical Constants and Conversion Tables

**TABLE 444  Metric Prefixes**

| Prefix | Symbol | Multiplication Factor |
|--------|--------|----------------------|
| tera | T | $1\ 000\ 000\ 000\ 000 = 10^{12}$ |
| giga | G | $1\ 000\ 000\ 000 = 10^{9}$ |
| mega | M | $1\ 000\ 000 = 10^{6}$ |
| kilo | k | $1000 = 10^{3}$ |
| hecto | h | $100 = 10^{2}$ |
| deka | da | $10 = 10^{1}$ |
| deci | d | $0.1 = 10^{-1}$ |
| centi | c | $0.01 = 10^{-2}$ |
| milli | m | $0.001 = 10^{-3}$ |
| micro | μ | $0.000\ 001 = 10^{-6}$ |
| nano | n | $0.000\ 000\ 001 = 10^{-9}$ |
| pico | p | $0.000\ 000\ 000\ 001 = 10^{-12}$ |
| femto | f | $0.000\ 000\ 000\ 000\ 001 = 10^{-15}$ |
| atto | a | $0.000\ 000\ 000\ 000\ 000\ 001 = 10^{-18}$ |

**TABLE 445  Metric Units**

| Quantity | Unit | Symbol | Formula |
|---|---|---|---|
| **Base units** | | | |
| Length | metre | m | |
| Mass | kilogram | kg | |
| Time | second | s | |
| Electric current | ampere | A | |
| Thermodynamic temperature | kelvin | K | |
| Plane angle | radian | rad | |
| Solid angle | steradian | sr | |
| **Derived units** | | | |
| Acceleration | metre per second squared | – | $m / s^2$ |
| Activity (of a radioactive source) | disintegration per second | – | (disintegration) / s |
| Angular acceleration | radian per second squared | – | $rad / s^2$ |
| Angular velocity | radian per second | – | rad / s |
| Area | square metre | – | $m^2$ |
| Density | kilogram per cubic metre | – | $kg / m^3$ |
| Electric capacitance | farad | F | A·s / V |
| Electric conductance | siemens | S | A / V |
| Electric field strength | volt per metre | – | V / m |
| Electric inductance | henry | H | W / A |
| Electric potential difference | volt | V | W / A |
| Electric resistance | ohm | Ω | V / A |
| Electromotive force | volt | V | W / A |
| Energy | joule | J | N·m |
| Force | newton | N | $kg·m / s^2$ |
| Power | watt | W | J / s |
| Pressure | pascal | Pa | $N / m^2$ |
| Quality of electricity | coulomb | C | A·s |
| Quality of heat | joule | J | N·m |
| Radiant intensity | watt per steradian | – | W / sr |
| Specific heat | joule per kilogram - kelvin | – | J / kg·K |
| Stress | pascal | Pa | $N / m^2$ |
| Thermal conductivity | watt per metre - kelvin | – | W / m·K |
| Velocity | metre per second | – | m / s |
| Viscosity, dynamic | pascal - second | – | Pa·s |
| Viscosity, kinematic | square metre per second | – | $m^2 / s$ |
| Voltage | volt | V | W / A |
| Volume | cubic metre | – | $m^3$ |
| Work | joule | J | N·m |

**TABLE 446   Chemical Elements**

| Element | Symbol | Atomic Weight | Valence |
|---|---|---|---|
| Aluminium | Al | 26.97 | 3 |
| Boron | B | 10.82 | 3 |
| Bromine | Br | 79.916 | 1,3,5,7 |
| Cadmium | Cd | 112.41 | 2 |
| Calcium | Ca | 40.08 | 2 |
| Carbon | C | 12.01 | 2,4 |
| Chlorine | Cl | 35.457 | 1,3,5,7 |
| Cobalt | Co | 58.94 | 2,3 |
| Copper | Cu | 63.57 | 1,2 |
| Fluorine | F | 19.00 | 1 |
| Hydrogen | H | 1.008 | 1 |
| Iodine | I | 126.92 | 1,3,5,7 |
| Iron | Fe | 55.85 | 2,3 |
| Lead | Pb | 207.21 | 2,4 |
| Magnesium | Mg | 24.32 | 2 |
| Manganese | Mn | 54.94 | 2,3 |
| Mercury | Hg | 200.61 | 1,2 |
| Molybdenum | Mo | 95.95 | 3,4,6 |
| Nickel | Ni | 58.69 | 2,3 |
| Nitrogen | N | 14.008 | 3,5 |
| Oxygen | O | 16 | 2 |
| Phosphorus | P | 30.98 | 3,5 |
| Potassium | K | 39.096 | 1 |
| Selenium | Se | 78.96 | 3 |
| Silicon | Si | 28.06 | 4 |
| Sodium | Na | 22.997 | 1 |
| Sulphur | S | 32.06 | 2,4,6 |
| Zinc | Zn | 65.38 | 2 |

**TABLE 447   Conversion Tables**

Example: to convert 10 miles to kilometres, find 1 mile in the Length table. Numbers on that same horizontal are equal units to 1 mile, therefore 1 mile = 1.6094 km; 10 miles = 16.094 km.

## 1. Length

| km | m | mm | mile | yard | ft | in |
|---|---|---|---|---|---|---|
| 1 | 1000 | $10^6$ | 0.6214 | 1094 | 3281 | $3.937 \times 10^4$ |
| $10^{-3}$ | 1 | 1000 | $6.214 \times 10^{-4}$ | 1.0936 | 3.281 | 39.370 |
| $10^{-6}$ | $10^{-3}$ | 1 | $6.214 \times 10^{-7}$ | $1.094 \times 10^{-3}$ | $3.281 \times 10^{-3}$ | $3.937 \times 10^{-2}$ |
| 1.6094 | 1609.4 | $1.609 \times 10^6$ | 1 | 1760 | 5280 | 63360 |
| $9.144 \times 10^{-4}$ | 0.9144 | 914.41 | $5.682 \times 10^{-4}$ | 1 | 3 | 36 |
| $3.048 \times 10^{-4}$ | 0.3048 | 304.8 | $1.894 \times 10^{-4}$ | 0.3333 | 1 | 12 |
| $2.54 \times 10^{-5}$ | 0.0254 | 25.4 | $1.578 \times 10^{-5}$ | $2.778 \times 10^{-2}$ | $8.333 \times 10^{-2}$ | 1 |

## 2. Area

| km² | m² | cm² | mm² | sq. mile | acre | yd² | ft² | in² |
|---|---|---|---|---|---|---|---|---|
| 1 | $10^{-6}$ | $10^{10}$ | $10^{12}$ | 0.38612 | 247.11 | $1.196 \times 10^6$ | $1.076 \times 10^7$ | $1.550 \times 10^9$ |
| $10^6$ | 1 | $10^4$ | $10^6$ | $3.86 \times 10^{-7}$ | $2.471 \times 10^{-4}$ | 1.1960 | 10.764 | 1550 |
| $10^{-10}$ | $10^{-4}$ | 1 | 100 | $3.86 \times 10^{-11}$ | $2.471 \times 10^{-8}$ | $1.196 \times 10^{-4}$ | $1.076 \times 10^{-3}$ | 0.1550 |
| $10^{-12}$ | $10^{-6}$ | $10^{-2}$ | 1 | $3.86 \times 10^{-13}$ | $2.47 \times 10^{-10}$ | $1.196 \times 10^{-6}$ | $1.076 \times 10^{-5}$ | $1.550 \times 10^{-3}$ |
| 2.590 | $2.59 \times 10^6$ | $2.59 \times 10^{10}$ | $2.59 \times 10^{12}$ | 1 | 639.96 | $3.097 \times 10^6$ | $2.788 \times 10^7$ | $4.01 \times 10^9$ |
| $4.047 \times 10^{-3}$ | 4047 | $4.047 \times 10^7$ | $4.047 \times 10^9$ | $1.563 \times 10^{-3}$ | 1 | 4840 | 43560 | $6.273 \times 10^{-6}$ |
| $8.36 \times 10^{-7}$ | 0.8361 | 8361 | $8.36 \times 10^5$ | $3.228 \times 10^{-7}$ | $2.066 \times 10^{-4}$ | 1 | 9 | 1296 |
| $9.29 \times 10^{-8}$ | $9.29 \times 10^{-2}$ | 929 | 92900 | $3.587 \times 10^{-8}$ | $2.296 \times 10^{-5}$ | 0.1111 | 1 | 144 |
| $6.45 \times 10^{-10}$ | $6.45 \times 10^{-4}$ | 6.4516 | 645.16 | $2.491 \times 10^{-10}$ | $1.594 \times 10^{-7}$ | $7.716 \times 10^{-4}$ | $6.944 \times 10^{-3}$ | 1 |

## 3. Volume

| m³ | dm³ / litre | cm³ / ml | yd³ | ft³ | in³ | UK gallon | US gallon |
|---|---|---|---|---|---|---|---|
| 1 | $10^3$ | $10^6$ | 1.3079 | 35.311 | 61013 | 219.97 | 264.17 |
| $10^{-3}$ | 1 | $10^3$ | $1.308 \times 10^{-3}$ | $3.531 \times 10^{-2}$ | 61.013 | 0.2200 | 0.2642 |
| $10^{-6}$ | $10^{-3}$ | 1 | $1.308 \times 10^{-6}$ | $3.531 \times 10^{-5}$ | $6.101 \times 10^{-2}$ | $2.199 \times 10^{-4}$ | $2.642 \times 10^{-4}$ |
| 0.7646 | 764.6 | $7.646 \times 10^5$ | 1 | 27 | 46650 | 168.19 | 201.99 |
| $2.832 \times 10^{-2}$ | 28.32 | $2.832 \times 10^4$ | $3.704 \times 10^{-2}$ | 1 | 1728 | 6.229 | 7.481 |
| $1.639 \times 10^{-5}$ | $1.639 \times 10^{-2}$ | 16.387 | $2.144 \times 10^{-5}$ | $5.787 \times 10^{-4}$ | 1 | $3.605 \times 10^{-3}$ | $4.329 \times 10^{-3}$ |
| $4.546 \times 10^{-3}$ | 4.546 | $4.546 \times 10^3$ | $5.946 \times 10^{-3}$ | 0.1605 | 277.42 | 1 | 1.2008 |
| $3.785 \times 10^{-3}$ | 3.785 | $3.785 \times 10^3$ | $4.951 \times 10^{-3}$ | 0.1337 | 231 | 0.8327 | 1 |

*NB* 1 UK gall = 8 pints = 4 quarts

## 4. Mass

| Tonne / Mg | kg | g | UK ton | US ton | cwt | lb | oz |
|---|---|---|---|---|---|---|---|
| 1 | 1000 | $10^6$ | 0.9832 | 1.1011 | 19.66 | $2.205 \times 10^3$ | $3.527 \times 10^4$ |
| $10^{-3}$ | 1 | 1000 | $9.832 \times 10^{-4}$ | $1.101 \times 10^{-3}$ | $1.966 \times 10^{-2}$ | 2.2046 | 35.274 |
| $10^{-6}$ | $10^{-3}$ | 1 | $9.832 \times 10^{-7}$ | $1.101 \times 10^{-6}$ | $1.966 \times 10^{-5}$ | $2.204 \times 10^{-3}$ | $3.527 \times 10^{-2}$ |
| 1.017 | 1017 | $1.017 \times 10^6$ | 1 | 1.12 | 20 | 2240 | 35840 |
| 0.9081 | 908.1 | $9.081 \times 10^5$ | 0.8928 | 1 | 17.856 | 2000 | 32000 |
| $5.085 \times 10^{-2}$ | 50.85 | $5.085 \times 10^4$ | 0.05 | 0.0560 | 1 | 112 | 1792 |
| $4.536 \times 10^{-4}$ | 0.4536 | 453.6 | $4.46 \times 10^{-4}$ | $5 \times 10^{-4}$ | $8.92 \times 10^{-3}$ | 1 | 16 |
| $2.835 \times 10^{-5}$ | $2.835 \times 10^{-2}$ | 28.349 | $2.79 \times 10^{-5}$ | $3.125 \times 10^{-5}$ | $5.580 \times 10^{-4}$ | $6.10 \times 10^{-2}$ | 1 |

*(continued over)*

**TABLE 447 (continued)  Conversion Tables**

**5. Density**

| Tonne/m³ Mg/m³ g/cm³ | kg/m³ | lb/in³ | UK ton/yd³ | US ton/yd³ | lb/ft³ |
|---|---|---|---|---|---|
| 1 | 1000 | 0.03613 | 0.75247 | 0.8428 | 62.423 |
| $10^{-3}$ | 1 | $3.613 \times 10^{-5}$ | $7.525 \times 10^{-4}$ | $8.428 \times 10^{-4}$ | $6.243 \times 10^{-2}$ |
| 27.680 | 27680 | 1 | 20.828 | 23.328 | $1.728 \times 10^3$ |
| 1.3289 | $1.328 \times 10^3$ | $4.801 \times 10^{-2}$ | 1 | 1.12 | 82.955 |
| 1.1865 | $1.186 \times 10^3$ | $4.287 \times 10^{-2}$ | 0.8929 | 1 | 74.074 |
| $1.602 \times 10^{-2}$ | 16.019 | $5.787 \times 10^{-4}$ | $1.205 \times 10^{-2}$ | $1.35 \times 10^{-2}$ | 1 |

**6. Force and weight**

| MN | kN | N | kgf | tonf | lbf |
|---|---|---|---|---|---|
| 1 | 1000 | $10^6$ | $1.0196 \times 10^5$ | 100.4 | $2.248 \times 10^5$ |
| $10^{-3}$ | 1 | $10^3$ | 101.96 | 0.1004 | 224.82 |
| $10^{-6}$ | $10^{-3}$ | 1 | 0.10196 | $1.004 \times 10^{-4}$ | 0.2248 |
| $9.807 \times 10^{-6}$ | $9.807 \times 10^{-3}$ | 9.807 | 1 | $9.842 \times 10^{-4}$ | 2.2048 |
| $9.964 \times 10^{-3}$ | 9.964 | 9964 | 1016 | 1 | 2240 |
| $4.448 \times 10^{-6}$ | $4.448 \times 10^{-3}$ | 4.448 | 0.45455 | $4.464 \times 10^{-4}$ | 1 |

## 7. Pressure, stress and moculus of elasticity

| MN/m² | kN/m² | kp | | | | | | | psi | |
|---|---|---|---|---|---|---|---|---|---|---|
| MPa | kPa | kgf/cm² | bar | atm | mH₂O | ftH₂O | mmHg | Ton/ft² | lbf/in² | lbf/ft² |
| 1 | 1000 | 0.197 | 10 | 9.869 | 102.2 | 335.2 | 7500.6 | 9.320 | 145.04 | 20886 |
| 0.001 | 1 | $.019 \times 10^{-2}$ | 0.0100 | $9.87 \times 10^{-3}$ | 0.1022 | 0.3352 | 7.5006 | 0.0093 | 0.14504 | 20.886 |
| $9.807 \times 10^{-2}$ | 98.07 | 1 | 0.9807 | 0.9678 | 10.017 | 32.866 | 735.56 | 0.9139 | 14.223 | 2048.1 |
| 0.100 | 100 | 1.0197 | 1 | 0.9869 | 10.215 | 33.515 | 750.06 | 0.9320 | 14.504 | 2088.6 |
| 0.1013 | 101.33 | 1.0332 | 1.0132 | 1 | 10.351 | 33.959 | 760.02 | 0.9444 | 14.696 | 2116.2 |
| $9.788 \times 10^{-3}$ | 9.7885 | $9.983 \times 10^{-2}$ | $9.789 \times 10^{-2}$ | $9.661 \times 10^{-2}$ | 1 | 3.2808 | 73.424 | $9.124 \times 10^{-2}$ | 1.4198 | 204.45 |
| $2.983 \times 10^{-3}$ | 2.9835 | $3.043 \times 10^{-2}$ | $2.984 \times 10^{-2}$ | $2.945 \times 10^{-2}$ | 0.3048 | 1 | 22.377 | $2.781 \times 10^{-2}$ | 0.43275 | 62.316 |
| $1.333 \times 10^{-4}$ | 0.1333 | $1.3595 \times 10^{-3}$ | $1.333 \times 10^{-3}$ | $1.315 \times 10^{-3}$ | $1.362 \times 10^{-2}$ | $4.469 \times 10^{-2}$ | 1 | $1.243 \times 10^{-3}$ | $1.934 \times 10^{-3}$ | 2.7846 |
| 0.1073 | 107.3 | 1.0942 | 1.0730 | 1.0589 | 10.960 | 35.960 | 804.78 | 1 | 15.562 | 2240.0 |
| $6.895 \times 10^{-3}$ | 6.895 | $7.031 \times 10^{-2}$ | $6.895 \times 10^{-2}$ | $6.805 \times 10^{-2}$ | 0.7043 | 2.3108 | 51.714 | $6.426 \times 10^{-2}$ | 1 | 144 |
| $4.788 \times 10^{-5}$ | $4.788 \times 10^{-2}$ | $4.883 \times 10^{-4}$ | $4.788 \times 10^{-4}$ | $4.725 \times 10^{-4}$ | $4.891 \times 10^{-3}$ | $1.605 \times 10^{-2}$ | 0.3591 | $4.464 \times 10^{-4}$ | $6.944 \times 10^{-3}$ | 1 |

## 8. Permeability

| m/s | cm/s | m/year | Darcy | ft/yr | ft/day |
|---|---|---|---|---|---|
| 1 | 100 | $3.156 \times 10^7$ | $1.04 \times 10^5$ | $1.035 \times 10^8$ | $2.835 \times 10^5$ |
| 0.01 | 1 | $3.156 \times 10^5$ | $1.04 \times 10^3$ | $1.035 \times 10^6$ | $2.834 \times 10^3$ |
| $3.169 \times 10^{-8}$ | $3.169 \times 10^{-6}$ | 1 | $3.28 \times 10^3$ | 3.281 | $8.982 \times 10^{-3}$ |
| $9.66 \times 10^{-6}$ | $9.66 \times 10^{-4}$ | 304 | 1 | 1000 | 2.74 |
| $9.658 \times 10^{-9}$ | $9.659 \times 10^{-7}$ | 0.3048 | $10^{-3}$ | 1 | $2.738 \times 10^{-3}$ |
| $3.527 \times 10^{-6}$ | $3.527 \times 10^{-4}$ | 111.33 | 0.365 | 365.25 | 1 |

*Source: Agronomics Catalogue* (1986) ELE International, Hemel Hempstead, Herts, UK.

## TABLE 448   Conversion Factors

### 1. Power, Energy, Heat, Thermal Transmission

| | |
|---|---|
| 1 watt (w) | $= 9.478 \times 10^{-4}$ BTU / s |
| | $= 0.2388$ calories/s |
| | $= 0.7376$ ft-lb (force) / s |
| 1 H.P. | $= 33\ 000$ lb lifted through 1 ft in 1 minute |
| 1 kilo watt (kw) | $= 1.34$ horse power |
| 1 000 watts | $= 1$ kw |
| 1 horse power | $= 0.746$ kw |
| Watts | $=$ volts x amps |
| 1 kilo watt-hour | $= 3600$ kJ $= 3410$ BTU |
| To raise 1 kg of water by 1°C requires | 4200 J of heat |
| 1 joule (J) | $= 0.24$ calorles |
| | $= 9.478 \times 10^{-4}$ BTU |
| | $= 0.7376$ ft-lb (force) |
| | $= 2.778 \times 10^{-7}$ kw-hr |
| 1 calorie | $= 4.2$ J |
| 1 kilojoule (kJ) | $= 240$ calories |
| 1 Therm | $= 105.5$ Mega Joules |
| 1 kwh | $= 3.6$ Mega Joules |
| 1 BTU | $= 1.055$ kilojoules |
| 1 J / m² | $= 8.806 \times 10^{-5}$ BTU / ft² |
| | $= 2.390 \times 10^{-5}$ calorie / cm² |
| 1 J / kg | $= 4.299 \times 10^{-4}$ BTU / lb (mass) |
| | $= 2.388 \times 10^{-4}$ calorie / g |
| BTU / lb x 0.556 | $=$ kcal / kg |
| kcal / kg x 1.79 | $=$ BTU / lb |
| 1 BTU / h | $= 0.293$ W (Joule / sec) |
| 1 kcal / h | $= 1.163$ W (Joule / sec) |

### 2. Calorific Value

| | |
|---|---|
| 1 BTU / gal | $= 0.232$ kJ / l |
| 1 BTU / ft³ | $= 37.26$ kJ / m³ |
| 1 kcal / lb | $= 9.23$ kJ / kg |
| 1 BTU / lb | $= 2.326$ kJ / kg |

### 3. Power Ratings

Brake horsepower (bhp) is the horsepower of an engine measured at the flywheel.

DIN bhp rates the horsepower of an engine in installed condition but without the alternator / dynamo actually charging. DIN is the rating commonly used by tractor manufacturers.

SAE bhp refers to a bare engine rating and gives a higher figure than DIN.

PTO power (kilowatts) $= \dfrac{\pi NT}{30\ 000}$

where N = rotational speed of PTO in rev / min

T = torque at the PTO in Nm

Hydraulic power (kilowatts) =

$\dfrac{\text{Pressure (kPa) x flow rate (l / s)}}{1000}$  or

$\dfrac{\text{Pressure (bars) x flow rate (l / s)}}{10}$

Conveyor and elevator powers (approx allowing for normal friction):– horizontal conveyor kilowatt power =

$\dfrac{\text{t / hr x distance conveyed (m)}}{400}$

– vertical elevator kilowatt power =

$\dfrac{\text{t / hr x height elevated (m)}}{80}$

Water power =

$\dfrac{\text{rate of flow (l / s) x total head (m) x 9.81}}{1000}$

Power of engine to drive water pump = Water power x efficiency (efficiency of pump and drive is normally about 50-60%)

### 4. Water Quality

1 mg / l = 1 ppm

= E.wt of ion $= \dfrac{\text{Atomic wt of ion}}{\text{Valency of ion}}$

me / l of ion $= \dfrac{\text{mg / l of ion (or ppm)}}{\text{E.wt of ion}}$

1 me / l x E.wt = ppm

EC x $10^3$ (millimhos / cm) x 1000 =

EC x $10^6$ (micromhos / cm)

1% $= \dfrac{1}{100}$ = 10 000 ppm

**TABLE 448** *(continued)*   **Conversion Factors**

$1 \mu s / cm = 1 \mu mho / cm$

| | |
|---|---|
| $1 \mu s / cm$ | *Approximations for most natural water in range* |
| $= 0.65$ mg / l | |
| $= 0.10$ meq / l of cations | 100 - 5 000 |
| | $\mu s$ / cm at 25 °C |

| | |
|---|---|
| $mol / m^3$ | $=$ me / l |
| Total cations OR anions (meq / l) $\simeq$ | |
| | 10 EC mmhos / cm |
| TDS mg / l | $\simeq 650$ x EC mmhos / cm |
| 1 dS / m | $= 1$ millimho / cm |

### 5. Area and volume calculations

| | |
|---|---|
| Area of circle | $= \pi r^2$ |
| Circumference of circle | $= 2 \pi r$ |
| Volume of cylinder | $= \pi r^2 h$ |
| Volume of sphere | $= 3$ x diameter x 0.5236 |
| | or $^4/_3 \pi r^2$ |
| Area of sphere | $=$ Diameter$^2$ x $\pi$ |
| Volume of cone | $= \dfrac{\pi r^2 h}{3}$ |
| | or base area x $\dfrac{h}{3}$ |
| Area of triangle | $= ^1/_2$ base x height |
| | $\pi = \dfrac{22}{7}$ or 3.142 |

### 6. Light

| | |
|---|---|
| 1 lux = 1 lumen / $m^2$ | $= 0.093$ foot candles |
| 1 foot candle | $= 10.764$ lux |

### 7. Temperature

Celsius (centigrade) to Fahrenheit conversion

$$C = \frac{5 (F - 32)}{9}$$

$$F = \frac{9C}{5} + 32$$

### 8. Velocity

| | |
|---|---|
| 1 m / s | $= 3.281$ ft / s $= 2.237$ mph |
| 1 ft / s | $= 0.305$ m / s |
| 1 km / hr | $= 0.621$ mph $= 0.539$ knots / hr |
| 1 mph | $= 1.609$ km / hr |
| 1 knot | $= 1$ nautical mile / hr |

### 9. Rate of flow

1 $m^3$ / s = 35.32 cu ft / s = $1.585$ x $10^4$ gall / min

| | |
|---|---|
| 1 cu ft / s | $= 0.028$ $m^3$ / s |
| 1 $m^3$ / day | $= 0.1834$ gall / min |
| gpm x 0.27 | $= m^3$ / hr |
| gpm x 0.076 | $= l / s$ |
| US gpm x 0.227 | $= m^3$ / h |
| 0.116 l / s | $\equiv 1$ mm / ha / day |

### 10. Consumption

| | |
|---|---|
| 1 mpg | $= 0.354$ km / l |
| 1 km / l | $= 2.825$ mpg |

### 11. Sundry field conversions

| | |
|---|---|
| 1 g / l | $= 0.160$ ounces / gall |
| 1 ounce / gall | $= 6.236$ g / l |
| 1 fluid ounce | $= 29.52$ ml |
| 1 ml / l | $= 0.160$ fluid ounces / gall |
| 1 fluid ounce / gall | $= 6.250$ ml / l |
| 1 l / ha | $= 0.712$ pints / acre |
| 1 gall / acre | $= 11.233$ l / ha |
| 1 kg / ha | $= 0.892$ pounds / acre |
| 1 pound / acre | $= 1.121$ kg / ha |
| 1 British bushel | $= 1.28$ $ft^3$ |
| 1 USA bushel | $= 0.969$ British bushels |
| 1 cord of wood | $= 128$ $ft^3$ |
| Hundred-weights / acre x 125 = kg / ha | |

*(continued over)*

**TABLE 448** *(continued)* **Conversion Factors**

### 12. Imperial measures

**Volume**

| | |
|---|---|
| 5 ounces of water | = 1 gill |
| 4 gills | = 1 pint |
| 2 pints | = 1 quart |
| 4 quarts | = 1 gallon (= 10 lbs of water) |
| 2 gallons | = 1 peck |
| 4 pecks | = 1 bushel |
| 8 bushels | = 1 quarter |
| 31.5 gallons | = 1 barrel |

### 13. Length

| | |
|---|---|
| 12 inches | = 1 foot |
| 3 feet | = 1 yard |
| 5.5 yards | = 1 pole (perch or rod) |
| 4 poles | = 1 chain |
| 22 yards | = 1 chain |
| 10 chains | = 1 furlong |
| 8 furlongs | = 1 mile |
| 1760 yards | = 1 mile |
| 6 feet | = 1 fathom |
| 3 miles | = 1 league |

### 14. Area

| | |
|---|---|
| 144 sq in | = 1 sq ft |
| 9 sq ft | = 1 sq yd |
| 30.25 sq yds | = 1sq pole |
| 40 sq poles | = 1 rood |
| 4 roods | = 1 acre |
| 1 acre | = 4840 sq yd |
| 640 acres | = 1 sq mile |
| 10 sq chains | = 1 acre |

### 15. Avoirdupois weight

| | |
|---|---|
| 27.34 grains | = 1 drachm |
| 16 drachms | = 1 ounce |
| 16 ounces | = 1 pound |
| 14 pounds | = 1 stone |
| 2 stones | = 1 quarter |
| 4 quarters | = 1 hundred-weight |
| 20 hundred-weight | = 1 ton |
| 2240 pounds | = 1 ton |

### 16. Concentration

| | | |
|---|---|---|
| 1% | = 10 000 ppm | = 10 g / l |
| 0.1% | = 1000 | = 1000 mg / l = 1 g / l |
| 0.01% | = 100 ppm | = 100 mg / l |
| 0.001% | = 10 ppm | = 10 mg / l |
| 0.0001% | = 1 ppm | = 1 mg / l |

### 17. Fertilizer Analyses

| | |
|---|---|
| P x 2.29 | = $P_2O_5$ |
| $P_2O_5$ x 0.436 | = P |
| K x 1.205 | = $K_2O$ |
| $K_2O$ x 0.830 | = K |
| Mg x 1.1658 | = MgO |
| MgO x 0.603 | = Mg |
| Ca x 1.399 | = CaO |
| CaO x 0.715 | = Ca |
| Na x 2.5 | = NaCl |

A 'unit' of fertilizer is 1% of one 'hundred-weight'
(112 lb) or 1.12 lb.
Units / acre of plant food x 1.25 =
kg / ha of plant food.

### 18. Feed energy

| | |
|---|---|
| kcal / lb x 2.2046 | = kcal / kg |
| kcal / kg x 0.4536 | = kcal / lb |
| kcal / kg x 0.004184 | = MJ / kg |
| MJ / kg x 239 | = kcal / kg |

# Appendix III Key Reference Texts

**TABLE 452   Key References**

The following references are a selection (not intended or compiled to be exclusive or exhaustive) which can be used for further more detailed and explanatory reading. Most, but not all, are referred to in the text for some data.

| Subject | Titles |
| --- | --- |
| **Agricultural Compendium** | ILACO (1985) *Agricultural Compendium for the Tropics and Subtropics.* |
| **Agricultural Database** | CAB International Information Services, Wallingford, Oxford, OX10 8DE. UK. Tx. 847964. Fax. 01491 33508. |
| **Agriculture, Tropical** | An extensive range of subject specific texts are produced in the Tropical Agriculture Series (Longman Group). |
| **Agriculture, Temperate** | Halley, R.J. & Saffe, R.J., eds. (1988) *The Agricultural Notebook.* |
| **Animal Feedstuffs** | FAO (1981) *Tropical Feeds.* Animal Production and Health Series No. 12. National Academy of Sciences (1971) *Atlas of Nutritional Data on United States and Canadian Feeds.* MAFF/ADAS (1986) *Feed Composition.* UK tables of feed composition and nutritive values for ruminants. |
| **Animal Nutrition** | McDonald, P., Edwards R.A., & Greenhalgh, J.F.D. (1988) *Animal Nutrition.* MAFF/ADAS (UK). (1987) *Energy Allowances and Feeding Systems for Ruminants.* Reference Book 433. National Research Council (USA). *Nutrient Requirements of Domestic Animals.* A series with individual texts on sheep, poultry, beef cattle, goats, swine, dairy cattle, horses, rabbits and others. |
| **Climate** | FAO (1987) *Agroclimatology Data.* Plant Production and Protection Series No. 25. |
| **Crops, General** | Whiteman, P.C. (1980) *Tropical Pasture Science.* Acland, J.D. (1972) *East African Crops.* Williams, C.N., Uzo J.O. & Peregrine, W.T.H. (1991) *Vegetable Production in the Tropics.* Samson, J.A. (1986) *Tropical Fruits.* Weiss, E.A. (1983) *Oilseed Crops.* |
| **Crop Protection** | Ivens, G.W. ed. (1981) *The UK Pesticide Guide.* British Crop Protection Council. Page, B.E. & Thomson, W.T. (1989) *The Quick Guide.* Hill, D.S. & Waller, J.M. (1988) *Pests and Diseases of Tropical Crops.* Vol. 2. Field Handbook. |

*(continued over)*

**TABLE 452** *(continued)*  **Key References**

| Subject | Titles |
|---|---|
| **Crop Taxonomy / Distribution** | de Rougemont, G.M. (1989) *Crops of Britain and Europe.*<br>FAO (1959) *Grasses in Agriculture.* Ag. Studies No. 42.<br>FAO (1973) *Legumes in Agriculture.* Ag. Studies No. 21.<br>Purseglove, J.W. (1976) *Tropical Crops. Monocotyledons.*<br>Purseglove, J.W. (1974) *Tropical Crops. Dicotyledons.* |
| **Crop Water Requirements** | FAO (1988) *Crop Water Requirements.*<br>    Irrigation and Drainage Paper No. 24.<br>FAO (1979) *Yield Response to Water.*<br>    Irrigation and Drainage Paper No. 33. |
| **Farm Machinery** | John Deere (1981) *Machinery Management.*<br>    *Fundamentals of Machinery Operation.*<br>Massey Ferguson, *Implement Guide.* |
| **Fertilizer Use** | de Geus, J.G. (1973) *Fertilizer Guide for the Tropics and Subtropics.* Centre d'Etude de l'Azote.<br>Cooke, G.W. (1982) *Fertilizing for Maximum Yields.*<br>FAO (1984) *Fertilizer and Plant Nutrition Guide.*<br>    Fertilizer and Plant Nutrition Bulletin No. 9.<br>MAFF/ADAS (1988) *Fertilizer Recommendations.* Ref. Book 209. |
| **Field Engineering** | Hudson, N.W. (1974)<br>    *Field Engineering for Agricultural Development.* |
| **Livestock Breeds** | Mason, I.L. (1969) *Livestock Breeds. A World Dictionary of Livestock Breeds, Types and Varieties.*<br>    CAB Technical Communication No. 8.<br>Maule, J.P. (1990) *The Cattle of The Tropics.*<br>    CTVM. University of Edinburgh. |
| **Livestock Housing** | Sainsbury, D. & Sainsbury, P. (1988)<br>    *Livestock Health and Housing.* |
| **Livestock General** | Wilkinson, J.M. & Stark, B.A. (1987)<br>    *Commercial Goat Production.*<br>Payne W.J.A. (1990)<br>    *An Introduction to Animal Husbandry in the Tropics.*<br>Gatenby, R.M. (1986)<br>    *Sheep Production in the Tropics and Sub-Tropics.*<br>Chamberlain, A. (1990)<br>    *Milk Production in the Tropics.* |

**TABLE 452**  *(continued)*  **Key References**

| Subject | Titles |
| --- | --- |
| **Management Handbooks** | Lorenz, O.A. & Maynard, D.N. (1980)<br>    *Knott's Handbook for Vegetable Growers.*<br>Nix, J. *Farm Management Pocketbook.*<br>    Wye College, University of London. (revised annually)<br>Agro Business Consultants (UK)<br>    *The Agricultural and Costing Book.* (revised monthly)<br>Scottish Agricultural College, Edinburgh.<br>    *The Scottish Agricultural Colleges Farm Management Handbook.* (revised annually) |
| **Micronutrients** | Mortvedt, J.J. *et al.* (1972) *Micronutrients in Agriculture.*<br>    Soil Science Society of America.<br>FAO (1983) *Micronutrients. Fertiliser and Plant Nutrition*<br>    Bulletin No. 7. |
| **Plant Analysis** | Reuter, D.J. & Robinson, J.B. (1986)<br>    *Plant Analysis. An Interpretation Manual.* |
| **Research Organisations**<br>**(addresses world-wide)** | *Agricultural Research Centres.* (1990) (Longman) |
| **Seed Suppliers** | FAO (1986) *World List of Seed sources*<br>    SAP: SIDP/86.2 Plant Production Division. |
| **Soils** | BAI (1984) *Booker Tropical Soil Manual.* (A second edition has<br>    become available since compilation of this text)<br>John Deere (1984)<br>    *Tillage. Fundamentals of Machinery Operation.* |
| **Units and Conversions** | Horvath, A.L. (1986)<br>    *Conversion Tables of Units for Science and Engineering.* |
| **Water Quality** | FAO (1985) *Water Quality For Agriculture.*<br>    Irrigation and Drainage Paper No. 29. |
| **Weed Control** | Boettger, M.S., ed. (1987)<br>    *Weed Control Manual and Herbicide Guide.* |
| **World Agriculture** | International Association of Agricultural Economists (1969)<br>    *World Atlas of Agriculture.*<br>van Royen, W (1954) *Agricultural Resources of the World.*<br>    University of Maryland/USDA. |

# Index